Plant Stress Tolerance

Plant Stress Tolerance: Molecular Mechanisms and Breeding Strategies, Volume One provides effective ways for organizing precision and sustainable agriculture. The methods include the use of advanced molecular techniques covering multiple omics, high-throughput technology, computational biology, epigenetic manipulation, and CRISPR genome editing. These methods can advance the development of high-yield, high-quality, and stress-resilient crops that meet the requirements for supporting global food and nutrition security.

The book proposes strategies for omics-assisted and speed breeding techniques, exploring molecular mechanisms of plant abiotic stress caused by temperature, drought, salinity, and various pollutants. These are uncovered by quantitative trait loci analysis and mapping, genomic selection, functional genomics, multiple omics, high-throughput sequencing, and high-throughput phenotyping, and are integrated into the various systems of crop improvement.

Plant Stress Tolerance: Molecular Mechanisms and Breeding Strategies, Volume One presents emerging and comprehensive knowledge and is an ideal reference for students, researchers, teachers, and professors. It inspires ideas for investigations into the fields of plant stress physiology, plant functional genomics, plant multiple omics, plant genetic engineering, systems biology, and crop breeding.

Plant Stress Tolerance

Molecular Mechanisms and Breeding Strategies

Volume One

Edited by
Jen-Tsung Chen

CRC Press
Taylor & Francis Group
Boca Raton London New York

CRC Press is an imprint of the
Taylor & Francis Group, an **informa** business

Designed cover image: Shutterstock

First edition published 2025
by CRC Press
2385 NW Executive Center Drive, Suite 320, Boca Raton FL 33431

and by CRC Press
4 Park Square, Milton Park, Abingdon, Oxon, OX14 4RN

CRC Press is an imprint of Taylor & Francis Group, LLC

ISBN: 9781032596495 (hbk)
ISBN: 9781032600314 (pbk)
ISBN: 9781003457237 (ebk)

DOI: 10.1201/9781003457237

Typeset in Times
by codeMantra

Contents

Editor

Dr. Jen-Tsung Chen is a professor of cell biology at the National University of Kaohsiung in Taiwan. He also teaches genomics, proteomics, plant physiology, and plant biotechnology. Dr. Chen's research interests include bioactive compounds, chromatography techniques, plant molecular biology, plant biotechnology, bioinformatics, and systems pharmacology. He became an active editor of academic books and journals in an effort to advance the exploration of multidisciplinary knowledge involving plant physiology, plant biotechnology, nanotechnology, ethnopharmacology, systems biology, and drug discovery. He serves as an editorial board member and guest editor on several reputed journals. Dr. Chen has published books in collaboration with international publishers on diverse topics such as drug discovery, herbal medicine, medicinal biotechnology, nanotechnology, bioengineering, plant functional genomics, plant speed breeding, CRISPR-based plant genome editing, and artificial intelligence. In 2023 and 2024, Dr. Chen was included in the World's Top 2% Scientists ranking by Stanford University.

Contributors

Puzhakkal Abhirami
Department of Genomic Science
Central University of Kerala
Kasaragod, India

Jothi Dheivasikamani Abidharini
Department of Human Genetics and Molecular Genetics
Bharathiar University
Coimbatore, India

Gerardo Acosta-García
Laboratorio de Biología Molecular
Departamento de Ingeniería Bioquímica, Tecnológico Nacional de México/ITCelaya
Celaya, México

Padmore Adu-Antwi
Council for Scientific and Industrial Research-Crops Research Institute (CSRI-CRI)
Kumasi, Ghana

Shreni Agrawal
Department of Biotechnology, Parul Institute of Applied Science
Parul University
Vadodara, India

Abdel Rahman Mohammad Al Tawaha
Department of Biological Sciences
Al Hussein Bin Talal University
Ma'an, Jordan

Maria Almira-Casellas
Lab. Plant Physiology, Department of Animal Biology, Plant Biology and Ecology
Autonomous University Barcelona
Barcelona, Spain

Priscilla Amponsah
Key Laboratory of Integrated Pest Management on Crops in East China, Ministry of Agriculture
Nanjing Agricultural University
Nanjing, China

Arumugam Vijaya Anand
Department of Human Genetics and Molecular Genetics
Bharathiar University
Coimbatore, India

Muthukrishnan Anand
Department of English
Ramco Institute of Technology
Rajapalayam, India

Augustine Antwi-Boasiako
National Center for Soybean Improvement, Key Laboratory of Biology and Genetics and Breeding for Soybean, Ministry of Agriculture, State Key Laboratory of Crop Genetics and Germplasm Enhancement
Nanjing Agricultural University
Nanjing, China
and
Council for Scientific and Industrial Research-Crops Research Institute (CSRI-CRI)
Fumesua, Kumasi, Ghana

Gunasekaran Arthi
Department of Human Genetics and Molecular Genetics
Bharathiar University
Coimbatore, India

Muthukrishnan Arun
Department of Biotechnology
Bharathiar University
Coimbatore, India

Asad Azeem
Department of Plant Breeding and Genetics
University of Agriculture
Faisalabad, Pakistan

Smiljana Goreta Ban
Institute of Agriculture and Tourism
Poreč, Croatia
Centre of Excellence for Biodiversity and Molecular Plant Breeding
Zagreb, Croatia

Juan Barceló
Laboratory of Plant Physiology, Department of Animal Biology, Plant Biology and Ecology
Autonomous University Barcelona
Barcelona, Spain

Angélica Barrales-López
Academia de la Ingeniería en Innovación Agrícola Sustentable
Instituto Tecnológico Superior de Libres Puebla
de Libres, Pue, México

Nataša Bauer
Department for Molecular Biology, Faculty of Science
University of Zagreb
Zagreb, Croatia

Naheeda Begum
State Key Laboratory of Biocontrol, School of Ecology
Sun Yat-sen University
Shenzhen, China

Divya Raj Bharti
Department of Genetics and Plant Breeding
Palli Siksha Bhavana (Institute of Agriculture)
Bolpur, West Bengal

Indrani Bhattacharya
Department of Biotechnology, Parul Institute of Applied Science
Parul University
Vadodara, India

Silvia Busoms
Laboratory of Plant Physiology, Department of Animal Biology, Plant Biology and Ecology
Autonomous University Barcelona
Barcelona, Spain

Pablo I. Calzadilla
Department of Earth and Environmental Sciences, Faculty of Science and Engineering
University of Manchester
Manchester, United Kingdom
and
Instituto de Fisiología Vegetal (INFIVE)
Universidad Nacional de La Plata – CONICET
Buenos Aires, Argentina

Jen-Tsung Chen
Department of Life Sciences
National University of Kaohsiung
Kaohsiung, Taiwan

Richa Das
Department of Biotechnology, Parul Institute of Applied Science
Parul University
Vadodara, India

Sandip Debnath
Department of Genetics and Plant Breeding
Palli Siksha Bhavana (Institute of Agriculture)
Bolpur, West Bengal

Chenniappan Anchana Devi
PG & Research Department of Biotechnology
Women's Christian College
Chennai, India

Vinay Kumar Dhiman
Department of Basic Sciences
Dr. Yashwant Singh Parmar University of Horticulture and Forestry
Nauni, India

Vivek Kumar Dhiman
Department of Biotechnology
Dr. Yashwant Singh Parmar University of Horticulture and Forestry
Nauni, India

Fatima El Amerany
Laboratory of Natural Resources and Environment, Polydisciplinary Faculty of Taza, Department of Biology, Chemistry, and Geology
Sidi Mohamed Ben Abdellah University
Fes, Morocco

Nizar El Mazouni
Center of Agrobiotechnology and Bioengineering, Centre AgroBiotech-URL-CNRST-05, Physiology of Abiotic Stresses Team
Cadi Ayyad University
Marrakesh, Morocco
and
Laboratory of Agro-Food, Biotechnologies and Valorization of Plant Bioresources (AGROBIOVAL), Faculty of Sciences Semlalia
Cadi Ayyad University
Marrakesh, Morocco
and
Laboratory of Materials Sciences and Process Optimization, Chemistry of Condensed Matter and Environment Team, Chemistry Department, Faculty of Sciences Semlalia
Cadi Ayyad University
Marrakesh, Morocco

Hassan Ragab El-Ramady
Faculty of Agriculture
Kafrelsheikh University
Kafr el-Sheikh, Egypt

Samia Elouali
Interdisciplinary Laboratory in Bio-Resources, Environment and Materials, Higher Normal School, Department of Biology
Cadi Ayyad University
Marrakesh, Morocco
and
Service of Polymeric and Composite Materials (SMPC)
Mons University
Mons, Belgium

Mohamed S. Elshikh
Department of Botany and Microbiology, College of Science
King Saud University
Riyadh, Saudi Arabia

Eduardo Espítia-Rangel
Instituto Nacional de Investigaciones Forestales, Agrícolas y Pecuarias (INIFAP)
Campo Experimental Valle de México
Texcoco, México

Jemaa Essemine
National Key Laboratory of Plant Molecular Genetics
CAS Center for Excellence in Molecular Plant Sciences, Chinese Academy of Sciences
Shanghai, China

Laura Fonseca-Mendoza
División de Ciencias de la Salud e Ingeniería
Universidad de Guanajuato
Guanajuato, México

Selene Fragoso-Iñiguez
Laboratorio de Biología Molecular
Departamento de Ingeniería Bioquímica
Tecnológico Nacional de México/ITCelaya
Celaya, México

Roberto Gamboa-Becerra
Instituto de Ecología
Xalapa, México

Silverio García-Lara
Escuela de Ingeniería y Ciencias
Tecnológico de Monterrey
Campus Monterrey, N.L., México

Karen Ghazaryan
Faculty of Biology
Yerevan State University
Yerevan, Armenia

Tzitziki González-Rodríguez
Escuela de Ingeniería y Ciencias
Tecnológico de Monterrey
Campus Monterrey, N.L., México

Sumbul Habib
Centre of Agricultural Biochemistry and Biotechnology (CABB)
University of Agriculture
Faisalabad, Pakistan

Youssef Ait Hamdan
Interdisciplinary Laboratory in Bio-Resources, Environment and Materials, Higher Normal School, Department of Biology
Cadi Ayyad University
Marrakech, Morocco

Karli Haritha
Department of Human Genetics and Molecular Genetics
Bharathiar University
Coimbatore, India

Maryan Hernández-Jiménez
Laboratorio de Biología Molecular, Departamento de Ingeniería Bioquímica
Tecnológico Nacional de México/ITCelaya
Celaya, México

Chayma Ikan
Center of Agrobiotechnology and Bioengineering, Research Unit labeled CNRST (Centre AgroBiotech-URL-CNRST-05), "Physiology of Abiotic Stresses" Team
Cadi Ayyad University
Marrakesh, Morocco
and
Laboratory of Agro-Food, Biotechnologies and Valorization of Plant Bioresources (AGROBIOVAL), Faculty of Sciences Semlalia
Cadi Ayyad University
Marrakesh, Morocco

Iman Janah
Laboratory of Agro-Food, Biotechnologies and Valorization of Plant Bioresources, Faculty of Science Semlalia, Department of Biology
Cadi Ayyad University
Marrakesh, Morocco

Sasikumar Janani
Department of Human Genetics and Molecular Genetics
Bharathiar University
Coimbatore, India

Nenad Jasprica
Institute for Marine and Coastal Research
University of Dubrovnik
Dubrovnik, Croatia

Giles N. Johnson
Department of Earth and Environmental Sciences, Faculty of Science and Engineering
University of Manchester
Manchester, United Kingdom

Faiz Ahmad Joyia
Centre of Agricultural Biochemistry and Biotechnology (CABB)
University of Agriculture
Faisalabad, Pakistan

Srinivasan Kameswaran
Department of Botany
Vikrama Simhapuri University College
Nellore, India

Muhammad Sarwar Khan
Centre of Agricultural Biochemistry and Biotechnology (CABB)
University of Agriculture
Faisalabad, Pakistan

Deepak Kumar
R&D Division
Nextnode Bioscience Pvt Ltd.
Gujrat, India

Manish Kumar
National Institute of Plant Biotechnology
IARI
New Delhi, India

Anupam Kumari
Department of Biotechnology, Parul Institute of Applied Science
Parul University
Vadodara, India

Rachid Lahlali
Phytopathology Unit, Department of Plant Protection
Ecole Nationale d'Agriculture de Meknès
Menkes, Morocco

Saglara Mandzhieva
Academy of Biology and Biotechnology
Southern Federal University
Rostov-on-Don, Russia

Rashmi Maurya
Department of Botany
Lucknow University
Lucknow, U. P., India

Abdelilah Meddich
Center of Agrobiotechnology and Bioengineering, Centre AgroBiotech-URL-CNRST-05, Physiology of Abiotic Stresses Team
Cadi Ayyad University
Marrakesh, Morocco
African Sustainable Agriculture Research Institute (ASARI)
University Mohammed VI Polytechnic (UM6P)
Laayoune, Morocco
Laboratory of Agro-Food, Biotechnologies and Valorization of Plant Bioresources (AGROBIOVAL), Faculty of Sciences Semlalia
Cadi Ayyad University
Marrakesh, Morocco

Tatiana Minkina
Department of Biology and Biotechnology
Southern Federal University
Rostov-on-Don, Russia

Sumithran Subhadevi Suhith Mithra
Department of Human Genetics and Molecular Genetics
Bharathiar University
Coimbatore, India

Josaphat Miguel Montero-Vargas
Instituto Nacional de Enfermedades Respiratorias (INER)
Mexico City, Mexico

Ghulam Mustafa
Centre of Agricultural Biochemistry and Biotechnology (CABB)
University of Agriculture
Faisalabad, Pakistan

Zhang NanNan
National Center for Soybean Improvement, Key Laboratory of Biology and Genetics and Breeding for Soybean, Ministry of Agriculture, State Key Laboratory of Crop Genetics and Germplasm Enhancement
Nanjing Agricultural University
Nanjing, China

Marimuthu Oviya
Department of Human Genetics and Molecular Genetics
Bharathiar University
Coimbatore, India

Himanshu Pandey
PG Department of Agriculture Khalsa College
Amritsar, Punjab

Aqsa Parvaiz
Department of Biochemistry and Biotechnology
The Women University Multan
Multan, Pakistan

Amitava Paul
Department of Genetics and Plant Breeding
Palli Siksha Bhavana (Institute of Agriculture)
Sriniketan, India

Charlotte Poschenrieder
Laboratory of Plant Physiology, Department of Animal Biology, Plant Biology and Ecology
Autonomous University Barcelona
Barcelona, Spain

Venkatachalam Sivaprakash
Department of Human Genetics and Molecular Genetics
Bharathiar University
Coimbatore, India

Biswajit Pramanik
Department of Genetics and Plant Breeding
Palli Siksha Bhavana (Institute of Agriculture)
Sriniketan, West Bengal

Amardeep Ray Preethi
Department of Human Genetics and Molecular Genetics
Bharathiar University
Coimbatore, India

Natarajan Pushpa
PG & Research Department of Microbiology
Cauvery College for Women
Tiruchirappalli, India

Mingnan Qu
Jiangsu Key Laboratory of Crop Genetics and Physiology
Agricultural College of Yangzhou University
Yangzhou, China

Vishnu D. Rajput
Academy of Biology and Biotechnology
Southern Federal University
Rostov-on-Don, Russia

Mandala Ramakrishna
Department of Botany
Vikrama Simhapuri University College
Nellore, India

Muhammad Khuram Razzaq
Faculty of Agriculture and Veterinary Sciences
Superior University
Lahore, Pakistan

Saravanan Renuka
Department of Chemistry and Biosciences
SASTRA Deemed to be University, Srinivasa Ramanujan Centre
Kumbakonam, India

Mohammed Rhazi
Interdisciplinary Laboratory in Bio-Resources, Environment and Materials, Higher Normal School, Department of Biology
Cadi Ayyad University
Marrakesh, Morocco

Aryadeep Roychoudhury
Discipline of Life Sciences, School of Sciences
Indira Gandhi National Open University
New Delhi, India

Rajaswini Rupali
Department of Biotechnology
Bharathiar University
Coimbatore, India

Branka Salopek-Sondi
Department for Molecular Biology
Ruđer Bošković Institute
Zagreb, Croatia

Santanu Samanta
Post Graduate Department of Biotechnology
St. Xavier's College (Autonomous)
Mumbai, India

Palanisamy Sampathkumar
Department of Chemistry and Biosciences
SASTRA Deemed to be University, Srinivasa Ramanujan Centre
Kumbakonam, India

Krishnagowdu Saravanan
Department of Biotechnology
Bharathiar University
Coimbatore, India

Md. Shamim
Dr Kalam Agricultural College
Bihar Agricultural University
Bihar, India

Avinash Sharma
Faculty of Agricultural Sciences
Arunachal University of Studies
Namsai, India

Neha Sharma
National Institute of Plant Biotechnology
IARI
New Delhi, India

Nimisha Sharma
Indian Agricultural Research Institute
New Delhi, India

Mohammed Haris Siddiqui
Department of Agriculture, Integral Institute of Agricultural Science and Technology
Integral University
Lucknow, India

Saba Siddiqui
Department of Agriculture, Integral Institute of Agricultural Science and Technology
Integral University
Lucknow, India

Guillermo Antonio Silva-Martínez
Laboratorio de Biología Molecular, Departamento de Ingeniería Bioquímica
Tecnológico Nacional de México/ITCelaya
Celaya, México

Abhishek Singh
Faculty of Biology
Yerevan State University
Yerevan, Armenia

Devendra Singh
Department of Basic Sciences
Dr. Yashwant Singh Parmar University of Horticulture and Forestry
Nauni, India

Nagendra Kumar Singh
National Institute of Plant Biotechnology
IARI
New Delhi, India

Sandeep Kumar Singh
Department of Agriculture, Integral Institute of Agricultural Science and Technology
Integral University
Lucknow, India

Ramalingam Sivakumar
Department of Chemistry and Biosciences
SASTRA Deemed to be University, Srinivasa Ramanujan Centre
Kumbakonam, India

Shivangee Solanki
Department of Biotechnology, Parul Institute of Applied Science
Parul University
Waghodia, India

Fatima Ezzahra Soussani
Center of Agrobiotechnology and Bioengineering, Centre AgroBiotech-URL-CNRST-05, Physiology of Abiotic Stresses Team
Cadi Ayyad University
Marrakesh, Morocco
and
Laboratory of Agro-Food, Biotechnologies and Valorization of Plant Bioresources (AGROBIOVAL), Faculty of Sciences Semlalia
Cadi Ayyad University
Marrakesh, Morocco
and
Laboratory of Plant, Animal, and Agro-Industry Productions, Faculty of Science
University Ibn Toufail
Kenitra, Morocco

Deepti Srivastava
Department of Agriculture, Integral Institute of Agricultural Science and Technology
Integral University
Lucknow, India

Battana Swapna
Department of Botany
Vikrama Simhapuri University College
Nellore, India

Lokesh Thakur
Department of Biotechnology
Dr. Yashwant Singh Parmar University of Horticulture and Forestry
Solan, India

Mirta Tkalec
Department for Molecular Biology, Faculty of Science
University of Zagreb
Zagreb, Croatia

Salma Toubali
Center of Agrobiotechnology and Bioengineering, Centre AgroBiotech-URL-CNRST-05, Physiology of Abiotic Stresses Team
Cadi Ayyad University
Marrakesh, Morocco
and
Laboratory of Agro-Food, Biotechnologies and Valorization of Plant Bioresources (AGROBIOVAL), Faculty of Sciences Semlalia
Cadi Ayyad University
Marrakesh, Morocco

Fabiola Estefania Tristan-Flores
Laboratorio de Biología Molecular, Departamento de Ingeniería Bioquímica
Tecnológico Nacional de México/ITCelaya
Celaya, México

Nandakumar Vidya
Department of Biotechnology
Bharathiar University
Coimbatore, Tamil Nadu, India

K. Vijayalakshmi
PG & Research Department of Microbiology
Jamal Mohamed College
Tiruchirappalli, India

Kannan Vijayarani
Department of Education
Bharathiar University
Coimbatore, India

Puliyampatti Gunasekaran Vishnu
Department of Human Genetics and Molecular Genetics
Bharathiar University
Coimbatore, India

Robert Winkler
Unidad de Genómica Avanzada (UGA) – Laboratorio Nacional de Genómica para la Biodiversidad (LANGEBIO)
Irapuato, México

Zhan Xu
State Key Laboratory for Conservation and Utilization of Subtropical Agro-Bioresources
South China Agricultural University
Guangzhou, China

Chunting Zhang
National Center for Soybean Improvement, Key Laboratory of Biology and Genetics and Breeding for Soybean, Ministry of Agriculture, State Key Laboratory of Crop Genetics and Germplasm Enhancement
Nanjing Agricultural University
Nanjing, China

Tuanjie Zhao
National Center for Soybean Improvement, Key Laboratory of Biology and Genetics and Breeding for Soybean, Ministry of Agriculture, State Key Laboratory of Crop Genetics and Germplasm Enhancement
Nanjing Agricultural University
Nanjing, China

Preface

To face stress, plants have adapted diverse tolerance mechanisms by regulating the effects of stressors originating from different sources. For decades, these mechanisms have been substantially uncovered, which led to dramatic progress in crop improvement and breeding programs. Under the scenario of global climate change, stress conditions can be worse, and it's getting critical to developing resilient or smart crops against single-type or complicated stresses. For organizing sustainable agriculture, breeding strategies based on advanced molecular techniques, such as multiple omics, high-throughput technology, computational biology, and CRISPR genome editing, hold promise to developing high-yield, high-quality, and stress-resilient crops that meet the requirements for global food and nutrition security.

This book explores knowledge of non-coding RNAs that are critical in regulating plant stress response through epigenetic mechanisms which are valuable in crop breeding programs to develop resilient crops against stress. In recent years, advanced approaches have been updated frequently such as high-throughput sequencing technologies, which can explore in-depth knowledge of epigenetics at the level of whole-genome sequencing. Additionally, transcriptomics uncovers the role of transcription factors, cis-acting elements, DNA methylation, histone modifications, and post-translational modifications in the epigenetic regulation of gene expression under stress conditions.

A "photosynthetic re-design" approach has been presented in this book, chiefly involving engineering the molecular and physiological pathways in photosynthesis to enhance light capture, maximize energy conversion, and, eventually, increase carbon fixation. An approach for balancing crop yield and stress tolerance with re-designing photosynthesis has been provided.

In recent years, computational biology has seen huge progress, chiefly benefiting from the achievements of computer science, particularly the technologies based on artificial intelligence (AI) such as machine learning, deep learning, and emerging generative AI models. It leads to the highly efficient unraveling of complicated biological networks for predicting key players, for instance, involving plant stress tolerance, which can promote crop resilience to climate change and diverse biological attacks of pathogens and pests. This book gives an overview of the use of AI-driven systems to organize digital plant omics and their applications in crop breeding.

In plants, metabolites are crucial for the resistance against both abiotic and biotic stresses. This book illustrates approaches involving the analysis of metabolite quantitative trait loci (mQTLs), which can identify genetic regions associated with plant stress resistance. Targeted breeding strategies based on mQTL analyses integrated with advanced omics technologies and high-throughput phenotyping were proposed.

Systematic refines of literature dealing with molecular mechanisms of plant abiotic stress caused by temperature (cold or heat), drought, salinity, and pollutants (heavy metals or metalloids) were provided. Additionally, the integration of the resulting knowledge into breeding programs has been proposed as a series of strategies including omics-assisted breeding and speed breeding.

Under global climate change, extreme temperatures are becoming a new threat to damaging the quality and yield of crops. For regulating cold stress, plants possess crucial signaling pathways, such as inducer of CBF expression (ICE)–C-repeat binding factor (CBF)–cold-regulated gene (COR) signaling, to sense, transduce, and respond, and thus can mitigate the negative effects. Such key players involved in signaling are ideal targets for genetic manipulation assisted by certain success in QTL analyses and potent genome editing technologies, most efficiently being the system of clustered regularly interspaced short palindromic repeats (CRISPR) with Cas9. To mitigate heat stress, the genetic manipulation of heat shock proteins and heat stress transcription factors plays a crucial role in developing heat-tolerant crops. Detailed molecular mechanisms and crop improvement strategies are proposed in this book.

Drought stress can induce the formation of reactive oxygen species (ROS), which attack photosynthetic machinery and a range of proteins or enzymes of crops, causing considerable quality and yield loss in agricultural production. Molecular approaches-based strategies for adaptation against drought stress, for instance, in cereals were summarized, highlighting QTL analysis and mapping, genomic selection, functional genomics, multiple omics, high-throughput sequencing, and high-throughput phenotyping, to manipulate the balance of ROS and the scavenging in cellular processes and toward resilience to drought stress.

Under arid and semi-arid conditions, there is high demand in breeding programs to develop salt-tolerant, high-yielding varieties for major crops. This book addresses advances in understanding the complicated physiological, molecular, and genetic mechanisms responsible for alkaline salinity tolerance in plants, and the applications and integrations into breeding programs and research directions are comprehensively discussed.

Heavy metals, such as cadmium, copper, zinc, and mercury, and metalloids, such as arsenic, germanium, and tellurium, are toxic to ecosystems and plants. Approaches such as QTL mapping and genome-wide association studies and breeding strategies, including gene silencing that can reduce the negative effects of heavy metals and metalloids, were summarized highlighting their applications in grain crops. Using genetic manipulation, several genes have been overexpressed in plants, which shows promising results in mitigating stress responses caused by aluminum or cadmium.

To provide updated and comprehensive knowledge, Volume One of *Plant Stress Tolerance: Molecular Mechanisms and Breeding Strategies* was collaboratively organized by diverse scientists and experts. To be an ideal reference for students, researchers, teachers, and professors, this book is expected to inspire new ideas for designing future investigations in the fields of plant stress physiology, plant functional genomics, and crop breeding. In the end, the editor would like to thank all contributors for their innovative chapters and the support from the publisher.

1 Re-Designing Photosynthesis
Balancing Yield and Stress Tolerance

Pablo I. Calzadilla and Giles N. Johnson

1.1 INTRODUCTION

Changing climates have been severely impacting crops in recent years due to the increased frequency of extreme weather events (Ray et al., 2019; Razzaq et al., 2021). Losses of up to 50% in grain yield have been reported due to drought stress in species such as maize, wheat, and soybean (Johari-Pireivatlou, 2010; Aydinsakir et al., 2013; Candogan et al., 2013). Extreme weather also alters crops indirectly, increasing their susceptibility to pathogens and consequently affecting their productivity (Chaloner et al., 2021). Moreover, the quality of crops has been reported to change due to the current climate crisis. Studies performed on tea plants (*Camelia sinensis*) show that the concentration of secondary metabolites (e.g. caffeine, phenolic compounds, or terpenoids) fluctuates due to shifts in weather parameters (Ahmed et al., 2019); while increases in temperature alter the carbohydrate composition of wheat and barley grains, and of beans (DaMatta et al., 2010; Asseng et al., 2019).

The impacts of climate change on crops are, in themselves, a major concern for global food security. They come however against a background of increasing food demand, due to a combination of rising populations and increased *per capita* consumption. This combination has created what has been described as a "perfect storm" in agriculture (Poppy et al., 2014). Not only do we need to mitigate the effects of increasingly stressful and unpredictable climates, but also to simultaneously *increase* productivity. New varieties of crops are required, with enhanced stress resilience and yield, while maintaining their quality, aiming to fulfill the needs of a world with a growing population (Dhankher and Foyer, 2018).

Different approaches have been proposed to improve stress tolerance and yield in plants. These include overexpressing transcription factors to modulate gene regulatory networks (Wang et al., 2016), engineering metabolic pathways toward the production of secondary metabolites (e.g. glycine betaine, proline, polyamines; Vendruscolo et al., 2007; Espasandin et al., 2018), or enhancing antioxidant responses (Liu et al., 2013; Li et al., 2015), among others. In recent years, one strategy that has been gaining particular attention in the search for increasing plant yield is photosynthetic re-design. This involves engineering the different physiological, biophysical, and biochemical components of the photosynthetic apparatus to enhance its metabolic output: organic carbon. A range of approaches have been proposed, and, in some cases, promising results have been obtained (Kumar et al., 2009; Kromdijk et al., 2016; Driever et al., 2017). However, the impact of modifying the photosynthetic apparatus on stress tolerance, and its consequences in the context of plant responses to fluctuating, often stressful, environments, has not been fully addressed (Table 1.1).

Photosynthesis is a sensor hub for environmental change (Foyer and Noctor, 2012). Short-term environmental changes (seconds to minutes) lead to short-term regulatory photosynthetic responses, while sustained changes (hours to days) in time lead to photosynthetic acclimation, involving changes in the composition of the photosynthetic apparatus and other metabolic pathways (Gjindali et al., 2021). These regulatory responses and acclimation processes are of fundamental importance for plants' survival in natural habitats and agricultural systems and, as such, must be addressed

DOI: 10.1201/9781003457237-1

when re-engineering photosynthesis. In this review, we present an overview of proposed strategies to increase crop yield through alterations of the photosynthetic pathway. We address how these manipulations might impact plant resilience to stress. Finally, we discuss where resilience and yield meet, in the generation of new varieties of crops for the future. Is there an optimal solution for increasing plant yield in a climate change scenario?

1.2 PHOTOSYNTHETIC RE-DESIGN: PATHWAYS AND STRATEGIES TO INCREASE YIELD IN THE REAL WORLD

Photosynthesis is the fundamental process through which plants fix carbon dioxide into sugars, transforming light into chemical energy. This depends on the primary reactions of two photosystems,

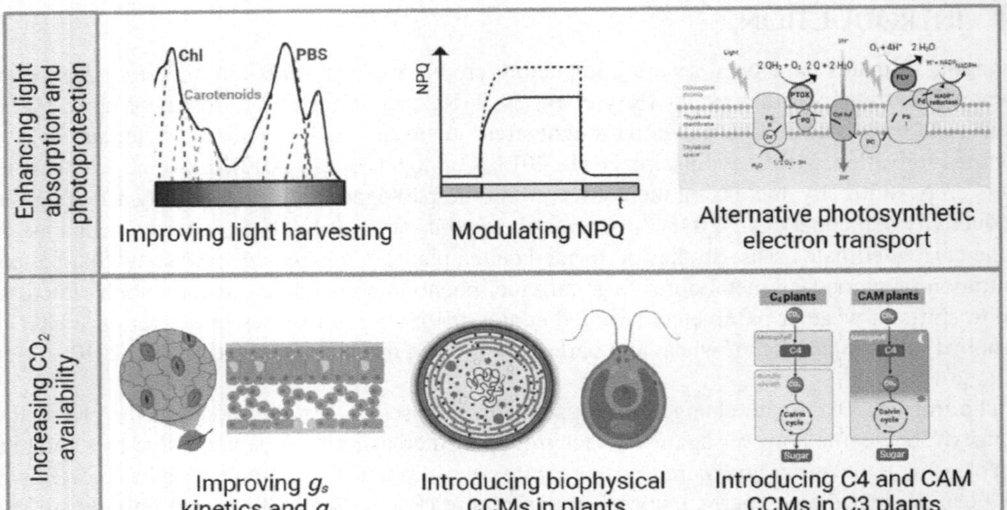

FIGURE 1.1 Strategies to increase yield and stress tolerance through re-designing photosynthesis. Created with BioRender.com. [CCM: Carbon Concentration Mechanisms; CBB cycle: Calvin-Benson-Bassham cycle; NPQ: Non-Photochemical Quenching; RuBP: ribulose-1,5-biphosphate; gs: stomatal conductance; gm: mesophyll conductance]

Photosystems I (PSI) and II (PSII), in the chloroplast thylakoid membranes. Each photosystem absorbs light energy using chlorophylls and uses that energy to drive electron transfer to electron acceptors. The electron flow from PSII to PSI, through the plastoquinone (PQ) pool and the cytochrome b_6f is defined as linear electron flow, generating ATP and reductant power (NADPH). These molecules are then used in the Calvin–Benson–Bassham (CBB) cycle, in which the enzyme Rubisco (ribulose-1,5-biphosphate carboxylase/oxygenase) fixes CO_2 in the chloroplast stroma, generating carbohydrates for wider metabolism and production of biomass.

As for any metabolic process, photosynthesis can be limited by substrate availability (light and CO_2 being the most relevant limiting substrates), the kinetic properties of their enzymatic steps, or damage to their molecular components. Photosynthesis is particularly vulnerable to the latter, due to side-reactions which produce reactive oxygen species (ROS; Roach and Krieger-Liszkay, 2014), and multiple complex systems exist to protect against this (collectively called photoprotection; Pinnola and Bassi, 2018). Re-engineering photosynthesis involves strategies to increase the availability of photosynthetic substrates, speed up their assimilation, or protect against photodamage. In the following sections, we will review approaches to re-design photosynthesis (Figure 1.1) and discuss their impact on stress responses under fluctuating environments.

1.2.1 Enhancing Light Absorption through Designing New Antenna Systems and Photoprotection

1.2.1.1 Improving Light Harvesting

The first step in photosynthesis is the absorption of light energy by pigments in the photosynthetic apparatus. Photosynthetic organisms have evolved different light-harvesting systems which absorb photons and trigger charge separation in the reaction centers of the photosystems. In plants and green algae, this system is made of membrane-bounds light-harvesting complexes (LHCs), which bind chlorophylls *a* and *b* and the xanthophylls lutein, neoxanthin, and the pigments of the xanthophyll cycle – violaxanthin, antheraxanthin, and zeaxanthin (Eskling et al., 1997). These antenna systems transfer absorbed energy ("excitons") to the reaction centers, which in plants bind chlorophyll *a* and β-carotene.

Photosynthetic eukaryotes mainly use light energy across the wavelength range 400–700 nm, with absorption being maximal between 430–470 and 650–670 nm (the absorption peaks for chlorophylls; Chen and Blankenship, 2011). The potential to increase absorption across a wider wavelength range has been discussed, with approaches ranging from expressing bacterial and algal light-harvesting proteins in plants to de novo design of photosynthetic systems (Ort et al., 2015; Hitchcock et al., 2022; Leister, 2023). The aim is to increase the use of incident light to drive photosynthesis (Blankenship and Chen, 2013). However, not all wavelengths of absorbed light are able to drive water splitting in the reaction centers of Chl *a* photosystems (Nürnberg et al., 2018; Mascoli et al., 2020). Although ambitious, the heterologous introduction of coupled antenna systems and reaction centers binding different pigments (e.g. chlorophyll *d*) has been proposed to solve this issue (Ort et al., 2015; Hitchcock et al., 2022).

Increasing the range of wavelengths of the incident light absorbed can be advantageous under limited light conditions. However, plants directly exposed to sunlight are usually light saturated. Since more photons are absorbed than can be used to drive photochemistry, this decreases photosynthetic efficiency and increases the probability of oxidative stress and photodamage (Roach and Krieger-Liszkay, 2014). As a result, photoprotective mechanisms are triggered under these conditions, demanding energy and resources which could be used for other metabolic processes. Thus, a different strategy to improve photosynthetic efficiency involves *decreasing* light-harvesting pigments per photosystem, to enhance the quantity of photons used in photochemistry (Ort et al., 2011; Blankenship and Chen, 2013; Table 1.1). This will also enhance light distribution within the canopy, resulting in higher integrated photosynthetic rates of cultivated crops.

Characterization of chlorophyll-deficient mutants in soybean and tobacco has shown that a reduction in PSII antenna can improve their photosynthetic efficiency when compared to wild-type plants (Kirst et al., 2017; Pettigrew et al., 1989; Table 1.1). The experiments performed on tobacco also showed that these plants can be cultivated at higher densities, minimizing the surface area required to produce a certain amount of biomass (Kirst et al., 2017). Furthermore, modeling experiments suggest that decreasing chlorophyll content can significantly improve nitrogen use efficiency (NUE; Song et al., 2017). Despite these advantages, a reduction in antenna size will also have an impact on the regulatory processes controlling the photosynthetic electron transport chain (PETC) and the concentration ratio of their photosynthetic components (e.g. PSII/PSI; Ghirardi and Melis, 1988; Kim et al., 2009). These can trigger negative effects during plant development or affect plant responses under different growth conditions (Kim et al., 2009; Leister, 2023; Mao et al., 2023). It has been shown that decreasing antenna size in mutants of the green alga *Chlamydomonas reinhardtii* paradoxically *increases* their sensitivity to photodamage under fluctuating light conditions (De Mooij et al., 2015). This effect can be explained by a possible impairment in photoprotection. *Arabidopsis* mutants deficient in chlorophyll *b* have also been described as more prone to photodamage (Kim et al., 2009; Calzadilla et al., 2024).

To date, approaches to engineer new light-harvesting systems have been most developed in cyanobacteria and microalgae (Kirst et al., 2014; Kumar et al., 2021; Hitchcock et al., 2022), while their successful implementation in plants remains challenging. New synthetic biology technologies will facilitate the introduction of alternative antennae in plants. Nevertheless, their impact on photoprotection and the regulation of photosynthetic electron transport might hinder their use to increase stress resilience and plant yield under fluctuating environments. Photoprotection is essential for the survival

TABLE 1.1
Enhancing Light Absorption through Designing New Antenna Systems and Photoprotection

	Article	Strategy	Studied Condition	Stress Tolerance Increased	Evaluation of Yield Impact	Species
Improving light harvesting	Pettigrew et al. (1989)	Decreased antenna size due to chlorophyll deficiency	Field conditions	Not assessed	Increase in photosynthetic rates	*Glycine max*
	Kirst et al. (2017)	Decreased antenna size due to chlorophyll deficiency	Greenhouse conditions	Not assessed	Decrease in photosynthetic rates but increase in biomass accumulation	*Nicotiana tabacum*
	Song et al. (2017)	Modeling decreased antenna size due to chlorophyll deficiency	Modeling inputs were measured under field conditions	Not assessed	Increase in canopy photosynthesis and nitrogen use efficiency	*Oryza sativa*
	Mao et al. (2023)	Decreased antenna size due to chlorophyll deficiency	Field conditions	Not assessed	Decrease in photosynthetic rates. No increase in biomass observed	*O. sativa*

(Continued)

TABLE 1.1 (*Continued*)
Enhancing Light Absorption through Designing New Antenna Systems and Photoprotection

	Article	Strategy	Studied Condition	Stress Tolerance Increased	Evaluation of Yield Impact	Species
Manipulating photoprotection	De Souza et al. (2022)	Accelerating NPQ recovery	Fluctuating light – field conditions	Not assessed	Seed yield increase	*G. max*
	Kromdijk et al. (2016)	Accelerating NPQ recovery	Fluctuating light – field and greenhouse conditions	Not assessed	Biomass increase	*N. tabacum*
	Hubbart et al. (2018)	Increasing NPQ by overexpressing PsbS	Fluctuating light – greenhouse conditions	Not assessed	Increase in biomass and grain yield	*O. sativa*
	Garcia-Molina and Leister (2020)	Accelerating NPQ recovery	Fluctuating light and abiotic stress – greenhouse and growth cabinet conditions	No-transgenic lines slightly under-performed	Growth rate and biomass accumulation decrease	*Arabidopsis thaliana*
	Lehretz et al. (2022)	Accelerating NPQ recovery	Fluctuating light and drought stress – greenhouse and growth cabinet conditions	No – acceleration of induction and relaxation of NPQ does not translate into higher stress tolerance. Drought tolerance was not increase	No improvement under control conditions, and tuber yield decreased under fluctuating light	*Solanum tuberosum (potato)*
	Tognetti et al. (2006)	Expressing cyanobacterial flavodoxin	Multiple stresses – high irradiation and temperature, drought	Yes	No	*N. tabacum*
	Tognetti et al. (2007)	Expressing cyanobacterial flavodoxin	Iron starvation	Yes	Biomass increases under stress	*N. tabacum*
	Li et al. (2017)	Expressing cyanobacterial flavodoxin	Oxidative stress, drought, high temperature, N starvation	Yes	Biomass decrease	*Agrostis stolonifera* (creeping bentgrass)
	Yamamoto et al. (2016)	Expressing cyanobacterial flavodoxin	Fluctuating light – growth cabinet conditions	Yes	Not assessed	*A. thaliana*
	Messant et al. (2023)	PTOX overexpression	High light	Yes-Photoprotection of PSI	Not assessed	*A. thaliana and Marchantia polymorpha*
	Calzadilla et al. (2024)	PTOX overexpression	High light	Yes-Photoprotection of PSII	Not assessed	*A. thaliana*

and growth of plants in natural conditions (Suorsa et al., 2012; Kromdijk et al., 2016; De Souza et al., 2022) and needs to be integral to the design of any synthetic biology light-harvesting systems.

1.2.1.2 Non-Photochemical Quenching

When plants experience a change in their environment, e.g., a sudden increase in light intensity induced by changes in cloud cover or a transition from shade to sun, a phenomenon termed non-photochemical quenching (NPQ) is induced. This process is measured using the technique of chlorophyll fluorescence analysis (Maxwell and Johnson, 2000) and reflects an increase in the conversion of absorbed light energy to heat with a concomitant decrease in the efficiency of exciton transfer to reaction centers. This reduces the "excitation pressure" on the reaction centers, mitigates over-reduction of electron carriers, and therefore protects against oxidative damage. Several molecular processes are known to contribute to NPQ, each having different activation and deactivation kinetics (Müller et al., 2001). In most plants, the main (and fastest) process involved is called high-energy state quenching (qE; Walters and Horton, 1991).

The molecular steps involved in qE have been studied over many years, resulting in some detailed models for the physical processes involved; however, two clear biochemical regulators are recognized (Müller et al., 2001). (1) Generation of qE involves the production of the xanthophyll zeaxanthin through the reversible two-step de-epoxidation of violaxanthin in the xanthophyll cycle. De-epoxidation is catalyzed by the enzyme violaxanthin de-epoxidase (VDE), while the reverse reaction is catalyzed by zeaxanthin epoxidase (ZEP). (2) qE is promoted by the protonation of PSII subunit S (PsbS), a polypeptide with homology to the LHC proteins. Both zeaxanthin formation and PsbS protonation are driven by the light-dependent acidification of the thylakoid lumen (ΔpH formation; Müller et al., 2001). Following a transition from dark to light, qE typically takes 5–10 minutes to be fully induced, with a similar time required for its relaxation. However, the kinetics of these reactions and their final extent vary substantially depending on species, physiological status, and environmental conditions (Johnson et al., 1993; Demmig-Adams et al., 2014).

In the last decade, different studies have focused on how alterations in qE kinetics could increase photosynthetic efficiency and crop yields in naturally fluctuating environments (Table 1.1). Hubbart et al. (2018) showed that overexpressing PsbS increases biomass and grain yield in rice plants, due to higher NPQ under fluctuating light conditions. Surprisingly, this increase in NPQ was not accompanied by a decrease in PSII electron transport rate, even though these processes should be in competition. It was suggested that the better performance of the transformed plants was related to reduced photo-oxidative stress (Hubbart et al., 2018). The kinetics of NPQ induction and relaxation were not analyzed with a high time resolution.

It has been shown that inefficiencies in switching off qE can lead to significant yield loss in fluctuating conditions, due to energy being lost as heat following transitions to low-light (Kromdijk et al., 2016). Kromdijk et al. (2016) produced tobacco plants which overexpressed both VDE and ZEP, in addition to PsbS (termed VPZ plants). These had accelerated activation and deactivation of qE following changes in light intensity, leading to higher integrated CO_2 fixation efficiency and an approximately 15% biomass increase under field conditions. Similar results were observed in soybean, where seed yield increased up to 33%, without affecting seed quantity, protein, or oil content (De Souza et al., 2022). However, there was a year-on-year variation in the impact of this alteration in soybeans (De Souza et al., 2022). Moreover, studies characterizing VPZ in *Arabidopsis* and potato plants showed no improvement in photosynthesis and an impairment in growth rate and biomass accumulation under both greenhouse and fluctuating light conditions (Garcia-Molina and Leister, 2020; Lehretz et al., 2022; Table 1.1). When the latter were tested under abiotic stress conditions, no improvement in their stress tolerance was observed, when compared to their wild type.

The contrasting effect of VPZ transformation in different plant species might be related to their specific trade-off between photoprotection and productivity (Leister, 2023). Higher NPQ is expected to result in less photodamage (higher photoprotection) at the expense of photochemistry and vice versa. Depending on the species-specific limitations on NPQ, the VPZ transformation may result in more or the same NPQ capacity as wild-type plants. Similarly, there is substantial natural variation in the kinetics of both induction and relaxation of NPQ (Johnson et al., 1993), and so the transformation of different species will impact this process differently.

The factors underlying the variation in NPQ kinetics between species are poorly understood but likely include differences in the rate of generation of a ΔpH across the thylakoid membrane. ΔpH generation depends on the regulation of electron transport and the kinetics of induction and deactivation of the CBB cycle, which will also impact the effectiveness of the VPZ transformation in a species-specific way. VPZ overexpression has also been reported to modify hormone levels (Kaiser et al., 2019; Lehretz et al., 2022), leading to alterations in plant growth, independently of their photosynthetic performance. The specific molecular mechanism through which VPZ plants enhance plant yield needs to be further elucidated before this approach can be routinely used to increase productivity in crop species.

1.2.1.3 Alternative Electron Transport Pathways

A different approach to modulating photoprotection involves the introduction of alternative electron transport routes (Table 1.1). These alternative pathways can divert electrons away from the PETC, reducing the probability of ROS formation. Alternative electron sinks have been proposed, which take electrons from either PSI or PSII. In plants, the terminal electron acceptor in the PETC is ferredoxin, which takes electrons from PSI and transfers them to NADP, a reaction catalyzed by ferredoxin NADP reductase. Photosynthetic bacteria and some algal species possess alternative terminal electron acceptors, which provide protection against stress (Zhang et al., 2009; Chaux et al., 2017). Flavodoxins and flavodiiron proteins are PSI electron acceptors, absent in angiosperms, and heterologous expression of these can increase stress tolerance in different plant species (Tognetti et al., 2006; Yamamoto et al., 2016; Li et al., 2017).

The heterologous expression of flavodoxin as an electron sink of PSI was characterized in tobacco and creeping bentgrass (*Agrostis stolonifera*; Li et al., 2017; Tognetti et al., 2006, 2007; Table 1.1). Tognetti et al. (2007) observed an increase in fresh weight and photosynthetic rates of tobacco overexpressing flavodoxin under iron deprivation conditions. No phenotypical differences were observed between transgenic and wild-type plants under normal growth conditions. Flavodoxin expression protects photosynthesis by maintaining electron flux through the PETC and helping ensure redox homeostasis in the chloroplast, when ferredoxin is lacking under iron deficiency (Tognetti et al., 2007). These plants also showed higher photosynthetic rates under a wide variety of stress conditions, although in these cases their impact on plant yield was not addressed (Tognetti et al., 2006; Table 1.1). The expression of flavodoxin in creeping bentgrass increases their performance under stress (Li et al., 2017; Table 1.1); however, these plants showed retardation in plant development and reproduction and a decrease in total biomass accumulation under control conditions when compared with the wild-type plants.

Similar to flavodoxin for PSI, the plastid terminal oxidase (PTOX) has been shown to act as an electron sink for PSII in some extremophile species (Streb et al., 2005; Stepien and Johnson, 2009; Savitch et al., 2010). PTOX has for many years been seen as an enigma in photosynthesis research (Nawrocki et al., 2015; Johnson and Stepien, 2016; Krieger-Liszkay and Feilke, 2016), as simple overexpression of the protein does not lead to significant activity or provide any photoprotection (Rosso et al., 2006; Heyno et al., 2009). However, Calzadilla et al. (2024) have recently confirmed PTOX's potential as a safety valve for PSII, showing that structural rearrangements of the thylakoid membrane during plant acclimation to stress can act as a regulatory step in its activation. These chloroplast ultrastructural modifications are sufficient to give rise to substantial rates of PTOX activity, with this activity being photoprotective. A role for PTOX in PSI photoprotection was also suggested in *Arabidopsis* and *Marchantia polymorpha*, under high light conditions (Messant et al., 2023). This photoprotective effect was related to PTOX's ability to oxidize the PQ pool, increase the donor-side limitation of PSI, and, as such, decrease the reduction of $P700^+$ under high light.

Based on these studies, PTOX can be used as a strategy to re-design photosynthesis and increase stress tolerance in plants. Nevertheless, data on its impact on plant yield are still currently missing (Table 1.1).

Altering the redox state of the PETC has been shown to affect gene expression and acclimation responses in plants, with the redox state of the PQ pool or of ferredoxin, linked to thioredoxin (Thormählen et al., 2017), both implicated in photosynthetic acclimation to the environment (Pfannschmidt et al., 1999; Frigerio et al., 2007). Oxidation of PQH_2 by PTOX may be expected to alter this signaling pathway though, to our knowledge, this has not been reported. Expressing flavodoxin, however, was shown to trigger changes in gene expression in creeping bentgrass (Li et al., 2017). This may explain the increased stress tolerance observed in these plants but also their growth impairment. Furthermore, the pleiotropic impact of flavodoxin expression on other chloroplast redox-regulated pathways cannot be excluded (Li et al., 2017). Since these phenotypical differences were not observed in the flavodoxin-expressing tobacco (Tognetti et al., 2006, 2007), further studies should be performed to address the impact of diverting electrons downstream of PSI and their consequences on chloroplast redox homeostasis.

In addition to the linear electron transport pathways described above, where electrons from water are passed through PSII and onto CO_2 or alternative electron acceptors, there is electron flow involving only PSI – cyclic electron flow (CEF; Johnson, 2011; Nawrocki et al., 2019). CEF is crucial not only for ensuring a balance between NADPH and ATP production by electron transport but also for generating the ΔpH needed for NPQ. As such, CEF plays a vital role both in optimizing photosynthesis and protection against environmental challenges, making it a potential target of manipulation to increase crop yields and stress tolerance.

In CEF, electrons passing through PSI are returned to the PETC, via one of at least two pathways. A protein complex, often termed the NDH complex due to its homology with mitochondrial complex I, functions as a ferredoxin plastoquinone oxidoreductase (Ma et al., 2021b). Only very low activity of this complex has been reported in C3 plants, though this may play an important role in C4 plants (Ishikawa et al., 2016; Ogawa et al., 2023). There is limited evidence that plants lacking the complex have compromised productivity under stress conditions (Munné-Bosch et al., 2005). The second pathway, sometimes referred to as the FQR (or the antimycin-sensitive) pathway, has not been fully resolved but involves the proteins PGR5 and PGRL1 (Munekage et al., 2002; DalCorso et al., 2008; Ma et al., 2021a). Knocking out this protein has a major impact on NPQ and the regulation of the ETC and results in plants having extreme sensitivity to environmental fluctuations (Gollan et al., 2017). However, whether this pathway is suitable for manipulation is unclear. Overexpression of the PGR5 protein was shown to have a negative effect on plant growth (Okegawa et al., 2007).

1.2.2 Improving Photosynthesis by Increasing CO_2 Availability

The efficiency of carbon fixation through the CBB cycle depends not only on the kinetics of the enzymes involved but also on the availability of CO_2. In C3 plants, this is limited by the gas diffusion rates from the bulk atmosphere, via the stomata and intercellular airspace of the leaves to the mesophyll chloroplast stroma, where it is fixed into an organic form by Rubisco. Limitations in the supply of CO_2 result in a restriction to the maximum rate of carbon fixation and an increase in the rate of the alternative Rubisco-catalyzed reaction, oxygenation, leading to a pathway termed photorespiration (see Section 1.2.3). Thus, increasing CO_2 availability for Rubisco carboxylase activity is a suitable strategy to maximize CO_2 fixation. This is being attempted by altering the behavior of stomata to improve CO_2 entry into the leaf, changing leaf architecture to improve CO_2 diffusion within a leaf, and inducing carbon-concentrating mechanisms that actively increase the CO_2 concentration around Rubisco (Table 1.2).

TABLE 1.2
Improving Photosynthesis by Increasing CO_2 Availability

	Article	Strategy	Studied Condition	Stress Tolerance Increased	Evaluation of Yield Impact	Species
Modulating stomata dynamics	Eyland et al. (2021)	Studying stomatal conductance (g_s) kinetics	Growth cabinet and greenhouse conditions – fluctuating light conditions	Not assessed	Stomatal conductance (g_s) kinetics limits photosynthetic capacity and water use efficiency (WUE)	*Musa* spp. (Banana)
	Faralli et al. (2022)	Exploring natural variation in stomatal dynamics	Growth cabinet and greenhouse conditions – heat stress	Stomatal dynamics (g_s) linked to higher heat stress tolerance	Stomatal dynamics (g_s) linked to higher WUE	*Vitis vinifera* (grapevine)
	Yang et al. (2023)	Characterizing stomatal behavior in response to drought priming	Drought stress and stress recovery	Faster stomatal dynamics associated with stress tolerance	Faster stomatal dynamics associated with higher photosynthetic rates and biomass	wheat *(Triticum aestivum)*
	Ozeki et al. (2022)	Dynamic photosynthesis model to evaluate the contribution of stomatal dynamics on photosynthesis and WUE	Growth cabinet conditions – fluctuating light. High and low nitrogen	Not assessed	Accelerating stomatal closure enhances WUE	C3 and C4 *Poaceae* species
	McAusland et al. (2016)	Characterizing photosynthesis and stomatal behavior under fluctuating light	Greenhouse conditions – fluctuating light conditions	Not assessed	Improving the rapidity of stomatal responses leads to improved productivity and intrinsic water use efficiency ($_i$WUE)	C3 and C4 species
	Kimura et al. (2020)	Characterization of stomatal mutants	Growth cabinet conditions – fluctuating and constant light	Not assessed	Faster stomatal dynamics lead to higher photosynthetic rates and biomass, with similar WUE than wild-type plants	*Arabidopsis thaliana*

(Continued)

TABLE 1.2 (*Continued*)
Improving Photosynthesis by Increasing CO_2 Availability

	Article	Strategy	Studied Condition	Stress Tolerance Increased	Evaluation of Yield Impact	Species
Improving mesophyll conductance (g_m)	Yamori et al. (2020)	Characterization of stomatal mutant	Growth cabinet conditions – fluctuating light	Not assessed	Increased stomatal conductance leads to a higher rate of photosynthetic induction	*O. sativa*
	Papanatsiou et al. (2019)	Expression of light-gated K^+ channel BLINK1 to enhance stomatal aperture	Growth cabinet conditions – fluctuating light	Not assessed	Faster stomatal dynamics increase photosynthetic rates and biomass and improve WUE	*A. thaliana*
	Horaruang et al. (2022)	Altering the regulation of the GORK K^+ channel to promote its activity	Growth cabinet conditions – fluctuating light	Not assessed	Faster stomatal dynamics increase growth and improve WUE	*A. thaliana*
	Hanba et al. (2004)	Overexpression of the barley aquaporin HvPIP2;1	Growth cabinet conditions	Not assessed	Higher CO_2 diffusion leads to higher CO_2 assimilation rates	*O. sativa*
	Flexas et al. (2006)	Overexpression of tobacco aquaporin NtAQP1	Growth cabinet and greenhouse conditions	Not assessed	Higher CO_2 diffusion leads to higher CO_2 assimilation rates	*Nicotiana tabacum*
	Gómez-Ocampo et al. (2021)	Overexpression of AtBBX21 transcription factor	Greenhouse conditions – water restriction conditions	Higher tolerance to water restrictions	Increase in photosynthetic rates and yield, improve WUE	*Solanum tuberosum*
	Lehmeier et al. (2017)	Altering the expression of cell cycle genes in the leaf mesophyll	Growth cabinet conditions	Not assessed	Changes in leaf airspace pattern and cell density lead to higher photosynthetic capacity	*A. thaliana*
	Atkinson et al. (2016)	Expression of Ci transporters LCIA and HLA3 (CCM from *Chlamydomonas reinhardtii*)	Ambient and low CO_2	Not assessed	No phenotypical difference with plants	*A. thaliana*

(*Continued*)

TABLE 1.2 (*Continued*)
Improving Photosynthesis by Increasing CO_2 Availability

	Article	Strategy	Studied Condition	Stress Tolerance Increased	Evaluation of Yield Impact	Species
Introducing carbon concentration mechanisms (CCMs)	Simkin et al. (2015)	Expression of cyanobacterial putative-inorganic carbon transporter B (*ictB*)	Greenhouse conditions	Not assessed	Photosynthetic rates and biomass increase	*N. tabacum*
	Lieman-Hurwitz et al. (2003)	Expression of cyanobacterial putative-inorganic carbon transporter B (*ictB*)	Limiting CO_2 concentration and low humidity	Not assessed	Photosynthetic rates increase	*A. thaliana* and *N. tabacum*
	Hay et al. (2017)	Expression of cyanobacterial putative-inorganic carbon transporter B (*ictB*)	Field and greenhouse conditions; drought; ambient and high CO_2	Yes – photosynthetic rates and biomass increase	Photosynthetic rates, biomass, and seed mass increase	*G. max*
	Gong et al. (2015)	Expression of cyanobacterial putative-inorganic carbon transporter B (*ictB*)	Field conditions	Not assessed	Photosynthetic rates and biomass increase	*O. sativa*
	Yang et al. (2008)	Expression of cyanobacterial putative-inorganic carbon transporter B (*ictB*)	Field and greenhouse conditions	Not assessed	Photosynthetic rates, biomass, and grain yield increase	*O. sativa*
	Chen et al. (2023)	Introduction of proteobacterium carboxysomes to plants	Growth cabinet conditions – ambient and high CO_2 concentrations	Not assessed	Slower growth at high CO_2, no growth at ambient CO_2	*N. tabacum*
	Lin et al. (2014)	Expression of cyanobacterial Rubisco with internal carboxysomal protein CcmM35	Ambient and high CO_2	Not assessed	Slower growth at high CO_2, no growth at ambient CO_2	*N. tabacum*

1.2.2.1 Modulating Stomata Dynamics

The relationship between stomatal responses (and morphology) and photosynthetic capacity has recently been explored as a strategy to improve crop performance (Lawson and Vialet-Chabrand, 2019; Harrison et al., 2020). Although the factors limiting photosynthesis can vary depending on the environmental condition or due to stress, stomatal conductance (g_s) has been identified as a major limitation for photosynthesis (Wang et al., 2022b). Since stomatal responses are slower than photosynthetic ones (Lawson and Blatt, 2014; McAusland et al., 2016), efforts have been made to study and characterize g_s kinetics to enhance plant yield and stress resilience. Stomata controls not only gaseous fluxes into the leaf (i.e. CO_2 uptake) but also water transpiration rates. Therefore, identifying faster g_s responses can also improve water use efficiency (WUE).

Fast stomatal dynamics were associated with better thermoregulation and WUE in grapevine and banana (Eyland et al., 2021; Faralli et al., 2022). In grapevine, a reduced stomatal response to changes in vapor pressure deficit can improve leaf cooling, leading to higher photosynthetic performance under heat stress (Faralli et al., 2022). This is predicated however on the assumption that plants are in a well-watered environment, which may well not be the case. In more arid environments, the loss of water from the leaves would result in a water deficit, severely impacting water availability if g_s does not respond accordingly. In this regard, faster stomatal closure under drought stress and faster re-opening during post-drought recovery were reported to be part of drought priming in wheat, linking stomatal responses to stress tolerance in these plants (Yang et al., 2023).

Plant productivity has been enhanced by faster stomatal responses in several plant species, under fluctuating light conditions (McAusland et al., 2016; Kimura et al., 2020; Yamori et al., 2020; Faralli et al., 2022; Ozeki et al., 2022). A study comparing C3 and C4 grasses showed that rapid stomatal closure under fluctuating light contributes to higher WUE in C4 crops when compared to C3 (Ozeki et al., 2022). These stomatal dynamics were influenced by stomata morphology and nitrogen availability, the contribution of the latter still being unclear. Stomatal morphology can impact gas exchange through their shape, size, density, or patterning (Harrison et al., 2020). Smaller and denser stomata were associated with faster stomatal responses, faster photosynthetic induction, and higher biomass accumulation under fluctuating light in *Arabidopsis* and rice (Sakoda et al., 2020; Xiong et al., 2022).

Accelerating stomatal closure under high-to-low-light transitions results in significant savings in water and a 40% increase in grain yield under drought stress in rice (Qu et al., 2020). This phenotype was strongly correlated with a single Na^+/H^+ tonoplastic antiporter (OsNHX1), showing the relevance of genetic variability in stomatal dynamics traits and the significant contribution of ion channels. Stomatal movements are highly coordinated with the transport of ions across the membranes of the guard cells (Jezek and Blatt, 2017). This ion traffic generates changes in water fluxes, which trigger turgor changes that open and close the pore. Recently, using optogenetics and synthetic biology approaches, Papanatsiou et al. (2019) introduced a blue-light responsive ion channel (BLINK1) into *Arabidopsis* stomata. BLINK1 enhanced K^+ conductance in the guard cells, accelerating stomatal opening and closure under fluctuating light. Plant overexpressing BLINK1 showed a more than twofold increase in biomass, under water-replete and water-deficit conditions, when compared with the wild type (Papanatsiou et al., 2019). Similar results were observed by Horaruang et al. (2022), when the regulation of the GORK K^+ channel was impaired to promote its activity.

1.2.2.2 Improving Mesophyll Conductance (g_m)

In addition to stomatal conductance and kinetics, mesophyll conductance (g_m) limits photosynthesis (Flexas et al., 2012; Lundgren and Fleming, 2020), with this limitation being as high as 35% during photosynthetic induction (Knauer et al., 2022; Liu et al., 2022). The parameter g_m comprises the structural and biochemical barriers that define the diffusion of gases within the leaf (Ellsworth et al., 2018; Liu et al., 2022). Since a higher CO_2 diffusion rate will increase CO_2 availability for the CBB cycle, modifying g_m has been proposed to enhance photosynthesis (Lundgren and Fleming, 2020). g_m has been positively correlated with photosynthetic capacity in species like cassava (*Manihot esculenta*), wheat, rice, and soybean (Barbour et al., 2016; Tomeo and Rosenthal, 2017; De Souza et al., 2020). However, due to the complexity of the genetic factors regulating CO_2 diffusion, the identification of useful g_m traits to improve photosynthetic efficiency has been hindered.

Mesophyll conductance is a dynamic trait and shows large plasticity in response to nutrient availability and environmental stress factors (Niinemets et al., 2009; Buckley and Warren, 2014). Nutrient deficiencies were reported to decrease photosynthesis due to a decrease in g_m (Xiong et al., 2015; Lu et al., 2016; Xie et al., 2020). Under such conditions, mesophyll conductance is mainly impaired by changes in anatomy and CO_2 membrane barriers, probably related to the synthesis cost of enzymes or membrane proteins. In this regard, nitrogen availability has been linked to fast g_m responses to environmental fluctuations due to aquaporin expression (Xiong et al., 2015). Under conditions of nitrogen or potassium deficiency, alterations of chloroplast cellular location, number, or structure also contribute to the g_m limitation of photosynthesis (Lu et al., 2016; Xie et al., 2020).

g_m can be affected by combined drought and heat stress, CO_2 concentration, and fluctuating light (Flexas et al., 2007; Campany et al., 2016; Zeng et al., 2023). Several studies address the contribution of g_m to drought acclimation in plants (Cano et al., 2014; Han et al., 2016; Olsovska et al., 2016; Ouyang et al., 2017). Its significance in regulating forest ecosystem productivity under drought has also been discussed (Keenan et al., 2010). Increasing g_m without affecting g_s can enhance photosynthesis without affecting WUE, making the uncoupling of g_m and g_s a valuable trait for breeding productive drought-resistant crops (Flexas et al., 2013). However, the relationship between g_m and g_s seems to be species dependent. While g_m is positively correlated with g_s in soybean (Tomeo and Rosenthal, 2017), no correlation of these parameters was found in wheat (Barbour et al., 2016).

Although the advantages of engineering mesophyll conductance to enhance photosynthesis might be clear, the genetic strategies to achieve an increased g_m are still poorly defined. A simple approach is to overexpress aquaporins to increase CO_2 diffusion across membranes as was attempted in rice and tobacco plants (Hanba et al., 2004; Flexas et al., 2006; Table 1.2). Overexpressing these membrane proteins increased internal CO_2 conductance and assimilation rates in both species. However, Hanba et al. (2004) reported that the overexpression of the barley aquaporin HvPIP2;1 also increases susceptibility to water stress due to higher transpiration rates. This agrees with Aharon et al. (2003), showing that the overexpression of PIP1b in tobacco enhances their susceptibility to drought and salt stress. A more complex approach will be to overexpress transcription factors controlling developmental networks (i.e. BBX21; Gómez-Ocampo et al., 2021), or cell cycle genes aiming to modify leaf architecture (Lehmeier et al., 2017). Although these approaches enhanced photosynthetic capacity through alterations of mesophyll morphology, the pleiotropic effects related to altering developmental patterns cannot be ignored and should be considered before these strategies are broadly used in crop species. In conclusion, it should be possible to increase g_m without negatively impacting overall water relations and therefore stress sensitivity, it is not yet clear how this can be achieved across different species. For a detailed review of how to enhance photosynthesis by manipulating mesophyll conductance, see Ren et al. (2019).

1.2.2.3 Introducing Carbon Concentration Mechanisms

Evolution has shaped different strategies to increase CO_2 availability for photosynthesis. In C4 and crassulacean acid metabolism (CAM) plants, the initial reaction of CO_2 capture is catalyzed not by Rubisco but by the enzyme phosphoenolpyruvate carboxylase (PEPC). Unlike Rubisco, PEPC is not sensitive to O_2. PEPC fixes HCO_3^- to form oxaloacetate, as the first step of a series of reactions to transport and/or store carbon, typically as malate (Edwards, 2019). Malate is converted back to pyruvate and CO_2 by NAD(P)-malic enzyme, CO_2 then being recaptured in the CBB cycle. While C4 metabolism separates PEPC and Rubisco activity spatially (between the mesophyll and bundle sheaths cells), CAM plants do this temporally (with the former being active at night and the latter during the daytime) (Edwards, 2019). CAM is primarily an adaptation to dry environments, with plants able to close their stomata during the day to minimize transpiration, while C4 is particularly seen as an adaptation to high temperatures.

The strategies for carbon fixation seen in C4 and CAM plants are examples of so-called inorganic carbon concentrating mechanisms (CCMs). Efforts have been made to engineer some of these CCM traits into C3 species, particularly the ones related to C4 photosynthesis (Schuler et al., 2016; Kubis and Bar-Even, 2019). Probably the highest profile is the *C4 Rice Project*, a global consortium aiming to introduce C4 traits in rice to increase photosynthetic efficiency (https://c4rice.com/). Although challenging, due to differences in leaf anatomy and morphology, biochemical pathways, and metabolite transport; the main enzymes of the C4 metabolism are already present in C3 plants and similar carbon shuttling mechanisms already exist (Schuler et al., 2016; Kubis and Bar-Even, 2019). Advances in synthetic biology and the increased number of C4 species established as model organisms for research may help shed light on this pathway (Hibberd et al., 2008; Schuler et al., 2016).

The introduction of so-called C2 CCMs traits, a form of C3–C4 intermediate with fewer leaf anatomical requirements, has been suggested as a simpler alternative to this strategy (Bellasio and Farquhar, 2019; Lundgren, 2020). In C2 plants, the enzyme glycine decarboxylase is localized to the bundle sheath. As a result, the decarboxylation step in photorespiration is concentrated within this tissue, allowing the plants to generate a locally high CO_2 concentration.

In addition to higher plants, other photosynthetic organisms possess CCMs to enhance the catalytic activity of Rubisco (Raven et al., 2008). These involve biophysical CCMs, which instead of concentrating CO_2 through organic compounds such as malate, concentrate HCO_3^- close to clustered Rubisco (Hennacy and Jonikas, 2020). The introduction of these biophysical CCMs – carboxysomes and pyrenoids (from cyanobacteria and green algae, respectively) – is being attempted to improve photosynthetic yields in C3 plants (Hanson et al., 2016; Hennacy and Jonikas, 2020; Adler et al., 2022). Significant results were recently obtained toward achieving this aim, through the assembly and condensation of CCMs in *Arabidopsis* and tobacco (Atkinson et al., 2020; Chen et al., 2023). Nevertheless, further optimization of these strategies is required before the complete introduction of CCMs succeeds in enhancing photosynthesis in plants (Hanson et al., 2016; Adler et al., 2022).

The expression of a putative cyanobacterial inorganic carbon transporter B (ictB) has been shown to enhance photosynthetic capacity and biomass in different plant species (Table 1.2). Although the precise role of ictB is unclear, its functionality in carbon uptake under ambient atmospheric conditions has been reported (Price et al., 2008). Transgenic ictB soybean plants showed higher photosynthetic rates and biomass accumulation under drought stress conditions, which was related to a higher CO_2 concentration around Rubisco even under conditions where stomatal conductivity (g_s) should be low (Hay et al., 2017). Interestingly, a higher g_m was also observed in ictB plants (Gong et al., 2015; Hay et al., 2017), which could justify their higher CO_2 availability for carboxylation and their higher photosynthetic rates.

Although technically challenging, engineering CO_2 concentration mechanisms seems to be a suitable strategy to tackle the objective of increasing both stress tolerance and photosynthesis in plants, thereby increasing yield across a range of conditions. There are, however, metabolic costs associated with these mechanisms, which may affect yield gains in some environments. Furthermore, the kinetic properties of photosynthetic enzymes and their activation states (particularly under stress conditions, e.g. heat) can still limit photosynthesis. In the following section, we will address current approaches to increasing photosynthetic capacity through manipulating carbon metabolism.

1.2.3 Re-designing Photosynthesis through Fine-Tuning Carbon Metabolic Reactions

1.2.3.1 Enhancing Rubisco Carboxylase Capacity and Activation State

Rubisco (EC4.1.1.39) catalyzes the primary photosynthetic CO_2 capture reaction, between CO_2 and ribulose-1,5-bisphosphate (RuBP) to form two molecules of 3-phosphoglycerate (3PG; Andersson and Backlund, 2008). This is the first reaction of the CBB cycle and a significant regulatory point for photosynthesis (Amaral et al., 2024). Rubisco is an inefficient enzyme with a slow catalytic rate and with a carboxylase activity in competition with a side reaction that binds O_2 instead of CO_2 in its active site. Consequently, C3 plants typically contain high Rubisco concentrations to achieve a high photosynthetic capacity (Sage, 2002).

Considering that Rubisco catalytic properties are a major limitation for photosynthesis, Rubisco becomes a relevant target for improving photosynthesis (Parry et al., 2013; Table 1.3). Overexpressing Rubisco has been suggested as a strategy to enhance photosynthesis in plants (Suzuki et al., 2009; Yoon et al., 2020; Suganami et al., 2021); however, an increase in carbon fixation and biomass accumulation was only observed under high nitrogen concentrations (Yoon et al., 2020; Suganami et al., 2021), since high Rubisco content compromises nitrogen availability for other metabolic processes. An alternative strategy involves increasing Rubisco K_{cat} and carboxylation to oxygenation ratio (Carmo-Silva et al., 2015). Increasing K_{cat} would require less investment in Rubisco content to support photosynthesis, leading to a higher NUE and availability (Zhu et al., 2007; Carmo-Silva et al., 2015).

TABLE 1.3
Re-designing Photosynthesis through Fine-Tuning Carbon Metabolic Reactions

	Article	Strategy	Studied Condition	Stress Tolerance Increased	Evaluation of Yield Impact	Species
Enhancing Rubisco carboxylase capacity and activation state	Suzuki et al. (2009)	Over-expression of Rubisco to increase Rubisco content	Greenhouse conditions	Not assessed	No increase in photosynthetic rates or biomass due to downregulation of Rubisco activation state	*Oryza sativa*
	Yoon et al. (2020)	Overexpression of RBCS to increase Rubisco content	Field conditions	Not assessed	Increase in photosynthetic rates and biomass accumulation with N fertilization	*O. sativa*
	Suganami et al. (2021)	Co-overexpression of both Rubisco and Rubisco activase	Growth cabinet conditions	Not assessed	Co-overexpression increases photosynthetic rates at moderately high temperatures	*O. sativa*
	Bi et al. (2017)	Overexpression of Rubisco activase	Growth cabinet and greenhouse conditions – low temperature and weak light stress	Increase tolerance to low temperature and weak light stress	Increase in photosynthetic rates and biomass accumulation	Cucumber (*Cucumis sativus*)
	Yamori et al. (2012)	Overexpression of Rubisco activase	Different temperature conditions	Not assessed	Higher photosynthetic rates under high temperature	*O. sativa*
	Fukayama et al. (2012)	Overexpression of Rubisco activase	Greenhouse conditions	Not assessed	Photosynthetic rates decrease	*O. sativa*
	Kumar et al. (2009)	Overexpression of thermostable Rubisco activase isoform	Heat stress	Yes	Increase in photosynthetic rates, biomass, and seed yield	*Arabidopsis thaliana*
	Kurek et al. (2007)	Overexpression of thermostable Rubisco activase isoform	Heat stress	Yes	Photosynthetic rates, biomass, and seed yield increase	*A. thaliana*
	Scafaro et al. (2018)	Overexpression of thermostable Rubisco activase isoform	Heat stress	Yes	Faster light induction of photosynthesis, growth, and yield when exposed to heat stress (45°C)	*O. sativa*

(Continued)

TABLE 1.3 (*Continued*)
Re-designing Photosynthesis through Fine-Tuning Carbon Metabolic Reactions

	Article	Strategy	Studied Condition	Stress Tolerance Increased	Evaluation of Yield Impact	Species
Re-engineering photorespiration	Carmo-Silva and Salvucci (2013)	Removing the redox-regulated inhibition of the α isoform of Rubisco activase	Growth cabinet – fluctuating light conditions	Not assessed	Higher photosynthetic induction and growth under fluctuating light	*A. thaliana*
	Harvey et al. (2022)	Reducing the redox-regulated α RCA isoform content	Growth cabinet and field conditions	Not assessed	No differences in photosynthetic induction rates either growth	*Glycine max*
	Gunn et al. (2020)	Heterologous expression of *Rhodobacter sphaeroides* Rubisco and Rubisco activase	Growth cabinet conditions	Not assessed	Higher photosynthetic rates and growth at elevated CO_2	*Nicotiana tabacum*
	López-Calcagno et al. (2018)	Accelerating photorespiration	Field and greenhouse conditions	Yes – under high photorespiratory conditions (low CO_2 and continuous light)	Increase in photosynthetic rates and biomass accumulation	*N. tabacum*
	Timm et al. (2015)	Accelerating photorespiration	Different photoperiod conditions	Not assessed	Photosynthetic rates increase. Biomass increase under short days	*A. thaliana*
	Timm et al. (2012)	Accelerating photorespiration	Growth cabinet conditions	Not assessed	Increase in photosynthetic rates and biomass accumulation	*A. thaliana*
	Kebeish et al. (2007)	Alternative photorespiratory pathway	Growth cabinet conditions	Not assessed	Increase in photosynthetic rates and biomass accumulation	*A. thaliana*
	Maier et al. (2012)	Alternative photorespiratory pathway	Growth cabinet conditions	Not assessed	Increase in photosynthetic rates and biomass accumulation	*A. thaliana*
	Shen et al. (2019)	Alternative photorespiratory pathway	Greenhouse and field conditions; high light	Higher photosynthetic capacity at higher light intensities	Photosynthetic rates, biomass, and nitrogen content increase. Seed yield increase under certain conditions	*O. sativa*

(*Continued*)

TABLE 1.3 (*Continued*)
Re-designing Photosynthesis through Fine-Tuning Carbon Metabolic Reactions

	Article	Strategy	Studied Condition	Stress Tolerance Increased	Evaluation of Yield Impact	Species
Modulating metabolic fluxes in the CBB cycle	Wang et al. (2020)	Alternative photorespiratory pathway	Greenhouse and field conditions	Not assessed	Increase in photosynthetic rates, growth, and grain yield	*O. sativa*
	Nayak et al. (2022)	Alternative photorespiratory pathway	Greenhouse conditions	Not assessed	Increase in photosynthetic rates, growth, and grain yield	*O. sativa*
	Dalal et al. (2015)	Alternative photorespiratory pathway	Growth cabinet conditions	Not assessed	Increase in photosynthetic rates, growth, and seed yield	*Camelina sativa*
	South et al. (2019)	Alternative photorespiratory pathway	Greenhouse and field conditions	Not assessed	Increase in photosynthetic rates and biomass accumulation	*N. tabacum*
	Yao et al. (2022)	Alternative photorespiratory pathway	Greenhouse conditions	Not assessed	Increase in photosynthetic rates, growth, and yield	*Solanum tuberosum*
	Chen et al. (2019)	Alternative photorespiratory pathway	Greenhouse conditions	Not assessed	Increase in photosynthetic rates, growth, and seed yield in a low CO_2 environment	*Cucumis sativus*
	Cavanagh et al. (2022)	Alternative photorespiratory pathway	Greenhouse and field conditions – heat stress	Yes	Increase in photosynthetic rates and biomass accumulation	*N. tabacum*
	Tamoi et al. (2006)	Overexpression of enzymes of the CBB (FBPase and SBPase)	Growth cabinet conditions	Not assessed	Increase in photosynthetic rates and biomass accumulation	*N. tabacum*
	Driever et al. (2017)	Overexpression of the CBB enzyme SBPase	Greenhouse conditions	Not assessed	Increase in photosynthetic rates, biomass, and seed yield	*Triticum aestivum*
	Lefebvre et al. (2005)	Overexpression of the CBB enzyme SBPase	Growth cabinet and greenhouse conditions	Not assessed	Photosynthetic rates and biomass increase under growth cabinet conditions	*N. tabacum*
	Uematsu et al. (2012)	Overexpression of the CBB enzyme FBPase	Ambient and high CO_2	Not assessed	Increase in photosynthetic rates and biomass accumulation	*N. tabacum*
	Simkin et al. (2015)	Overexpression of enzymes of the CBB (FBPase and SBPase)	Greenhouse conditions	Not assessed	Increase in photosynthetic rates and biomass accumulation	*N. tabacum*

(*Continued*)

TABLE 1.3 (*Continued*)
Re-designing Photosynthesis through Fine-Tuning Carbon Metabolic Reactions

	Article	Strategy	Studied Condition	Stress Tolerance Increased	Evaluation of Yield Impact	Species
	Gong et al. (2015)	Overexpression of enzymes of the CBB (FBPase and SBPase)	Field conditions	Not assessed	Increase in photosynthetic rates and biomass accumulation	*O. sativa*
	Rosenthal et al. (2011)	Overexpression of the CBB enzyme SBPase	Field conditions under elevated CO_2	Not assessed	Increase in photosynthetic rates and biomass accumulation at elevated CO_2	*N. tabacum*
	Ding et al. (2016)	Overexpression of the CBB enzyme SBPase	Growth cabinet and greenhouse conditions – chilling stress	Yes	Increase in photosynthetic rates and biomass accumulation	*Solanum lycopersicum*
	Feng et al. (2007a)	Overexpression of the CBB enzyme SBPase	Greenhouse conditions – salt stress	Yes	Increase in photosynthetic rates and biomass accumulation	*O. sativa*
	Feng et al. (2007b)	Overexpression of the CBB enzyme SBPase	Greenhouse conditions – heat stress	Yes	Increase in photosynthetic rates and biomass accumulation	*O. sativa*
	López-Calcagno et al. (2020)	Co-expression of cyanobacterial bifunctional FBP/SBPase or SBPase, together with cytochrome c_6	Greenhouse and field conditions	Not assessed	Increase in photosynthetic rates, biomass, and iWUE	*N. tabacum*

Rubisco catalytic turnover rate and substrate specificity are generally tightly linked, with specificity achieved at a cost to rate, hindering the possibility of molecular improvement (South et al., 2018). Hence, while attempts are being made to introduce Rubisco with a high K_{cat} into plants, this is generally part of a strategy that is combined with the introduction of a CCM (Lin et al., 2014; Orr et al., 2020). New isoforms of Rubisco with better catalytic properties are being explored, from wild and ancestral varieties of crops as well as from other photosynthetic organisms (Lin et al., 2022; Orr et al., 2016; Prins et al., 2016; Chen et al. 2023; Table 1.3). Directed evolution has also been used to search for improved Rubisco isoforms, with the generation of mutant libraries and their corresponding screening for improvements in catalytic properties (Mueller-Cajar and Whitney, 2008; Zhou and Whitney, 2019).

Several abiotic stresses can trigger inhibition of photosynthesis due to a reduction in the activation state of Rubisco, induced by inhibition of its active site or a decrease in its protein concentration (Galmés et al., 2013). Reactivation of Rubisco active sites requires the participation of Rubisco activase (RCA). Two RCA isoforms have been described in plants, α and β. These differ in size, with the former being subject to redox regulation, via reduction/oxidation of a disulfide bond (Portis, 2003). Moreover, RCA is a thermolabile protein whose activity decreases with temperature. RCA overexpression leads to increased photosynthetic rates under high and low temperatures in rice and cucumber, respectively (Bi et al., 2017; Scafaro et al., 2018; Table 1.3). However, a negative correlation between RCA and Rubisco content was reported, which can hinder the improvement of photosynthesis induced by RCA overexpression (Fukayama et al., 2012, 2018). Suganami et al. (2020) recently showed that this correlation also depends on nitrogen availability and that the negative effect of RCA overexpression on Rubisco content is reduced with fertilization.

Increasing RCA thermostability enhances photosynthesis and growth under heat stress in *Arabidopsis* (Kurek et al., 2007; Kumar et al., 2009; Table 1.3). These plants achieve higher seed yield than wild types when exposed to heat stress. Similarly, the overexpression of a thermotolerant RCA variant in rice increases light induction of photosynthesis and seed yield at 45°C growth temperature (Scafaro et al., 2018). A different approach in engineering RCA was performed by Carmo-Silva and Salvucci (2013), removing the redox-regulated inhibition of the α isoform in *Arabidopsis*. These plants showed a faster photosynthetic induction but had reduced growth compared to the wild type. They did, however, grow better under fluctuating than continuous irradiance, implying a benefit of faster induction in natural environments. By contrast RCA-α knockdown lines in soybean showed no improvement in either photosynthetic activity or growth, when compared to their respective wild types (Harvey et al., 2022).

The differences observed between Carmo-Silva and Salvucci (2013) and Harvey et al. (2022) can be accounted for by a different RCA isoform ratio between species, soybean expressing less RCA-α than *Arabidopsis* (Harvey et al., 2022). A better understanding of RCA thermo-sensitivity and regulatory pathways, under different environmental conditions and in different plant species, should be developed for further implementation of this approach (Wijewardene et al., 2021; Sparrow-Muñoz et al., 2023). Moreover, transformations of both Rubisco and its cognate RCA should be considered, since it was shown that simultaneous expression of these proteins can further enhance photosynthetic capacity (Gunn et al., 2020). Although there remain several challenges, engineering Rubisco and RCA seems to be a promising approach to enhance plant yield, reducing the impact of environmental stress, particularly at non-optimal temperatures.

1.2.3.2 Re-engineering Photorespiration

Photorespiration is a metabolic pathway triggered by an oxygenase reaction occurring instead of carboxylation at the active site of Rubisco (Peterhansel and Maurino, 2011). The carboxylase and oxygenase capacity of this enzyme compete, but under certain circumstances (e.g. drought or heat stress), the oxygenase activity becomes dominant, leading to the production of 2-phosphoglycolate (2PG), a toxic byproduct. A coordinated network of enzymatic reactions is required for 2PG recycling, with the peroxisomes, mitochondria, and chloroplasts all being involved (Smith et al., 2023). Photorespiration is a highly energy-demanding process, and a major cause of carbon loss in plants, making it a suitable target for improvement (Hagemann and Bauwe, 2016). Photorespiration has been suggested to play an important role in photoprotection under stress conditions (Eisenhut et al., 2017; Shi et al., 2022; Villalobos-González et al., 2022).

In the absence of mechanisms to suppress the oxygenase reaction of Rubisco (e.g. CCMs or improving Rubisco specificity), it is essential that any 2PG produced is recycled to minimize carbon loss and avoid the accumulation of toxic molecules. There is potential for increasing 2PG recycling to minimize carbon loss and avoid toxic effects. This includes accelerating overall flux (Timm et al., 2015; López-Calcagno et al., 2019) or bypassing photorespiratory metabolic reactions (Maier et al., 2012; Shen et al., 2019; Table 1.3). Optimizing photorespiration will also accelerate carbon recapture through the regeneration of RuBP, leading to higher plant productivity (Maier et al., 2012; López-Calcagno et al., 2019).

Overexpressing proteins of the glycine cleavage system, needed for recycling glycolate to CO_2, increased photosynthetic rates and plant growth in *Arabidopsis* and tobacco plants (Timm et al., 2012, 2015; López-Calcagno et al., 2019). In *Arabidopsis*, the overexpression of glycine decarboxylase (GDC) was correlated with decreased concentration of photorespiratory metabolic intermediates, which inhibit the CBB cycle (Timm et al., 2012, 2015). In tobacco, the acceleration of photorespiration decreased photoinhibition of PSII under conditions of low CO_2 and continuous light (López-Calcagno et al., 2019). Although overexpressing the L- and H-proteins of GDC led to photosynthetic improvement, regulation of their expression levels is required, since high levels of expression result in growth impairment in tobacco (López-Calcagno et al., 2019). Alteration in carbon allocation between starch and soluble sugars was also reported in the same study.

Several strategies for bypassing photorespiratory metabolic reactions have been gaining attention to increase carbon assimilation in plants, most of them involving modifications of the glycolate catabolic pathway (Wang et al., 2022a). Moreover, with the use of modeling approaches, the potential for applying synthetic biology to develop new metabolic routes and the potential impact of these *in planta* can be analyzed (Basler et al., 2016; Khurshid et al., 2020; Osmanoglu et al., 2021). Studies have shown that bypassing photorespiration increases photosynthesis and biomass production not only in *Arabidopsis* (Kebeish et al., 2007; Maier et al., 2012) but in different crop species (Dalal et al., 2015; Chen et al., 2019; Shen et al., 2019; South et al., 2019; Nayak et al., 2022; Yao et al., 2022).

Increasing fluxes through the photorespiratory pathway, if they increase the overall capacity for metabolism to act as a sink for photosynthesis, are expected to reduce vulnerability to environmental stress, especially when that stress is associated with decreased entry of CO_2 into the leaf. Indeed, in some cases, there is evidence that modifying photorespiration leads to increased stress tolerance (e.g. heat stress and high light; Shen et al., 2019; Cavanagh et al., 2022) and higher seed yield (Dalal et al., 2015; Wang et al., 2020; Nayak et al., 2022). In rice, higher seed yield was accompanied by better grain quality (Zhang et al., 2022), although with a decrease in seed setting rate (Shen et al., 2019; Wang et al., 2020). The latter was suggested to be related to alterations in the source-to-sink flow, due to increased photosynthetic rates without a concomitant increase in sugar transport (Wang et al., 2020). This emphasizes the need to consider strategies for improving photosynthesis in combination (see Section 1.3).

1.2.3.3 Modulating Metabolic Fluxes in the CBB Cycle

Several enzymes of the CBB cycle have been reported to be limiting for carbon fixation, particularly sedoheptulose-1,7-bisphosphatase (SBPase) and fructose-1,6-bisphosphatase (FBPase; Raines et al., 2022). These potential metabolic bottlenecks were identified using modeling approaches and metabolic control analysis, which allow the study of the role of individual enzymes in controlling overall fluxes (Pettersson and Ryde-Pettersson, 1988; Poolman et al., 2000). Predictions were confirmed using knockdown transgenic plants with reduced expression of single enzymes of the CBB cycle, which were impaired in carbon fixation (Harrison et al., 2001; reviewed by Raines, 2003). Consequently accelerating RuBP regeneration by overexpressing these enzymes becomes suitable for enhancing photosynthesis (Raines, 2003).

Overexpression of SBPase or FBPase alone increases photosynthetic capacity and growth in several plant species (Lefebvre et al., 2005; Tamoi et al., 2006; Rosenthal et al., 2011; Uematsu et al., 2012; Driever et al., 2017; Table 1.3), with a positive impact in grain yield reported for wheat (Driever et al., 2017). Moreover, increasing SBPase content also increases salt and high temperature resilience in rice (Feng et al., 2007b, 2007a) and improves chilling tolerance in tomato plants (Ding et al., 2016). The latter was related to reduced chilling-induced oxidative stress in the SBPase-overexpressing plants, due to higher photosynthetic activity and metabolic flux (Ding et al., 2017).

Simultaneous transgenic manipulation of multiple enzymes of the CBB cycle has also been performed, leading to further photosynthetic improvement (Simkin et al., 2015). The effect of this is however species-specific, since the positive effect of co-overexpressing SBPase and FBPase in tobacco was not reproduced in *Arabidopsis* (Simkin et al., 2017). Other strategies involve the co-overexpression of enzymes of the CBB cycle, together with proteins aiming to stimulate electron transport (López-Calcagno et al., 2020), increase CO_2 availability (Gong et al., 2015; Simkin et al., 2015), or accelerate photorespiration (Simkin et al., 2017). In all cases, a synergistic effect on photosynthetic capacity was observed due to co-expression. Some of these strategies involved the heterologous expression of cyanobacterial photosynthetic proteins (Gong et al., 2015; Simkin et al., 2015; López-Calcagno et al., 2020).

Although increasing the abundance of CBB cycle enzymes has yielded promising results toward improving yield and stress resilience, these approaches were mainly performed in C3 plants. Recently, Ermakova et al. (2023) demonstrated that overexpressing SBPase in the bundle sheath of a C4 model species *Setaria vidiris* does not enhance photosynthetic capacity, probably due to differences between C3 and C4 metabolism. In C4 plants, carbon metabolism is distributed between different cell types (mesophyll and bundle sheaths), involving different shuttles and metabolite transporters, while in C3 all carbon metabolism occurs in similar cell types. These differences might limit the effect of overexpressing SBPase in increasing carbon fluxes and overall photosynthetic capacity (Ermakova et al., 2023). Many important crops are C4 plants (e.g. maize, sorghum, and sugarcane), and, therefore, further research is needed to address the impact of modulating metabolic fluxes through the CBB cycle in these species. Moreover, the effect of overexpressing CBB cycle enzymes will be affected by developmental and environmental conditions (Zhao et al., 2021), hindering the possibility of generalizing their impact on photosynthesis.

Overall, strategies intended to increase the capacity of the CBB cycle are likely to have a positive impact on both yield and stress tolerance, where stresses result in limitations of carbon assimilation or excesses in electron transport. As with altering primary reactions of CO_2 fixation, there remains a need to ensure that different reactions are kept in balance across a range of environmental conditions.

1.2.4 Mining Natural Photosynthetic Variability to Identify Higher-Yield Photosynthetic Traits

Nature offers a plethora of adaptative traits that can be exploited for crop improvement, making germplasm banks a vital resource for the discovery of new physiological processes and the acceleration of crop breeding programs (Sharwood et al., 2022). The selection of new materials from crop wild relatives has been advanced with the development of 'omic technologies' (from transcriptomics to phenomics) and the analysis of high-throughput data, aiming to increase our understanding of the relationship between genotype and phenotype (van Bezouw et al., 2019; Calzadilla et al., 2022). Genome-wide association studies (GWAS) allow the identification of the genetic architecture of a desirable trait/s in a given population (Korte and Farlow, 2013).

GWAS has been successfully used in the search for novel photosynthetic genes involved in NPQ (Wang et al., 2017; Rungrat et al., 2019), phosphorus deficiency responses (Lü et al., 2018), and even stomatal dynamics (Pignon et al., 2021; Table 1.4). GWAS was also employed in the search for 'stay-green' traits (Zhao et al., 2019). Stay-green phenotypes can maintain carbon assimilation and photosynthetic capacity, delaying the senescence process under stress conditions (Thomas

TABLE 1.4
Mining Natural Photosynthetic Variability to Identify Higher-Yield Photosynthetic Traits

Article	Strategy	Studied Condition	Stress Tolerance Increased	Evaluation of Yield Impact	Species
Orr et al. (2016)	Exploring Rubisco thermotolerance	Greenhouse conditions	Not assessed	Not assessed	Plant species from different taxa
Prins et al. (2016)	Exploring Rubisco thermotolerance	Greenhouse conditions	Not assessed	Potential increase in photosynthetic rates (modeling results)	*Triticum aestivum*
Lin et al. (2022)	Exploring ancient Rubisco from Solanaceae family	In silico analysis and in vitro biochemical characterization	Not assessed	Not assessed	–
Christopher et al. (2016)	Exploring stay-green traits	Field conditions – water stress environments	Not assessed	Stay-green traits correlated with higher yield	*T. aestivum*
Zhang et al. (2019)	Exploring stay-green traits – QTL analysis	Field conditions	Not assessed	Biomass, seed yield, and nitrogen accumulation increase	*Zea mays*
Zhao et al. (2019)	Exploring stay-green traits – GWAS	Field conditions	Not assessed	Not assessed	*Oryza sativa*
Qu et al. (2020)	Exploring genes related to stomatal dynamics – GWAS	Field and growth cabinet conditions – drought stress	Yes	Biomass and grain yield increase under drought	*O. sativa*
Pignon et al. (2021)	Exploring genes related to stomatal dynamics – GWAS/TWAS	Greenhouse conditions	Not assessed	Not assessed	*Sorghum bicolor*
Barbour et al. (2016)	Exploring natural variation in mesophyll conductance (g_m)	Growth cabinet conditions	Not assessed	Mesophyll conductance positively correlated with photosynthetic rates	*T. aestivum*
Tomeo and Roshental (2017)	Exploring natural variation in mesophyll conductance (g_m)	Field and growth cabinet conditions	Not assessed	Mesophyll conductance positively correlated with photosynthetic rates	*Glycine max*
Wang et al. (2017)	Exploring non-photochemical quenching capacity – GWAS	Field conditions	Not assessed	Not assessed	*O. sativa*
Rungrat et al. (2019)	Exploring non-photochemical quenching capacity – GWAS	Growth cabinet conditions	Not assessed	Not assessed	*Arabidopsis thaliana*

(Continued)

TABLE 1.4 (*Continued*)
Mining Natural Photosynthetic Variability to Identify Higher-Yield Photosynthetic Traits

Article	Strategy	Studied Condition	Stress Tolerance Increased	Evaluation of Yield Impact	Species
Lü et al. (2018)	Exploring the photosynthetic response to P stress – GWAS	Growth cabinet conditions	Not assessed	Not assessed	*Glycine max*
Mantilla-Perez et al. (2020)	Exploring leaf angle as a smart canopy trait – GWAS	Field conditions	Not assessed	Not assessed	*S. bicolor*
Natukunda et al. (2022)	Exploring leaf angle as a smart canopy trait – RNAseq	Greenhouse conditions	Not assessed	Not assessed	*S. bicolor*

and Ougham, 2014). Thus, stay-green plants of certain crops (e.g. maize and wheat) have longer grain-filling and, consequently, higher yields (Christopher et al., 2016; Zhang et al., 2019). Since nitrogen allocation and remobilization can impact grain yield, nitrogen availability must be considered when growing stay-green varieties (Liu et al., 2021b). Furthermore, the photosynthetic capacity of the leaves is tightly linked to the sink-source nitrogen balance of the entire plant, and delayed leaf senescence can impact this relationship (Hikosaka, 2005).

Mining photosynthetic variability also involves the identification of traits that can enhance photosynthetic capacity in the entire cultivation system, a concept defined as a 'smart canopy'. Smart canopies refer to a group of plants that interact cooperatively in the field to maximize their light harvesting and biomass production capacity (Ort et al., 2015). This implies changes in the architecture, physiology, or biochemistry of the plant to adjust to the different microclimates inside the canopy (Song et al., 2016). Leaf angle is one of the most discussed traits in the generation of smart canopies, since it defines light interception, photosynthetic capacity, and, therefore, crop productivity (Mantilla-Perez and Salas Fernandez, 2017).

Ideally, in a smart canopy, leaves will be more vertical in the upper canopy but horizontal in the lower canopy, allowing a better distribution of sunlight and ensuring high photosynthesis in the entire plant. The search for genes controlling leaf angle has been challenging since this characteristic is controlled by multiple loci. Nevertheless, the utilization of technologies such as RNAseq, GWAS, and 3D canopy architecture acquisition is helping to identify the quantitative trait loci (QTL) contributing to this trait (Liu et al., 2021a; Mantilla-Perez et al., 2020; Natukunda et al., 2022). Although promising for increasing yield, the relationship between smart canopies and stress responses has not to our knowledge been addressed. The use of systems biology and new modeling approaches will allow a better understanding of the impact of 're-designing' canopies in fluctuating environments (Zhu et al., 2012; Song et al., 2016; Rodrigues et al., 2021).

1.3 BALANCING YIELD AND RESILIENCE IN RE-DESIGNING PHOTOSYNTHESIS

As discussed in the preceding sections, the objective of enhancing photosynthesis as a solution to increasing food security is being explored extensively. The various strategies proposed will have implications for stress tolerance, which have been considered to different degrees in different studies. At the same time, there has been discussion from the alternative direction, about how we might enhance the stress resilience of photosynthesis, minimizing reactive oxygen production and photodamage. Photosynthetic ROS production lies at the heart of plant stress responses and any manipulation of the photosynthetic apparatus is therefore liable to impact the stress resilience of the plant. Of the strategies outlined in this review (Figure 1.1), some are explicitly linked to stress responses, others less directly, but all need to be viewed through the lens of how they impact plants' tolerance to a fluctuating environment.

Many strategies can be seen as dual purpose in terms of productivity and stress resilience. C4 metabolism is an adaptation that increases both the potential rate of photosynthesis, by increasing Rubisco's substrate concentration, and WUE. Accelerating NPQ induction and relaxation allows plants to maintain an optimal level of heat dissipation across changing environments, while increasing integrated carbon fixation. However, it must be remembered that each strategy comes at a metabolic cost and what works out as a net gain in one environment may not be of benefit in another (De Souza et al., 2020). For example, the cost of accelerating NPQ may not be recovered by the benefits of accelerating its relaxation (i.e. increased carbon capture). Moreover, while VPZ plants showed high yield under fluctuating light conditions (Kromdijk et al., 2016; De Souza et al., 2022), improved performance was not observed under other environmental constraints (e.g. drought; Garcia-Molina and Leister, 2020; Lehretz et al., 2022; Table 1.1). This approach may therefore need tuning to both the crop species and the local environment. Selection of appropriate promoters may also be important in ensuring that genes of interest are only expressed where and when they are beneficial.

Other interventions aiming to increase photosynthesis may not have obvious direct effects on stress tolerance but may impact them indirectly (either positively or negatively). For example, attempts to enhance the CBB cycle may increase photosynthesis; however, this may need to be accompanied by changes to downstream metabolism. Many stress conditions can result in sink limitation, resulting in the accumulation of photosynthate in the leaf, which in turn causes feedback inhibition of photosynthesis. Such an effect not only means that gains in photosynthesis are not realized but may also induce ROS production and leaf senescence (Kumar et al., 2023). Leaf metabolic balance is also known to be important in regulating the acclimation responses of plants (Saunders et al., 2022; Gjindali and Johnson, 2023), so changing the rate of primary carbon fixation may result in unforeseen and suboptimal acclimation responses.

A particular stress that can hinder the benefits obtained by re-designing photosynthesis is nutrient deficiency. Some of the strategies previously discussed demand high nitrogen availability to enhance photosynthetic capacity. For example, Rubisco overexpression leads to increased carbon fixation only under non-limiting nitrogen concentrations (Yoon et al., 2020; Suganami et al., 2021), while, to overcome the negative effect of overexpressing RCA on Rubisco, nitrogen is also required (Suganami et al., 2020). Moreover, g_s and g_m properties will also depend on nitrogen and potassium levels (Xiong et al., 2015; Lu et al., 2016; Ozeki et al., 2022). Given that, in the wider context of climate change, there is a need to reduce agricultural inputs (Mahmud et al., 2021); the use of strategies of photosynthetic re-design requiring high nutrient availability may not be a suitable solution to tackle the challenges facing food production.

The simultaneous application of multiple approaches to re-design photosynthesis might be essential for the generation of high-yield resilient plants. Although not tested under stress, synergistic effects on yield were observed when accelerating the CBB flux together with increased capacity for photorespiration, photosynthetic electron transport, or increases in CO_2 concentration, giving higher rates of carboxylation (Gong et al., 2015; Simkin et al., 2015, 2017; López-Calcagno et al., 2020). New combinations of strategies can be implemented to fine-tune the entire photosynthetic pathway, targeting both photoprotection and carbon fixation (e.g. the introduction of alternative electron pathways and new RCA isoforms). In this regard, the development of the CRISPR-Cas9 technology is a promising resource for plant transformation, particularly in crops and non-model organisms (Schindele et al., 2018).

Several challenges remain in the field of re-designing photosynthesis. First, many of the strategies discussed in this review involve the incorporation or manipulation of multigenic (i.e. QTL) traits, such as CCM strategies or g_m characteristics. In addition to the technical difficulties, the lack of knowledge of the genes regulating some of these traits hinders their use to improve photosynthesis. Second, it was demonstrated that the strategies that worked in one species cannot simply be assumed to work in another (e.g. accelerating NPQ or overexpressing enzymes of the CBB cycle). Thus, further studies are needed to understand the broad impact of engineering the photosynthetic pathway in different plant species. In this regard, the differences between photosynthetic pathways (C3 vs C4) or plant functional types need to be considered, in terms of physiology, anatomy, and biochemistry (Knauer et al., 2022; Ermakova et al., 2023).

1.4 CONCLUSION – IS THERE AN OPTIMAL SOLUTION TO BALANCE YIELD AND STRESS TOLERANCE IN A CLIMATE CHANGE SCENARIO?

Although considerable progress has been made in recent years, a suitable and unique strategy to increase both yield and stress tolerance by re-designing photosynthesis remains elusive. The impact of the different strategies can be species- or stress-dependent, increasing the complexity of finding an optimal solution. However, based on the approaches discussed above, fine-tuning carbon metabolic reactions seems to be the most promising for improving photosynthetic capacity in a fluctuating environment. Accelerating carbon metabolic flux has successfully enhanced photosynthetic capacity and stress resilience in several plant species under different conditions, while other approaches have only been shown to increase one or the other (e.g. NPQ acceleration or introducing alternative electron pathways). Moreover, improving stomatal dynamics and mesophyll conductance is tightly linked to the physiology, anatomy, and morphology of each species and, as such, is difficult to consider as a general strategy. For now, other approaches, like improving antenna systems, introducing Carbon Concentration Mechanisms (CMM)s, or designing smart canopies, require further investigation before their implementation leads to photosynthetic improvement in plants. The development of novel in silico techniques and genetic transformation technologies, together with the simultaneous combination of different strategies to re-design photosynthesis, may pave the way for the generation of improved crops to face the current climate crisis.

REFERENCES

Adler, L., Díaz-Ramos, A., Mao, Y., Pukacz, K. R., Fei, C., & McCormick, A. J. (2022). New horizons for building pyrenoid-based CO_2-concentrating mechanisms in plants to improve yields. *Plant Physiology, 190*, 1609–1627.

Aharon, R., Shahak, Y., Wininger, S., Bendov, R., Kapulnik, Y., & Galili, G. (2003). Overexpression of a plasma membrane aquaporin in transgenic tobacco improves plant vigor under favorable growth conditions but not under drought or salt stress. *Plant Cell, 15*, 439–447.

Ahmed, S., Griffin, T. S., Kraner, D., Schaffner, M. K., Sharma, D., Hazel, M., Leitch, A. R., Orians, C. M., Han, W., & Stepp, J. R. (2019). Environmental factors variably impact tea secondary metabolites in the context of climate change. *Frontiers in Plant Science, 10*, 939.

Amaral, J., Lobo, A. K. M., & Carmo-Silva, E. (2024). Regulation of Rubisco activity in crops. *New Phytologist, 241*, 35–51.

Andersson, I., & Backlund, A. (2008). Structure and function of Rubisco. *Plant Physiology and Biochemistry, 46*, 275–291.

Asseng, S., Martre, P., Maiorano, A., Rötter, R. P., O'Leary, G. J., Fitzgerald, G. J., Girousse, C., Motzo, R., Giunta, F., & Babar, M. A. (2019). Climate change impact and adaptation for wheat protein. *Global Change Biology, 25*, 155–173.

Atkinson, N., Feike, D., Mackinder, L. C. M., Meyer, M. T., Griffiths, H., Jonikas, M. C., Smith, A. M., & McCormick, A. J. (2016). Introducing an algal carbon-concentrating mechanism into higher plants: Location and incorporation of key components. *Plant Biotechnology Journal, 14*, 1302–1315.

Atkinson, N., Mao, Y., Chan, K. X., & McCormick, A. J. (2020). Condensation of Rubisco into a proto-pyrenoid in higher plant chloroplasts. *Nature Communications, 11*, 6303.

Aydinsakir, K., Erdal, S., Buyuktas, D., Bastug, R., & Toker, R. (2013). The influence of regular deficit irrigation applications on water use, yield, and quality components of two corn (Zea mays L.) genotypes. *Agricultural Water Management*, *128*, 65–71.

Barbour, M. M., Bachmann, S., Bansal, U., Bariana, H., & Sharp, P. (2016). Genetic control of mesophyll conductance in common wheat. *New Phytologist*, *209*, 461–463.

Basler, G., Küken, A., Fernie, A. R., & Nikoloski, Z. (2016). Photorespiratory bypasses lead to increased growth in Arabidopsis thaliana: Are predictions consistent with experimental evidence? *Frontiers in Bioengineering and Biotechnology*, *4*, 31.

Bellasio, C., & Farquhar, G. D. (2019). A leaf-level biochemical model simulating the introduction of C2 and C4 photosynthesis in C3 rice: Gains, losses and metabolite fluxes. *New Phytologist*, *223*, 150–166.

van Bezouw, R. F. H. M., Keurentjes, J. J. B., Harbinson, J., & Aarts, M. G. M. (2019). Converging phenomics and genomics to study natural variation in plant photosynthetic efficiency. *The Plant Journal*, *97*, 112–133.

Bi, H., Liu, P., Jiang, Z., & Ai, X. (2017). Overexpression of the rubisco activase gene improves growth and low temperature and weak light tolerance in Cucumis sativus. *Physiologia Plantarum*, *161*, 224–234.

Blankenship, R. E., & Chen, M. (2013). Spectral expansion and antenna reduction can enhance photosynthesis for energy production. *Current Opinion in Chemical Biology*, *17*, 457–461.

Buckley, T. N., & Warren, C. R. (2014). The role of mesophyll conductance in the economics of nitrogen and water use in photosynthesis. *Photosynthesis Research*, *119*, 77–88.

Calzadilla, P. I., Carvalho, F. E. L., Gomez, R., Neto, M. C. L., & Signorelli, S. (2022). Assessing photosynthesis in plant systems: A cornerstone to aid in the selection of resistant and productive crops. *Environmental and Experimental Botany*, *201*, 104950.

Calzadilla, P. I., Song, J., Gallois, P., & Johnson, G. N. (2024). Proximity to Photosystem II is necessary for activation of Plastid Terminal Oxidase (PTOX) for photoprotection. *Nature Communicaions*, *15*, 287.

Campany, C. E., Tjoelker, M. G., von Caemmerer, S., & Duursma, R. A. (2016). Coupled response of stomatal and mesophyll conductance to light enhances photosynthesis of shade leaves under sunflecks. *Plant Cell & Environment*, *39*, 2762–2773.

Candogan, B. N., Sincik, M., Buyukcangaz, H., Demirtas, C., Goksoy, A. T., & Yazgan, S. (2013). Yield, quality and crop water stress index relationships for deficit-irrigated soybean [Glycine max (L.) Merr.] in sub-humid climatic conditions. *Agricultural Water Management*, *118*, 113–121.

Cano, F. J., Lopez, R., & Warren, C. R. (2014). Implications of the mesophyll conductance to CO_2 for photosynthesis and water-use efficiency during long-term water stress and recovery in two contrasting Eucalyptus species. *Plant Cell & Environment*, *37*, 2470–2490.

Carmo-Silva, E., & Salvucci, M. E. (2013). The regulatory properties of rubisco activase differ among species and affect photosynthetic induction during light transitions. *Plant Physiology*, *161*, 1645–1655.

Carmo-Silva, E., Scales, J. C., Madgwick, P. J., & Parry, M. A. J. (2015). Optimizing Rubisco and its regulation for greater resource use efficiency. *Plant Cell & Environment*, *38*, 1817–1832.

Cavanagh, A. P., South, P. F., Bernacchi, C. J., & Ort, D. R. (2022). Alternative pathway to photorespiration protects growth and productivity at elevated temperatures in a model crop. *Plant Biotechnology Journal*, *20*, 711–721.

Chaloner, T. M., Gurr, S. J., & Bebber, D. P. (2021). Plant pathogen infection risk tracks global crop yields under climate change. *Nature Climate Change*, *11*, 710–715.

Chaux, F., Burlacot, A., Mekhalfi, M., Auroy, P., Blangy, S., Richaud, P., & Peltier, G. (2017). Flavodiiron proteins promote fast and transient O_2 photoreduction in Chlamydomonas. *Plant Physiology*, *174*, 1825–1836.

Chen, M., & Blankenship, R. E. (2011). Expanding the solar spectrum used by photosynthesis. *Trends in Plant Science*, *16*, 427–431.

Chen, T., Hojka, M., Davey, P., Sun, Y., Dykes, G. F., Zhou, F., Lawson, T., Nixon, P. J., Lin, Y., & Liu, L.-N. (2023). Engineering α-carboxysomes into plant chloroplasts to support autotrophic photosynthesis. *Nature Communications*, *14*, 2118.

Chen, Z., Kang, X., Nie, H., Zheng, S., Zhang, T., Zhou, D., Xing, G., & Sun, S. (2019). Introduction of exogenous glycolate catabolic pathway can strongly enhances photosynthesis and biomass yield of cucumber grown in a low-CO_2 environment. *Frontiers in Plant Science*, *10*, 702.

Christopher, J. T., Christopher, M. J., Borrell, A. K., Fletcher, S., & Chenu, K. (2016). Stay-green traits to improve wheat adaptation in well-watered and water-limited environments. *Journal of Experimental Botany*, *67*, 5159–5172.

Dalal, J., Lopez, H., Vasani, N. B., Hu, Z., Swift, J. E., Yalamanchili, R., Dvora, M., Lin, X., Xie, D., & Qu, R. (2015). A photorespiratory bypass increases plant growth and seed yield in biofuel crop Camelina sativa. *Biotechnology for Biofuels*, *8*, 1–22.

DalCorso, G., Pesaresi, P., Masiero, S., Aseeva, E., Schünemann, D., Finazzi, G., Joliot, P., Barbato, R., & Leister, D. (2008). A complex containing PGRL1 and PGR5 is involved in the switch between linear and cyclic electron flow in Arabidopsis. *Cell*, *132*, 273–285.

DaMatta, F. M., Grandis, A., Arenque, B. C., & Buckeridge, M. S. (2010). Impacts of climate changes on crop physiology and food quality. *Food Research International*, *43*, 1814–1823.

Demmig-Adams, B., Koh, S.-C., Cohu, C. M., Muller, O., Stewart, J. J., & Adams, W. W. (2014). Non-photochemical fluorescence quenching in contrasting plant species and environments. In B. Demmig-Adams, G. Garab, W. Adams III, & G. Govindjee (Eds.), *Non-Photochemical quenching and energy dissipation in plants, algae and cyanobacteria* (pp 531–552). Springer.

Dhankher, O. P., & Foyer, C. H. (2018). Climate resilient crops for improving global food security and safety. *Plant Cell & Environment*, *41*, 877–884.

Ding, F., Wang, M., & Zhang, S. (2017). Overexpression of a Calvin cycle enzyme SBPase improves tolerance to chilling-induced oxidative stress in tomato plants. *Scientia Horticulturae*, *214*, 27–33.

Ding, F., Wang, M., Zhang, S., & Ai, X. (2016). Changes in SBPase activity influence photosynthetic capacity, growth, and tolerance to chilling stress in transgenic tomato plants. *Scientific Reports*, *6*, 32741.

Driever, S. M., Simkin, A. J., Alotaibi, S., Fisk, S. J., Madgwick, P. J., Sparks, C. A., Jones, H. D., Lawson, T., Parry, M. A. J., & Raines, C. A. (2017). Increased SBPase activity improves photosynthesis and grain yield in wheat grown in greenhouse conditions. *Philosophical Transactions of the Royal Society B*. https://doi.org/10.1098/rstb.2016.0384

Edwards, E. J. (2019). Evolutionary trajectories, accessibility and other metaphors: The case of C4 and CAM photosynthesis. *New Phytologist*, *223*, 1742–1755.

Eisenhut, M., Bräutigam, A., Timm, S., Florian, A., Tohge, T., Fernie, A. R., Bauwe, H., & Weber, A. P. M. (2017). Photorespiration is crucial for dynamic response of photosynthetic metabolism and stomatal movement to altered CO_2 availability. *Molecular Plant*, *10*, 47–61.

Ellsworth, P. V., Ellsworth, P. Z., Koteyeva, N. K., & Cousins, A. B. (2018). Cell wall properties in Oryza sativa influence mesophyll CO_2 conductance. *New Phytologist*, *219*, 66–76.

Ermakova, M., Lopez-Calcagno, P. E., Furbank, R. T., Raines, C. A., & von Caemmerer, S. (2023). Increased sedoheptulose-1,7-bisphosphatase content in Setaria viridis does not affect C4 photosynthesis. *Plant Physiology*, *191*, 885–893.

Eskling, M., Arvidsson, P., & Åkerlund, H. (1997). The xanthophyll cycle, its regulation and components. *Physiologia Plantarum*, *100*, 806–816.

Espasandin, F. D., Calzadilla, P. I., Maiale, S. J., Ruiz, O. A., & Sansberro, P. A. (2018). Overexpression of the arginine decarboxylase gene improves tolerance to salt stress in Lotus tenuis plants. *Journal of Plant Growth Regulation*, *37*, 156–165.

Eyland, D., van Wesemael, J., Lawson, T., & Carpentier, S. (2021). The impact of slow stomatal kinetics on photosynthesis and water use efficiency under fluctuating light. *Plant Physiology*, *186*, 998–1012.

Faralli, M., Bontempo, L., Bianchedi, P. L., Moser, C., Bertamini, M., Lawson, T., Camin, F., Stefanini, M., & Varotto, C. (2022). Natural variation in stomatal dynamics drives divergence in heat stress tolerance and contributes to seasonal intrinsic water-use efficiency in Vitis vinifera (subsp. sativa and sylvestris). *Journal of Experimental Botany*, *73*, 3238–3250.

Feng, L., Han, Y., Liu, G., An, B., Yang, J., Yang, G., Li, Y., & Zhu, Y. (2007a). Overexpression of sedoheptulose-1,7-bisphosphatase enhances photosynthesis and growth under salt stress in transgenic rice plants. *Functional Plant Biology*, *34*, 822–834.

Feng, L., Wang, K., Li, Y., Tan, Y., Kong, J., Li, H., Li, Y., & Zhu, Y. (2007b). Overexpression of SBPase enhances photosynthesis against high temperature stress in transgenic rice plants. *Plant Cell Reports*, *26*, 1635–1646.

Flexas, J., Barbour, M. M., Brendel, O., Cabrera, H. M., Carriquí, M., Díaz-Espejo, A., Douthe, C., Dreyer, E., Ferrio, J. P., & Gago, J. (2012). Mesophyll diffusion conductance to CO_2: An unappreciated central player in photosynthesis. *Plant Science*, *193*, 70–84.

Flexas, J., Diaz-Espejo, A., Galmes, J., Kaldenhoff, R., Medrano, H., & Ribas-Carbo, M. (2007). Rapid variations of mesophyll conductance in response to changes in CO_2 concentration around leaves. *Plant Cell & Environment*, *30*, 1284–1298.

Flexas, J., Niinemets, Ü., Gallé, A., Barbour, M. M., Centritto, M., Diaz-Espejo, A., Douthe, C., Galmés, J., Ribas-Carbo, M., & Rodriguez, P. L. (2013). Diffusional conductances to CO_2 as a target for increasing photosynthesis and photosynthetic water-use efficiency. *Photosynthesis Research*, *117*, 45–59.

Flexas, J., Ribas-Carbó, M., Hanson, D. T., Bota, J., Otto, B., Cifre, J., McDowell, N., Medrano, H., & Kaldenhoff, R. (2006). Tobacco aquaporin NtAQP1 is involved in mesophyll conductance to CO_2 in vivo. *The Plant Journal*, *48*, 427–439.

Foyer, C. H., & Noctor, G. (2012). Managing the cellular redox hub in photosynthetic organisms. *Plant Cell & Environment*, *35*, 199–201.

Frigerio, S., Campoli, C., Zorzan, S., Fantoni, L. I., Crosatti, C., Drepper, F., Haehnel, W., Cattivelli, L., Morosinotto, T., & Bassi, R. (2007). Photosynthetic antenna size in higher plants is controlled by the plastoquinone redox state at the post-transcriptional rather than transcriptional level. *Journal of Biological Chemistry*, *282*, 29457–29469.

Fukayama, H., Mizumoto, A., Ueguchi, C., Katsunuma, J., Morita, R., Sasayama, D., Hatanaka, T., & Azuma, T. (2018). Expression level of Rubisco activase negatively correlates with Rubisco content in transgenic rice. *Photosynthesis Research*, *137*, 465–474.

Fukayama, H., Ueguchi, C., Nishikawa, K., Katoh, N., Ishikawa, C., Masumoto, C., Hatanaka, T., & Misoo, S. (2012). Overexpression of Rubisco activase decreases the photosynthetic CO_2 assimilation rate by reducing Rubisco content in rice leaves. *Plant and Cell Physiology*, *53*, 976–986.

Galmés, J., Aranjuelo, I., Medrano, H., & Flexas, J. (2013). Variation in Rubisco content and activity under variable climatic factors. *Photosynthesis Research*, *117*, 73–90.

Garcia-Molina, A., & Leister, D. (2020). Accelerated relaxation of photoprotection impairs biomass accumulation in Arabidopsis. *Nature Plants*, *6*, 9–12.

Ghirardi, M. L., & Melis, A. (1988). Chlorophyll b deficiency in soybean mutants. I. Effects on photosystem stoichiometry and chlorophyll antenna size. *Biochimica et Biophysica Acta - Bioenergetics*, 932, 130–137.

Gjindali, A., Herrmann, H. A., Schwartz, J.-M., Johnson, G. N., & Calzadilla, P. I. (2021). A holistic approach to study photosynthetic acclimation responses of plants to fluctuating light. *Frontiers in Plant Science*, *12*, 668512.

Gjindali, A., & Johnson, G. N. (2023). Photosynthetic acclimation to changing environments. *Biochemical Society Transactions*, *51*, 473–486.

Gollan, P. J., Lima-Melo, Y., Tiwari, A., Tikkanen, M., & Aro, E.-M. (2017). Interaction between photosynthetic electron transport and chloroplast sinks triggers protection and signalling important for plant productivity. *Philosophical Transactions of the Royal Society B*, *372*, 20160390.

Gómez-Ocampo, G., Ploschuk, E. L., Mantese, A., Crocco, C. D., & Botto, J. F. (2021). BBX21 reduces abscisic acid sensitivity, mesophyll conductance and chloroplast electron transport capacity to increase photosynthesis and water use efficiency in potato plants cultivated under moderated drought. *The Plant Journal*, *108*, 1131–1144.

Gong, H. Y., Li, Y., Fang, G., Hu, D. H., Bin, J. W., Wang, Z. H., & Li, Y. S. (2015). Transgenic rice expressing Ictb and FBP/Sbpase derived from cyanobacteria exhibits enhanced photosynthesis and mesophyll conductance to CO_2. *PLoS One*, *10*, e0140928.

Gunn, L. H., Martin Avila, E., Birch, R., & Whitney, S. M. (2020). The dependency of red Rubisco on its cognate activase for enhancing plant photosynthesis and growth. *Proceedings of the National Academy of Sciences of the United States of America*, *117*, 25890–25896.

Hagemann, M., & Bauwe, H. (2016). Photorespiration and the potential to improve photosynthesis. *Current Opinion in Chemical Biology*, *35*, 109–116.

Han, J.-M., Meng, H.-F., Wang, S.-Y., Jiang, C.-D., Liu, F., Zhang, W.-F., & Zhang, Y.-L. (2016). Variability of mesophyll conductance and its relationship with water use efficiency in cotton leaves under drought pretreatment. *Journal of Plant Physiology*, *194*, 61–71.

Hanba, Y. T., Shibasaka, M., Hayashi, Y., Hayakawa, T., Kasamo, K., Terashima, I., & Katsuhara, M. (2004). Overexpression of the barley aquaporin HvPIP2; 1 increases internal CO_2 conductance and CO_2 assimilation in the leaves of transgenic rice plants. *Plant and Cell Physiology*, *45*, 521–529.

Hanson, M. R., Lin, M. T., Carmo-Silva, A. E., & Parry, M. A. J. (2016) Towards engineering carboxysomes into C3 plants. *The Plant Journal*, *87*, 38–50.

Harrison, E. L., Arce Cubas, L., Gray, J. E., & Hepworth, C. (2020). The influence of stomatal morphology and distribution on photosynthetic gas exchange. *The Plant Journal*, *101*, 768–779.

Harrison, E. P., Olcer, H., Lloyd, J. C., Long, S. P., & Raines, C. A. (2001). Small decreases in SBPase cause a linear decline in the apparent RuBP regeneration rate, but do not affect Rubisco carboxylation capacity. *Journal of Experimental Botany*, *52*, 1779–1784.

Harvey, C. M., Cavanagh, A. P., Kim, S. Y., Wright, D. A., Edquilang, R. G., Shreeves, K. S., Perdomo, J. A., Spalding, M. H., Ort, D. R., & Bernacchi, C. J. (2022). Removal of redox-sensitive Rubisco activase does not alter Rubisco regulation in soybean. *Photosynthesis Research*, *154*, 169–182.

Hay, W. T., Bihmidine, S., Mutlu, N., Le Hoang, K., Awada, T., Weeks, D. P., Clemente, T. E., & Long, S. P. (2017). Enhancing soybean photosynthetic CO_2 assimilation using a cyanobacterial membrane protein, ictB. *Journal of Plant Physiology*, *212*, 58–68.

Hennacy, J. H., & Jonikas, M. C. (2020). Prospects for engineering biophysical CO_2 concentrating mechanisms into land plants to enhance yields. *Annual Review of Plant Biology*, *71*, 461–485.

Heyno, E., Gross, C. M., Laureau, C., Culcasi, M., Pietri, S., & Krieger-Liszkay, A. (2009). Plastid alternative oxidase (PTOX) promotes oxidative stress when overexpressed in tobacco. *Journal of Biological Chemistry*, *284*, 31174–31180.

Hibberd, J. M., Sheehy, J. E., & Langdale, J. A. (2008). Using C4 photosynthesis to increase the yield of rice – Rationale and feasibility. *Current Opinion in Plant Biology*, *11*, 228–231.

Hikosaka, K. (2005). Leaf canopy as a dynamic system: Ecophysiology and optimality in leaf turnover. *Annals of Botany*, *95*, 521–533.

Hitchcock, A., Hunter, C. N., Sobotka, R., Komenda, J., Dann, M., & Leister, D. (2022). Redesigning the photosynthetic light reactions to enhance photosynthesis–The PhotoRedesign consortium. *The Plant Journal*, *109*, 23–34.

Horaruang, W., Klejchová, M., Carroll, W., Silva-Alvim, F. A. L., Waghmare, S., Papanatsiou, M., Amtmann, A., Hills, A., Alvim, J. C., & Blatt, M. R. (2022). Engineering a K^+ channel 'sensory antenna'enhances stomatal kinetics, water use efficiency and photosynthesis. *Nature Plants*, *8*, 1262–1274.

Hubbart, S., Smillie, I. R. A., Heatley, M., Swarup, R., Foo, C. C., Zhao, L., & Murchie, E. H. (2018). Enhanced thylakoid photoprotection can increase yield and canopy radiation use efficiency in rice. *Communications Biology*, *1*, 22.

Ishikawa, N., Takabayashi, A., Noguchi, K., Tazoe, Y., Yamamoto, H., Von Caemmerer, S., Sato, F., & Endo, T. (2016). NDH-mediated cyclic electron flow around photosystem I is crucial for C4 photosynthesis. *Plant and Cell Physiology*, *57*, 2020–2028.

Jezek, M., & Blatt, M. R. (2017). The membrane transport system of the guard cell and its integration for stomatal dynamics. *Plant Physiology*, *174*, 487–519.

Johari-Pireivatlou, M. (2010). Effect of soil water stress on yield and proline content of four wheat lines. *African Journal of Biotechnology*, *9*, 36–40.

Johnson, G. N. (2011). Physiology of PSI cyclic electron transport in higher plants. *Biochimica et Biophysica Acta (BBA) - Bioenergetics*, *1807*, 384–389.

Johnson, G. N., & Stepien, P. (2016). Plastid terminal oxidase as a route to improving plant stress tolerance: Known knowns and known unknowns. *Plant and Cell Physiology*, *57*, 1387–1396.

Johnson, G. N., Young, A. J., Scholes, J. D., & Horton, P. (1993). The dissipation of excess excitation energy in British plant species. *Plant Cell & Environment*, *16*, 673–679.

Kaiser, E., Correa Galvis, V., & Armbruster, U. (2019). Efficient photosynthesis in dynamic light environments: A chloroplast's perspective. *Biochemical Journal*, *476*, 2725–2741.

Kebeish, R., Niessen, M., Thiruveedhi, K., Bari, R., Hirsch, H.-J., Rosenkranz, R., Stäbler, N., Schönfeld, B., Kreuzaler, F., & Peterhänsel, C. (2007). Chloroplastic photorespiratory bypass increases photosynthesis and biomass production in Arabidopsis thaliana. *Nature Biotechnology*, *25*, 593–599.

Keenan, T., Sabate, S., & Gracia, C. (2010). The importance of mesophyll conductance in regulating forest ecosystem productivity during drought periods. *Global Change Biology*, *16*, 1019–1034.

Khurshid, G., Abbassi, A. Z., Khalid, M. F., Gondal, M. N., Naqvi, T. A., Shah, M. M., Chaudhary, S. U., & Ahmad, R. (2020). A cyanobacterial photorespiratory bypass model to enhance photosynthesis by rerouting photorespiratory pathway in C3 plants. *Scientific Reports*, *10*, 20879.

Kim, E. H., Li, X. P., Razeghifard, R., Anderson, J. M., Niyogi, K. K., Pogson, B. J., & Chow, W. S. (2009). The multiple roles of light-harvesting chlorophyll a/b-protein complexes define structure and optimize function of Arabidopsis chloroplasts: A study using two chlorophyll b-less mutants. *Biochimica et Biophysica Acta (BBA) - Bioenergetics*, *1787*, 973–984.

Kimura, H., Hashimoto-Sugimoto, M., Iba, K., Terashima, I., & Yamori, W. (2020). Improved stomatal opening enhances photosynthetic rate and biomass production in fluctuating light. *Journal of Experimental Botany*, *71*, 2339–2350.

Kirst, H., Formighieri, C., & Melis, A. (2014). Maximizing photosynthetic efficiency and culture productivity in cyanobacteria upon minimizing the phycobilisome light-harvesting antenna size. *Biochimica et Biophysica Acta (BBA) – Bioenergetics*, *1837*, 1653–1664.

Kirst, H., Gabilly, S. T., Niyogi, K. K., Lemaux, P. G., & Melis, A. (2017). Photosynthetic antenna engineering to improve crop yields. *Planta*, *245*, 1009–1020.

Knauer, J., Cuntz, M., Evans, J. R., Niinemets, Ü., Tosens, T., Veromann-Jürgenson, L., Werner, C., & Zaehle, S. (2022). Contrasting anatomical and biochemical controls on mesophyll conductance across plant functional types. *New Phytologist*, *236*, 357–368.

Korte, A., & Farlow, A. (2013). The advantages and limitations of trait analysis with GWAS: A review. *Plant Methods*, *9*, 1–9.

Krieger-Liszkay, A., & Feilke, K. (2016). The dual role of the plastid terminal oxidase PTOX: Between a protective and a pro-oxidant function. *Frontiers in Plant Science*, *6*, 2015–2017.

Kromdijk, J., Głowacka, K., Leonelli, L., Gabilly, S. T., Iwai, M., Niyogi, K. K., & Long, S. P. (2016). Improving photosynthesis and crop productivity by accelerating recovery from photoprotection. *Science*, *354*, 857–861.

Kubis, A., & Bar-Even, A. (2019). Synthetic biology approaches for improving photosynthesis. *Journal of Experimnetal Botany*, *70*, 1425–1433.

Kumar, A., Li, C., & Portis, A. R. (2009). Arabidopsis thaliana expressing a thermostable chimeric Rubisco activase exhibits enhanced growth and higher rates of photosynthesis at moderately high temperatures. *Photosynthesis Research*, *100*, 143–153.

Kumar, R., Brar, M. S., Kunduru, B., Ackerman, A. J., Yang, Y., Luo, F., Saski, C. A., Bridges, W. C., de Leon, N., & McMahan, C. (2023). Genetic architecture of source–sink-regulated senescence in maize. *Plant Physiology*, *193*, 2459–2479.

Kumar, V., Sharma, N., Jaiswal, K. K., Vlaskin, M. S., Nanda, M., Tripathi, M. K., & Kumar, S. (2021). Microalgae with a truncated light-harvesting antenna to maximize photosynthetic efficiency and biomass productivity: Recent advances and current challenges. *Process Biochemistry*, *104*, 83–91.

Kurek, I., Chang, T. K., Bertain, S. M., Madrigal, A., Liu, L., Lassner, M. W., & Zhu, G. (2007). Enhanced thermostability of Arabidopsis Rubisco activase improves photosynthesis and growth rates under moderate heat stress. *Plant Cell*, *19*, 3230–3241.

Lawson, T., & Blatt, M. R. (2014). Stomatal size, speed, and responsiveness impact on photosynthesis and water use efficiency. *Plant Physiology*, *164*, 1556–1570.

Lawson, T., & Vialet-Chabrand, S. (2019). Speedy stomata, photosynthesis and plant water use efficiency. *New Phytologist*, *221*, 93–98.

Lefebvre, S., Lawson, T., Fryer, M., Zakhleniuk, O. V., Lloyd, J. C., & Raines, C. A. (2005). Increased sedoheptulose-1,7-bisphosphatase activity in transgenic tobacco plants stimulates photosynthesis and growth from an early stage in development. *Plant Physiology*, *138*, 451–460.

Lehmeier, C., Pajor, R., Lundgren, M. R., Mathers, A., Sloan, J., Bauch, M., Mitchell, A., Bellasio, C., Green, A., & Bouyer, D. (2017). Cell density and airspace patterning in the leaf can be manipulated to increase leaf photosynthetic capacity. *The Plant Journal*, *92*, 981–994.

Lehretz, G. G., Schneider, A., Leister, D., & Sonnewald, U. (2022). High non-photochemical quenching of VPZ transgenic potato plants limits CO_2 assimilation under high light conditions and reduces tuber yield under fluctuating light. *Journal of Integrative Plant Biology*, *64*, 1821–1832.

Leister, D. (2023). Enhancing the light reactions of photosynthesis: Strategies, controversies, and perspectives. *Molecular Plant*, *16*, 4–22.

Li, Z., Yuan, S., Jia, H., Gao, F., Zhou, M., Yuan, N., Wu, P., Hu, Q., Sun, D., & Luo, H. (2017). Ectopic expression of a cyanobacterial flavodoxin in creeping bentgrass impacts plant development and confers broad abiotic stress tolerance. *Plant Biotechnology Journal*, *15*, 433–446.

Li, Z. Q., Li, J. X., Li, H. J., Shi, Z. H., & Zhang, G. F. (2015). Overexpression of TsApx1 from Thellungiella salsuginea improves abiotic stress tolerance in transgenic Arabidopsis thaliana. *Biologia Plantarum*, *59*, 497–506.

Lieman-Hurwitz, J., Rachmilevitch, S., Mittler, R., Marcus, Y., & Kaplan, A. (2003). Enhanced photosynthesis and growth of transgenic plants that express ictB, a gene involved in HCO_3^- accumulation in cyanobacteria. *Plant Biotechnology Journal*, *1*, 43–50.

Lin, M. T., Occhialini, A., Andralojc, P. J., Parry, M. A. J., & Hanson, M. R. (2014). A faster Rubisco with potential to increase photosynthesis in crops. *Nature*, *513*, 547–550.

Lin, M. T., Salihovic, H., Clark, F. K., & Hanson, M. R. (2022). Improving the efficiency of Rubisco by resurrecting its ancestors in the family Solanaceae. *Science Advances*, *8*, eabm6871.

Liu, D., Liu, Y., Rao, J., Wang, G., Li, H., Ge, F., & Chen, C. (2013). Overexpression of the glutathione S-transferase gene from Pyrus pyrifolia fruit improves tolerance to abiotic stress in transgenic tobacco plants. *Molecular Biology*, *47*, 515–523.

Liu, F., Song, Q., Zhao, J., Mao, L., Bu, H., Hu, Y., & Zhu, X. (2021a). Canopy occupation volume as an indicator of canopy photosynthetic capacity. *New Phytoloogist*, *232*, 941–956.

Liu, T., Barbour, M. M., Yu, D., Rao, S., & Song, X. (2022). Mesophyll conductance exerts a significant limitation on photosynthesis during light induction. *New Phytologist*, *233*, 360–372.

Liu, Z., Hu, C., Wang, Y., Sha, Y., Hao, Z., Chen, F., Yuan, L., & Mi, G (2021b). Nitrogen allocation and remobilization contributing to low-nitrogen tolerance in stay-green maize. *Field Crops Research*, *263*, 108078.

López-Calcagno, P. E., Brown, K. L., Simkin, A. J., Fisk, S. J., Vialet-Chabrand, S., Lawson, T., & Raines, C. A. (2020). Stimulating photosynthetic processes increases productivity and water-use efficiency in the field. *Nature Plants*, *6*, 1054–1063.

López-Calcagno, P. E., Fisk, S., Brown, K. L., Bull, S. E., South, P. F., & Raines, C. A. (2019). Overexpressing the H-protein of the glycine cleavage system increases biomass yield in glasshouse and field-grown transgenic tobacco plants. *Plant Biotechnology Journal*, *17*, 141–151.

Lü, H., Yang, Y., Li, H., Liu, Q., Zhang, J., Yin, J., Chu, S., Zhang, X., Yu, K., & Lv, L. (2018). Genome-wide association studies of photosynthetic traits related to phosphorus efficiency in soybean. *Frontiers in Plant Science*, *9*, 1226.

Lu, Z., Lu, J., Pan, Y., Lu, P., Li, X., Cong, R., & Ren, T. (2016). Anatomical variation of mesophyll conductance under potassium deficiency has a vital role in determining leaf photosynthesis. *Plant Cell & Environment*, *39*, 2428–2439.

Lundgren, M. R. (2020). C2 photosynthesis: A promising route towards crop improvement? *New Phytologist*, *228*, 1734–1740.

Lundgren, M. R., & Fleming, A. J. (2020). Cellular perspectives for improving mesophyll conductance. *The Plant Journal*, *101*, 845–857.

Ma, M., Liu, Y., Bai, C., Yang, Y., Sun, Z., Liu, X., Zhang, S., Han, X., & Yong, J. W. H. (2021a). The physiological functionality of PGR5/PGRL1-dependent cyclic electron transport in sustaining photosynthesis. *Frontiers in Plant Science*, *12*, 702196.

Ma, M., Liu, Y., Bai, C., & Yong, J. W. H. (2021b). The significance of chloroplast NAD(P)H dehydrogenase complex and its dependent cyclic electron transport in photosynthesis. *Frontiers in Plant Science*, *12*, 661863.

Mahmud, K., Panday, D., Mergoum, A., & Missaoui, A. (2021). Nitrogen losses and potential mitigation strategies for a sustainable agroecosystem. *Sustainability*, *13*, 2400.

Maier, A., Fahnenstich, H., Von Caemmerer, S., Engqvist, M. K. M., Weber, A. P. M., Flügge, U.-I., & Maurino, V. G. (2012). Transgenic introduction of a glycolate oxidative cycle into A. thaliana chloroplasts leads to growth improvement. *Frontiers in Plant Science*, *3*, 38.

Mantilla-Perez, M. B., Bao, Y., Tang, L., Schnable, P. S., & Salas-Fernandez, M. G. (2020). Toward "smart canopy" sorghum: Discovery of the genetic control of leaf angle across layers. *Plant Physiology*, *184*, 1927–1940.

Mantilla-Perez, M. B., & Salas Fernandez, M. G. (2017). Differential manipulation of leaf angle throughout the canopy: Current status and prospects. *Journal of Experimental Botany*, *68*, 5699–5717.

Mao, L., Song, Q., Li, M., Liu, X., Shi, Z., Chen, F., Chen, G., Zheng, H., & Zhu, X.-G. (2023). Decreasing photosystem antenna size by inhibiting chlorophyll synthesis: A double-edged sword for photosynthetic efficiency. *Crop and Environment*, *2*, 46–58.

Mascoli, V., Bersanini, L., & Croce, R. (2020). Far-red absorption and light-use efficiency trade-offs in chlorophyll f photosynthesis. *Nature Plants*, *6*, 1044–1053.

Maxwell, K., & Johnson, G. N. (2000). Chlorophyll fluorescence – A practical guide. *Journal of Experimental Botany*, *51*, 659–668.

McAusland, L., Vialet-Chabrand, S., Davey, P., Baker, N. R., Brendel, O., & Lawson, T. (2016). Effects of kinetics of light-induced stomatal responses on photosynthesis and water-use efficiency. *New Phytologist*, *211*, 1209–1220.

Messant, M., Hani, U., Lai, T., Wilson, A., Shimakawa, G., & Krieger-Liszkay, A. (2023). Plastid terminal oxidase (PTOX) protects photosystem I and not photosystem II against photoinhibition in Arabidopsis thaliana and Marchantia polymorpha. *The Plant Journal*, *117*, 669–678.

De Mooij, T., Janssen, M., Cerezo-Chinarro, O., Mussgnug, J. H., Kruse, O., Ballottari, M., Bassi, R., Bujaldon, S., Wollman, F.-A., & Wijffels, R. H. (2015). Antenna size reduction as a strategy to increase biomass productivity: A great potential not yet realized. *Journal of Applied Phycology*, *27*, 1063–1077.

Mueller-Cajar, O., & Whitney, S. M. (2008). Directing the evolution of Rubisco and Rubisco activase: First impressions of a new tool for photosynthesis research. *Photosynthesis Research*, *98*, 667–675.

Müller, P., Li, X.-P., & Niyogi, K. K. (2001). Non-photochemical quenching. A response to excess light energy. *Plant Physiology*, *125*, 1558–1566.

Munekage, Y., Hojo, M., Meurer, J., Endo, T., Tasaka, M., & Shikanai, T. (2002). PGR5 is involved in cyclic electron flow around photosystem I and is essential for photoprotection in Arabidopsis. *Cell*, *110*, 361–371.

Munné-Bosch, S., Shikanai, T., & Asada, K. (2005). Enhanced ferredoxin-dependent cyclic electron flow around photosystem I and α-tocopherol quinone accumulation in water-stressed ndhB-inactivated tobacco mutants. *Planta, 222*, 502–511.

Natukunda, M. I., Mantilla-Perez, M. B., Graham, M. A., Liu, P., & Salas-Fernandez, M. G. (2022). Dissection of canopy layer-specific genetic control of leaf angle in Sorghum bicolor by RNA sequencing. *BMC Genomics, 23*, 95.

Nawrocki, W. J., Bailleul, B., Picot, D., Cardol, P., Rappaport, F., Wollman, F.-A., & Joliot, P. (2019). The mechanism of cyclic electron flow. *Biochimica et Biophysica Acta (BBA) - Bioenergetics, 1860*, 433–438.

Nawrocki, W. J., Tourasse, N. J., Taly, A., Rappaport, F., & Wollman, F.-A. (2015). The plastid terminal oxidase: Its elusive function points to multiple contributions to plastid physiology. *Annual Review of Plant Biology, 66*, 49–74.

Nayak, L., Panda, D., Dash, G. K., Lal, M. K., Swain, P., Baig, M. J., & Kumar, A. (2022). A chloroplast Glycolate catabolic pathway bypassing the endogenous photorespiratory cycle enhances photosynthesis, biomass and yield in rice (Oryza sativa L.). *The Plant Science, 314*, 111103.

Niinemets, Ü., Díaz-Espejo, A., Flexas, J., Galmes, J., & Warren, C. R. (2009). Role of mesophyll diffusion conductance in constraining potential photosynthetic productivity in the field. *Journal of Experimental Botany, 60*, 2249–2270.

Nürnberg, D. J., Morton, J., Santabarbara, S., Telfer, A., Joliot, P., Antonaru, L. A., Ruban, A. V., Cardona, T., Krausz, E., & Boussac, A. (2018). Photochemistry beyond the red limit in chlorophyll f–containing photosystems. *Science, 360*, 1210–1213.

Ogawa, T., Kobayashi, K., Taniguchi, Y. Y., Shikanai, T., Nakamura, N., Yokota, A., & Munekage, Y. N. (2023). Two cyclic electron flows around photosystem I differentially participate in C4 photosynthesis. *Plant Physiology, 191*, 2288–2300.

Okegawa, Y., Long, T. A., Iwano, M., Takayama, S., Kobayashi, Y., Covert, S. F., & Shikanai, T. (2007). A balanced PGR5 level is required for chloroplast development and optimum operation of cyclic electron transport around photosystem I. *Plant and Cell Physiology, 48*, 1462–1471.

Olsovska, K., Kovar, M., Brestic, M., Zivcak, M., Slamka, P., & Shao, H. B. (2016). Genotypically identifying wheat mesophyll conductance regulation under progressive drought stress. *Frontiers in Plant Sciene, 7*, 1111.

Orr, D. J., Alcântara, A., Kapralov, M. V., Andralojc, P. J., Carmo-Silva, E., & Parry, M. A. J. (2016). Surveying Rubisco diversity and temperature response to improve crop photosynthetic efficiency. *Plant Physiology, 172*, 707–717.

Orr, D. J., Worrall, D., Lin, M. T., Carmo-Silva, E., Hanson, M. R., & Parry, M. A. J. (2020). Hybrid cyanobacterial-tobacco Rubisco supports autotrophic growth and procarboxysomal aggregation. *Plant Physiology, 182*, 807–818.

Ort, D. R., Merchant, S. S., Alric, J., Barkan, A., Blankenship, R. E., Bock, R., Croce, R., Hanson, M. R., Hibberd, J. M., & Long, S. P. (2015). Redesigning photosynthesis to sustainably meet global food and bioenergy demand. *Proceedings of the National Academy of Sciences of the United States of America, 112*, 8529–8536.

Ort, D. R., Zhu, X., & Melis, A. (2011). Optimizing antenna size to maximize photosynthetic efficiency. *Plant Physiology, 155*, 79–85.

Osmanoglu, Ö., Khaled AlSeiari, M., AlKhoori, H. A., Shams, S., Bencurova, E., Dandekar, T., & Naseem, M. (2021). Topological analysis of the carbon-concentrating CETCH cycle and a photorespiratory bypass reveals boosted CO_2-sequestration by plants. *Frontiers in Bioengineering and Biotechnology, 9*, 708417.

Ouyang, W., Struik, P. C., Yin, X., & Yang, J. (2017). Stomatal conductance, mesophyll conductance, and transpiration efficiency in relation to leaf anatomy in rice and wheat genotypes under drought. *Journal of Experimental Botany, 68*, 5191–5205.

Ozeki, K., Miyazawa, Y., & Sugiura, D. (2022). Rapid stomatal closure contributes to higher water use efficiency in major C4 compared to C3 Poaceae crops. *Plant Physiology, 189*, 188–203.

Papanatsiou, M., Petersen, J., Henderson, L., Wang, Y., Christie, J. M., & Blatt, M. R. (2019). Optogenetic manipulation of stomatal kinetics improves carbon assimilation, water use, and growth. *Science, 363*, 1456–1459.

Parry, M. A. J., Andralojc, P. J., Scales, J. C., Salvucci, M. E., Carmo-Silva, A. E., Alonso, H., & Whitney, S. M. (2013). Rubisco activity and regulation as targets for crop improvement. *Journal of Experimental Botany, 64*, 717–730.

Peterhansel, C., & Maurino, V. G. (2011). Photorespiration redesigned. *Plant Physiology, 155*, 49–55.

Pettersson, G., & Ryde-Pettersson, U. (1988). A mathematical model of the Calvin photosynthesis cycle. *European Journal of Biochemistry, 175*, 661–672.

Pettigrew, W. T., Hesketh, J. D., Peters, D. B., & Woolley, J. T. (1989). Characterization of canopy photosynthesis of chlorophyll-deficient soybean isolines. *Crop Science*, *29*, 1025–1029.

Pfannschmidt, T., Nilsson, A., & Allen, J. F. (1999). Photosynthetic control of chloroplast gene expression. *Nature*, *397*, 625–628.

Pignon, C. P., Fernandes, S. B., Valluru, R., Bandillo, N., Lozano, R., Buckler, E., Gore, M. A., Long, S. P., Brown, P. J., & Leakey, A. D. B. (2021) Phenotyping stomatal closure by thermal imaging for GWAS and TWAS of water use efficiency-related genes. *Plant Physiology*, *187*, 2544–2562.

Pinnola, A., & Bassi, R. (2018). Molecular mechanisms involved in plant photoprotection. *Biochemical Society Transactions*, *46*, 467–482.

Poolman, M. G., Fell, D. A., & Thomas, S. (2000). Modelling photosynthesis and its control. *Journal of Experimental Botany*, *51*, 319–328.

Poppy, G. M., Chiotha, S., Eigenbrod, F., Harvey, C. A., Honzák, M., Hudson, M. D., Jarvis, A., Madise, N. J., Schreckenberg, K., & Shackleton, C. M. (2014). Food security in a perfect storm: Using the ecosystem services framework to increase understanding. *Philosophical Transactions of the Royal Society B*, *369*, 20120288.

Portis, Jr., A. R. (2003). Rubisco activase–Rubisco's catalytic chaperone. *Photosynthesis Research*, *75*, 11–27.

Price, G. D., Badger, M. R., Woodger, F. J., & Long, B. M. (2008). Advances in understanding the cyanobacterial CO_2-concentrating-mechanism (CCM): Functional components, Ci transporters, diversity, genetic regulation and prospects for engineering into plants. *Journal of Experimental Botany*, *59*, 1441–1461.

Prins, A., Orr, D. J., Andralojc, P. J., Reynolds, M. P., Carmo-Silva, E., & Parry, M. A. J. (2016). Rubisco catalytic properties of wild and domesticated relatives provide scope for improving wheat photosynthesis. *Journal of Experimental Botany*, *67*, 1827–1838.

Qu, M., Essemine, J., Xu, J., Ablat, G., Perveen, S., Wang, H., Chen, K., Zhao, Y., Chen, G., & Chu, C. (2020). Alterations in stomatal response to fluctuating light increase biomass and yield of rice under drought conditions. *The Plant Journal*, *104*, 1334–1347.

Raines, C. A. (2003). The Calvin cycle revisited. *Photosynthesis Research*, *75*, 1–10.

Raines, C. A., Cavanagh, A. P., & Simkin, A. J. (2022). Improving carbon fixation. In A. Ruban, C. H. Foyer, & E. H. Murchie (Eds.), *Photosynthesis in action* (pp 175–192). Elsevier.

Raven, J. A., Cockell, C. S., & De La Rocha, C. L. (2008). The evolution of inorganic carbon concentrating mechanisms in photosynthesis. *Philosophical Transactions of the Royal Society B*, *363*, 2641–2650.

Ray, D. K., West, P. C., Clark, M., Gerber, J. S., Prishchepov, A. V., & Chatterjee, S. (2019). Climate change has likely already affected global food production. *PLoS One*, *14*, e0217148.

Razzaq, A., Wani, S. H., Saleem, F., Yu, M., Zhou, M., & Shabala, S. (2021). Rewilding crops for climate resilience: Economic analysis and de novo domestication strategies. *Journal of Experimental Botany*, *72*, 6123–6139.

Ren, T., Weraduwage, S. M., & Sharkey, T. D. (2019). Prospects for enhancing leaf photosynthetic capacity by manipulating mesophyll cell morphology. *Journal of Experimental Botany*, *70*, 1153–1165.

Roach, T., & Krieger-Liszkay, A. (2014). Regulation of photosynthetic electron transport and photoinhibition. *Current Protein & Peptide Science*, *15*, 351–362.

Rodrigues, A. F., de Mello, C. R., Nehren, U., de Coimbra Ribeiro, J. P., Mantovani, V. A., & de Mello, J. M. (2021). Modeling canopy interception under drought conditions: The relevance of evaporation and extra sources of energy. *Journal of Environmental Management*, *292*, 112710.

Rosenthal, D. M., Locke, A. M., Khozaei, M., Raines, C. A., Long, S. P., & Ort, D. R. (2011). Over-expressing the C3 photosynthesis cycle enzyme sedoheptulose-1–7 bisphosphatase improves photosynthetic carbon gain and yield under fully open air CO_2 fumigation (FACE). *BMC Plant Biology*, *11*, 1–12.

Rosso, D., Ivanov, A. G., Fu, A., Geisler-Lee, J., Hendrickson, L., Geisler, M., Stewart, G., Krol, M., Hurry, V., & Rodermel, S. R. (2006). IMMUTANS does not act as a stress-induced safety valve in the protection of the photosynthetic apparatus of Arabidopsis during steady-state photosynthesis. *Plant Physiology*, *142*, 574–585.

Rungrat, T., Almonte, A. A., Cheng, R., Gollan, P. J., Stuart, T., Aro, E., Borevitz, J. O., Pogson, B., & Wilson, P. B. (2019). A genome-wide association study of non-photochemical quenching in response to local seasonal climates in Arabidopsis thaliana. *Plant Direct*, *3*, e00138.

Sage, R. F. (2002). Variation in the k cat of Rubisco in C3 and C4 plants and some implications for photosynthetic performance at high and low temperature. *Journal of Experimental Botany*, *53*, 609–620.

Sakoda, K., Yamori, W., Shimada, T., Sugano, S. S., Hara-Nishimura, I., & Tanaka, Y. (2020). Higher stomatal density improves photosynthetic induction and biomass production in Arabidopsis under fluctuating light. *Frontiers in Plant Science*, *11*, 589603.

Saunders, H. A., Calzadilla, P. I., Schwartz, J.-M., & Johnson, G. N. (2022). Cytosolic fumarase acts as a metabolic fail-safe for both high and low temperature acclimation of Arabidopsis thaliana. *Journal of Experimental Botany, 73*, 2112–2124.

Savitch, L. V., Ivanov, A. G., Krol, M., Sprott, D. P., Quist, G., & Huner, N. P. A. (2010). Regulation of energy partitioning and alternative electron transport pathways during cold acclimation of Lodgepole pine is oxygen dependent. *Plant and Cell Physiology, 51*, 1555–1570.

Scafaro, A. P., Atwell, B. J., Muylaert, S., Van Reusel, B., Ruiz, G. A., Van Rie, J., & Gallé, A. (2018). A thermotolerant variant of Rubisco activase from a wild relative improves growth and seed yield in rice under heat stress. *Frontiers in Plant Science, 9*, 1663.

Schindele, P., Wolter, F., & Puchta, H. (2018). Transforming plant biology and breeding with CRISPR/Cas9, Cas12 and Cas13. *FEBS Letters, 592*, 1954–1967.

Schuler, M. L., Mantegazza, O., & Weber, A. P. M. (2016). Engineering C4 photosynthesis into C3 chassis in the synthetic biology age. *The Plant Journal, 87*, 51–65.

Sharwood, R. E., Quick, W. P., Sargent, D., Estavillo, G. M., Silva-Perez, V., & Furbank, R. T. (2022). Mining for allelic gold: Finding genetic variation in photosynthetic traits in crops and wild relatives. *Journal of Experimental Botany, 73*, 3085–3108.

Shen, B.-R., Wang, L.-M., Lin, X.-L., Yao, Z., Xu, H.-W., Zhu, C.-H., Teng, H.-Y., Cui, L.-L., Liu, E.-E., & Zhang, J.-J. (2019). Engineering a new chloroplastic photorespiratory bypass to increase photosynthetic efficiency and productivity in rice. *Molecular Plant, 12*, 199–214.

Shi, Q., Sun, H., Timm, S., Zhang, S., & Huang, W. (2022). Photorespiration alleviates photoinhibition of photosystem I under fluctuating light in tomato. *Plants, 11*, 195.

Simkin, A. J., Lopez-Calcagno, P. E., Davey, P. A., Headland, L. R., Lawson, T., Timm, S., Bauwe, H., & Raines, C. A. (2017). Simultaneous stimulation of sedoheptulose 1,7-bisphosphatase, fructose 1, 6-bisphophate aldolase and the photorespiratory glycine decarboxylase-H protein increases CO_2 assimilation, vegetative biomass and seed yield in Arabidopsis. *Plant Biotechnology Journal, 15*, 805–816.

Simkin, A. J., McAusland, L., Headland, L. R., Lawson, T., & Raines, C. A. (2015). Multigene manipulation of photosynthetic carbon assimilation increases CO_2 fixation and biomass yield in tobacco. *Journal of Experimental Botany, 66*, 4075–4090.

Smith, E. N., van Aalst, M., Tosens, T., Niinemets, Ü., Stich, B., Morosinotto, T., Alboresi, A., Erb, T., Gómez-Coronado, P. A., & Tolleter, D. (2023). Improving photosynthetic efficiency toward food security: Strategies, advances, and perspectives. *Molecualr Plant, 16*, 1547–1563.

Song, Q., Chu, C., Parry, M. A. J., & Zhu, X. (2016). Genetics-based dynamic systems model of canopy photosynthesis: The key to improve light and resource use efficiencies for crops. *Food and Energy Security, 5*, 18–25.

Song, Q., Wang, Y., Qu, M., Ort, D. R., & Zhu, X. (2017). The impact of modifying photosystem antenna size on canopy photosynthetic efficiency – Development of a new canopy photosynthesis model scaling from metabolism to canopy level processes. *Plant and Cell Environment, 40*, 2946–2957.

South, P. F., Cavanagh, A. P., Liu, H. W., & Ort, D. R. (2019). Synthetic glycolate metabolism pathways stimulate crop growth and productivity in the field. *Science, 363*, eaat9077.

South, P. F., Cavanagh, A. P., Lopez-Calcagno, P. E., Raines, C. A., & Ort, D. R. (2018). Optimizing photorespiration for improved crop productivity. *Journal of Integrative Plant Biology, 60*, 1217–1230.

De Souza, A. P., Burgess, S. J., Doran, L., Hansen, J., Manukyan, L., Maryn, N., Gotarkar, D., Leonelli, L., Niyogi, K. K., & Long, S. P. (2022). Soybean photosynthesis and crop yield are improved by accelerating recovery from photoprotection. *Science, 377*, 851–854.

De Souza, A. P., Wang, Y., Orr, D. J., Carmo-Silva, E., & Long, S. P. (2020). Photosynthesis across African cassava germplasm is limited by Rubisco and mesophyll conductance at steady state, but by stomatal conductance in fluctuating light. *New Phytologist, 225*, 2498–2512.

Sparrow-Muñoz, I., Chen, T. C., & Burgess, S. J. (2023). Recent developments in the engineering of Rubisco activase for enhanced crop yield. *Biochemical Society Transactions, 51*, 627–637.

Stepien, P., & Johnson, G. N. (2009). Contrasting responses of photosynthesis to salt stress in the glycophyte Arabidopsis and the halophyte Thellungiella: Role of the plastid terminal oxidase as an alternative electron sink. *Plant Physiology, 149*, 1154–1165.

Streb, P., Josse, E., Gallouet, E., Baptist, F., Kuntz, M., & Cornic, G. (2005). Evidence for alternative electron sinks to photosynthetic carbon assimilation in the high mountain plant species Ranunculus glacialis. *Plant Cell & Environment, 28*, 1123–1135.

Suganami, M., Suzuki, Y., Kondo, E., Nishida, S., Konno, S., & Makino, A. (2020). Effects of overproduction of Rubisco activase on Rubisco content in transgenic rice grown at different N levels. *International Journal of Molecular Sciences, 21*, 1626.

Suganami, M., Suzuki, Y., Tazoe, Y., Yamori, W., & Makino, A. (2021). Co-overproducing Rubisco and Rubisco activase enhances photosynthesis in the optimal temperature range in rice. *Plant Physiology, 185*, 108–119.

Suorsa, M., Järvi, S., Grieco, M., Nurmi, M., Pietrzykowska, M., Rantala, M., Kangasjärvi, S., Paakkarinen, V., Tikkanen, M., & Jansson, S. (2012). PROTON GRADIENT REGULATION5 is essential for proper acclimation of Arabidopsis photosystem I to naturally and artificially fluctuating light conditions. *Plant Cell, 24*, 2934–2948.

Suzuki, Y., Miyamoto, T., Yoshizawa, R., Mae, T., & Makino, A. (2009). Rubisco content and photosynthesis of leaves at different positions in transgenic rice with an overexpression of RBCS. *Plant Cell & Environment, 32*, 417–427.

Tamoi, M., Nagaoka, M., Miyagawa, Y., & Shigeoka, S. (2006). Contribution of fructose-1,6-bisphosphatase and sedoheptulose-1,7-bisphosphatase to the photosynthetic rate and carbon flow in the Calvin cycle in transgenic plants. *Plant and Cell Physiology, 47*, 380–390.

Thomas, H., & Ougham, H. (2014). The stay-green trait. *Journal of Experimental Botany, 65*, 3889–3900.

Thormählen, I., Zupok, A., Rescher, J., Leger, J., Weissenberger, S., Groysman, J., Orwat, A., Chatel-Innocenti, G., Issakidis-Bourguet, E., & Armbruster, U. (2017). Thioredoxins play a crucial role in dynamic acclimation of photosynthesis in fluctuating light. *Molecular Plant, 10*, 168–182.

Timm, S., Florian, A., Arrivault, S., Stitt, M., Fernie, A. R., & Bauwe, H. (2012). Glycine decarboxylase controls photosynthesis and plant growth. *FEBS Letters, 586*, 3692–3697.

Timm, S., Wittmiß, M., Gamlien, S., Ewald, R., Florian, A., Frank, M., Wirtz, M., Hell, R., Fernie, A. R., & Bauwe, H. (2015). Mitochondrial dihydrolipoyl dehydrogenase activity shapes photosynthesis and photorespiration of Arabidopsis thaliana. *Plant Cell, 27*, 1968–1984.

Tognetti, V. B., Palatnik, J. F., Fillat, M. F., Melzer, M., Hajirezaei, M.-R., Valle, E. M., & Carrillo, N. (2006). Functional replacement of ferredoxin by a cyanobacterial flavodoxin in tobacco confers broad-range stress tolerance. *Plant Cell, 18*, 2035–2050.

Tognetti, V. B., Zurbriggen, M. D., Morandi, E. N., Fillat, M. F., Valle, E. M., Hajirezaei, M.-R., & Carrillo, N. (2007). Enhanced plant tolerance to iron starvation by functional substitution of chloroplast ferredoxin with a bacterial flavodoxin. *Proceedings of the National Academy of Sciences of the United States of America, 104*, 11495–11500.

Tomeo, N. J., & Rosenthal, D. M. (2017). Variable mesophyll conductance among soybean cultivars sets a tradeoff between photosynthesis and water-use-efficiency. *Plant Physiology, 174*, 241–257.

Uematsu, K., Suzuki, N., Iwamae, T., Inui, M., & Yukawa, H. (2012). Increased fructose 1,6-bisphosphate aldolase in plastids enhances growth and photosynthesis of tobacco plants. *Journal of Experimental Botany, 63*, 3001–3009.

Vendruscolo, E. C. G., Schuster, I., Pileggi, M., Scapim, C. A., Molinari, H. B. C., Marur, C. J., & Vieira, L. G. E. (2007). Stress-induced synthesis of proline confers tolerance to water deficit in transgenic wheat. *Journal of Plant Physiology, 164*, 1367–1376.

Villalobos-González, L., Alarcón, N., Bastías, R., Pérez, C., Sanz, R., Peña-Neira, Á., & Pastenes, C. (2022). Photoprotection is achieved by photorespiration and modification of the leaf incident light, and their extent is modulated by the stomatal sensitivity to water deficit in grapevines. *Plants, 11*, 1050.

Walters, R. G., & Horton, P. (1991). Resolution of components of non-photochemical chlorophyll fluorescence quenching in barley leaves. *Photosynthesis Research, 27*, 121–133.

Wang, H., Wang, H., Shao, H., & Tang, X. (2016). Recent advances in utilizing transcription factors to improve plant abiotic stress tolerance by transgenic technology. *Frontiers in Plant Science, 7*, 67.

Wang, L.-M., Shen, B.-R., Li, B.-D., Zhang, C.-L., Lin, M., Tong, P.-P., Cui, L.-L., Zhang, Z.-S., & Peng, X.-X. (2020). A synthetic photorespiratory shortcut enhances photosynthesis to boost biomass and grain yield in rice. *Molecular Plant, 13*, 1802–1815.

Wang, Q., Yang, H., Cao, P., Chen, F., & Zhao, L. (2022a). Biosynthetic approaches to efficient assimilation of CO_2 via photorespiration modification in plant chassis. *Frontiers in Bioengineering and Biotechnology, 10*, 979627.

Wang, Q., Zhao, H., Jiang, J., Xu, J., Xie, W., Fu, X., Liu, C., He, Y., & Wang, G. (2017). Genetic architecture of natural variation in rice nonphotochemical quenching capacity revealed by genome-wide association study. *Frontiers in Plant Science, 8*, 1773.

Wang, Y., Wang, Y., Tang, Y., & Zhu, X.-G. (2022b). Stomata conductance as a goalkeeper for increased photosynthetic efficiency. *Current Opinion in Plant Biology, 70*, 102310.

Wijewardene, I., Shen, G., & Zhang, H. (2021). Enhancing crop yield by using Rubisco activase to improve photosynthesis under elevated temperatures. *Stress Biology, 1*, 2.

Xie, K., Lu, Z., Pan, Y., Gao, L., Hu, P., Wang, M., & Guo, S. (2020). Leaf photosynthesis is mediated by the coordination of nitrogen and potassium: The importance of anatomical-determined mesophyll conductance to CO_2 and carboxylation capacity. *The Plant Science*, *290*, 110267.

Xiong, D., Liu, X. I., Liu, L., Douthe, C., Li, Y., Peng, S., & Huang, J. (2015). Rapid responses of mesophyll conductance to changes of CO_2 concentration, temperature and irradiance are affected by N supplements in rice. *Plant Cell & Environment*, *38*, 2541–2550.

Xiong, Z., Dun, Z., Wang, Y., Yang, D., Xiong, D., Cui, K., Peng, S., & Huang, J. (2022). Effect of stomatal morphology on leaf photosynthetic induction under fluctuating light in rice. *Frontiers in Plant Science*, *12*, 754790.

Yamori, W., Masumoto, C., Fukayama, H., & Makino, A. (2012). Rubisco activase is a key regulator of non-steady-state photosynthesis at any leaf temperature and, to a lesser extent, of steady-state photosynthesis at high temperature. *The Plant Journal*, *71*(6), 871-880

Yamamoto, H., Takahashi, S., Badger, M. R., & Shikanai, T. (2016). Artificial remodelling of alternative electron flow by flavodiiron proteins in Arabidopsis. *Nature Plants*, *2*, 1–7.

Yamori, W., Kusumi, K., Iba, K., & Terashima, I. (2020). Increased stomatal conductance induces rapid changes to photosynthetic rate in response to naturally fluctuating light conditions in rice. *Plant Cell & Environment*, *43*, 1230–1240.

Yang, M., He, J., Sun, Z., Li, Q., Cai, J., Zhou, Q., Wollenweber, B., Jiang, D., & Wang, X. (2023). Drought priming mechanisms in wheat elucidated by in-situ determination of dynamic stomatal behavior. *Frontiers in Plant Science*, *14*, 1138494.

Yang, S.-M., Chang, C.-Y., Yanagisawa, M., Park, I., Tseng, T.-H., & Ku, M. S. B. (2008). Transgenic rice expressing cyanobacterial bicarbonate transporter exhibited enhanced photosynthesis, growth and grain yield. In: Allen, J.F., Gantt, E., Golbeck, J.H., Osmond, B. (eds) *Photosynthesis energy from the sun: 14th International congress on photosynthesis* (pp. 1243–1246). Springer.

Yao, Z., Rao, Z., Hou, S., Tian, C., Liu, C.-Y., Yang, X., & Zhu, G. (2022). The appropriate expression and coordination of glycolate oxidase and catalase are vital to the successful construction of the photorespiratory metabolic pathway. *Frontiers in Plant Science*, *13*, 999757.

Yoon, D.-K., Ishiyama, K., Suganami, M., Tazoe, Y., Watanabe, M., Imaruoka, S., Ogura, M., Ishida, H., Suzuki, Y., & Obara, M. (2020). Transgenic rice overproducing Rubisco exhibits increased yields with improved nitrogen-use efficiency in an experimental paddy field. *Nature Food*, *1*, 134–139.

Zeng, Z.-L., Wang, X.-Q., Zhang, S.-B., & Huang, W. (2023). Mesophyll conductance limits photosynthesis in fluctuating light under combined drought and heat stresses. *Plant Physiology*, *194*, 1498–1511.

Zhang, C., Zhong, X., Lin, D., Wu, K., Wu, Z., Zhang, Z., & Peng, X. (2022). Grain quality affected by introducing photorespiratory bypasses into rice. *Agronomy*, *12*, 566.

Zhang, J., Fengler, K. A., Van Hemert, J. L., Gupta, R., Mongar, N., Sun, J., Allen, W. B., Wang, Y., Weers, B., & Mo, H. (2019). Identification and characterization of a novel stay-green QTL that increases yield in maize. *Plant Biotechnology Journal*, *17*, 2272–2285.

Zhang, P., Allahverdiyeva, Y., Eisenhut, M., & Aro, E.-M. (2009). Flavodiiron proteins in oxygenic photosynthetic organisms: Photoprotection of photosystem II by Flv2 and Flv4 in Synechocystis sp. PCC 6803. *PLoS One*, *4*, e5331.

Zhao, H., Tang, Q., Chang, T., Xiao, Y., & Zhu, X.-G. (2021). Why an increase in activity of an enzyme in the Calvin–Benson cycle does not always lead to an increased photosynthetic CO_2 uptake rate? – a theoretical analysis. *In Silico Plants*, *3*, diaa009.

Zhao, Y., Qiang, C., Wang, X., Chen, Y., Deng, J., Jiang, C., Sun, X., Chen, H., Li, J., & Piao, W. (2019). New alleles for chlorophyll content and stay-green traits revealed by a genome wide association study in rice (Oryza sativa). *Scientific Reports*, *9*, 2541.

Zhou, Y., & Whitney, S. (2019). Directed evolution of an improved Rubisco; in vitro analyses to decipher fact from fiction. *International Journal of Molecular Sciences*, *20*, 5019.

Zhu, X.-G., Song, Q., & Ort, D. R. (2012). Elements of a dynamic systems model of canopy photosynthesis. *Current Opinion in Plant Biology*, *15*, 237–244.

Zhu, X.-G., De Sturler, E., & Long, S. P. (2007). Optimizing the distribution of resources between enzymes of carbon metabolism can dramatically increase photosynthetic rate: A numerical simulation using an evolutionary algorithm. *Plant Physiology*, *145*, 513–526.

2 Plant Tolerance to Alkaline Salinity
A Gordian Knot to Untie

Silvia Busoms, Maria Almira-Casellas, Juan Barceló, and Charlotte Poschenrieder

2.1 INTRODUCTION

Multiple concurrent stress factors are continuously challenging plant performance (Zandalinas and Mittler, 2022). Plant responses to complex or combined stresses have been mainly reviewed for drought and heat (Rizhsky et al., 2004) and for abiotic–biotic stress interactions (Atkinson and Urwin, 2012). Salinity caused by high NaCl concentrations is a complex stress factor combining ion toxicity and osmotic threats. The mechanisms of injury and tolerance in plants exposed to high NaCl conditions have been a research focus for more than a hundred years (Hill, 1908). Currently, we already have an advanced picture of these response mechanisms to NaCl stress at both the physiological and genetic-molecular levels (e.g. Shabala and Munns, 2017; Busoms et al., 2018; Ali et al., 2023). Nonetheless, the impact of this progress on the actual improvement of crop yields under saline conditions has been rather limited. Major reasons for this failure are the multi-trait characteristic of plants' salinity tolerance mechanisms and the complexity of stress factors that in addition to high Na^+ limit crop performance on salinized soils.

Drought and alkalinity are major constraints in large areas of salinized crop lands. Breeding strategies based mainly on Na^+ exclusion are insufficient or even can be disadvantageous for crop productivity under these multifactorial stress conditions (Pérez-Martín et al., 2022; Chen et al., 2024). Although the severe damage that the combination of salinity and alkalinity stress causes to crop plants has also been documented for more than a hundred years (e.g. Harris, 1920), the underlying mechanisms are less explored. However, the fast progress of alkaline salinization of crop soils and the increasing food demand have strongly activated research on alkaline salinity in plants during the past two decades (Fang et al., 2021).

2.1.1 Characteristics and Origin of Alkaline Salinity

The US Salinity Laboratory (1947) defines saline soils as soils 'containing sufficient exchangeable sodium to interfere with the growth of crop plants and containing appreciable quantities of soluble salts'. Based on the salt sensitivity of most crops, the usually considered threshold for saline soils is an electrical conductivity of the saturation extract (ECe) of 4 dS/m or greater, which is equivalent to approximately 40 mM of sodium chloride (NaCl) exerting an osmotic pressure of 0.2 MPa (Munns and Tester, 2008). However, more sensitive plants can even be damaged by ECe values between 2 and 4 dS/m and vegetation changes due to salinity can occur at ECe below 4 dS/m (Bui and Henderson, 2003).

In soils with pH 7 or lower, NaCl and Na_2SO_4 mostly contribute to salinity. Contrastingly, under alkaline conditions, $NaHCO_3$ and Na_2CO_3 are becoming the predominant salts causing further increases in soil pH (Al-Busaidi and Cookson, 2003; Fang et al., 2021). In alkaline-saline soils, the percentage of exchangeable sodium (the degree of saturation of the soil exchange complex with sodium, defined as [(exchangeable sodium/cation exchange capacity) $\times$ 100]) is established to be

DOI: 10.1201/9781003457237-2

above 15, the exchangeable sodium adsorption rate (SARe) remains below 6, the ECe criterion is above 4 dS/m, and the pH of the saturated soil usually raises to 8.5. Posterior classifications further differentiated salt-alkali conditions in soils according to salt content and pH range, from mild (salt content up to 3‰ and pH between 7.1 and 8.5), to moderate (salt content between 3‰ and 6‰ and pH between 8.5 and 9.5), or severe (salt content exceeding 6‰, and pH value over 9.5) (Fang et al., 2021). The extreme conditions found in highly alkaline-sodic soils with SARe higher than 6 and pH higher than 9 affect crop yield not only due to the extreme salinity and alkalinity stress but additionally due to the bad soil structure, leading to poor aeration and occasional waterlogging (Rengasamy, 2016; Zewd and Siban, 2021).

Saline-alkaline soils are typically found in arid and semi-arid regions, where low precipitation and high evaporation lead to incomplete leaching and salt accumulation on the soil surface (Li, 2010). Changes in climate, changes in the chemical properties of major rock-forming materials, and geologic changes such as erosion, weathering, and deposition are primary contributors to the occurrence of alkaline-saline soils over time. Moreover, saline alkalinization is accelerated by anthropogenic sources, mainly due to irrigation mismanagement since irrigation waters frequently carry significant amounts of dissolved salts. However, alkaline salinity can also occur in inland marsh soils under dry climates (Sun et al., 2000) and occasionally even under tropical climate conditions (Oliveira Junior et al., 2017).

2.1.2 Implications for Agricultural Production

Soil alkaline salinization is of global concern as this problem has become a major limiting factor for crop production (Shabala, 2013). According to different estimations, more than 833 million hectares of arable land are affected by salinity (FAO, 2021). Although we have still a fragmentary picture of the distribution of saline soils in certain regions (Ivushkin et al., 2019) salinity can be considered one of the major causes of crop land loss worldwide (Hussain et al., 2019). Of particular concern is the fact that soil salinization is progressing quickly at a rate of 1.5×10^6 hm^2 per year (FAO, 2024). Sodic soils make up approximately 60% of all saline soils (Szabolcs, 1989). There is no exact estimation of the world area of alkaline-saline soils. Alkaline-saline soils can be found all over the world in Asia, Africa, South America, Australia, and the Mediterranean regions (Daliakopoulus et al., 2016). Alkaline salinization of soils is strongly favored under semi-arid and arid climates due to the movement of soluble ions like Na^+, Ca^{2+}, Mg^{2+}, K^+, CO_3^{2-} and HCO_3^- to the superficial soil layers produced by the high evapotranspiration rates (Singh, 2022). Climate change causing temperature rise and enhanced evapotranspiration exacerbates these processes directly due to enhanced soil drying and indirectly due to higher irrigation demands and the progressive salinization of irrigation water (Al-Najar and Ashour, 2013; Issanova et al., 2022). Mid-term and long-term predictions regarding soil salinization under the climate change scenario foresee an aggravation of soil salinization especially in dry lands of South America, South Africa, and Australia, among others, while a decrease of salinization is forecasted for Eastern Europe. Within Europe, the Mediterranean area is the most affected region suffering from enhanced soil salinization and alkalinization in both the mid- and long-term (2036–2060 and 2071–2100, respectively) (Hassani et al., 2021). Despite the efforts to reduce greenhouse gas emissions, climate change is already causing heat waves and drought exacerbating saline-alkaline conditions. This scenario will be further aggravated in the coming years, and both heat waves and drought will worsen soil salinization and alkalinization even if an occasional increase of flash floods may temporally reduce the salt load of soils locally.

The development of adequate land use strategies is crucial to palliate the problems of cropland scarcity (Yang et al., 2022). This must be necessarily addressed from the perspective of adequate soil management (Minhas et al., 2021), the breeding of tolerant crops, and the cultivation of halophytes for soil restoration, grazing, energy crops, and dietary use (Ventura et al., 2015; Liu and Wang, 2021). These actuations should be framed in the context of climate-smart agriculture, aiming to increase the resilience of agricultural systems to the impacts of climate change (Lipper et al.,

2014). The design of efficient breeding strategies for crops in semi-arid and arid scenarios requires a deep understanding of the genetic-molecular and physiological traits underlying both plant injury responses and tolerance mechanisms to alkaline-saline stress.

2.2 PLANT RESPONSES TO ALKALINE SALINITY

2.2.1 Plant Injury: Differences between Neutral and Alkaline Salinity

The mechanisms of salinity stress injury and tolerance have been extensively explored under slightly acid or neutral pH conditions. Under these circumstances, NaCl salinity is the major stress component. Currently, significant progress has been made in the identification of key genes and processes relevant to yield maintenance under conditions of NaCl salinity (Kumar et al., 2022; Liu et al., 2019; Zörb et al., 2018). Both the osmotic stress imposed by the high salt concentration that causes drought stress and the toxicity of specific ions, primarily Na^+ and to a lesser extent Cl^-, Mg^{2+}, and/or SO_4^{2-}, are mainly responsible for the reduction of crop yield under neutral salinity. Yield-depleting secondary effects are caused by a reduction in photosynthesis, hormone and nutrient imbalances, oxidative stress, and the energy costs for maintenance of ion homeostasis (Cabot et al., 2014; Zörb et al., 2018; Liu et al., 2019; Munns et al., 2020; Tomaz et al., 2020).

In alkaline-saline soils, $NaHCO_3$ instead of NaCl is the main challenge for plant performance. Relevant stress factors shared with neutral saline soils are Na^+ toxicity and the low soil osmotic potential. Important additional stress factors are the high pH of the alkaline-saline soils and the carbonate and bicarbonate anions. The first fundamental question is whether the common stress factors, Na^+ toxicity and low osmotic potential, are determinants for the selection of tolerance to both types of soil salinity or if the specific stress factors of saline-alkaline soils are relevant yield-limiting factors and are drivers of specific tolerance mechanisms.

Several studies reveal that alkaline salinity is more stressful to crops than neutral salinity (Figure 2.1a) (Shi and Sheng, 2005; Gong et al., 2013; Guo et al., 2017; Cao et al., 2020; Wang et al., 2020a, c). As an example, 100 mM of $NaHCO_3$ or Na_2CO_3 inhibited canola seed germination by almost 80%, while 250 mM NaCl had to be reached to reduce the germination rate by only 40% (Figure 2.1a) (Wang et al., 2022b). Comparisons under equally low osmotic potentials also demonstrated stronger inhibition of root elongation by $NaHCO_3$ than by NaCl (Figure 2.1b) (Hazman et al., 2016).

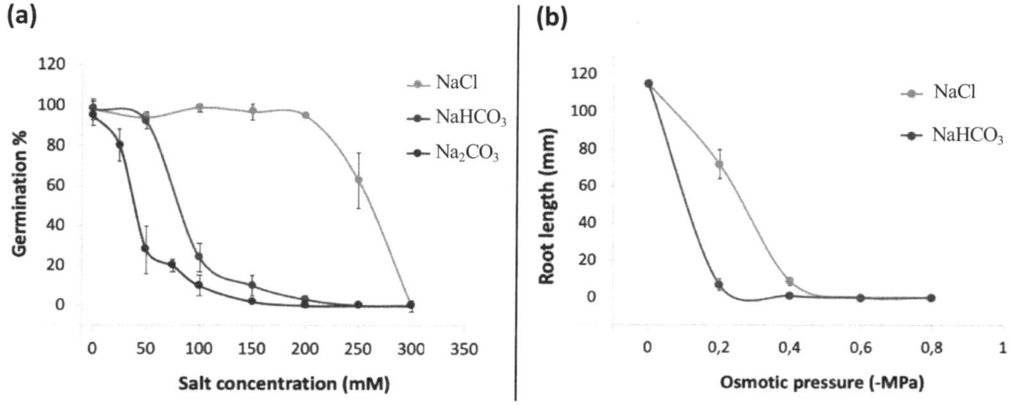

FIGURE 2.1 (a) Germination percentage of canola seeds under varying levels of neutral salinity (light grey line: NaCl, pH ≈ 6) and alkaline salinity (black line: $NaHCO_3$, pH ≈ 8.5; or dark grey line: Na_2CO_3, pH ≈ 11). Data from Wang et al. (2022). (b) Dose–response curves for the root length of rice seedlings submitted to osmotic pressures between 0 and −0.8 MPa by the addition of NaCl (light grey line) or $NaHCO_3$ (black line). Data from Hazman et al. (2016).

Under saline-alkali stress, plants experience not only the ion toxicity and osmotic stress imposed by the presence of a high Na^+ concentration but also the disruption of cell pH stability by HCO_3^- and CO_3^{2-} (Cao et al., 2022) and the consequences of impaired micronutrient acquisition (Hazman et al., 2016). The exploration of several *Arabidopsis thaliana* populations showed that naturally evolved tolerance to neutral salinity does not confer tolerance to alkaline salinity (Pérez-Martín et al., 2022). Adaptation to neutral salinity was found to be insufficient – even detrimental – for plant fitness under alkaline salinity. Moreover, alkaline salinity exerts stronger transcriptome alterations than neutral salinity (Pérez-Martín et al., 2021) (see Section 2.2.3).

Taken together, these findings support the hypothesis that in addition to Na^+ toxicity and osmotic constraints, either or both the bicarbonate anion and the high soil pH are compelling stress components of alkaline-saline soils. Bicarbonate buffers the pH of the soil solution within the range of around 7.5–8.5. It is well established that such alkaline pH values hamper the availability of essential micronutrients like iron and zinc; the chalk chlorosis syndrome is a frequent consequence even in non-saline carbonate-rich calcareous soils (Tagliavini and Rombola, 2001). Then, is alkaline salinity more stressful than neutral salinity simply because of the higher soil pH, or is the bicarbonate anion more toxic than the hydroxyl and chloride anions?

Bicarbonate seems to be more phytotoxic than chloride. Bicarbonate at 10 mM is toxic to most crop plants (Alhendawi et al., 1997). Highly sensitive acidophilous species like *Sphagnum* mosses or *Rhododendron* are damaged even at HCO_3^- concentrations between 0.8 and 2 mM (Chaanin and Preil, 1994; Koks et al., 2022). Contrastingly, most chloride-sensitive lemon trees suffer a 20% yield reduction at 15 mM Cl^-. In Cl^- tolerant crops like maize and barley, such yield penalties are caused by Cl^- concentrations of 70 and 90 mM, respectively (Bergmann, 1993). Moreover, the toxicity threshold for substrate bicarbonate depends on the Zn and or Fe efficiency of the genotype (Hajiboland et al., 2003; Pérez-Martín et al., 2021). The higher toxicity of HCO_3^- than of OH^- has been shown comparing bicarbonate versus non-bicarbonate alkalinity under non-saline conditions (Zhao et al., 2023).

An internal toxicity threshold concentration for bicarbonate has not been established so far. Early experiments using labeled HCO_3^- have tried to determine the influence of bicarbonate uptake on both the internal cation/anion balance and the pH homeostasis (Overstreet et al., 1940). The labeled carbon from roots supplied with $H^{14}CO_3^-$ can be found in leaf sugars and proteins (Stringer and Kimmerer, 1993). The dark fixation of HCO_3^--derived inorganic carbon may compensate for the lower photosynthetic leaf CO_2 assimilation under water deficit (Rao and Wu, 2017). However, due to the carbonic anhydrase activity, the intracellular presence of labeled carbon is not proof of the plasma membrane uptake and transport of HCO_3^- inside the plant. In fact, up to date, no specific bicarbonate transporter has been identified in higher land plants, although HCO_3^- may enter the symplasm through unspecific anion transporters (Poschenrieder et al., 2018).

2.2.1.1 Alkaline pH and Bicarbonate as Plant Stress Factors: Commons and Differences

Experimental designs for addressing response differences of plants to alkaline pH per se and bicarbonate usually cultivate plants in nutrient solutions with the same initial alkaline pH, containing or not bicarbonate. The bicarbonate-containing solution maintains a quite stable alkaline pH because of the buffering capacity of the $H_2CO_3/HCO_3^-/CO_2$ system. Contrastingly, if the pH of the reference solution without bicarbonate is adjusted to alkaline values simply by KOH or NaOH, the pH is unstable because plant roots quickly acidify such solutions. The stability of the alkaline pH can be provided by continuous titration, ion exchange resins, and organic buffers like TES, BTP, or inorganic buffers. Using the continuous titration system to separate nutrient deficiency from alkaline pH effects, it has been shown that OH^- is toxic to the root growth of mung beans at pH $\geq$ 8.5 (Kopittke and Menzies, 2004). The concentration of bicarbonate in solutions is highest between pH 8 and 8.5. However, bicarbonate toxicity can even be detected at lower pH. Concentrations between 10 and 20 mM HCO_3^- are mainly used in toxicity/tolerance studies. Thus, when applying alkaline salts

($NaHCO_3$, Na_2CO_3), the association between observed phenotypes and high pH should be taken with caution. Recent research clearly demonstrates that high bicarbonate concentrations damage plants more than high pH (Zhao et al. 2023). Bicarbonate caused more intense growth reduction in plants when compared to (1) neutral salinity, (2) high pH, or (3) neutral salts combined with high pH. For instance, in *Brassica napus* L., alkaline salts inhibited seedling growth more than neutral salt. Moreover, the enhanced growth inhibition was irrevocably linked to the presence of Na_2CO_3 rather than to high pH alone (Wang et al., 2022b).

Both alkaline pH per se and bicarbonate challenge cell pH homeostasis and micronutrient acquisitions. The first fundamental question is how plants achieve pH homeostasis under these alkaline conditions. The pH in the plant cell cytoplasm is usually maintained around 7.5, while apoplastic pH can reach low values between 4 and 5 in growing cells. The concentration of intracellular calcium, $[Ca]_{cyt}$ (Behera et al., 2018), proton pumping (Cosse and Seidel, 2021), proton-coupled transporters (Schuhmacher, 2014), and metabolic processes (Sakano, 2001) regulate pH homeostasis inside the cells. Multiple metabolic reactions are involved in plant H^+ homeostasis. A key component is plant elicitor peptide (PEP) carboxylase, the alkaline-pH-sensitive trigger in plants. The enzyme's activity is favored by alkaline pH. Bicarbonate and pyruvate are the substrates, and oxalacetate is the product, respectively. Thus, PEP carboxylase can metabolize HCO_3^- either entering the cells or formed from CO_2 by carbonic anhydrase. Further metabolization to malate, by malate dehydrogenase, will have a strong influence on the entire organic acid (OA) and respiratory metabolism (Sakano, 2001).

Classically, enhanced production of OAs under bicarbonate exposure has been pointed out as a possible factor involved in the low biological availability of Fe and Zn inside plant cells. In fact, root growth inhibition under bicarbonate exposure is paralleled by a substantial increase in OA biosynthesis (Yang et al., 1994). Citrate is the main OA accumulating not only in roots but also in the xylem, in leaves, and in the apoplastic fluid (Abadía et al., 2002). While a high rate of citrate exudation from roots into the soil can favor iron solubilization in the rhizosphere and the root apoplast, a high internal citrate to iron molar ratio may inhibit the activity of iron chelate reductase (González-Vallejo et al., 1999) (see also Section 2.2.2.3).

Apoplastic pH homeostasis is achieved by collaboration between the plasma membrane H^+-ATPase and homogalacturonan, the negatively charged cell wall pectin that can bind protons (Xu and Yu, 2023). Plants may perceive pH changes already in the apoplastic space due to specific receptors. Recent investigations have shown that the root surface pH is acidic in the elongation zone, while alkalinization occurs in the transition zone. The acidic pH in the elongation zone is clearly related to the proton pumping activity mediated by proton ATPase (AHA). Alkaline pH negatively regulates AHA activity through the calcium-sensor CBL10 (Xie et al., 2022). Contrastingly, the alkaline zone in the root tip transition zone is regulated by a system that quickly responds to auxin. The system comprises three components: an auxin influx carrier, an auxin receptor, and a calcium channel (Serre et al., 2023).

Stress can cause the alkalinization of the apoplast (Geilfuss, 2017). Root growth inhibition due to apoplastic alkalinization by biotic stress has deserved much attention. Pattern-triggered immunity is the first line of defense of plants against pathogens (Shu et al., 2023). Without stress, the acidic pH of the apoplast (pH 6) favors the protonated state and thus the interaction between the root meristem growth factor 1 (RGF1) and its receptor (RGFR1) through a sulfotyrosine residue. The acid pH sensor is functioning, and root growth is promoted. At pH 8 the unprotonated state predominates and no hydrogen bonds are formed; root growth is inhibited (Liu et al., 2022). PEPs involved in multiple plant immune responses bind to their specific receptors in an alkaline-dependent manner mediated by the Glu/Asp pH sensor. In the experiment by Liu et al. (2022), alkalinization was induced using an H^+-ATPase inhibitor; whether pH rise induced by other stress factors like bicarbonate exposure, drought, or salinity is sensed in the same way has not been explored so far.

Surprisingly, the overall root and leaf apoplastic pH values remain quite constant in the acidic range even in external media with an alkaline pH of up to 8 (Martinière et al., 2018) or bicarbonate exposure (Nikolic and Römheld, 2002). The ability to maintain the AHA activity under stress conditions is crucial for pH homeostasis and root growth. Tolerance to carbonate-rich soils is intimately related to AHA activity. Apoplastic acidification due to proton pumping is required for both root growth and the activity of ferric chelate reductase, which is highest at low pH (see below). Contrastingly, the pH of the xylem sap increases within 5 minutes after root exposure to bicarbonate (Wegner and Zimmermann, 2004). The authors claim that this observation supports the hypothesis that bicarbonate-induced alkalinization of the apoplast causes precipitation of Fe in the leaf apoplastic space, making it unavailable for biochemical processes. This hypothesis, however, is contradicted by studies with sunflower and grape plants where no increase in apoplastic Fe concentration was found under bicarbonate exposure (Nikolic and Römheld, 2002). It has been argued that in chlorotic leaves of plants exposed to bicarbonate microsites of alkalinization in leaf apoplast may drastically reduce the iron reductase activity causing Fe deficiency. Bulk tissue analyses may not detect these locally restricted events (Kosegarten et al., 1999). Iron translocation to the leaves can further be inhibited by the binding of a large proportion of Fe to hemicellulose in the root cell walls (Wang et al., 2020b).

Summarizing, alkalinization by bicarbonate is more stressful than high pH alone mainly because bicarbonate exacerbates Fe deficiency strongly interfering with the processes of uptake and transport in plants. The mechanisms behind still need research efforts that should focus on the identification of bicarbonate transport mechanisms and the regulation of apoplastic pH at the tissular and subtissular level in relation to Fe-chelate reductase activity. Salinity, the increasingly relevant stress factor for crop yields, aggravates these knowledge gaps. Below we highlight the current knowledge on how Na^+ salinity in combination with bicarbonate alkalinity affects plants, how tolerant plants manage this complex stress situation, and how these tolerance mechanisms can be used for the breeding of improved crops.

2.2.2 Physiological and Molecular Responses to Alkaline Salinity

Alkaline salinity stress alters multiple physiological and biochemical processes including water relations (Lambers and Oliveira, 2019; Hailu and Mehari, 2021), accumulation of compatible solutes (Ahmad et al., 2014; Mehta and Vyas, 2023), ion transport regulation (Wang et al., 2022b; Li and Yang 2023), photosynthesis efficiency (Yang et al., 2011; Shamsabad et al., 2022), antioxidant responses (Ahmad et al., 2014; Liu et al., 2015b), phytohormonal signaling (Rao et al., 2023), and mineral nutrition (Li et al., 2016; Almira-Casellas et al., 2023). In saline-alkali soils, plants absorb large amounts of Na^+ and Cl^- while the absorption of K^+, Ca^{2+}, Mg^{2+}, and other essential micronutrients is inhibited. This ion imbalance leads to accelerated chlorophyll degradation and the reduction of chlorophyll content. In addition to the reduction of growth, the stability of cell membranes can be harmed. The relative electrical conductivity of cells reflects the permeability change and damage degree of cell membrane under adverse conditions, and malonaldehyde (MDA) is a product of lipid peroxidation. MDA accumulates under alkali-saline stress; thus, cell membrane stability is one of the most used criteria to evaluate the salinity tolerance of plants. Reactive oxygen species (ROS) are a byproduct of normal cell metabolism of plants. However, abiotic stresses such as alkaline salinity break the balance between the production and elimination of ROS, causing oxidative damage and apoptosis in plant cells. When plants are exposed to salt-alkaline damage, intracellular ROS will increase and react with macromolecules, leading to active oxygen imbalance. Protective enzymes and antioxidants act against membrane lipid peroxidation, which can remove excessive free radicals under adverse conditions, thus playing a role in protecting the cell membrane structure (Dumanović et al., 2021).

2.2.2.1 Ion Homeostasis and Membrane Transport

Under alkaline salinity, CO_3^{2-} and HCO_3^- ions buffer the protons released for apoplast and rhizosphere acidification. This not only inhibits Fe uptake from the apoplast into the cytosol of root cells, due to a reduced Fe-reductase activity, but also weakens Na^+ extrusion and compartmentalization, processes that are H^+-gradient-dependent. This is the main reason why, at the same pH level (pH 8), plants on alkaline Na^+-rich soils accumulate significantly higher Na^+ and suffer from more intense K^+ and Fe depletion than plants on neutral saline soils (Paz et al., 2012; Wang et al., 2021).

The increase of intracellular Ca^{2+} is a further major ionic stress response under alkaline salinity, as Na^+ and bicarbonate ions exacerbate Ca^{2+} displacement from the cell walls to a higher degree than observed under neutral salinity or high pH alone (Wang et al., 2021). Under alkaline salinity root Na^+ efflux is also regulated by SOS1 Na^+/H^+ antiporter in many plant species (Cao et al., 2022). The increase of cytosolic Ca^{2+} in the cell is sensed by the Ca^{2+}-binding proteins SOS3 and SCaBP8/CBL10, which form a complex with SOS2. The activated SOS2 phosphorylates the Na^+/H^+ antiporter SOS1, promoting Na^+ efflux (Qiu et al., 2002). Besides the SOS pathway, the implication of other PM Na^+/H^+ antiporters (NHAs), vacuolar Na^+/H^+ and Ca^{2+}/H^+ exchangers (NHXs and CAXs), and K^+ transporters (HAK/KUP/HKTs) in reducing Na^+/K^+ ratios has been confirmed specifically under alkaline salinity (Fang et al., 2021). Maintenance of K^+ homeostasis is a key trait for alkaline-saline tolerance. This is further supported by the observation that $NaHCO_3$, but not NaCl, strongly enhanced the expression of HAK, NHX, and AKT gene families in rice. The overexpression of OsHAKs and OsAKT1 contributes to enhanced K^+ homeostasis in rice under alkaline salinity (Yang et al., 2012). Furthermore, lower K^+ efflux may indirectly benefit plant performance under alkaline salinity as the maintenance of K^+ homeostasis is also an important factor in the amelioration of Fe deficiency (Ye et al., 2019) (see also Section 2.2.3.2).

Alkaline salinity not only hampers the homeostasis of cationic nutrients like Ca, K, Fe, Zn, and Cu but also has a deep impact on nutrients that plants take up in anionic form. Tissue concentrations of N, P, S, and Mo can also be substantially decreased (Zhang et al., 2020; Hu et al., 2022). A high proportion of the differentially expressed genes in *Leymus chinensis* under alkaline salinity were genes related to nutrient transport including not only those for alkaline cations, alkaline earth cations, Fe, and Zn but also those coding for nitrate, phosphate, sulfate, and molybdate (Hu et al., 2022). This broad influence of alkaline salinity on the entire mineral nutrient profile of plants is a highly relevant trait to be considered in both breeding programs for tolerance to alkaline salinity and the management of crops on alkaline-saline soils.

2.2.2.2 Increased Cytosolic and Organelle Accumulation of Compatible Solutes

The roles of proline and betaine in osmoregulation under neutral salinity are well established (Singh et al., 2015). However, osmoregulation by compatible solutes under alkaline salinity seems to be more intricate. Enhanced accumulation of both Pro and GB under alkaline salinity has been observed in different species. In mulberry (Ahmad et al., 2014), but not in rice (Lv et al., 2014), Pro and GB were found to be good indicators for sodium carbonate-induced alkalinity tolerance. At equimolar Na^+ concentrations, Pro and GB shoot concentrations are higher under alkaline than under neutral salinity (Lv et al., 2013). These observations point to high pH (or bicarbonate ion) rather than Na^+ as a pivotal factor for Pro and GB accumulation under alkaline salinity under similar or even slightly lower osmotic stress (Yang et al., 2008).

Moreover, under alkaline salinity, other osmoregulators may play a role. In bean, root and shoot Pro and GB displayed higher accumulation under alkaline than under neutral salinity, but their proportion among all osmoregulators was small. Their contribution to the total molarity of osmotically active substances decreased under alkaline salinity due to an enhancing effect on the OA concentrations in shoots and, most of all, in roots (Yu et al., 2019). Likewise, in *Suaeda glauca*, OAs were found to be the dominant factor in maintaining ionic equilibrium under alkaline salinity, with a 74% contribution to the total produced anions (Yang et al., 2008).

2.2.2.3 OA Production

Several alkali-tolerant halophytes exhibit elevated OA concentrations in shoots when subjected to alkaline salinity (Shi et al., 2002; Yang et al., 2007). This shoot accumulation of OA may compensate for the imbalance in the inorganic cation/anion ratio. In roots enhancement of both biosynthesis and secretion of OA may contribute to rhizosphere acidification and micronutrient mobilization in the alkali-tolerant halophyte *Chloris virgata* (Yang et al., 2010). Contrastingly, in alkaline-sensitive plants bicarbonate-induced OA accumulation in roots can be a sign of detrimental response (Yang et al., 1994).

Dissecting alkaline salinity into pH, bicarbonate, or Na^+ toxicity in sensitive plants has revealed a bicarbonate-specific elicitation of the OAs production in roots (Hajiboland et al., 2005) and shoots (Yang et al., 2008; Shi and Sheng, 2005), coupled with a lower root OAs exudation (Sagervanshi et al., 2021). In such a case, Fe mobilization is not favored, and the high tissue OA concentrations may hamper Fe acquisition by the leaf mesophyll (Abadía et al., 2002). Clearly, the role of bicarbonate-induced enhancement of the dark fixation of inorganic carbon and other sources of enhanced OA production under alkaline-saline stress deserves further investigation.

2.2.2.4 Hormone Signaling and Other Plant Growth Regulators

An increasing number of studies reveal an involvement of stress-related genes associated with plant hormone biosynthesis and signal transduction under alkaline salinity (Zhang et al., 2022a; Wang et al., 2023a). However, there is no clear trend of phytohormone accumulation and related gene expression patterns in plants upon alkaline salinity. Their endogenous concentrations in tissues are extremely low and variable depending on the tissue, species, and time point. Exposure to alkaline salinity results in positive regulation of abscisic acid (ABA) through both biosynthesis and the ABA signal transduction pathway: ABA receptors PYL/PYRs and ABA-responding element binding factors (ABFs). Under alkaline salinity, ABA biosynthesis not only reduces K^+ efflux, membrane injury, and ROS accumulation (Li et al., 2022) but also activates the PM H^+-ATPase by inhibiting PP2Cs. Generally, an increase of indole acetic acid, jasmonic acid (JA), and the ethylene precursor under alkaline salinity is also reported, whereas JA–salicylic acid (SA) and ABA–SA antagonism is found. Small auxin-up RNAs (*SAURs*) are auxin-responding genes with a promising role in the plant alkaline salinity response mechanisms, for contributing to PP2Cs inhibition and PM H^+-ATPase activation together with ABA. Whereas the activation of auxin-responsive families of GH3 and SAUR in the plant's early response to alkaline salinity is reported (Zhang et al., 2022a), the repression of SAURs due to a sustained ABA and JA accumulation upon long-term exposure to alkaline salinity needs to be addressed, as it potentially constitutes an ABA/auxin-mediated PM H^+-ATPase regulation mechanism. To date, special emphasis has been placed on the involvement of strigolactone, a carotenoid-derived phytohormone, in the ABA-dependent and ABA-independent regulation of expression of H^+-ATPase genes and induction of OAs upon alkaline salinity (Ma et al., 2022a).

Furthermore, the polyamines (PAs) putrescine (Put), spermidine (Spd), and spermine (Spm) accumulate in response to alkaline salinity (Do et al., 2014). PA degradation is one synthesis pathway of the non-protein amino acid γ-aminobutyric acid (GABA). Under alkaline salinity both PAs and GABA increase. (Xu et al., 2019). Nitric oxide induced by ROS accumulation and acting downstream to GABA may regulate the PM H^+-ATPase under alkaline salinity (Xu et al., 2021b).

2.2.3 Molecular Genetics of Plant Tolerance to Alkaline Salinity

2.2.3.1 Osmoregulation

The regulation of Pro and GB metabolic pathways is essential for the accumulation of these osmolytes under abiotic stress. Δ1-pyrroline-5-carboxylate synthase (P5CS), pyrroline-5-carboxylate reductase (P5CR), ornithine transaminase (δ-OAT), proline dehydrogenase (PDH), pyrroline-5 carboxylate dehydrogenase (P5CDH), and proline transporter (ProT) are the key enzymes in the proline metabolic pathway. The induction of Pro synthesis genes and the inhibition of Pro catabolic genes as a plant response to alkaline salinity have been described in different species (Lv et al., 2014; Min et al., 2023). Indeed, in the *Ms*P5CS1a-RNAi mutant root growth rates decreased compared to the empty vector line under saline-alkali stress (Min et al., 2023), indicating that proline metabolism-related genes play an important role in saline-alkali stress tolerance.

Increased betaine aldehyde dehydrogenase (BADH) activity catalyzes the production of betaine from choline and implies GB accumulation, protecting the activity of intracellular proteins and metabolic enzymes, stabilizing the cell membrane, and reducing the ROS accumulation (Tian et al., 2017). Besides *BADH*, genes encoding IMP biosynthetic process, sucrose synthase (*SUS*), tyrosine metabolic process, sucrose phosphatase, and annexin have been reported among the differentially expressed genes related to osmoregulation under saline-alkali stress in *Trollius chinensis* seedlings (Hou et al., 2023).

Further candidate genes involved in the improvement of the plant osmotic balance for enhancing tolerance to alkaline salinity have been identified through molecular mapping and validation. These include the aquaporin CrPIP2.3 from the alkaline-halophyte *Canavalia rosea* (Zheng et al., 2021), the transcriptional repressor GmMYB3A of *Glycine max* (He et al., 2018), and the S-phase kinase-associated GsSKP21 from the ancestral soybean *Glycine soja* (Liu et al., 2015a) (Table 2.1). The generation of overexpression lines for each target gene in *A. thaliana* and soybean has revealed how the maintenance of the cell water status is achieved through the modulation of the ABA-response signaling pathway under alkaline salinity treatment. Enhanced expression of PIP2 and SKP21 led to major activation of transcription factors (TFs) regulating the ABA-mediated dehydration responses, like RD26, ANAC019, and NCED3, and the ABA-inducible gene *RD29A*, leading to increased tolerance to alkaline salinity. Likewise, *GmMYB3a* activation revealed an increase in the sensitivity to alkaline salinity due to the enhanced repression of the ABA regulator ABI5, responsible for ROS-scavenging processes among others.

2.2.3.2 Regulation of Ion Transport

We already have an advanced picture of the mechanisms underlying tolerance to either neutral salinity (Zelm et al., 2020) or micronutrient deficiency associated with non-saline-alkaline soils (Vélez-Bermúdez and Schmidt, 2023). However, under the combination of salinity and alkaline stress, we need still many answers to the key questions regarding how bicarbonate affects the tolerance to toxic Na^+ concentrations, and vice versa, how high Na^+ concentrations affect the ability of plants to maintain micronutrient acquisition and internal pH homeostasis.

Tolerance to alkaline salinity is associated with increasing transcript levels of PM-H^+-ATPases, for maintaining root activity and reducing Na^+ uptake. This increase in gene expression is mediated by Ca^{2+}-mediated degradation of PM-H^+-ATPase repressors. The specific induction under $NaHCO_3$ of some NHX members is an important mechanism of ion and pH homeostasis under alkaline salinity. Compartmentalization of Na^+ into vacuoles by NHXs is coupled with H^+ extrusion and provides a source of cytosolic H^+ for posterior H^+ efflux (by PM-H^+-ATPases) or H^+ recirculation into the vacuoles (by V-H^+-ATPases and H^+-PPases). The fine-tuning of proton extrusion and proton tonoplast import is, therefore, crucial in the plant's response to alkaline salinity. Moreover, NHXs cation selectivity for Na^+ or K^+ is regulated through vacuolar Ca^{2+} signaling, where the binding of calmodulin (CaM)-like protein 15 (AtCAM15 in *Arabidopsis*) to AtNHX1 decreases its Na^+/H^+ exchange activity. Interestingly, CAM15-NHX1 interaction decreases dramatically under high pH (Yamaguchi et al., 2005). The release of AtCaM15 from the C-terminus of AtNHX1 increases Na^+/H^+ exchange activity and favors higher vacuolar Na^+ accumulation and H^+ cytosolic pool. Therefore, pH-dependent tonoplast Na^+/H^+ antiporters are key players for ionic balance under alkaline salinity, favored by the vacuolar alkalinization occurring upon stress exposure and modulated by the activity of V-H^+-ATPases and H^+-PPases for vacuolar re-acidification.

Potassium homeostasis, K^+ transporters (HAKs/KUPs/HKTs), and K^+ channels (AKTs) suffer severe activity decrease when the proton motive force (pmf) is not maintained. K^+ efflux is strongly enhanced and favored by both the consequent change in the electrochemical gradient and the activation of the outward-rectifying K^+ channels and non-specific cation channels (Falhof et al., 2016). The major plant growth restriction under alkaline salinity has indeed been linked to a strong impairment in potassium influx. Similarly, proton gradient restriction under alkaline salinity hampers solubilization and uptake of iron in species from both Strategy I (relying on PM H^+-ATPase

TABLE 2.1
Key Genes for Tolerance to Alkaline Salinity Validated through the Generation of Mutant Lines, Complementation Studies, or Gene Expression Analysis

Gene Symbol	Species	Treatment	Tissue	Role	Validation Method	References
CrPIP2.3	*Canavalia rosea*	150 mM $NaHCO_3$ (pH 8.2)	Leaf	Enhancing water management functions	Overexpression lines in *A. thaliana*	Zheng et al. (2021)
OsPPa6	*Oryza sativa*	50 mM $NaHCO_3$/Na_2CO_3 (pH 9.2)	Leaf	Regulating osmotic balance	CRISPR/Cas9-generated mutants in rice	Wang et al. (2019a)
GmMYB3A	*Glycine soja*	100 mM Na_2CO_3	Leaf	Negatively regulating stress-related genes DREB, ABF4, GmABI5, AOS, MYC, and PAL	Overexpression lines in soybean	He et al. (2018)
GsSKP21	*G. soja*	50 mM $NaHCO_3$	Root and leaf	Positively regulating the ABA-mediated stress response	Overexpression lines in *A. thaliana*	Liu et al. (2015a)
GmNHX6	*Glycine max*	90 mM $NaHCO_3$	Root and leaf	Maintaining K^+ homeostasis	Overexpression lines in *A. thaliana*	Jin et al. (2022a)
OsDMI3	*Oryza sativa*	75 mM $NaHCO_3$ (pH 8.0)	Root	Regulating OsSOS1, OsA3, and OsA8 expression	Knockout and overexpression lines in rice	Ni et al. (2021)
MdTyDc	*Malus domestica*	1 M $NaHCO_3$: $NaCO_3$ (1:1) (pH 9.1)	Root and leaf	Increasing dopamine content and tissue N, maintaining root activity and reducing tissue [Na^+]	Overexpression lines in apple tree	Liu et al. (2021b)
ZmNSA1	*Zea mays*	100 mM $NaHCO_3$	Root	Mediating the transcriptional upregulation of MHAs	UniformMu mutant and overexpression lines in maize	Cao et al. (2020)
BOR2	*G. soja*	50 mM $NaHCO_3$	Root	Potential B efflux transporter with HCO_3^- affinity	Knockout and transgenic lines in *A. thaliana*	Duan et al. (2018b)
SLAH3	*G. soja*	50 mM $NaHCO_3$	Root, hypocotyl	Facilitating root-to-shoot nitrate transport	Knockout and overexpression lines in *A. thaliana*	Duan et al. (2018a)
ERF71	*G. soja*	50 mM $NaHCO_3$ (pH 8.5)	Root	Increasing AHA2 expression	Overexpression lines in *A. thaliana*	Yu et al. (2017)

(Continued)

TABLE 2.1 (*Continued*)
Key Genes for Tolerance to Alkaline Salinity Validated through the Generation of Mutant Lines, Complementation Studies, or Gene Expression Analysis

Gene Symbol	Species	Treatment	Tissue	Role	Validation Method	References
ERF6	*G. soja*	50 and 100 mM $NaHCO_3$ (pH 8.5)	Root	Increasing NADP-ME, H^+-Ppase, and H^+-ATPase genes; inducing ABA and ET signaling	Overexpression lines in *A. thaliana*	Yu et al. (2016)
GsPPCK3	*G. soja*	50 mM $NaHCO_3$ (pH 8.5)	Root and leaf	Inducing expression of plant elicitor peptide carboxylase, CS, H^+-ATPase, and NADP-ME genes	Overexpression lines in alfalfa	Sun et al. (2014)
JAZ2	*G. soja*	50 mM $NaHCO_3$	Root and leaf	Positively regulating the expression of NHX1, NADP-ME and H^+-Ppase	Overexpression lines in *A. thaliana*	Zhu et al. (2012)
SbAT1	*Sorghum bicolor*	62.5 mM $NaHCO_3$ + 12.5 mM Na_2CO_3 (pH = 9.3)	Root and leaf	Negatively regulating phosphorylation of aquaporin PIP2s, increasing H_2O_2 levels in cells	GWAS and NILs in sorghum, knockout, and overexpression lines in sorghum, millet, rice, and maize	Zhang et al. (2023)
GmSNF1	*Glycine max*	50 mM $NaHCO_3$	Root and leaf	Interacting with GmPKS4, increasing CAT activity, reducing H_2O_2	Overexpression lines in *Arabidopsis* and soybean, gene-silenced lines in soybean	Lu et al. (2023)
AhALDH3H1	*Arachis hypogaea*	40 mM NaCl:$NaHCO_3$:Na_2CO_3 (1:10:1)	Root	Maintaining cell wall structure and metabolite transport. Increasing ALDH and SOD activity, reducing MDA	Overexpression lines in soybean	Cao et al. (2023)
AmPAL	*Astragalus membranaceus*	50, 60, 100, and 150 mM $NaHCO_3$	Root and leaf (cytoplasmic membrane)	Protective role. Increasing PAL, SOD, and Pro, reducing MDA	Overexpression lines in tobacco and maize	Fan et al. (2022)

(*Continued*)

TABLE 2.1 (*Continued*)
Key Genes for Tolerance to Alkaline Salinity Validated through the Generation of Mutant Lines, Complementation Studies, or Gene Expression Analysis

Gene Symbol	Species	Treatment	Tissue	Role	Validation Method	References
RBOH1	*Solanum lycopersicum*	15 mM NaCl + 15 mM $NaHCO_3$ + 15 mM Na_2CO_3 (pH 8.9)	Leaf	Increasing SOD, CAT, APX, and GR activity	Agrobacterium-mediated transgenic lines	Xu et al. (2021a)
SlWRKY28	*Salix linearistipularis*	100 mM $NaHCO_3$	Root and leaf	Inducing PtAPX, PtEnolase, PtSOD, and PtSPDS gene expression. Increasing APX activity, reducing MDA	Overexpression lines in *Populus davidiana* × *Populus bolleana*	Wang et al. (2020c)
MsSiR	*M. sativa*	150 mM $NaHCO_3$ (pH 8.5)	Leaf	Improving antioxidant activity, reducing H_2O_2 and O2, increasing GSH, Cys, and GSSG	Overexpression lines in alfalfa	Sun et al. (2020)
MdATG18a	*Malus domestica*	1 M $NaHCO_3$:$NaCO_3$ (1:1) (pH 9.1)	Leaf	Positively regulating autophagic activity. Increasing CAT, SOD, POD, and APX activities, enhancing AsA, DHA, GSH, GSSG, Glu, and GABA, reducing H_2O_2 and O_2^-	Overexpression lines in apple tree	Li et al. (2020c)
MsGSTU8	*M. sativa*	100 mM Na_2CO_3: $NaHCO_3$, 1:2	Leaf	Decreasing the accumulation of ROS and increasing the levels of antioxidant enzymes	Overexpression lines in tobacco	Du et al. (2019)
GSNOR	*S. lycopersicum*	100 mM $NaHCO_3$	Root and leaf	Regulating redox signaling (ROS and RNS)	Suppression and overexpression lines in tomato	Gong et al. (2015)

(*Continued*)

TABLE 2.1 (*Continued*)
Key Genes for Tolerance to Alkaline Salinity Validated through the Generation of Mutant Lines, Complementation Studies, or Gene Expression Analysis

Gene Symbol	Species	Treatment	Tissue	Role	Validation Method	References
MsNF-YC2	*M. sativa*	150 mM $NaHCO_3$ (pH 8.5)	Root and leaf	Increasing RWC, SOD, and POD activities and Pro. Reducing MDA, H_2O_2, and O_2^-; putatively modulating phytohormone and photosynthesis signaling	Overexpression lines in alfalfa	Yu et al. (2022)
VaERF3	*Vigna angularis*	50 and 100 mM $NaHCO_3$	Root and leaf	Positively regulating ERF expression. High levels of Pro, low levels of MDA and ROS	Overexpression lines in *A. thaliana*	Li et al. (2020a)
LcCHI2	*Leymus chinensis*	100 mM Na_2CO_3	Leaf	Inducing chitinase activity, bacterial and fungal resistance. Reducing shoot Na^+, MDA, and CEC	Overexpression lines in tobacco and maize	Liu et al. (2020b)
MsCBL4	*M. sativa*	100 mM $NaHCO_3$	Root and leaf	Enhancing Ca^{2+} transport and accumulation, increasing CAT, SOD, and POD activity, reducing MDA,H_2O_2,andO_2^-	Overexpression lines in tobacco	An et al. (2020)
GsbZIP67	*M. sativa*	50, 100, and 150 mM $NaHCO_3$	Root and leaf	Inducing MtNADP-ME and MtH^+PP-ase gene expression. Increasing POD activity, reducing MDA and ion leakage	Overexpression lines in alfalfa	Wu et al. (2018)
SsMT2	*Suaeda salsa*	400 and 500 mM $NaHCO_3$	Root and leaf	Improving ROS-scavenging ability. Putatively interacting with Na^+ transporters for improved Na^+ extrusion	Yeast transformation; Agrobacterium-mediated transgenic lines of *A. thaliana*	Jin et al. (2017)

(Continued)

TABLE 2.1 (*Continued*)
Key Genes for Tolerance to Alkaline Salinity Validated through the Generation of Mutant Lines, Complementation Studies, or Gene Expression Analysis

Gene Symbol	Species	Treatment	Tissue	Role	Validation Method	References
MhANR	*Mallus halliana*	100 mM 1:1 NaCl:$NaHCO_3$ (pH 8)	Leaf and calli	Negatively regulating ion homeostasis, osmotic, and antioxidant responses	Overexpression lines in *A. thaliana*, tobacco, and apple calli	Wang et al. (2023c)
MhZDS	*M. halliana*	101 mM 1:1 NaCl:$NaHCO_3$ (pH 8)	Leaf and calli	Positively regulating ion homeostasis, osmotic, and antioxidant responses	Overexpression lines in *Arabidopsis*, tobacco, and apple calli	Wang et al. (2023c)
GmPKS4	*G. max*	90 mM NaCl + 40 mM $NaHCO_3$	Root and leaf	Inducing AtCBL1, AtNHX1, GmCBL1, GmNHX1, and GmCAX1 gene expression. Increasing Pro and RWC and CAT, POD, and SOD activity, reducing MDA	Overexpression lines in *A. thaliana* and soybean	Ketehouli et al. (2021)
NINJA/YUC8	*Arabidopsis thaliana*	75 mM NaCl + 15 mM $NaHCO_3$ (pH 8.3)	Root and leaf	NA	GWAS and gene expression analyses	Almira-Casellas et al. (2023)
GmDNAJC7	*G. max*	90 mM $NaHCO_3$	Root and leaf	Inducing ER stress response, H_2O_2 response, and cellulose biosynthesis gene expression (in silico)	Knockout and overexpression lines of *Arabidopsis*, soybean, and tobacco	Jin et al. (2022b)
LpCPC	*Lilium pumilum*	500 mM $NaHCO_3$	Root and leaf	Enhancing lignin biosynthesis and soluble phenols, increasing NHX1 and SOS1 gene expression, reducing membrane damage	Overexpression lines in tobacco	Dong et al. (2022)
OsFBDUF54	*O. sativa*	40 mM $NaHCO_3$	Root	NA	GWAS and CRISPR/Cas9-generated mutants in rice	Li et al. (2020b)

(*Continued*)

TABLE 2.1 (*Continued*)
Key Genes for Tolerance to Alkaline Salinity Validated through the Generation of Mutant Lines, Complementation Studies, or Gene Expression Analysis

Gene Symbol	Species	Treatment	Tissue	Role	Validation Method	References
GsNAC019	*G. soja*	50 mM $NaHCO_3$ (pH 8.5)	Root and leaf	Inducing AHA2, KIN, and RD29A gene expression. Putatively modulating the ABA signaling pathway	Overexpression lines in *A. thaliana* and soybean	Cao et al. (2017)
AKT1	*Puccinellia tenuiflora*	150–200 mM Na_2CO_3 (pH 11)	Root	Induced under alkaline salinity	RNA sequencing	Zhao et al. (2016)
Gshdz4	*G. soja*	125 mM $NaHCO_3$ (pH 8.5)	Root	Inducing bicarbonate-responsive genes NADP-ME and H^+-PPase. Putatively inducing GsRD29B via GsNAC019 regulation. Increasing POD activity, reducing MDA	Overexpression lines in *A. thaliana*	Cao et al. (2016)
GsCML27	*G. soja*	50 mM $NaHCO_3$	Root	Increasing germination rates specifically under alkaline salinity	Overexpression lines in *A. thaliana*	Chen et al. (2016)
osa-MIR396c	*O. sativa*	75–100 mM $NaHCO_3$	Leaf	Negatively regulating target GRF and other regulatory proteins	Suppression and overexpression lines in *A. thaliana*, rice, and tobacco	Gao et al. (2010)
osa-MIR393	*O. sativa*	75–100 mM $NaHCO_3$	Leaf	Negative regulator to target alkali response genes	Overexpression lines in rice and *A. thaliana*	Gao et al. (2010)

activity) and Strategy II (relying on phytosiderophore secretion). Altered expression levels of the ferric chelate reductase FRO2 (FERRIC REDUCTASE OXIDASE 2), which reduces Fe(III) to Fe(II), the iron transporter IRT1 (IRON-REGULATED TRANSPORTER 1), which transports Fe(II) into root epidermal cells, and other Fe-acquisition related genes (*IRO2*, *YSL2*, and *YSL15*) are reported as part of the plant response to alkaline salinity stress, and their upregulation correlates with increased Fe uptake efficiency under alkaline salinity conditions. Likewise, the upregulation of genes coding for key phytosiderophore biosynthetic enzymes (*NAS1-NAS2*) and coumarin secretion (*PDR9*, *BGLU42*) enhances the response of plants under iron-limiting conditions as occurring in alkaline-saline soils (Kobayashi et al., 2015; Li et al., 2016; Waters et al., 2018).

Nitrate is the primary N source in soils and, apart from being an essential nutrient, serves as a signaling molecule in various biological processes that regulate abiotic stresses. Most Arabidopsis SLAC/SLAHs are reported to be nitrate conductive, and the overexpression of *AtSLAH1/3* increases the shoot NO_3^-/Cl^- ratio under high Cl^- levels (Qiu et al., 2016). In soya, *GsSLAH3* is upregulated after alkaline salinity treatment in both roots (Zhang et al., 2016) and shoots (Duan et al., 2018a). *A. thaliana GsSLAH3*-overexpressing lines accumulated more NO_3^- than the wild type and exhibited major tolerance, suggesting that *GsSLAH3* is a positive bicarbonate-responsive gene that increases resistance to alkaline salinity (Table 2.1).

The generation of knockout, suppression, and overexpression mutants of gene variants detected specifically under alkaline salinity has confirmed the involvement of an extensive number of novel candidates for the improvement of ionic homeostasis. It is the case of DMI3 and PpA6 (in rice), TyDC (in apple), and ERFs 6–71, PPCK3, and JAZ2 (in soybean), all observed to enhance proton pump transcript expression under alkaline salinity, promoting the overall plant tolerance. Moreover, the phosphorylation of H^+-ATPase repressors mediated by Ca^{2+}-binding proteins has been depicted in maize for NSA1, a negative regulator of MHA2 and MHA4, hampering root Na^+ efflux under alkaline salinity. The bicarbonate-specific induction of GsBOR2, a putative boron efflux transporter, has been linked to a potential role of BOR2 in bicarbonate transport and enhanced plant tolerance to alkaline salinity (Table 2.1).

2.2.3.3 ROS Scavenging

When plants are subjected to saline-alkali stress, ROS can accumulate in large amounts, usually boosting O_2^- at early stages and then converting it into H_2O_2 for further accumulation (Hou et al., 2023). As a stress signal, the increase of ROS can trigger protecting enzymes such as superoxide dismutase (SOD), peroxidase (POD), ascorbate peroxidase (APX), and/or glutathione reductase (GR). SOD as the first line of defense catalyzes the conversion of O_2^- into H_2O_2, and POD catalyzes the conversion of H_2O_2 into O_2 and H_2O. AsA and GSH are both antioxidants aiding membranes against oxidation by acting as buffers in redox processes. AsA serves as an electron donor for APX to scavenge H_2O_2. GR is a rate-limiting enzyme in the ROS-scavenging system, which can increase and maintain enhanced GSH concentrations (Foyer and Noctor, 2011). As an example, the activities of AsA, APX, GSH, and GR were maintained or increased in *T. chinensis* seedlings exposed to saline-alkali stress. The increase in GR activity elevated the GSH content, which was then reduced to AsA through the AsA–GSH cycle (Hou et al., 2023). The content of AsA and APX activity were still significantly higher than the control although they were slightly decreased under high saline-alkali concentration.

NADPH oxidases play a fundamental role in ROS signaling of salt stress (Liu et al., 2020a). Arabidopsis mutants *rbohD* that are unable to produce H_2O_2 when exposed to NaCl stress accumulate higher Na^+ concentrations and are more salt sensitive than wild type (Wang et al., 2019c). Rice knockout mutants, *OsRbohA*, are more NaCl sensitive due to worse K^+ homeostasis (Wang et al., 2022a). Contrastingly, NADPH oxidase mutants of rice, *osrbohd* or *osrbohh*, are more tolerant to alkaline salinity than wild-type rice with intact capacity to generate superoxide radicals (Shen et al., 2023). These contrasting findings may be due to the large structural and functional differences among the multiple NADPH oxidases in plants (Liu et al., 2020a).

Furthermore, in rice and other cereals, ATI1 (Alkaline Tolerance 1) has been found to cause sensitivity to alkaline salinity (Zhang et al., 2023). ATI1, by interacting with Gβ, inhibits the phosphorylation of PIP2, an aquaporin able to transport H_2O_2 (Table 2.1). The consequent intracellular accumulation of H_2O_2 enhances the sensitivity to alkaline salinity. Other novel candidates for alkaline salinity tolerance have been validated by their involvement in enhancing the plant redox status and are listed in Table 2.1. Among them, the salt-alkali tolerance function of GmSNF1 (a Ser/Thr protein kinase) has been demonstrated to rely on its interaction with a salt-alkaline tolerance gene (*PKS4*) and the enhancement of antioxidant enzyme activity, reflected by increased CAT activity and decreased H_2O_2 tissue accumulation.

2.3 AGRONOMICAL SOLUTIONS FOR CULTIVATION IN SALINE-ALKALINE SOILS

2.3.1 Chasing Natural Variation in Wild Crop Relatives

The degree of salt and alkaline tolerance largely differs among and within plant species based on their genetic variations. Exploration of more than 300 sorghum accessions revealed varieties tolerating alkaline soils up to pH 10 (Zhang et al., 2023). GWAS analysis in sorghum allowed us to identify AT1 as a major player in saline-alkaline tolerance, as mentioned above. Nonetheless, variability may be poor in other cultivated plants. Crop wild relatives (CWR), instead, have higher genetic and phenotypic variability and are sources of desirable tolerance genes. The wide genetic pool of CWRs has allowed the identification of candidate genes for salinity tolerance and transcription factors associated with salt stress responsive genes (see Abdul Aziz and Masmoudi, 2023; Busoms et al., 2023; Kapazoglou et al., 2023). For example, Orsini et al. (2010) described the distinctive traits and physiological responses of halophytic *Arabidopsis*-relative model species under salt stress to establish the basis for salinity tolerance genetic research and exploration. Later, several studies used *Eutrema salsugineum* as a model for deciphering mechanisms of salt and other abiotic stresses (see Li et al., 2022). In rice, 22 wild *Oryza* relatives have been recognized as important genetic resources, highlighting *Oryza rufipogon*, *Oryza coarctat*, *Oryza latifolia*, and *Oryza alta* as best germplasms for the development of salt-tolerant cultivated rice (Solis et al., 2020).

The sources of the Na^+ exclusion mechanism among wheat wild relatives such as *Aegilops* spp., *Thinopyrum* spp., *Elytrigia elongata*, or *Hordeum* spp. were identified (Colmer et al., 2006) and have been successfully implemented in molecular breeding programs (Naeem et al., 2020). The sequencing and construction of a recombinant inbred population of wild soybeans allowed the identification of multiple QTLs relevant to traits of interest in agriculture such as the ion transporter *GmCHX1* (Qi et al., 2014). CHXs are members of the CPA2 subfamily, are involved in the exchange and transport of Na^+ and H^+, and confer salt tolerance in several species (Isayenkov et al., 2020). In millets, whole-genome sequence information of wild relatives such as sorghum, pearl millet, or foxtail millets is available and has been used efficiently to identify candidate genes for abiotic stress tolerance and for advancing breeding strategies (Satyavathi et al., 2019).

The Solanaceae family includes commercially important vegetable crops (tomato, pepper, eggplant) characterized by their relative sensitivity to salinity. One of the most promising species for introgression breeding in eggplant is *Solanum insanum* because it grows in a wide range of soil conditions and interspecific hybrids between *Solanum melongena* and *S. insanum* are easily obtained. The higher salt tolerance of *S. insanum* seems to rely mostly on its ability to accumulate higher concentrations of proline (Brenes et al., 2020), but the genomic bases have not yet been studied. In addition, transcriptome sequencing studies of wild relatives of potato (*Ipomoea imperati*), sweet potato (*Ipomoea pes-caprae*), tomato (*Solanum chilense*), cotton (*Gossypium aridum*), among others have been performed to identify high salt-tolerant genes (Razzaq et al., 2021).

All these important studies are mainly focused on neutral salinity, and there are few CWRs well-characterized under salt-alkali conditions. Wild soybean (*Glycine soja*) is an exception, and efforts have been allocated to characterize genes involved in ion homeostasis and ROS damage regulation under salt-alkaline stress (see Cai et al., 2022). Recently, Zhou et al. (2023) compared the proteomic landscape of two contrasted wild soybean strains under saline-alkaline stress and concluded that tonoplast intrinsic protein (GsTIP2-1 and GsTIP2-2), inositol transporter (GsINT1), sucrose transport protein (GsSUC4), and autoinhibited Ca^{2+}-ATPase (GsACA11) can enhance saline-alkaline tolerance by relieving the inhibition of protein synthesis and amino acid metabolism.

Cao et al. (2020) showed that natural maize inbred lines harbor large genetic variations of shoot Na^+ exclusion under saline-alkaline stress. They detected a 4-bp deletion in the *ZmNSA1* gene and

TABLE 2.2
Saline-Alkaline-Tolerant Plant Species with Potential for Commercial and Conservational Uses

Species	Potential Use	References
Miscanthus hybrid *(Miscanthus sinensis × Miscanthus lutarioriparius)*	Biomass and soil remediation	Xu et al. (2021c)
Leymus multicaulis (Kar. and Kir.) Tzvelev	Fodder and cover crop	Majerus (1996)
Leymus angustus (Trin.) Pilg.	Fodder and cover crop	Majerus (1996)
*Leymus chinensis (*Trin.) Tzvelev	Fodder and cover crop	Liu et al. (2015b)
Thinopyrum ponticum Podp.	Fodder and cover crop	Majerus (1996)
Agropyrum cristatum L.	Fodder and cover crop	Majerus (1996)
Agropyrum desertorum (Fisch. ex Link) Schult.	Fodder and cover crop	Majerus (1996)
Agropyrum sibericum (Willd.) Beauv.	Fodder and cover crop	Majerus (1996)
Lepidium latifolium L.	Biomass and cover crop	Hajiboland et al. (2018)
Lolium sp. inoculated with *Fusarium incarnatum*	Remediation of Zn polluted saline-alkaline soil	Zhang et al. (2022b)
Tripolium pannonicum (Jacq.) Dobrocz.	Culinary use and soil remediation	Jekabsone et al. (2023)
Cochlearia officinalis L.	Culinary use and soil remediation	Jekabsone et al. (2023)
Lobularia maritima (L.) Desv.	Culinary use and soil remediation	Jekabsone et al. (2023)

observed that the lack of ZmNSA1 promoted shoot Na^+ exclusion and saline-alkaline tolerance. Therefore, ZmNSA1 was proposed as a target gene for improving maize saline-alkaline tolerance by gene editing or marker-assisted selection strategies.

Highly stress-tolerant but low-fruit-quality genotypes can contribute to improving yield under saline conditions by grafting. This surgical means of joining a scion and rootstock of two different genotypes with the desired traits offers an alternative to breeding and biotechnological approaches to salt tolerance. Grafting high-yielding but salt-susceptible cultivars onto salt-resistant/tolerant rootstocks is already an established procedure for sustainable yield improvement of tomatoes and other vegetables under neutral salinity (Singh et al., 2020). However, grafting can also be useful under alkaline salinity. In this sense, better performance, enhanced leaf iron concentrations, and increased OA root exudation have been observed when watermelon, a crop susceptible to alkaline salinity, was grafted onto an alkaline salinity-tolerant pumpkin (Colla et al., 2010).

Moreover, the use of wild halophytic species as forage resources in saline environments has gained increasing attention. Many wild species have forage potential or can be used for the rehabilitation of problematic soils, biofuel production, or intercropping. Table 2.2 gives some examples of saline-alkaline species with potential practical use. However, adequate management is required to avoid salt enrichment of surface soil by favoring the transport from deep soil layers and the replacement of valuable native species by invasive plants (Hooks et al., 2018).

2.3.2 Genomics-Assisted Breeding and Genome Editing

Genomics-assisted breeding is an advanced approach where both genomic and phenotypic information are considered for selecting the best phenotypes (Satyavathi et al., 2019). It is well known that traditional breeding is time-consuming, laborious, and sometimes inefficient. Instead, genetic engineering, which involves directly introducing genes conferring tolerance to alkaline salinity, is a powerful tool for breeding salt-alkaline tolerant cultivars. Although transgenic technology shortens the time required for breeding, public acceptance is low; thus, researchers have preferred to focus on mutagenesis or genome-editing strategies that are more accepted. For example, Zhang et al. (2022c) used heavy-ion-beam mutagenesis in rice to induce broad-spectrum mutations at high frequencies and produce salt-alkaline tolerant rice mutants. Indeed, the grain yield of mutant line M89 was higher than the local cultivar (TH899) when cultivated in soda saline-alkaline soils. Transgenic Kitaake rice plants generated with high-throughput activation tagging-based T-DNA insertion mutagenesis and overexpressing *OsSDG721* showed tolerance to alkaline salinity by the upregulation of *OsHKT1;5*, along with increased leaf angle and advanced heading and ripening dates (Liu et al., 2021a).

In forest trees, a collection of genes related to saline-alkali tolerance have been cloned and functionally localized, and later some transgenic lines were generated by genetic transformation to increase the resilience of Chinese forest from salinized lands (Liu et al., 2012). However, poplar is still the most transformed tree. For example, the TF WRKY (*SlWRKY28* from the *Salix linearistipularis*) was transferred to *Populus davidiana*×*Populus bolleana*, and the significant overexpression of the gene under alkaline salinity alleviated the symptoms of stress by increasing ROS scavenging (Wang et al., 2020c).

2.3.3 Amendments

The combination of traditional amelioration strategies with novel biotechnological and ecoengineering tools offers some of the most advanced and effective solutions against salinity and alkalinity.

Biochar is a solid carbonaceous material derived from pyrolyzing biomass at 350–700°C under oxygen-free or oxygen-limiting conditions (Brassard et al., 2019). The carbonaceous material can be obtained from agricultural and forestry waste, sewage sludge, manure, etc. Application of biochar as a soil amendment can enhance crop yield by improving physical, chemical, and biological soil properties, and by ameliorating the effects of abiotic stresses in plants (Das et al., 2021). However, the positive effect of biochar application in ameliorating salt-affected soils depends on its own structural and compositional characteristics (Zang et al., 2022). Ineffective improvement of soil quality and plant growth in saline-alkaline soils has been reported. The main reason for such negative outcomes is the alkalinization effect of some biochars. Certain types of biochars contain soluble alkali ions (e.g. Na^+, K^+, Mg^{2+}, Ca^{2+}) that can increase soil salinity. Moreover, some have carbonates and oxygen-containing functional groups (e.g. $-COO^-$, $-O^-$) that also can boost soil alkalinity (Yuan et al., 2011; Fidel et al., 2017). Therefore, the usage of alkaline biochar must be avoided as such materials can increase the degradation of soil texture, enhance soil salt alkalization, and stunt plant growth. Nowadays, modified biochar, derived from physicochemical and biological modifications of pristine biochar, presents an effective approach for saline-alkali soil reclamation (Li et al., 2023). Compared to other modification methods of biochar, ball milling is an efficient, economical, and environmentally friendly method to modify biochar for loading phosphate on the surface. Red P, an elemental form of P that can slowly be converted into phosphate in the presence of water and oxygen, releases a large amount of acid in the conversion process and could effectively reduce the alkalinity of saline-alkaline soils (Zang et al., 2022).

Despite there is not a complete understanding of the interaction of nanoparticles (NPs) and the intracellular mechanisms of plants, some studies have confirmed the positive effects of nanotechnology in the cell recovery of salt-stressed plants (see Etesami et al., 2021). The biological function of NPs depends on their physicochemical properties, the concentration used, the application method, and the developmental status of the plant. In general terms, NPs interact at cellular and subcellular levels after entry through roots or leaves, promoting molecular, biochemical, and physiological changes. When applied at specific concentrations (Etesami et al., 2021) carbon nanomaterials, ZnO-, CeO-, TiO_2-, PAO-NPs, and silicon NPs, among others, can enhance plants' tolerance to salinity stress. This is achieved by multiple mechanisms: changing DNA methylation, gene expression of transcription factors and genes associated with ion homeostasis and hormonal responses, soluble sugars, the antioxidant defense system, and photosynthetic parameters. There are limited studies establishing the specific application rates for saline-alkaline soils, but some progress has been done in alfalfa, where the supply of 5 g/kg of graphene (Nano-C) promoted plant growth under saline-alkali stress (Chen and Wang, 2021), and in maize, where applying silicon nanoparticles through seed priming mitigated the effects of salt-alkaline stress (Alsamadany et al., 2023).

Other amendments successfully applied in saline-alkaline soils include coumarins and flue-gas desulfurization (FGD) gypsum. In particular, coumarin application restored ion homeostasis in *Moricandia charantia* plants submitted to alkaline salinity, augmenting osmotic adjustment and intensifying oxidative defense (Iram et al., 2023). In alkaline, salinized, and secondary salinized soils, FGD gypsum alone or combined with a K–Zn–Mn fertilizer increased the yield of rice by 21.2% and 60.4%, the yield of sunflower by 2.4% and 23.6%, and the yield of medlar by 18.81% and 20.78%, respectively (Wang et al., 2023b).

2.3.4 Plant–Microbial Associations

The increase of alkaline salinity in crop soils not only has a direct impact on plant growth but also affects the activity and community composition of soil microbes. Many studies have shown that plants' resistance or tolerance to abiotic stresses is not exclusively derived from the plant genome itself; the associated microbiome such as mycorrhizal fungi, rhizobia, and other beneficial microorganisms also have a positive impact on the host's tolerance, becoming an effective extension of the plant genome (see Pantigoso et al., 2022). For example, rhizospheric microorganisms regulate plant metabolism by stimulating the release of specific root exudates. On the one hand, the diffusion, transformation, and decomposition of root exudates increase the complexity of the microecological circulation system in the rhizosphere (Vives-Peris et al., 2020). On the other hand, roots can secrete bioactive substances into the rhizosphere soils, attracting beneficial microorganisms, and thus affecting the species and quantity of rhizosphere microorganisms (Ma et al., 2022b). Halophyte species such as *Arthrocnemum* spp. or *Salicornia* spp. can shift their root exudate composition to adapt to severe saline-alkali stress, thus promoting the growth of salt-tolerant microorganisms (Ghanem et al., 2021).

Among the environmental factors conditioning the microbiome structure and composition, the pH is a primary driving force of bacterial distribution in the soil (Wang et al., 2019b). Even though soil salinization and alkalinization frequently co-occur, the meta-analyses of plant–microbiome interactions have mainly been conducted in saline soils omitting the pH factor. Soil microorganisms participate in the adjustment of ecological processes in multiple ways, degradation of organic matter, nutrient element transformations, enzyme production, etc. However, some of the plant growth-promoting (PGP) activities beneficial in alkaline conditions could be detrimental in saline conditions or vice versa. Understanding how plant-associated bacteria communities respond to alkaline salinity may enable the identification of specific bacterial taxa that promote vegetation growth and contribute to the remediation of severe saline-alkali regions. Among the predominant phyla associated with halophytes on alkaline-saline soils were Proteobacteria (Halobacteria, Pseudomonas), Actinobacteria, Firmicutes, Chloroflexi, Gemmatimonadetes, Acidobacteria, and Bacteroidetes (Wang et al., 2021; Yin et al., 2021; Nan et al., 2022); thus, they have become targets for future bioinoculant formulations.

To highlight some examples, the synergic effect of silicon and the PGP strain *Enterobacter* sp. FN0603 mitigated saline-alkaline stress in wheat (Xu et al., 2023). Kumar et al. (2022) focused on the screening and isolation of nitrogen-fixing actinobacteria that, combined with the cultivation of salt-tolerant tree species (*Casuarina* and *Acacia*), were able to restore salt-alkaline soils and allowed subsequent cultivation of rice. Moreover, the active substances (Gutierrez et al., 2020), exopolysaccharides (Gan et al., 2020), and enzymes (Chen et al., 2022) produced by members of Halomonas could play important functions under high salt and high pH pressure, which is of great significance for the restoration of saline-alkali land.

2.4 CONCLUDING REMARKS

Plants must overcome multiple stress factors to perform on alkaline-saline soils. It is becoming evident that mechanisms conferring Na^+ tolerance are required but insufficient under alkaline salinity.

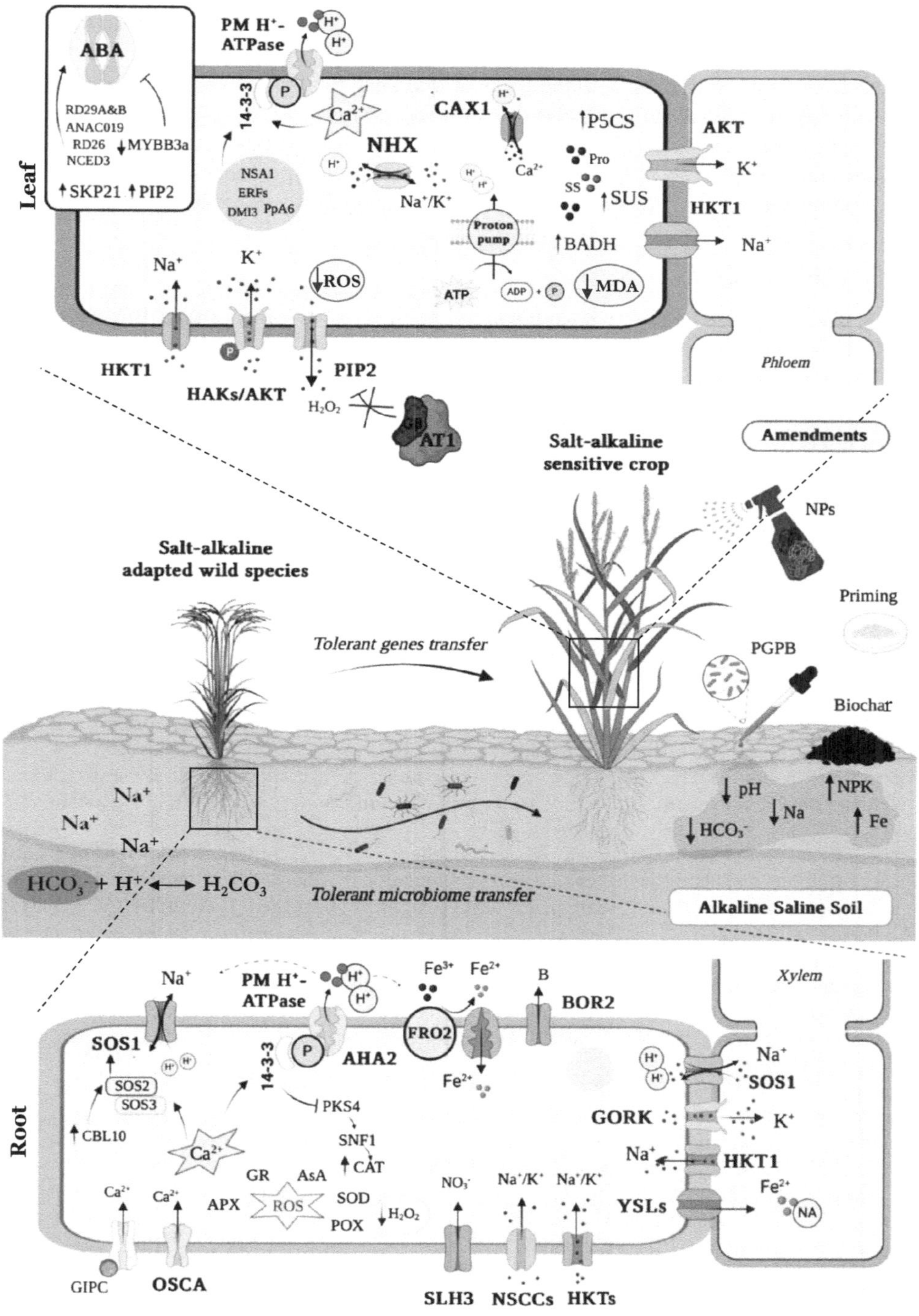

FIGURE 2.2 Overview of agronomical strategies for cultivation in saline-alkaline soils, including the key genes involved in tolerance to alkaline salinity. Gene symbols are shown in bold letters. Ion fluxes are indicated with black arrows; gene activation/repression and molecule increase/decrease are indicated with black arrows. Star-framed symbols denote signaling pathways; attenuated signaling responses are circled in grey. Some ubiquitous organelles and transport and signaling processes are presented in only one scheme section (leaf/root cell) for space distribution purposes.

Strategies that allow plants to handle not only bicarbonate and alkaline pH but also alkalinity-induced micronutrient deficiency under conditions of concurrent Na^+ ion toxicity and low osmotic potential are required. The task of untangling these multiple interconnected tolerance mechanisms is of great complexity. Naturally selected traits of tolerance to alkaline salinity in adapted species are of invaluable help in selecting genetic traits for successful breeding. Nonetheless, this Gordian knot will not be solved by a single stroke as Alexander the Great did; i.e. not a single, but multiple approaches combining basic molecular genetics, physiological approaches, breeding efforts, and different crop management strategies are required to efficiently solve this complex problem.

ACKNOWLEDGMENTS

This work was supported by the Spanish Research Agency (AEI) project PID2019–104000RB-I00. Silvia Busoms was supported by a Maria Zambrano scholarship from the Spanish Ministry of Universities.

REFERENCES

Abadía, J., López-Millán, A. F., Rombolà, A., & Abadía, A. (2002). Organic acids and Fe deficiency: A review. *Plant and Soil*, *241*, 75–86. https://doi.org/10.1023/A:1016093317898

Abdul Aziz, M., & Masmoudi, K. (2023). Insights into the transcriptomics of crop wild relatives to unravel the salinity stress adaptive mechanisms. *International Journal of Molecular Sciences*, 24, 9813. https://doi.org/10.3390/ijms24129813

Ahmad, P., Ozturk, M., Sharma, S., & Gucel, S. (2014). Effect of sodium carbonate-induced salinity-alkalinity on some osmoprotectants, protein profile, antioxidant enzymes, and lipid peroxidation in two mulberry (*Morus alba* L.) cultivars. *Journal of Plant Interactions*, 9, 460–467. https://doi.org/10.1080/17429145.2013.855271

Al-Busaidi, A., & Cookson, P. (2003). Salinity-pH relationships in calcareous soils. *Journal of Agriculture and Marine Sciences*, *8*, 41–46. https://doi.org/10.24200/JAMS.VOL8ISS1PP41-46

Al-Najar, H., & Ashour, E. K. (2013). The impact of climate change and soil salinity in irrigation water demand on the Gaza strip. *Journal of Water and Climate Change*, *4*, 11–130. https://doi.org/10.2166/wcc.2013.142

Alhendawi, R. A., Römheld, V., Kirkby, E. A., & Marschner, H. (1997). Influence of increasing bicarbonate concentrations on plant growth, organic acid accumulation in roots and iron uptake by barley, sorghum, and maize. *Journal of Plant Nutrition*, *20*, 1731–1753. https://doi.org/10.1080/01904169709365371

Ali, A., Petrov, V., Yun, D.-J., & Gechev, T. (2023). Revisiting plant salt tolerance: Novel components of the SOS pathway. *Trends in Plant Science*, *28*, 1060–1069. https://doi.org/10.1016/j.tplants.2023.04.003

Almira-Casellas, M. J., Pérez-Martín, L., Busoms, S., Boesten, R., Llugany, M., Aarts, M. G. M., & Poschenrieder, C. (2023). A genome-wide association study identifies novel players in Na and Fe homeostasis in *Arabidopsis thaliana* under alkaline-salinity stress. *Plant Journal*, *113*, 225–245. https://doi.org/ 10.1111/tpj.16042

Alsamadany, H., Alharby, H., Ahmad, Z., Al-Zaharani, H. S., Alzaharani, Y. M., & Almaghamsi, A. (2023). Improving alkaline stress tolerance in maize through seed priming with silicon nanoparticles: A comprehensive investigation of growth, photosynthetic pigments, antioxidants, and ion balance. *Silicon*, *2023*, 2233–2244. https://doi.org/10.1007/s12633-023-02833-5

An, Y., Yang, X. X., Zhang, L., Zhang, J., Du, B., Yao, L., Li, X.-T., & Guo, C. (2020). Alfalfa MsCBL4 enhances calcium metabolism but not sodium transport in transgenic tobacco under salt and saline–alkali stress. *Plant Cell Reports*, *39*, 997–1011. https://doi.org/10.1007/s00299-020-02543-x

Atkinson, N. J., & Urwin, P. E. (2012). The interaction of plant biotic and abiotic stresses: From genes to the field. *Journal of Experimental Botany*, *63*, 3523–3544. https://doi.org/10.1093/jxb/ers100

Behera, S., Xu, Z., Luoni, L., Bonza, M. C., Doccula, F. G., De Michelis, M. I., Morris, R. J., Schwarzländer, M., & Costa, A. (2018). Cellular Ca^{2+} signals generate defined pH signatures in plants. *The Plant Cell*, *30*, 2704–2719. https://doi.org/10.1105/tpc.18.00655

Bergmann, W. (1993). *Ernährungsstörungen bei Kulturpflanzen*. Gustav Fischer Verlag, Jena. ISBN: 3334604144

Brassard, P., Godbout, S., Lévesque, V., Palacios, J. H., Raghavan, V., & Ahmed, A. (2019). Biochar for soil amendment. In M. Jeguirim & L. Limousy (Eds.), *Char and carbon materials derived from biomass* (pp 109–146). Elsevier Inc. ISBN: 978-0-12-814893-8. https://doi.org/10.1016/C2017-0-02406-0

Brenes, M., Solana, A., Boscaiu, M., Fita, A., Vicente, O., Calatayud, A., Prohens, J., & Plazas, M. (2020). Physiological and biochemical responses to salt stress in cultivated eggplant (*Solanum melongena* L.) and in *S. insanum* L., a close wild relative. *Agronomy, 10*, 651. https://doi.org/10.3390/agronomy10050651

Bui, E. N., & Henderson, B. L. (2003). Vegetation indicators of salinity in northern Queensland. *Austral Ecology, 28*, 539–552. https://doi.org/10.1046/j.1442-9993.2003.01311.x

Busoms, S., Paajen, P., Marburger, S., Bray, S., Huang, X.-Y., Poschenrieder, C., Yant, L., & Salt, D. E. (2018). Fluctuating selection on migrant adaptive sodium transporter alleles in coastal *Arabidopsis thaliana. Proceedings of the National Academy of Sciences, USA, 115*(52), E12443–E12452. https://doi.org/10.1073/pnas.1816964115

Busoms, S., Fischer, S., & Yant, L. (2023). Chasing the mechanisms of ecological adaptive salinity tolerance. *Plant Communications, 4*, 100571. https://doi.org/10.1016/j.xplc.2023.100571

Cabot, C., Sibole, J. V., Barceló, J., & Poschenrieder, C. (2014). Lessons from crops struggling with salinity. *Plant Science, 226*, 2–13. https://doi.org/10.1016/j.plantsci.2014.04.013

Cai, X., Jia, B., Sun, M., & Sun, X. (2022). Insights into the regulation of wild soybean tolerance to salt-alkaline stress. *Frontiers in Plant Science, 13*, 1002302. https://doi.org/10.3389/fpls.2022.1002302

Cao, L., Yu, Y., Duanmu, H., Chen, C., Duan, X., Zhu, P., Chen, R., Li, Q., Zhu, Y., & Ding, X. (2016). A novel *Glycine soja* homeodomain-leucine zipper (HD-Zip) I gene, *Gshdz4*, positively regulates bicarbonate tolerance and responds to osmotic stress in Arabidopsis. *BMC Plant Biology, 16*, 184. https://doi.org/10.1186/s12870-016-0872-7

Cao, L., Yu, Y., Ding, X., Zhu, D., Yang, F., Liu, B., Liu, B., Sun, X., Duan, X., Yin, K., & Zhu, Y. (2017). The *Glycine soja* NAC transcription factor *GsNAC019* mediates the regulation of plant alkaline tolerance and ABA sensitivity. *Plant Molecular Biology, 95*, 253–268. https://doi.org/10.1007/s11103-017-0643-3

Cao, Y., Zhang, M., Liang, X., Li, F., Shi, Y., Yang, X., & Jiang, C. (2020). Natural variation of an EF-hand Ca^{2+}-binding-protein coding gene confers saline-alkaline tolerance in maize. *Nature Communications, 11*, 186. https://doi.org/10.1038/s41467-019-14027-y

Cao, Y., Song, H., & Zhang, L. (2022). New insight into plant saline-alkali tolerance mechanisms and application to breeding. *International Journal of Molecular Sciences, 23*, 16048. https://doi.org/10.3390/ijms232416048

Cao, Y., Wang, J., Zhao, S., Fang, Q., Ruan, J., Li, S., Liu, T., Qi, Y., Zhang, L., Zhang, X., & Meng, F. (2023). Overexpression of the aldehyde dehydrogenase AhALDH3H1 from *Arachis hypogaea* in soybean increases saline-alkali stress tolerance. *Frontiers in Plant Science, 14*, 1165384. https://doi.org/10.3389/fpls.2023.1165384

Chaanin, A., & Preil, W. (1994). Influence of bicarbonate on iron deficiency chlorosis in Rhododendron. *Acta Horticulturae, 364*, 71–78. https://doi.org/10.17660/ActaHortic.1994.364.8

Chen, C., Sun, X., Duanmu, H., Zhu, D., Yu, Y., Cao, L., Liu, A., Jia, B., Xiao, J., & Zhu, Y. (2016). *GsCML27*, a gene encoding a calcium-binding Ef-hand protein from *Glycine soja*, plays differential roles in plant responses to bicarbonate, salt and osmotic stresses. *PLoS One, 10*, e0141888. https://doi.org/10.1371/journal.pone.0141888

Chen, G. Q., Zhang, X., Liu, X., Huang, W., Xie, Z., Han, J., Xu, T., Mitra, R., Zhou, C., Zhang, J., & Chen, T. (2022). *Halomonas spp.*, as chassis for low-cost production of chemicals. *Applied Microbiology and Biotechnology, 106*, 6977–6992. https://doi.org/10.1007/s00253-022-12215-3

Chen, X., Zhao, C., Yun, P., Zhou, M., Chen, Z.-H., & Shabala, S. (2024). Climate-resilient crops: Lessons from xerophytes. *The Plant Journal, 117*, 1815–1835. https://doi.org/10.1111/tpj.16549

Chen, Z., & Wang, Q. (2021). Graphene ameliorates saline-alkaline stress-induced damage and improves growth and tolerance in alfalfa (*Medicago sativa*). *Plant Physiology and Biochemistry, 163*, 128–138. https://doi.org/10.1016/j.plaphy.2021.03.039

Colla, G., Rouphael, Y., Cardarelli, M., Salerno, A., & Rea, E. (2010). the effectiveness of grafting to improve alkalinity tolerance in watermelon. *Environmental Experimental Botany, 68*, 283–291. https://doi.org/10.1016/j.envexpbot.2009.12.005

Colmer, T. D., Flowers, T. J., & Munns, R. (2006). Use of wild relatives to improve salt tolerance in wheat. *Journal of Experimental Botany, 57*, 1059–1078. https://doi.org/10.1093/jxb/erj124

Cosse, M., & Seidel, T. (2021). Plant proton pumps and cytosolic pH-homeostasis. *Frontiers in Plant Science, 12*, 672873. https://doi.org/10.3389/fpls.2021.672873

Daliakopoulus, I. N., Tsanis, I. K., Koutroulis, A., Kourgialas, N. N., Varouchakis, A. E., Karatzas, G. P., & Ritsema, C. J. (2016). The threat of soil salinity: A European scale review. *Science of the Total Environment*, *573*, 727–739. https://doi.org/10.1016/j.scitotenv.2016.08.177

Das, S., Mohanty, S., Sahu, G., Rana, M., & Pilli, K. (2021). Biochar: A sustainable approach for improving soil health and environment. Soil erosion-current challenges and future perspectives in a changing world. In A. Vieira & S. C. Rodrigues (Eds.), *Soil erosion – Current challenges and future perspectives in a changing world* (pp. 1–18). Intech Open, ISBN: 978-1-83962-304-2.

Do, P. T., Drechsel, O., Heyer, A. G., Hincha, D. K., & Zuther, E. (2014). Changes in free polyamine levels, expression of polyamine biosynthesis genes, and performance of rice cultivars under salt stress: A comparison with responses to drought. *Frontiers in Plant Science*, *5*, 182. https://doi.org/10.3389/fpls.2014.00182

Dong, Y., Zhang, L., Chang, X., Wang, X., Li, G., Chen, S., & Jin, S. (2022). Overexpression of LpCPC from *Lilium pumilum* confers saline-alkali stress ($NaHCO_3$) resistance. *Plant Signaling & Behavior*, *17*(1), 2057723. https://doi.org/10.1080/15592324.2022.2057723

Du, B., Zhao, W., An, Y., Li, Y., Zhang, X., Song, L., & Guo, C. (2019). Overexpression of an alfalfa glutathione S-transferase gene improved the saline-alkali tolerance of transgenic tobacco. *Biology Open*, *8*, bio 043505. https://doi.org/10.1242/bio.043505

Duan, X., Yu, Y., Duanmu, H., Chen, C., Sun, X., Cao, L., Li, Q., Ding, X., Liu, B., & Zhu, Y. (2018a). GsSLAH3, a *Glycine soja* slow type anion channel homolog, positively modulates plant bicarbonate stress tolerance. *Physiologia Plantarum*, *164*, 145–162. https://doi.org/10.1111/ppl.12683

Duan, X., Yu, Y., Chen, C., Duanmu, H., Cao, L., Sun, M., Sun, X., & Zhu, Y. (2018b). A potential efflux boron transporter gene *GsBOR2*, positively regulates Arabidopsis bicarbonate tolerance. *Plant Science*, *274*, 284–292. https://pubmed.ncbi.nlm.nih.gov/30080614/

Dumanović, J., Nepovimova, E., Natić, M., Kuča, K., & Jaćević, V. (2021). The significance of reactive oxygen species and antioxidant defense system in plants: A concise overview. *Frontiers in Plant Science*, *11*, 552969. https://doi.org/10.3389/fpls.2020.552969

Etesami, H., Fatemi, H., & Rizwan, M. (2021). Interactions of nanoparticles and salinity stress at physiological. Biochemical and molecular levels in plants: A review. *Ecotoxicology and Environmental Safety*, *225*, 112769. https://doi.org/10.1016/j.ecoenv.2021.112769

Falhof, J., Pedersen, J. T., Fuglsang, A. T., & Palmgren, M. (2016). Plasma membrane H^+-ATPase regulation in the center of plant physiology. *Molecular Plant*, *9*, 323–337. https://doi.org/10.1016/j.molp.2015.11.002

Fan, L., Shi, G., Yang, J., Liu, G., Niu, Z., Ye, W., Wu, S., Wang, L., & Guan, Q. (2022). A protective role of phenylalanine ammonia-lyase from *Astragalus membranaceous* saline-alkali stress. *International Journal of Molecular Sciences*, *23*, 15686. https://doi.org/10.3390/ijms232415686

Fang, S., Hou, X., & Liang, X. (2021). Response mechanisms of plants under saline-alkali stress. *Frontiers in Plant Science*, *12*, 667458. https://doi.org/10.3389/fpls.2021.667458

FAO. (2021). World map of salt-affected soils launched at virtual conference. Global Soil Partnership. Food and Agriculture Organization of the United Nations (fao.org)

FAO. (2024). 20th Working session of the intergovernmental technical panel on soils. https://www.fao.org/fileadmin/user_upload/GSP/ITPS/20th_ITPS/20_ITPS_INSAS.pdf

Fidel, R. B., Laird, D. A., Thompson, M. L., & Lawrinenko, M. (2017). Characterization and quantification of biochar alkalinity. *Chemosphere*, *167*, 367–373. https://doi.org/10.1016/j.chemosphere.2016.09.151

Foyer, C., & Noctor, G. (2011). Ascorbate and glutathione: The heart of the redox hub. *Plant Physiology*, *155*, 2–18. https://doi.org/10.1104/pp.110.167569

Gan, L., Li, X., Zhang, H., Zhang, R., Wang, H., Xu, Z., Peng, B., & Tian, Y. (2020). Preparation, characterization and functional properties of a novel exopolysaccharide produced by the halophilic strain *Halomonas saliphila* LCB169^T. *International Journal of Biological Macromolecules*, *156*, 372–380. https://doi.org/10.1016/j.ijbiomac.2020.04.062

Gao, P., Bai, X., Yang, L., Lv, D., Li, Y., Cai, H., Ji, W., Guo, D., & Zhu, Y. (2010). Over-expression of *osa-MIR396c* decreases salt and alkali stress tolerance. *Planta*, *231*, 991–1001. https://doi.org/10.1007/s00425-010-1104-2

Geilfuss, C.-M. (2017). The pH of the apoplast: Dynamic factor with impact under stress. *Molecular Plant*, *10*, 1371–1386. https://doi.org/10.1016/j.molp.2017.09.018

Ghanem, A. F. M., Mohamed, E., Kasem, A. M. M. A., & El-Ghamery, A. A. (2021). Differential salt tolerance strategies in three halophytes from the same ecological habitat: Augmentation of antioxidant enzymes and compounds. *Plants*, *10*, 1100. https://doi.org/10.3390/plants10061100

Gong, B., Wen, D., Van den Langenberg, K., Wei, M., Yang, F., Sho, Q., & Wang, X. (2013). Comparative effects of NaCl and $NaHCO_3$ stress on photosynthetic parameters, nutrient metabolism, and the antioxidant system in tomato leaves. *Scientia Horticulturae*, *157*, 1–12. https://doi.org/10.1016/j.scienta.2013.03.032

Gong, B., Wen, D., Wang, X., Wei, M., Yang, F., Li, Y., & Shi, Q. (2015). S-nitrosoglutahione reductase-modulated redox signaling controls sodic alkaline stress responses in *Solanum lycopersicum* L. *Plant and Cell Physiology*, *56*, 790–802. https://doi.org/10.1093/pcp/pcv007

González-Vallejo, E. B., González-Reyes, J. A., Abadía, A., López Millán, A. F., Yunta, F., Lucena, J. J., & Abadía, J. (1999). Reduction of ferric chelates by leaf plasma membrane preparations from Fe-deficient and Fe-sufficient sugar beet. *Australian Journal of Plant Physiology*, *26*, 601–611. https://doi.org/10.1071/PP99002

Guo, R., Shi, L. X., Yan, C., Zhong, X., Gu, F. X., Liu, Q., Xia, X., & Li, H. (2017). Ionomic and metabolic responses to neutral salt or alkaline salt stresses in maize (*Zea mays* L.) seedlings. *BMC Plant Biology*, *17*, 41. https://doi.org/10.1186/s12870-017-0994-6

Gutierrez, T., Morris, G., Ellis, D., Mulloy, B., & Aitken, M. D. (2020). Production and characterization of a marine Halomonas surface-active exopolymer. *Applied Microbiology and Biotechnology*, *104*, 1063–1076. https://doi.org/10.1007/s00253-019-10270-x

Hailu, B., & Mehari, H. (2021). Impacts of soil salinity/sodicity on soil-water relations and plant growth in dry land areas: A review. *Journal of Natural Sciences Research*, *112*, 1–10. https://doi.org/10.7176/JNSR/12-3-01

Hajiboland, R., Yang, X. E., & Römheld, V. (2003). Effects of bicarbonate and high pH on growth of Zn-efficient and Zn-inefficient genotypes of rice, wheat and rye. *Plant and Soil*, *250*, 349–357. https://doi.org/10.1023/A:1022862125282

Hajiboland, R., Yang, X. E., Römheld, V., & Neumann, G. (2005). Effect of bicarbonate on elongation and distribution of organic acids in root and root zone of Zn-efficient and Zn-inefficient rice (*Oryza sativa*) genotypes. *Environmental and Experimental Botany*, *54*, 163–173. https://doi.org/10.1016/j.envexpbot.2004.07.001

Hajiboland, R., Bahrami-Rad, S., Akhani, H., & Poschenrieder, C. (2018). Salt tolerance mechanisms in three Irano-Turanian Brassicaeae halophytes relatives of Arabidopsis thaliana. *Journal of Plant Research*, *131*, 1029–1046. https://doi.org/10.1007/s10265-018-1053-6

Harris, F. S. (1920). *Circular No. 41-Soil Alkali. UAES Circulars. Paper*, *36*. https://digitalcommons.usu.edu/uaes_circulars/36

Hassani, A., Azapagic, A., & Shokri, N. (2021). Global predictions of primary soil salinization under changing climate in the 21st century. *Nature Commununications*, *12*, 6663. https://doi.org/10.1038/s41467-021-26907-3

Hazman, M., Hause, B., Eiche, E., Riemann, M., & Nick, P. (2016). Different forms of osmotic stress evoke qualitatively different responses in rice. *Journal of Plant Physiology*, *202*, 45–56. https://doi.org/10.1016/j.jplph.2016.05.027

He, Y., Yang, X., Xu, C., Guo, D., Niu, L., Wang, Y., Li, J., Yan, F., & Wang, Q. (2018). Overexpression of a novel transcriptional repressor GmMYB3a negatively regulates salt-alkali tolerance and stress-related genes in soybean. *Biochemical and Biophysical Research Communications*, *498*, 586–591. https://doi.org/10.1016/j.bbrc.2018.03.026

Hill, T. G. (1908). Observations on the osmotic properties of the root hairs of certain salt marsh plants. *New Phytologist*, *7*, 133–142. https://www.jstor.org/stable/2427254

Hooks, T. N., Picchioni, G. A., Schutte, B. J., Shukla, M. K., & Daniel, D. L. (2018). Sodium chloride effects on seed germination, growth, and water use of *Lepidium alyssoides, L. draba, and L. latifolium*: Traits of resistance and implications for invasiveness on saline soils. *Rangeland Ecology and Management*, *71*, 433–442. https://doi.org/10.1016/j.rama.2018.04.001

Hou, R., Yang, L., Wuyun, T., Chen, S., & Zhang, L. (2023). Genes related to osmoregulation and anti-oxidation play important roles in the response of *Trollius chinensis* seedlings to saline-alkaline stress. *Frontiers in Plant Science*, *14*, 1080504. https://doi.org/10.3389/fpls.2023.1080504

Hu, X., Wang, D., Ren, S., Feng, S., Zhang, H., Zhang, J., Kun, Q., & Aimin, Z. (2022). Inhibition of root growth by alkaline salts due to disturbed ion transport an accumulation in *Leymus chinensis*. *Environmental and Experimental Botany*, *200*, 104907. https://doi.org/10.1016/j.envexpbot.2022.104907

Hussain, S., Shaukat, M., Ashraf, M., Zhu, C., Jin, Q., & Zhang, J. (2019). Salinity stress in arid and semi-arid climates: Effects and management in field crops. In S. Hussain (Ed.), *Climate change and agriculture* (Vol. 13, pp. 201–226). Intech Open, ISBN: 978-1-83880-039-0.

Iram, K., Ashraf, M. A., Ibrahim, S. M., Rasheed, R., & Ali, S. (2023). Coumarin regulated redox homeostasis to facilitate phytoremediation of saline and alkaline soils by bitter gourd (*Momordica charantia* L.). *Environmental Science and Pollution Research*, *30*, 99584–99604. https://doi.org/10.1007/s11356-023-29360-5

Isayenkov, S., Dabravoski, S., Pan, T., & Shabala, S. (2020). Phylogenetic diversity and physiological roles of plant monovalent cation/H^+ antiporters. *Frontiers in Plant Science*, *11*, 573564. https://doi.org/10.3389/fpls.2020.573564

Issanova, G., Abuduwaili, J., Tynybayeva, K., Kaldybayev, A., Kaldybayev, A., Tanirbergenov, S., & Ge, Y. (2022). Assessment of the soil cover in the dried Aral seabed in Kazakhstan and climate change in the region. *Water Air and Soil Pollution*, *233*, 525. https://doi.org/10.1007/s11270-022-05966-2

Ivushkin, K., Bartholomeus, H., Bregt, A. K., Pulatov, A., Kempen, B., & de Sousa, L. (2019). Global mapping of soil salinity change. *Remote Sensing of Environment*, *231*, 1112602019. https://doi.org/10.1016/j.rse.2019.111260

Jekabsone, A., Kulika, J., Romanovs, M., Andersone-Ozola, U., & Ievinsh, G. (2023). Salt tolerance and ion accumulation in several halophytic plant species depending on the type of anion. *International Journal of Plant Biology*, *14*, 1131–1154. https://doi.org/10.3390/ijpb14040083

Jin, S., Xu, C., Li, G., Sun, D., Li, Y., Wang, X., & Liu, S. (2017). Functional characterization of a type 2 metallothionein gene, SsMT2, from alkaline-tolerant *Suaeda salsa*. *Scientific Report*, *7*, 17914. https://doi.org/10.1038/s41598-017-18263-4

Jin, T., Jiaxin, A., Xu, H., Chen, J., Pan, L., Zhao, R., Wang, N., Gai, J., & Li, Y. (2022a). A soybean sodium/hydrogen exchanger GMNHX6 confers plant alkaline salt tolerance by regulating Na^+/K^+ homeostasis. *Frontiers in Plant Science*, *13*, 9386365. https://doi.org/10.3389/fpls.2022.938635

Jin, T., Shan, Z., Zhou, S., Yang, Q., Junyi, G., & Li, Y. (2022b). *GmDNAJC7* form soybean is involved in plant tolerance to alkaline-salt, salt, and drought stresses. *Agronomy*, *12*, 1419. https://doi.org/10.3390/agronomy12061419

Kapazoglou, A., Gerakar, M., Lazaridi, E., Kleftogianni, K., Sarri, E., Tani, E., & Bebeli, P. J. (2023). Crop wild relatives: A valuable source of tolerance to various abiotic stresses. *Plants*, *12*, 328. https://doi.org/10.3390/plants12020328

Ketehouli, T., Zhou, Y. G., Dai, S. Y., Carther, K. F. I., Sun, D. Q., Li, Y., Nguyen, Q. V. H., Xu, H., Wang, F.-W., Liu, W.-C., Li, X.-W., & Li, H.-Y. (2021). A soybean calcineurin B-like protein-interacting protein kinase, *GmPKS4*, regulates plant responses to salt and alkali stresses. *Journal of Plant Physiology*, *256*, 153331. https://doi.org/10.1016/j.jplph.2020.153331

Kobayashi, S., Satone, H., Tan, E., Kurokochi, H., Asakawa, S., Liu, S., & Takano, T. (2015). Transcriptional responses of a bicarbonate-tolerant monocot, *Puccinellia tenuiflora*, and a related bicarbonate-sensitive species, *Poa annua*, to $NaHCO_3$ stress. *International Journal of Molecular Sciences*, *16*, 496–509. https://doi.org/10.3390/ijms16010496

Koks, A. H. W., Fritz, C., Smolders, A. J. P., Rehlmeyer, K., Elzenga, J. T. M., Krosse, S., Lamers, L. P. M., & van Dijk, G. (2022). Sphagnum bleaching: Bicarbonate "toxicity" and tolerance for seven Sphagnum species. *Plant Biology*, *24*, 780–790. https://doi.org/10.1111/plb.13423

Kopittke, P. M., & Menzies, N. (2004). Control of nutrient solutions for studies at high pH. *Plant and Soil*, *266*, 343–354. https://doi.org/10.1007/s11104-005-2658-9

Kosegarten, H. U., Hoffmann, B., & Mengel, K. (1999). Apoplastic pH and Fe^{3+} reduction in intact sunflower leaves. *Plant Physiology*, *121*, 1069–1079. https://doi.org/10.1104/pp.121.4.1069

Kumar, P., Choudhary, M., Halder, T., Prakash, N. R., Singh, V., Vineeth, T. V., Sheoran, S., Ravikiran, K. T., Longmei, N., Rakshit, S., & Siddique, K. H. M. (2022). Salinity stress tolerance and omics approaches: Revisiting the progress and achievements in major cereal crops. *Heredity*, *128*, 497–518. https://doi.org/10.1038/s41437-022-00516-2

Lambers, H., & Oliveira, R. S. (2019). *Plant physiological ecology* (3rd ed.). Cham Switzerland: Springer Nature. ISBN: 978-3-030-29638-4

Li, B. (2010). Soil salinization. In L. Ci & X. Yang (Eds.), *Desertification and its control in China*. Springer. https://doi.org/10.1007/978-3-642-01869-5_6

Li, C., Duan, C., Zhang, H., Zhao, Y., Meng, Z., Zhao, Y., & Zhang, Q. (2022). Adaptive mechanisms of halophytic *Eutrema salsugineum* encountering saline environment. *Frontiers in Plant Science*, *13*, 909527. https://doi.org/10.3389/fpls.2022.909527

Li, C., Wang, Z., Xu, Y., Sun, J., Ruan, X., Mao, X., Hu, X. & Liu, P. (2023). Analysis of the effect of modified biochar on saline–alkali soil remediation and crop growth. *Sustainability*, *15*(6), 5593. https://doi.org/10.3390/su15065593

Li, J., & Yang, Y. (2023). How do plants maintain pH and ion homeostasis under saline-alkaline stress? *Frontiers in Plant Science*, *14*, 1217193. https://doi.org/10.3389/fpls.2023.1217193

Li, Q., Yang, A., & Zhang, W. H. (2016). Efficient acquisition of iron confers greater tolerance to saline-alkaline stress in rice (*Oryza sativa* L.). *Journal of Experimental Botany*, *67*, 6431–6444. https://doi.org/10.1093/jxb/erw407

Li, W., Wang, C., Shi, H., Wang, B., Wang, J., Liu, Y., Ma, J.-Y., Tian, S.-Y., & Zhang, Y.-W. (2020a). Genome-wide analysis of ethylene response factor family in adzuki bean and functional determination of VaERF3 under saline alkaline stress. *Plant Physiology and Biochemistry*, *147*, 215–222. https://doi.org/10.1016/j.plaphy.2019.12.019

Li, X., Zheng, H., Wu, W., Liu, H., Wang, J., Jia, Y., Li, J., Yang, L., Lei, L., Zou, D., & Zhao, H. (2020b). QTL mapping and candidate gene analysis for alkali tolerance in Japonica rice at the bud stage based on linkage mapping and genome-wide association study. *Rice*, *13*, 48. https://doi.org/10.1186/s12284-020-00412-5

Li, Y., Liu, C., Sun, X., Liu, B., Zhang, X., Liang, W., Huo, L., Wang, P., Ma, F., & Li, C. (2020c). Overexpression of MdATG18a enhances alkaline tolerance and GABA shunt in apple through increased autophagy under alkaline conditions. *Tree Physiology*, *40*, 1509–1519. https://doi.org/10.1093/treephys/tpaa075

Li, Y., Zeng, H., Xu, F., Yan, F., & Xu, W. (2022). H^+-ATPases in plant growth and stress responses. *Annual Review of Plant Biology*, *73*, 495–521. https://doi.org/10.1146/annurev-arplant-102820-114551

Lipper, L., Thornton, P., Campbell, B., Baedekar, T., Braimoh, A., Bwalya, M., Caron, P., Cattaneo, A., Garrity, D., Henry, K., Hottle, R., Jackson, L., Jarvis, A., Kossam, F., Mann, W., McCarthy, N., Meybeck, A., Neufeldt, H., Remington, T., … Torquebiau, E. F. (2014). Climate-smart agriculture for food security. *Nature Climate Change*, *4*, 1068–1072. https://doi.org/10.1038/nclimate2437

Liu, A., Yu, Y., Duan, X., Sun, X., Duanmu, H., & Zhu, Y. (2015a). *GsSKP21*, a glycine soja S-phase kinase-associated protein, mediates the regulation of plant alkaline tolerance and ABA sensitivity. *Plant Molecular Biology*, *87*, 111–124. https://doi.org/10.1007/s11103-014-0264-z

Liu, B., Kang, C., Wang, X., & Bao, G. (2015b). Tolerance mechanisms of *Leymus chinensis* to salt-alkaline stress. *Acta Agriculturae Scandinavica, Section B – Soil & Plant Science*, *65*(8), 723–734. https://doi.org/10.1080/09064710.2015.1054867

Liu, C., Chen, K., Zhao, X., Wang, X., Shen, C., Zhu, Y., Dai, M., Qiu, X., Yang, R., Xing, D., Pang, Y., & Xu, J. (2019). Identification of genes for salt tolerance and yield-related traits in rice plants grown hydroponically and under saline field conditions by genome-wide association study. *Rice*, *12*, 88. https://doi.org/10.1186/s12284-019-0349-z

Liu, L., & Wang, B. (2021). Protection of halophytes and their use for cultivation of saline-alkali soil in China. *Biology*, *10*, 353. https://doi.org/10.3390/biology10050353

Liu, L., Song, W., Huang, S., Jiang, K., Moriwak, Y., Wang, Y., Men, Y., Zhang, D., Wen, X., Han, Z., Chai, J., & Guo, H. (2022). Extracellular pH sensing by plant cell-surface peptide receptor complexes. *Cell*, *185*, 3341–3355. https://doi.org/10.1016/j.cell.2022.07.012

Liu, M., Yu, H., Ouyang, B., Shi, C., Demidchik, V., Hao, Z., Yu, M., & Shabala, S. (2020a). NADPH oxidases and the evolution of plant salinity tolerance. *Plant, Cell and Environment*, *43*, 2957–2968. https://doi.org/10.1111/pce.13907

Liu, Y., Chen, X., Xue, S., Quan, T., Cui, D., Han, L., Cong, W., Li, M., Yun, D.-J., Liu, B., & Xu, Z.-Y. (2021a). SET DOMAIN GROUP 721 protein functions in saline–alkaline stress tolerance in the model rice variety Kitaake. *Plant Biotechnology Journal*, *19*, 2576–2588. https://doi.org/10.1111/pbi.13683

Liu, X., Yu, Y., Deng, S., Jin, X., Yin, Y., Guo, J., Li, N., Liu, Y., Han, S., Wang, C., & Hao, D. (2020b). A Na_2CO_3-responsive chitinase gene from *Leymus chinensis* improves pathogen resistance and saline-alkali stress tolerance in transgenic tobacco and maize. *Frontiers in Plant Science*, *11*, 504. https://doi.org/10.3389/fpls.2020.00504

Liu, X., Jin, Y., Tan, K., Zheng, J., Gao, T., Zhang, Z., Zhao, Y., Ma, F., & Li, C. (2021b). MdTyDc overexpression improves alkalinity tolerance in *Malus domestica*. *Frontiers in Plant Science*, *12*, 625890. https://doi.org/10.3389/fpls.2021.625890

Liu, Z., Zhang, H., Yang, X., & Liu, T. (2012). Research advances in gene and genetic engineering for saline-alkaline tolerance of forest trees. *World Forestry Research*, *25*, 11–17.

Lu, P., Dai, S. Y., Yong, L. T., Zhou, B. H., Wang, N., Dong, Y. Y., Liu, W.-C., Wang, F.-W., Yang, H.-Y., & Li, X.-W. (2023). A soybean sucrose non-fermenting protein kinase 1 gene *GmSNF1*, positively regulates plant response to salt and salt-alkali stress in transgenic plants. *International Journal of Molecular Sciences*, *24*, 12482. https://doi.org/10.3390/ijms241512482

Lv, B. S., Wei, X. W., Ma, H. Y., Sun, Y., Wei, L. X., Jiang, C. J., & Liang, Z. W. (2013). Differences in growth and physiology of rice in response to different saline-alkaline factors. *Agronomy Journal, 105*, 1119–1128. https://doi.org/10.2134/agronj2013.0017

Lv, B. S., Li, X. W., Ma, H. Y., Yang, H. Y., Wei, L. X., Lv, H. Y., Jiang, C.-J., & Liang, Z.-W. (2014). Different modes of proline accumulation in response to saline alkaline stress factors in rice (*Oryza sativa* L.). *Research on Crops, 15*, 14–21. https://dx.doi.org/10.5958/j.2348-7542.15.1.002

Ma, C., Bian, C., Liu, W., Sun, Z., Xi, X., Guo, D., Liu, X., Tian, Y., Wang, C., & Zheng, X. (2022a). Strigolactone alleviates the salinity-alkalinity stress of *Malus hupehensis* seedlings. *Frontiers in Plant Science, 13*, 901782. https://doi.org/10.3389/fpls.2022.901782

Ma, W., Tan, S., Dengzeng, Z., Zhang, D., Zhang, T., & Ma, X. (2022b). Root exudates contribute to belowground ecosystem hotspots: A review. *Frontiers in Microbiology, 13*, 937940. https://doi.org/10.3389/fmicb.2022.937940

Majerus, M. (1996). *Plant materials for saline-alkaline soils*. USDA Natural Resources Conservation Service, Bridger Plant Materials Center. https://animalrangeextension.montana.edu/forage/saline_alkaline_soils.html

Martinière, A., Gibrat, R., Sentenac, H., Dumont, X., Gaillard, I., & Paris, N. (2018). Uncovering pH at both sides of the root plasma membrane interface using noninvasive imaging. *Proceedings of the National Academy of Sciences, USA, 115*, 6488–6493. https://doi.org/10.1073/pnas.1721769115

Mehta, D., & Vyas, S. (2023). Comparative bio-accumulation of osmoprotectants in saline stress tolerating plants: A review. *Plant Stress, 9*, 100177. https://doi.org/10.1016/j.stress.2023.100177

Min, Y., Yu, D., Yang, J., Zhao, W., Zhang, L., Bai, Y., & Guo, C. (2023). Bioinformatics and expression analysis of proline metabolism-related gene families in alfalfa under saline-alkaline stress. *Plant Physiology and Biochemistry, 205*, 108182. https://doi.org/10.1016/j.plaphy.2023.108182

Minhas, P. S., Yadav, R. K., & Sharma, P. C. (2021). *Managing salt affected soils for sustainable agriculture*. Directorate of Knowledge Management in Agriculture Publ. ISBN: 978-81-7164-196-3.

Munns, R., & Tester, M. (2008). Mechanisms of salinity tolerance. *Annual Review of Plant Biology, 59*, 651–681. https://doi.org/10.1146/annurev.arplant.59.032607.092911

Munns, R., Day, D. D., Fricke, W., Watt, M., Arsova, B., Barkla, B. J., Bose, J., Byrt, C. S., Chen, Z.-H., Foster, K. J., Gilliham, M., Henderson, S. W., Jenkins, C. L. D., Kronzucker, H. J., Miklavcic, S. J., Plett, D., Roy, S. J., Shabala, S., Shelden, M. C., … Tyerman, S. D. (2020). Energy costs of salt tolerance in crop plans. *New Phytologist, 225*, 1072–1090. https://doi.org/10.1111/nph.15864

Naeem, M., Iqbal, M., Shakeel, A., Ul-Allah, S., Hussain, M., Rehman, A., Zafar, Z. U., Athar, H., & Ashraf, M. (2020). Genetic basis of ion exclusion in salinity stresses wheat: Implications in improving crop yield. *Plant Growth Regulation, 92*, 479–496. https://doi.org/10.1007/s10725-020-00659-4

Nan, L., Guo, Q., Cao, S., & Zhan, Z. (2022). Diversity of bacterium communities in saline-alkali soil in arid regions of Nortwest China. *BMC Microbiology, 22*, 11. https://doi.org/10.1186/s12866-021-02424-7

Ni, L., Wang, Q., Chen, C., Wang, S., Shen, T., Jian, J., Cui, Z., Li, K., Yang, Q., & Jiang, M. (2022). OsDMI3-mediated OsUXS3 phosphorylation improves oxidative stress tolerance by modulating OsCATB protein abundance in rice. *Journal of Integrative Plant Biology, 64*, 1087–1101. https://doi.org/10.1111/jipb.13255

Nikolic, M., & Römheld, V. (2002). Does high bicarbonate supply to roots change availability of iron in the leaf apoplast. *Plant and Soil, 241*, 67–74. https://doi.org/10.1023/A:1016029024374

Oliveira Junior, J. C., Chiapini, M., Nascimento, A. F., Couto, E. G., Beirigo, R. M., & Vidal-Torrado, P. (2017). Genesis and classification of sodic soils in the northern Pantanal. *Revista Brasileira Ciencia do Solo, 41*, e0170015. https://doi.org/10.1590/18069657rbcs20170015

Orsini, F., Paino D'Urzo, M., Inan, G., Serra, S., Oh, D. H., Mickelbart, M. V., Consiglio, F., Li, X., Jeong, J. C., Yun, D.-J., Bohnert, H. J., Bressan, R. A., & Maggio, A. (2010). A comparative study of salt tolerance parameters in 11 wild relatives of *Arabidopsis thaliana*. *Journal of Experimental Botany, 61*, 3787–3798. https://doi.org/10.1093/jxb/erq188

Overstreet, R., Ruben, S., & Broyer, T. C. (1940). The absorption of bicarbonate ion by barley plants as indicated by studies with radioactive carbon. *Proceedings of the National Academy of Sciences, USA, 26*, 688–695. https://doi.org/10.1073/pnas.26.12.688

Pantigoso, H. A., Newberger, D., & Vivanco, J. M. (2022). The rhizosphere microbiome: Plant–microbial interactions for resource acquisition. *Journal of Applied Microbiology, 133*, 2864–2876. https://doi.org/10.1111/jam.15686

Paz, R. C., Rocco, R. A., Reinoso, H., Menéndez, A. B., Piekenstain, F. L., Ruiz, O. A. (2012). Comparative study of alkaline, saline, and mixed saline-alkaline stresses with regard to their effects on growth, nutrient accumulation, and root morphology of *Lotus tenuis*. *Journal of Plant Growth Regulation, 31*, 448–459. https://doi.org/10.1007/s00344-011-9254-4

Pérez-Martín, L., Busoms, S., Tolrà, R., & Poschenrieder, C. (2021). Transcriptomics reveals fast changes in salicylate and jasmonate signaling pathways in shoots of carbonate-tolerant Arabidopsis thaliana under bicarbonate exposure. *International Journal of Molecular Sciences, 22*, 1226. https://doi.org/10.3390/ijms22031226

Pérez-Martín, L., Busoms, S., Almira, M. J., Azagury, N., Terés, J., Tolrà, R., Poschenrieder, C., & Barcelo, J. (2022). Evolution of salt tolerance in *Arabidopsis thaliana* on siliceous soils does not confer tolerance to saline calcareous soils. *Plant and Soil, 476*, 455–475. https://doi.org/10.1007/s11104-022-05439-9

Poschenrieder, C., Fernández, J. A., Rubio, L., Pérez, L., Terés, J., & Barceló, J. (2018). Transport and use of bicarbonate in plants: Current knowledge and challenges ahead. *International Journal of Molecular Sciences, 19*, 1352; 25 pp. https://doi.org/10.3390/ijms19051352

Qi, X., Li, M. W., Xie, M., Liu, X., Ni, M., Shao, G., Yim, A. K.-Y., Tao, Y., Wong, F.-L., Isobe, S., Wong, C.-F., Wong, K.-S., Xu, C., Li, C., Wang, Y., Guan, R., Sun, F., Fan, G., … Lam, H.-M. (2014). Identification of a novel salt tolerance gene in wild soybean by whole-genome sequencing. *Nature Communications, 5*, 4340. https://doi.org/10.1038/ncomms5340

Qiu, J., Henderson, S. W., Tester, M., Roy, S. J., & Gilliham, M. (2016). SLAH1, a homologue of the slow type anion channel SLAC1, modulates shoot Cl^- accumulation and salt tolerance in *Arabidopsis thaliana*. *Journal of Experimental Botany, 67*, 4495–4505. https://doi.org/10.1093/jxb/erw237

Qiu, Q. S., Guo, Y., Dietrich, M. A., Schumaker, K. S., & Zhu, J. K. (2002). Regulation of SOS1, a plasma membrane Na^+/H^+ exchanger in *Arabidopsis thaliana*, by SOS2 and SOS3. *Proceedings of the National Academy of Sciences, USA, 99*, 8436–8441. https://doi.org/10.1073/pnas.122224699

Rao, S., & Wu, Y. (2017). Root-derived bicarbonate assimilation in response to variable water deficit in *Camptotheca acuminate* seedlings. *Photosynthesis Research, 134*, 59–70. https://doi.org/10.1007/s11120-017-0414-7

Rao, Y., Peng, T., & Xue, S. (2023). Mechanisms of plant saline-alkaline tolerance. *Journal of Plant Physiology, 281*, 153916. https://doi.org/10.1016/j.jplph.2023.153916

Razzaq, A., Saleem, F., Wani, S. H., Alyousf, H. A., Abdelbacki, A. M. M., Tamam, N., & Elansary, H. O. (2021). De-novo domestication for improving salt tolerance in crops. *Frontiers in Plant Science, 12*, 1623. https://dx.doi.org/10.3389/fpls.2021.681367

Rengasamy, P. (2016). Soil chemistry factors confounding crop salinity tolerance – A review. *Agronomy, 6*, 53. https://doi.org/10.3390/agronomy6040053

Rizhsky, L., Liang, H., Shuman, J., Shulaev, V., Davletova, S., & Mittler, R. (2004). When defense pathways collide. The response of Arabidopsis to a combination of drought and heat stress. *Plant Physiology, 134*, 1683–1696. https://doi.org/10.1104/pp.103.033431

Sagervanshi, A., Naeem, A., Kaiser, H., Pitann, B., & Mühling, K. H. (2021). Early growth reduction in *Vicia faba* L. under alkali salt stress is mainly caused by excess bicarbonate and related to citrate and malate over accumulation. *Environmental and Experimental Botany, 192*, 104636. https://doi.org/10.1016/j.envexpbot.2021.104636

Sakano, K. (2001). Metabolic regulation of pH in plant cells: Role of cytoplasmic pH in defense reaction and secondary metabolism. *International Review of Cytology, 206*, 1–44. https://doi.org/10.1016/s0074-7696(01)06018-1

Satyavathi, C. T., Solanki, R. K., Kakani, R. K., Bharadwaj, C., Singhal, T., Padaria, J., Khandelwal, V., Srivastava, R. V., Tomar, R. S., & Iqubal, M. A. (2019). Genomics assisted breeding for abiotic stress tolerance in millets. In V. R. Rajpal,, A. Kumar, D. Sehgal, & S. N. Raina (Eds.), *Genomics assisted breeding of crops for abiotic stress tolerance* (Vol. II, pp. 241–255). Springer. https://doi.org/10.1007/978-3-319-99573-1_13

Schuhmacher, K. (2014). pH in the endomembrane system—An import and export buisness. *Current Opinion in Plant Biology, 22*, 71–76. https://pubmed.ncbi.nlm.nih.gov/25282587/#:~:text=10.1016/j.pbi,2014%20Oct%202

Serre, N. B. C., Wernervá, D., Vittal, P., Dubey, S. M., Medvecá, E., Jelínková, A., Petrášek, J., Grossmann, G., & Fendrych, M. (2023). The AUX1-AFB1-CNGC14 module establishes a longitudinal root surface profile. *eLife, 12*, e85193. https://doi.org/10.7554/eLife.85193

Shabala, S. (2013). Learning from halophytes: Physiological basis and strategies toimprove abiotic stress tolerance in crops. *Annals of Botany, 112*, 1209–1221. https://doi.org/10.1093%2Faob%2Fmct205

Shabala, S., & Munns, R. (2017). Salinity stress: Physiology and adpative mechanisms. In S. Shabala (Ed.), *Plant stress physiology* (2nd ed., pp 46–108). CABI.

Shamsabad, M. R. M., Esmaeilizadeh, M., Roosta, H. R., Deghani, M. R., Dabrowski, P., & Kalaji, H. M. (2022). The effect of supplementary light on the photosynthetic apparatus of strawberry plants under salinity and alkalinity stress. *Scientific Reports, 12*, 13257. https://doi.org/10.1038/s41598-022-17377-8

Shen, T., Yan, R., Xu, F., Wang, Q., Chen, D., Li, K., Ni, L., & Jiang, M. (2023). The NADPH oxidase OsRbohD and OsRbohH negatively regulate saline alkaline tolerance in rice. *Environmental and Experimental Botany, 213*, 105445. https://doi.org/10.1016/j.envexpbot.2023.105445

Shi, D., & Sheng, Y. (2005). Effect of various salt-alkaline mixed stress conditions on sunflower seedlings and analysis of their stress factors. *Environmental and Experimental Botany, 54*, 8–21. https://doi.org/10.1016/j.envexpbot.2004.05.003

Shi, D. C., Yin, S. J., Yang, G. H., & Zhao, K. F. (2002). Citric acid accumulation in an alkali-tolerant plant *Puccinellia tenuiflora* under alkaline stress. *Journal of Integrative Plant Biology, 44*, 537540,. https://www.jipb.net/EN/Y2002/V44/I5/537

Shu, L. J., Kahlon, P. S., & Ranf, S. (2023). The power of patterns: New insights into pattern-triggered immunity. *New Phytologist, 240*, 960–967. https://doi.org/10.1111/nph.19148

Singh, A. (2022). Soil salinity: A global threat to sustainable development. *Soil Use and Management, 38*, 39–61. https://doi.org/10.1111/sum.12772

Singh, H., Kumar, P., Kumar, A., Kyriacou, M. C., Colla, G., & Rouphael, Y. (2020). Grafting tomato as a tool to improve salt tolerance. *Agronomy, 10*, 263. https://doi.org/10.3390/agronomy10020263

Singh, M., Kumar, J., Singh, S., Singh, V. P., & Prasad, S. M. (2015). Roles of osmoprotectants in improving salinity and drought tolerance in plants: A review. *Reviews in Environmental Science/Biotechnology, 14*, 407–426. https://doi.org/10.1007/s11157-015-9372-8

Solis, C., Yong, M., Vinarao, R., Jena, K., Holford, P., Shabala, L., Zhou, M., Shabala, S., & Chen, Z.-H. (2020). Back to the wild on a quest for donors toward salinity tolerant rice. *Frontiers in Plant Science, 11*, 323. https://doi.org/10.3389/fpls.2020.00323

Stringer, J. W., & Kimmerer, T. W. (1993). Refixation of xylem sap CO_2 in *Populus deltoides*. *Physiologia Plantarum, 89*, 243–251. https://doi.org/10.1111/j.1399-3054.1993.tb00150.x

Sun, G., Luo, X., Yi, F., & Zhang, X. (2000). Concept, distribution law and formation mechanism of inland saline alkaline wetland – taking Songliao plain as an example. *Chinese Geographical Science, 3*, 44–47. https://doi.org/10.1007/s11769-000-0009-6

Sun, M., Sun, X., Zhao, Y., Zhao, C., Duanmu, H., Yu, Y., Ji, W., & Zhu, Y. (2014). Ectopic expression of *GsPPCK3* and *SCMRP* in *Medicago sativa* enhances plant alkaline stress tolerance and methionine content. *PLoS ONE, 25*, e89578. https://doi.org/10.1371/journal.pone.0089578

Sun, N., Song, T., Ma, Z., Dong, L., Zhan, L., Xing, Y., Liu, J., Song, J., Wang, S., & Cai, H. (2020). Overexpression of MsSiR enhances alkali tolerance in alfalfa (*Medicago sativa* L.) by increasing the glutathione content. *Plant Physiology and Biochemistry 154*, 538–546. https://doi.org/10.1016/j.plaphy.2020.07.001

Szabolcs, I. (1989). *Salt-affected soils* (274 pp). CRC Press. ISBN: 978-0-8493-4818-1

Tagliavini, M., & Rombolà, A. D. (2001). Iron deficiency and chlorosis in orchard and vineyard ecosystems. *European Journal of Agronomy, 15*, 71–92. https://doi.org/10.1016/S1161-0301(01)00125-3

Tian, F., Wang, W., Liang, C., Wang, X., Wang, G., & Wang, W. (2017). Overaccumulation of glycine betaine makes the function of the thylakoid membrane better in wheat under salt stress. *The Crop Journal, 5*, 73–82. https://doi.org/10.1016/j.cj.2016.05.008

Tomaz, A., Palma, P., Alvarenga, P., & Gonçalves, M. C. (2020). Soil salinity risk in a climate change scenario and its effect on crop yield. In M. N. V. Prasad & M. Pietrzykowski (Eds.), *Climate change and soil interactions* (pp. 351–396). Elsevier. ISBN: 978-0-12–818032-7. https://doi.org/10.1016/B978-0-12-818032-7.00013-8

Vélez-Bermúdez, I. C., & Schmidt, W. (2023). Plant strategies to mine iron from alkaline substrates. *Plant and Soil, 483*, 1–25. https://doi.org/10.1007/s11104-022-05746-1

Ventura, Y., Eshel, A., Pasternak, D., & Sagi, M. (2015). The development of halophyte-based agriculture: Past and present. *Annals of Botany, 115*, 529–540. https://doi.org/10.1093/aob/mcu173

Vives-Peris, V., de Ollas, C., Gómez-Cadenas, A., & Pérez-Clemente, R. M. (2020). Root exudates: From plant to rhizosphere and beyond. *Plant Cell Reports, 39*, 3–17. Springer. https://doi.org/10.1007/s00299-019-02447-5

Wang, B., Xie, G., Liu, Z., He, R., Han, J., Huang, S., Liu, L., & Cheng, X. (2019a). Mutagenesis reveals that the OsPPa6 gene is required for enhancing the alkaline tolerance in rice. *Frontiers in Plant Science, 10*, 759. https://doi.org/10.3389/fpls.2019.00759

Wang, C., Zhou, X., Guo, D., Zhao, J., Li, Y., Feng, G., Gao, Q., Yu, H., & Zhao, L.-P. (2019b). Soil pH is the primary factor driving the distribution and function of microorganisms in farmland soils in northeastern China. *Annals of Microbiology, 69*, 1461–1473. https://doi.org/10.1007/s13213-019-01529-9

Wang, F., Chen, Z. H., Liu, X., Shabala, L., Yu, M., Zhou, M., Salih, A., & Shabala, S. (2019c). The loss of RBOHD function modulates root adaptive responses to combined hypoxia and salinity stress in Arabidopsis. *Environmental and Experimental Botany, 158*, 125–135. https://doi.org/10.1016/j.envexpbot.2018.11.020

Wang, H., Ye, L., Zhou, L., Yu, J., Pang, B., Zuo, D., Gu, L., Zhu, B., Du, X., & Wang, H. (2023a). Co-expression network analysis of the transcriptome identified hub genes and pathways responding to saline-alkaline stress in *Sorghum bicolor* L. *International Journal of Molecular Sciences*, *24*, 16831. https://doi.org/10.3390/ijms242316831

Wang, J., Zhang, Y., Yan, X., & Guo, J. (2020a). Physiological and transcriptomic analyses of yellow horn (*Xanthoceras sorbifolia*) provide important insights into salt and saline-alkali stress tolerance. *PLoS One*, *15*(12), e0244365. https://doi.org/10.1371/journal.pone.0244365

Wang, J., Zhao, A., Ma, F., Liu, J., Xiao, G., & Xu, X. (2023b). Amendment of saline–alkaline soil with flue-gas desulfurization gypsum in the Yinchuan Plain, Northwest China. *Sustainability*, *15*, 8658. https://doi.org/10.3390/su15118658

Wang, N., Dong, X., Chen, Y., Ma, B., Yao, C., Ma, F., & Liu, Z. (2020b). Direct and bicarbonate-induced iron deficiency differently affect iron translocation in Kiwifruit roots. *Plants*, *9*, 1578. https://doi.org/10.3390/plants9111578

Wang, Q., Ni, L., Cui, Z., Jiang, J., Chen, C., & Jiang, M. (2022a). The NADPH oxidase OsRbohA increases salt tolerance by modulating K^+ homeostasis in rice. *The Crop Journal*, *10*, 1611–1622. https://doi.org/10.1016/j.cj.2022.03.004

Wang, S., Sun, L., Narsing Rao, M. P., Wang, L., Wang, Y., & Li, W. J. (2021). Insights into the microbial diversity in saline-alkaline soils of China. In D. Egamberdieva, N. K. Birkeland, W. J. Li, & H. Panosyan (Eds.), *Microbial communities and their interactions in the extreme environment. Microorganisms for sustainability* (Vol. 32). Springer. https://doi.org/10.1007/978-981-16-3731-5_2

Wang, W., Zhang, F., Sun, L., Yang, L., Yang, Y., Wang, Y., Siddique, K. H. M., & Pang, J. (2022b). Alkaline salt inhibits seed germination and seedling growth of canola more than neutral salt. *Frontiers in Plant Science*, *13*, 814755. https://doi.org/10.3389/fpls.2022.814755

Wang, W. X., Zhang, Z. X., Wang, X., Han, C., Dong, Y. J., & Wang, Y. X. (2023c). Functional identification of ANR genes in apple (*Malus halliana*) that reduce saline-alkali stress tolerance. *Plant Biology*, *25*, 892–901. https://doi.org/10.1111/plb.13559

Wang, X., Ajab, Z., Liu, C., Hu, S., Liu, J., & Guan, Q. (2020c). Overexpression of transcription factor SlWRKY28 improved the tolerance of *Populus davidiana* × *P. bolleana* to alkaline salt stress. *BMC Genetics*, *21*, 1–13. https://doi.org/10.1186/s12863-020-00904-9

Waters, B. M., Amundsen, K., & Graef, G. (2018). Gene expression profiling of iron deficiency chlorosis sensitive and tolerant soybean indicates key roles for phenylpropanoids under alkalinity stress. *Frontiers in Plant Science*, *9*, 10. https://doi.org/10.3389/fpls.2018.00010

Wegner, L. H., & Zimmermann, U. (2004). Bicarbonate-induced alkalinization of the xylem sap in intact maize seedlings as measured in situ with a novel xylem pH probe. *Plant Physiology*, *136*, 3469–3477. https://doi.org/10.1104/pp.104.043844

Wu, S., Zhu, P., Jia, B., Yang, J., Shen, Y., Cai, X., Sun, X., Zhu, Y., & Sun, M. (2018). A *Glycine soja* group S2 bZIP transcription factor GsbZIP67 conferred bicarbonate alkaline tolerance in *Medicago sativa*. *BMC Plant Biology*, *18*, 234. https://doi.org/10.1186/s12870-018-1466-3

Xie, Q., Yang, Y., Wang, Y., Pan, C., Hong, S., Wu, Z., Song, J., Zhou, Y., & Jiang, X. (2022). The calcium sensor CBL10 negatively regulates plasma membrane H^+-ATPase activity and alkaline stress response in Arabidopsis. *Environmental and Experimental Botany*, *194*, 104752. https://doi.org/10.1016/j.envexpbot.2021.104752

Xu, F., & Yu, F. (2023). Sensing and regulation of plant extracellular pH. *Trends in Plant Science*, *28*, 1422–1437. https://doi.org/10.1016/j.tplants.2023.06.015

Xu, F., Liang, Y., Wang, X., Guo, Y., Tang, K., & Feng, F. (2023). Synergic mitigation of saline-alkaline stress in wheat plant by silicon and *Enterobacter sp.* FN0603. *Frontiers in Microbiology*, *13*, 1100232. https://doi.org/10.3389/fmicb.2022.1100232

Xu, J., Liu, T., Yang, S., Jin, X., Qu, F., Huang, N., & Hu, X. (2019). Polyamines are involved in GABA-regulated salinity-alkalinity stress tolerance in muskmelon. *Environmental Experimental Botany*, *164*, 181–189. https://doi.org/10.1016/j.envexpbot.2019.05.011

Xu, J., Kang, Z., Zhu, K., Zhao, D., Yuan, Y., Yang, S., Zhen, W., & Hu, X. (2021a). RBOH1-dependent H_2O_2 mediates spermine-induced antioxidant enzyme system to enhance tomato seedling tolerance to salinity-alkalinity stress. *Plant Physiology and Biochemistry*, *164*, 237–246. https://doi.org/10.1016/j.plaphy.2021.04.017

Xu, J., Liu, T., Qu, F., Jin, X., Huang, N., Wang, J., & Hu, X. (2021b). Nitric oxide mediates γ-aminobutyric acid-enhanced muskmelon tolerance to salinity–alkalinity stress conditions. *Scientia Horticulturae*, *286*, 110229. https://doi.org/10.1016/j.scienta.2021.110229

Xu, Y., Zheng, C., Liang, L., Yi, Z., & Xue, S. (2021c). Quantitative assessment of the potential for soil improvement by planting Miscanthus on saline-alkaline soil and the underlying microbial mechanisms. *GCB Bioenergy*, *13*, 1191–1205. https://doi.org/10.1111/gcbb.12845

Yamaguchi, T., Aharon, G. S., Sottosanto, J., & Blumwald, E. (2005). Vacuolar Na^{+}/H^{+} antiporter cation selectivity is regulated by calmodulin from within the vacuole in a Ca^{2+} and pH-dependent manner. *Proceedings of the National Academy of Sciences*, USA, *102*, 16107–16112. https://doi.org/10.1073/pnas.0504437102

Yang, C., Chong, J., Kim, C., Li, C., Shi, D., & Wang, D. (2007). Osmotic adjustment and ion balance traits of an alkali resistant halophyte *Kochia sieversiana* during adaptation to salt and alkali conditions. *Plant and Soil*, *294*, 263–276. https://doi.org/10.1007/s11104-007-9251-3

Yang, C., Shi, D., & Wang, D. (2008). Comparative effects of salt and alkali stresses on growth, osmotic adjustment, and ionic balance of an alkali-resistant halophyte *Suaeda glauca* (Bge.). *Plant Growth Regulation*, *56*, 179–190. https://doi.org/10.1007/s10725-008-9299-y

Yang, C., Guo, W., & Shi, D. (2010). Physiological roles of organic acids in alkali-tolerance of the alkali-tolerant halophyte *Chloris virgata*. *Agronomy Journal*, *102*, 1081–1089. https://doi.org/10.2134/agronj2009.0471

Yang, C., Zhao, N., Xu, C., Liu, B., & Shi, D. (2012). Regulation of ion homeostasis in rice subjected to salt and alkali stresses. *Australian Journal of Crop Science*, *6*, 724–731. https://search.informit.org/doi/10.3316/informit.362941256372230

Yang, J. Y., Zheng, W., Tian, Y., Wu, Y., & Zhou, D. W. (2011). Effects of various mixed salt-alkaline stresses on growth, photosynthesis, and photosynthetic pigment concentrations of *Medicago ruthenica* seedling. *Photosynthetica*, *49*, 275–284. https://doi.org/10.1007/s11099-011-0037-8

Yang, S., Hao, X., Xu, Y., Yang, J., & Su, D. (2022). Meta-analysis of the effect of saline-alkali land improvement and utilization on soil organic carbon. *Life*, *12*, 1870 https://doi.org/10.3390/life12111870

Yang, X., Römheld, V., & Marschner, H. (1994). Effect of bicarbonate on root growth and accumulation of organic acids in Zn-inefficient and Zn-efficient rice cultivars (*Oryza sativa* L.). *Plant and Soil*, *164*, 1–7. https://doi.org/10.1007/BF00010104

Ye, Y. Q., Luo, H. Y., Li, M., Zhang, J. J., Cao, G. Q., Lin, K. M., Lin, S. Z., Xu, S. S. (2019). Potassium ameliorates iron deficiency by facilitating the remobilization of iron from root cell walls and promoting its translocation from roots to shoots. *Plant and Soil, 440,* 507–521. https://doi.org/10.1007/s11104-019-04111-z

Yin, F., Zhang, F., & Wang, H. (2021). Rhizosphere bacteria community and functions under typical natural halophyte communities in North China salinized areas. *PLoS ONE*, *16*, e0259515. https://doi.org/10.1371/journal.pone.0259515

Yu, J., Yuan, Y., Zhang, W., Song, T., Hou, X., Kong, L., & Cui, G. (2022). Overexpression of an NF-YC2 gene confers alkali tolerance to transgenic alfalfa (*Medicago sativa* L.). *Frontiers in Plant Science*, *13*, 960160. https://doi.org/10.3389/fpls.2022.960160

Yu, Y., Liu, A., Duan, X., Wang, S., Sun, X., Duanmu, H., Duanmu, H., Zhu, D., Chen, C., Cao, L., Xiao, J., Li, Q., Nisa, Z. U., Zhu, Y., & Ding, X. (2016). *GsERF6*, an ethylene-responsive factor from *Glycine soja*, mediates the regulation of plant bicarbonate tolerance in Arabidopsis. *Planta*, *244*, 681–698. https://dx.doi.org/10.1007/s00425-016-2532-4

Yu, Y., Duan, X., Ding, X., Chen, C., Zhu, D., Yin, K., Cao, L., Song, X., Zhu, P., Li, Q., Nisa, Z. U., Yu, J., Du, J., Song, Y., Li, H., Liu, B., & Zhu, Y. (2017). A novel AP2/ERF family transcription factor form Glycine soja, GsERF71, is a DNA binding protein that positively regulates alkaline stress tolerance in Arabidopsis. *Plant Molecular Biology*, *94*, 509–530. https://doi.org/10.1007/s11103-017-0623-7

Yu, S., Yu, L., Hou, Y., Zhang, Y., Guo, W., & Xue, Y. (2019). Contrasting effects of NaCl and NaHCO3 stresses on seed germination, seedling growth, photosynthesis, and osmoregulators of the common bean (*Phaseolus vulgaris* L.). *Agronomy, 9*, 409. https://doi.org/10.3390/agronomy9080409

Yuan, J. H., Xu, R. K., & Zhang, H. (2011). The forms of alkalis in the biochar produced from crop residues at different temperatures. *Bioresource Technology*, *102*, 3488–3497. https://doi.org/10.1016/j.biortech.2010.11.018

Zandalinas, S. I., & Mittler, R. (2022). Plant responses to multifactorial stress combinations. *New Phytologist*, *234*, 1161–1167. https://doi.org/10.1111/nph.18087

Zang, P., Bing, X., Xiao, H., Li, B., & Sun, H. (2022). Amelioration effects of coastal saline-alkali soil by ball-milled red phosphorus-loaded biochar. *Chemical Engineering Journal*, *431*, 133904. https://doi.org/10.1016/j.cej.2021.133904

Zelm, E., Zhang, Y., & Testernik, C. (2020). Salt tolerance mechanisms of plants. *Annual Review of Plant Biology*, *72*, 403–433. https://doi.org/10.1146/annurev-arplant-050718-100005

Zewd, I., & Siban, M. (2021). The effects of alkalinity on physical and chemical properties of soil. *Journal of Plant Biology and Agriculture Science*, *3*, 1–5 https://www.pulsus.com/scholarly-articles/the-effects-of-alkalinity-on-physical-and-chemical-properties-of-soil-9338.html

Zhang, C., Meng, W., Wang, Y., Zhou, Y., Wang, S., Qi, F., Wang, N., & Ma, J. (2022a). Comparative analysis of physiological, hormonal and transcriptomic responses reveal mechanisms of saline-alkali tolerance in autotetraploid rice (*Oryza sativa* L.). *International Journal of Molecular Sciences*, *23*, 16146. https://doi.org/10.3390/ijms232416146

Zhang, H., Yu, F., Xie, P., Sun, S., Qiao, X., Tang, S., Chen, C., Yang, S., Mei, C., Yang, D., Wu, Y., Xia, R., Li, X., Lu, J., Liu, Y., Xie, X., Ma, D., Xu, X., Liang, Z., … Xie, Q. (2023). A Gγ protein regulates alkaline sensitivity in crops. *Science*, *379*, 1204. https://doi.org/10.1126/science.ade8416

Zhang, J., Wang, J., Jiang, W., Liu, J., Yang, S., Gai, J., & Li, Y. (2016). Identification and analysis of $NaHCO_3$ stress responsive genes in wild soybean (*Glycine soja*) roots by RNA-seq. *Frontiers in Plant Science*, *7*: 1842. https://doi.org/10.3389/fpls.2016.01842

Zhang, J., Fan, X., Wang, X., Tang, Y., Zhang, H., Yuan, Z., Zhou, J., Han, Y., & Li, T. (2022b). Bioremediation of saline-alkali soil polluted with Zn using ryegrass associated with *Fusarium incarnatum*. *Environmental Pollution*, *312*, 119929. https://doi.org/10.1016/j.envpol.2022.119929

Zhang, K., Tang, J., Wang, Y., Kang, H., & Zeng, J. (2020). The tolerance to saline-alkaline stress was dependent on the roots in wheat. *Physiology and Molecular Biology of Plants*, *26*, 947–954. https://doi.org/10.1007/s12298-020-00799-x

Zhang, X., Yang, F., Ma, H., & Li, J. (2022c). Evaluation of the saline–alkaline tolerance of rice (*Oryza sativa* L.) mutants induced by heavy-ion beam mutagenesis. *Biology*, *11*, 126. https://doi.org/10.3390/biology11010126

Zhao, Y., Chen, Y., Liu, S., Li, F., Sun, M., Liang, Z., Sun, Z., Yu, F., Rengel, Z., & Li, H. (2023). Bicarbonate rather than high pH in growth medium induced Fe-deficiency chlorosis in dwarfing rootstock quince A (*Cydonia oblonga* Mill.) but did not impair Fe nutrition of vigorous rootstock *Pyrus betulifolia*. *Frontiers in Plant Science*, *14*, 1237327. https://doi.org/10.3389/fpls.2023.1237327

Zheng, J., Lin, R., Pu, L., Wang, Z., Mei, Q., Zhang, M., & Jian, S. (2021). Ectopic expression of CrPIP2;3 a plasma membrane intrinsic protein gene from halophyte *Canavalia rosea*, enhances drought and salt-alkali stress tolerance in Arabidopsis. *International Journal of Molecular Sciences*, *22*, 565. https://doi.org/10.3390/ijms22020565

Zhao, Q., Suo, J., Chen, S., Jin, Y., Ma, X., Yin, Z., Zhang, Y., Wang, T., Luo, J., Jin, W., Zhang, X., Zhou, Z., & Dai, S. (2016). Na2CO3-responsive mechanisms in halophyte Puccinellia tenuiflora roots revealed by physiological and proteomic analyses. *Scientific Reports, 6*, 32717. https://doi.org/10.1038/srep32717

Zhou, Y., Xu, K., Gao, H., Yao, W., Zhang, Y., Zhang, Y., Hussain, M. A., Wang, F., Yang, X., & Li, H. (2023). Comparative proteomic analysis of wild soybean (*Glycine soja*) genotypes reveals positive regulation of saline-alkaline stress tolerance by tonoplast transporters. *Journal of Agriculture and Food Chemistry*, *71*, 14109–14124. https://doi.org/10.1021/acs.jafc.3c02111

Zhu, D., Cai, H., Luo, X., Bai, X., Deyholos, M. K., Chen, Q., Ji, W., & Zhu, Y. (2012). Over–expression of a novel JAZ family gene from *Glycine soja*, increases salt and alkali stress tolerance. *Biochemical and Biophysical Research Communications*, *426*, 273–279. https://doi.org/10.1016/j.bbrc.2012.08.086

Zörb, C., Geilfus, C. M., & Dietz, K. J. (2018). Salinity and crop yield. *Plant Biology*, *21*, 31–38. https://doi.org/10.1111/plb.12884

3 Metabolic Markers and Metabolite Quantitative Trait *Loci* (mQTLs) for the Breeding of Crop Quality and Resistance

Tzitziki González-Rodríguez, Josaphat Miguel Montero-Vargas, Roberto Gamboa-Becerra, Silverio García-Lara, and Robert Winkler

3.1 INTRODUCTION

Plants are the primary source of human and animal nutrition. The chemical composition of plant materials is highly informative for food quality, safety, and authenticity (Nollet and Winkler 2022). At the same time, the metabolic profile of plant tissue represents a precise phenotype, which is the endpoint of physiological activities during the genotype/phenotype interaction (Figure 3.1; Johannsen 1911; Crick 1970).

In the past, farmers selected the grains for the next cultivation cycle based on the harvested food's productivity, taste, color, and other desirable qualities. However, new technologies have emerged to select food crops more efficiently based on chemical fingerprints and metabolic biomarkers. In addition, the fast climate change forces the accelerated breeding of plants capable of coping with increased abiotic and biotic stresses.

Thus, metabolomics became an essential tool to assist in breeding programs. Metabolomics is a branch of 'omics', integrating biochemistry, instrumental analysis, and bioinformatics and aiming

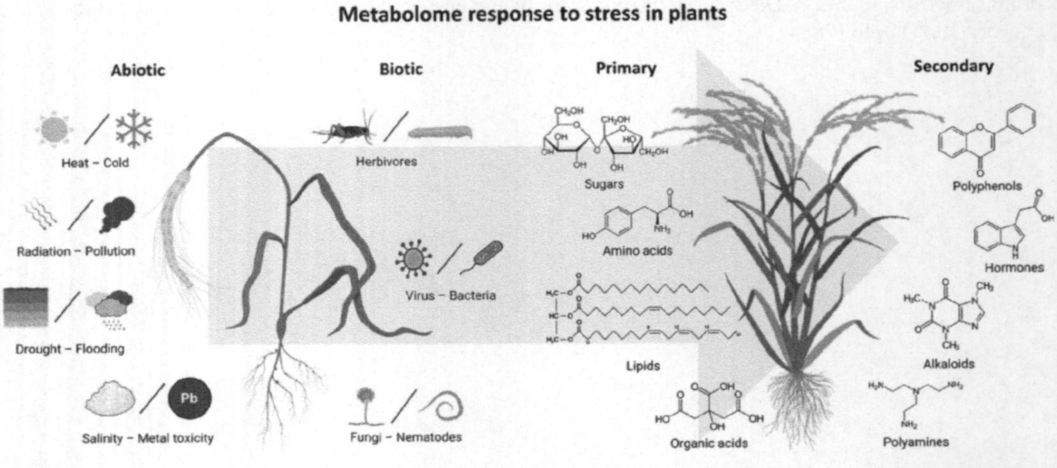

FIGURE 3.1 Metabolic profiles of plants are the result of their genome, their developmental stage, and environmental factors. Created with BioRender.com.

 DOI: 10.1201/9781003457237-3

toward identifying and quantifying all metabolites of an organism (Shen et al. 2023). Food crops contain many compounds, including sugars, amino acids, organic acids, lipids, starch, lignin, alkaloids, flavonoids, polyamines, hormones, glucosinolates, polyphenols, and terpenes. All these compounds are directly affected by biotic and abiotic stresses, and plant researchers can assess plant's response to these alterations by analyzing shifts in metabolite profiles across different stages of crop development. Integrating metabolomics with quantitative genetics, transcriptomics, and genetic manipulation has proven essential in advancing crop improvement strategies (Manickam et al. 2023). This integration may facilitate metabolic editing or engineering by simultaneously targeting multiple genes that govern a specific metabolic response (Razzaq et al. 2022).

Addressing the complete chemical composition of complete metabolomes is not trivial. Currently, two techniques dominate metabolite profiling studies: liquid chromatography coupled to mass spectrometry (LC–MS) and nuclear magnetic resonance (NMR) (Fernie and Schauer 2009). Two approaches are used: (1) 'targeted' analysis refers to studies of a predefined set of metabolites, using directed extraction methods and delivering accurate profiles of defined compounds. Alternatively, (2) 'untargeted' metabolomics focuses on the global detection and relative quantification of metabolites present in a sample (Johnson et al. 2016).

Large-scale metabolomics usually uses MS because of its high sensitivity, selectivity, and broad applicability to theoretically any molecule. For analyzing complex mixtures, MS analyzers are typically coupled to chromatography for pre-separating compounds. Gas chromatography coupled to mass spectrometry (GC–MS), liquid chromatography coupled to mass spectrometry (LC–MS), and capillary-based MS are the most frequently used analytical strategies (Razzaq et al. 2019; Rama and Vinutha 2019). In addition, processing and interpreting metabolomics data is challenging (Winkler 2020). Conventional statistics, machine learning, and artificial intelligence are used. Open-source web platforms, databases, and software facilitate the setup of custom workflows and complete data analysis pipelines (Berthold et al. 2009; Röst et al. 2016; Xia et al. 2009; Pang et al. 2021; Chong et al. 2019; Smith et al. 2006; Tautenhahn et al. 2012; Pluskal et al. 2010; Schmid et al. 2023).

Despite being at an earlier stage of development compared to genomics, transcriptomics, and proteomics, metabolomics offers a unique advantage in studying cellular entities with the most significant impact on the final phenotype (Fiehn 2002). This discipline typically focuses on a specific cell, tissue, or organism. In recent years, considerable advancements have been made in metabolomics by developing new instruments and software. These advancements have enabled the analysis of the entire metabolome of plant species using high-throughput methods (Plumb et al. 2023). Furthermore, integrating multi-omics methodologies has accelerated the screening process and reduced the time required to develop improved crop varieties (Zhu et al. 2023). In addition, novel approaches such as metabolite quantitative trait *loci* (mQTLs) mapping and metabolite genome-wide association studies (GWAS) support the identification of genetic variants associated with metabolic traits (Kumar et al. 2021).

Thus, 'metabolic breeding' strategies promise to push forward the agile development of stress-tolerant plants that are productive in hostile environmental conditions while maintaining and improving their nutritional and sensory quality as food sources.

3.2 METABOLIC MARKERS FOR EVALUATING PLANTS AND PLANT PRODUCTS

Today, it would not be easy to understand crop improvement without using metabolomics and other omics data (genomics, transcriptomics, and proteomics). Significantly, applying these advanced tools has fostered the success of crop improvement programs by elevating the quality characteristics, encompassing flavonoids, folate, and protein composition.

A biomarker is a quantifiable characteristic for predicting or objectively evaluating plant performance (Fernandez et al. 2016). Unlike stable genotypic markers, phenotypic biomarkers are multi-dimensionally regulated in terms of genetics (plant genotype), temporal factors (different development stages, diurnal patterns), spatial regulation (specific organs, tissues, and cell types), and environmental responses (Kooke and Keurentjes 2012). Consequently, utilizing these metabolic traits as biomarkers demands thoughtful consideration of such multifaceted factors.

In plant sciences, metabolic markers find primary utility in three key domains: as traits of agricultural significance, as diagnostic indicators, and as genotype performance markers. The objective for traits of agricultural importance involves directly measuring specific molecules or their precursors, targeting germplasm assessment, and determining desirable or undesirable metabolic markers (Pissard et al. 2013; Ruiz-García et al. 2014). As a diagnostic tool, individual metabolic markers have been suggested to gauge the plant resistance or severity of specific biotic and biotic stressors (Roessner et al. 2000; Korn et al. 2008; Hayat et al. 2012).

Genotype performance markers selection explores the relationships between performance traits (biomass yield and crop quality) of a specific set of genotypes and metabolic markers, with the fundamental assumption being that these predictive markers hold for any genotypes cultivated under different environmental conditions, offering promising prospects for improving plant performance (Meyer et al. 2007).

Plant chemical diversity is vast, and plant-based foods offer essential human health nutrients, such as amino acids, organic acids, vitamins, sugars, and fatty acids, among many other health-promoting phytochemicals (Hall et al. 2008). These compounds serve as precursors for secondary metabolites, including volatiles, phenylpropanoids, and terpenoid compounds, contributing to plant-based foods' color, aroma, and nutritional characteristics (Romero et al. 2021).

Plant metabolomics and chemometrics lifted the comprehensive analysis of nutritional metabolites to a new level. Given the often low nutrient concentrations in crop-based foods, metabolome-assisted breeding techniques utilizing metabolic markers linked to nutritional metabolite production represent a valuable tool to enhance crop nutritional quality and sensory quality because sensory perception is pivotal for consumer acceptance (Fang et al 2019; Jacobs et al. 2021).

Table 3.1 lists representative examples of studies that identified biomarkers associated mainly with the quality of plant-based products and for which different metabolomics analytical platforms and appropriate advanced statistical analysis were used to identify the quality biomarkers.

Metabolomic studies for the identification of quality biomarkers have benefited not only model crops such as tomato and rice but also other major and minor crops, including banana, barley, coffee, grape, kiwi, lettuce, tangerine, mango, potato, walnut, among others, to support the demands of the food industry to provide plant-based products with desirable characteristics.

Understanding the complex relationships between quality biomarkers and attributes is challenging. For such projects, optimized workflows (sample preparation, data mining, annotation, statistical analysis, biomarker identification, and validation) are central for obtaining helpful information (Jacobs et al. 2021; Razzaq et al. 2022).

Since metabolism is under considerable genetic control, knowledge of metabolites' natural variation and heritability is of utmost relevance because it informs about the conservation of advantageous properties in the next generation, assuming similar environmental conditions (Soltis and Kliebenstein 2015). Heritability of metabolites in plants refers to the degree to which the variation in metabolite levels among individuals can be attributed to genetic factors. Studies of heritability of qualitative and quantitative traits have served as the basis for improving desirable plant characteristics such as better plant performance and crop quality.

Knowing the levels and modes of heritability of plant metabolites is essential for plant breeding and crop improvement programs. Identifying the genetic factors contributing to desirable metabolite profiles assists plant scientists and breeders in developing crops with improved quality, nutritional content, stress tolerance, and other essential traits (Razzaq et al. 2022).

The nutritional value of food crops is altered by nutrient supply during plant growth, and the effects of the availability of critical elements such as nitrogen, carbon, phosphorous, and sulfur have been evaluated. For example, a nitrogen deficiency has accumulated metabolites like threonate, glycerate, and inositol in barley (Comadira et al. 2015).

TABLE 3.1
Metabolite Markers Associated with Plant-Based Products Identified Using Different Analytical Platforms and Advanced Statistical Approaches

Crop/Plant-Based Product	Equipment	Quality Trait Studied	Statistical Determination of Biomarker	References
Apple	GC–MS; LC-UV/Vis–MS	Post-storage changes/metabolite levels associated with ethylene synthesis, flavonoid pigment synthesis, and fruit texture	PCA-VIP scores	Rudell et al. (2008)
	UHPLC–DAD–HRMS	Postharvest quality/postharvest blue mold decay resistance	PCA-VIP scores	Sun et al. (2017)
	^{1}H NMR; GC–MS; HS–SPME-GC–MS	Post-storage metabolite changes under low oxygen-controlled atmosphere storage	PCA-VIP scores; PLS	Brizzolara et al. (2017)
Banana	^{1}H NMR	Postharvest quality/changes in metabolite profile of banana during postharvest senescence	PCA-VIP scores	Yuan et al. (2017)
Coffee	UPLC–ESI–MS	Coffee cup quality traits/correlation between highly heritable metabolites and sensory traits of coffee beverage	Spearman's rank correlation	Gamboa-Becerra et al. (2019)
Dark black walnut (*Juglans nigra*)	HS–SPME–GC–MS	Aromatic quality/volatiles associated with light, medium, and dark black walnut	PLS	Lee et al. (2011)
Grapevine	HPLC–IT–MS	Wine quality/biomarkers associated with grapevine berry development and withering	O2PLS-DA	Zamboni et al. (2010)
Grapefruit	HPLC–PAD; GC–MS	Postharvest quality/association between volatile emission and carotenoid content, with chilling injury development sensitivity	PCA-VIP scores	Lado et al. (2019)
Kiwi fruit	GC–ISQ–MS	Postharvest fruit quality/metabolite changes in response to ethylene-induced kiwifruit ripening	PCA; PLS-DA	Lim et al. (2017)
Lettuce	HPLC–PDA–UV/Vis; HPLC–LTQ–Orbitrap FTMS	Phytochemicals associated with crop quality and nutritional quality of lettuce	Random forest model; correlation analysis	Treuren et al. (2018)

(Continued)

TABLE 3.1 (*Continued*)
Metabolite Markers Associated with Plant-Based Products Identified Using Different Analytical Platforms and Advanced Statistical Approaches

Crop/Plant-Based Product	Equipment	Quality Trait Studied	Statistical Determination of Biomarker	References
Malting	UPLC–MS	Malt quality traits/grain metabolites as functional markers of quality traits	PCA-VIP scores; Spearman's rank correlation	Heuberger et al. (2014)
Mandarin	^{1}H NMR	Compounds associated with sensory traits	PLS1 and PLS2; Pearson correlation	Simons et al. (2019)
Mango	UPLC–TOF-MS	Metabolite levels in leave tissues associated with increased tolerance to fruit malformation linked to *Fusarium* infection	PCA; OPLS-DA	Augustyn et al. (2014)
Potato	GC–TOF-MS	Biomarkers associated with chip quality (browning of potato chip during frying) and potato quality traits (susceptibility to black spot bruising)	PCA-VIP scores; PLS; Spearman's rank correlation; prediction models	Steinfath et al. (2010)
Rice	HS–SPME/ GC–TOF-MS	Aromatic quality attributes/volatile differentiating between aromatic and non-aromatic rice grains	PCA-eigenvalues	Setyaningsih et al. (2019)
	SPME–GC-qMS; GC–TOF-MS; ^{1}H NMR; DI–ESI-MS	Post-storage changes/ metabolic changes between fragrant and non-fragrant rice in aged grains	HCA; PLS-DA	Mumm et al. (2016)
Tomato	^{1}H NMR; GC–MS	Agronomic traits and fruit quality/shelf life and firmness	HCA; neuronal networks (SOM)	López et al. (2015)

DA, data analysis; DAD, diode array detector; DI, direct injection; ESI, electrospray ionization; FTMS, Fourier transform mass spectrometry; GC, gas chromatography; HCA, hierarchical cluster analysis; HRMS, high-resolution mass spectrometry; HS, headspace; ISQ, (Thermo) single quadrupole; IT, ion trap; LC, liquid chromatography; LTQ, linear ion trap (Thermo); MS, mass spectrometry; NMR, nuclear magnetic resonance; O(2)PLS, (two-way) orthogonal partial least square; PCA, principal component analysis; PDA, photodiode array; PLS, partial least square; SOM, self-organizing map; SPME, solid-phase microextraction; TOF, time-of-flight; U(H)PLC, ultra-high-performance liquid chromatography; UV, ultraviolet; VIP, variable importance on projection; vis, visible.

In a wheat metabolome characterization during grain filling employing an integrated GC-TOF-MS technique, elevated fructose, ribulose, and lyxose levels were detected under nitrogen stress (Heyneke et al. 2017). Other examples used GC–MS and LC–MS technologies for wheat metabolomic profiling revealed enhanced synthesis of tyrosine, lysine, allo-inositol, and L-ascorbic acid in wheat under nitrogen stress (Khan et al. 2019).

An un-targeted GC–MS-MS study identified metabolites with therapeutic significance in Mappillai Samba, a traditional rice variety renowned for its medicinal properties (Rajagopalan et al. 2022).

Employing an un-targeted analysis, various metabolites, including fructose, tyrosine, glucose, glutamine, threonine, serine, and valine, have been recognized as diagnostic biomarkers for post-harvest quality trait identification in potato tubers (Steinfath et al. 2010).

The chemical variation is influenced by genotype and market class, with notable profiles observed, for example, in yellow-flesh potatoes (higher carotenoids) and specialty potatoes (higher chlorogenic acid), revealing that variation in health-associated molecules among market classes, including vitamins, bioactive, medicinals, and minerals are essential for potatoes (Chaparro et al. 2018).

Merging of metabolomics and association-mapping methodologies has demonstrated connections between genomic regions in maize and characteristics like kernel composition, potato starch content, tomato pigment content, and maize pro-vitamin A (Fernie and Schauer 2009).

Recently, a combination of transcriptome and metabolome studies on *Camellia reticulata* concluded that flavonoids, especially anthocyanins, play a significant role in flower color, resulting in potential applications for developing *Camellia* varieties with specific flower colors and qualities (Geng et al. 2022).

Metabolomics has accelerated phenotyping, reducing the time needed to screen varieties resistant to biotic and abiotic influences or with nutrition quality traits. This strategy was applied to evaluate the natural variation in the metabolism of rice that identified 36 genes that modulate the levels of metabolites with potential physiological and nutritional importance, such as alanine, threonine, leucine, and histidine (Chen et al. 2014).

In rice, flavonoids and amino acids are associated with phenotypic traits such as grain color, size, and weight (Chen et al. 2016). Diverse winter wheat lines were analyzed with targeted metabolomics, identifying oxalic acid, ornithine, L-arginine, pentose alcohol III, L-tyrosine, and a sugar oligomer (oligo II) as a metabolic marker associated with genome variation (Matros et al. 2017). These findings underscore the growing feasibility of marker-assisted breeding through metabolites for crop improvement.

A metabolic phenotyping strategy was used to assess the physiological state of maize plants under various stress conditions (García-Flores et al. 2012). An evaluation of maize hybrids grown in distinct agroecologies in Mexico showed that phytochemical profiles are relevant to nutraceutical effects and antioxidative capacity according to each location (Montero-Vargas et al. 2020).

Related to biotic stress, un-targeted metabolomics analysis using LC–MS in maize, metabolites associated with pathogen response, such as benzoxazinoids and flavonoids, were proposed for marker-assisted breeding to cultivate resistant maize lines (Zhou et al. 2019).

The metabolomic profiling of 266 maize inbred lines under salinity stress with LC–MS identified several metabolic markers associated with salt tolerance, such as terpenoids, benzoxazinoids, lipids, amino acids, and flavonoids (Liang et al. 2021).

In another vital food crop, hundreds of varieties of soybean lines were analyzed by a targeted metabolomics approach, allowing the identification of metabolites related to lipid metabolism and oil-related traits and amino acids such as alanine, arginine, asparagine, aspartic acid, and daizein (Liu et al. 2020).

A targeted analysis of tomatoes identified 980 putative metabolites in the ripened tomato pericarp. Amino acids, alkaloids, vitamins, polyamines, and polyphenols are involved in the effects of domestication and breeding on the metabolomic features of tomatoes (Zhu et al. 2018).

All these studies suggest using metabolic profiles for the precise phenotyping of plants and evaluating food materials. The compounds present show the nutritional and sensory quality and presence of possibly hazardous contaminants. At the same time, chemical markers inform about the plant's health status and possible resistance mechanisms.

3.3 CONCEPT OF USING MQTLS IN PLANT BREEDING

Crop improvement has been practiced for thousands of years, but the technology has recently emerged to simultaneously select and improve various traits of agronomic interest. The rapid development of high-density genotyping platforms and diverse phenotyping tools has allowed comprehensive approaches, including different omics, to discover genetic variants controlling known and unknown biochemical pathways in various plants. These studies entail metabolic quantitative trait *loci* (mQTLs), chromosomic regions containing genetic variation responsible for differences in chemical composition, and the number of metabolites between individuals (Swain-Lenz et al. 2017).

Researchers require segregating populations, molecular markers, and phenotypic measurements to identify QTLs. Mapping populations are created through the genetic variance between parents concerning a specific trait of interest. These parents undergo a crossbreeding process, yielding heterozygous organisms (F_1). Subsequently, these can be further crossed utilizing various schemes until a homozygous population is achieved. While the total number of individuals within mapping populations may vary, a higher count enhances the precision of the analysis. These populations undergo genotyping using molecular markers and DNA sequences that indicate locating a gene, whether within or in proximity. Such markers, like single-nucleotide polymorphisms (SNPs), restriction fragment length polymorphisms, and microsatellites, offer the advantage of being unaffected by environmental factors, thus ensuring the integrity of the analysis (Collard et al. 2005).

Phenotypic traits, including morphological characteristics, metabolite levels, and agronomic traits, are contingent upon the study. Therefore, it is imperative to devise an appropriate experimental design facilitating trait quantification and assigning individual values within the population. Pioneering research focused on finding mQTLs on a large scale was performed in the model plant *Arabidopsis thaliana*. In this work, an un-targeted metabolic profiling approach from leaves of 14

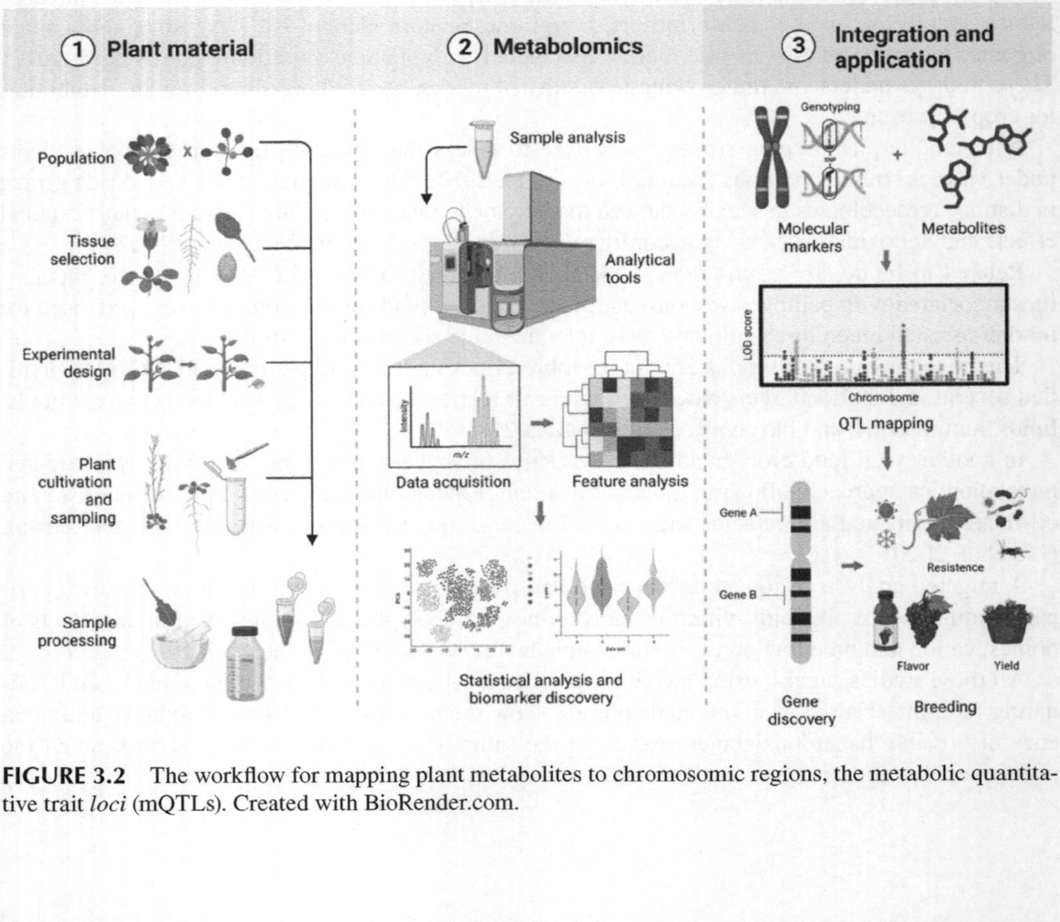

FIGURE 3.2 The workflow for mapping plant metabolites to chromosomic regions, the metabolic quantitative trait *loci* (mQTLs). Created with BioRender.com.

accessions was obtained using LC–MS, revealing the natural chemical diversity within a single plant species. A recombinant inbred line (RIL) mapping population was generated between the most chemically distant accessions to explore the origin of this variation. Certain genomic regions were associated with metabolite profiles, leading to the identification of genes and biosynthetic pathways. This study demonstrated for the first time that metabolic variation can be mapped and established that metabolites are heritable, showing that mQTL distribution is non-uniform throughout the genome. These findings have opened up new avenues for studying genetics' role in plant metabolism and its relation to complex traits such as stress tolerance (Keurentjes et al. 2006) (Figure 3.2).

Following, there has been an increase in the exploration of mQTLs in plants (Table 3.2). Several studies have demonstrated the existence of genomic hotspots that regulate plant metabolism (Keurentjes et al. 2006; Lisec et al. 2008; Khan et al. 2012; Matsuda et al. 2012; Wahyuni et al. 2014; Knoch et al. 2017; Li et al. 2019; Maharijaya et al. 2019; Teh et al. 2019; Pott et al. 2020). This phenomenon indicates a single polymorphism influencing widespread changes in downstream gene expression, with the expression mapping to the same genomic locus (Breitling et al. 2008). Such hotspots for mQTLs suggest the presence of regulatory genes controlling the synthesis or breakdown of groups of metabolites at that map position (Knoch et al. 2017; Zhu et al. 2018).

Our understanding of the genetic regulation of metabolism continues to expand. It is now evident that metabolic composition is regulated in a tissue-specific manner linked to the developmental stage (Gong et al. 2013; Knoch et al. 2017; Garbowicz et al. 2018; Maharijaya et al. 2019; González-Rodríguez et al. 2023). Also, different modes of inheritance between tissues have been reported for mQTLs (Toubiana et al. 2012).

Multiple studies have reported several mQTLs associated with diverse plant origins and biochemical pathways (Table 3.2). By identifying these mQTLs, breeders can select specific alleles related to a particular metabolic profile without affecting other important traits, such as the agronomic yield. Therefore, mQTL-assisted breeding promises to develop plants endowed with enhanced nutritional value, improved volatile profiles, and increased resilience, among other desirable attributes.

Many fruits are sources of health-beneficial compounds, and their consumption can prevent diseases. Examples include phenolic compounds present in strawberries (Pott et al. 2020), grapevine (*Vitis* spp.) (Teh et al. 2019), and apples (Khan et al. 2012). Still, the genetic basis of the production of these has been limited. Through a mQTL study made in apple (*Malus domestica*), it was found a mQTL hotspot in linkage group 16 (LG16) for mQTLs for metabolites such as procyanidins, quinic acid, phenolic esters, coumaroyl hexoside, kaempferol glycosides, etc., all from the phenylpropanoid pathway (PPP). In LG16, they also found structural genes from the PPP, like the *leucoanthocyanidin reductase* (*LAR1*) and other important transcription factors such as *MYB*, *bHLH*, *AP2*, and *bZIP* (Khan et al. 2012).

By analyzing the correlation between the genetic markers and metabolite levels, mQTLs can be identified. One of the significant advantages of studying mQTLs is that they link genotype and phenotype at the molecular level. This information can help to understand the underlying genetic mechanisms that control metabolic pathways and their response to different environmental conditions, including plant stress. However, mQTL studies focused on plant stress are still limited.

3.4 EXPLORING MQTLS UNDER STRESS CONDITIONS

As sessile organisms, plants have evolved unique mechanisms to defend and adapt to adverse environmental conditions. Generally, there are two types of plant stresses: biotic and abiotic. Biotic stress is produced by biological agents like herbivores or pathogens. In contrast, abiotic stress is caused by environmental factors such as drought, low and high temperatures, nutrient limitation, high radiation, salinity, and heavy metals.

The mechanisms involved in the plant's response to stress are multifaceted and involve diverse players and processes (Gou et al. 2023). Metabolites, including primary and secondary, play crucial roles in plant defense. There is plenty of evidence of its impact and roles against stress (Erb and

TABLE 3.2
mQTL Research in Plants

Plant	Populations	Number of mQTLs	Measured Traits/ Result	Approach/ Technique	References
Apple	F_1 (Prima×Fiesta)	669 (488 in the peel and 181 in the flesh)	418 metabolites in peel and 254 in flesh. mQTL hotspots that mainly regulate thephenylpropanoid pathway.	Untargeted LC–ESI-Q-TOF	S. A. Khan et al. (2012)
Arabidopsis	RIL and IL (Col-0×C24)	157	Identification of genes related to the control of primary metabolism.	Untargeted GC–MS	Lisec et al. (2008)
	RIL (Bay×Sha)	11 clusters	557 metabolites. Epistatic network that controls central metabolism.	Untargeted GC-TOF-MS	Rowe et al. (2008)
	RIL and IL (Col-0×C24)	153 for augmented additive and 83 for dominance effects	181 metabolites. Identification of the mode of gene action for heterosis in primary metabolism.	Untargeted GC–MS	Lisec et al. (2009)
	RIL (Col-0×C24)	786	311 polar primary metabolites. Finding master regulatory locus of seed metabolism.	Untargeted GC–MS	Knoch et al. (2017)
Buckwheat	Diverse genotypes	27	Among 24 metabolites identified in seed endosperm, 18 were associated with mQTLs.	Targeted HPLC–DAD–MS	Zargar et al. (2023)
Grape	F_2	19	Detection of 1,317 ions and identification of a major locus on chromosome 18 related to disease-resistance motifs.	Untargeted LC–Q-Orbitrap	Teh et al. (2019)
Maize	RIL (B73×Mo17)	31 in leaves and six in seeds	236 ions from leaves and 71 from seeds. Identification of *loci* associated with multiple metabolite biosynthetic pathways through a new technical perspective.	Untargeted DLI–MS	González-Rodríguez et al. (2023)

(*Continued*)

TABLE 3.2 (*Continued*)
mQTL Research in Plants

Plant	Populations	Number of mQTLs	Measured Traits/ Result	Approach/ Technique	References
	RIL (By804×B73)	297	Detection of 69 metabolites in seedling leaves, 64 in reproductive stage leaves, and 41 in kernel. Description of the genetic basis of primary metabolism and key candidates' genes.	Untargeted GC–TOF-MS	Wen et al. (2015)
Pepper	F_2 (*Capsicum annuum*×*Capsicum chinense*)	279	Detection of 542 metabolites, 52 annotated. Two mQTL hotspots in chromosome 9.	Untargeted LC–PDA–QTOF-MS	Wahyuni et al. (2014)
Potato	C (*Solanum phureja*×*Solanum tuberosum*)×E (*Solanum vernei*×*S. tuberosum*)	87	139 polar metabolites related to primary metabolism.	Untargeted GC–TOF-MS	Carreno-Quintero et al. (2012)
Rapeseed	DH (EXPRESS×SWU07)	4	Identification of genetic control of glucosinolates content to environmental variation.	Targeted NIRS	He et al. (2018)
	DH (Tapidor×Ningyou7)	105	Dissection of the genetic basis of glucosinolate biosynthesis.	Targeted HPLC	Feng et al. (2012)
Rice	RIL ('Lemont'×'Teqing')	1,797	512 metabolites. Identification of two hotspots, C74A on chromosome 3 and CDO497 on 7, opposite effects on carbon and nitrogen metabolites.	Untargeted GC–TOF	Li et al. (2016)
	RIL (ZS97×MH63)	1,884 in flag leaves and 937 in seeds	Identification of 24 genes for flavonoids and functional validation of three.	Untargeted LC–MS/MS	Gong et al. (2013)
	BC (Sasanishiki×Habatak)	802	Large-scale analysis to dissect the genetic architecture of metabolome and in silico analysis of candidates' genes.	Untargeted CE-MS, GC–MS, and LC–MS	Matsuda et al. (2012)

(*Continued*)

TABLE 3.2 (*Continued*)
mQTL Research in Plants

Plant	Populations	Number of mQTLs	Measured Traits/ Result	Approach/ Technique	References
Strawberry	F_1 ('232'×'1,392')	Seven for total phenolic content, nine for antioxidant capacity, and 309 for the polar secondary metabolites	77 polar secondary metabolites. Genetic control of phenylpropanoid compounds and identification of hotspots and candidate genes for the flavonoid pathway.	Targeted and Untargeted UPLC–Orbitrap-MS/MS	Pott et al. (2020)
	F_1 ('232'×'1,392')	133	50 primary metabolites. Complementary analysis of RNA-seq reveals 826 differentially expressed genes and one responsible for variation in ascorbic acid.	Targeted GC-TOF-MS	Vallarino et al. (2019)
Tomato	ILs (*Solanum lycopersicum* 'M82'×*Solanum pennellii*)and the recurrent parent 'M82'	338	Profiling of specialized metabolites, including flavonoids and steroidal glycoalkaloids. Validation of some mQTLs through fine mapping and overexpression of genes in *Nicotiana*.	Untargeted LC–MS	Alseekh et al. (2020)
	ILs and BILs	1,528 in fruit and 428 in leaves	Lipidomic profiling. Identification of candidate genes and a major mQTL on a chromosome that regulates levels of diacylglycerols and triacylglycerols.	Untargeted GC–TOF-MS and UPLC–FT-MS	Garbowicz et al. (2018)

(*Continued*)

TABLE 3.2 (*Continued*)
mQTL Research in Plants

Plant	Populations	Number of mQTLs	Measured Traits/ Result	Approach/ Technique	References
	IL4-4×'M82'	72	327 metabolites, including lipids, volatile organic compounds, and primary and secondary metabolites. Identification of a major locus encoding 20 genes that influence multiple pathways.	Targeted and Untargeted HPLC, GC–MS, and LC–MS	Liu et al. (2016)
	ILs	679	Identification of the mode of inheritance of primary and secondary metabolites. Comparison with morphological traits and description of transcriptional mechanisms for some pathways.	Untargeted UPLC–Orbitrap	Alseekh et al. (2015)
	ILs (*S. lycopersicum* ('M82')×*S. pennellii*)	2,820	Identification of QTL for health-related antioxidant carotenoids and tocopherols and molecular signatures for over 2,000 compounds.	Targeted and Untargeted NMR, H PLC-PDA, GC–MS, and LC–MS.	Perez-Fons et al. (2014)
	ILs	46	The finding of strong post-transcriptional regulation of seed metabolism. The identification of genetic factors significantly affecting amino and organic acids was confirmed in two consecutive seasons.	Untargeted GC–MS	Toubiana et al. (2012)

(*Continued*)

TABLE 3.2 (*Continued*)
mQTL Research in Plants

Plant	Populations	Number of mQTLs	Measured Traits/ Result	Approach/ Technique	References
	IL and ILH	174 are dominantly inherited, 61 are additively inherited, and 80 are recessively inherited	Primary metabolites and comparison on the modes of inheritance of the mQTLs.	Untargeted GC–MS	Schauer et al. (2008)
	ILs	889	74 metabolites, including amino-, organic-, and fatty acids, sugars, and vitamins. Integration of metabolic and whole-plant phenotypic traits.	Untargeted GC–MS	Schauer et al. (2006)
Wheat	RIL (KN9204×J411)	1,005	1,260 metabolic features, and 24 candidate genes related to grain traits. Association and prediction of agronomic traits.	Targeted LC–MS/MS	Shi et al. (2020)
	DH (Excalibur×Kukri)	238	558 features, including alkaloids and flavonoids. Identification of phenology-related loci and potential targets for breeding in chromosome 7A.	LC–MS	Hill et al. (2015)

Populations: BIL, backcrossed inbred lines; DH, double haploids; IL, introgression lines; RIL, recombinant inbred lines. Techniques: DAD, diode array detector; DLI, direct liquid introduction; ESI, electrospray ionization; FT, Fourier transform; GC, gas chromatography; HPLC, high-performance liquid chromatography; LC, liquid chromatography; MS, mass spectrometry; NIRS, near-infrared spectroscopy; NMR, nuclear magnetic resonance; Q, quadrupole; TOF, time-of-flight; UPLC, ultra-performance liquid chromatography.

Kliebenstein 2020). As mentioned, these are controlled by multiple QTLs, reflecting the polygenic nature of these traits. Previous QTL studies of agronomic traits have needed more information on the biochemical pathways underlying those traits and their responses to stress. Studies focused on mQTLs related to biotic and abiotic stresses are summarized in Table 3.3.

A wheat study investigated the relation of metabolic and 29 agronomic traits in a double haploid line 'Excalibur' × 'Kukri' population. It was documented that a single genomic interval on chromosome 7A affected 12 agronomic traits and also co-localized with mQTLs for several metabolites. This chromosome was found to impact agronomic and metabolic traits substantially, indicating its importance in influencing multiple aspects of wheat physiology and productivity under drought conditions (Hill et al. 2013).

Studies in barley (*Hordeum vulgare* L.) have identified mQTLs related to abiotic stress. Flag leaves from Mediterranean and German accessions were subjected to a combination of heat and drought stress, revealing mQTLs for γ-tocopherol, glutathione, and succinate content, which co-localize with genes encoding enzymes of the pathways producing these antioxidant metabolites (Templer et al. 2017). Another research identified 98 metabolites that significantly changed their levels in response to water deficit, revealing differences among groups between RILs and parental varieties at the early and late stages of the stress. The study also identified mQTL hotspots in the barley genome, highlighting the linkage groups 1H.1 and 2H, revealing regions and candidate genes for regulating metabolite levels in drought conditions (Piasecka et al. 2017).

In biotic stress, investigations on pepper (*Capsicum*) leaves detected 674 metabolites, revealing mQTLs differing between tissues (fruits and leaves). They found metabolites and mQTLs associated with thrips resistance, such as two diterpene glycosides and a flavonoid compound (Maharijaya et al. 2019). Additional studies in rice revealed metabolic traits associated with resistance to white-backed planthopper. Here, the authors identified flavanone 3-hydroxylase (F3H), a secondary metabolite-related gene, as a significant contributor to WBPH resistance in rice (Kim et al. 2021).

Proteomics has also contributed to these studies, as evidenced in a wheat analysis. In this research, it was possible to identify putatively resistance-related metabolites, including plant-signaling molecules, jasmonic acid and salicylic acid, and terpenoids, in rachis and spikelets of a resistant NIL upon *Fusarium graminearum* inoculation. It identified that the resistance in the QTL 'Fhb1' is mainly associated with cell wall thickening due to the deposition of specific compounds, such as hydroxycinnamic acid amides, phenolic glucosides, and flavonoids (Gunnaiah et al. 2012).

These findings highlight the complexity and interconnection of plant metabolic and genetic responses to various environmental challenges, providing insight into their ability to adapt and survive in changing environments.

While these studies did not specifically target stress, they have unearthed traits associated with stress by examining mQTLs. An extensive analysis was conducted in field conditions to explore the genomic regions linked to secondary metabolism in the pericarp of tomato fruit. The study found that environmental conditions had a more significant impact on mQTLs associated with primary metabolites than those related to secondary metabolites in these tomato introgression lines. The research also identified substantial genetic effects underlying multiple mQTL hotspots on chromosomes 6 and 10, significantly influencing the fruit's nutritional properties (Alseekh et al. 2015). These findings suggest that targeted-breeding strategies could be employed to improve the nutritional quality of tomato fruit.

3.5 CHALLENGES AND PROSPECTS OF MQTLS IN PLANT BREEDING

QTL analysis presents challenges that impact its applicability in plant breeding. One of the main limitations is generating the populations needed for QTL mapping, which often requires labor-intensive and time-consuming processes such as crossing and genotyping (Takagi et al. 2013). This challenge can be addressed through the generation of double haploid lines, which can significantly reduce the time and cost of generation of the mapping population (Maqbool et al. 2020).

Another constraint is the construction of linkage maps, which involves large populations and extensive phenotypic data availability. QTL analysis often requires large sample sizes to achieve statistical power and accurately detect QTL. Small sample sizes can lead to false positives or false negatives (Broman and Sen 2009; Kassie et al. 2023).

Another area for improvement is that QTL analysis identifies chromosomal regions associated with traits but needs more precision in locating specific genes. The density and resolution of genetic markers used in QTL analysis impact mapping precision; low marker density can result in broad QTL regions, complicating the identification of individual genes (Piasecka et al. 2017). Even when

TABLE 3.3
Stress-Related mQTLs Studies in Plants

Plant	Type of Stress	Population	Result	Approach/ Technique	References
			Abiotic		
Barley	Drought and heat	A diverse set of barley genotypes	Identification of 13 mQTLs involved in metabolite content and antioxidative defense.	Targeted LC–MS	Templer et al. (2017)
	Drought	RIL (Maresi×CamB)	Detection of 138 mQTL for 98 traits, some related to genes involved in stress responses to cold, heat, and oxidative stress.	Targeted LC–MS	Piasecka et al. (2017)
Wheat	Drought	DH (Excalibur×Kukri)	Identified 112 mQTLs for 95 metabolic features.	Untargeted GC–MS	Hill et al. (2013)
			Biotic		
Pepper	Thrips (*Frankliniela occidentalis*)	F_2 (*Capsicum annuum*×*Capsicum chinense*)	304 mQTLs of the 674 metabolites detected. Novel metabolites correlated with thrips resistance and co-localized with the major resistance QTL on chromosome 6.	Untargeted LC–MS	Maharijaya et al. (2019)
Wheat	*Fusarium graminearum*	NIL-R and NIL-S	Identification of putatively resistance-related metabolites, including plant-signaling molecules and terpenoids.	Untargeted LC-LTQ-Orbitrap	Gunnaiah et al. (2012)

QTL regions are identified, the biological relevance of the associated genes may remain unclear (Broman and Sen 2009).

The potential of metabolomics in crop genetic improvement will continue to increase with new technologies that allow the measurement and analysis of a more significant number and variety of metabolites. New technical perspectives must be addressed to decrease cost and time in mQTL analysis (González-Rodríguez et al. 2023).

Beyond the challenges inherent in QTL analysis, a crucial aspect lies in the subsequent exploration of identified QTL regions. Further studies become imperative to unravel the intricate mechanisms underlying trait variation. Additionally, validating mQTLs and associated genes is paramount (Garbowicz et al. 2018; Alseekh et al. 2020). The necessity for complementary studies becomes evident, emphasizing the need for confirmation through advanced biotechnological approaches. Techniques such as the generation of transgenic lines and the application of CRISPR/CAS for genome editing offer invaluable tools for delving into the functions of the identified genes. This comprehensive approach strengthens the reliability of QTL findings and provides a deeper understanding of the genetic factors influencing traits, bridging the gap between identification and functional comprehension.

The knowledge obtained from the mQTL studies has to be used for breeding new cultivars with increased levels of these putatively beneficial metabolites through application in marker-assisted breeding.

3.6 CONCLUSIONS

Metabolic traits serve both for plant characterization and as breeding markers. Using mass spectrometry methods provides rich information, which, combined with adequate statistical analyses, could drastically speed up the generation of plant varieties with the desired properties.

Metabolic QTLs play a crucial role in this avenue for understanding the genetic basis of plant stress responses. They can provide valuable insights into the underlying mechanisms of metabolic adaptation and molecular networks that regulate developmental and physiological processes.

Evaluating and predicting chemical signatures instead of genes is a new paradigm in plant breeding. Mapping high-resolution mass spectrometry data to genomic regions provides new perspectives for the rational design of improved crops. Still, it needs research with large plant populations and detailed molecular and metabolic characterization.

However, it has become clear that the prediction of metabolic profiles and, therefore, promoting plants with robust characteristics against environmental perturbations is still a long way to go.

Future challenges will include the use of cutting-edge techniques such as CRISPR/CAS and double haploids to bridge the gap between the identification and comprehension of mQTLs.

REFERENCES

Alseekh, S., Ofner, I., Liu, Z., Osorio, S., Vallarino, J., Last, R. L., Zamir, D., Tohge, T., & Fernie, A. R. (2020). Quantitative trait loci analysis of seed-specialized metabolites reveals seed-specific flavonols and differential regulation of glycoalkaloid content in tomato. *The Plant Journal*, *103*(6), 2007–24. https://doi.org/10.1111/tpj.14879

Alseekh, S., Tohge, T., Wendenberg, R., Scossa, F., Omranian, N., Li, J., Kleessen, S., Giavalisco, P., Pleban, T., Mueller-Roeber, B., Zamir, D., Nikoloski, Z., & Fernie, A. R. (2015). Identification and mode of inheritance of quantitative trait loci for secondary metabolite abundance in tomato. *Plant Cell*, *27*(3), 485–512. https://doi.org/10.1105/tpc.114.132266

Augustyn, W. A., Regnier, T., Combrinck, S., & Botha, B. M. (2014). Metabolic profiling of mango cultivars to identify biomarkers for resistance against fusarium infection. *Phytochemistry Letters*, *10*(December), civ–cx. https://doi.org/10.1016/j.phytol.2014.05.014

Berthold, M. R., Cebron, N., Dill, F., Gabriel, T. R., Kötter, T., Meinl, T., Ohl, P., Thiel, K., & Wiswedel, B. (2009). KNIME - The Konstanz information miner: Version 2.0 and beyond. *SIGKDD Exploration Newsletter*, *11*(1), 26–31. https://doi.org/10.1145/1656274.1656280

Breitling, R., Li, Y., Tesson, B. M., Fu, J., Wu, C., Wiltshire, T., Gerrits, A., Bystrykh, L. V., de Haan, G., Su, A. I., & Jansen, R. C. (2008). Genetical genomics: Spotlight on QTL hotspots. *PLoS Genet*, *4*(10), e1000232. https://doi.org/10.1371/journal.pgen.1000232

Brizzolara, S., Santucci, C., Tenori, L., Hertog, M., Nicolai, B., Stürz, S., Zanella, A., & Tonutti, P. (2017). A metabolomics approach to elucidate apple fruit responses to static and dynamic controlled atmosphere storage. *Postharvest Biology and Technology*, *127*(May), 76–87. https://doi.org/10.1016/j.postharvbio.2017.01.008

Broman, K. W., & Sen, S. (2009). *A guide to QTL mapping with R/Qtl*. Statistics for biology and health. Springer. https://doi.org/10.1007/978-0-387-92125-9

Carreno-Quintero, N., Acharjee, A., Maliepaard, C., Bachem, C. W. B., Mumm, R., Bouwmeester, H., Visser, R. G. F., & Keurentjes, J. J. B. (2012). Untargeted metabolic quantitative trait loci analyses reveal a relationship between primary metabolism and potato tuber quality. *Plant Physiology*, *158*(3), 1306–1318. https://doi.org/10.1104/pp.111.188441

Chaparro, J. M., Holm, D. G., Broeckling, C. D., Prenni, J. E., & Heuberger, A. L. (2018). Metabolomics and ionomics of potato tuber reveals an influence of cultivar and market class on human nutrients and bioactive compounds. *Frontiers in Nutrition*, *5*, 36. https://doi.org/10.3389/fnut.2018.00036

Chen, W., Gao, Y., Xie, W., Gong, L., Lu, K., Wang, W., Li, Y., Liu, X., Zhang, H., Dong, H., Zhang, W., Zhang, L., Yu, S., Wang, G., Lian, X., & Luo, J. (2014). Genome-wide association analyses provide genetic and biochemical insights into natural variation in rice metabolism. *Nature Genetics*, *46*(7), 714–721. https://doi.org/10.1038/ng.3007

Chen, W., Wang, W., Peng, M., Gong, L., Gao, Y., Wan, J., Wang, S., Shi, L., Zhou, B., Li, Z., Peng, X., Yang, C., Qu, L., Liu, X., & Luo, J. (2016). Comparative and parallel genome-wide association studies for metabolic and agronomic traits in cereals. *Nature Communications*, *7*(1), 12767. https://doi.org/10.1038/ncomms12767

Chong, J., Yamamoto, M., & Xia, J. (2019). MetaboAnalystR 2.0: From raw spectra to biological insights. *Metabolites*, *9*(3), 57. https://doi.org/10.3390/metabo9030057

Collard, B. C. Y., Jahufer, M. Z. Z., Brouwer, J. B., & Pang, E. C. K. (2005). An introduction to markers, quantitative trait loci (QTL) mapping and marker-assisted selection for crop improvement: The basic concepts. *Euphytica*, *142*(1), 169–196. https://doi.org/10.1007/s10681-005-1681-5

Comadira, G., Rasool, B., Karpinska, B., Morris, J., Verrall, S. R., Hedley, P. E., Foyer, C. H., & Hancock, R. D. (2015). Nitrogen deficiency in barley (Hordeum vulgare) seedlings induces molecular and metabolic adjustments that trigger aphid resistance. *Journal of Experimental Botany*, *66*(12), 3639–3655. https://doi.org/10.1093/jxb/erv276

Crick, F. (1970). Central dogma of molecular biology. *Nature*, *227*(5258), 561–563. https://doi.org/10.1038/227561a0

Erb, M., & Kliebenstein, D. J. (2020). Plant secondary metabolites as defenses, regulators, and primary metabolites: The blurred functional trichotomy. *Plant Physiology*, *184*(1), 39–52. https://doi.org/10.1104/pp.20.00433

Fang, C., Luo, J., & Wang, S. (2019). The diversity of nutritional metabolites: Origin, dissection, and application in crop breeding. *Frontiers in Plant Science*, *10*, 1028. https://www.frontiersin.org/articles/10.3389/fpls.2019.01028

Feng, J., Long, Y., Shi, L., Shi, J., Barker, G., & Meng, J. (2012). Characterization of metabolite quantitative trait loci and metabolic networks that control glucosinolate concentration in the seeds and leaves of Brassica napus. *New Phytologist*, *193*(1), 96–108. https://doi.org/10.1111/j.1469-8137.2011.03890.x

Fernandez, O., Urrutia, M., Bernillon, S., Giauffret, C., Tardieu, F., Gouis, J. L., Langlade, N., Charcosset, A., Moing, A., & Gibon, Y. (2016). Fortune telling: Metabolic markers of plant performance. *Metabolomics*, *12*(10), 158. https://doi.org/10.1007/s11306-016-1099-1

Fernie, A. R., & Schauer, N. (2009). Metabolomics-assisted breeding: A viable option for crop improvement? *Trends in Genetics*, *25*(1), 39–48. https://doi.org/10.1016/j.tig.2008.10.010

Fiehn, O. (2002). Metabolomics – The link between genotypes and phenotypes. *Plant Molecular Biology*, *48*(1–2), 155–171. https://doi.org/10.1023/A:1013713905833

Gamboa-Becerra, R., Hernández-Hernández, M. C., González-Ríos, O., Suárez-Quiroz, M. L., Gálvez-Ponce, E., Ordaz-Ortiz, J. J., & Winkler, R. (2019). Metabolomic markers for the early selection of Coffea canephora plants with desirable cup quality traits. *Metabolites*, *9*(214), 1–19. https://doi.org/10.3390/metabo9100214

Garbowicz, K., Liu, Z., Alseekh, S., Tieman, D., Taylor, M., Kuhalskaya, A., Ofner, I., Zamir, D., Klee, H. J., Fernie, A. R., & Brotman, Y. (2018). Quantitative trait loci analysis identifies a prominent gene involved in the production of fatty acid-derived flavor volatiles in tomato. *Molecular Plant*, *11*(9), 1147–1165. https://doi.org/10.1016/j.molp.2018.06.003

García-Flores, M., Juárez-Colunga, S., Montero-Vargas, J. M., López-Arciniega, J. A. I., Chagolla, A., Tiessen, A., & Winkler, R. (2012). Evaluating the physiological state of maize (Zea mays L.) plants by direct-injection electrospray mass spectrometry (DIESI-MS). *Molecular BioSystems*, *8*(6), 1658–1660. https://doi.org/10.1039/C2MB25056J

Geng, F., Nie, R., Yang, N., Cai, L., Hu, Y. C., Chen, S., Cheng, X., Wang, Z., & Chen, L. (2022). Integrated transcriptome and metabolome profiling of Camellia reticulata reveal mechanisms of flower color differentiation. *Frontiers in Genetics*, *13*, 1059717. https://www.frontiersin.org/articles/10.3389/fgene.2022.1059717

Gong, L., Chen, W., Gao, Y., Liu, X., Zhang, H., Xu, C., Yu, S., Zhang, Q., & Luo, J. (2013). Genetic analysis of the metabolome exemplified using a rice population. *Proceedings of the National Academy of Sciences of the United States of America*, *110*(50), 20320–20325. https://doi.org/10.1073/pnas.1319681110

González-Rodríguez, T., Pérez-Limón, S., Peniche-Pavía, H., Rellán-Álvarez, R., Sawers, R. J. H., & Winkler, R. (2023). Genetic mapping of maize metabolites using high-throughput mass profiling. *Plant Science*, *326*(January), 111530. https://doi.org/10.1016/j.plantsci.2022.111530

Gou, M., Balint-Kurti, P., Xu, M., & Yang, Q. (2023). Quantitative disease resistance: Multifaceted players in plant defense. *Journal of Integrative Plant Biology*, *65*(2), 594–610. https://doi.org/10.1111/jipb.13419

Gunnaiah, R., Kushalappa, A. C., Duggavathi, R., Fox, S., & Somers, D. J. (2012). Integrated metabolo-proteomic approach to decipher the mechanisms by which wheat QTL (Fhb1) contributes to resistance against Fusarium graminearum. *PLoS One*, *7*(7), e40695. https://doi.org/10.1371/journal.pone.0040695

Hall, R. D., Brouwer, I. D., & Fitzgerald, M. A. (2008). Plant metabolomics and its potential application for human nutrition. *Physiologia Plantarum*, *132*(2), 162–175. https://doi.org/10.1111/j.1399-3054.2007.00989.x

Hayat, S., Hayat, Q., Alyemeni, M. N., Wani, A. S., Pichtel, J., & Ahmad, A. (2012). Role of proline under changing environments: A review. *Plant Signaling Behavior*, *7*(11), 1456–1466. https://doi.org/10.4161/psb.21949

He, Y., Fu, Y., Hu, D., Wei, D., & Qian, W. (2018). QTL mapping of seed glucosinolate content responsible for environment in Brassica napus. *Frontiers in Plant Science*, *9*(June), 891. https://doi.org/10.3389/fpls.2018.00891

Heuberger, A. L., Broeckling, C. D., Kirkpatrick, K. R., & Prenni, J. E. (2014). Application of nontargeted metabolite profiling to discover novel markers of quality traits in an advanced population of malting barley. *Plant Biotechnology Journal*, *12*(2), 147–160. https://doi.org/10.1111/pbi.12122

Heyneke, E., Watanabe, M., Erban, A., Duan, G., Buchner, P., Walther, D., Kopka, J., Hawkesford, M. J., & Hoefgen, R. (2017). Characterization of the wheat leaf metabolome during grain filling and under varied N-supply. *Frontiers in Plant Science*, *8*, 2048. https://www.frontiersin.org/articles/10.3389/fpls.2017.02048

Hill, C. B., Taylor, J. D., Edwards, J., Mather, D., Bacic, A., Langridge, P., & Roessner, U. (2013). Whole-genome mapping of agronomic and metabolic traits to identify novel quantitative trait loci in bread wheat grown in a water-limited environment. *Plant Physiology*, *162*(3), 1266–1281. https://doi.org/10.1104/pp.113.217851

Hill, C. B., Taylor, J. D., Edwards, J., Mather, D., Langridge, P., Bacic, A., & Roessner, U. (2015). Detection of QTL for metabolic and agronomic traits in wheat with adjustments for variation at genetic loci that affect plant phenology. *Plant Science*, *233*(April), 143–154. https://doi.org/10.1016/j.plantsci.2015.01.008

Jacobs, D. M., van den Berg, M. A., & Hall, R. D. (2021). Towards superior plant-based foods using metabolomics. *Current Opinion in Biotechnology*, *70*(August), 23–28. https://doi.org/10.1016/j.copbio.2020.08.010

Johannsen, W. (1911). The genotype conception of heredity. *The American Naturalist*, *45*(531), 129–159. https://doi.org/10.1086/279202

Johnson, C. H., Ivanisevic, J., & Siuzdak, G. (2016). Metabolomics: Beyond biomarkers and towards mechanisms. *Nature Reviews Molecular Cell Biology*, *17*(7), 451–459. https://doi.org/10.1038/nrm.2016.25

Kassie, F. C., Nguepjop, J. R., Ngalle, H. B., Assaha, D. V. M., Gessese, M. K., Abtew, W. G., Tossim, H.-A., Sambou, A., Seye, M., Rami, J.-F., Fonceka, D., & Bell, J. B. (2023). An overview of mapping quantitative trait loci in peanut (Arachis hypogaea L.). *Genes*, *14*(6), 1176. https://doi.org/10.3390/genes14061176

Keurentjes, J. J. B., Fu, J., Ric de Vos, C. H., Lommen, A., Hall, R. D., Bino, R. J., van der Plas, L. H. W., Jansen, R. C., Vreugdenhil, D., & Koornneef, M. (2006). The genetics of plant metabolism. *Nature Genetics*, *38*(7), 842–849. https://doi.org/10.1038/ng1815

Khan, F., Fuentes, D., Threthowan, R., Mohammad, F., & Ahmad, M. (2019). Comparative metabolite profiling of two wheat genotypes as affected by nitrogen stress at seedling stage. *Journal of Animal and Plant Sciences*, *29*(February), 260–268.

Khan, S. A., Chibon, P.-Y., De Vos, R. C. H., Schipper, B. A., Walraven, E., Beekwilder, J., Van Dijk, T., Finkers, R., Visser, R. G. F., van de Weg, E. W., Bovy, A., Cestaro, A., Velasco, R., Jacobsen, E., & Schouten, H. J. (2012). Genetic analysis of metabolites in apple fruits indicates an mQTL hotspot for phenolic compounds on linkage group 16. *Journal of Experimental Botany*, *63*(8), 2895–2908. https://doi.org/10.1093/jxb/err464

Kim, E.-G., Yun, S., Park, J.-R., & Kim, K.-M. (2021). Identification of F3h, major secondary metabolite-related gene that confers resistance against whitebacked planthopper through QTL mapping in rice. *Plants*, *10*(1), 81. https://doi.org/10.3390/plants10010081

Knoch, D., Riewe, D., Meyer, R. C., Boudichevskaia, A., Schmidt, R., & Altmann, T. (2017). Genetic dissection of metabolite variation in arabidopsis seeds: Evidence for mQTL hotspots and a master regulatory locus of seed metabolism. *Journal of Experimental Botany*, *68*(7), 1655–1667. https://doi.org/10.1093/jxb/erx049

Kooke, R., & Keurentjes, J. J. B. (2012). Multi-dimensional regulation of metabolic networks shaping plant development and performance. *Journal of Experimental Botany*, *63*(9), 3353–3365. https://doi.org/10.1093/jxb/err373

Korn, M., Peterek, S., Mock, H.-P., Heyer, A. G., & Hincha, D. K. (2008). Heterosis in the freezing tolerance, and sugar and flavonoid contents of crosses between Arabidopsis thaliana accessions of widely varying freezing tolerance. *Plant Cell & Environment*, *31*(6), 813–827. https://doi.org/10.1111/j.1365-3040.2008.01800.x

Kumar, R., Sharma, V., Suresh, S., Ramrao, D. P., Veershetty, A., Kumar, S., Priscilla, K., Hangargi, B. S., Narasanna, R., Pandey, M. K., Naik, G. R., Thomas, S., & Kumar, A. (2021). Understanding omics driven plant improvement and de Novo Crop domestication: Some examples. *Frontiers in Genetics*, *12*(April), 637141. https://doi.org/10.3389/fgene.2021.637141

Lado, J., Gurrea, A., Zacarías, L., & Rodrigo, M. J. (2019). Influence of the storage temperature on volatile emission, carotenoid content and chilling injury development in Star Ruby red grapefruit. *Food Chemistry*, *295*(October), 72–81. https://doi.org/10.1016/j.foodchem.2019.05.108

Lee, J., Vázquez-Araújo, L., Adhikari, K., Warmund, M., & Elmore, J. (2011). Volatile compounds in light, medium, and dark black walnut and their influence on the sensory aromatic profile. *Journal of Food Science*, *76*(2), C199–C204. https://doi.org/10.1111/j.1750-3841.2010.02014.x

Li, B., Zhang, Y., Mohammadi, S. A., Huai, D., Zhou, Y., & Kliebenstein, D. J. (2016). An integrative genetic study of rice metabolism, growth and stochastic variation reveals potential C/N partitioning loci. *Scientific Reports*, *6*(1), 30143. https://doi.org/10.1038/srep30143

Li, K., Wen, W., Alseekh, S., Yang, X., Guo, H., Li, W., Wang, L., et al. (2019). Large-scale metabolite quantitative trait locus analysis provides new insights for high-quality maize improvement. *The Plant Journal*, *99*(2), 216–230. https://doi.org/10.1111/tpj.14317

Liang, X., Liu, S., Wang, T., Li, F., Cheng, J., Lai, J., Qin, F., Li, Z., Wang, X., & Jiang, C. (2021). Metabolomics-driven gene mining and genetic improvement of tolerance to salt-induced osmotic stress in maize. *New Phytologist*, *230*(6), 2355–2370. https://doi.org/10.1111/nph.17323

Lim, S., Lee, J. G., & Lee, E. J. (2017). Comparison of fruit quality and GC–MS-based metabolite profiling of kiwifruit 'Jecy Green': Natural and exogenous ethylene-induced ripening. *Food Chemistry*, *234*(November), 81–92. https://doi.org/10.1016/j.foodchem.2017.04.163.

Lisec, J., Meyer, R. C., Steinfath, M., Redestig, H., Becher, M., Witucka-Wall, H., Fiehn, O., Törjék, O., Selbig, J., Altmann, T., & Willmitzer, L. (2008). Identification of metabolic and biomass QTL in Arabidopsis thaliana in a parallel analysis of RIL and IL populations. *The Plant Journal*, *53*(6), 960–972. https://doi.org/10.1111/j.1365-313X.2007.03383.x

Lisec, J., Steinfath, M., Meyer, R. C., Selbig, J., Melchinger, A. E., Willmitzer, L., & Altmann, T. (2009). Identification of heterotic metabolite QTL in Arabidopsis thaliana RIL and IL populations. *The Plant Journal*, *59*(5), 777–788. https://doi.org/10.1111/j.1365-313X.2009.03910.x

Liu, J.-Y., Li, P., Zhang, Y.-W., Zuo, J.-F., Li, G., Han, X., Dunwell, J. M., & Zhang, Y.-M. (2020). Three-dimensional genetic networks among seed oil-related traits, metabolites and genes reveal the genetic foundations of oil synthesis in soybean. *The Plant Journal*, *103*(3), 1103–1124. https://doi.org/10.1111/tpj.14788

Liu, Z., Alseekh, S., Brotman, Y., Zheng, Y., Fei, Z., Tieman, D. M., Giovannoni, J. J., Fernie, A. R., & Klee, H. J. (2016). Identification of a solanum pennellii chromosome 4 fruit flavor and nutritional quality-associated metabolite QTL. *Frontiers in Plant Science*, *7*(November), 1671. https://doi.org/10.3389/fpls.2016.01671

López, M. G., Zanor, M. I., Pratta, G. R., Stegmayer, G., Boggio, S. B., Conte, M., Bermúdez, L., et al. (2015). Metabolic analyses of interspecific tomato recombinant inbred lines for fruit quality improvement. *Metabolomics*, *11*(5), 1416–1431. https://doi.org/10.1007/s11306-015-0798-3

Maharijaya, A., Vosman, B., Pelgrom, K. T. B., Wahyuni, Y., de Vos, R. C. H., & Voorrips, R. E. (2019). Genetic variation in phytochemicals in leaves of pepper (Capsicum) in relation to thrips resistance. *Arthropod-Plant Interactions*, *13*(1), 1–9. https://doi.org/10.1007/s11829-018-9628-7

Manickam, S., Rajagopalan, V. R., Kambale, R., Rajasekaran, R., Kanagarajan, S., & Muthura, R. (2023). Plant metabolomics: Current initiatives and future prospects. *Current Issues in Molecular Biology*, *45*(11), 8894–8906. https://doi.org/10.3390/cimb45110558

Maqbool, M. A., Beshir, A., & Khokhar, E. S. (2020). Doubled haploids in maize: Development, deployment, and challenges. *Crop Science*, *60*(6), 2815–2840. https://doi.org/10.1002/csc2.20261

Matros, A., Liu, G., Hartmann, A., Jiang, Y., Zhao, Y., Wang, H., Ebmeyer, E., Korzun, V., Schachschneider, R., Kazman, E., Schacht, J., Longin, F., Reif, J. C., & Mock, H.-P. (2017). Genome–metabolite associations revealed low heritability, high genetic complexity, and causal relations for leaf metabolites in winter wheat (Triticum aestivum). *Journal of Experimental Botany*, *68*(3), 415–428. https://doi.org/10.1093/jxb/erw441

Matsuda, F., Okazaki, Y., Oikawa, A., Kusano, M., Nakabayashi, R., Kikuchi, J., Yonemaru, J.-I., Ebana, K., Yano, M., & Saito, K. (2012). Dissection of genotype–phenotype associations in rice grains using metabolome quantitative trait loci analysis. *The Plant Journal*, *70*(4), 624–636. https://doi.org/10.1111/j.1365-313X.2012.04903.x

Meyer, R. C., Steinfath, M., Lisec, J., Becher, M., Witucka-Wall, H., Törjék, O., Fiehn, O., Eckardt, A., Willmitzer, L., Selbig, J., & Altmann, T. (2007). The metabolic signature related to high plant growth rate in Arabidopsis thaliana. *Proceedings of the National Academy of Sciences*, *104*(11), 4759–4764. https://doi.org/10.1073/pnas.0609709104

Montero-Vargas, J. M., Ortíz-Islas, S., Ramírez-Sánchez, O., García-Lara, S., & Winkler, R. (2020). Prediction of the antioxidant capacity of maize (Zea mays) hybrids using mass fingerprinting and data mining. *Food Bioscience*, *37*(October), 100647. https://doi.org/10.1016/j.fbio.2020.100647

Mumm, R., Hageman, J. A., Calingacion, M. N., de Vos, R. C. H., Jonker, H. H., Erban, A., Kopka, J., Hansen, T. H., Laursen, K. H., Schjoerring, J. K., Ward, J. L., Beale, M. H., Jongee, S., Rauf, A., Habibi, F., Indrasari, S. D., Sakhan, S., Ramli, A., Romero, M., … Hall, R. D. (2016). Multi-platform metabolomics analyses of a broad collection of fragrant and non-fragrant rice varieties reveals the high complexity of grain quality characteristics. *Metabolomics*, *12*(2), 38. https://doi.org/10.1007/s11306-015-0925-1

Nollet, L. M. L., & Winkler, R. (2022). *Mass spectrometry in food analysis*. Boca Raton: CRC Press. https://doi.org/ 10.1201/9781003091226

Pang, Z., Chong, J., Zhou, G., de Lima Morais, D. A., Chang, L., Barrette, M., Gauthier, C., Jacques, P.-É., Li, S., & Xia, J. (2021). MetaboAnalyst 5.0: Narrowing the gap between raw spectra and functional insights. *Nucleic Acids Research*, *49*(W1), W388–W396. https://doi.org/10.1093/nar/gkab382

Perez-Fons, L., Wells, T., Corol, D. I., Ward, J. L., Gerrish, C., Beale, M. H., Seymour, G. B., Bramley, P. M., & Fraser, P. D. (2014). A genome-wide metabolomic resource for tomato fruit from Solanum pennellii. *Scientific Reports*, *4*(1), 3859. https://doi.org/10.1038/srep03859

Piasecka, A., Sawikowska, A., Kuczyńska, A., Ogrodowicz, P., Mikołajczak, K., Krystkowiak, K., Gudyś, K., Guzy-Wróbelska, J., Krajewski, P., & Kachlicki, P. (2017). Drought-related secondary metabolites of barley (*Hordeum vulgare* L.) leaves and their metabolomic quantitative trait loci. *The Plant Journal*, *89*(5), 898–913. https://doi.org/10.1111/tpj.13430

Pissard, A., Pierna, J. A. F., Baeten, V., Sinnaeve, G., Lognay, G., Mouteau, A., Dupont, P., Rondia, A., & Lateur, M. (2013). Non-destructive measurement of vitamin c, total polyphenol and sugar content in apples using near-infrared spectroscopy. *Journal of the Science of Food and Agriculture*, *93*(2), 238–244. https://doi.org/10.1002/jsfa.5779

Plumb, R. S., Gethings, L. A., Rainville, P. D., Isaac, G., Trengove, R., King, A. M., & Wilson, I. D. (2023). Advances in high throughput LC/MS based metabolomics: A review. *TrAC Trends in Analytical Chemistry*, *160*(March): 116954. https://doi.org/10.1016/j.trac.2023.116954

Pluskal, T., Castillo, S., Villar-Briones, A., & Oresic, M. (2010). MZmine 2: Modular framework for processing, visualizing, and analyzing mass spectrometry-based molecular profile data. *BMC Bioinformatics*, *11*(July), 395. https://doi.org/10.1186/1471-2105-11-395

Pott, D. M., Vallarino, J. G., Cruz-Rus, E., Willmitzer, L., Sánchez-Sevilla, J. F., Amaya, I., & Osorio, S. (2020). Genetic analysis of phenylpropanoids and antioxidant capacity in strawberry fruit reveals mQTL hotspots and candidate genes. *Scientific Reports*, *10*(1), 20197. https://doi.org/10.1038/s41598-020-76946-x

Rajagopalan, V. R., Manickam, S., & Muthura, R. (2022). A comparative metabolomic analysis reveals the nutritional and therapeutic potential of grains of the traditional rice variety Mappillai Samba. *Plants*, *11*(4), 543. https://doi.org/10.3390/plants11040543

Rama, P. G., & Vinutha, T. (2019). Metabolomic profiling of plants to understand reasons for plant stress resilience to abiotic stress. In S. H. Wani (Ed.), *Recent approaches in omics for plant resilience to climate change* (pp. 57–74). Springer International Publishing. https://doi.org/10.1007/978-3-030-21687-0_3

Razzaq, A., Sadia, B., Raza, A., Hameed, M. K., & Saleem, F. (2019). Metabolomics: A way forward for crop improvement. *Metabolites*, *9*(12), 303. https://doi.org/10.3390/metabo9120303

Razzaq, A., Wishart, D. S., Wani, S. H., Hameed, M. K., Mubin, M., & Saleem, F. (2022). Advances in metabolomics-driven diagnostic breeding and crop improvement. *Metabolites*, *12*(6), 511. https://doi.org/10.3390/metabo12060511

Roessner, U., Wagner, C., Kopka, J., Trethewey, R. N., & Willmitzer, L. (2000). Simultaneous analysis of metabolites in potato tuber by gas chromatography–Mass spectrometry. *The Plant Journal*, *23*(1), 131–142. https://doi.org/10.1046/j.1365-313x.2000.00774.x

Romero, H., Pott, D. M., Vallarino, J. G., & Osorio, S. (2021). Metabolomics-based evaluation of crop quality changes as a consequence of climate change. *Metabolites*, *11*(7), 461. https://doi.org/10.3390/metabo11070461

Röst, H. L., Sachsenberg, T., Aiche, S., Bielow, C., Weisser, H., Aicheler, F., Andreotti, S., Ehrlich, H.-C., Gutenbrunner, P., Kenar, E., Liang, X., Nahnsen, S., Nilse, L., Pfeuffer, J., Rosenberger, G., Rurik, M., Schmitt, U., Veit, J., Walzer, M., Wojnar, D., … Kohlbacher, O. (2016). OpenMS: A flexible open-source software platform for mass spectrometry data analysis. *Nature Methods*, *13*(9), 741–748. https://doi.org/10.1038/nmeth.3959

Rowe, H. C., Hansen, B. G., Halkier, B. A., & Kliebenstein, D. J. (2008). Biochemical networks and epistasis shape the Arabidopsis thaliana metabolome. *The Plant Cell*, *20*(5), 1199–1216. https://doi.org/10.1105/tpc.108.058131

Rudell, D. R., Mattheis, J. P., & Curry, E. A. (2008). Prestorage ultraviolet-white light irradiation alters apple peel metabolome. *Journal of Agricultural and Food Chemistry*, *56*(3), 1138–1147. https://doi.org/10.1021/jf072540m

Ruiz-García, L., Hellín, P., Flores, P., & Fenoll, J. (2014). Prediction of Muscat aroma in table grape by analysis of rose oxide. *Food Chemistry*, *154*(July), 151–157. https://doi.org/10.1016/j.foodchem.2014.01.005

Schauer, N., Semel, Y., Balbo, I., Steinfath, M., Repsilber, D., Selbig, J., Pleban, T., Zamir, D., & Fernie, A. R. (2008). Mode of inheritance of primary metabolic traits in tomato. *The Plant Cell*, *20*(3), 509–523. https://doi.org/10.1105/tpc.107.056523

Schauer, N., Semel, Y., Roessner, U., Gur, A., Balbo, I., Carrari, F., Pleban, T., Perez-Melis, A., Bruedigam, C., Kopka, J., Willmitzer, L., Zamir, D., & Fernie, A. R. (2006). Comprehensive metabolic profiling and phenotyping of interspecific introgression lines for tomato improvement. *Nature Biotechnology*, *24*(4), 447–454. https://doi.org/10.1038/nbt1192

Schmid, R., Heuckeroth, S., Korf, A., Smirnov, A., Myers, O., Dyrlund, T. S., Bushuiev, R., Murray, K. J., Hoffmann, N., Lu, M., Sarvepalli, A., Zhang, Z, Fleischauer, M., Dührkop, K., Wesner, M., Hoogstra, S. J., Rudt, E., Mokshyna, O., Brungs, C., … Pluskal, T. (2023). Integrative analysis of multimodal mass spectrometry data in MZmine 3. *Nature Biotechnology*, 41(4), 447–449. https://doi.org/10.1038/s41587-023-01690-2

Setyaningsih, W., Majchrzak, T., Dymerski, T., Namieśnik, J., & Palma, M. (2019). Key-marker volatile compounds in aromatic rice (*Oryza sativa*) grains: An HS-SPME extraction method combined with GC×GC-TOFMS. *Molecules*, *24*(22), 4180. https://doi.org/10.3390/molecules24224180

Shen, S., Zhan, C., Yang, C., Fernie, A. R., & Luo, J. (2023). Metabolomics-centered mining of plant metabolic diversity and function: Past decade and future perspectives. *Molecular Plant*, *16*(1), 43–63. https://doi.org/10.1016/j.molp.2022.09.007

Shi, T., Zhu, A., Jia, J., Hu, X., Chen, J., Liu, W., Ren, X., Sun, D., Fernie, A. R., Cui, F., & Chen, W. (2020). Metabolomics analysis and metabolite-agronomic trait associations using kernels of wheat (Triticum aestivum) recombinant inbred lines. *The Plant Journal*, *103*(1), 279–292. https://doi.org/10.1111/tpj.14727

Simons, T. J., McNeil, C. J., Pham, V. D., Suh, J. H., Wang, Y., Slupsky, C. M., & Guinard, J.-X. (2019). Evaluation of California-grown blood and cara cara oranges through consumer testing, descriptive analysis, and targeted chemical profiling. *Journal of Food Science*, *84*(11), 3246–3263. https://doi.org/10.1111/1750-3841.14820

Smith, C. A., Want, E. J., O'Maille, G., Abagyan, R., & Siuzdak, G. (2006). XCMS: Processing mass spectrometry data for metabolite profiling using nonlinear peak alignment, matching, and identification. *Analitcal Chemistry*, *78*(3), 779–787. https://doi.org/10.1021/ac051437y

Soltis, N. E., & Kliebenstein, D. J. (2015). Natural variation of plant metabolism: Genetic mechanisms, interpretive caveats, and evolutionary and mechanistic insights. *Plant Physiology*, *169*(3), 1456–1468. https://doi.org/10.1104/pp.15.01108

Steinfath, M., Strehmel, N., Peters, R., Schauer, N., Groth, D., Hummel, J., Steup, M., Selbig, J., Kopka, J., Geigenberger, P., & Van Dongen, J. T. (2010). Discovering plant metabolic biomarkers for phenotype prediction using an untargeted approach. *Plant Biotechnology Journal*, *8*(8), 900–911. https://doi.org/10.1111/j.1467-7652.2010.00516.x

Sun, J., Janisiewicz, W. J., Nichols, B., Jurick II, W. M., & Chen, P. (2017). Composition of phenolic compounds in wild apple with multiple resistance mechanisms against postharvest blue mold decay. *Postharvest Biology and Technology*, *127*(May), 68–75. https://doi.org/10.1016/j.postharvbio.2017.01.006

Swain-Lenz, D., Nikolskiy, I., Cheng, J., Sudarsanam, P., Nayler, D., Staller, M. V., & Cohen, B. A. (2017). Causal genetic variation underlying metabolome differences. *Genetics*, *206*(4), 2199–2206. https://doi.org/10.1534/genetics.117.203752

Takagi, H., Abe, A., Yoshida, K., Kosugi, S., Natsume, S., Mitsuoka, C., Uemura, A., Utsushi, H., Tamiru, M., Takuno, S., Innan, H., Cano, L. M., Kamoun, S., & Terauchi, R. (2013). QTL-Seq: Rapid mapping of quantitative trait loci in rice by whole genome resequencing of DNA from two bulked populations. *The Plant Journal*, *74*(1), 174–183. https://doi.org/10.1111/tpj.12105

Tautenhahn, R., Patti, G. J., Rinehart, D., & Siuzdak, G. (2012). XCMS online: A web-based platform to process untargeted metabolomic data. *Analytical Chemistry*, *84*(11), 5035–5039. https://doi.org/10.1021/ac300698c

Teh, S. L., Rostandy, B., Awale, M., Luby, J. J., Fennell, A., & Hegeman, A. D. (2019). Genetic analysis of stilbenoid profiles in grapevine stems reveals a major mQTL hotspot on chromosome 18 associated with disease-resistance motifs. *Horticulture Research*, *6*(1), 1–11. https://doi.org/10.1038/s41438-019-0203-x

Templer, S. E., Ammon, A., Pscheidt, D., Ciobotea, O., Schuy, C., McCollum, C., Sonnewald, U., Hanemann, A., Förster, J., Ordon, F., von Korff, M., & Voll, L. M. (2017). Metabolite profiling of barley flag leaves under drought and combined heat and drought stress reveals metabolic QTLs for metabolites associated with antioxidant defense. *Journal of Experimental Botany*, *68*(7), 1697–1713. https://doi.org/10.1093/jxb/erx038

Toubiana, D., Semel, Y., Tohge, T., Beleggia, R., Cattivelli, L., Rosental, L., Nikoloski, Z., Zamir, D., Fernie, A. R., & Fait, A. (2012). Metabolic profiling of a mapping population exposes new insights in the regulation of seed metabolism and seed, fruit, and plant relations. *PLoS Genetics*, *8*(3), e1002612. https://doi.org/10.1371/journal.pgen.1002612

Treuren, R. V., van Eekelen, H. D. L. M., Wehrens, R., & de Vos, R. C. H. (2018). Metabolite variation in the lettuce gene pool: Towards healthier crop varieties and food. *Metabolomics*, *14*(11), 146. https://doi.org/10.1007/s11306-018-1443-8

Vallarino, J. G., Pott, D. M., Cruz-Rus, E., Miranda, L., Medina-Minguez, J. J., Valpuesta, V., Fernie, A. R., Sánchez-Sevilla, J. F., Osorio, S., & Amaya, I. (2019). Identification of quantitative trait loci and candidate genes for primary metabolite content in strawberry fruit. *Horticulture Research*, *6*(1), 4. https://doi.org/10.1038/s41438-018-0077-3

Wahyuni, Y., Stahl-Hermes, V., Ballester, A.-R., De Vos, R. C. H., Voorrips, R. E., Maharijaya, A., Molthoff, J., Zamora, M. V., Sudarmonowati, E., Arisi, A. C. M., Bino, R. J., & Bovy, A. G. (2014). Genetic mapping of semi-polar metabolites in pepper fruits (Capsicum Sp.): towards unravelling the molecular regulation of flavonoid quantitative trait loci. *Molecular Breeding: New Strategies in Plant Improvement*, *33*(3), 503–518. https://doi.org/10.1007/s11032-013-9967-0

Wen, W., Li, K., Alseekh, S., Omranian, N., Zhao, L., Zhou, Y., Xiao, Y., et al. (2015). Genetic determinants of the network of primary metabolism and their relationships to plant performance in a maize recombinant inbred line population. *The Plant Cell*, *27*(7), 1839–1856. https://doi.org/10.1105/tpc.15.00208

Winkler, R., (Ed.). 2020. *Processing metabolomics and proteomics data with open software: A practical guide* (1st ed.). New Developments in Mass Spectrometry 8. Royal Society of Chemistry.

Xia, J., Psychogios, N., Young, N., & Wishart, D. S. (2009). MetaboAnalyst: A web server for metabolomic data analysis and interpretation. *Nucleic Acids Research*, *37*(Suppl 2), W652–W660. https://doi.org/10.1093/nar/gkp356

Yuan, Y., Zhao, Y., Yang, J., Jiang, Y., Lu, F., Jia, Y., & Yang, B. (2017). Metabolomic analyses of banana during postharvest senescence by 1h-high resolution-NMR. *Food Chemistry*, *218*(March), 406–412. https://doi.org/10.1016/j.foodchem.2016.09.080

Zamboni, A., Di Carli, M., Guzzo, F., Stocchero, M., Zenoni, S., Ferrarini, A., Tononi, P., Toffali, K., Desiderio, A., Lilley, K. S., Enrico Pè, M., Benvenuto, E., Delledonne, M., & Pezzotti, M. (2010). Identification of putative stage-specific grapevine berry biomarkers and omics data integration into networks. *Plant Physiology*, *154*(3), 1439–1459. https://doi.org/10.1104/pp.110.160275

Zargar, S. M., Manzoor, M., Bhat, B., Wani, A. B., Sofi, P. A., Sudan, J., Ebinezer, L. B., Dall'Acqua, S., Peron, G., & Masi, A. (2023). Metabolic-GWAS provides insights into genetic architecture of seed metabolome in buckwheat. *BMC Plant Biology*, *23*(1), 373. https://doi.org/10.1186/s12870-023-04381-x

Zhou, S., Kremling, K. A., Bandillo, N., Richter, A., Zhang, Y. K., Ahern, K. R., Artyukhin, A. B., Hui, J. X., Younkin, G. C., Schroeder, F. C., Buckler, E. S., & Jander, G. (2019). Metabolome-scale genome-wide association studies reveal chemical diversity and genetic control of maize specialized metabolites. *The Plant Cell*, *31*(5), 937–955. https://doi.org/10.1105/tpc.18.00772

Zhu, F., Wen, W., Cheng, Y., Alseekh, S., & Fernie, A. R. (2023). Integrating multiomics data accelerates elucidation of plant primary and secondary metabolic pathways. *aBIOTECH*, *4*(1), 47–56. https://doi.org/10.1007/s42994-022-00091-4

Zhu, G., Wang, S., Huang, Z., Zhang, S., Liao, Q., Zhang, C., Lin, T., Qin, M., Peng, M., Yang, C., Cao, X., Han, X., Wang, X., van der Knaap, E., Zhang, Z., Cui, X., Klee, H., Fernie, A. R., Luo, J., & Huang, S. (2018). Rewiring of the fruit metabolome in tomato breeding. *Cell*, *172*(1), 249–261.e12. https://doi.org/10.1016/j.cell.2017.12.019

4 Plants' Abiotic Stress Responses

Molecular Mechanisms of Stress Tolerance and Breeding Strategies

Jemaa Essemine, Zhan Xu, Jen-Tsung Chen, and Mingnan Qu

4.1 INTRODUCTION

Warming events and climate change are expected to intensify and exacerbate severe environmental conditions, including extreme weather waves and enhanced average temperatures (Lämke and Bäurle, 2017). In their natural habitat, as sessile organisms, plants are subjected to a plethora of abiotic stress including salt, heat, low temperature, drought, and heavy metals. To withstand the various prevailing stressful conditions and overcome their deleterious effects, plants developed various responses to mitigate or avoid their harmful effects. In this regard, several abiotic stress-regulatory mechanisms could be established and synergistically operate, in an orchestrated manner, to foster plant stress tolerance and promote breeding strategies. Abiotic stress may disrupt and dramatically affect several processes within the plant such as physiology, biochemistry, molecular, and even genetic and epigenetics, leading, thereby, to a regression in growth, development, crop yield, and plant fitness in general. Abiotic stress may also affect the plant's innate immunity, making it more vulnerable to disease and other pathogen invasions such as pests and other pathogen attacks (Dietz and Vogelsang, 2023). Interestingly, plants respond differently to various abiotic constraints, especially at the molecular, genetic, and epigenetic levels. This suggests a versatility in the breeding strategies between plant species, especially large crops such as wheat, barely, and rice, in response to multiple challenging conditions (biotic and abiotic stresses). Globally, the exclusion of starvation and malnutrition from society represents a major challenge for scientists and agricultural stakeholders worldwide. To feed an ever-increasing global population (worldwide) under challenging environmental conditions such as global warming and climate changes, scientists thought to incorporate various interconnected plant research realms, and accordingly, researchers have to be well-equipped or skilled and endowed with strong bioinformatics insights and tools to be able handling big datasets (Essemine et al., 2024). In the same trend, based on the notification (or statement) released by the Economic and Social Affairs Department of the United Nations, UN (https://www.un.org/en/desa, March 2023), the worldwide population is expected to attain and may even exceed 10 billion by 2050. According to the global Agricultural Productivity Index 2022 (http://www.globalagriculturalproductivity.org/, March 2023), the worldwide agricultural output and yield were found to drop below the levels required for the sustainable development of agricultural productivity. Furthermore, the Food and Agriculture Organization (FAO) of the UN displayed in a recent statistical yearbook that hunger and malnutrition are still threatening the five world's continents, but to different extents, and even increase to reach globally 770 million people undernourished in 2021. Hence, in 20 years (2000–2020) the arable land decreased by around 134 million ha (FAO, 2017) due to water scarcity and flooding and/or waterlogging. When exposed to severe environmental hurdles, plants evolve various strategies to overcome the harmful effects of stress, including escape, resistance, and avoidance. In line with that, rice constitutes the main staple food and most nutritionally-rich crop for more

 DOI: 10.1201/9781003457237-4

than half of the worldwide population (Sasaki and Burr, 2000) which deserves to be investigated to safeguard food security through various breeding strategies to create new varieties more tolerant to abiotic stress. Taking into consideration that abiotic stress threatens and causes substantial and deep damage to rice yield improvement and sustainable production (Usman et al., 2023).

Using new biotechnology techniques (NBTs) such as third-generation sequencing (long-reads sequencing), next-generation sequencing (short-reads sequencing), and CRISPR/Cas9-based gene-editing system, scientists unearthed regulatory molecular markers (genes, enzymes, transcription factors [TFs], Cis-regulatory elements, proteins, etc.) that could be responsible for certain crucial agronomic traits through the effective use of genome-wide association studies (GWAS) and integrative omics and genomics strategies, which was confirmed significant in rice breeding programs (Huang et al., 2010; Chen et al., 2016). This gene-editing method is characterized by its simplicity, versatility, and robustness in identifying candidate causal genes associated with master traits in crop breeding strategies to enhance crop productivity and boost plant resilience to abiotic challenges. In the last years, GWAS has been considered, by several scientists, as a potent and promising approach in crop breeding (Zhou et al., 2016b; Li et al., 2017; Singh et al., 2018). Nevertheless, crop enhancement in quantity and quality and breeding new varieties endowed with certain specific characteristics such as stress resilience and high yield require the production (or creation) of plants with transgenerational heritable traits (plant cell memory). The stochastic nature of certain hybridization methods such as the off-target effects (inaccuracy) sometimes of CRISPR gene-editing technology in plants represents also a major challenge, but such mutation generated through off-target effects can be readily screened out (eliminated or removed) through successive genetic crosses (Tang et al., 2019).

In brief, addressing food security, under challenging environmental conditions and the ever-increasing global population, is an endless and challenging task. Thus, tedious work and random breeding programs cannot abide by the unexpected demands for high crop production. Alternative sophisticated gene-editing technologies through targeted mutations, such as CRISPR, are so imperatively needed. In this chapter, we compile (gather or review) the prominent most recent achievements in crops, particularly rice, genetics, and breeding strategies, in the face of challenging environmental conditions, such as climate change. We would like, as well, to highlight the paramount importance of regulatory epigenetics, regulon (union of TFs and *Cis*-regulatory elements), and omics in fostering rice breeding strategies. A key concern in using CRISPR in agricultural genotype screening is to select (or define) plants that are transgene-free and particularly that have been edited with steady heritable (or stably transmissible) traits across generations (Gao and Zhao, 2014). The aim of crop improvement, through breeding programs, is to generate (or create) new plant varieties via targeted-promising agronomic traits that are responsible for improved productivity and enhanced plant adaptability or resilience to abiotic stress. These traits can be defined through GWAS and the interplay of several interconnected plant science realms such as epigenetics, Synbio, and molecular biology. We, eventually, stressed that the regulatory mechanistic responses of crop plants to abiotic stresses can be unraveled and precisely dissected genetically and epigenetically.

4.2 HIGHER PLANT INTERACTION WITH ABIOTIC STRESS IN THEIR HABITAT

Abiotic stress responses are crucial for plants, as immobile organisms, since they are unable to survive unless they can withstand severe surrounding challenges (biotic and abiotic) by developing suitable (relevant) and intricate mechanisms. These mechanisms span various aspects including physiological, molecular, genetics, omics, and even above-genetics (epigenetics). Furthermore, Synbio represents an advanced plant science topic integrating biology and engineering to, simultaneously, foster plant breeding and resilience to stress. Synbio may also curatively serve humanity through discovering, from plants such as tobacco and other medicinal plants, new drugs for chronic diseases such as cancer and Alzheimer's disease. It is expected that the abiotic stress will be exacerbated (intensified) sooner (near future) due to the earth's continuous warming attributable to climate change and extensive release of CO_2 in the air, as reported by the Intergovernmental Panel on Climate Change (IPCC; http:// www.ipcc.ch). Thus, in the European heatwave of 2003, crop production was diminished by around 30% (Ciais et al., 2005). Thus, comprehending plant–environment interactions constitutes nowadays the most important, urgent, and challenging plant science topic. Thus, disentangling the mechanism of key stress-responsive genes mediating the plant adaptation to the fluctuating environment is of great interest to enhance plant tolerance and adaptability. This would help in deciphering the potential scenarios of adaptation, expanding the genetic variation (segregation) of abiotic stress-responsive genes to promote and extrapolate the extent of stress resistance or vulnerability conferred in natural ecosystems (Singhal et al., 2016).

4.3 COMPARISON OF BREEDING METHODS USED IN MODERN AGRICULTURE

Three standard and main breeding strategies have been extensively used in modern agriculture (Moda). These are three breeding approaches, which are crossing, mutational, and transgenic. The three methods are essentially used for crop improvement in Moda (Chen et al., 2019). Thus, the first approach (crossbreeding) necessitates several years to incorporate the desired alleles and to enhance segregation by genetic recombination (Scheben et al., 2017). The second one (mutation breeding) shows broader (larger) genetic change (segregation or variation) through integrating (creating) stochastic (or random) mutations using specific mutagens (chemically) or physical irradiation (Pacher and Puchta, 2017). However, these breeding approaches are highly restricted due to their stochastic aspect (or nature), and thus it remains challenging and complicated to generate and screen large numbers of mutants (Chen et al., 2019). The restrictions in these techniques are the fact that they are time-consuming (10–12 years), laborious, un-targeted breeding programs (random mutation) and cannot satisfy the increasing requirements for improved crop productivity or yield, even if genome-assisted breeding (MAB) strategy was implemented using marker-assisted selection (MAS) technique to promote selection efficiency (Scheben et al., 2017). However, transgenic breeding creates the desired traits via the transfer of foreigners' genes or genetic elements (TFs or enzymes) into elite background lines, to foster certain traits such as plant resilience to abiotic stress and/or enhance yield quality and quantity, which may bypass the complexities of reproductive isolation. However, a recently emerged genetic biotechnological tool to breed crops is the CRISPR/Cas approach known for its time-saving, simplicity (not laborious), versatility, power, and use of targeted-breeding programs. The CRISPR/Cas method involves the modification of endogenous genes through the nucleotide base insertion–deletion (Indel) technique (Chen et al., 2019; Qu et al., 2020a, b; Essemine et al., 2024).

4.4 MOLECULAR MECHANISMS OF CROP RESPONSES TO ABIOTIC STRESS

As a fixed organism, plants evolved various strategies to withstand abiotic stress effects. At the physiological level, plants develop a first line of defense, which includes escape, avoidance, and adaptation. However, at the molecular level, plants may react in different ways including synthesis and sequestration of the harmful reactive oxygen species (ROS) such as hydrogen peroxide (H_2O_2) and hydroxyl radicals (HO), which possess the highest electron reduction potential, and they (•HO) are mostly responsible for the cytotoxic effect in the aerobic organism. Besides, the accumulation of organic compounds (or osmolytes) such as proline, glycine betaine (Mata et al., 2016; Raza, 2020), and abscisic acid (ABA) helps crops species to overcome deleterious effects of abiotic stress through functioning as safety valves to scavenge (or sequestre) reactive elements. At the genomic level, plant reactions to environmental abiotic constraints include the upregulation of stress-responsive genes that are identified to be immediately and/or directly involved in the plant protection against the environmental hurdles either biotic or abiotic (Flowers and Yeo, 1995;

Chinnusamy et al., 2004; Krasensky and Jonak, 2012). For instance, under drought, researchers observed an increase in the expression of ABA biosynthetic genes and upregulation of the ABA-responsive genes (Onaga and Wydra, 2016). However, at the metabolomics level, crops respond to stress by upregulation of certain metabolites-related genes and downregulation or even silencing of others; however, at the transcriptomic level, plants react to stress by monitoring the transcripts' abundance of some genes and regulons including TFs, *Cis*-regulatory elements, putative expressed genes, and transposons. This interconnected regulatory network (genes, TFs, proteins, etc.) operates in tandem to enhance plant resistance to stress and maintain crops with high productivity to secure urgent and huge food demands for a large population under challenging environmental circumstances.

4.5 GWAS FUNCTIONS AS A GENE EDITOR VIA THE CRISPR/CAS SYSTEM IN PLANT BREEDING STRATEGIES

Gene engineering based on the CRISPR/Cas gene-editing system constitutes a potent method to efficiently disperse (diffuse and propagate) genetic variants (or elements) across populations via cross-reproduction (duplication or amplification) and sexual breeding reproduction. In Moda (modern agriculture), a CRISPR/Cas-based gene-editing system can be used to destroy (or alter) and eventually remove invasive pathogenic elements (organisms or factors), such as pests and weeds. It (CRISPR/Cas) could be harnessed (or utilized) as well to degrade pathogens and incorporate new traits and/or phenotype into the pre-existing population genome. Thus, the CRISPR/Cas approach ensures genetic recombination and genetic variation by creating new varieties endowed with high productivity and/or high resilience to abiotic challenges. For instance, pigweed (*Amaranthus*) could be designed by gene drive technique to turn vulnerable to an extensively used herbicide glyphosate. Because the CRISPR/Cas-induced gene drive system has the power (potential) to affect the whole agricultural ecosystem genome, any thought regarding its eventual application should pass by an extensive and accurate evaluation including, for ethical and fairness concerns, the safety and values of both natural and social sciences. The facility, diversity, and potential of CRISPR/Cas technology pinpoint (or define) genome editing as a robust tool for precise and fastidious (or squeamish) crop improvement via gene deletion, insertion, replacement, point mutations, fine-tuning of functional regulation, and other manipulations for any gene locus or biomarkers, such TFs or *Cis*-acting elements, in crops. However, substantial technical issues remain yet to be conquered for wider implementation of the CRISPR/Cas-editing system in a precise and effective way across crop species. Among these improvement procedures, the main thing would be accomplishing an efficient nuclease reagent distribution and substantially increasing the frequency of homology-directed repair (HDR) for DNA damage repair in plants. Eventually, the CRISPR/Cas system functions concomitantly as an eminent writer, reader, eraser, and editor in creating gene mutations through indels, replacement, base editing, and so on of tools to introduce a desired targeted-trait(s) of potentially great interest in crops resilience to environmental changes and improved productivity. Gene-editing system (CRISPR) can be harnessed as well to eliminate unwanted agronomic traits or transfer genes from one species to another through non-homologous end-joining or double-strand break (DSB) DNA damage repair. Thus, the term 'editor' encompasses various functionalities that can be fulfilled by the gene-editing system to enhance crop quality and quantity. Hence, CRISPR/Cas operates simultaneously as a writer, reader, and erasing to incorporate any preferred trait into a targeted gene.

4.6 CRISPR/CAS9 GENE-EDITING TECHNOLOGIES AS A PROMISING TOOL FOR CROP BREEDING STRATEGIES

Earlier, researchers have used the so-called zinc finger nucleases (ZFNs) and transcription activator-like effector nucleases (TALENs) in breeding strategies. However, the construction complexity and the high cost and skill required for the analysis of raw datasets have limited their use by the scientific community in Moda. Alternatively, GWAS has been extensively used in the agriculture field to promote crop breeding strategies for the sake of improving crop productivity and plant resilience to severe environmental conditions. In the last decade (since 2013), several research groups have developed a CRISPR (clustered regularly interspaced short palindromic repeat) system for use in rice (*Oryza sativa*), wheat (*Triticum aestivum*), *Nicotiana benthamiana*, and *Arabidopsis thaliana* (Li et al., 2013; Nekrasov et al., 2013; Shan et al., 2013). This method has also improved hybrid-breeding strategies and eliminated undesired traits or integrated preferred ones (traits) to crop elite cultivars, which represents nowadays a direct approach and is easily applicable and feasible, rendering crop traits to be accurately and suitably manipulated, even within one generation (short time). Accordingly, CRISPR/Cas has the power (or ability) to enhance food production and thus global alimentary security through continuous cropping and sustainable developmental agriculture. Eventually, gene-editing systems (CRISPR-Ca9, CRISPR-Cpf1, base editing) represent the most evolved breeding approach in Moda and are also recognized by their high-throughput (HTP) yield, reliability, fast run time (shorter time by half compared to other methods), and easy application. This recently emerged molecular genetic method empowered by various sequencing methodologies (next-generation sequencing [NGS] and third-generation sequencing [TGS]), thanks to the sophisticated and fine technologies including PacBio and Oxford Nanopore Technologies (ONT), can improve crop productivity and adaptability to challenging environmental conditions and speed up hybridization and crop breeding strategies.

4.7 CRISPR SYSTEMS FOR PLANT GENOME EDITING

The delivery of CRISPR constructs a product to plant cells, and the generation of editing events (introducing mutations through indels) constitutes the main phases (or steps) in the genome-editing process. Thus, CRISPR-induced editing elements, encompassing DNA, RNA, and RNPs (ribonucleoproteins), can be carried and delivered into the plant cell via protoplast transfection (transient process), *Agrobacterium*-mediated transfer DNA (T-DNA) transformation, or particle bombardment. The first technique is mainly used for transient gene expression; however, the two other types of transformation represent the major constructs delivery (carriers) technique for the generation of an edited system (Chen et al., 2019). During the last decade, scientists have applied three CRSIPR-editing systems on crops, mainly on rice and tomato, due to their small genome size, and it is also genetically easy to deal with small chromosome numbers (12 chromosomes for each) during any genetic mapping or model prediction building. These three kinds are CRISPR/Cas9, CRIPR/Cpf1, and base-editing systems (Chen et al., 2013). Where the first is CRIPR-associated protein 9 (Cas9), the second refers to CRISPR from *Prevoltella* and *Francisella* (Cpf1), and the third one represents a base-editing genome technique discovered by Gaudelli and co-workers (2017). Recognition enzymes based on the two first types (Cas9 and Cpf1) differ in many features. CRISPR/Cas9 is known by its long guide RNA (gRNA, ~80 bp), G-rich protospacer-adjacent motif (PAM, NGG), the splitting (or cleavage) site being positioned proximal and upstream of the PAM (around 5 pb upstream PAM), and cleavage of the DNA in a blunt-ends manner. However, the gRNA for CRISPR-Cpf1 is shorter (~43 bp), with a T-rich PAM sequence, Cpf1 cleavage site being situated distal and downstream of the PAM sequence, and cleavage of the DNA in a staggered way, generating thus a 5-nucleotide (nts) 5′ overhang commencing at 18 nts 3′ end of the PAM sequence (Zetsche et al., 2015). The base-editing methodologies allow the precise and direct replacement (or conversion) of one DNA-targeted nucleotide by another independently of the random nature of DSB known to be formed in the aforementioned two CRISPR (Cas9 and Cpf1). For example, it has been reported that two types of enzyme-editing tools can lead to two kinds of base-editing conversion: notably the CBE-mediated C-to-T base-editing replacement and strategy and the ABE-mediated A-to-G base-editing tool (Gaudelli et al., 2017; Chen et al., 2019).

As reported in previous studies, the CRISPR-editing system has the great potential to foster breeding programs for crops improvement (Li et al., 2017, 2018; Rao and Wang, 2021), and it can also enhance plant tolerance to environmental biotic and abiotic constraints (Shi et al., 2017;

Wang et al., 2017; Rahman et al., 2022; Li et al., 2022; Esmaeili et al., 2022; Angon et al., 2023). Genome-editing technology using CRISPR/Cas RNPs generates a DNA-free and low off-target plant. CRISPR/Cas editing of *Cis*-acting factors, such as promoters and upstream open reading frames, enables the fine-tuning of quantitative trait variation and post-transcriptional upregulation of translation in eukaryotic genes.

4.8 APPLICATION OF CRISPR/CAS FOR THE PRECISION OF PLANT BREEDING

Based on revolutionized and advanced genomic and omics methods, scientists thought to dissect plant genomes and pinpoint eventually candidate biomarkers (genes or *Cis*-acting elements) and/or TFs genes family that can be related to certain physiological and/or agronomic traits using featured smart technologies such as NGS, TGS, and GWAS using CRISPR-Cas9-based genome-editing system. This method efficiently (or reliably) identifies crucial biomarkers (gene(s) or TFs) controlling crop production (quality and/or quantity)-related traits including grain size, grain width, and grain weight. Through mutating and manipulating the candidate genes (knockout, Indel, overexpression, and near-isogenic lines, NILs), followed by specific validating wet lab techniques including qRT-PCR, transcriptomics analysis, and metabolomics profiling, scientists and stockholders might synergistically cooperate to improve crops yield through targeted-breeding programs. Thus, breeders look forward to ameliorate plant plasticity (adaptation or acclimatization) to withstand severe stress conditions, which may lead to more resilient and productive plants (Essemine et al., 2024). CRISPR/Cas approach offers easy, valuable, reliable, and diversified tools to study plant molecular genetics and robust and fruitful plant breeding strategies. CRISPR/Cas-gene-editing systems, through knockout, insertion/deletion (Indel), and mutation, are being extensively used in attempts to improve crop yield, quality, disease resistance, and stress tolerance traits, which may promote (or foster) hybrid breeding and crop domestication.

4.9 PREDOMINANT GENETIC ANALYSIS IN CROP RESPONSES TO ABIOTIC STRESS

In this context, our concluding observations were elaborated based on a bibliometric analysis performed by different research groups using computational sweeping and imputation with the help of various specific and professional software. In this regard, Nascimento and co-workers revealed that, in several previous reports, the term CRISPR/Cas was found to be the highly investigated technique of editing genes tool for the tolerance to water and salt stress in *Arabidopsis* and rice in the past seven years (Nascimento et al., 2023). Similar bibliometric imputation (attribution) results were attained by Hamdan and collaborators in checking the presence of terms and keywords analogous (comparable or identical) to those used by Nascimento and co-workers, although Hamdan et al.'s analysis was implemented with publications related to CRISPR/Cas stored and retrieved (or extracted) data only from SCOPUS repository database in the past ten years (Hamdan et al., 2022). Thus, in crops, CRISPR/Cas can be employed to investigate gene function through mutation generation (loss of function) by direct knockout (Ma and Liang, 2021). Furthermore, the CRISPR/Cas-editing system constitutes a powerful tool to identify stress-responsive genes and thereby may foster crop tolerance to the concerned abiotic stress. Due to its high efficiency, simplicity, short processing time, and low cost, CRISPR/Cas technology has expanded (diffused or widespread) rapidly among breeders and scientists, during the last decade, as a potent method of crop breeding compared to the other traditional (standard) approaches.

4.10 TRANSCRIPTOMIC REGULATION AND PROFILING IN CROP IN RESPONSE TO ABIOTIC STRESS

Owing to their immobile nature, plants can frequently be subjected to multiple biotic and abiotic stresses during their lifecycle (Jha et al., 2020). However, plants developed intricate mechanisms involving the interplay of different interconnected biological systems to cope with and tolerate the challenging effects caused (or exerted) by environmental constraints (biotic and abiotic). In this context, long non-coding RNAs (lncRNAs), a diverse category of RNAs containing more than 200 nucleotides (nts), were identified to play a crucial importance in plant resilience to diverse abiotic hurdles (Jha et al., 2020). Thus, as revealed in a number of studies, lncRNAs mediate various regulatory functions through an orchestrated interplay between different interconnected biological pathways and/or systems, which becomes, nowadays, an emerging research topic to improve plant resistance to abiotic stress and crop productivity (Nejat and Mantri, 2018; Budak et al., 2020; Jha et al., 2020).

For instance, among the crucial target genes detected, using GWAS screening, to be specifically related to amino acid content in maize kernels were an arogenate dehydratase, which is responsible for phenylalanine accumulation, and opaque2 (O2), a basine leucine zipper (bZIP) TF that acts as a prominent regulator for the expression levels of many genes during the kernel developmental stage and identified to be associated with lysine/total amino acid ratio (Deng et al., 2017; Wen et al., 2015, 2018; Pott et al., 2021). In the same trend, a recent report combining transcriptomic, metabolomic, and genomics datasets displayed the improvement and breeding of certain agronomic traits such as crop color, flavor, and enhanced amino acid content (Pott et al., 2021).

A synergistically integrated genomics and transcriptomics analysis has ushered by a novel period (new era) of dissecting crop genomes for the sake of improving their resistance to abiotic challenges that might be caused by climate changes and global warming. On the one hand, transcriptomic approaches have revolutionized our understanding and knowledge regarding gene regulation under abiotic stress. On the other hand, traditional methods such as northern blotting, microarray analyses, and future-generation sequencing have paved the way toward advanced technologies including RNA-seq data analysis through TGS using sophisticated sequencing platforms such as Pacific Bioscience (PacBio) and ONT, which allow scientists to simultaneously analyze large genome size (thousands of genes) and facilitate the detection of both known and novel transcripts (long- and short-reads) reported after retrieval, assembly and ligation as putative expressed genes or cis-regulatory elements with still unknown functions. The recent surge in TGS (known also as long-read sequencing) technologies, although it has few inconveniences such as high assembly error rates and low-throughput sequencing, has substantially influenced the field of transcriptomics, as it offers several advantages over the traditional sequencing methods (such as sanger). Among these advantages, we mention that the long-read products (results) of TGS make easier de novo assembly over even the closest sequencing method of second-generation sequencing known also NGS (Athanasopoulou et al., 2021). These few inconveniences of TGS can be conquered (overcome) because this new technology is still undergoing concurrent and continuous development. TGS platforms, including PacBio and ONT, elaborate longer reads and thus can sequence RNA molecules.

4.11 TRANSCRIPTOMIC DISSECTION FOSTERS ABIOTIC STRESS TOLERANCE AND CROP BREEDING

Transcriptomics signifies the expressive (or significant) analysis of total RNAs transcribed by a cell or tissue, including both coding and non-coding RNA, among them lncRNAs, at a particular functional stage. It focuses on investigating the nature, structural features, functional role, and regulation of gene transcription (Nejat et al., 2018). It offers precise and valuable insights through the quantitative analysis of the modifications in plant gene expressions and transcriptome profiling. Thus, transcriptomics allows delineating (tracking or conceiving) the gene regulatory network

(GRN) and achieving the whole-genome sequencing, thereafter plotting (imagining and predicting) the expression patterns (or models), which will help in revealing new stress-resistant genes in crops (Wang et al., 2020). In a recent report, authors have shown that the integration of omics (metabolomics and transcriptomics) and mGWAS constitutes a key and potent method to comprehend the metabolic and quality modifications of fruits that can happen along the domestication and breeding programs (Zhu et al., 2018). Further, transcriptomics helps uncover the plant response to abiotic stress by analyzing gene expression levels and the transcripts' abundance of particular genes associated with certain stress response compounds such as osmo-regulators, ROS, and heat shock proteins (HSPs). The HSPs are known to be synthesized under heat stress as a molecular mechanism of heat stress resistance and defense adopted by the plant. Hence, transcriptomics is a revolutionized analysis approach of RNAs to understand the gene regulations through transcripts abundance and/or expression levels of such a targeted or defined candidate gene.

4.12 REGULON INVOLVEMENT IN CONTROLLING ABIOTIC STRESS-RESPONSIVE GENES IN CROP

In this subject, it has been shown that a single TF can monitor the expression of several downstream genes concomitantly via signal transduction of genetic information and/or tight physical interaction or association between the TF to *Cis*-acting elements (form a module) embedded (or integrated) in the promoter regions of their corresponding target genes, forming thereby a transcriptional cluster or unit named 'Regulon' (Nakashima and Yamaguchi-Shinozaki, 2006). Notably, several regulons were defined in the gene responses of plants to challenging environmental conditions (Marco et al., 2015). In relation to the transcript abundance analysis, the TFs were pinpointed to possess paramount importance (crucial role) in adjusting crop resistance to abiotic constraints as well. For instance, the expression levels of genes encoding dehydration-responsive element binding TFs were reported to increase (upregulation) under water deficit, suggesting the potential role of these genes in controlling drought-tolerance-associated genes (Zhu et al., 2013). Furthermore, using expressed sequence tag and microarray test in cotton uncovered several genes modulating abiotic stress responses, including MYB-related genes, such as FAR1, bHLH, bZIP, MADS (Zhou et al., 2016a) and RTNLB5, and PRA1 (Tahmasebi et al., 2019).

4.13 OMICS APPROACHES TO DEVELOP ABIOTIC STRESS-RESILIENT CROPS

Although the availability of several recent studies performed on the role of omics in plant abiotic stress tolerance in crops and fruits, we limited this report to corroborate briefly the importance of integrative analysis of transcriptomics and metabolomics in mitigating stress plants. Notably, various fruitful and valuable reports were stressed in the realm of agriculture and recognized to unravel the potential implication of genomic and omics (transcriptomic, metabolomic, proteomic, etc.) in mitigating abiotic stress effects and enhancing crop plants' resilience with improved productivity (Yang et al., 2020; Makhumbila et al., 2022). Thus, an integrated genomics and transcriptomics analysis was reported to play pivotal roles in crop improvement by promoting its resilience and offering a wealth of valuable insights regarding the genetic rearrangement, epigenome makeup, metabolomic profiling of crucial metabolites, and the gene expression patterns of key genes and stress-responsive gene families (Yang et al., 2021). In other studies, metabolomic profiling of key metabolites likely to be involved in abiotic stress resilience was also performed (Raza et al., 2020; Van Esse et al., 2020; Litvinov et al., 2021). Several earlier reports asserted (corroborated) the participation of certain metabolic pathways, enzymes, and/or proteins in refining (refreshing or updating) plant resistance to environmental hurdles such as heat (Alaux et al., 2018; Marko et al., 2019; Andrási et al., 2021) and drought (Delauney and Verma, 1993; Akbudak et al., 2018; Zhao et al., 2020; Schaarschmidt et al., 2020).

4.14 METABOLOMIC INVOLVEMENT IN STRESS TOLERANCE AND CROP BREEDING PROGRAMS

Understanding the function and structure of genes involved (or engaged) in the main metabolic pathways would certainly help in promoting crop resistance to abiotic stress such as heat and drought, improving food quality, and enhancing crop yield (Gong et al., 2013). Thus, the identification of genes mediating and regulating the crucial metabolic pathways makes the breeding method refinement (or amendment) and crop biofortification, which may serve as a master approach to struggle against society malnutrition and starvation (Galili and Amir, 2013; Galili et al., 2016; Deng et al., 2017; Yang et al., 2020). Thus, metabolomics and its integration with one or more 'omics' realm(s) constitute a purposeful approach in elucidating the significantly relevant information of plant biochemistry and molecular mechanisms that can be employed in the breeding of continuous cropping and improving yields. To date, the most successful results are derived (extracted or obtained) from metabolomic datasets on plant stress response (Raza, 2020). Thus, understanding plant metabolomics under certain stress conditions may help pinpoint the crucial stress-responsive genes (or enzymes, proteins, TFs gene family). In this regard, the gained insights would be of great interest to breeders to engineer and redesign metabolic pathways integrating these newly detected features and insights to boost plant resilience toward improving crop breeding and productivity to preserve food security for society and feed animals.

4.15 EPIGENETIC REGULATORY MECHANISMS IN CROP RESPONSES TO ABIOTIC STRESS

Epigenetics (above-genetics) contributes to gene regulation without changing the gene structure. For instance, epigenetic regulatory modules (repressor and activator) may influence the plant resistance to climate change. Thus, chromatin changes, nucleosome placement, and DNA methylation have been shown as eminent epigenetic marks (epimarks) in plant adaptation to challenging conditions. Thus, the epimark regulation of gene expressions depends not only on the type of modification but also on their location (or placement) on the genes. This means the epimark functions depend on their positions on the gene, i.e., if they are situated on the promoter region or gene body (Miryeganeh, 2021). For instance, histone change and DNA methylation were shown to alter the expressions of stress-responsive genes at the transcriptional and post-transcriptional steps by modifying the chromatin status of those genes (Luo and He, 2020). However, TFs and RNA-directed DNA methylation (RdDM) play a key role in the regulation of gene expression in plants under abiotic stress. Certain crucial heat stress-responsive genes are regulated through RdDM pathway-mediated DNA methylation (Popova et al., 2013). Mostly following stress termination, the epigenetic modifications return to their prestressed state (initial static state); however, it has been reported that the epigenetic modifications (or, at least, a part of them) remain even after stress wave cessation and may act, in this case, as stress memory permanent motifs for the plants (Friedrich et al., 2019; Srikant and Drost, 2020; Miryeganeh, 2021) to recognize the same type of stress during the next generation. With time, the plants inherit these epigenetic changes, which may grant the plant an innate immunity, in the future, against exclusively that type of stress. However, the epigenetic marks respond differently to each kind of stress and may lead to repression or activation of certain stress-responsive genes to boost the plant's ability to withstand the prevailing stressful conditions and even set (establish) specific biosensors to detect such types of stress during the next generation of growth.

Consistently, it has been reported, in this regard, that DNA methylation can be involved in the transgenerational inheritance (known as cell memory) of the response to heavy metals in rice (Cong et al., 2019). Further to DNA methylation dynamics, histone modifications were also characterized to be directly implicated in intergenerational and/or intragenerational stress memory in plants (Ashapkin et al., 2020). In addition, recent reports have shown as well that various epigenetic regulatory modules of histone (de)acetylation act as potent modulators of abiotic stress response (Asensi-Fabado et al., 2016; Luo et al., 2017).

4.16 FOSTERING CROP RESILIENCE TO ABIOTIC STRESS

Genetic engineering of plant tolerance to abiotic stress helps pinpoint the induced metabolic pathways in response to one or several abiotic stress(es) (Debnath et al., 2011). Genetically, promoting crop resilience to abiotic stress requires the engineering of transgenic plants endowed with a high capacity to withstand severe abiotic challenges. This may occur through overexpressing genes from a resistant variety (stress resistant) to an elite variety (stress susceptible) to obtain a transgenic line with an enhanced stress tolerance and high yield productivity (Chen et al., 2019). However, plant resilience to abiotic stress might be reinforced through epigenetic and Synbio technologies. Notably, to enhance crop yield (or productivity), it is required (recommended) to comprehend beforehand the interplay of various interconnected biological systems (omics, genetics, epigenetics, etc.) in the plant responses to challenging environments. At the genomic level, the plant response to abiotic stress might be orchestrated through both synergistic and sequential events of a number of responsible or key genes. Hence, the stress-responsive genes can be classified into two classes: early and delayed stress-responsive genes. Thus, the former class (early) can be induced quickly and transiently, whereas the genes of the latter group are mostly more slowly activated and their expressions can be maintained for longer. Interestingly, crop plants manage abiotic stress through an intricate mechanism involving the interplay of interconnected omics realms such as genomics and omics (transcriptomics, metabolomics, interactomics, etc.). The function of this GRN can be coordinated by intermediates such as *Cis*-acting elements, transposons, transcriptions factors (TFs), and putative expressed genes still with accurately unknown functions, except those that operate (or serve) as scaffold and/or transposon elements.

4.17 POTENTIALS AND CHALLENGES IN CROP BREEDING

Crop breeding programs meet a series of challenges and hurdles before being successfully implemented. Nevertheless, crop breeding offers several advantages for agriculture, stockholders, and society. Among the criteria that need to be considered before the release of new crop cultivars to the market, based on breeding strategies, we report the higher productivity potential, nutritional value, and improved resilience to environmental challenges, considering both biotic and abiotic factors (Breseghello and Coelho, 2013). Crop breeding has the power to reinforce global food safety and sustainable agricultural development in the face of continuous increases in the entire world population and decreased area of arable land. Particularly, the integration of Synbio in breeding programs offers advantages over the ability to synthesize a diverse range of useful (utile) secondary metabolites for medicinal and industrial purposes. For instance, the NBTs, such as the CRISPR/Cas approach, have the great potential to prevent (and even exclude) off-target effects within timelines that are yet highly limited compared to the standard crossbreeding crop development method. Furthermore, the gene-editing system reduces the time of the breeding program to almost half (4–6 years) compared to the standards three other methods aforementioned, which are crossbreeding (8–10 years), mutation breeding (8–10 years), and transgene breeding (8–12 years) (Scheben et al., 2017; Chen et al., 2019).

However, the key challenges in crop breeding strategies are the availability of sophisticated and suitable facilities and well-trained and skilled staff in the field (protocols, techniques, technologies, etc.) empowered by powerful breeding methodologies to apply successful and purposeful protocols and incorporate evolved and advanced technology in the design and implementation of their breeding programs and the need for long-term financial support owing to the uncertainties of breeding experiments that need to be reproduced several times and over several generations to eliminate (or suppress) free transgene generated by off-target effects. This explains the time-consuming procedures and high-cost requirements of the breeding techniques. In addition, climate changes have entrained challenges for the crop breeding strategies employed to ensure global population food security. The cramped genetic diversity of elite varieties has further restricted our ability and exacerbated our burden to enhance genetic success through the common (or ordinary) breeding protocol. Eventually, some breeding procedures (cross- and mutation breeding) are limited by their random nature (or aspect), where creating (then screening) an elevated number of mutants is uneasy (complicated) and tedious to implement. This procedure is time-consuming and laborious; the stochastic nature of mutations may slow down the breeding pipeline, which cannot keep pace with the enhanced demands for crop yield, even if GAB methods are applied to increase selection effectiveness (Scheben et al., 2017).

4.18 GENERAL OVERVIEW OF SYNBIO AND ITS APPLICATION IN PLANTS

Synbio integrates several research branches or bifurcations of biology including biology, bioinformatics, computational biology, and engineering. Therefore, Synbio offers potential in the development of sustainable fuels, with crops being an ideal feedstock. Synbio provides deep learning of complex living organisms at the molecular and genetic levels, giving scientists the required knowledge to build up or strip down biological systems to their basic constructing bulk. With an in-depth understanding of the genetic structure of plants, Synbio enables the development and generation of new varieties from elite lines by changing big clusters of genes and/or gene parts, rather than modifying just one or a few genes as in the case of genetic engineering (or genetic modifying [GM]) approach. For example, better domesticated and sophisticated crops could be able to grow efficiently under changing climates, such as crops that produce more nutritious grains and crops able to cope with both drought and freezing. The creation of crops that sense and respond to their surrounding environmental factors (heat, salt, drought, etc.) could also be possible, through integrated biosensors (biomarkers) and genes that turn on or off in response to the prevalent particular conditions. Synbio constitutes an awesome less frequent (or adopted) avenue for creating artificial life as it poses complex social and ethical fears and anxieties. Accordingly, the application of Synbio in agriculture may face social and ethical challenges because of the general public's limited knowledge of technology and its beneficial impacts on life in general. Hence, the use of Synbio in crop improvement will face social and public acceptance problems as a legacy of GM technologies in several countries (Fesenko and Edwards, 2014).

4.19 SYNBIO ROLE IN CROP BREEDING AND PLANT ABIOTIC STRESS TOLERANCE

Synbio can be successfully and fruitfully implemented (or used) in the photosynthesis research field to improve crop yield and quality. To optimize and modulate the pre-existing photosynthetic apparatus and electron transporter super-complexes in crop Synbio, gene-editing systems (CRISPR, base editing, etc.) or genome design and reconstruction may constitute a powerful and purposeful alternative directive (or new avenue) in crop breeding via Synbio technology. Thus, crop Synbio constitutes, nowadays, an inclusive (or universal) and explicit incorporative method in crop-engineered key traits to promote breeding strategies and improve output, which not only inserts the foreign DNA template (the introduced DNA fragment can be obtained from plants of the same or different species or artificially engineered and synthesized) into the receiver plant genome, identical to the usually (or standard) adopted transgenic plants technique (or principle) but also facilitates the manipulation (such as editing, damage repair, and indels) of the genome itself of any species. This aforementioned might be implemented (or excuted) thanks to genome-editing (or -tailoring), engineering, and redesigning the synthesis pathway (anabolism) of targeted-key traits (for instance, metabolites) identified beforehand (or earlier) to possess a potent (or great) impact on the crop yield (Kaul et al., 2021; Rao et al., 2021). Interestingly, many believe (thought) that Synbio would be considered among the creative and transformative technologies that could be fundamental (or essential) to combat (or struggle) challenging environmental conditions, lack of energy (paucity), global food safety problems, and scarcity of water resources due to drought and warming issues.

In the same context, it has been stressed that Synbio can play an eminent role in plant stress response. This may occur by inducing (through encoding) genes involved in the synthetic signal transduction pathway with a specific biomarker sensor (or modular) receptor. For instance, plants can be planned (engineered or programmed) to recognize and then react to a wide range of environmental challenges such as nutrient deficiency (or depletion), abiotic stress, and other external constraints and hurdles (Pouvreau et al., 2018). This may happen, for instance, through integrating specific synthetic *Cis*-acting elements to their appropriate positions on the gene to easily manipulate the targeted gene known earlier to be associated with a particular type of stress. Integrating Synbio and epigenetics can certainly shape (or guide) the epigenetic marks functions to the desired goal that the researcher wants to reach. In addition, Synbio has great ability (potential) to generate genetically modified organisms (plants) with an increased drought resistance capacity (Borland et al., 2014; De Paoli et al., 2014; Llorente et al., 2018).

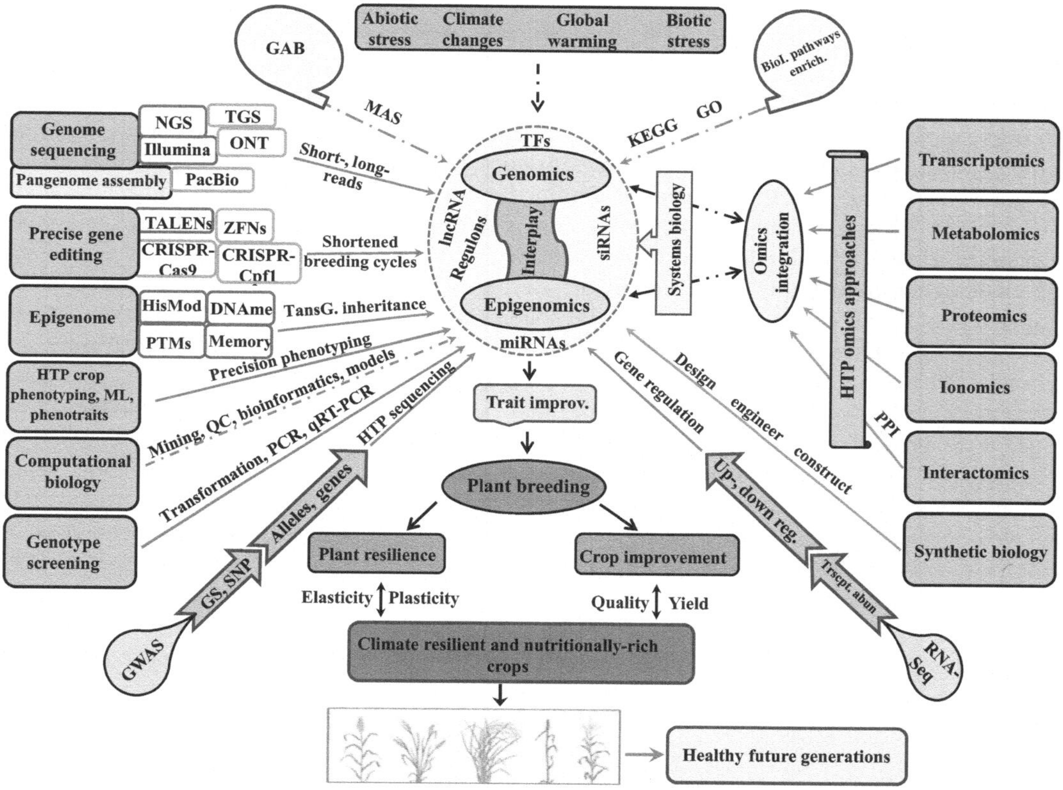

FIGURE 4.1 A representative schematic diagram summarizing the interplay of various interconnected biological systems and omics pathways to improve crop productivity and enhance plant resilience to challenging conditions. Abbreviations: NGS, next-generation sequencing; TGS, third-generation sequencing; PacBio, Pacific Bioscience; ONT, oxford nanopore technology; GAB, genomics-assisted selection; MAS, marker-assisted selection; TALENs, transcription activator-like effector nucleases; ZFNs, zinc finger nucleases; CRISPR/Cas9, clustered regularly interspaced short palindromic repeats; Cas9; CRISPR-associated (Cas) proteins 9; CRISPR-Cpf1, CRISPR from *Prevotella* and *Francisella* 1; DNAme, DNA methylation; HisMod., histone modifications; PTMs, post-translational modifications; HTP, high throughput; ML, machine learning; QC, quality control; qRT-PCR; quantitative real-time polymerase chain reaction; PCR, polymerase chain reaction; TFs, transcription factors; lncRNAs, long non-coding RNAs; miRNAs, micro RNAs; siRNAs, small interfering RNAs; Bio. Pthwys, biological pathways; KEEG, Kyoto Encyclopedia of genes and genomes; GO, gene ontology; PPI, protein–protein interactions; Trscpt. abun., transcripts abundance; Up-, Down reg., upregulation, and downregulation, respectively; RNA-seq, RNA sequencing; GS, genome sequencing; SNP, single-nucleotide polymorphism; trait improv., trait improvement; Regulon, union of TFs with *Cis*-regulatory elements.

A promising global (universal) strategy to exploit (or harness) the power of Synbio will deliver the future generation of synthetically improved (or genetically modified) crops. The potential of specifically improving yield and nutritional value of crops will address the impending global challenges such as the increased demand for food and fiber owing to the rising (increasing) global population growth , the expected deleterious repercussions (effects) of climate changes (global warming and stresses) on large crops, and the emerging doubts and fears among stockholders, consumers, and the entire society about the health, origin (provenance), continuous clean bio-agriculture, environmental safety, global priority for sustainable development, and eventually the security and ethical food production.

4.20 CONCLUSIONS

In this chapter, we shed light on the recent progress made essentially in the technologies and methodologies of molecular biology and the versatility of plant breeding strategies to enhance crop yield and quality (Figure 4.1). This book chapter also highlights the interplay of different interconnected plant science realms including genetics, epigenetics, molecular biology, omics, and Synbio to improve plant resilience to abiotic stresses and foster crop breeding programs toward improved crop productivity and/or generation of new varieties endowed with high-stress tolerance ability (Figure 4.1). Globally, the avoidance (or elimination) of starvation (or hunger) and malnutrition from society constitutes the main goal of scientists (breeders and geneticists) and farmers around the world. Hence, ensuring food security and safety for the worldwide population has never been so challenging, particularly under continuous decrease in cultivated and arable land due to the expansion of urban construction and water resources (due to salinity, pollution, and drought), an ever-growing worldwide population and environmental challenges (including global warming and climate changes). So, scientists have resorted to developing new crop varieties with more tolerance to climate change and water deficits through assisted-smart agriculture and applying developed technologies to promote crop breeding strategies (Figure 4.1). Notably, researchers use HTP technologies to genetically pinpoint stress-responsive genes and genetic markers including genes and regulons (TFs and *Cis*-regulatory elements) by proceeding to a high sequencing coverage rate employing very sophisticated platforms such as PacBio and ONT. Thus, advanced sequencing technologies (TGS and NGS) hold promising prospects to solve all transcriptome-associated modifications. Herein, the word 'transcriptome' reflects all the RNA transcripts, including coding and non-coding RNAs, because the term can be used sometimes to refer to either the total RNAs in the plant cell or just mRNA. Transcriptomics technologies represent a treasure of valuable and robust insights essential for mediating and/or engineering gene expression regulation. Eventually, our current report emphasized the role of integration of metabolomics and Synbio technologies in pinpointing the key metabolic pathways-related genes, enzymes, and/or proteins that might be involved in abiotic stress tolerance and hence may constitute valuable biomarkers for crop breeding efforts to enhance productivity or create new varieties.

REFERENCES

Akbudak, M. A., Filiz, E., & Kontbay, K. (2018). DREB2 (dehydration-responsive element-binding protein 2) type transcription factor in sorghum (*Sorghum bicolor*): Genome-wide identification, characterization and expression profiles under cadmium and salt stresses. *3 Biotech*, *8*(10), 426. https://doi.org/10.1007/s13205-018-1454-1

Alaux, M., Rogers, J., Letellier, T., Flores, R., Alfama, F., Pommier, C., Mohellibi, N., Durand, S., Kimmel, E., Michotey, C., Guerche, C., Loaec, M., Lainé, M., Steinbach, D., Choulet, F., Rimbert, H., Leroy, P., Guilhot, N., Salse, J., & Quesneville, H. (2018). Linking the international wheat genome sequencing consortium bread wheat reference genome sequence to wheat genetic and phenomic data. *Genome Biology*, *19*, 111.

Andrási, N., Pettkó-Szandtner, A., & Szabados, L. (2021). Diversity of plant heat shock factors: Regulation, interactions, and functions. *Journal of Experimental Botany, 72*, 1558–1575.

Angon, P. B., Mondal, S., Akter, S., Sakil, M. A., & Jalil, M. A. (2023). Roles of CRISPR to mitigate drought and salinity stresses on plants. *Plant Stress*, 8, 100169. https://doi.org/10.1016/j.stress.2023.100169.

Asensi-Fabado, M. A., Amtmann, A., & Perrella, G. (2016). Plant responses to abiotic stress: The chromatin context of transcriptional regulation. *Biochimica et Biophysica Acta*, 1860(1), 106–122. https://doi.org/10.1016/j.bbagrm.2016.07.015.

Ashapkin, V. V., Kutueva, L. I., Aleksandrushkina, N. I., & Vanyushin, B. F. (2020). Epigenetic mechanisms of plant adaptation to biotic and abiotic stresses. *International Journal of Molecular Sciences, 21*, 7457. https://doi.org/10.3390/ijms21207457.

Athanasopoulou, K., Boti, M. A., Adamopoulos, P. G., Skourou, P. C., & Scorilas, A. (2021). Third-generation sequencing: The spearhead towards the radical transformation of modern genomics. *Life*, 12, 30.

Borland, A. M., Hartwell, J., Weston, D. J., Schlauch, K. A., Tschaplinski, T. J., Tuskan, G. A., Yang, X., & Cushman, J. C. (2014). Engineering crassulacean acid metabolism to improve water-use efficiency. *Trends in Plant Science, 19*, 327–338. https://doi.org/10.1016/j.tplants.2014.01.006

Breseghello, F., & Coelho, A. S. G. (2013). Traditional and modern plant breeding methods with examples in rice (*Oryza sativa* L.). *Journal of Agricultural and Food Chemistry, 61*(35), 8277–8286. https://doi.org/10.1021/jf305531j.

Budak, H., Kaya, S. B., & Cagirici, H. B. (2020). Long non-coding RNA in plants in the era of reference sequences. *Frontiers in Plant Science, 11*, 276. https://doi.org/10.3389/fpls.2020.00276

Chen, B., Gilbert, L. A., Cimini, B. A., Schnitzbauer, J., Zhang, W., Li, G.-W., Park, J., Blackburn, E. H., Weissman, J. S., Qi, L. S., & Huang, B. (2013). Dynamic imaging of genomic loci in living human cells by an optimized CRISPR/Cas system. *Cell*, 155, 1479–1491. https://doi.org/10.1016/j.cell.2013.12.001

Chen, K., Wang, Y., Zhang, R., Zhang, H., & Gao, C. (2019). CRISPR/Cas genome editing and precision plant breeding in agriculture. *Annual Review of Plant Biology, 70*, 28.1–28.31. https://doi.org/10.1146/annurev-arplant-050718–100049

Chen, W., Wang, W., Peng, M., Gong, L., Gao, Y., Wan, J., Wang, S., Shi, L., Zhou, B., & Li, Z. (2016). Comparative and parallel genome-wide association studies for metabolic and agronomic traits in cereals. *Nature Communications, 7*, 12767. https://doi.org/10.1038/ncomms12767

Chinnusamy, V., Schumaker, K., & Zhu, J. K. (2004). Molecular genetic perspectives on cross-talk and specificity in abiotic stress signaling in plants. *Journal of Experimental Botany, 55*, 225–236. https://doi.org/10.1093/jxb/erh005

Ciais, P., Reichstein, M., Viovy, N., Granier, A., Ogée, J., Allard, V., Aubinet, M., Buchmann, N., Bernhofer, C., Carrara, A., Chevallier, F., Noblet, N. D., Friend, A. D., Friedlingstein, P., Grünwald, T., Heinesch, B., Keronen, P., Knohl, A., Krinner, G., … Valentini, R. (2005). Europe-wide reduction in primary productivity caused by the heat and drought in 2003. *Nature*, 437(7058), 529–533. https://doi.org/10.1038/nature03972

Cong, W., Miao, Y., Xu, L., Zhang, Y., Yuan, C., Wang, J., Zhuang, T., Lin, X., Jiang, L., Wang, N., Ma, J., Sanguinet, K. A., Liu, B., Rustgi, S., & Ou, X. (2019). Transgenerational memory of gene expression changes induced by heavy metal stress in rice (*Oryza sativa* L.). *BMC Plant Biology, 19*, 282. https://doi.org/10.1186/s12870-019-1887-7

De Paoli, H. C., Borland, A. M., Tuskan, G. A., Cushman, J. C., & Yang, X. (2014). Synthetic biology as it relates to CAM photosynthesis: Challenges and opportunities. *Journal of Experimental Botany, 65*, 3381–3393. https://doi.org/10.1093/jxb/eru038

Debnath, M., Pandey, M., & Bisen, P. S. (2011). An omics approach to understand the plant abiotic stress. *OMICS: A Journal of Integrative Biology, 15*(11), 739–762. https://doi.org/10.1089/omi.2010.0146.

Delauney, A. J., & Verma, D. P.S. (1993). Proline biosynthesis and osmoregulation in plants. *The Plant Journal, 4*, 215–223. https://doi.org/10.1046/j.1365-313X.1993.04020215.x

Deng, M., Li, D., Luo, J., Xiao, Y., Liu, H., Pan, Q., Zhang, X., Jin, M., Zhao, M., & Yan, J. (2017). The genetic architecture of amino acids dissection by association and linkage analysis in maize. *Plant Biotechnology Journal, 15*(10), 1250–1263. https://doi.org/10.1111/pbi.12712

Dietz, K. J., & Vogelsang, L. A. (2023). General concept of quantitative abiotic stress sensing. *Trends in Plant Science, 29*, 319–328. https://doi.org/10.1016/j.tplants.2023.07.006

Esmaeili, N., Shen, G., & Zhang, H. (2022). Genetic manipulation for abiotic stress resistance traits in crops. *Frontiers in Plant Science, 13*, 1011985. https://doi.org/10.3389/fpls.2022.1011985

Essemine, J., Guerfel, M., & Qu, M. (2024). Editorial: Genetic and epigenetic regulatory mechanisms in higher plants in response to abiotic stress. *Frontiers in Plant Science*, *15*, 1374289. https://doi.org/10.3389/fpls.2024.1374289

FAO (2017). *The future of food and agriculture-trends and challenges* (pp. 1–163). Rome, Italy: Food and Agriculture Organization of the United Nations.

Fesenko, E., & Edwards, R. (2014). Plant synthetic biology: A new platform for industrial biotechnology. *Journal of Experimental Botany*, *65*(8), 1927–1937. https://doi.org/10.1093/jxb/eru070.

Flowers, T. J., & Yeo, A. R. (1995). Breeding for salinity resistance in crop plants-where next? *Australian Journal of Plant Physiology*, *22*, 875–884. https://doi.org/10.1071/PP9950875

Friedrich, T., Faivre, L., Bäurle, I., & Schubert, D. (2019). Chromatin-based mechanisms of temperature memory in plants. *Plant Cell Environment*, *42*(3), 762–770. https://doi.org/10.1111/pce.13373

Galili, G., & Amir, R. (2013). Fortifying plants with the essential amino acids lysine and methionine to improve nutritional quality. *Plant Biotechnology*, *1*(2), 211–222. https://doi.org/10.1111/pbi.12025

Galili, G., Amir, R., & Fernie, A. R. (2016). The regulation of essential amino acid synthesis and accumulation in plants. *Annual Review of Plant Biology*, *67*, 153–178. https://doi.org/10.1146/annurev-arplant-043015–112213.

Gao, Y., & Zhao, Y. (2014). Self-processing of ribozyme-flanked RNAs into guide RNAs in vitro and in vivo for CRISPR-mediated genome editing. *Journal of Integrative Plant Biology*, *56*, 343–349. https://doi.org/10.1111/jipb.12152.

Gaudelli, N. M., Komor, A. C., Rees, H. A., Packer, M. S., Badran, A. H., Bryson, D. I., & Liu, D. R. (2017). Programmable base editing of A$_*$T to G$_*$C in genomic DNA without DNA cleavage. *Nature*, 551(7681), 464–471. https://doi.org/10.1038/nature24644

Gong, L., Chen, W., Gao, Y., Liu, X., Zhang, H., Xu, C., Yu, S., Zhang, Q., & Luo, J. (2013). Genetic analysis of the metabolome exemplified using a rice population. *Proceedings of the National Academy of Sciences of the United States of America*, 110, 20320. https://doi.org/10.1073/pnas.1319681110

Hamdan, M. F., Karlson, C. K. S., Teoh, E. Y., Lau, S. E., & Tan, B. C. (2022). Genome editing for sustainable crop improvement and mitigation of biotic and abiotic stresses. *Plants*, 11, 2625. https://doi.org/10.3390/plants11192625

Huang, X., Sang, T., Zhao, Q., Feng, Q., Zhao, Y., Li, C., Zhu, C., Lu, T., Zhang, Z., & Li, M. (2010). Genome-wide association studies of 14 agronomic traits in rice landraces. *Nature Genetics*, *42*, 961–967. https://doi.org/10.1038/ng.695

Jha, U. C., Nayyar, H., Jha, R., Khurshid, M., Zhou, M., Mantri, N., & Siddique, K. H. M. (2020). Long non-coding RNAs: Emerging players regulating plant abiotic stress response and adaptation. *BMC Plant Biology*, *20*, 466. https://doi.org/10.1186/s12870-020-02595-x

Kaul, S., Choudhary, M., Gupta, S., & Dhar, M. K. (2021). Engineering host microbiome for crop improvement and sustainable agriculture. *Frontiers in Microbiology*, *12*, 635917. https://doi.org/10.3389/fmicb.2021.635917

Krasensky, J., & Jonak, C. (2012). Drought, salt, and temperature stress-induced metabolic rearrangements and regulatory networks. *Journal of Experimental Botany*, *63*, 1593–1608. https://doi.org/10.1093/jxb/err460

Li, J., Manghwar, H., Sun, L., Wang, P., Wang, G., Sheng, H., Zhang, J., Liu, H., Qin, L., Rui, H., Li, B., Lindsey, K., Daniell, H., Jin, S., & Zhang, X. (2018). Whole genome sequencing reveals rare off-target mutations and considerable inherent genetic or/and somaclonal variations in CRISPR/Cas9-edited cotton plants. *Plant Biotechnology Journal*, *17*, 858–868. https://doi.org/10.1111/pbi.13020

Li, J., Zhang, H., Si, X., Tian, Y., Chen, K., Liu, J., Chen, H., & Gao, C. (2017). Generation of thermosensitive male-sterile maize by targeted knockout of the ZmTMS5 gene. *Journal of Genetics and Genomics*, *44*(9), 465–468. https://doi.org/10.1016/j.jgg.2017.02.002

Li, J. F., Norville, J. E., Aach, J., McCormack, M., Zhang, D., Bush, J., Church, G. M., & Sheen, J. (2013). Multiplex and homologous recombination-mediated genome editing in Arabidopsis and *Nicotiana benthamiana* using guide RNA and Cas9. *Nature Biotechnology*, *31*, 688–691. https://doi.org/10.1038/nbt.2654

Li, Y., Wu, X., Zhang, Y., & Zhang, Q. (2022). CRISPR/Cas genome editing improves abiotic and biotic stress tolerance of crops. *Frontiers in Genome Editing*, *4*, 987817. https://doi.org/10.3389/fgeed.2022.987817

Litvinov, D. Y., Karlov, G. I., & Divashuk, M. G. (2021). Metabolomics for crop breeding: General considerations. *Genes*, *12*(1602), 1–24. https://doi.org/10.3390/genes12101602

Llorente, B., Williams, T. C., & Goold, H. D. (2018). The multiplanetary future of plant synthetic biology. *Genes*, *9*, b348. https://doi.org/10.3390/genes9070348

Luo, M., Cheng, K., Xu, Y., Yang, S., & Wu, K. (2017). Plant responses to abiotic stress regulated by histone deacetylases. *Frontiers in Plant Science*, *8*, 2147. https://doi.org/10.3389/fpls.2017.02147

Luo, X., & He, Y. (2020). Experiencing winter for spring flowering: A molecular epigenetic perspective on vernalization. *Journal of Integrative Plant Biology*, *62*, 104–117. https://doi.org/10.1111/jipb.12896

Ma, L., & Liang, Z. (2021). CRISPR technology for abiotic stress resistant crop breeding. *Plant Growth Regulation*, *94*, 115–129. https://doi.org/10.1007/s10725-021-00704-w

Makhumbila, P., Rauwane, M., Muedi, H., & Figlan, S. (2022). Metabolome profiling: A breeding prediction tool for legume performance under biotic stress conditions. *Plants*, 11, 1756. https://doi.org/10.3390/plants11131756

Marco, F., Bitrián, M., Carrasco, P., Rajam, M. V., Alcázar, R., Antonio, F., & Tiburcio, A. F. (2015). Genetic engineering strategies for abiotic stress tolerance in plants. In B. Bahadur, M. V. Rajam, L. Sahijram, & K. V. Krishnamurthy (Eds.), *Plant biology and biotechnology volume II: Plant genomics and biotechnology*. Springer Nature. https://doi.org/10.1007/978-81-322-2283-5_29

Marko, D., El-shershaby, A., Carriero, F., Summerer, S., Petrozza, A., Iannacone, R., Schleiff, E., & Fragkostefanakis, S. (2019). Identification and characterization of a thermotolerant TILLING allele of heat shock binding protein 1 in tomato. *Genes*, *10*(7), 516. https://doi.org/10.3390/genes10070516

Mata, A. T., Jorge, T. F., Pires, M. V., & Antonio, C. (2016). Drought stress tolerance in plants: Insights from metabolomics. In M. Anwar Hossain, S. Hussain Wani, S. Bhattacharjee, D. J. Burritt, & L.- S. Phan Tran (Eds.), *Drought stress tolerance in plants*, vol 2: *molecular and genetic perspectives* (pp. 187–216). Springer International.

Miryeganeh, M. (2021). Plants' epigenetic mechanisms and abiotic stress. Genes, 12(8), 1106. https://doi.org/10.3390/genes12081106.

Nakashima, K., & Yamaguchi-Shinozaki, K. (2006). Regulons involved in osmotic stress-responsive and cold stress-responsive gene expression in plants. *Physiologia Plantarum*, *126*, 62–71.

Nascimento, F. D. S., Rocha, A. D. J., Soares, J. M. D. S., Mascarenhas, M. S., Ferreira, M. D. S., Lino, L. S. M., de Souza Ramos, A. P., Diniz, L. E. C., de Oliveira Mendes, T. A., Ferreira, C. F., dos Santos-Serejo, J. A. & Amorim, E. P. (2023). Gene editing for plant resistance to abiotic factors: A systematic review. *Plants*, 12, 305. https://doi.org/10.3390/plants12020305

Nejat, N., & Mantri, N. (2018). Emerging roles of long non-coding RNAs in plant response to biotic and abiotic stresses. *Critical Reviews in Biotechnology*, *38*(1), 93–105. https://doi.org/10.1080/07388551.2017.1312270

Nejat, N., Ramalingam, A., & Mantri, N. (2018). Advances in transcriptomics of plants. In R. K. Varshney, M. K. Pandey, & A. Chitikineni (Eds.), *Plant genetics and molecular biology* (pp. 161–185). Springer International.

Nekrasov, V., Staskawicz, B., Weigel, D., Jones, J. D., & Kamoun, S. (2013). Targeted mutagenesis in the model plant *Nicotiana benthamiana* using Cas9 RNA-guided endonuclease. *Nature Biotechnology*, *31*, 691–693. https://doi.org/10.1038/nbt.2655

Onaga, G., & Wydra, K. (2016). Advances in plant tolerance to abiotic stresses. *Plant Genome*, *10*, 229–272. https://doi.org/10.5772/64350

Pacher, M., & Puchta, H. (2017). From classical mutagenesis to nuclease-based breeding-directing natural DNA repair for a natural end-product. *The Plant Journal*, *90*, 819–833. https://doi.org/10.1111/tpj.13469

Popova, O. V., Dinh, H. Q., Aufsatz, W., & Jonak, C. (2013). The RdDM pathway is required for basal heat tolerance in Arabidopsis. *Molecular Plant*, *6*, 396–410.

Pott, D. M., Durán, S. V., Osorio, S., & Vallarino, J. G. (2021). Combining metabolomic and transcriptomic approaches to assess and improve crop quality traits. *CABI Agriculture and Bioscience*, *2*(1), 1–20. https://doi.org/10.1186/s43170-020-00021-8

Pouvreau, B., Vanhercke, T., & Singh, S. (2018). From plant metabolic engineering to plant synthetic biology: The evolution of the design/build/test/learn cycle. *Plant Science*, *273*, 3–12. https://doi.org/10.1016/j.plantsci.2018.03.035

Qu, M., Essemine, J., Li, M., Chang, S., Chang, T., Chen, G. Y., & Zhu, X. G. (2020a). Genome-wide association study unravels LRK1 as a dark respiration regulator in rice (*Oryza sativa* L.). *International Journal of Molecular Sciences*, *21*, 4930. https://doi.org/10.3390/ijms21144930

Qu, M., Essemine, J., Xu, J., Ablat, G. J., Perveen, S., Wang, H., Chen, K., Zhao, Y., Chen, G., Chu, C., & Zhu, X. (2020b). Alterations in stomatal response to fluctuating light increase biomass and yield of rice under drought conditions. *The Plant Journal*, *104*, 1334–1347. https://doi.org/10.1111/tpj.15004

Rahman, M., Zulfiqar, S., Raza, M. A., Ahmad, N., & Zhang, B. (2022). Engineering abiotic stress tolerance in crop plants through CRISPR genome editing. *Cell*, *11*(22), 3590. https://doi.org/10.3390/cells11223590

Rao, M. J., & Wang, L. (2021). CRISPR/Cas9 technology for improving agronomic traits and future prospective in agriculture. *Planta*, *254*, 68. https://doi.org/10.1007/s00425-021-03716-y

Raza, A. (2020). Metabolomics: A systems biology approach for enhancing heat stress tolerance in pants. *Plant Cell Reports*, *41*, 741–763. https://doi.org/10.1007/s00299-020-02635-8

Raza, A., Ashraf, F., Zou, X., Zhang, X., & Tosif, H. (2020). Plant adaptation and tolerance to environmental stresses: Mechanisms and perspectives. In M. Hasanuzzaman (Ed.), *Plant ecophysiology and adaptation under climate change: Mechanisms and perspectives I: General consequences and plant responses* (pp. 114–117). Springer.

Sasaki, T., & Burr, B. (2000) International rice genome sequencing project: The effort to completely sequence the rice genome. *Current Opinion in Plant Biology*, *3*(2): 138–141. doi: 10.1016/s1369-5266(99)00047-3.

Schaarschmidt, S., Fischer, A., Lawas, L. M. F., Alam, R., Septiningsih, E. M., Bailey-Serres, J., Jagadish, S. V. K., Huettel, B., Hincha, D. K., & Zuther, E. (2020). Utilizing PacBio Iso-Seq for novel transcript and gene discovery of abiotic stress responses in *Oryza sativa* L. *International Journal of Molecular Sciences*, *21*(21), 8148. https://doi.org/10.3390/ijms21218148

Scheben, A., Wolter, F., Batley, J., Puchta, H., & Edwards, D. (2017). Towards CRISPR/Cas crops-bringing together genomics and genome editing. *New Phytologist*, *216*(3), 682–698. https://doi.org/10.1111/nph.14702

Shan, Q., Wang, Y., Li, J., Zhang, Y., Chen, K., Liang, Z., Zhang, K., Liu, J., Xi, J. J., Qiu, J.-L., & Gao, C. (2013). Targeted genome modification of crop plants using a CRISPR-Cas system. *Nature Biotechnology*, *31*(8), 686–688. https://doi.org/10.1038/nbt.2650

Shi, D., Zhuang, K., Xia, Y., Zhu, C., Chen, C., Hu, Z., & Shen, Z. (2017). Hydrilla verticillata employs two diferent ways to affect DNA methylation under excess copper stress. *Aquatic Toxicology*, *193*, 97–110. https://doi.org/10.1016/j.aquatox.2017.10.007

Singh, M., Kumar, M., Albertsen, M. C., Young, J. K., & Cigan, A. M. (2018). Concurrent modifications in the three homeologs of Ms45 gene with CRISPR-Cas9 lead to rapid generation of male sterile bread wheat (*Triticum aestivum* L.). *Plant Molecular Biology*, *97*, 371–383. https://doi.org/10.1007/s11103-018-0749-2

Singhal, P., Jan, A. T., Azam, M., & Rizwanul-Haq, Q. M. (2016). Plant abiotic stress: A prospective strategy of exploiting promoters as alternative to overcome the escalating burden. *Frontiers in Life Science*, *9*(1), 52–63. https://doi.org/10.1080/21553769.2015.1077478

Srikant, T., & Drost, H. G. (2020). How stress facilitates phenotypic innovation through epigenetic diversity. *Frontiers in Plant Science*, *11*, 606800. https://doi.org/10.3389/fpls.2020.606800

Tahmasebi, A., Ashrafi-Dehkordi, E., Shahriari, A. G., Mazloomi, S. M., & Ebrahimie, E. (2019). Integrative meta-analysis of transcriptomic responses to abiotic stress in cotton. *Progress in Biophysics and Molecular Biology*, *146*, 112–122. https://doi.org/10.1016/j.pbiomolbio.2019.02.005

Tang, J., Chen, L., & Liu, Y. G. (2019). Off-target effects and the solution. *Nature Plants*, *5*, 341–342. https://doi.org/10.1038/s41477-019-0406-z.

Usman, B., Derakhshani, B., & Jung, K. H. (2023). Recent molecular aspects and integrated omics strategies for understanding the abiotic stress tolerance of rice. *Plants*, *12*(10), 2019. https://doi.org/10.3390/plants12102019

Van Esse, H. P., Reuber, T. L., & van der Does, D. (2020). Genetic modification to improve disease resistance in crops. *New Phytologist*, *225*, 70–86. https://doi.org/10.1111/nph.15967

Wang, D., Lu, X., Chen, X., Wang, S., Wang, J., Guo, L., Yin, Z., Chen, Q., & Ye, W. (2020). Temporal salt stress-induced transcriptome alterations and regulatory mechanisms revealed by PacBio long-reads RNA sequencing in Gossypium hirsutum. *BMC Genomics*, *21*(1), 838. https://doi.org/10.1186/s12864-020-07260-z

Wang, P., Sun, X., Jia, X., & Ma, F. (2017). Apple autophagy-related protein MdATG3s afford tolerance to multiple abiotic stresses. *Plant Science*, *256*, 53–64. https://doi.org/10.1016/J.PLANTSCI.2016.12.003.

Wen, W., Jin, M., Li, K., Liu, H., Xiao, Y., Zhao, M., Alseekh, S., Li, W., Lima, F. D. A. E., Brotman, Y., Willmitzer, L., Fernie, A. R., & Yan, J. (2018). An integrated multi-layered analysis of the metabolic networks of different tissues uncovers key genetic components of primary metabolism in maize. *The Plant Journal*, *93*(6), 1116–1128. https://doi.org/10.1111/tpj.13835

Wen, W., Li, K., Alseekh, S., Omranian, N., Zhao, L., Zhou, Y., Xiao, Y., Jin, M., Yang, N., Liu, H., Florian, A., Li, W., Pan, Q., Nikoloski, Z., Yan, J., & Fernie, A. R. (2015). Genetic determinants of the network of primary metabolism and their relationships to plant performance in a maize recombinant inbred line population. *Plant Cell*, *27*(7), 1839–1856. https://doi.org/10.1105/tpc.15.00208

Yang, X., Cushman, J. C., Borland, A. M., & Liu, Q. (2020). Editorial: Systems biology and synthetic biology in relation to drought tolerance or avoidance in plants. *Frontiers in Plant Science*, *11*, 394. https://doi.org/10.3389/fpls.2020.00394

Yang, Q., Zhao, D., & Liu, Q. (2020). Connections between amino acid metabolisms in plants: Lysine as an example. *Frontiers in Plant Science*, *11*, 928. https://doi.org/10.3389/fpls.2020.00928

Zetsche, B., Gootenberg, J. S., Abudayyeh, O. O., Slaymaker, I. M., Makarova, K. S., Essletzbichler, P., Volz, S. E., Joung, J., van der Oost, J., Regev, A., Koonin, E. V., & Zhang, F. (2015). Cpf1 is a single RNA-guided endonuclease of a class 2 CRISPR-Cas system. *Cell*, *163*(3), 759–771. https://doi.org/10.1016/j.cell.2015.09.038

Zhao, T., Wu, T., Zhang, J., Wang, Z., Pei, T., Yang, H., Li, J., & Xu, X. (2020). Genome-wide analyses of the genetic screening of C_2H_2-type zinc finger transcription factors and abiotic and biotic stress responses in tomato (*Solanum lycopersicum*) based on RNA-seq data. *Frontiers in Genetics*, *11*, 540. https://doi.org/10.3389/fgene.2020.00540

Zhou, B., Zhang, L., Ullah, A., Jin, X., Yang, X., & Zhang, X. (2016a). Identification of multiple stress responsive genes by sequencing a normalized cDNA library from sea-land cotton (*Gossypium barbadense* L.). *PLoS One*, *11*(3), e0152927. https://doi.org/10.1371/journal.pone.0152927

Zhou, H., He, M., Li, J., Chen, L., Huang, Z., Zheng, S., Zhu, L., Ni, E., Jiang, D., Zhao, B., & Zhuang, C. (2016b). Development of commercial thermo-sensitive genic male sterile rice accelerates hybrid rice breeding using the CRISPR/Cas9-mediated TMS5 editing system. *Scientific Reports*, *6*, 37395. https://doi.org/10.1038/srep37395

Zhu, G., Wang, S., Huang, Z., Zhang, S., Liao, Q., Zhang, C., Lin, T., Qin, M., Peng, M., Yang, C., Cao, X., Han, X., Wang, X., van der Knaap, E., Zhang, Z., Cui, X., Klee, H., Fernie, A. R., Luo, J., & Huang, S. (2018). Rewiring of the fruit metabolome in tomato breeding. *Cell*, *172*, 249–261. https://doi.org/10.1016/j.cell.2017.12.019

Zhu, Y. N., Shi, D.-Q., Ruan, M. B., Zhang, L. L., Meng, Z.-H., Liu, J., & Yang, W. C. (2013). Transcriptome analysis reveals crosstalk of responsive genes to multiple abiotic stresses in cotton (*Gossypium hirsutum* L.). *PLoS One*, *8*(11), e80218. https://doi.org/10.1371/journal.pone.0080218

5 Integration of Artificial Intelligence (AI) and Machine Learning (ML) in Plant Breeding

Neha Sharma, Manish Kumar, Nimisha Sharma, and Nagendra Kumar Singh

5.1 INTRODUCTION

An increase in population growth with each passing day increases overall pressure on the resources available, and hence it may lead to excessive consumption of natural resources and depletion of the ecosystem, which ultimately causes poverty and inequality among human races, and by the year 2050 the situation would become worse [1]. Plant breeding approaches serve as an important strategy to ensure food security among people to reduce world hunger and hence offer enough food for everyone. With the advancement in technologies, plant breeding has shown great potential to develop different and improved new varieties of all crops that are more adaptable to biotic and abiotic stresses and can produce higher and better yields [2,3], resulting in increased food production. Due to many fundamental advancements in recent years, plant breeding has evolved, with a variety of new technologies and techniques, such as speed breeding, better hybridization, precision breeding, targeted genetic engineering, molecular marker-based breeding, and omics selection to develop superior and more productive varieties. The outcome of agricultural biotechnology has increased manifold with the different applications and advancements of omics. The new era of omics addresses large datasets, and their speedy analysis gives a plant breeder an extra edge to either develop new varieties or improve yield and combat various biotic and abiotic stresses [4]. Breeders create new trait varieties concerning various stresses by selective breeding approaches like gene editing, e.g. CRISPR/Cas9 [5,6] or gene pyramiding. For instance, plant breeders need better algorithms for the analysis and interpretation of complex datasets generated via different omics technologies. Artificial intelligence (AI), and machine learning (ML) provide those targeted insights into all the complex data generated for efficient breeding for better traits [7].

5.2 WHAT IS ML AND AI?

ML is the learning of machines from data already provided to the machine, and now it has been opted by several areas of the agriculture industry for improved cultivars and designing better plant breeding strategies [8]. Recently, ML or we can say AI has evolved in such a way that now its applications have increased manifolds in science, particularly in plant genomics. With the help of ML, more resistant varieties can be developed which are better suited for any abiotic and biotic stress. The integration of AI and ML for plant breeding in the agricultural industry would provide efficient and cost-effective varieties [9]. These technological developments have enabled breeders and researchers to study large and complex datasets to find meaningful patterns in their test plants. But where there are benefits of course there are also challenges as these ML algorithms need

 DOI: 10.1201/9781003457237-5

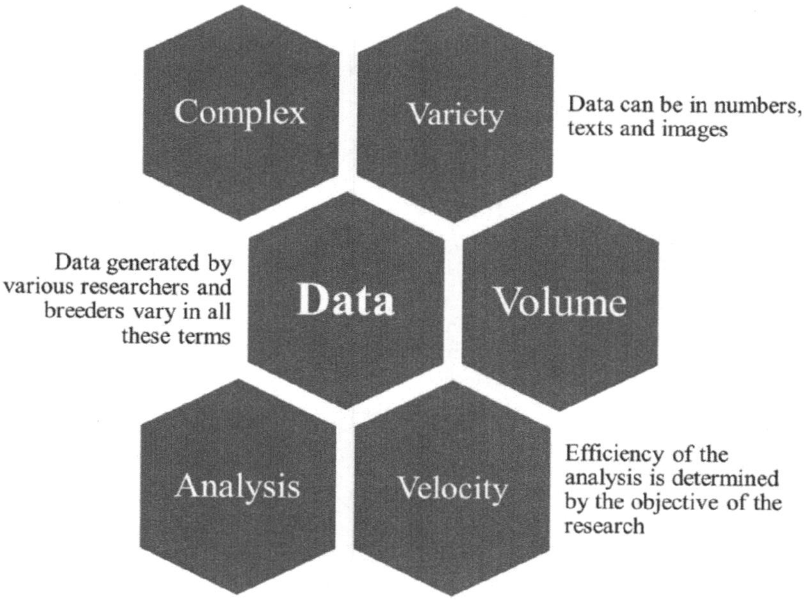

FIGURE 5.1 Characteristics of the data resulted from different studies.

high-quality datasets, large servers, and big computing resources, and sometimes a single algorithm is not suitable for all [10]. But, in a nutshell, AI and ML have the potential to increase the probability of identifying a true genotype by ultimately focusing on desired traits. Nowadays, multinational seed enterprises (MSEs) have opted for AI and ML or similar approaches for predictive breeding for the majority of crops. It is also assumed that shortly AI will create new opportunities for plant breeding by incorporating an integrated genomic-environs prediction for selection via combining multi-omics [11] (Figure 5.1).

ML algorithms build predictive precise models, recognize the objects in image forms and videos, and channelize the robotic systems. The results obtained from different omics technologies include massive volumes of data that may be structured (viz., numerical data) or unstructured (viz., texts and images) and usually expressed in terms of bytes, kilobytes, megabytes, gigabytes, and terabytes. Due to the high volume (in terms of the amount of data received), velocity (efficiency of analysis), and variety (refers to diversity; i.e. differences in the types), data plant breeders or researchers have to collect big volumes of data concerning the environmental interaction of the same. This big data then would be used to improve desirable traits, such as resistance of plants to different stress environments and production of better and higher yields. In addition, big complex data obtained in plant molecular breeding can explore new insights into desired precise traits, the better and improved yield of a crop under different environmental conditions, or unique genes that can be used to develop new improved varieties with desired traits/characteristics [12]. Initially in plant breeding, it was difficult to detect the transcription factors, but now different TF–DNA databases, namely PlantPAN [13] and PlnTFDB [14], are currently available in plants. Similarly, shortly AI and ML have opted for many plant breeding strategies, and now there are hundreds of databases for rapid breeding methods; these databases are depicted to improve the upcoming omics analysis and that would ultimately produce efficient and successful crops, which would give food security to entire populations and prevent global poverty globally (Table 5.1).

Therefore, in plant breeding, AI technologies analyze large amounts of result data by developing algorithms that can grasp knowledge and learn from past data, make decisions, and ultimately

TABLE 5.1
Different AI-Generated Databases for Plant Studies

S. No.	Study	Database	Crop/Method	References
1.	Protein–protein interaction	Predicted Rice Interactome Network	Rice	Gu et al. [15]
2.	Metabolic interaction	AraCyc databases	*Arabidopsis*	Mueller et al. [16]
3.	Protein–protein interaction	PPIM	Maize	Zhu et al. [17]
4.	Plant metabolic pathway stoichiometric analysis	PlantCyc	Seed development	Grafahrend et al. [18]
5.	Optimization of disinfection protocols	AI- or ML-mediated disinfection and development	*Cannabis* seeds	Pepe et al. [19]
6.	AI- and ML-assisted in vitro rooting	Optimized environment creation	*Passiflora caerulea*	Jafari et al. [20]
7.	Genome-wide association studies	Yield-associated hyperspectral reflectance control	Soybean	Yoosefzadeh-Najafabadi et al. [21]
8.	Multi-omics	ZEAMAP	Maize	Gui et al. [22]

help in selecting plant varieties with improved desired characteristics concerning biotic and abiotic stresses. In fact, AI and ML robotics and computer vision are also gaining attention and also offer large applications in plant breeding. For example, a robot named BoniRob V2 was created and designed as a crop scout, which has the potential to check soil attributes and characterization of plants concerning their morphological and genotypic data using multi-sensor applications and control weeds [23,24]. For pre-processing phenotypic data, the AllInOne R-shiny package was developed, which is an open-source, licensed, highly sophisticated analytical R package and is user-friendly as no coding skills are required for the same [25].

5.3 FEATURES OF AI OR ML

- Data-driven technology
- Learn from past data and automatically detect different patterns and conclude the program for desired known output

5.4 RELATION OF PLANT BREEDING AND ML OR AI

In traditional plant breeding, the very first stage is selection followed by experimental design and then statistical analysis. Marker-assisted breeding is an extra edge to the breeders to cut short the breeding cycle concerning the trait of interest. Smart breeding effectively uses both traditional and marker-assisted breeding for the analysis of big data or extensive datasets generated in the omics era. Basically, ML brings computer science and statistics to one common platform for creating prediction models (Figure 5.2).

In modern breeding, the techniques of AI and ML have been used to design different programs for selection, to protect endangered germplasms, to assess the genetic variations among varieties and cultivars, and for multi-environmental trials for important crops. With AI's help, frequently larger datasets comprising G-P-E (genotype–phenotype–environment) with replications with large sample sizes and many populations or breeding lines could be assessed [26]. ML can be incorporated at around each step of plant breeding, i.e. first for the selection of appropriate parents for any

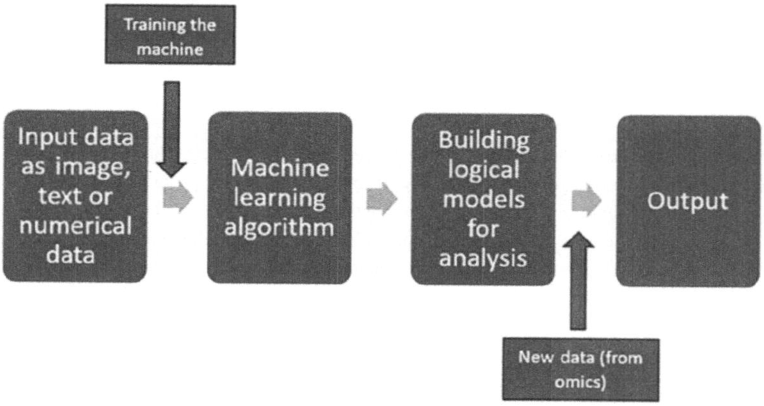

FIGURE 5.2 Pipeline of machine learning or artificial intelligence.

cross to check the progeny breeding lines across various environments. In fact, over the coming years with the advent of new technologies, the researchers or breeders will get high-throughput measurement data so the role of AI and ML will become too large.

5.5 ALGORITHMS OF AI AND ML FOR PLANT BREEDING

The term 'machine learning' was first introduced by Arther Samuel in 1959 [27]. ML enables a machine to grasp knowledge automatically from the data provided to build mathematical models based on various algorithms and make predictions and decisions. Different AI and ML algorithms have been frequently opted and used in molecular plant breeding for the past few years. Currently, AI and ML are being used for image recognition, text recognition, speech recognition, Facebook auto-tagging, and many more. But before using AI and ML in plant breeding, we need to know that each algorithm has proved to have its advantages and disadvantages. If we express ML in terms of gardening, seeds are algorithms, data plays nutrients, plants are programs, and we researchers are gardeners. Induction of decision trees (predictive model to conclude a set of observations), artificial neural networks (modeled like the function of the human brain; derive conclusions from big, complex, and unrelated set of information), hidden Markov model (for modeling linkage disequilibrium), clustering (mapping to nearby slots for closely related data), support vector machines (used for classification and regression analysis), and simulated annealing are different strategies that are being used in different algorithms of AI and ML (Table 5.2).

Overall, further research and in-depth knowledge are necessary to assess the actual efficacy and output of AI and ML algorithms for predicting traits and characteristics of interest via high-throughput omics datasets in plant breeding.

5.6 CHALLENGES FACED IN ADOPTING AI AND ML

AI or ML would guarantee progress in plant sub-atomic examinations by laying out speedy associations among genotypes and aggregates and their regarded expansive affiliations. Concentrates on plant and microbe communications, plant and stress resistance, different biochemical ordered trials, and top-to-bottom omics could depend on simulated AI and ML. It has been expected that AI and ML will soon suggest better alternatives in plant research via accurate decision-making to get the desired characteristics in plants by predicting the exact location where any particular genetic modification is to be made. The analysis of the crop performance under different environmental conditions adds to a future opportunity to check the optimum temperature, pressure, etc., for desired

TABLE 5.2
Advantages and Characteristics of Different Algorithms

S. No.	Algorithm	Helps in	How It Works	Advantages	References
1.	Multivariate adaptive regression splines (MARS)	Shoot quality assessment, multiplication, and leaf color responses	It captures the basic significant factors and their inter- and intra-relationships needed to predict the optimal nutrient levels for the crop	Improved accuracy, better stability, and potential to work with large datasets	Akin et al. [28]
2.	Random forest	Remote sensing	Random selection so the accuracy varies depending on the data	Can handle complex varietal datasets without tweaking Disadvantage: Slow and difficult to interpret	Belgiu and Dragu [29]
3.	Support vector machines (SVM)	Differentiation of patterns of spectral signatures of various potato plants infected with (PVY) Potato Virus Y	The utilization of near-infrared rays and shortwave infrared wavelength rays for classification and characterization of PVY	Robust, consistent, scalable, and accurate	Griffel et al. [30]
4.	Artificial neural network (ANN)	Diversity assessment	It ensures great differentiation, even in close datasets	Accurate and able to classify all genotypes correctly and computationally intensive	Silva et al. [31]; Brasileiro et al. [32]; Sant'Anna et al. [33]
5.	Ensemble learning	Classification, regression, and clustering	Combines the effect of multiple algorithms to design and create much more robust, precise, and powerful models	More accurate and robust predictions	Dietterich [34]; Polikar [35]

growth. The data obtained from *n* number of experiments is very complex and shows variability so there must be different models and equations for such different types of heterogeneous data [36]. Also, a few models created via AI and ML showing good results in one approach may not work in relatively closer research operations at differential geographic locations. Even after offering many advantages, AI and ML and their driven technological models are not opted by all. Here are a few challenges:

1. As we know, when it comes to food-related things, people generally do not accept the same very fast. Genetically engineered products are still not able to make its place. So similarly, it is difficult for common people to adopt AI and follow the decisions of genotypic or phenotypic prediction based on their model analysis.

2. The cost of AI-driven technology is also one of the major factors, and sometimes the estimates or predictions provided by the technology do not meet the user's needs. Also sometimes the estimates are not good enough for the actual validation.
3. A skilled person may be needed for the interpretation of the confidence of estimates and predictions.
4. Compared to the traditional statistical approaches, AI technology needs all the information before making the machine learn first. So the output should be known in advance so that the problem can be solved. Hence AI and ML require more effort compared to the conventional approach.
5. Finding more innovative AI and ML models for everyday problems is mandatory and challenging.

Even though there are many challenges, AI and ML offer multi-dimensional tools to researchers or plant breeders that allow them to work on their complex data and allow fast interpretation of the data utilizing models and decision trees. With the progression in innovation, specifically, through man-made reasoning and AI, it has become truly simple to gauge plant genotypes and aggregates, yet that prompts the creation of large, complex informational indexes. In rearing, both AI and ML center on the expanded pressure resilience in plants and examinations of those plants at various genotypic, phenotypic, and biochemical levels and thus to have upgraded yields of those harvests. Both AI and ML give significant examples in plant information, which eventually offers simple techniques for specialists and researchers. Eventually, we can say AI and ML would really overcome any barrier between genotype and breeding approaches and subsequently prompt the advancement of further developed cultivars. Also, based on the changing environments, the technology of AI and ML has the potential to ensure global food security and address poverty for the growing population.

5.7 FUTURE OPPORTUNITIES

Shortly, AI and ML would support the process of plant breeding via rapid decision-making, suggesting the genome-editing process to be more precise to ensure much-improved crop performance in all the soil attributes concerning all the abiotic and biotic factors. Soon the process of plant breeding will be cut short by the newly emerged technologies and algorithms of AL or ML. The newly emerged science of AI or ML would ensure progress in plant molecular studies by establishing quick connections between genotype and phenotype and their respective genome-wide associations. Studies on plant and pathogen interactions, plant and stress tolerance, various biochemical experimentations, and in-depth omics could simply rely on AI and ML [37].

Therefore, we can say computer-based intelligence and ML will uphold the course of plant molecular breeders by recommending the genome-altering cycle to be more exact to guarantee significantly better harvest execution concerning all the abiotic and biotic variables.

5.8 CONCLUSION

Future plant breeding will be more accurate, precise, and speedy with a bigger sample size where the implementation of AI and ML will address the efficiency, better predictability, and more economic gain as well as genetic gain. AI and ML offer a variety of opportunities for plant breeding. Integration of more innovation in breeding techniques concerning genome editing, transgenics, and marker-assisted breeding and selection would benefit humankind and prevent hunger. The databases generated will aid in the establishment of better plant breeding platforms for improved breeding initiatives with more genetic and economic gain in the plant breeding pipelines for global and national programs of small- and medium-type breeding experiments carried out by researchers and breeders. AI and ML offer several applications in plant breeding, genotyping, high-throughput phenotyping, genetic transformation, and genome editing. In general, the generation of huge, complex high-quality datasets is an essential requirement in all future breeding approaches and for the successful analysis of those datasets' implementation of ML is a prerequisite. Such big datasets can provide a way for new forward-looking requirements of AI- or ML-mediated plant breeding, and that could hopefully overcome many numbers of the routine challenges faced by plant breeders and researchers and ultimately benefit humankind by overcoming hunger and poverty.

REFERENCES

1. Coole, D. (2018). *Should we control world population?* John Wiley & Sons.
2. Yoosefzadeh-Najafabadi, M., Rajcan, I., & Eskandari, M. (2022). Optimizing genomic selection in soybean: An important improvement in agricultural genomics. *Heliyon*, *8*, e11873.
3. Najafabadi, M. Y., Soltani, F., Noory, H., & Díaz-Pérez, J. C. (2018). Growth, yield and enzyme activity response of watermelon accessions exposed to irrigation water déficit. *International Journal of Vegetable Science*, *24*, 323–337.
4. Li, Q., & Yan, J. (2020). *Sustainable agriculture in the era of omics: Knowledge-driven crop breeding* (Vol. 21, pp. 1–5). Springer.
5. Pramanik, D., Shelake, R. M., Kim, M. J., & Kim, J.-Y. (2021). CRISPR-mediated engineering across the central dogma in plant biology for basic research and crop improvement. *Molecular Plant*, *14*, 127–150.
6. Hesami, M., Najafabadi, M. Y., Adamek, K., Torkamaneh, D., & Jones, A. M. P. (2021). Synergizing off-target predictions for in silico insights of CENH3 knockout in Cannabis through CRISPR/CAS. *Molecules*, *26*, 2053.
7. Yoosefzadeh-Najafabadi, M., Torabi, S., Tulpan, D., Rajcan, I., & Eskandari, M. (2021). Genome-wide association studies of soybean yield-related hyperspectral reflectance bands using machine learning-mediated data integration methods. *Frontiers in Plant Science*, *12*, 777028.
8. Jafari, M., Daneshvar, M. H., Jafari, S., & Hesami, M. (2022). Machine learning-assisted in vitro rooting optimization in Passiflora caerulea. *Forests*, 13, 2020.
9. Yoosefzadeh-Najafabadi, M., Torabi, S., Tulpan, D., Rajcan, I., & Eskandari, M. (2021). Genome-wide association studies of soybean yield-related hyperspectral reflectance bands using machine learning-mediated data integration methods. *Frontiers in Plant Science*, *12*, 777028.
10. Zhang, N., Gupta, A., Chen, Z., & Ong, Y.-S. (2021). Evolutionary machine learning with minions: A case study in feature selection. *IEEE Transactions on Evolutionary*, *26*, 130–144.
11. Lee, T. G. (2021). *Artificial intelligence for breeding better crops*. https://vscnews.com/artificial-intelligence-crop-breeding/
12. Najafabadi, M. Y., & Rajcan, I. (2022). Six decades of soybean breeding in Ontario, Canada: A tradition of innovation. *Canadian Journal of Plant Science*, 103, 333–352.
13. Chow, C.-N., Zheng, H.-Q., Wu, N.-Y., Chien, C.-H., Huang, H.-D., Lee, T.-Y., Chiang-Hsieh, Y.-F., Hou, P.-F., Yang, T.-Y., & Chang, W.-C. (2016). PlantPAN 2.0: An update of plant promoter analysis navigator for reconstructing transcriptional regulatory networks in plants. *Nucleic Acids Research*, *44*, D1154–D1160.
14. Riaño-Pachón, D. M., Ruzicic, S., Dreyer, I., & Mueller-Roeber, B. (2007). PlnTFDB: An integrative plant transcription factor database. *BMC Bioinformatics*, *8*, 42.
15. Gu, H., Zhu, P., Jiao, Y., Meng, Y., & Chen, M. (2011). PRIN: A predicted rice interactome network. *BMC Bioinformatics*, *12*, 161.
16. Mueller, L. A., Zhang, P., & Rhee, S. Y. (2003). AraCyc: A biochemical pathway database for Arabidopsis. *Plant Physiology*, *132*, 453–460.
17. Zhu, G., Wu, A., Xu, X.-J., Xiao, P.-P., Lu, L., Liu, J., Cao, Y., Chen, L., Wu, J., & Zhao, X.-M. (2016). PPIM: A protein-protein interaction database for maize. *Plant Physiology*, *170*, 618–626.
18. Grafahrend-Belau, E., Junker, B. H., & Schreiber, F. (2012). Plant metabolic pathways: Databases and pipeline for stoichiometric analysis. In G. K. Agrawal, & R. Rakwal (Eds.), *Seed development: OMICS technologies toward improvement of seed quality and crop yield* (pp. 345–366). Springer.
19. Pepe, M., Hesami, M., & Jones, A. M. P. (2021). Machine learning-mediated development and optimization of disinfection protocol and scarification method for improved *in vitro* germination of Cannabis seeds. *Plants*, *10*, 2397.
20. Jafari, M., Daneshvar, M. H., Jafari, S., & Hesami, M. (2022). Machine learning-assisted *in vitro* rooting optimization in *Passiflora caerulea*. *Forests*, *13*, 2020.

21. Yoosefzadeh-Najafabadi, M., Torabi, S., Tulpan, D., Rajcan, I., & Eskandari, M. (2021). Genome-wide association studies of soybean yield-related hyperspectral reflectance bands using machine learning-mediated data integration methods. *Frontiers in Plant Science*, *12*, 777028.
22. Gui, S., Yang, L., Li, J., Luo, J., Xu, X., Yuan, J., Chen, L., Li, W., Yang, X., & Wu, S. (2020). ZEAMAP, a comprehensive database adapted to the maize multi-omics era. *IScience*, *23*, 101241.
23. Bangert, W., Kielhorn, A., Rahe, F., Albert, A., Biber, P., Grzonka, S., Haug, S., Michaels, A., Mentrup, D., & Hänsel, M. (2013). Field-robot based agriculture: "Remote Farming. 1" and "BoniRob-Apps". *VDI-Berichte*, *2193*, 2.1.
24. Falk, K. G., Jubery, T. Z., Mirnezami, S. V., Parmley, K. A., Sarkar, S., Singh, A., Ganapathysubramanian, B., & Singh, A. K. (2020). Computer vision and machine learning enabled soybean root phenotyping pipeline. *Plant Methods*, *16*, 1–19.
25. Yoosefzadeh-Najafabadi, M., Heidari, A., & Rajcan, I. (2022). *AllInOne*. https://github.com/MohsenYN/AllInOne/wiki
26. Xu, Y., Zhang, X., Li, H., Zheng, H., Zhang, J., Olsen, M. S., Varshney, R. K., Prasanna, B. M., & Qian, Q. (2022). Smart breeding driven by big data, artificial intelligence, and integrated genomic-enviromic prediction. *Molecular Plant*, *15*, 1664–1695.
27. Samuel, A. L. (2000). Some studies in machine learning using the game of checkers. *IBM Journal of Research and Development*, *44*, 206–226. https://doi.org/10.1147/rd.441.0206
28. Akin, M., Eyduran, S. P., Eyduran, E., & Reed, B. M. (2020). Analysis of macro nutrient related growth responses using multivariate adaptive regression splines. *Plant Cell, Tissue and Organ Culture (PCTOC): Journal of Plant Biotechnology*, *140*, 661–670.
29. Belgiu, M., & Draguţ, L. (2016). Random forest in remote sensing: A review of applications and future directions. *ISPRS Journal of Photogrammetry and Remote Sensing*, *114*, 24–31.
30. Griffel, L., Delparte, D., & Edwards, J. (2018). Using support vector machines classification to differentiate spectral signatures of potato plants infected with Potato Virus Y. *Computers and Electronics in Agriculture*, *153*, 318–324.
31. Silva, G. N., Tomaz, R. S., Sant'Anna, I. D. C., Nascimento, M., Bhering, L. L., & Cruz, C. D. (2014). Neural networks for predicting breeding values and genetic gains. *Scientia Agricola*, *71*, 494–498.
32. Brasileiro, B. P., Marinho, C. D., Costa, P. M. D. A., Cruz, C. D., Peternelli, L. A., & Barbosa, M. H. P. (2015). Selection in sugarcane families with artificial neural networks. *Crop Breeding and Applied Biotechnology*, *15*, 72–78.
33. Sant'Anna, I., Tomaz, R. S., Silva, G. N., Nascimento, M., Bhering, L. L., & Cruz, C. D. (2015). Superiority of artificial neural networks for a genetic classification procedure. *Genetics and Molecular Research*, *14*, 9898–9906.
34. Dietterich, T. G. (2002). Ensemble learning. In M. A. Arbib (Ed.), *The handbook of brain theory and neural networks* (Vol. 2, pp. 110–125). MIT Press.
35. Polikar, R. (2012). Ensemble learning. In *Ensemble machine learning* (pp. 1–34). Springer.
36. Azodi, C. B., Tang, J., & Shiu, S.-H. (2020). Opening the black box: Interpretable machine learning for geneticists. *Trends in Genetics*, *36*, 442–455.
37. Wang, H., Cimen, E., Singh, N., & Buckler, E. (2020). Deep learning for plant genomics and crop improvement. *Current Opinion in Plant Biology*, 54, 34–41.

6 CRISPR-Based Precision Breeding for Crop Improvement

Current Achievements and Future Directions

Divya Raj Bharti, Biswajit Pramanik, Amitava Paul, and Sandip Debnath

6.1 INTRODUCTION

Global food security has been threatened by a number of significant challenges, including rapid population expansion, climate change, and environmental risks (Sundström et al. 2014). Achieving food sufficiency requires increasing agricultural production. Therefore, it is crucial to produce crops that are resistant to disease, have a high yield, and are climate-adaptable to increase crop output. Over the years, traditional breeding techniques including hybridization and mutation breeding have been essential in raising agricultural yields. However, the output of crops has been greatly impacted by the steady decline in the natural genetic variety of agricultural plants (Govindaraj et al. 2015). Using transgenic or genetically modified (GM) crops is one example of how molecular breeding technology has been applied over the past few decades, and it has shown great promise in increasing agricultural output and overcoming problems with traditional breeding approaches. Still, a lot of effort and time is required. Various political and environmental issues regarding the social acceptance of these GM crops remain a challenging endeavor (Lusser et al. 2012).

Gene modification has recently become a revolutionary tool that successfully overcomes the drawbacks of traditional breeding methods by introducing mutations in selected genomic portions of the crops with great proficiency and accuracy using sequence-specific nucleases (SSNs). Zinc finger nucleases (ZFNs), transcriptional activator-like effector nucleases (TALENs), and the CRISPR/Cas9 endonuclease are examples of artificially engineered SSNs that have demonstrated remarkable efficacy in specifically designed mutagenesis in a range of model plants and crops. In the target area of the DNA, the designed nucleases can result in double-strand breaks (DSBs), which are then healed by the integrated cellular repair process, involving either non-homologous end-joining (NHEJ) or homology directed repair (HDR). Whereas the precise repair mechanism of the latter replaces the DNA residing genes with the foreign one following DSB, the former erroneous pathway follows knockout and frameshift mutations in the selected genes.

While HDR is more accurate than NHEJ, the incidence of illicit recombination in NHEJ is higher. Compared to NHEJ, relatively little gene substitution or gene targeting is done by HDR. While insertions and deletions (indels), arbitrary point mutations, or gene addition transmit many of the important traits, gene knockouts, and loss-of-function mutations are still not very useful in crop development. CRISPR-Cpf1, or *Prevotella* and *Francisella* I-derived CRISPR, may have been discovered recently as a novel and efficient gene-editing technique. It is more precise, efficient, and useful than that of CRISPR/Cas9, as reported by Zaidi et al. (2017). This contemporary technique

 DOI: 10.1201/9781003457237-6

to become popular is base editing, which allows G–C to A–C base pairs conversion without the need for a donor template or DSB. These developments in genome editing have the potential to speed up agricultural innovation by facilitating the efficient and precise creation of new characteristics. Numerous insightful evaluations exist about genome-editing techniques and their potential for improving crops. Van de Wiel et al. (2017) conducted a comprehensive study of SDN-1 (site-directed nuclease-1) and noted its advancements and novel versions, which include using genome-editing tools to create tiny indels at chosen targets. This chapter provides a thorough analysis of the most recent advancements in the implementations of genome modification technologies for crop improvement, involving base editing as well as CRISPR/Cpf1 tool. This also devises the challenges and possible applications of this newly evolved technology to the development of agriculture.

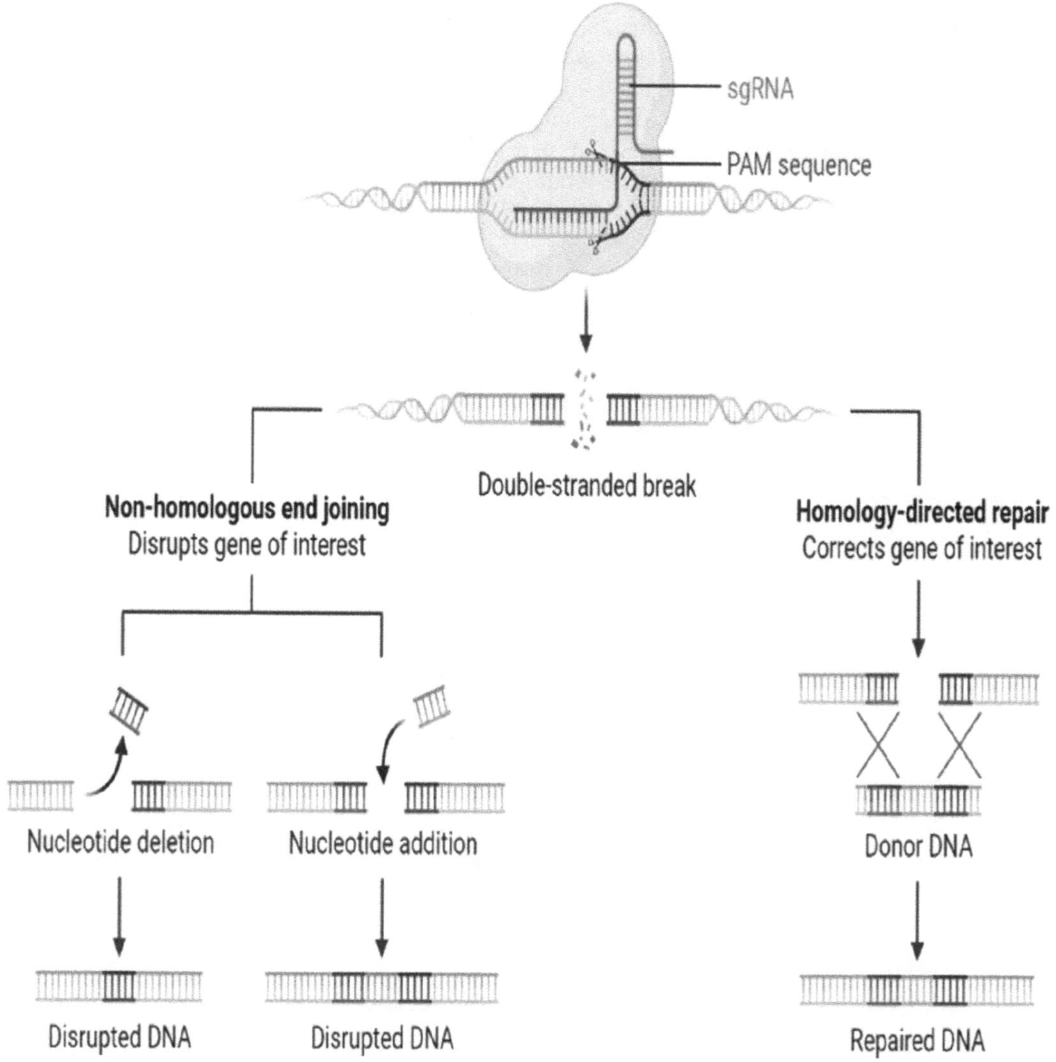

FIGURE 6.1 DNA repair mechanisms followed in the genome-editing process. Created in BioRender: https://biorender.com.

6.2 AN OVERVIEW OF THE GENE-EDITING TOOLS

The SSNs creating DSBs at precise locus have set the core foundation of all genome-editing techniques. The cellular endogenous repair mechanisms then mend the breaks using either HDR or NHEJ (Figure 6.1). By introducing locus-specific point mutations, these repair processes change the coding frames and produce chromosomal rearrangements by knocking out, inserting, or replacing several genes (Sar et al. 2022).

6.2.1 Meganucleases

The first enzymes to be utilized for genome editing were mega nucleases (Figure 6.2). The four primary families into which the mega nucleases are divided are LAGLIDADG, GIY-YIG, His-Cys box, and HNH. The former family includes the well-known mega nucleases, namely I-CreI and I-SceI. Typically, they can identify to cleave 12–40 bp long DNA sequences. Although this tool was predominantly developed for both human and animal systems, this has also been of great use for plants, such as *Zea mays* and *Arabidopsis thaliana*, for the aim of genome editing (Antunes et al. 2012). The maize MS26 gene was targeted with an altered mega nuclease that was generated from the I-CreI homing nuclease. Upon knocking out the genes, male sterility was induced (Djukanovic et al. 2013). However, it is challenging to design homing endonucleases to target distinct DNA locations because their nuclease and DNA-binding domains overlap, and their usage has also been restricted by patent disputes.

6.2.2 ZFNs

It is believed that zinc finger nucleases were the first generation of engineered endonucleases that detect a particular DNA sequence in the genome using specially created zinc finger domains (ZFPs) (Figure 6.3). Three to six ZFPs make up a typical Cys2-His2 zinc finger protein, and each ZFP can recognize a target DNA sequence that is three base pairs long. The non-specific nuclease domain FokI is associated with the zinc finger protein. Upon attachment of two ZFNs with different selected DNA sequences, the oppositely directional arrangement can be observed

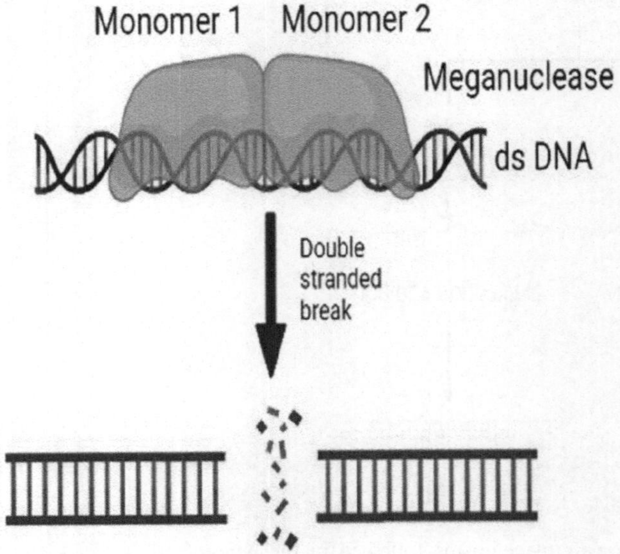

FIGURE 6.2 Schematic representation of meganucleases. Created in BioRender: https://biorender.com.

between them. The FokI nuclease domain becomes active and can form DSBs between recognized sequences when it dimerizes. The prone NHEJ pathway is typically used to repair DSBs.

The first business to create ZFNs was Sangamo Biosciences (Scott 2005). While vectors with matching DNA were widely accessible from Sigma-Aldrich, the implementation of ZFNs in large, publicly financed molecular facilities has been restricted by licensing agreements with two industries, namely Sangamo Biosciences and Dow Agro Sciences. Plant genome editing has made use of oligomerized pool engineering. Alternative strategies for producing ZFNs include bacterial one-hybrid screening of zinc finger libraries and context-dependent assembly, which employs previously selected two-finger units (Sander et al. 2011). The editing tool may be made simpler by certain advancements in ZFN production, although off-target cleavage with ZFNs is frequently caused by the random binding of the zinc finger motifs.

The first two plants in which ZFN was used for plant genome editing were tobacco and *Arabidopsis*, respectively (Wright et al. 2005; Lloyd et al. 2005). *Arabidopsis*, canola, maize, tobacco, soybean, and other plants have all had successful genome-editing reports to date (Mishra and Zhao 2018). ZFNs were able to effectively target a tobacco gene sequence called endochitinase-50 (CHN50) for herbicide resistance and site-directed DNA integration (Cai et al. 2009). *Agrobacterium* effectively delivered the donor extracellular DNA, which included both ZFNs and a contributor DNA assembly with a PAT-herbicide tolerance gene cassette fringed by brief homology having CHN50 gene homology. This resulted in a maximum of 10% precise homology-mediated transgene incorporation. In the same correspondence, Shukla et al. (2009) discovered that one of the maize

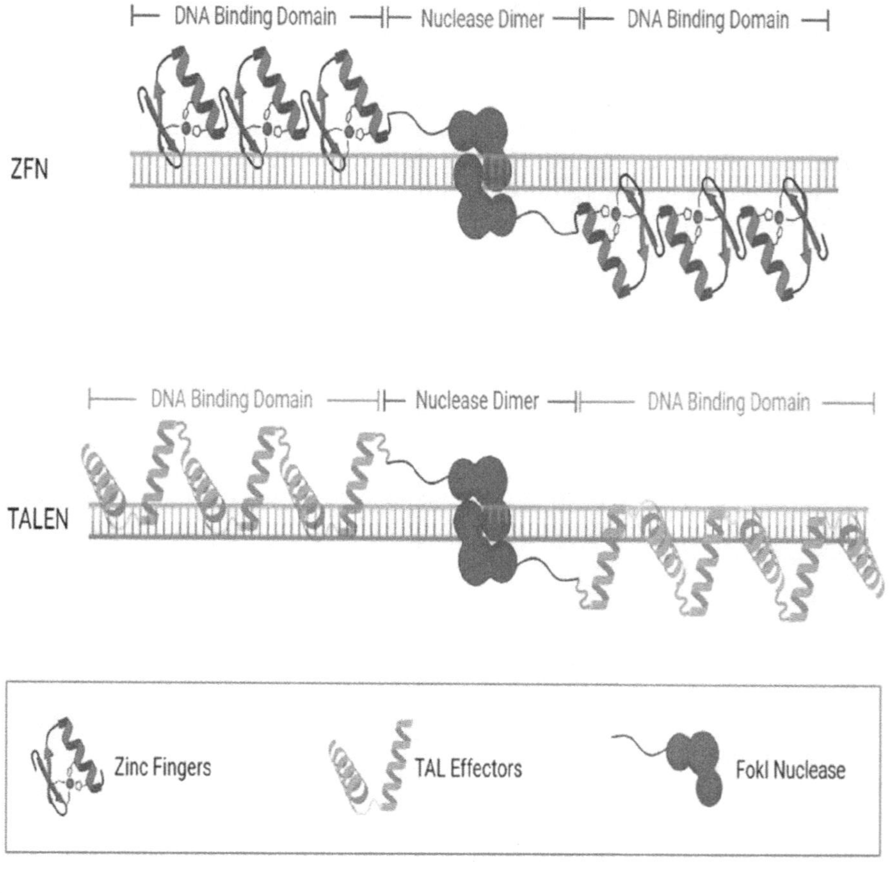

FIGURE 6.3 Schematic representation of ZFNs and TALENs. Created in BioRender: https://biorender.com.

genes involved in phytic acid production, Inositol-1,3,4,5,6-Pentakisphosphate kinase 1 (IPK1), was targeted for cleavage. In addition to a 2A 'stutter' region encircled by almost 815 bp similarity to the IPK1, the donor template DNA was co-delivered by *Agrobacterium* and contained a promoter-less PAT gene. ZFNs were effective in directing site-targeted mutagenesis toward the following genes: the *Arabidopsis* ABA-insensitive-4 (ABI4) gene (Osakabe et al. 2010), the soybean DCL4a and DCL4b gene-like genes (Curtin et al. 2012), and the *Arabidopsis* alcohol dehydrogenase and chalcone synthase genes (Zhang et al. 2010).

6.2.3 TALENs

To activate the transcription of particular target genes that promote plant disease, a class of secreted proteins known as transcription activator-like effectors (TALE) of *Xanthomonas*, a well-renowned genus of plant pathogenic bacteria, are typically injected into the phyto-cellular system via the type III secretion system. The core DNA-binding domain (with around 35 amino acid repeats) is present in every TALE, except variations of amino acids at the 12th and 13th positions, respectively. These are well denoted by repeat variable residues (RVDs). The proper combination of repeats and RVDs establishes the perfect targeted sequence of TALE in DNA. Single-nucleotide recognition is followed by each RVD (Moscou and Bogdanove 2009). The DNA-binding domain of TALE and the nuclease domain of FokI combine to form the chimeric protein known as TALEN. Similar to ZFNs, two TALENs work together to cut a desired area of DNA. The NHEJ or HDR pathway is subsequently used to fix the DSBs. The efficiency of TALENs is way more intense than that of ZFNs when the effects on off-targets are considered (Figure 6.3).

Li et al. (2011) reported the first effective application of TALENs for gene editing in yeast. As of right now, practically all significant model and crop plants, including rice, wheat, maize, tomato, potato, barley, *Arabidopsis*, tobacco, and many more, have been successfully edited genomes using TALENs (Mishra and Zhao 2018). Acetolactate synthase (ALS), the target gene, was mutated in tobacco (Zhang et al. 2013) and potatoes (Nicolia et al. 2015) using the TALEN system. To produce persistent and heritable mutations, TALENs have been used to modify the glossy2 (gl2) and the PROCERA gene of maize and tomato, respectively (Lor et al. 2014; Char et al. 2015). The development of aroma (Shan et al. 2015), bacterial blight, and blast resistance (Li et al. 2012; Wang et al. 2017) in rice – all are examples of the effective application of TALEN technology. These studies show how crucial the TALEN system is for enhancing crop characteristics.

6.2.4 CRISPR/Cas System

The most recently created genome-modifying tool is the type II CRISPR/Cas9 system (Figure 6.4), discovered from *Streptococcus pyogenes*, which is well-liked for its efficiency, friendliness, ease of use, and simplicity. This is, basically, a bacterial defensive system against bacteriophages. The origins of CRISPR started in the 1980s when Ishino and colleagues were working on the iap gene when they found 29-nucleotide repeats in the genome of *Escherichia coli*. Following the first discoveries that sparked interest in CRISPR research, the pace of the field picked up speed as several papers elucidated various facets of the CRISPR system. The major three components of this system are Cas9 endonuclease, CRISPR RNA (crRNA), and trans-activating crRNA (tracrRNA), respectively. The functionality of Cas9 endonuclease relies on the fragmentation of the phage DNA to be incorporated within the CRISPR array as a spacer. That array is then used to transcribe crRNA and a corresponding tracrRNA, which combine to produce a dsRNA structure that attracts Cas proteins for the process of cleavage. Protospacer-adjacent motif or protospacer-adjacent motif (PAM) sequence is the prime prerequisite for target DNA binding and cleavage. This sequence (5′-NGG-3′) can be detected downstream of the specific target DNA. On the contrary, Cas9 endonuclease has two grooves namely, NUC and REC lobe for binding nucleic acid. The NUC lobe is made up of three components: PAM-interacting domains, HNH, and RuvC. In contrast, the latter is a functional

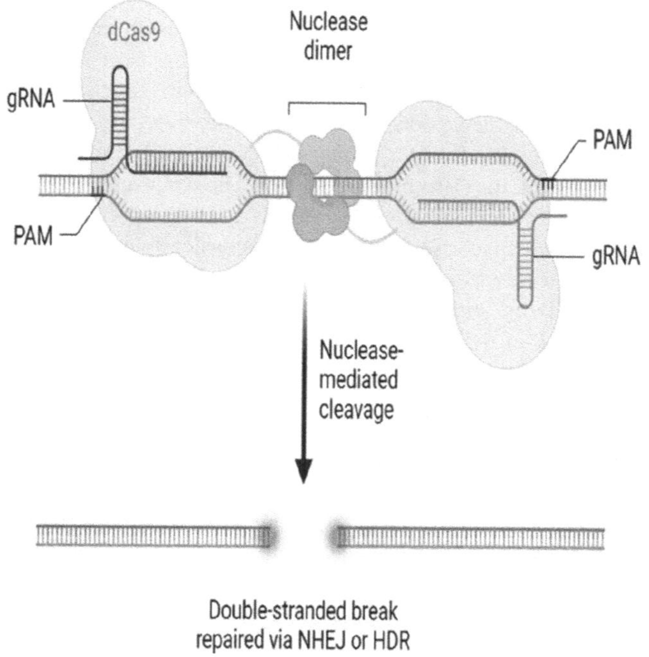

FIGURE 6.4 Schematic representation of CRISPR/Cas9. Created in BioRender: https://biorender.com.

domain exclusive to Cas9 (Nishimasu et al. 2014). RuvC and HNH's homology aids in the prediction of their nuclease domain structures. When sgRNA gets into touch with the Cas9 enzyme at its REC lobe, the enzyme becomes active. The RNA–DNA hybrid is subsequently cut by the HNH nuclease domain, and the Cas9-sgRNA complex looks for PAM sites in the double-stranded DNA while RuvC cleaves the alternative strand. After that, the highly accurate HDR pathways or the prone-to-mistake NHEJ pathways are used to repair the DSBs. While HDR uses gene replacement or insertion techniques to more accurately repair DSBs, NHEJ typically introduces indel mutations (Shukla et al. 2009).

This advanced system can be constructed without the need for intricate protein engineering procedures, in contrast to ZFNs and TALENs. By simultaneously introducing DSBs at various loci, the CRISPR/Cas system also enables simultaneous editing of several genes. These benefits have enabled the system to surpass its forerunners to gain widespread recognition within the scientific community. *A. thaliana* and *N. benthamiana*, as well as the rice crop, were the subjects of the first set of investigations that showed how the CRISPR/Cas system might be applied to plants. Following that, a number of publications were released showcasing the CRISPR system's potential in numerous crops, including tomato, wheat, rice, and maize. In rice, genome modification via this tool has been shown to increase grain weight, improve disease resistance, and increase grain tolerance to herbicides (Mishra and Zhao 2018).

6.2.4.1 Discovery and Insights

The discovery of repetitive DNA in *Escherichia coli* marked the beginning of the CRISPR/Cas9 system investigation, although Barrangou and colleagues later verified their function. They ward against viruses that invade bacteria or archaea. These sections match the bacteriophage's unicellular bacteria's or archaea's cellular mechanism. It is primarily composed of two components. The former consists of regularly spaced, palindromic short DNA sequences (20–40 bp long), which are identical. Another hyper-variably spaced DNA is the second component, which is made up of individual segments that are exactly matched to the viral DNA, particularly that of bacteriophages, also known as protospacers. Numerous other genes known as 'Cas genes' are also linked to CRISPR (Hsu et al. 2014). Cas gene-produced Cas proteins generally act as nucleases or helicases for either cleaving or unwinding DNA. They belong to the class 1 and class 2 major classes of CRISPR systems, which are distinguished by distinctions in their constituent parts. In the latter systems (consisting of types II, V, and VI), cleavage of the invasive genetic material is mediated by a single RNA-guided endonuclease (e.g., Cas9 in type II). In class 1 systems (i.e. types I, III, and IV), cleavage of the target needs a complex of several proteins (Brouns et al. 2008). The bacterial system transcribes as well as translates Cas proteins and repeats DNA to make crRNAs when it recognizes foreign DNA. The 5′ end of the crRNA consists of a spacer sequence, which is basically a smaller RNA segment outside the genetic element. On the other side, i.e. in the 3′ end, a portion of the CRISPR repeat sequence can be found. After hybridizing with the crRNA, it is adjoined with Cas proteins, which subsequently break down the DNA of the bacteriophage. Thus, the infection has basically halted before it ever begins. Both the base sequence and its cleavage are recognized by crRNA and tracer RNA, and more cleavage is caused by Cas. The mature crRNA-effector complex detects DNA targets to remove similar sequences in alien nucleic acids. This is a novel characteristic through which the entire system of CRISPR/Cas can be detected. Notably, in most CRISPR/Cas systems, target DNA selection and degradation are critically dependent on a crRNA-juxtaposed 2–5 bp long conserved sequence, known as PAM. Upon lacking the spacer DNA possessing a sequence similarity with that of the bacteriophage, a cell forms class 1 Cas protein. This protein digests the DNA and inserts the desirable copy into the CRISPR system. Thus, the spacer serves as a record of all previous infections to prevent new infections in the cell (Bhaya et al. 2011).

Cas proteins and CRISPR repeat-spacer arrays are components of the system. Based on their Cas genes and the kind of interference complex, CRISPR/Cas systems are categorized into two groups, which are further subdivided into six varieties. While class 2 CRISPR/Cas systems also employ complicated single-effector C proteins to interfere, class 1 CRISPR/Cas systems use interference multi-cases of proteins. RNA-led DNA interference prepares the basic foundation of this innovative tool. The Cas9 protein is gaining prominence in the gene modification field among a wide range of organisms due to the single protein requirement, especially in the type II among the class 2 system. This feature provides an easier way of manipulation than that of the other one. Sequence-specific gene editing has become simpler with the advent of Cas9 as a tool since it is an RNA endonuclease that can be readily trained to target new places by changing its guide RNA sequence. Basically, sgRNA is made up of two components: crRNA, which binds to a particular target, and tracrRNA, or trans-activating crRNA, which attaches to Cas9 (Doudna and Charpentier 2014).

DNA cleavage and DNA repair are the two processes that make up Cas9-mediated editing. When Cas9 is administered to a particular genomic region under gRNA (guide RNA) in the presence of PAM, it causes a DSB that initiates cellular repair processes through HDR and NHEJ (Sar et al. 2022). A random indel mutation at the DSB site in NHEJ might result in gene knockout if it modifies the gene's reading frame or if it affects a crucial area of the encoded protein. In contrast, homologous recombination driven by a contributor DNA template can be utilized in HDR to produce a desirable sequence alteration or replacement at the DSB location, leading to mutagenesis, insertion, deletion, or correction in the genomic sequences. This system acts as a potent tool for sequence-dependent gene modifications, which include knocking in, knocking out of genes, or gene-specific repairing and mutagenesis.

6.3 IMPLEMENTATIONS OF CRISPR TOOLS FOR CROP ADVANCEMENT

Genome editing has emerged as a very crucial tool for achieving a betterment in the quality, yield, and climate-resilience in a crop keeping the factors of sustainable agriculture in mind. This section discusses the use of CRISPR/Cas systems to boost food production and address malnutrition-related problems. Several approaches to crop enhancement with CRISPR-based gene-editing technologies will also be discussed in this section. In short, the safe use of this novel technology in crop development can assist researchers in creating climate-smart crop types.

6.3.1 Development of Biotic Stress Resistance

Herbivores like insects also directly harm plants and spread disease, resulting in a considerable reduction in output. Plant pathogens, including bacteria, fungi, and viruses, are responsible for a variety of plant illnesses. The shifting environmental conditions pose a negative impact on the crop's ability to withstand stresses. It is believed that 20–40% of crop losses are directly caused by biotic stressors. An array of articles was found claiming the possible and effective implications of CRISPR technology against biotic factors of crop loss, but to demonstrate the CRISPR/Cas system's potential for accomplishing SDG2, a few genome alterations done to crops, especially food crops, were critically examined in this section.

Using breeding or transgenic technologies to introduce dominant resistance (R) genes is one of the main strategies for crop protection. R genes are often known for their target specificity and avirulence. Because of this, R gene-mediated techniques may not always provide long-term resistance with a wide-spectrum resistance, which eventually leads to encouraging the target pathogen to evolve resistance. Another tactic is to aim for the susceptibility (Su) genes inside the plant genome using CRISPR/Cas. The pathogen employs the factors encoded by these genes to support infection. Through targeted mutagenesis, researchers are able to delete Su genes, reducing the vulnerability to illness. Targeting the Su genes that encode sucrose transporters is one way to increase disease resistance. Under ideal circumstances, the causal organism of bacterial blight, *Xanthomonas oryzae* pv *oryzae*, significantly reduces rice output. TALEs are released by the bacterial pathogen during infection, and they promote the expression of genes that code for several sugar transporters in the cell, which include OsSWEET11, OsSWEET13, OsSWEET14, etc. These SWEET genes are actually one of the most renowned Su genes for causing bacterial blight, as they possess a greater ability to promote bacterial growth in plant tissues. Scientists were able to generate bacterial blight-resistant lines of rice by omitting the link between SWEET genes and TALE by applying the CRISPR system (Oliva et al. 2019). To execute this objective, they altered the TALE binding sites in the SWEET promoters, mentioned above, due to which the resultant plants showed increased resistance against the pathogens.

Magnaporthe oryzae, a fungus, causes blast disease, which has an impact on rice production worldwide. Certain rice Su genes, namely OsERF92231 and Pi21.32, have been knocked out to create rice blast-resistant lines. The Mildew Locus O (MLO) gene-targeted application of CRISPR/Cas also improved wheat's ability to withstand powdery mildew. Furthermore, to protect plants from viral infections, 33 genes linked to plant Su have been modified. A serious danger to rice production is the tungro disease, which is cumulatively caused by both the rice tungro bacilliform virus and the rice tungro spherical virus. The translation initiation factor 4 gamma gene (eIF4G) controls resistance against this disease. Tungro virus-resistant rice lines were developed via CRISPR/Casing eIF4G. Eliminating Su loci becomes more challenging in polyploid crops like wheat; however, CRISPR/Cas may be used to simultaneously target several homologs, or copies of a gene found on different copies of the genome. In fact, employing CRISPR/Cas to knock out the Su gene has enhanced wheat resilience to several diseases. For instance, CRISPR/Cas9 was used to simultaneously target all three homologs of MLO33 to generate powdery mildew-resistant wheat lines. Interestingly, the resistant plants also possessed well-developed yield-attributing characteristics, showing that using CRISPR technology to create disease-resistant plants might increase output and help accomplish the objective of zero hunger.

Additionally, researchers have employed this system for modifying pathogen genes encouraging disease development in the host. To improve rice's resistance to the targeted pathogen, for example, CRISPR/Cas has been used to knock off several fungal genes. Those include RSY1,36,

ALB1 (ALBINA 1), UvSLT2,37, and USTA. This has stopped the spread of false smut disease and rice blast. Another way to generate virus resistance is to directly target the viral genome using CRISPR/Cas systems. Different DNA or RNA genomes of viruses are targeted by various Cas variants. For instance, wheat dwarf virus resistance was developed by preventing the replication process of geminivirus in barley (*Hordeum vulgare*), notably through CRISPR-associated tools (Kis et al. 2019). Furthermore, gene alteration or disruption of the microbial genome through implicating CRISPR/Cas developed resistance to various diseases in several fruits and vegetable crops like cucumber (*Cucumis sativus*), tomato (*Solanum lycopersicum*), watermelon (*Citrullus lanatus*), and many more (Table 6.1).

TABLE 6.1
Development of CRISPR-Mediated Disease-Resistant Food Crops

Food Crop Species	Biotic Stress	Targeted Gene(s)	Gene Function	Transgene-Free Plants
Barley	DNA viral disease	MP, CP, Rep/Rep, IR/virus genome	Spreading of virus	No
Cucumber	RNA viral disease	*elf4E*/host S-gene	Disease transmission	Yes
	Powdery mildew	*TaEDR1*/host S-gene		No
Rice	Bacterial blight	*Xa13*/host S-gene		Yes
		OsSWEET13/host S-gene		No
		OsSWEET14/host S-gene		Yes
	Blast	*OsERF922*/host S-gene		Yes
		Pi21/host S-gene		Yes
		ALB1, *RSY1*/fungal gene	Facilitate pathogen growth	Yes
	False smut	*USTA*, *UvSLT2*/fungal gene		Yes
	Rice tungro disease	*eIF4G*	Disease transmission	Yes
Soybean	Soybean mosaic virus	*GmF3H1*, *GmF3H2*, and *GmFNSII-1*	Biosynthesis of isoflavonoid	Yes
Tomato	Bacterial speck	*SlDMR6-1*/host S-gene	Disease transmission	No
	Phytophthora blight	*SlDMR6-1*/host S-gene		No
	Bacterial spot	*SlDMR6-1*/host S-gene		No
	Bacterial speck	*SlJAZ2*/host S-gene		–
	Powdery mildew	*SlMlo1*/host S-gene		Yes
		PMR4/host S-gene		No
	Botrytis cinerea	*SlMYC2*/host S-gene	Disease inhibition	Yes
	TYLCV	CP, Rep/virus genome	Spreading of virus	No
	PVX, TMV	DCL2/virus genome		No
	Pepper mottle virus	*eIF4E1*/host S-gene	Viral infection	Yes
	Yellow leaf curl virus	IR and CP/virus genome		Yes
Wheat	Powdery mildew	*TaMLO*/host S-gene	Disease facilitation	Yes

Source: Adapted from Ahmad et al. (2021b).
PVX, potato virus X; TMV, tomato mosaic virus; TYLCV, tomato yellow leaf curl virus.

Enhancing disease resistance and hence lowering crop losses can be achieved by engineering biotic stress tolerance. For instance, a well-adaptable and long-term resistance may be offered by the CRISPR-directed viral genetic system. Targeting Su genes, however, requires delayed detection of Su genes specified to the host-pathogen relationship, although, genomic advancements may accelerate the entire process. Notable to mention, an array of further research will be needed to minimize growth-defense understanding for acquired resistance against diseases.

6.3.2 Introduction of Abiotic Stress Tolerance

Plants are exposed to a wide range of abiotic stresses, in addition to biotic ones, including drought, heavy metals, salt, and temperature. These can obstruct agriculture development, which eventually, sometimes causes significantly disastrous output depletion at a global level. Plant genes, designated as sensitivity (Se) genes, amp up the harmful effects of abiotic stressors. To boost the tolerance against several abiotic stressors in a variety of plant species, genome-modifying techniques, especially CRISPR/Cas, have been employed to disrupt these particular genes (Ahmad et al. 2021a). For instance, rice's ability to withstand cold is influenced by the B-Amylase (BMY) gene. They control the breakdown of starch and the buildup of maltose to prevent cold stress. When OsMYB30 and OsJAZ9, two cold-responsive R2R3-type MYB transcription factors, bind to the promoter regions of BMY genes, BMY function is compromised. Scientists disrupted OsMYB30 using CRISPR/Cas to create rice lines that could withstand low temperatures (Zeng et al. 2020). CRISPR/Cas9 method was also employed to alter the drought and salt tolerance (DST) gene in MTU1010, a cultivar that belongs to Indica rice. To enhance the efficiency of light utilization and water retention, mutant lines containing a large range of deletion (around 366 bp) in the exons have enlarged leaf areas and decreased stomatal density, making them resistant to salt, osmotic, and drought stressors. Heavy metal pollution in soils, such as cadmium and arsenic, may be extremely harmful to plants and humans alike since the metals can build up in grain. However, by implementing this technique to silence the heavy metal transporter genes, plants may be cultivated safely in polluted soil. For instance, using the system to alter the rice gene that codes for the cadmium 42 transporter gene natural resistance-associated macrophage protein 5 (OsNRAMP5) reduced the accumulated Cd amount and enhanced biological progression of the plant under circumstances of Cd adequacy (Khan et al. 2020). Researchers have improved drought resistance in maize by disrupting the Auxin-Regulated Gene Involved in Organ Size 8 (ARGOS8) using CRISPR/Cas9 (Yang et al. 2019). Orthologues of Mitogen-Activated Protein Kinase 3 (SlMAPK3) and Agamous-Like 6 (SlAGL6) in tomato, when undergoes CRISPR-oriented alteration, improved its tolerance to temperature stress, especially heat, whereas ADP-ribosylation factor 4 (SlARF4) enhanced the same for salt stresses. These CRISPR-edited plants exhibited improved agronomic traits and resistance to abiotic stresses (Bouzroud et al. 2020). Table 6.2 provides an overview of several uses of CRISPR/Cas to strengthen plant resistance to abiotic stressors. Furthermore, to the best of our ability to explore the literature, we were unable to locate any examples of the CRISPR/Cas system being used to enhance wheat's resistance to abiotic stress. Therefore, studies on CRISPR/Cas wheat genome editing to produce high-yielding, abiotic stress-tolerant wheat lines may raise food security and contribute to the goal of eradicating hunger. Herbicides are used by farmers not just to target crops directly but also to lessen weed competition for nutrients, water, and light. Herbicide-tolerant crops have therefore shown to be particularly beneficial for contemporary agriculture. In the past, considerable mutagenesis or transgenic techniques were needed to create a new herbicide-tolerant cultivar. In recent times, acetolactate synthase genes' (ZmALS1 and ZmALS2, respectively) modification were created through CRISPR-sourced base modifications and prime alteration, which eventually enhanced herbicide withstanding ability in maize (Nuccio et al. 2021). Herbicide resistance in rice was demonstrated by HDR-based gene repairing, specifically of the OsALS gene. This process was facilitated by the integration of Cas9 with VirD2. Herbicide resistance in rice has also grown as a result of precision OsALS gene nicking along with HDR-intended repair sponsored through prime

TABLE 6.2
Generation of Abiotic Stress-Resistant Food Crops via CRISPR/Cas

Food Crop Species	Abiotic Stress	Targeted Gene(s)	Transgene-Free Plants	Outcomes
Chickpea	Drought	*Coumarate ligase (4CL), Reveille 7 (RVE7)*	Yes	Enhanced drought tolerance
Maize	Drought	*ARGOS8*	No	
	Herbicide	*ZmALS1, ZmALS2*	Yes	Herbicide resistance
		ZmALS1, ZmALS2	Yes	Sulfonylurea herbicide-resistant plant
		ZmALS2	No	Herbicide resistance
Oilseed rape	Herbicide	*BnALS1*	Yes	
Rice	Cold	*TIFY1a*	No	Cold adaptation
		TIFY1b	No	
		OsMYB30	No	Improved cold tolerance
	Drought	*OsDST*	No	Moderate level tolerance of water stress
		OsEBP89	No	Enhanced tolerance to drought and direct seeding on wet land
		OsSAPK2	No	Enhanced drought tolerance
		OsSRL1, OsSRL2	Yes	
	Salinity	*OsRR22*	Yes	Enhanced salinity tolerance
		OsDST	No	
		qSOR1	No	Better root structure for saline soils
	Heavy metal	*OsNramp5*	No	Decreased Cd accumulation
		OsLCT1, OsNramp5	Yes	
		OsNramp5	Yes	
	Herbicide	*OsALS*	Yes	Herbicide tolerance
		OsALS (novel allele G628W)	Yes	
		OsALS	Yes	
		OsALS1	No	
		OsALS	Yes	
		OsALS	No	
Tomato	Heat	*SlMAPK3*	No	Enhanced heat tolerance
		SlAGL6	No	
	Salinity	*SlARF4*	No	Enhanced salinity tolerance
	Herbicide	*SlALS*	Yes	Herbicide tolerance development

Source: Adapted from Ahmad et al. (2021b).

editing (Ali et al. 2020). Furthermore, by modifying ALS1 through CRISPR-based gene editing, tomatoes and oilseed rape (*Brassica napus*) were able to withstand herbicides more effectively.

Furthermore, agricultural yields are frequently reduced by natural calamities like floods. Despite being seen as a crop that can withstand flooding, rice is significantly impacted by floods. No instances of using CRISPR-based editing to create rice or other crop types that are flood-tolerant were found. Submergence-tolerant rice varieties were produced, nonetheless, by introducing Submergence 1 A-1 (SUB1A-1) into HYVs of rice via MAB i.e. marker-assisted breeding. Thus, CRISPR-originated gene knocking in approach was suggested to introduce these genes – trehalose-6-phosphate phosphatase-7 (TPP7), SUB1A-1, semi-dwarfing 1 (SD1), and shikimate kinase 1 and 2 (SK1 and SK2), and – to enhance crop submergence tolerance.

6.3.3 Grain Yield and Plant Architecture

The QTLs regulate grain production, a complex trait, comprising several characters such as grain number in each panicle, number of panicles in each plant, florets count per panicle, and test weight. As of right now, 19 rice QTLs controlling the size of grains have been cloned and studied. A small number of important QTLs, including OsGS3 (encoding the heterotrimeric G-protein's Gγ-subunit) and GL3.1 (encoding a candidate serine/threonine phosphatase OsPPKL1 with a kelch-like repeat motif), adversely affect the grain production. Grain size was enhanced by CRISPR-originated alteration of OsGS3 and OsGL3.1. These eventually raised the test weight and total plant-wise production (Yuyu et al. 2020). Multiplex gene remodeling of three genes (i.e. GS3, GW2, and Gn1a) also improved the entire yield when comparing modified cultivated plants to the wild ones. In another research work, panicle length regulating gene OsPIN5b and drought resistivity maintaining genes GS3 and OsMYB30 were simultaneously knocked down using this tool to develop HYV of rice with drought resistance.

Furthermore, by deleting specific abominable genes or their regulators of transcription factors (TFs) regulators, this technique has been utilized to upgrade plant development. Several plant architecture-enhancing characteristics consist of the entire height of the plant, total flag leaf area, leaf angle, leaf length, average distance between nodes, tiller numbers, tiller angle, structure of inflorescence as well as root, and its growth. Additionally, this system was involved in disrupting grain size-associated genes. By modifying SLR1, which codes for a DELLA protein, dwarf lines of rice have been created (Yuyu et al. 2020). The selected interference in the activity of carotenoid cleavage dioxygenase 7 (CCD7) by CRISPR/Cas9 caused a significant hike in tillering and depletion in height. However, this could be avoided by providing the artificial strigolactone (SL) counterpart GR24 (Butt et al. 2018). The practice of multiplex gene modification has raised a double mutant of rice by altering MIR396e and MIR396f, well-known microRNA genes. This mutant exhibited higher levels of mevalonic acid, which in turn led to enhanced grain output and flag leaf area (Yuyu et al. 2020). Furthermore, negative regulation of root development, tiller number, and higher rice yield is observed when OsLOGL5, a cytokinin-activation enzyme, is edited using CRISPR-based technology (Miao et al. 2020). Pod cracking is a significant problem that jeopardizes crop productivity in grain crops. CRISPR/Cas9 was recently used to modify qSH1, producing rice lines that are resistant to breaking. Higher yields were seen in CRISPR-edited homozygous lines-derived hybrids than that of the parents as well as wild genotypes (Sheng et al. 2020). Disrupting the genes that adversely affect yield and yield-related features has also dramatically enhanced yield in various grain, vegetable, and fruit crops (Table 6.3).

6.3.4 Photosynthetic Improvement

For many years, researchers studying plants have been trying to increase agricultural productivity through photosynthesis by increasing the area of flag leaves, minimizing photorespiration, boosting flux through the Calvin cycle, and improving the ability to absorb CO_2 (Wang et al. 2020).

TABLE 6.3
An Overview of the CRISPR/Cas System's Impact on Improving Grain Yield, Quality, and Plant Architecture

Applications	Food Crop Species	Traits	Targeted Gene(s)	Gene Function	Transgene-Free Plant	Outcomes
Yield improvement	Barley	Development and yield	*HvCKX1*	Cytokinin metabolism	No	Increased grain yield
	Rice	Grain size	*OsGS3*	Negative regulation on the size of the grains	Yes	Increased grain size
		Grain width and weight	*OsGW2*, *OsGW5*, *TGW6*	Negative regulation on grain width and weight	Yes	Increased grain width and weight
		Grain number	*OsGn1a*	Adverse regulation on grain number	Yes	Increased grain number
		Grain yield	*OsDEP1*	Inflorescence architecture and plant height	Yes	Increased yield
		Plant growth and grain yield	*PYL1–PYL6* and *PYL12* *(gp-1)*, *PYL7–PYL11* and *PYL13* *(gp-2)*	Improved grain productivity and growth	No	
		Grain yield	*OsPIN5b*	Negatively operates yield parameters	Yes	
			OsLOGL5	Negative regulation on the growth of root as well as plant and the number of emerged tillers	No	
		Tiller numbers	*OsAAP3*	Negatively regulates the number of plant tillers	No	Improved production through an enhanced number of tillers
	Tomato	Fruit size	*SlCLV3/ Promoter*	Signaling peptide gene	Yes	Enhanced fruit size
			SlENO	Elevated AP2/ERF transcription factor to regulate floral meristem activity	No	
	Wheat	Grain size	*TaGS5-3A*	Negative regulation on the grain size	Yes	Larger kernel size and yield

(Continued)

TABLE 6.3 (*Continued*)
An Overview of the CRISPR/Cas System's Impact on Improving Grain Yield, Quality, and Plant Architecture

Applications	Food Crop Species	Traits	Targeted Gene(s)	Gene Function	Transgene-Free Plant	Outcomes
		Grain number	*TaCKX2-D1*	Negative control on spikelet-wise grain count	No	Boost grain number per spikelet
		Grain yield	*TaDEP1, TaGASR7, TaLOX2, TaNAC2, TaPIN1, TaGW2, TaGLW7, TaGW2, and TaGW8*	Inflorescence architecture and plant height regulator, lipoxygenase, negative regulator of grain weight	Yes	Boost grain yield
Quality improvement	Maize	Cooking/ baking quality	*ZmGBSSI (Wx gene)*	Biosynthesis of amylose	Yes	Waxy corn hybrids
	Potato	Cooking and eating quality	*StGBSSI, StSBE1, StSBE2*		Yes	Waxy potato
	Rice	Cooking and eating quality	*OsAAP6, OsAAP10*	Amino acid transporter controlling grain protein content (GPC)	No	Increasing eating and cooking quality (ECQ), reducing GPC
			OsAAP10	AGPase for starch biosynthesis	No	Starch depletion in leaves
			OsWaxy	GBSS for synthesis of amylose	Yes	Decrease in amylose content (glutinous rice production)
			OsBEI and *OsBEIIb*	Starch branching enzyme	Yes	High amylose content
			ISA1	Starch (isoamylase-type) debranching enzymes	Yes	Reduced amylose content, increased total soluble sugar (TSS)
Quality improvement			*BADH2*	Betaine aldehyde dehydrogenase (fragrant rice)	Yes	Enhanced fragrance

(Continued)

TABLE 6.3 (*Continued*)
An Overview of the CRISPR/Cas System's Impact on Improving Grain Yield, Quality, and Plant Architecture

Applications	Food Crop Species	Traits	Targeted Gene(s)	Gene Function	Transgene-Free Plant	Outcomes
			OsGBSSI	Amylose synthesis	No	Low amylose content
			OsGBSSI/ Promoter		–	
			OsGBSSI/ Promoter and *5′UTR*		–	
			OsPLDα1		–	
		Nutritional quality	Insertion of 5.2 kb carotenoid cassette	β-carotene content	Yes	Increased β-carotene content
			OsFAD2-1	Oleic acid contents	–	Increased oleic acid proportion
			OsPLDα1	Phytic acid synthesis	–	lower accumulation of phytic acid
	Sweet potato	Starch quality	*IbGBSSI*	Starch quality	No	Improved starch quality
	Wheat	Allergies (bakers' asthma)	α-amylase/ trypsin inhibitors	Wheat allergies (bakers' asthma)	No	Depleted number of potential allergens
Plant architecture	Rice	Root architecture	*qSOR1*	Angle of root growth	No	Better root structure for saline soils
		Plant height	*SLR1*	Control plant height	No	Dwarf plants
		Grain size, plant architecture	*MIR396e and MIR396f*	Negative regulator of mevalonic acid	No	Increased grain size and temper shoot architecture

Source: Adapted from Ahmad et al. (2021b).

Gene editing with CRISPR has also been used to enhance photosynthesis because of its critical role in biomass production, grain and fruit output, and plant growth and development. For instance, a negative Regulator of Photosynthesis 1 (NRP1) was edited using CRISPR to enhance the photosynthetic efficiency, along with the yield of grain and biomass of the entire rice plant (Voesenek and Bailey-Serres 2015). Ribulose-1,5-bisphosphate carboxylase/oxygenase (Rubisco) is a crucial enzyme for photosynthesis as it significantly fixes CO_2, which makes it an important target for improving photosynthesis. However, given the current environmental circumstances, Rubisco's meager catalytic rate (K_{cat}), weak affinity (K_c), and specificity toward CO_2 ($S_{C/O}$) restrict its action. It is made up of two subunits: small (14–15 kDa) and large (52 kDa), which are encoded by a multigene family (RbcS) in the nucleus and a single gene (RbcL) in the chloroplast genome, respectively. Rubisco frequently shows a lower K_{cat} in C_3 plants as opposed to C_4 plants. This CRISPR system has provided scientists the opportunity to substitute the RbcS gene of rice with that of sorghum (*Sorghum bicolor*). The resultant rice plants demonstrated higher crop output in addition to an enhanced rate of photosynthetic ability comparable to that of C_4 plants. Plants with homozygous mutants had increased K_{cat}, increased K_c, and decreased $S_{C/O}$. Other facets of C_4 metabolism inside the C_3 crops have been engineered by numerous researchers.

Researchers have developed both polyploid and diploid agricultural plants' photosynthetic mechanisms by applying CRISPR-oriented editing technology. In tobacco (*Nicotiana tabacum*), multiple homologs of rbcS were deleted simultaneously, and the resulting mutants exhibited a higher rate of photosynthetic activity than the wild-type plants (Lacchini et al. 2020). Therefore, it is possible to design diploid as well as polyploid animals with extremely effective photosynthetic networks using the CRISPR-based gene-editing technology.

6.3.5 Nutrient Use Efficiency (NUE) Enhancement

The extensive application of fertilizers during the era of the green revolution raised yields, but it also inflated costs along with environmental nuisance caused by fertilizer runoff. Increased nitrogen use efficiency in crops will result in lower fertilizer usage, higher farmer income, and less pollution. Greater nutrient use efficiency (NUE) can result from enhanced plant growth and productivity, improved absorption of micro- and macronutrients from soil, and more efficient nutrient consumption and transport toward plant systems. Certain genes cause plants to absorb nutrients less efficiently, which has an adverse effect on NUE. Scientists have modified crops to increase NUE, which has boosted nutrient uptake, growth, and production (Tiwari et al. 2020). For example, the CRISPR-based cytosine base-editing system (CBE) was used to alter the rice nitrate transporter gene NRT1.1B. The resultant mutant exhibited better NUE in comparison to rice of the natural type (Yuyu et al. 2020). The amino acid Thr at position 327 was altered to Met by the CBE, resulting in a C $\rightarrow$ T replacement in NRT1.1B. Furthermore, NUE was greater in mutant rice plants with NRT1.1B insertions/deletions than in wild-type rice plants. It is possible that CRISPR-based technologies may be used to increase the NUE of more fruit, vegetable, and grain crops, hence increasing crop yield.

6.3.6 Grain Quality Enhancement

The nutritional value, physical attributes, milling characteristics, and cooking and eating properties of grains are used to gauge their overall quality. The Waxy gene-coding enzyme granule-bound starch synthase 1 (GBSS1) controls the manufacture of amylose, which is what dictates how well a grain cooks and tastes when eaten. By using CRISPR/Cas9 to disable GBSS1, researchers have created exceptional rice64 and maize65 lines with low amylose levels. Similarly, low amylose-containing rice lines were produced by modifying isoamylase 1 (ISA1). This gene codes for an isoamylase-type debranching enzyme. On the contrary, high amylose-packed rice lines were produced by the alteration process of starch branching enzymes, namely SBEI and SBEIIb. The biofortification of grains and other crops is an additional strategy for enhancing human nutrition. For instance, the CRISPR/Cas system was used for knocking in the genomic cassettes of golden rice, viz. GR1 and GR2, which eventually presented golden rice enriched with carotenoids (Ahmad et al. 2021b).

6.4 REGULATIONS REGARDING CRISPR-MEDIATED CROPS

The regulation of crops with GM traits is a topic of global discussion. Regulating GM crops is currently uneven among countries, albeit. Regarding the control of GM crops, there are two main models. The first model, which is implemented in India, Europe, New Zealand, and Australia, controls the method of acquiring the harvest. According to this approach, the processes utilized in the creation of novel breeding methods' offspring will determine how they are governed. This places

GM crops under similar rigid regulations to genetically modified organisms (GMOs). The second, less restrictive regulatory criterion places more emphasis on the extreme result than the method by which gene-edited crops are developed (Das et al. 2023). The United States, Argentina, and Canada adhere to this paradigm, which is case-by-case and focuses on the originality of the characteristic in issue, regardless of the technology utilized in its production. If the final product contains no foreign DNA, this strategy allows gene-edited crops to be registered as non-GMO.

International companies that prioritize research are relocating their operations to countries like the United States that have end-product-based regulations, which is impeding the progress of gene-edited crops in the European Union due to process-based regulation. Several countries in the European Union, such as France and Sweden, have expressed their discontent with the framework of process-oriented laws. Major economies – the United States and Canada, for instance – already-treated GMOs and gene-edited crops differently, and the European Union could soon be adjusting its regulatory approach. A commission was established in China in 2016 to design regulations for the handling of GM crops; these regulations are now being discussed. Regarding research and development efforts and financial support for CRISPR initiatives, the global community anticipates a lenient stance toward genetically engineered crops, with few restrictions. These advancements indicate that GM crops have a bright future and hold promise for achieving the objective of eradicating hunger.

6.5 FUTURE OUTLOOK

Gene-editing technology has revolutionized fundamental and applied plant research. Gene-editing methods have advanced via a succession of steps, i.e. a long run from the development of mega nucleases to the advancement of CRISPR-based primary modifiers. Improved agronomic features and high nutritional value crop types have been produced by CRISPR-based precision gene-editing technologies. We propound that the widespread use of CRISPR/Cas technologies in agriculture, coupled with their significant potential for crop improvement, will enable us to meet SDG2 by 2030. Apart from utilizing the CRISPR/Cas system to enhance the quality of grain and protect crops from biotic and abiotic stressors, this technology has assisted in the rapid ideotyping of various crop varieties through the enhancement of several features like photosynthetic proficiency and the entire architecture of the plant, which ultimately led to the eradication of hunger and malnutrition. Organelles with their own genomes, such as mitochondria and chloroplasts, include numerous significant genes that adversely control the most crucial physiological processes of plants, viz. photosynthesis and respiration, apart from the nuclear genes. Since these genomes have proven difficult for OBTs to target, there is tremendous promise for creating more productive food crop variants. Targeted crops' genomes may be genetically modified using CRISPR to enhance the photosynthetic apparatus and, eventually, boost biomass and crop output. In the meantime, CRISPR/Cas technology may soon be able to transform C_3 food crops (wheat, rice, and barley) into C_4-like plants, which have greater NUE, water usage efficiency, and photosynthetic rates. Furthermore, the CRISPR/Cas technology improves the likelihood of accurately and effectively breeding crops to meet the aim of eradicating hunger.

Using CRISPR to create two-line hybrids would not only reduce time but may also result in a yield boost of multiple times. Growing a segregated F_2 population comes at a very high expense. CRISPR can sustain hybrid vigor, which will change the way hybrids evolve and their breeding programs to address this issue. Employing CRISPR as a novel breeding tool to create smart and productive crops, such as climate-resilient crops, economically significant food crops that have self-compatibility induction, and material preparation for developing species hybrids, could ultimately help us address the global food security challenge being the little time-consumer, less costly, and energy saver.

Natural selection cannot accomplish the objective of eradicating hunger. Several years are needed to acquire the desired new features. The latest discovery of CRISPR-guided emergence of specified genes, which produces highly active genes in considerably smaller durations, can overcome this. Natural selection can also be circumvented by CRISPR/Cas-dependent directed evolution. With the aid of this technique, several crucial plant genes may be evolved to maximize productivity and accomplish the objective of zero hunger (Ahmad et al. 2021b).

It might take several years to domesticate a crop through the lengthy process of traditional crop domestication. Similar to how African rice landraces were domesticated, this process can be significantly accelerated using the CRISPR-based multiplex genome editing (MGE) technology. To increase the amount of food available, it is also possible to modify orphan crops, such as cowpea (*Vigna unguiculate*), pigeon pea (*Cajanus cajan*), millet (*Panicum miliaceum*), and cassava (*Manihot esculenta*), through the use of MGE systems. By using CRISPR to domesticate these crops from scratch, they might be in a better position to help end poverty and ensure global food security.

Plant transformation is one of the main methods that make gene editing possible. Researchers still need to develop more efficient delivery mechanisms since, despite considerable advancements in CRISPR/Cas system insertion techniques, transformation efficiency remains low or non-existent in complex genomic plants viz. *Brassica*, wheat, and cotton. Furthermore, tissue culture is needed for CRISPR-based genetic modifications to regenerate plants from callus, a tedious and time-consuming process. Even using well-established techniques, it may take 6–12 months for mutant rice and tomato plants to convert and regrow. For wheat or cotton, the procedure can take many months. On the contrary, a few plant species, including the chili pepper (*Capsicum rutescens*), continue to be resistant to tissue culture methods. Consequently, the need arises for tissue culture-independent gene alteration techniques, such as those that use RNA virus technology to deliver gRNA. A variety of genetic information manipulation techniques, i.e. knock-in and knockout, gene repression or activation, epigenome engineering, base editing, nuclear rearrangements, imaging, and many more, may be applied to any type of genetic material (DNA and RNA) using CRISPR systems. These systems' adaptability broadens the range of applications for crop development. Specifically, future optimization will present a great chance to comprehend and regulate the biological processes occurring in plants by breaking down the complications of epigenomes using CRISPR-mediated techniques.

Although CRISPR/Cas is a favorite biotechnological tool among academics, there are several issues with the way its products are regulated in the market. Among the principal issues are the off-target repercussions. According to certain studies, CRISPR systems come with several effects on the off-targets that might have adverse consequences on the goods, which eventually may create safety concerns. Correspondingly, it's imperative to generate more precise and well-structured tactics, viz. impartial and biased off-target identifying techniques, gRNA engineering and their modifications, improvised forms of Cas protein, effective delivery techniques, creation of anti-CRISPR proteins, and efficacious base modifications and prime editing structures. The latter ones are relatively new and less hazardous methods with insignificant off-target impacts on the biological systems, which makes it easier to produce GM food crops without risk. This helps us reach the objective of ending world hunger. Many genes govern many features in plants, and significant and related genes result in phenotypes.

6.6 CONCLUSION

Crossbreeding, mutation breeding, and transgenic breeding are the primary methods used in modern agriculture to improve crops. The world's expanding food needs cannot be met by breeding programs since they are labor-intensive, time-consuming, and lack concentration. Marker-assisted breeding and transgenic techniques have been used to address this issue and improve crop selection efficiency. These methods produce desirable features by transforming foreign organisms into elite cultivars. These genome-editing techniques are fantastic instruments that can pinpoint the precise plant molecular pathways for crop improvement and offer quick, targeted mutagenesis. Next-generation breeding techniques have completely changed crop breeding. Because genome-editing technologies are simple, efficient, highly selective, and amenable to multiplexing, they offer several benefits over conventional agricultural procedures. It has been concluded that plant biologists can further their study in the area of crop enhancement by utilizing genetic resources and technologies in conjunction with speed breeding.

Over the past 5 years, genome editing has become a thriving field of study and technology that has transformed plant crop development and functional genomics. Because of its diversity, tolerated off-target effects, simplicity, and ease of application, CRISPR/Cas9 has surfaced as the optimistic approach. The genome modification technology has the potential to produce crop types that will benefit farmers and consumers by having new agronomic features, increased disease resistance, and better yield and quality. Targeted mutagenesis has been effectively applied to different crops and a variety of model organisms using this method. CRISPR-Cpf1 has emerged as a novel and substitute technique for genome editing in plants. It can surmount some constraints of CRISPR/Cas9, such as the need for a PAM site, hence expanding the potent applications of genome alterations in crops. However, a growing number of research studies indicate that it is a productive as well as widely utilized technique that may expedite fundamental and practical research aimed at improving crops.

REFERENCES

Ahmad, S., Shahzad, R., Jamil, S., Tabassum, J., Chaudhary, M. A. M., Atif, R. M., Iqbal, M. M., Monsur, M. B., Lv, Y., Sheng, Z., Ju, L., Wei, X., Hu, P., & Tang, S. (2021a). Regulatory aspects, risk assessment, and toxicity associated with RNAi and CRISPR methods. In K. A. Abd-Elsalam & K.-T. Lim (Eds.), *CRISPR and RNAi systems* (pp. 687–721). Elsevier.

Ahmad, S., Tang, L., Shahzad, R., Mawia, A. M., Rao, G. S., Jamil, S., Wei, C., Sheng, Z., Shao, G., Wei, X., Hu, P., Mahfouz, M. M., Hu, S., & Tang, S. (2021b). CRISPR-based crop improvements: A way forward to achieve zero hunger. *Journal of Agricultural and Food Chemistry*, *69*(30), 8307–8323.

Ali, Z., Shami, A., Sedeek, K., Kamel, R., Alhabsi, A., Tehseen, M., Hassan, N., Butt, H., Kababji, A., Hamdan, S. M., & Mahfouz, M. M. (2020). Fusion of the Cas9 endonuclease and the VirD2 relaxase facilitates homology-directed repair for precise genome engineering in rice. *Communications Biology*, *3*(1), 44.

Antunes, M. S., Smith, J. J., Jantz, D., & Medford, J. I. (2012). Targeted DNA excision in Arabidopsis by a re-engineered homing endonuclease. *BMC Biotechnology*, *12*, 1–12.

Bhaya, D., Davison, M., & Barrangou, R. (2011). CRISPR-Cas systems in bacteria and archaea: Versatile small RNAs for adaptive defense and regulation. *Annual Review of Genetics*, *45*, 273–297.

Bouzroud, S., Gasparini, K., Hu, G., Barbosa, M. A. M., Rosa, B. L., Fahr, M., Bendaou, N., Bouzayen, M., Zsögön, A., Smouni, A., & Zouine, M. (2020). Down regulation and loss of auxin response factor 4 function using CRISPR/Cas9 alters plant growth, stomatal function and improves tomato tolerance to salinity and osmotic stress. *Genes*, *11*(3), 272.

Brouns, S. J., Jore, M. M., Lundgren, M., Westra, E. R., Slijkhuis, R. J., Snijders, A. P., Dickman, M. J., Makarova, K. S., Koonin, E. V., & Van Der Oost, J. (2008). Small CRISPR RNAs guide antiviral defense in prokaryotes. *Science*, *321*(5891), 960–964.

Butt, H., Jamil, M., Wang, J. Y., Al-Babili, S., & Mahfouz, M. (2018). Engineering plant architecture via CRISPR/Cas9-mediated alteration of strigolactone biosynthesis. *BMC Plant Biology*, *18*(1), 1–9.

Cai, C. Q., Doyon, Y., Ainley, W. M., Miller, J. C., DeKelver, R. C., Moehle, E. A., Rock, J. M., Lee, Y.-L., Garrison, R., Schulenberg, L., Blue, R., Worden, A., Baker, L., Faraji, F., Zhang, L., Holmes, M. C., Rebar, E. J., Collingwood, T. N., Rubin-Wilson, B., … Petolino, J. F. (2009). Targeted transgene integration in plant cells using designed zinc finger nucleases. *Plant Molecular Biology*, *69*, 699–709.

Char, S. N., Unger-Wallace, E., Frame, B., Briggs, S. A., Main, M., Spalding, M. H., Vollbrecht, E., Wang, K., & Yang, B. (2015). Heritable site-specific mutagenesis using TALENs in maize. *Plant Biotechnology Journal*, *13*(7), 1002–1010.

Chen, Y. Y., Zhu, A. K., Xue, P., Wen, X. X., Cao, Y. R., Wang, B. F., Zhang, Y., Shah, L., Cheng, S. H., Cao, L. Y., & Zhang, Y. X. (2020). Effects of GS3 and GL3.1 for grain size editing by CRISPR/Cas9 in rice. *Rice Science*, *27*(5), 405–413.

Curtin, S. J., Voytas, D. F., & Stupar, R. M. (2012). Genome engineering of crops with designer nucleases. *The Plant Genome*, *5*(2), 42.

Das, A., Mahanta, M., Pramanik, B., & Purkaystha, S. (2023). Genetically modified crops and crop species adapted to global warming in dry regions. In A. Naorem & D. Machiwal (Eds.), *Enhancing resilience of dryland agriculture under changing climate* (pp. 385–409). Springer.

Djukanovic, V., Smith, J., Lowe, K., Yang, M., Gao, H., Jones, S., Nicholson, M. G., West, A., Lape, J., Bidney, D., Falco, S. C., Jantz, D., & Alexander Lyznik, L. (2013). Male-sterile maize plants produced by targeted mutagenesis of the cytochrome P450-like gene (MS26) using a re-designed I–C reI homing endonuclease. *The Plant Journal*, *76*(5), 888–899.

Doudna, J. A., & Charpentier, E. (2014). The new frontier of genome engineering with CRISPR-Cas9. *Science*, *346*(6213), 1258096.

Govindaraj, M., Vetriventhan, M., & Srinivasan, M. (2015). Importance of genetic diversity assessment in crop plants and its recent advances: An overview of its analytical perspectives. *Genetics Research International*, 2015, 1–14.

Hsu, P. D., Lander, E. S., & Zhang, F. (2014). Development and applications of CRISPR-Cas9 for genome engineering. *Cell*, *157*(6), 1262–1278.

Khan, H., McDonald, M. C., Williams, S. J., & Solomon, P. S. (2020). Assessing the efficacy of CRISPR/Cas9 genome editing in the wheat pathogen Parastagonspora nodorum. *Fungal Biology and Biotechnology*, *7*, 1–8.

Kis, A., Hamar, É., Tholt, G., Bán, R., & Havelda, Z. (2019). Creating highly efficient resistance against wheat dwarf virus in barley by employing CRISPR/Cas9 system. *Plant Biotechnology Journal*, *17*(6), 1004.

Lacchini, E., Kiegle, E., Castellani, M., Adam, H., Jouannic, S., Gregis, V., & Kater, M. M. (2020). CRISPR-mediated accelerated domestication of African rice landraces. *PLoS One*, *15*(3), e0229782.

Li, T., Huang, S., Jiang, W. Z., Wright, D., Spalding, M. H., Weeks, D. P., & Yang, B. (2011). TAL nucleases (TALNs): Hybrid proteins composed of TAL effectors and FokI DNA-cleavage domain. *Nucleic Acids Research*, *39*(1), 359–372.

Li, T., Liu, B., Spalding, M. H., Weeks, D. P., & Yang, B. (2012). High-efficiency TALEN-based gene editing produces disease-resistant rice. *Nature Biotechnology*, *30*(5), 390–392.

Lloyd, A., Plaisier, C. L., Carroll, D., & Drews, G. N. (2005). Targeted mutagenesis using zinc-finger nucleases in Arabidopsis. *Proceedings of the National Academy of Sciences*, *102*(6), 2232–2237.

Lor, V. S., Starker, C. G., Voytas, D. F., Weiss, D., & Olszewski, N. E. (2014). Targeted mutagenesis of the tomato PROCERA gene using transcription activator-like effector nucleases. *Plant Physiology*, *166*(3), 1288–1291.

Lusser, M., Parisi, C., Plan, D., & Rodríguez-Cerezo, E. (2012). Deployment of new biotechnologies in plant breeding. *Nature Biotechnology*, *30*(3), 231–239.

Miao, C., Wang, D., He, R., Liu, S., & Zhu, J. K. (2020). Mutations in MIR 396e and MIR 396f increase grain size and modulate shoot architecture in rice. *Plant Biotechnology Journal*, *18*(2), 491–501.

Mishra, R., & Zhao, K. (2018). Genome editing technologies and their applications in crop improvement. *Plant Biotechnology Reports*, *12*, 57–68.

Moscou, M. J., & Bogdanove, A. J. (2009). A simple cipher governs DNA recognition by TAL effectors. *Science*, *326*(5959), 1501–1501.

Nicolia, A., Proux-Wéra, E., Åhman, I., Onkokesung, N., Andersson, M., Andreasson, E., & Zhu, L. H. (2015). Targeted gene mutation in tetraploid potato through transient TALEN expression in protoplasts. *Journal of Biotechnology*, *204*, 17–24.

Nishimasu, H., Ran, F. A., Hsu, P. D., Konermann, S., Shehata, S. I., Dohmae, N., Ishitani, R., Zhang, F., & Nureki, O. (2014). Crystal structure of Cas9 in complex with guide RNA and target DNA. *Cell*, *156*(5), 935–949.

Nuccio, M. L., Claeys, H., & Heyndrickx, K. S. (2021). CRISPR-Cas technology in corn: A new key to unlock genetic knowledge and create novel products. *Molecular Breeding*, *41*(2), 11.

Oliva, R., Ji, C., Atienza-Grande, G., Huguet-Tapia, J. C., Perez-Quintero, A., Li, T., Eom, J.-S., Li, C., Nguyen, H., Liu, B., Auguy, F., Sciallano, C., Luu, V. T., Dossa, G. S., Cunnac, S., Schmidt, S. M., Slamet-Loedin, I. H., Cruz, C. V., Szurek, B., … Yang, B. (2019). Broad-spectrum resistance to bacterial blight in rice using genome editing. *Nature Biotechnology*, *37*(11), 1344–1350.

Osakabe, K., Osakabe, Y., & Toki, S. (2010). Site-directed mutagenesis in Arabidopsis using custom-designed zinc finger nucleases. *Proceedings of the National Academy of Sciences*, *107*(26), 12034–12039.

Sander, J. D., Cade, L., Khayter, C., Reyon, D., Peterson, R. T., Joung, J. K., & Yeh, J. R. J. (2011). Targeted gene disruption in somatic zebrafish cells using engineered TALENs. *Nature Biotechnology*, *29*(8), 697–698.

Sar, P., Pramanik, B., & Debnath, S. (2022). CRISPR/Cas: A new horizon in crop improvement. In S. Debnath, I. K. Manjula, K. R. Shah, & S. Bhattacharjee (Eds.), *Futuristic trends in biotechnology* (pp. 9–27). IIP International Publishers.

Scott, C. T. (2005). The zinc finger nuclease monopoly. *Nature Biotechnology, 23*(8), 915–919.

Shan, Q., Zhang, Y., Chen, K., Zhang, K., & Gao, C. (2015). Creation of fragrant rice by targeted knockout of the OsBADH2 gene using TALEN technology. *Plant Biotechnology Journal, 13*(6), 791–800.

Sheng, X., Sun, Z., Wang, X., Tan, Y., Yu, D., Yuan, G., Yuan, D., & Duan, M. (2020). Improvement of the rice "easy-to-shatter" trait via CRISPR/Cas9-mediated mutagenesis of the qSH1 gene. *Frontiers in Plant Science, 11*, 619.

Shukla, V. K., Doyon, Y., Miller, J. C., DeKelver, R. C., Moehle, E. A., Worden, S. E., Mitchell, J. C., Arnold, N. L., Gopalan, S., Meng, X., Choi, V. M., Rock, J. M., Wu, Y.-Y., Katibah, G. E., Zhifang, G., McCaskill, D., Simpson, M. A., Blakeslee, B., Greenwalt, S. A., … Urnov, F. D. (2009). Precise genome modification in the crop species Zea mays using zinc-finger nucleases. *Nature, 459*(7245), 437–441.

Sundström, J. F., Albihn, A., Boqvist, S., Ljungvall, K., Marstorp, H., Martiin, C., Karin, N., Ivar, V., Jonathan, Y., & Magnusson, U. (2014). Future threats to agricultural food production posed by environmental degradation, climate change, and animal and plant diseases–A risk analysis in three economic and climate settings. *Food Security, 6*, 201–215.

Tiwari, J. K., Buckseth, T., Singh, R. K., Kumar, M., & Kant, S. (2020). Prospects of improving nitrogen use efficiency in potato: Lessons from transgenics to genome editing strategies in plants. *Frontiers in Plant Science, 11*, 597481.

Van de Wiel, C. C. M., Schaart, J. G., Lotz, L. A. P., & Smulders, M. J. M. (2017). New traits in crops produced by genome editing techniques based on deletions. *Plant Biotechnology Reports, 11*, 1–8.

Voesenek, L. A., & Bailey-Serres, J. (2015). Flood adaptive traits and processes: An overview. *New Phytologist, 206*(1), 57–73.

Wang, C., Wang, G., Gao, Y., Lu, G., Habben, J. E., Mao, G., Chen, G., Wang, J., Yang, F., Zhao, X., Zhang, J., Mo, H., Qu, P., Liu, J., & Greene, T. W. (2020). A cytokinin-activation enzyme-like gene improves grain yield under various field conditions in rice. *Plant Molecular Biology, 102*(4–5), 373–388.

Wang, F. J., Wang, C. L., Zheng, C. K., Qin, T. F., Ying, G. A. O., Liu, P. Q., & Zhao, K. J. (2017). Creation of gene-specific rice mutants by AvrXa23-based TALENs. *Journal of Integrative Agriculture, 16*(2), 424–434.

Wright, D. A., Townsend, J. A., Winfrey Jr, R. J., Irwin, P. A., Rajagopal, J., Lonosky, P. M., Hall, B. D., Jondle, M. D., & Voytas, D. F. (2005). High-frequency homologous recombination in plants mediated by zinc-finger nucleases. *The Plant Journal, 44*(4), 693–705.

Yang, C. H., Zhang, Y., & Huang, C. F. (2019). Reduction in cadmium accumulation in japonica rice grains by CRISPR/Cas9-mediated editing of OsNRAMP5. *Journal of Integrative Agriculture, 18*(3), 688–697.

Zaidi, S. S. E. A., Mahfouz, M. M., & Mansoor, S. (2017). CRISPR-Cpf1: A new tool for plant genome editing. *Trends in Plant Science, 22*(7), 550–553.

Zeng, Y., Wen, J., Zhao, W., Wang, Q., & Huang, W. (2020). Rational improvement of rice yield and cold tolerance by editing the three genes OsPIN5b, GS3, and OsMYB30 with the CRISPR–Cas9 system. *Frontiers in Plant Science, 10*, 1663.

Zhang, F., Maeder, M. L., Unger-Wallace, E., Hoshaw, J. P., Reyon, D., Christian, M., Li, X., Pierick, C. J., Dobbs, D., Peterson, T., Joung, J. K., & Voytas, D. F. (2010). High frequency targeted mutagenesis in Arabidopsis thaliana using zinc finger nucleases. *Proceedings of the National Academy of Sciences, 107*(26), 12028–12033.

Zhang, Y., Zhang, F., Li, X., Baller, J. A., Qi, Y., Starker, C. G., Bogdanove, A. J., & Voytas, D. F. (2013). Transcription activator-like effector nucleases enable efficient plant genome engineering. *Plant Physiology, 161*(1), 20–27.

7 Applications of CRISPR on Improving Plant Growth and Development through Secondary Metabolism

Amardeep Ray Preethi, Sasikumar Janani, Pulliyampatti Gunasekaran Vishnu, Karli Haritha, Sumithran Subhadevi Suhith Mithra, Kannan Vijayarani, Puzhakkal Abhirami, and Arumugam Vijaya Anand

7.1 INTRODUCTION

Genome editing is an approach to creating alteration in the genetic material of living organisms to improve some characteristics or to correct some disorders by insertion, omission, or substitution of DNA sequences (Zhang et al., 2015). Sequence-specific modification of the genome by targeted nucleases opens a vast area for researchers to virtually manipulate the sequences in recent years. Clustered regularly interspaced short palindromic repeats (CRISPR), transcription activator-like effector nucleases (TALENs), and zinc finger nucleases (ZFNs) are the most generally operated core technologies; among them CRISPR stands first. CRISPR was foremost found by Ishino in 1987 (Osaka University, Japan) from DNA sequences of *Escherichia coli*. In 1995, Francisco Mojica (University of Alicante, Spain) identified an identical structure in the genome of *Haloferax mediterranei* (archaea), bringing an effective breakthrough in comprehending the biological process of CRISPR (Mojica et al., 1993).

Mojica was one of the first scientists to hypothesize that the unusual repeats recreate a part of the immune system of archaea and bacteria. In 2007 two French nutritionists, Rodolphe Barrangou and Phillippe Horvath provided the foremost empirical knowledge on the action of CRISPR (Barrangou et al., 2007). CRISPR/Cas is an adaptive defense prokaryotes to eliminate foreign hereditary material (Koonin et al., 2017). Guide RNA provides particularity for DNA/RNA targets and contains memory of one-time infection (Gasiunas et al., 2012). Based on the architectonics of effector complexes, CRISPR can be classified into two categories (Makarova et al., 2020). Class I: DNA interference is mediated by CRISPR-related complex for antiviral defense via Cas3 helicase/nuclease subunit, and it is most abundant in prokaryotes (Brouns et al., 2008).

The cascade complex can be employed for genome editing like the introduction of new genome sequences, regulation of gene transcription, and site site-specific deletion. Class II: single subunit effector nucleases with highly diverse and distinct functioning groups (Harrington et al., 2018) have nuclease activities against ssDNA, dsDNA, or RNA. In class II, type II and V are RNA-guided DNA endonucleases families including Cas9 and Cas12. Cas9 has two different nuclease domains to form a double-strand break (DSB), the HNH domain to cleave the gRNA targeted strand and the RuvC domain to cleave non-targeted DNA (Gasiunas et al., 2012). The put-out action of both domains of Cas9 will create catalytically dead Cas9 (dCas9) and the render inactive of one of the domains creates nickase Cas9 (nCas9) and these are the formation blocks for DSB-free genome editors (Qi et al., 2013). Cas12 is widely utilized in genome editing, and it includes variants like Cas12a, Cas12b,

DOI: 10.1201/9781003457237-7

Cas12c, Cas12d, Cas12e, Cas12f, Cas12j, Cas12h, and Cas12i. Certain Cas12 nucleases can be used in vivo genome editing using viral genome due to their small size (Kim et al., 2022). Comfortable design for any genomic targets, multiplexing, and uncomplicated prediction about off-target sites are advantages over ZFNs and TALENs. Recently, a family of positively great transposons encoded RNA-guided system OMEGA was found in prokaryotes and eukaryotes by rebuilding the expansion of Cas8 and Cas12 endonucleases (Altae-Tran et al., 2021).

7.2 PRINCIPLE AND MECHANISM OF CRISPR SYSTEM

Through evolution, prokaryotes have achieved many protective measures to safeguard themselves from pathogens. CRISPR and its related proteins (Cas) work as an adaptive immune procedure in prokaryotes by memorizing prior conditions by combining brief sequences of intruding genomes called spacer into the CRISPR locus. These spacers are represented as small guide CRISPR RNAs (crRNAs), which are regulated by Cas proteins. They desired viral sequences following a recurring disease. The capability of the CRISPR/Cas9 approach to pinpoint DNA arrangements employing synthetic guide RNAs has allowed genome editing in a combination of cellular and organismic. It has a wide array of uses in therapeutics. Even though several types of research have been studied based on the biochemical functions after CRISPR/Cas systems use indications and also multiple facets of the immunity stages, it still lacks adequate comprehension (Hille and Charpentier, 2016). The basic biological characteristics of CRISPR/Cas are that it helps in comprehension of the evolution, antiviral activity, host–parasite reciprocal evolution, self-versus non-self-differentiation, and adaptation mechanisms. The most well-known example of Lamarckian evolution – the production of inheritable, adaptable genetic alterations in reaction to interactions with outside stimuli, in this instance exogenous genetic material – is provided by the CRISPR/Cas approach. CRISPR/Cas procedure uses many ways to differentiate between endogenous and exogenous, what the immune system normally does. The biological expense of CRISPR/Cas systems seems to be dictated by their obstructive influence on lateral gene transfer, restricting evolutionary breakthrough. Apoptosis and microbial dormancy are two inevitable features of the CRISPR/Cas system. It also takes part in a specific form of defense called altruistic means the shutdown of cellular processes results in the conservation of adjacent cells (Koonin, 2019).

Resulting of the increased flexibility, sensitivity, and specificity considerable CRISPR/Cas systems have been utilized as diagnostic devices. Cas9 variants have been used as RNA-guided targeted distinction effectors in a range of diagnostic uses, including imaging evaluation, bio-sensing tests, and target improvement for next-generation sequencing (NGS), which comes under adaptable and reasonably priced tests. The subsequent identification of Cas proteins with collateral division activity has also made it easier to construct a variety of diagnostic instruments for quick and portable detection, and these instruments have a lot of promise for use in point-of-care environments. Nevertheless, the representative evaluations that have been suggested on this subject are mostly limited to traditional bio-sensing applications; as a result, a thorough and organized characterization of this rapidly evolving field is needed (Wang et al., 2020).

The straightforward structure of the direction RNAs that help the Cas9 target DNA locus leads to the tremendous victory of CRISPR/Cas9 genome editing. Numerous recent research that effectively arranged disease-causing alleles ex vivo in bodily and induced pluripotent stem cells and in vivo in animal representatives utilizing CRISPR/Cas9. This increases the hope of genome editing in therapeutics (Zehra and Tunali, 2016). The small RNAs which help in CRISPR/Cas immunity are produced by the Cas6 superfamily, the host RNase III endoribonucleases, and the Cas5d subclass. New understandings of the fundamentals of crRNA processing and their assemblages are given by the structures of Cas6 and Cas5d (Li, 2015).

CRISPR/Cas is a complex machinery that works by integrating the elements of foreign genetic material, generally from transportable genetic materials, into CRISPR arrays present in microbe genomes. The incorporated segment undergoes translation called the spacers and they work as guide RNAs. The major function of the guide RNA is they act as the identifier and inactivator of the related targets. Modes:2 The CRISPR/Cas methods comprise different transformation and effector segments whose evolutional directions seem to be slightly partly separated. Comparative sequence analysis demonstrates the source of the transformation module from transposons, a different kind of transposable elements, which utilize a homolog of Cas1 protein, the integrating enzyme reliable for the spacer fusion into CRISPR arrays, as the transposase.

There are mainly two types of CRISPR approaches: Class 1 is characterized by multi-element effectors, and class 2 is characterized by effectors having large, single peptides. The type 2 effectors are produced from nucleases encoded by further MG. It is possible to theoretically reconstruct how the class 1 effector complexes evolved later on by sequential genomic duplication and substitution, mainly involving genetic blueprint encoding polypeptides with RNA identification motif domains. Apart from MGE's numerous contributions to the expansion of CRISPR/Cas, an equally noteworthy information flow is the incorporation of minimalist CRISPR/Cas system variations by MGE for purposes that are still unclear (Koonin, 2019).

The small RNAs which perform in the CRISPR/Cas9 immunity are produced by the Cas6 superfamily, the host RNase III endoribonucleases, and the Cas5d subclass. Current understandings of the basic regulations of crRNA processing and the structural aspects that connect crRNA processing to blocking have been provided by new advances in the structural biology of Cas6 and Cas5d and their associations with RNA molecules. The unique simple characteristic, the broad substrate ranges, and kinetic features make them very suitable for investigating molecular transition, protein–RNA association, and biotechnology uses (Li, 2015). The double RNA-guided DNA endonuclease Cas9 is used by several bacterial clustered regularly interspaced short palindromic repeats CRISPR-associated (Cas) systems to rescue target DNA from phages and fused plasmids by creating site-explicit DSBs. The desired DNA is distinguished by the existence of a short protospacer-adjacent motif adjoining the target location. The guide RNA and desired DNA base pairs complementarily which forms a cas9–gene interaction. It causes conformational changes and leads to R-loop arrangement and strand scission. By artificially creating a double trans-activating CRISPR RNA (tracrRNA)-CRISPR RNA (crRNA) system, it is easier to perform CRISPR/Cas9 as an RNA-programmable DNA editing and targeting foundation (Jiang et al., 2013a).

The significance of molecular dynamics simulations in identifying the allosteric communication mechanism can be observed with CRISPR/Cas9. The process primarily includes three steps that come under the function of CRISPR/Cas9: DNA recognition helps in nucleic acid cleavage and avoid off-target activity. The discovery of these 'allosteric transducers' led to the creation of unknown Cas9 protein forms with enhanced selectivity, creating a fresh method for regulating CRISPR/Cas9 activity. This established the basic principles behind the allosterism of CRISPR/Cas9, which helped in better genome editing (Nierzwicki et al., 2021). Nowadays many studies are focusing on CRISPR/Cas9, the specific details of DNA cleavage, and its kinetic process. In *Streptococcus pyogenes*, the detachment from a pre-existing trimeric structure of Cas9, guide RNA, and DNA occurs suddenly, during the initial catalytic cycle. Whereas the DNA association with the rate of progression is constrained by the Cas9-guide–RNA binary complex. Even though these much studies are conducted, the kinetic mechanism is still unknown (Raper et al., 2018)

7.3 CRISPR/CAS IN SECONDARY METABOLISM

A study in *Arabidopsis* shows that the MYB and bHLH transcription factors (TFs) are accountable for the control of secondary metabolic pathways (Shkryl et al., 2021). The appearance of the CRISPR/Cas9 system of gene modification technology revolutionized the exploitation and finding of secondary metabolites from filamentous fungi. CRISPR/Cas9 and CRISPRa showed improvement in the secondary metabolites content and also discovered other biologically active molecules in filamentous fungi (Jiang et al., 2021). Several studies related to the loss and gain of function in the deleted and activated mutants in polynucleotide/non-ribosomal peptide confirm that there are 22 different secondary metabolites present that are associated with BCGs. This was done with the help of the CRAGE-CRISPR system. Also, this demonstrated the profiling of the genotype–phenotype relationship in bacteria (Ke et al., 2022).

Scientists can control the bioproduction process of the metabolic products produced by plants (phytometabolites) through the miRNA-mediated CRISPR/Cas9 system. DNA editing in plant sciences is increasing these days, and it is easier to find and manipulate phyometabolites using the CRISPR techniques. These are of high medicinal value. Genome editing in phytometabolites plays a significant part in drug discovery (Sabzehzari et al., 2020a). In *Aspergillus niger*, the CRISPR/Cas9 technique efficiently reprogrammed the epigenetic modification in target genes related to secondary metabolites. The histone acetylase p300 was located on ATG adjoining parts of the secondary biosynthesis gene brief precisely, which activated the transcribing process. There the histone deacetylases HosA-dCas9 gene repressed transcription and initiated the synthesis of melanin. This was the foremost period that CRISPR/Cas9 was involved in regulating the expression of a secondary metabolite (Li et al., 2021).

Recent studies on the biosynthetic gene cluster (BGC) have shown that performing genetic and genomic technology such as CRISPR/Cas9 in the BGC regions results in the identification and investigation of more commercially viable secondary products. This CRISPR/Cas9 system enabled genetic engineering in fungi and aids in the finding of bioactive substances which are useful in therapy, farming, human wellness, and other applications (Wang et al., 2023). Employing the CRISPR/Cas9 genome modification system, a biallelic mutant strain named Hlmed5a/med5b was generated

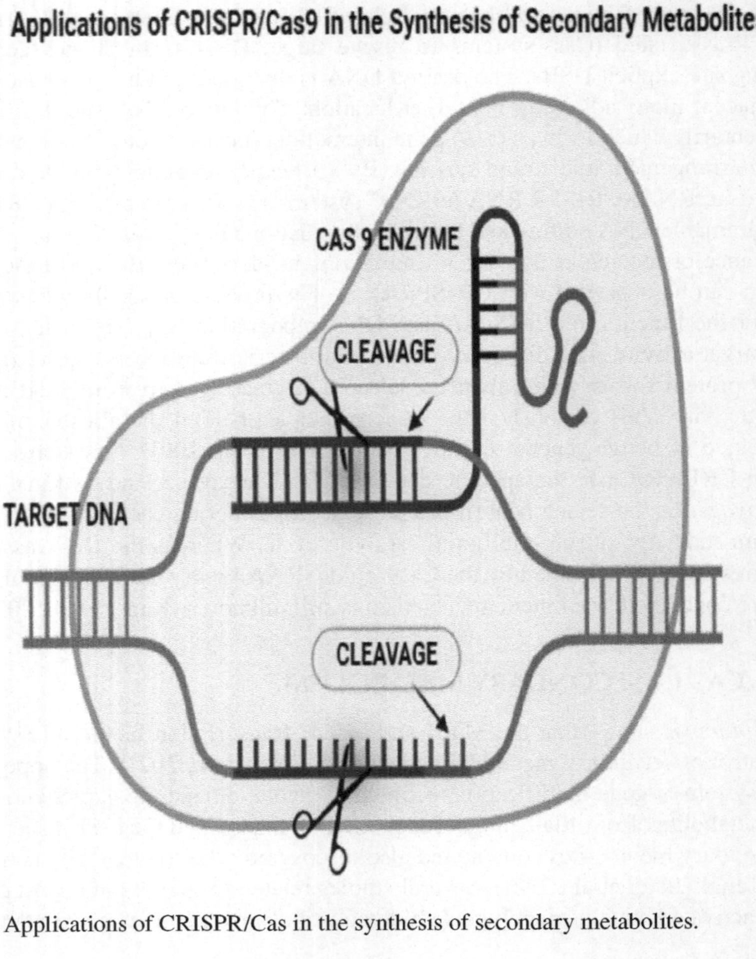

FIGURE 7.1 Applications of CRISPR/Cas in the synthesis of secondary metabolites.

in hop (*Humulus lupulus*), a critical marketable crop that is famous for the generation of specific secondary products in glandular trichomes (lupulin glands). Mainly the secondary metabolites produced by hop are used in the brewing industry. The mutation in the Hlmed5a and med5b genes results in the declined representation of the structural genes of flavonoids, humulone, and terpenoid metabolism (Awasthi et al., 2023). In plant sciences (Figure 7.1), the CRISPR/Cas system is the most contemporary and cutting-edge genetic engineering mechanism known because genome editing for secondary metabolites has a higher importance today because of the need for nutritious and therapeutic value (Rajyaguru et al., 2021). Using in vitro CRISPR/Cas9 in *Aspergillus wentii* the negative transcriptional regulator mrcA has been deleted for activating the silenced BGC regions which are capable of producing secondary metabolites. This resulted in the production of 17 different secondary metabolites. They are grown on potato dextrose media. Among 15 secondary metabolites, 9 secondary metabolites were characterized successfully (Yuan et al., 2022).

7.4 CRISPR-GENOME EDITING

For a long time, the methods for productive gene editing were controlled by particular animals (Sander and Joung, 2014). CRISPR is a genome editing system that was naturally seen in bacteria as an immune defense, and this mechanism was adopted by the scientific world. CRISPR is simpler and more efficient for designing RNA-guided molecules than TALENS and ZFN (Friedland et al., 2013). Hence, within a short period, it was rapidly undertaken as a genome editing system and involved all sorts of organisms (Hruscha et al., 2013). Cas9 and Cas12a, single protein nucleases, are most commonly used and can be efficiently reprogrammed to desired DNA by altering the guide RNA (Mali et al., 2013). An upgraded CRISPR/Cas 9 involves a Cas9/sgRNA–target DNA complex, which delivers a competent molecular pathway (Slaymaker et al., 2016). The tyrosinemia was foremost rectified using CRISPR in mice, which plays an essential step toward using CRISPR in humans. The lack of fumarylacetoacetate hydrolase causes the accumulation of cytotoxic metabolites in hepatocytes and leads to cell death, which was corrected by delivering specific sgRNA and vector encoding Cas9, together with DNA oligo for HDR. This corrected the allele mutation of fumarylacetoacetate hydrolase and rescued the mice by reducing the toxicity in hepatic cells (Yin et al., 2014).

Beta thalassemia, the hereditary blood disease, was remedied using CRISPR/Cas9 in induced pluripotent stem cells by transfecting the CRISPR/Cas9-mediated vectors for allele-specific targeting of the disease-generating gene (Xie et al., 2014). Single point mutation causing sickle cell disease can be reversed by CRISPR/Cas9 template-directed gene replacement. Another strategy involves the 'knockout' editing of TF BCL11a, which enables the reactivation of fetal globin and reduces symptoms of SCD (Hossain and Bungert, 2017). For harnessing HIV, CRISPR/Cas9-based modification is being explored. HIV is incorporated into host immune cells, where it works as an expression model of the virus. The integrated HIV genetic material originating from the genome of T cells infected with HIV was eradicated by CRISPR/Cas9 nucleases (Liao et al., 2015).

CRISPR is also widely used in agriculture and livestock, enabling scientists to create animals and plants with desirable traits. By complete elimination of CD163 polypeptides or omission of the region of CD163 gene which encodes SRC5, Cluster of Differentiation 163 (CD163)-inactivated pigs were developed, which have immunity against porcine respiratory and reproductive syndrome (Yang et al., 2018). Type 1V CRISPR/Cas system contains enzymatically inactive Cas13 to swap adenosine to inosine deaminase in mammal body cells. This is RNA editing for Programmable A to I Replacement (REPAIR) with the advantage of correcting full-length transcripts possessing pathogenic modifications and without sequence restrictions (Kim, 2018). Ex vivo methods of CRISPR-based editing involve the collection of cells from the body, fulfilling CRISPR-based editing, and putting back edited cells by autologous transplantation (Hoggat, 2016; Satish et al., 2021).

In immune therapy, regulatory T cells are modified outside the body to suppress the secretion of proinflammatory cytokines in rheumatoid arthritis. A CRISPR-driven clinical trial of chimeric antigen receptor (CAR) T-cell treatment concerns the removal of T cells and biological engineering to express CAR consisting of extracellular targeting compound and intracellular signaling moiety, and it was the first commercially approved gene transfer treatment by the US Food and Drug Administration (June et al., 2018). CRISPR enzymes can be used to detect sensitive nucleic acids. For that crRNA guide has to be designed, which is complementary with the nucleic acid of interest, joins up with the Cas enzyme, and provides a reporter DNA or RNA (a tailored genetic sequence with a fluorescent probe appended at one extremity and a quenching module attached at the opposite extremity). Extracted nucleic acid from biological samples is combined with the preparation; the Cas enzyme will cleave the reporter after the detection of the target nucleic acid and release fluorophore from the fluorophore–quencher complex, permitting it to fluorescence. The resulting signal is quantified, and changing the reporter will develop an antibody-responsive signal, which enables the optical readout on the test strip (Myhrvold et al., 2018).

7.5 PHYTOMETABOLIC EDITING BY CRISPR

CRISPR has revolutionized life science studies since 2013. CRISPR/Cas9 and Cpf1 are the most commonly used in plants (Gao, 2018). This enzyme can be delivered as protein–RNA, RNA, or DNA; cleaves the desired sequence; and makes a double helix break in DNA, later assembled as site-directed nucleases (Yin et al., 2017). Different genome-editing events like insertion/deletion, nucleotide substitution, and open reading frame inactivation take place when cells try to correct the DSBs (Shan et al., 2014). The biosynthetic pathway of phytometabolite can be manipulated by CRISPR. Controlling of TFs using CRISPR prevails over flux bottleneck by activating or repressing particular effectors in the biosynthetic pathway of secondary metabolites (Grotewold, 2018). *Arabidopsis thaliana*, *Nicotina benthamina*, and *Oryza sativa* are taken as model plants to employ CRISPR/Cas9-based induction of gene transcription. To designate Cas9-mediated transcriptional activation, the C end of inactivated pcoCas9 was converted to VP64 transcriptional activator region. AtPAP1 gene and miR319 genes were edited in *Arabidopsis* to specify the effectiveness of protein as well as non-protein-coding genes, among which AtPAP1 gene-edited plants disclose gene stimulations of two to seven times and miR319 gene-edited plants show overexpression of gene compared to control. To suppress the protein-encoding and non-coding genetic material, an artificial transcriptional inhibitor PCO-dCas9-3X is utilized in *Arabidopsis*. AtCSTF64 (gene encoding protein), miR159B, and miR159A (non-coding) were targeted, and results show that the repression of gene takes place in AtCSTF64, where transcript level reduced by 60%, and in miR159B/miR159A, the transcript level was halved (Lowder et al., 2015). Glutamate decarboxylase plays a pivotal role in the synthetic pathway of γ-aminobutyric acid (GABA) and contains an autoinhibitory region found in SlGAD3 and SlGAD2, which is excised using CRISPR/Cas9 methodology, thereby augmenting the accrual of GABA in tomatoes (Bouché and Fromm, 2004). The GABA bypass is adjusted, focusing on five vital genes SSADH, CAT9, GABA-TP3, GABA-TP2, and GABA-TP2 using pYL-CRISPR/Cas9, leading to a remarkable surge in GABA levels in both fruits and foliage (Li et al., 2018). Silencing the 4OMT2 gene, which controls the production of benzylisoquinoline in *Papaver somniferum*, decreases the level of S-reticuline, which has a role in the biosynthesis of morphine. In edited plants, the presence of uncharacterized alkaloids was detected, which shows the ability of CRISPR to detect unique secondary metabolites. SmRAS gene, which synthesizes rosmarinic acid synthetase enzyme, which recreates an essential function in the synthesis of rosmarinic acid in *Savia miltiorrhiza*, is manipulated by CRISPR, increasing the content of 3,4-dihydroxy phenyl lactic acid (precursor of RA) (Zhou et al., 2018).

FAD2-1B and FAD2-1A genes help to convert oleic acid to linoleic acid in the soybean (Tang et al., 2005; Haun et al., 2014), and mutation can be induced in this gene by CRISPR/Cpf1 genome editing, which results in the increased level of oleic acid in glycine max oil, having great nutritive belongings. The SmCPS1 gene, which produces diterpene synthase involved in the biosynthesis of tanshinone, is silenced by CRISPR/Cas9, resulting in the complete missing of tanshinone.

Disrupting the biosynthetic pathway of tanshinone in post-geranylgeranyl pyrophosphate (a common precursor for tanshinone and taxol) increases substrate for taxol (Sabzehzari et al., 2020b). Alpha-1,3-fucosyltransferase and beta (1,2)-xylosyltransferase genes were deactivated by employing CRISPR/Cas9 in BY-2 cells of *Nicotiana tabacum* to produce glycoproteins excluding plant-specific glycans (Mercx et al., 2017). FAD2 genes were inactivated by CRISPR/Cas9-based editing to enhance oleic acid composition in *Camelina sativa* plants, which have antioxidant and anti-cancerous properties (Jiang et al., 2017).

7.6 CRISPR IN MEDICINAL PLANTS FOR NATURAL PRODUCT BIOSYNTHESIS

CRISPR is used in diverse metabolic engineering tools that are effectively revealed in the control of secondary metabolite pathways in flora and fauna. Taxadiene formation in *Escherichia coli* was elevated through engineering methylerythritol-phosphate and terpenoid-forming pathways. This also led to minimal indole accumulation and P450-mediated 5α-oxidation of taxadiene to taxadien-5α-ol, resulting in the synthesis of taxol (Dey, 2021). The rosmarinic acid synthase gene (Sm*RAS*) in the water-soluble phenolic acid biosynthesis pathway of *Salvia miltiorrhiza* is modified using the CRISPR/Cas9 system. Zhou et al. (2018) and Guo et al. (2022) created a single-guide RNA (sgRNA) to edit the significant SmRAS gene.

CRISPR/Cas9 tools are mainly used in crop plants for harvest and stress forbearance, the enormous possibility of this technique has not been entirely operated in therapeutic plants in adjusting secondary metabolic routes to generate medicinally potent phytochemicals against cancer and other ailments. CRISPR/Cas9 serves as a good genetic modification instrument in plants and botanical-derived natural substances with cancer-preventing and other healing uses. Furthermore, it is involved in current patents on the medicinal utilizations of CRISPR-CAS systems for all mortal medical needs (Dey and Nandy, 2021). CRISPR/Cas platform is used for modifying DNAs in therapeutic plants as a gene-editing process for the production of biomolecular activators in therapeutic plants; it is employed to examine genome functionality, genetic enhancement, bioengineering, and alteration of germplasm in medicinal flora. This technology augments the therapeutic potency and agricultural significance of plants (Jamwal et al., 2023).

CRISPR/Cas9 technology expedites the progress of crop strains and reduces the duration of crop breeding cycles. By changing the size and GC composition of sgRNA and modifying the attention of the Cas9/sgRNA complex to mitigate the off-target impact of the CRISPR/Cas9 technique, the intended genes can be exploited with greater accuracy (Wan et al., 2022). The genomic accessibility of various food yields and therapeutic herbs signifies a notable benefit for employing CRISPR instruments in molecular biology to augment Plant secondary metabolites (PSM) levels. Incorporating prophylactic (nutraceutical) and remedial (therapeutic) metabolic profiles via plant genetic engineering offers a viable resolution to boost human well-being, specifically metabolic pathways, synthesizing botanical secondary metabolites (PSMs) influenced by genetic and ecological elements. Humans have relied on PSM-enriched botanical resources for thousands of years in healthcare and dietary supplements (Shelake et al., 2023).

7.7 REGULATORY AND ETHICAL ISSUES OF CRISPR EDITING IN PLANTS

The Animal and Plant Health Inspection Service's (APHIS), an associate of the USDA, reviews are used for investigation on GM crops. APHIS has shown that consequences from CRISPR/Cas that only *omit* a gene, in most circumstances, can't be controlled because no unique DNA is inserted into the host genome. In comparison to the United States, the European Union (EU) maintains a significantly more rigorous regulatory framework for genetically altered yields in agricultural practices.

It needs a comprehensive threat examination by the European Food Safety Authority (EFSA) before the EC determines whether to accept or withhold approval for use in the EU. EU regulation believes all genetically altered crops as transgenic irrespective of whether this contains any integration of foreign genes or direct gene modification and therefore subjected to ordinance and hazard estimation (Caplan et al., 2015). The CRISPR/Cas approach is a positively used gene refining tool both for the genetic modification of plants and creatures and for the medical research of human inborn disorders; it also creates possible threats in specific circumstances. The CRISPR/Cas system is the chance of unlawful investigation and transgression of CRISPR/Cas technology (Zhang et al., 2020). Advocates for CRISPR-altered plants assert that these plants and their derivatives lack any exogenous genetic material, while detractors regard crop cultivars produced through these methodologies as genetically engineered, potentially falling under GMO oversight (Ahmad et al., 2021a). The CRISPR/Cas system, being employed techniques for enhancing crops, facilitates rapid,

TABLE 7.1
Target Genes, Cas9 Version with Its Mutation Frequency

Plant	Target Gene	Cas9 Version	Mutation Frequency	References
Arabidopsis thaliana Glycine max	CHLI1 and 2, TT4	Human codon-enhanced Cas9	10–89	Huang and Li (2009)
	ADH1, RTEL1	*Arabidopsis* codon-enhanced Cas9	2.5–70.0	Jacobs et al. (1988)
	BRI1, GAI	Human codon-enhanced Cas9	30–84	Feng et al. (2013)
	TRY, ETC2	Maize codon-enhanced Cas9	42–90	Xing et al. (2014)
	PYR1, PYL1 and 2	Human codon-enhanced Cas9	13–93	Gonzalez-Guzman et al. (2012)
	FT, SPL4	Human codon-enhanced Cas9	10.00–84.78	Hyun et al. (2015)
	Bar, GmFEI1 and 12	Plant codon-enhanced Cas9	10.0–93.3	Meyer et al. (2009)
	DD20 and 43	Soybean codon-enhanced Cas9	59–76	Li (2015)
	GmPDS11 and 18	Plant codon-enhanced Cas9	11.7–48.1	Jacobs et al. (2015)
Citrus sinensis	CsPDS	Human codon-enhanced Cas9	3.2–3.9	Jia and Wang (2014)
Nicotiana benthamiana	GFP	*Chlamydomonas*	N/A	Jiang et al. (2013b)
Medicago truncatula	GUS	Soybean codon-enhanced Cas9	N/A	Michno et al. (2015)
Nicotiana tabacum	NtPDS	Plant codon-enhanced Cas9	81.8–87.5	Gao et al. (2015)
	mCherry	Plant codon-enhanced Cas9	N/A	Mercx et al. (2016)

robust, and precise genetic modification, prompting significant discussions regarding the biosafety, commercial application, and oversight of novel strains of genetically modified crops (Globus and Qimron, 2018).

Environmental concerns have also been identified as per CRISPR-mediated modifications in plants such as crop improvement. It is very important to mark the regulatory norms for the application of CRISPR. Safety and security measures should be developed to control any organism that can cause environmental damage (Shinwari et al., 2018). The matter of genome alteration falling under GMO regulations and broader legislation concerning risk assessment and oversight is identified. The topic of intellectual property rights for genetic modification technologies within the legal frameworks of the EU and the United States, as well as under global regulations, is examined. Disputes over patents among research groups involved in the creation of the CRISPR/Cas9 system are also analyzed. The chances of getting patent protection for plants created by genome editing are assessed (Medvedieva and Blume, 2018).

The CRISPR/Cas system has revealed significant promise for enhancing crops, yet the oversight of genome editing remains in the developmental phase. Methodological, biosafety, and societal apprehensions persist regarding the creation of genetically edited crops by humans and other organisms (Fiaz et al., 2021). To satisfy the increasing need for food globally, genetically engineered (GE) crops offer a robust substitute for conventionally bred plants. However, the societal embrace of GE crops hinges on their ratio of advantages to drawbacks exceeding unity (Munawar et al., 2024). Table 7.1 shows the target genes, the Cas9 version with its mutation frequency.

7.8 CRISPR APPLICATION IN PLANTS

7.8.1 Strategies for Boosting CRISPR Transcriptional Activation

Genome editing is revolutionized by CRISPR/Cas because of its flexible and intuitive system. Cas variations that are catalytically inactive (dCas) allow for precise genetic research without breaking DNA strands. Combining several effectors with dCas yields a versatile platform for sequence-specific imaging and gene regulation. Transcriptional repression, activation, and epigenetic alterations are examples of recent developments (Pan et al., 2021). Plants can have fine control over gene expression thanks to the use of synthetic promoters and modified TFs. Effective modulation is provided by synthetic promoters including modified cis-regulatory elements. dCas9, zinc fingers, and TALEs provide customized TFs for specific regulation, particularly in conjunction with activators or repressors. Through stress-tolerant inducible platforms, combinatorial CRISPR-dCas9-based cis-trans engineering provides a tool for regulating numerous genes and improving plant adaptation (Shrestha et al., 2018).

Plant transcriptional activation and multiplexed orthogonal genome editing are made possible by the CRISPR-Combo system, which uses only one Cas9 protein. This technique improves heritable mutations in rice without the utilization of external hormones and makes genome engineering in *Arabidopsis* more efficient for transgene-free, rapid breeding. Comprehensive guidelines encompass sgRNA design, hormone-free rice regeneration, *Agrobacterium*-mediated transformation, and observable detection of genetically unmodified *Arabidopsis* plants (Pan and Qi, 2023). New developments in gene-directing instruments have concentrated on CRISPR/Cas techniques, namely CRISPR-dCas, which makes use of an acceleratively nonfunctioning Cas protein. This platform attaches modular effector domains to enable precise epigenetic alterations and gene regulation. Summing up various DNA targeting technologies in plant biology, innovations include multi-effector targeting, transgenerational epigenetic modifications, and synergistic tactics for better plant applications (Gardiner et al., 2022).

Plant gene targeting (GT), which is essential for breeding and biotechnology, has low effectiveness. DSBs are induced using CRISPR/Cas systems, which show promise, although refinements are necessary. GT efficiency is enhanced by methods such as manipulating the DNA repair process, self-amplified GT vectors, and cell-type-specific Cas expression. Inspired by research on yeast and mammals, this development could lead to increased food yields and environmentally sustainable farming practices (Huang and Puchta, 2019). Improving plant resistance to abiotic stressors is critical for handling concerns connected to international nourishment safety in the context of weather shifts. Promising paths have the introduction of stress-related genes, the insertion of TFs, and the benefit of gene modification technologies like CRISPR/Cas9. The exploitation of hormone pathways, antioxidant defenses, and stress signaling is investigated. There is a discussion of the difficulties and opportunities in this area for food security and sustainable agriculture (Khokhar Voytas et al., 2023).

The genome engineering method known as CRISPR/Cas9 has spread like wildfire in the scientific world. Plant scientists now favor the CRISPR/Cas9 technology for genome editing, primarily because of its effectiveness, specificity, and adaptability. Most of the early CRISPR/Cas9 work in plants was restricted to functional research in model systems and proof of concept. Numerous CRISPR/Cas9 applications beyond single gene editing have been developed as a result of these and other biological research projects. These tools include site-specific transgene insertion, multiplexing to induce numerous cleavage events, and gene expression control. Plant scientists are using CRISPR/Cas9-gene-editing technologies (Figure 7.2) to improve crop traits now that a large portion of the conceptual work has been completed (Schaeffer and Nakata, 2015). The global food security is seriously threatened by plant viruses, which calls for creative solutions. Crops resistant to viruses

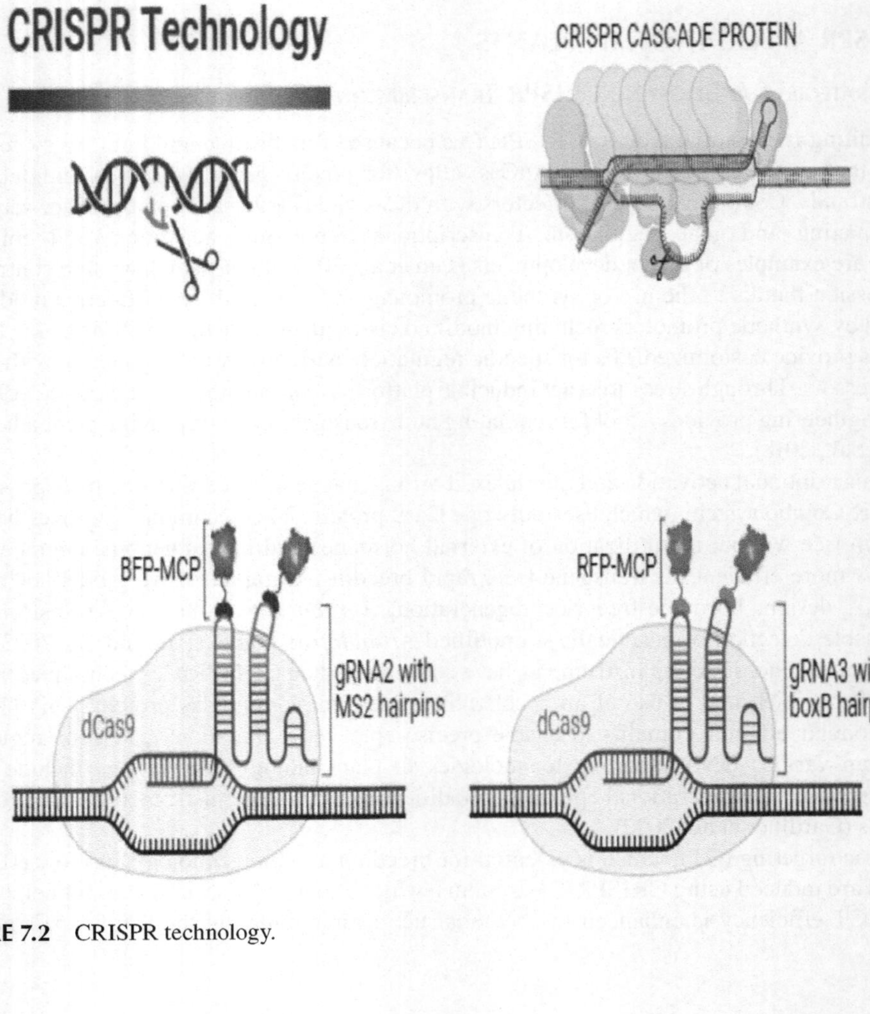

FIGURE 7.2 CRISPR technology.

are developed more quickly thanks to the CRISPR/Cas9 system, which is renowned for its specific genome editing. It either targets viral genomes directly (virus-mediated) or modifies host components essential for infection (plant-mediated) to achieve resistance (Jeyaraj et al., 2023).

The plant stress response is largely dependent on TFs, which is aided by an understanding of the structure–function relationships of TFs such as bZIP, WRKY, HD-Zip, MYB, DREB/CBF, SHN, and WXPL. Variant TF design is guided by insights into their three-dimensional structure, which discloses information on oligomerization, post-translational modifications, protein interactions, and DNA recognition. This information improves our capacity to modify TFs for better functional states (Hrmova and Hussain, 2021).

7.9 FUTURISTIC PROSPECTS OF CRISPR

Healthcare professionals have embraced the use of biosensor-based devices to detect diseases and evaluate patient signals in recent years. Developing smart and automated molecular diagnostic tools is a viable path toward illness identification and surveillance. These tools have the potential to completely transform the healthcare industry since they combine cloud computing, big data analysis in biomedicine, and medical artificial intelligence. CRISPR biosensors that are developed on microchips and make use of machine learning techniques are promising for the future. It is expected that wireless signal transmission via the Internet of Things and the cloud will improve decision-making abilities. The presentation addressed the discovery and use of CRISPR, emphasizing its role as a gene-editing tool and the high sensitivity of CRISPR-based biosensors, specifically Attomolar, Femtomolar, and Picomolar (Ibrahim et al., 2022).

Advances in CRISPR/Cas have led to revolutionary improvements in synthetic biology, metabolic engineering, medicines, and diagnostics. In the realm of genome manipulation, CRISPR/Cas systems have demonstrated greater precision, swiftness, cost-effectiveness, and efficacy compared to previous methodologies such as transcription activator-like effector nucleases (TALEN) and zinc finger nucleases (ZFNs). CRISPR/Cas systems provide difficulties such as off-target activity, delivery mechanism problems, and ethical and legal dilemmas despite their potential (Gohil et al., 2021). For CRISPR technologies to be broadly adopted, it is essential to develop amiable, widely acceptable solutions to these problems. The promise of integrating CRISPR/Cas with cutting-edge nanotechnology in the area of nutrition safety has been impressive. When paired with CRISPR/Cas-based observation, nanomaterials greatly enhance implementation and present a viable option for quick and accurate food safety detection (Zhao et al., 2023).

The benefit of CRISPR/Cas9 in horticulture, especially with dicot crops like tomatoes, was also discussed in the talk. Plant biology has altered dramatically as a consequence of genome-editing tools, which enable precise adjustments that foretell the permanent inheritance of epigenetic changes. Tomatoes with CRISPR/Cas9-based gene changes may have higher yields and productivity due to optimal gene discovery and trait enhancement (Naz et al., 2022). Along with agriculture, CRISPR/Cas9 has been used in phytoremediation technologies to solve environmental pollution issues. The ability of the technique to precisely alter the genome improves plants' capacity for bioremediation, which supports sustainable development (Saxena et al., 2020).

The unique blend of CRISPR/Cas and nucleic acid isothermal amplification offers a solution to worldwide public health issues about food safety. The detection process's sensitivity, specificity, adaptability, and dependability are all improved by this integration (Qiao et al., 2023). While the focus of current phytoremediation research is on finding new super-accumulators with strong chelation effects, molecular tools such as CRISPR/Cas9 are essential for comprehending plant tolerance and detoxification processes. With each new development in molecular research, our understanding of plants' capacity for adaptive phytoremediation in the face of the challenges posed by global warming grows (Naz et al., 2022). Using CRISPR/Cas approaches in electrochemical biosensors offers huge possible benefits, including enhanced adaptability, greater keenness and explicitness, shorter recuperation times, and the capacity to detect and determine analytes at lower attention.

The fundamental concepts of CRISPR/Cas systems and electrochemical biosensors were discussed in this work, along with the most current advancements in CRISPR/Cas system-based electrochemical biosensors for the identification of bacteria, viruses, and cancer (Ma et al., 2023).

The capacity to modify and regulate gene expression as desired has been enhanced by the CRISPR-associated system. Direct nuclease-dependent editing is now receiving more attention in biomedical science investigation because of gene-correcting techniques. Although most of the research to date has focused on editing the genome at the DNA level, newly developed methods are starting to gain traction because of their ability to induce differences in the RNA sequence (Satish et al., 2021).

7.10 CONCLUSION

CRISPR/Cas9 is an encouraging tool for plant synthetic biology due to its adaptability to genetic circuit design and gene stacking. The CRISPR/Cas toolkit has been further extended by the discovery of C2c2 class bacterial enzymes, which presents prospects for site-specific RNA modifications in eukaryotes. This breakthrough makes it easier to analyze gene expression at the RNA level, a domain that was previously difficult to explore. With its easy manipulation, high efficiency, and broad applicability, the CRISPR/Cas field has evolved quickly, allowing for its quick incorporation into a variety of molecular biology applications. Completely sequenced genomes of medicinal plants such as *Dendrobium officinale* and *Salvia miltiorrhiza* demonstrate the possibility of CRISPR/Cas9 to alter target genes, investigate the formation of phytoconstituents, and increase the yield of valuable ingredients while decreasing toxicity. Targeted gene editing has a framework thanks to the large-scale CRISPR screens in plants that have demonstrated the method's potential for synthetic biology and functional genomics. There are still difficulties, though, especially with keeping full plant mutant collections intact over several generations, which takes a lot of effort and money. Despite these difficulties, the CRISPR/Cas system is a better option for genome-wide knockout screening due to its capacity to produce precise targeted alterations in comparison to RNA interference. The benefit of CRISPR/Cas9 in bananas, which achieved 100% editing efficiency, demonstrates the technology's effectiveness in several plant species.

However, several delivery mechanisms such as pollen magnetoreception-mediated and nanoparticle-mediated ways are being explored because the effectiveness of these treatments depends on aspects like tissue type and regeneration capacity. By omitting time-consuming tissue culture procedures, these advances seek to speed up the process. Plant genome-editing technologies controlled by CRISPR/Cas have advanced remarkably in the past two years. The field of highly efficient sequences that can be targeted in the genome has increased with the development of both natural and synthetic nucleases. With the development of chromosome engineering and base editing, the toolkit for causing genomic alterations has expanded dramatically. It is predicted that, in the forthcoming years, significant progress will be made, providing breeders with the means to produce crops that are more resilient to pests, more disease-resistant, and more tolerant of the environmental pressures brought on by global warming. Although the use of CRISPR/Cas9 in agricultural crop genetic engineering is growing, there are still difficulties, especially in crops with little information available about their genomic sequences. As evidenced by the significance of target gene sequence knowledge for knockdown techniques, extensive genomic data continues to be required for the proper enactment of CRISPR/Cas9 strategies.

Altogether, plant molecular biology is entering a new era of possibilities thanks to CRISPR/Cas genome-editing technology. It is a vital instrument for advancing synthetic biology, functional genomics, and plant domestication due to its clarity of operation and high precision. CRISPR/Cas9 is positioned to play a key role in transforming agriculture as genomic sequencing efforts proceed, allowing the production of crops with improved characteristics, decreased environmental impact, and increased resistance to new challenges. The field of CRISPR/Cas research is anticipated to witness substantial advancements in the years to come due to its dynamic landscape.

ACKNOWLEDGMENT

All the authors are thankful to their respective universities and institutions for their valuable support. The corresponding author is thankful to Skill Development Centre, BEICH, RUSA2.0 (SDC Project order No. BU/RUSA/ SDC/BEICH 2020/15; 28.12.2020), Bharathiar University, for providing financial support.

DATA AVAILABILITY

The authors confirm that the data supporting the findings of this study are available within the book chapter.

REFERENCES

Ahmad, A., Ghouri, M. Z., Munawar, N., Ismail, M., Ashraf, S., & Aftab, S. O. (2021b). Regulatory, ethical, and social aspects of CRISPR crops. In A. Ahmad, S. H. Khan, & Z. Khan (Eds.), *CRISPR crops* (pp. 261–287). Springer.

Ahmad, A., Munawar, N., Khan, Z., Qusmani, A. T., Khan, S. H., Jamil, A., Ashraf, S., Ghouri, M. Z., Aslam, S., Mubarik, M. S., Munir, A., Sultan, Q., Abd-Elsalam K. A., & Qari, S. H. (2021a). An outlook on global regulatory landscape for genome-edited crops. *International Journal of Molecular Sciences*, *22*(21), 11753.

Altae-Tran, H., Kannan, S., Demircioglu, F. E., Oshiro, R., Nety, S. P., McKay, L. J., Dlakić, M., Inskeep, W. P., Makarova, K. S., Macrae, R. K., Koonin, E. V., & Zhang, F. (2021). The widespread IS200/IS605 transposon family encodes diverse programmable RNA-guided endonucleases. *Science*, *374*, 57–65.

Awasthi, P., Mishra, A. K., Kocábek, T., Nath, V. S., Mishra, S., Hazzouri, K. M., Sudalaimuthuasari, N., Stajner, N., Jakše, J., Krofta, K., Hájek, T., & Amiri, K. M. (2023). CRISPR/Cas9-mediated mutagenesis of the mediator complex subunits MED5a and MED5b genes impaired secondary metabolite accumulation in hop (Humulus lupulus). *Plant Physiology and Biochemistry*, *201*, 107851.

Barrangou, R., Fremaux, C., Deveau, H., Richards, M., Boyaval, P., Moineau, S., Romero, D. A., & Horvath, P. (2007). CRISPR provides acquired resistance against viruses in prokaryotes. *Science*, *315*, 1709–1712.

Bouché, N. & Fromm, H. G. (2004). ABA in plants: Just a metabolite? *Trends Plant Sciences*, *9*, 110–115.

Brouns, S. J., Jore, M. M., Lundgren, M., Westra, E. R., Slijkhuis, R. J., Snijders, A. P., Dickman, M. J., Makarova, K. S., Koonin, E. V., & van der Oost, J. (2008). Small CRISPR RNAs guide antiviral defense in prokaryotes. *Science*, *321*, 960–964.

Caplan, A. L., Parent, B., Shen, M., & Plunkett, C. (2015). No time to waste-the ethical challenges created by CRISPR: CRISPR/Cas, being an efficient, simple, and cheap technology to edit the genome of any organism, raises many ethical and regulatory issues beyond the use to manipulate human germ line cells. *EMBO Reports*, *16*(11), 1421–1426.

Dey, A. (2021). CRISPR/Cas genome editing to optimize pharmacologically active plant natural products. *Pharmacological Research*, *164*, 105359.

Dey, A., & Nandy, S. (2021). CRISPER/Cas in plant natural product research: Therapeutics as anticancer and other drug candidates and recent patents. *Recent Patents on Anti-Cancer Drug Discovery*, *16*(4), 460–468

Feng, Z., Zhang, B., Ding, W., Liu, X., Yang, D. L., Wei, P., Cao, F., Zhu, S., Zhang, F., Mao, Y., & Zhu, J. K. (2013). Efficient genome editing in plants using a CRISPR/Cas system. *Cell Research*, *23*(10), 1229–1232.

Fiaz, S., Khan, S. A., Noor, M. A., Ali, H., Ali, N., Alharthi, B., Qayyum, A., & Nadeem, F. (2021). CRISPR/Cas9 regulations in plant science. In A. Ahmad, S. H. Khan, & Z. Khan (Eds.), *CRISPR and RNAi systems* (pp. 33–45). Amsterdam, Netherlands: Elsevier.

Friedland, A. E., Tzur, Y. B., & Esvelt, K. M. (2013). Heritable genome editing in C. elegans via a CRISPR-Cas9 system. *Nature Methods*, *10*, 741–743.

Gao, C. (2018). The future of CRISPR technologies in agriculture. *Nature Reviews Molecular Cell Biology*, *19*(5), 275–276.

Gao, J., Wang, G., Ma, S., Xie, X., Wu, X., Zhang, X., Wu, Y., Zhao, P., & Xia, Q. (2015). CRISPR/Cas9-medialted targeted mutagenesis in Nicotiana tabacum. *Plant Molecular Biology*, *87*, 99–110

Gardiner, J., Ghoshal, B., Wang, M., & Jacobsen, S. E. (2022). CRISPR-Cas-mediated transcriptional control and epi-mutagenesis. *Plant Physiology, 188*(4), 1811–1824.

Gasiunas, G., Barrangou, R., Horvath, P., & Siksnys, V. (2012). Cas9-crRNA Ribonucleoprotein complex mediates specific DNA cleavage for adaptive immunity in bacteria. *Proceedings of the National Academy of Sciences of the United States of America, 109*, E2579–E2586.

Globus, R., & Qimron, U. (2018). A technological and regulatory outlook on CRISPR crop editing. *Journal of Cellular Biochemistry, 119*(2), 1291–1298.

Gohil, N., Bhattacharjee, G., Lam, N. L., Perli, S. D., & Singh, V. (2021). CRISPR-Cas systems: Challenges and future prospects. *Progress in Molecular Biology and Translational Science, 180*, 141–151.

Gonzalez-Guzman, M., Pizzio, G. A., Antoni, R., Vera-Sirera, F., Merilo, E., Bassel, G. W., Fernández, M. A., Holdsworth, M. J., Perez-Amador, M. A., Kollist, H., & Rodriguez, P. L. (2012). Arabidopsis PYR/PYL/RCAR receptors play a major role in quantitative regulation of stomatal aperture and transcriptional response to abscisic acid. *The Plant Cell, 24*(6), 2483–2496.

Grotewold, E. (2008). Transcription factors for predictive plant metabolic engineering: Are we there yet? *Current Opinion in Biotechnology, 19*(2):138–44.

Guo, M., Chen, H., Dong, S., Zhang, Z., & Luo, H. (2022). CRISPR-Cas gene editing technology and its application prospect in medicinal plants. *Chinese Medicine, 17*(1), 33.

Harrington, L. B., Burstein, D., Chen, J. S., Paez-Espino, D., Ma, E., Witte, I. P., Cofsky, J. C., Kyrpides, N. C., Banfield, J. F., & Doudna, J. A. (2018). Programmed DNA destruction by miniature CRISPR-Cas14 enzymes. *Science, 362*, 839–842.

Haun, W., Coffman, A. Clasen, B. M., Demorest, Z. L., Lowy, A., Ray, E., Retterath, A., Stoddard, T., Juillerat, A., Cedrone, F., Mathis. L., Voytas, D. F., & Zhang, F. (2014). Improved soybean oil quality by targeted mutagenesis of the fatty acid desaturase 2 gene family. *Plant Biotechnology Journal, 12*, 934–940.

Hille, F., & Charpentier, E. (2016). CRISPR-Cas: Biology, mechanisms and relevance. *Philosophical Transactions of the Royal Society B: Biological Sciences, 371*(1707), 20150496.

Hoggatt, J. (2016). Gene therapy for "Bubble Boy" disease. *Cell, 166*, 263.

Hossain, M. A., & Bungert, J. (2017). Genome editing for sickle cell disease: A little BCL11A goes a long way. *Molecular Theraphy, 25*, 561–562.

Hrmova, M., & Hussain, S. S. (2021). Plant transcription factors involved in drought and associated stresses. *International Journal of Molecular Sciences, 22*(11), 5662.

Hruscha, A., Krawitz, P., & Rechenberg, A. (2013). Efficient CRISPR/Cas9 genome editing with low off-target effects in zebrafish. *Development, 140*, 4982–4987.

Huang, T. K., & Puchta, H. (2019). CRISPR/Cas-mediated gene targeting in plants: Finally a turn for the better for homologous recombination. *Plant Cell Reports, 38*, 443–453.

Huang, Y. S., & Li, H. M. (2009). Arabidopsis CHLI2 can substitute for CHLI1. *Plant Physiology, 150*(2), 636–645.

Hyun, Y., Kim, J., Cho, S. W., Choi, Y., Kim, J. S., & Coupland, G. (2015). Site-directed mutagenesis in Arabidopsis thaliana using dividing tissue-targeted RGEN of the CRISPR/Cas system to generate heritable null alleles. *Planta, 241*, 271–284.

Ibrahim, A. U., Al-Turjman, F., Sa'id, Z., & Ozsoz, M. (2022). Futuristic CRISPR-based biosensing in the cloud and internet of things era: An overview. *Multimedia Tools and Applications, 81*(24), 35143–35171.

Jacobs, M., Dolferus, R., & Van Den Bossche, D. (1988). Isolation and biochemical analysis of ethyl methanesulfonate-induced alcohol dehydrogenase null mutants of Arabidopsis thaliana (L.) Heynh. *Biochemical Genetics, 26*, 105–122.

Jacobs, T. B., LaFayette, P. R., Schmitz, R. J., & Parrott, W. A. (2015). Targeted genome modifications in soybean with CRISPR/Cas9. *BMC Biotechnology, 15*, 1–10.

Jamwal, M., Ghora, B., Kumar, S., & Gupta, A. P. (2023). Applications of CRISPR/Cas9 in the synthesis of bioactive compounds from medicinal plants. In N. Kumar & R. S. Singh (Eds.), *Biosynthesis of bioactive compounds in medicinal and aromatic plants: Manipulation by conventional and biotechnological approaches* (pp. 81–97). Springer Nature.

Jeyaraj, G., Alphonse, V., Jayanthi, P., Angelin, N., & Govindan, G. (2023). Harnessing the potential of CRISPR/Cas system for enhancing virus resistance in plants: Targets, strategies, and challenges. *Physiological and Molecular Plant Pathology, 129*, 102202.

Jia, H., & Wang, N. (2014). Targeted genome editing of sweet orange using Cas9/sgRNA. *PloS One, 9*(4), e93806.

Jiang, C., Lv, G., Tu, Y., Cheng, X., Duan, Y., Zeng, B., & He, B. (2021). Applications of CRISPR/Cas9 in the synthesis of secondary metabolites in filamentous fungi. *Frontiers in Microbiology, 12*, 638096.

Jiang, W., Bikard, D., Cox, D., Zhang, F., & Marraffini, L. A. (2013a). RNA-guided editing of bacterial Genomes using CRISPR-Cas systems. *Nature Biotechnology*, *31*, 233–239.

Jiang, W., Zhou, H., Bi, H., Fromm, M., Yang, B., & Weeks, D. P. (2013b). Demonstration of CRISPR/Cas9/sgRNA-mediated targeted gene modification in Arabidopsis, tobacco, sorghum and rice. *Nucleic Acids Research*, *41*(20), e188–e188

Jiang, W. Z., Henry, I. M., Lynagh, P. G., Comai, L., Cahoon, E. B., & Weeks, D. P. (2017). Significant enhancement of fatty acid composition in seeds of allohexaploid, Camelina sativa, using CRISPR/Cas9 gene editing. *Plant Biotechnology Journal*, *15*(5), 648–657.

June, C. H., O'Connor, R. S., Kawalekar, O. U., Ghassemi, S., & Milone, M. C. (2018). CAR T cell Immunotherapy for human cancer. *Science*, *359*, 1361–1365.

Ke, J., Robinson, D., Wu, Z. Y., Kuftin, A., Louie, K., Kosina, S., Northen, T., Cheng, J.-F., & Yoshikuni, Y. (2022). CRAGE-CRISPR facilitates rapid activation of secondary metabolite biosynthetic gene clusters in bacteria. *Cell Chemical Biology*, *29*(4), 696–710.

Khokhar Voytas, A., Shahbaz, M., Maqsood, M. F., Zulfiqar, U., Naz, N., Iqbal, U. Z., Sara, M., Aqeel, M., Khalid, N., Ali, N., Zulfiqa, F., Al Syaad, K. M., & AlShaqhaa, M. A. (2023). Genetic modification strategies for enhancing plant resilience to abiotic stresses in the context of climate change. *Functional & Integrative Genomics*, *23*(3), 283.

Kim, D. Y., Lee, J. M., Moon, S. B., Chin, H. J., Park, S., Lim, Y., Kim, D., Koo, T., Ko, J.-H., & Kim, Y. S. (2022). Efficient CRISPR editing with a hypercompact Cas12f1 and engineered guide RNAs delivered by adeno-associated virus. *Nature Biotechnology*, *40*(1), 94–102.

Kim, V. N. (2018). RNA-targeting CRISPR comes of Age. *Nature Biotechnology*, *36*, 44.

Koonin, E. V. (2019). CRISPR: A new principle of genome engineering linked to conceptual shifts in evolutionary biology. *Biology & Philosophy*, *34*(1), 9.

Koonin, E. V., Makarova, K. S., & Zhang, F. (2017). Diversity, classification and evolution of CRISPR-Cas systems. *Current Opinion in Microbiology*, *37*, 67–78.

Li, H. (2015). Structural principles of CRISPR RNA processing. *Structure*, *23*(1), 13–20.

Li, R., Li, R., Li, X., Fu, D., Zhu, B., Tian, H., Luo, Y., & Zhu, H. (2018). Multiplexed CRISPR/Cas9-mediated metabolic engineering of γ-aminobutyric acid levels in Solanum lycopersicum. *Plant Biotechnology Journal*, *16* (2), 415–427.

Li, X., Huang, L., Pan, L., Wang, B., & Pan, L. (2021). CRISPR/dCas9-mediated epigenetic modification reveals differential regulation of histone acetylation on Aspergillus niger secondary metabolite. *Microbiological Research*, *245*, 126694.

Liao, H.-K., Gu, Y., & Diaz, A. (2015). Use of the CRISPR/Cas9 system as an intracellular defense Against HIV-1 infection in human cells. *Nature Communications*, *6*, 6413.

Lowder, L. G., Zhang, D., Baltes, N. J., Paul, J. W., Tang, X., Zheng, X., Voytas, D. F., Hsieh, T.-F., Zhang, Y., & Qi, Y. (2015). A CRISPR/Cas9 toolbox for multiplexed plant genome editing and transcriptional regulation. *Plant Physiology*, *169*, 971–985.

Ma, J., Li, X., Lou, C., Lin, X., Zhang, Z., Chen, D., & Yang, S. (2023). Utility of CRISPR/Cas mediated electrochemical biosensors. *Analytical Methods*, *15*, 3785–3801.

Makarova, K. S., Wolf, Y. I., Iranzo, J., Shmakov, S. A., Alkhnbashi, O. S., Brouns, S. J. J., Charpentier, E., Cheng, D., Haft, D. H., Horvath, P., Moineau, S., Mojica, F. J. M., Scott, D., Shah, S. A., Siksnys, V., Terns, M. P., Venclovas, Č., White, M. F., Yakunin, A. F., … Koonin, E. V. (2020). Evolutionary classification of CRISPR-Cas systems: A burst of class 2 and derived variants. *Nature Reviews Microbiology*, *18*, 67–83.

Mali, P., Esvelt, K. M., & Church, G. M. (2013). Cas9 as a versatile tool for engineering biology. *Nature Methods*, *10*, 957–963.

Medvedieva, M. O., & Blume, Y. B. (2018). Legal regulation of plant genome editing with the CRISPR/Cas9 technology as an example. *Cytology and Genetics*, *52*, 204–212.

Mercx, S., Smargiasso, N., Chaumont, F., De Pauw, E., & Boutry, C. (2017). Navarre Inactivation of the β(1,2)-xylosyltransferase and the α(1,3)-fucosyltransferase genes in Nicotiana tabacum BY-2 cells by a multiplex CRISPR/Cas9 strategy results in glycoproteins without plant-specific glycans. *Frontiers in Plant Science*, *8*, 403.

Mercx, S., Tollet, J., Magy, B., Navarre, C., & Boutry, M. (2016). Gene inactivation by CRISPR-Cas9 in Nicotiana tabacum BY-2 suspension cells. *Frontiers in Plant Science*, *7*, 40.

Meyer, J. D., Silva, D. C., Yang, C., Pedley, K. F., Zhang, C., van de Mortel, M., Hill, J. H., Shoemaker, R. C., Abdelnoor, R. V., Whitham, S. A., & Graham, M. A. (2009). Identification and analyses of candidate genes for Rpp4-mediated resistance to Asian soybean rust in soybean. *Plant Physiology*, *150*(1), 295–307.

Michno, J. M., Wang, X., Liu, J., Curtin, S. J., Kono, T. J., & Stupar, R. M. (2015). CRISPR/Cas mutagenesis of soybean and Medicago truncatula using a new web-tool and a modified Cas9 enzyme. *GM Crops & Food*, *6*(4), 243–252.

Mojica, F. J. M., Juez, G., & Rodriguez-Valera, F. (1993). Transcription at different salinities of Haloferax mediterranei sequences adjacent to partially modified PstI sites. *Molecular Microbiology*, *9*, 613–621.

Munawar, N., Ahsan, K., & Ahmad, A. (2024). CRISPR-edited plants' social, ethical, policy, and governance issues. In K. A. Abd-Elsalam & A. Ahmad (Eds.), *Global regulatory outlook for CRISPRized plants* (pp. 367–396). Switzerland: Academic Press.

Myhrvold, C., Freije, C. A., Gootenberg, J. S., Abudayyeh, O. O., Metsky, H. C., Durbin, A. F., Kellner, M. J., Tan, A. L., Paul, L. M., Parham, L. A., Garcia, K. F., Barnes, K. G., Chak, B., Mondini, A., Nogueira, M. L., Isern, S., Michael, S. F., Lorenzana, I., Yozwiak, N. L., … Sabeti, P. C. (2018). Field-deployable viral diagnostics using CRISPR-Cas13. *Science*, *360*, 444–448.

Naz, M., Benavides-Mendoza, A., Tariq, M., Zhou, J., Wang, J., Qi, S., Dai, Z., & Du, D. (2022). CRISPR/Cas9 technology as an innovative approach to enhancing the phytoremediation: Concepts and implications. *Journal of Environmental Management*, *323*, 116296.

Nierzwicki, Ł., Arantes, P. R., Saha, A., & Palermo, G. (2021). Establishing the allosteric mechanism in CRISPR-Cas9. *Wiley Interdisciplinary Reviews: Computational Molecular Science*, *11*(3), e1503.

Pan, C., & Qi, Y. (2023). CRISPR-combo-mediated orthogonal genome editing and transcriptional activation for plant breeding. *Nature Protocols*, *18*(6), 1760–1794.

Pan, C., Sretenovic, S., & Qi, Y. (2021). CRISPR/dCas-mediated transcriptional and epigenetic regulation in plants. *Current Opinion in Plant Biology*, *60*, 101980.

Qi, L. S., Larson, M. H., Gilbert, L. A., Doudna, J. A., Weissman, J. S., Arkin, A. P., & Lim, W. A. (2013). Repurposing CRISPR as an RNA-guided platform for sequence-specific control of gene expression. *Cell*, *152*, 1173–1183

Qiao, J., Zhao, Z., Li, Y., Lu, M., Man, S., Ye, S., Zhang, Q., & Ma, L. (2023). Recent advances of food safety detection by nucleic acid isothermal amplification integrated with CRISPR/Cas. *Critical Reviews in Food Science and Nutrition*, *10*, 1–22.

Rajyaguru, R., Maheshala, N., Mootapally, C., Nathani, N., Tomar, R., Bhalani, H., & Sharma, P. (2021). CRISPR-Cas9 approaches to enhance contents of plant secondary metabolites. In M. Shahnawaz (Ed.), *Biotechnological approaches to enhance plant secondary metabolites: Recent trends and future prospects* (p. 133). CRC Press.

Raper, A. T., Stephenson, A. A., & Suo, Z. (2018). Functional insights revealed by the kinetic mechanism of CRISPR/Cas9. *Journal of the American Chemical Society*, *140*(8), 2971–2984.

Sabzehzari, M., Hoveidamanesh, S., Modarresi, M. & Mohammadi, V. (2020a). Morphological, anatomical, physiological, and cytological studies in diploid and tetraploid plants of Ispaghul (Plantago ovata Forsk.). *Genetic Resources and Crop Evolution*, *67*, 129–137.

Sabzehzari, M., Zeinali, M., & Naghavi, M. R. (2020b). CRISPR-based metabolic editing: Next-generation metabolic engineering in plants. *Gene*, *759*, 144993.

Sander, J. D., & Joung, J. K. (2014). CRISPR-Cas systems for editing, regulating and targeting Genomes. *Nature Biotechnology*, *32*, 347–355.

Satish, L., Lavanya, G., Kasthuri, T., Kalaivaani, A., Shamili, S., Muthuramalingam, P., Gowrishankar, S., Pandian, S. K., Singh, V., Sitrit, Y., & Kushmaro, A. (2021). CRISPR based development of RNA editing and the diagnostic platform. *Progress in Molecular Biology and Translational Science*, *179*, 117–159.

Saxena, P., Singh, N. K., Singh, A. K., Pandey, S., Thanki, A., & Yadav, T. C. (2020). Recent advances in phytoremediation using genome engineering CRISPR-Cas9 technology. *Bioremediation of Pollutants*, *2020*, 125–141.

Schaeffer, S. M., & Nakata, P. A. (2015). CRISPR/Cas9-mediated genome editing and gene replacement in plants: Transitioning from lab to field. *Plant Science*, *240*, 130–142.

Shan, Q., Wang, Y., & Li, C. (2014). Gao genome editing in rice and wheat using the CRISPR/Cas system. *Nature Protocols*, *9*(10), 2395–2410.

Shelake, R. M., Jadhav, A. M., Bhosale, P. B., & Kim, J. Y. (2023). Unlocking secrets of nature's chemists: Potential of CRISPR/Cas-based tools in plant metabolic engineering for customized nutraceutical and medicinal profiles. *Plant Physiology and Biochemistry*, *203*, 108070.

Shinwari, Z. K., Tanveer, F., & Khalil, A. T. (2018). Ethical issues regarding CRISPR mediated genome editing. *Current Issues in Molecular Biology*, *26*(1), 103–110.

Shkryl, Y., Yugay, Y., Avramenko, T., Grigorchuk, V., Gorpenchenko, T., Grischenko, O., & Bulgakov, V. (2021). CRISPR/Cas9-mediated knockout of HOS1 reveals its role in the regulation of secondary metabolism in Arabidopsis thaliana. *Plants*, *10*(1), 104.

Shrestha, A., Khan, A., & Dey, N. (2018). Cis-trans engineering: Advances and perspectives on customized transcriptional regulation in plants. *Molecular Plant*, *11*(7), 886–898.

Slaymaker, I. M., Gao, L., & Zetsche, B. (2016). Rationally engineered Cas9 nucleases with improved specificity. *Science*, *351*(6268), 84–88.

Tang, G. Q., Novitzky, W. P., Carol Griffin, H., Huber, S. C., & Dewey, R. E. (2005). Oleate desaturase enzymes of soybean: Evidence of regulation through differential stability and phosphorylation. *Plant Journal*, *44*, 433–446.

Wan, H. F., Han, W. J., Zhou, L., Wang, S., & Sui, C. (2022). New advances of CRISPR/Cas9 technique and its application in disease treatment and medicinal plants research. *Current Pharmaceutical Biotechnology*, *23*(14), 1678–1690.

Wang, D., Jin, S., Lu, Q., & Chen, Y. (2023). Advances and challenges in CRISPR/Cas-based fungal genome engineering for secondary metabolite production: A review. *Journal of Fungi*, *9*(3), 362.

Wang, M., Zhang, R., & Li, J. (2020). CRISPR/Cas systems redefine nucleic acid detection: Principles and methods. *Biosensors and Bioelectronics*, *165*, 112430.

Xie, F., Ye, L., & Chang, J. C. (2014). Seamless gene correction of β-thalassemia mutations in patient-specific iPSCs using CRISPR/Cas9 and piggyBac. *Genome Research*, *24*, 1526–1533.

Xing, H. L., Dong, L., Wang, Z. P., Zhang, H. Y., Han, C. Y., Liu, B., Wang, X.-C., & Chen, Q. J. (2014). A CRISPR/Cas9 toolkit for multiplex genome editing in plants. *BMC Plant Biology*, *14*(1), 1–12.

Yang, H., Zhang, J., Zhang, X., Shi, J., Pan, Y., Zhou, R., Li, G., Li, Z., Cai, G., & Wu, Z. (2018). CD163 knockout pigs are fully resistant to highly pathogenic porcine reproductive and respiratory syndromevirus. *Antiviral Research*, *151*, 63–70.

Yin, H., Xue, W., & Chen, S. (2014). Genome editing with Cas9 in adult mice corrects a disease mutation and phenotype. *Nature Biotechnology*, *32*, 551–553.

Yin, K., Gao, C., & Qiu, J. L. (2017). Progress and prospects in plant genome editing. *Nature Plants*, *3*, 17107.

Yuan, B., Keller, N. P., Oakley, B. R., Stajich, J. E., & Wang, C. C. (2022). Manipulation of the global regulator mcrA upregulates secondary metabolite production in Aspergillus wentii using CRISPR-Cas9 with *in vitro* assembled Ribonucleoproteins. *ACS Chemical Biology*, *17*(10), 2828–2835.

Zehra, G. Ü. N., & Tunali, B. Ç. (2016). Biology, mechanism and applications of CRISPR-Cas immune system. *International Journal of Engineering Research and Development*, *8*(2), 11–21.

Zhang, D., Hussain, A., Manghwar, H., Xie, K., Xie, S., Zhao, S., Larkin, R. M., Qing, P., Jin, S., & Ding, F. (2020). Genome editing with the CRISPR-Cas system: An art, ethics and global regulatory perspective. *Plant Biotechnology Journal*, *18*(8), 1651–1669.

Zhang, N. J., Baltes, J. W., Paul, X., Tang, X., Zheng, D. F., Voytas, T. F., Hsieh, T.-F., Zhang, Y., & Zhang, Y. (2015). Qi A CRISPR/Cas9 toolbox for multiplexed plant genome editing and transcriptional regulation. *Plant Physiology*, *169*(2), 971–985.

Zhao, Z., Lu, M., Wang, N., Li, Y., Zhao, L., Zhang, Q.,Man, S., Ye, S., & Ma, L. (2023). Nanomaterials-assisted CRISPR/Cas detection for food safety: Advances, challenges and future prospects. TrAC. *Trends in Analytical Chemistry*, *2023*, 117269.

Zhou, Z., Tan, H., Li, Q., Chen, J., Gao, S., Wang, Y., Chen, W., & Zhang, L. (2018). CRISPR/Cas9-mediated efficient targeted mutagenesis of RAS in Salvia miltiorrhiza. *Phytochemistry*, *148*, 63–70.

8 Plant Biochemical and Molecular Responses to Extreme Temperature

Santanu Samanta and Aryadeep Roychoudhury

8.1 INTRODUCTION

Plants have a sessile lifestyle that makes them vulnerable to changes in their surroundings. One of these environmental changes is the extreme temperature conditions caused by fluctuating climate parameters, which are one of the most important ecological concerns nowadays. Additionally, extreme temperature events are expected to be more intense and more frequent in the near future (Meehl et al. 2007). Extreme temperature can be of two types: (1) high temperature or heat stress and (2) low temperature or chilling/freezing stress in plants. Heat stress is often defined as the rise in temperature beyond a threshold level (>25°C). In contrast, low temperature can be defined as subzero (freezing, i.e., <0°C) to chilling (0–15°C) temperature, which are common in nature

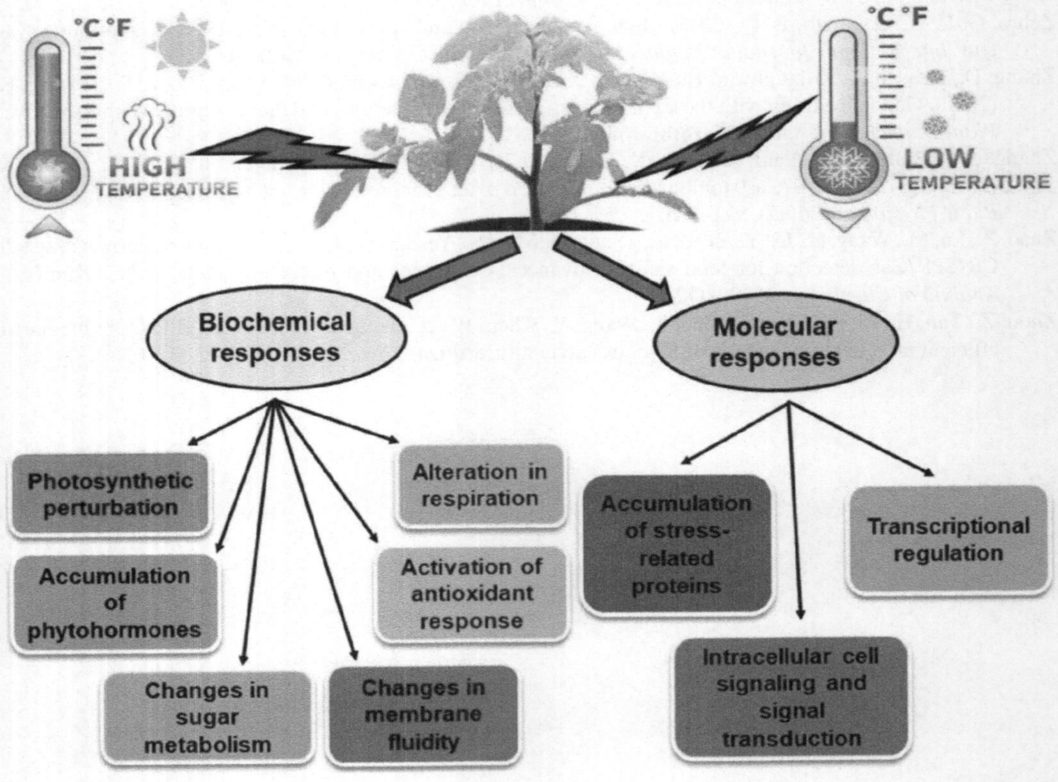

FIGURE 8.1 Generalized plant responses during extreme temperature conditions with sustainable yield production.

DOI: 10.1201/9781003457237-8

(Burke and Wanjura 2010; Hatfield and Prueger 2015). Extreme temperature affects the geographical distribution of plants and reduces plant productivity, thus having large societal and economic consequences (Porter 2005; Arun-Chinnappa et al. 2017). Both high- and low-temperature episodes have a wide range of negative effects on plant physiology and overall productivity since they occur at different temporal scales and with various levels of intensity and duration (Stewart et al. 2021). Plants experience temperature stress in different ways and adaptation or acclimation to extreme temperature conditions occur over different time scales and levels of plant organization.

Since temperature is an uncontrollable climatic factor, understanding plant responses and defense mechanisms is imperative to alleviate the adverse effects of extreme temperature for the development of plants with improved thermotolerance/cold resistance for ensuring food security. To maintain growth and productivity under extreme temperature regimes, plants must adapt to stress conditions and develop an extensive array of adaptive measures at various levels of organization to counter high- or low-temperature stresses. The ability of plant modifications for enhanced tolerance is mostly based on the ability to perceive heat/cold signals and transduce them into the downstream molecules that trigger the appropriate defense pathways (Zandalinas et al. 2020). The complex plant response to both high- and low-temperature conditions which involves biochemical and molecular mechanisms are schematically elucidated in Figure 8.1. In this chapter, we briefly consider how extreme temperature limits the physiological and metabolic processes of plants along with biochemical and molecular responses to promote plant acclimation, tolerance, and survival under such harsh environmental conditions. From this perspective, this chapter provides a closer insight into strategies, which could lead to the generation of rational designing of novel cultivars.

8.2 PLANT RESPONSES TO EXTREME TEMPERATURE

Extreme temperature, such as both heat and cold stress, may hinder different aspects of plant growth and development in several ways. The plant responses during thermal or cold stress can be broadly divided into two general types: (1) biochemical responses and (2) molecular responses. Some of these common responses are widely discussed in the underlying sections.

8.2.1 Biochemical Responses

8.2.1.1 Membrane Dynamics

Temperature is a physical cue, and it directly influences the state of the plasma membrane and the remodeling of membrane microdomains. Membranes, particularly plasma and chloroplast membranes, act as a first responder during temperature fluctuation; modifications in membranes may be a prelude to several metabolic alterations in cells. During acclimation in response to extreme temperature conditions, the changes in membrane lipid composition are crucial for maintaining integrity for plant survival. Rising temperature leads to increased fluidity of the membrane as well as ion permeability. To counteract heat stress-incited increased fluidity, membranes accumulate high concentration of unsaturated triacylglycerols to tackle increased viscosity under thermal stress (Higashi et al. 2015; Mueller et al. 2017). A rapid accumulation of phosphatidylinositol-4,5-bisphosphate (PIP_2) and phosphatidic acid and the activation of phospholipase D and phosphatidylinositol phosphate kinase are among the earliest events following the onset of thermal stress (Mishkind et al. 2009). The contrasting events occurred during chilling temperature, where the rigidity of the membrane is induced, ultimately causing the loss of the bilayer structure, membrane leakiness and perturbed transport and metabolic activity, which have dramatic effects on cellular metabolism (Heidarvand and Amiri 2010). The degree of fatty acid saturation reduction during cold environment is a determining factor for plant survival under such conditions (Uemura et al. 2006). Degradation of membrane phospholipids by phospholipase D has been observed in cortical cells of poplar trees under freezing conditions (Voshida and Sakai 1974). Freeze-induced breakdown of membrane phospholipids is a common phenomenon because of physical pressure exerted by extracellular ice formation, thereby disrupting normal functions of the plasma membrane. H^+-ATPase is a membrane-localized primary transporter that plays an indispensable role in physiological functions such as maintaining intracellular pH balance and electrical potential gradient across the plasma membrane and tonoplast. Substantial reports have shown that the impairment of H^+-ATPase is considered as one of the initial physiological responses of cells during low-temperature conditions (Ahn et al. 2000; Janicka-Russak et al. 2012; Muzi et al. 2016).

8.2.1.2 Photosynthesis

Photosynthesis is highly sensitive to extreme temperature conditions, especially during elevated temperature regimes, and is often inhibited before other cellular functions are impaired. The photosynthetic apparatus is highly sensitive to extreme temperature, which can respond to an imbalance in cellular energy and redox status (Mathur et al. 2014). The impairment of chlorophyll biosynthesis is a common phenomenon in plastids during exposure to both high- and low-temperature stress. Lesser accumulation of chlorophyll in heat-stressed plants results from diminished enzyme activities involved in chlorophyll biosynthesis or its accelerated degradation or a combination of both (Dutta et al. 2009). 5-Aminolevulinic acid dehydratase, involved in the pyrrole biosynthetic pathway, was recorded to be downregulated under heat stress (Mohanty et al. 2006; Tiwari and Tripathi 1998). In addition, protochlorophyllide (Pchlide) formation, as well as Pchlide oxidoreductase and porphobilinogen deaminase, was analogously affected in heat-stressed wheat seedlings (Tiwari and Tripathi 1998). The impact of chilling stress was also documented with sharp decrement in maximum quantum yield of photosystem II (PSII) (F_v/F_m), efficiency of photosystem I (PSI) and PSII (PI_{ABS}), which was accompanied by reduction in net photosynthesis (P_N), intracellular CO_2, water use efficiency in nine cultivars of *Stevia rebaudiana* (a sweet medicinal herb) (Hajihashemi et al. 2018). During exposure to high-temperature stress, the electron transport system in PSII gets perturbed owing to an increment in the fluidity of the thylakoid membranes, which causes the dislodging of PSII light-harvesting complexes from thylakoid membrane (Srivastava et al. 1997; Baker and Rosenqvist 2004; Janka et al. 2013). In addition, the dissociation of extrinsic 18, 24, and 33 kDa polypeptides and loss of divalent cations like Ca^{2+} and Mn^{2+} and Cl^- anion from the PSII pigment–protein complex from the thylakoid membranes are also the prime reasons behind the heat-inactivated PSII (Havaux 1993; Wise et al. 2004). Like PSII, PSI is also affected during thermal stress conditions. Several researchers have shown that PSI activity is much more thermostable than PSII (Havaux 1993). An increased PSI activity has been observed under moderate high-temperature conditions, leading to more adenosine triphosphate (ATP) production, which ultimately results in a higher nicotinamide adenine dinucleotide phosphate hydrogen/adenosine triphosphate (NADPH/ATP) ratio, favoring the non-photochemical reduction of plastoquinone pool (Mathur et al. 2014). Similarly, there are numerous studies of cold-induced photoinhibition, subsequently leading to a reduction in photosynthetic activity under low-temperature stress (Hou et al. 2016). Low temperature causes stomatal closure by affecting guard cell functioning that directly or indirectly alters internal CO_2 concentration, leading to the reduction of ribulose 1,5-bisphosphate carboxylase oxygenase (Rubisco) activity (Allen and Ort 2001). Previously, Cvetkovic et al. (2017) found overexcitation of PSII under cold stress-triggered energy dissipation through non-radiative reactions. The maximum efficiency of PSII was attenuated by 18% after 1 day of exposure to chilling stress (Venzhik et al. 2011). A 45% decreased photosynthetic rate in primary leaves was reported upon imposing cold stress (day: night, 5°C:5°C), as compared to control (day: night, 20°C:16°C) (Leonardos et al. 2003). Thus, low temperature-incited diminished photosynthetic activity reduced the photo-assimilation or transportation of assimilates, ultimately causing significant yield losses in plants.

8.2.1.3 Respiration

Respiration is instrumental for the survival and growth of plants, which are highly temperature sensitive; even small temperature fluctuation affects the whole-plant respiration cycle and carbon balance of plants. Generally, the rate of respiration is triggered exponentially with short-term increments in temperature (Armstrong et al. 2006), which is primarily driven by an increase in the demand for cellular maintenance against heat stress-mediated damages. However, long-term warming of the temperature regime may downgrade plant respiration (non-photorespiratory mitochondrial CO_2 release). In an experiment, moderately high-temperature regimes (27°C and 30°C) led to increased respiration rates by 49% and 56%, respectively, with respect to the control seedlings (Loka and Oosterhuis 2010). The multiprotein complexes involved in mitochondrial respiration like NADH dehydrogenase (complex I), succinate dehydrogenase (complex II), cytochrome bc1 (complex III), cytochrome c oxidase (complex IV), ATPase (complex V), and alternate oxidase (AOX) are severely affected in response to both high- and low-temperature stress (Yang et al. 2011; Rashid et al. 2020), thereby influencing the ATP production. Furthermore, a rise in atmospheric temperature induces leakage of protons from mitochondrial inter-membrane space to the mitochondrial matrix, which results in the loss of proton motive force and ATP synthase activity (Brookes 2005). The impeded ATP production also causes the accumulation of adenosine diphosphate (ADP) and a lower ATP/ADP ratio. The rate of respiration is increased in a wide range of Antarctic species, evergreen shrubs, trees, cereals, forbs, and grasses, acclimated to low temperature (Bravo et al. 2007; Campbell et al. 2007). The enhanced respiration rate, induced by low temperature, was correlated with higher metabolic demand for the production of energy and carbon skeletons for biosynthesis. The mitochondrial electron transport chain is sensitive to chilling stress due to the inactivation of several respiratory enzymes. Plants possess two respiratory electron transport pathways: salicylhydroxamic acid (SHAM: selective competitive inhibitor of the alternate oxidase in the mitochondrial electron transport pathway)-resistant, cytochrome pathway and cyanide-resistant, alternate oxidase (AOX) pathway. Low-temperature stress repressed the activity of the cytochrome pathway; however, the alternate pathway, escalated under cold conditions, is attributed to induced synthesis along with the increased activity of AOX to mitigate low-temperature stress (Miroslava and Otília 1999; Grabelnych et al. 2004). The AOX pathway mediates the integration of respiration and other metabolic pathways by generating H_2O molecules from the intracellular oxygen–water concentration and preventing the blockade of electron flow to maintain metabolic homeostasis during cold stress (Wang et al. 2011; Vanlerberghe 2013).

8.2.1.4 Sugar Metabolism

Sugars (primary metabolites) act as prime carbon and energy sources for most of the cell types. Being the major source of carbon, they are important for energy metabolism and act as substrates for plant growth and development (Sami et al. 2016). The accumulation of sugars is fundamental for membrane stabilization, osmotic adjustment, and protection of proteins from desiccation during temperature acclimation via maintaining cellular water balance and buffering intracellular redox balance (Durand et al. 2017). It has been observed that the concentration of soluble sugar is rapidly induced under extreme temperature conditions as an early metabolic response (Yang et al. 2009; Wang et al. 2018). In addition, the dramatic induction of sugar might result from reduced sink activity due to growth retardation in response to cold stress (Huner et al. 1998). An amplified sugar accumulation, correlated with freezing tolerance of natural accessions of *Arabidopsis* and alfalfa (*Medicago sativa*), has been documented (Klotke et al. 2004; Hannah et al. 2006; Bertrand et al. 2017). Sucrose is the most common and most abundant sugar found to be accumulated during and after exposure to low temperature (non-freezing) stress (Steponkus and Lanphear 1968; Kaurin et al. 1981; Guy et al. 1992) along with the increased activity of sucrose phosphate synthase (SPS; responsible for sucrose synthesis from fructose and uridine diphosphate [UDP]-glucose). However, other forms of sugar like fructan have also been reported to accumulate in temperate grasses under low-temperature environment. Studies have shown that heat stress exposure also promotes the rise of soluble sugar levels to combat stress-mediated damage (Lokesha et al. 2019; Zheng et al. 2023). Previously, Zhao et al. (2013) illustrated how heat stress affected sugar accumulation and related enzyme activity during the grain-filling stage in sweet corn. The intracellular level of soluble sugar and sucrose initially increased and then decreased, which was corroborated by the lowered activity of SPS and increased activity of sucrose synthase (SS; responsible for sucrose catabolism and interconversion) that led to a lowering of sucrose content and a reduction of kernel quality. A similar accumulation of soluble sugars was evident in sugarcane sprouts upon heat stress treatment, which contributed significant defense responses against heat damages for short-term heat stress (up to 25 hours) conditions (Gomathi et al. 2013). However, an opposite scenario was observed where heat stress declined the concentration of sugars such as fructose, sugar nucleotides, and hexose phosphate in wheat seedlings, which might be related to increased assimilate utilization under thermal stress conditions (Jenner 1991). Sugar molecules, serving as primary signaling molecules, also mediate the source–sink relationship under heat stress at the anthesis stage (Zhang et al. 2018).

8.2.1.5 Antioxidant Response

When the environmental temperature overpasses the physiological threshold of a plant, the concentration of reactive oxygen species (ROS) and reactive nitrogen species (RNS) are elevated inside the cells. Temperature stress favors excess ROS and RNS production that causes the loss of redox homeostasis in tissues, resulting in oxidative injuries in plants (Munné-Bosch and Alegre 2002). In normal conditions, basal levels of ROS and RNS are tightly regulated and are necessary for numerous signaling purposes in plants. However, the balance between ROS and RNS production and subsequent scavenging is disrupted under stressed environment like temperature extremes. At moderate to high concentrations, such toxic active oxygen species are deleterious for cellular macromolecules. Plants have developed sophisticated antioxidant machinery primarily composed of enzymatic and non-enzymatic compounds. Enzymatic antioxidants include superoxide dismutase (SOD), catalase (CAT), guaiacol peroxidase (GPOX), glutathione reductase (GR), monodehydroascorbate reductase (MDHAR), dehydroascorbate reductase (DHAR), and ascorbate peroxidase (APX). Non-enzymatic antioxidants comprise glutathione (GSH), ascorbate (ASC), carotenoids, xanthophyll, and flavonoids. Short-term (for 10 hours) of three levels of heat exposure (37°C, 42°C, and 47°C) significantly induced both enzymatic antioxidants (CAT, SOD, POD, APX, GR) and non-enzymatic antioxidants (ASC and GSH), which scavenged ROS in lily (*Lilium longiflorum*) seedlings (Yin et al. 2008). Similarly, thermal stress-triggered antioxidant defense responses (APX, SOD, and CAT) alleviated heat-induced excess ROS scavenging for maintaining the stability of the photosynthetic apparatus in *Jatropha curcas* (Silva et al. 2017). In another report, Hameed et al. (2012) reported that CAT activity was diminished upon exposure to high temperature, although peroxidase, SOD, and APX activity was induced in six genotypes of wheat seedlings. The heat-induced depletion in CAT activity was compensated by parallel induction in peroxidases to quench the toxic ROS molecules. The induction of antioxidative enzyme activities such as SOD and peroxidases along with the upregulation of transcript levels of *Mn-SOD*, *Cu/Zn-SOD*, *POD*, and *APX* has been recorded to counteract the cold stress-mediated repercussions in bermudagrass (*Cynodon dactylon*) (Fan et al. 2014). The protection from chilling injury could be induced by chilling acclimation, which was partly mediated by increased enzyme activities as well as isozyme profiling of CAT and APX in leaves and SOD, CAT, APX, and GR in roots in chilling-susceptible rice plants (Kuk et al. 2003). The differential antioxidant response was recorded in two different cultivars of *Brassica* seedlings. The activity of the enzymes related to the ASC-GSH cycle such as APX, MDHAR, DHAR, and GR, together with content of ASC and GSH, were recorded higher in WS-1 (chilling-tolerant genotype) compared to Ta2 (chilling-susceptible genotype) in wucai (*Brassica campestris*) against chilling stress (Wang et al. 2020). Therefore, tolerance to chilling in WS-1 is well confirmed by the enhancement of antioxidant machinery under chilling environment.

8.2.1.6 Phytohormones

Phytohormones are the endogenous biological messengers for coordinating a myriad of phenomena associated with plant growth and development. Several previous reports provided solid evidence that phytohormones are actively involved in plant responses to extreme temperature conditions (Jha et al. 2022; Lado et al. 2023). Both high and low temperature alters phytohormone homeostasis, endogenous level, biosynthesis, stability, and compartmentalization in plants. Heat stress-mediated depletion of endogenous auxin levels in the developing anthers of barley, associated with the down-regulation of *YUCCA* (auxin biosynthetic genes), was reported by Sakata et al. (2010). However, the abscisic acid (ABA) level was triggered in response to heat stress, which was correlated with the induced transcript level of *zeaxanthin epoxidase* (*ABA1/ZEP*) as well as the rate-limiting enzyme, *9-cisepoxycarotenoid dioxygenase* (*NCED2*, *NCED5*, and *NCED9*). Contrastingly, the gibberellic acid (GA) level was lowered, together with the suppressed gene expression of *GA20 oxidase* genes (*GA20ox1*, *GA20ox2*, and *GA20ox3*) and *GA3 oxidase* genes (*GA3ox1*, *GA3ox2*) in *Arabidopsis* seeds (Toh et al. 2008). Interestingly, Todorova et al. (2005) revealed that *Arabidopsis* seedlings exposed to heat stress exhibited decreased cytokinin content, while cold stress treatment resulted in elevated total cytokinin levels. Sometimes, these phytohormones work coordinately to modulate stress responses. An increment in ABA level was documented in green algae under heat stress in the presence of brassinosteroids (Bajguz 2009). A complex phytohormonal response during cold acclimation in two different wheat cultivars (winter Samanta and spring Sandra) was uncovered by Kosová et al. (2012). A rapid elevation of ABA level has been observed in both cultivars, with faster and stronger response in winter Samanta with a simultaneous decrease in growth-promoting phytohormones (auxin, bioactive cytokinins, and GA) during the alarm phase (1 day). After 3–7 days of cold exposure, the levels of growth-promoting phytohormones were restored with a decrease in ABA and the level of other stress phytohormones (jasmonic acid [JA] and salicylic acid [SA]) began to rise. After prolonged cold treatment (21 days: resistance phase), JA and SA levels were gradually elevated, with a decline in bioactive cytokinins and auxin, which indicated an acclimation period. Recent evidence supported the involvement of JA in leaf senescence and tolerance to cold stress via positively regulating C-repeat binding factor (CBF) transcriptional pathway to upregulate downstream cold-responsive genes (Hu et al. 2017). ABA also plays an important role in cold stress tolerance. The biosynthesis of ABA is not an early event in response to cold stress; rather, it functions late in the development of cold-induced metabolic alterations (Shi and Yang 2014).

8.2.2 Molecular Responses

8.2.2.1 Stress-Related Proteins

To cope with temperature stress, plants have developed several defense mechanisms. At the molecular level, plants deploy a series of defense responses during exposure to temperature extremes. Some of these defense components are discussed briefly in the underlying sections.

8.2.2.1.1 Heat Shock Proteins

One strategy is inducing the synthesis of a small group of evolutionarily conserved polypeptides known as heat shock proteins (HSPs) in response to high temperature (Parsell and Lindquist 1993; Morimoto and Westerheide 2010). They are not only synthesized in response to short-term heat exposure, but their production is also a necessary step for heat acclimation. According to molecular weight, amino acid sequences, and functions, five major families of HSPs are synthesized in plants: the HSP100 (Clp) family, the HSP90 family, the HSP70 (DnaK) family, the chaperonins (GroEL and HSP60), and the low-molecular-weight HSPs or small HSPs (~17–30 kDa) family. The evolutionary conservation of heat shock responses and HSPs with cellular resistance to high temperature has led to the long-standing hypothesis that HSPs protect from detrimental effects of elevated temperature, and the accumulation of HSPs leads to increased thermotolerance. It has been suggested that the general role of HSPs is to act as molecular chaperones that bind to partially folded or heat-denatured substrate proteins and prevent irreversible aggregation or promote correct folding of a substrate during thermal stress. In a study, creeping bentgrass (*Agrostis stolonifera*) was subjected to 38°C; the results showed that HSP101, HSP90, HSP70, and small HSPs were found to be upregulated (Wang et al. 2014a). HSP101 proteins are abundantly available in grass species, such as rice, wheat, and maize, as well as perennial grass, *Dichanthelium lanuginosum* (Young et al. 2001;

Campbell et al. 2001; Al-Niemi and Stout 2002; Agarwal et al. 2003a). Increased evidence established that, among the five conserved families of HSPs, small HSPs are the most dominant and prevalent proteins expressed in higher plants upon heat exposure (Harndahl et al. 1999; Sun et al. 2002). For instance, a significantly higher thermotolerance was observed by overexpressing HSP18.1, which conferred protection against heat shock in carrots (Malik et al. 1999). The *Hsp20* genes, identified in bread wheat, *Triticum aestivum*, were upregulated during high-temperature stress. Recent research has found another class of stress-related proteins, viz., the SP family of proteins that have some homology with small HSPs, with additional characteristics and functions (Wang et al. 2002).

8.2.2.1.2 Antifreeze Proteins

Plants differ in their resistance to chilling and freezing temperature. Plants can avoid freezing stress either by surviving in a dormant state (avoidance) or by readjusting their metabolism (tolerance). Freezing-tolerant plants (those able to thrive in extracellular freezing) can enhance the capacity of their tolerance and survive the injury caused by the formation of extracellular ice in their tissues at subzero temperature (Antikainen and Griffith 1997). Freezing injury usually arises from cellular dehydration via the removal of water from the cytoplasm to the growing ice crystals. Moreover, growing ice undergoes a spontaneous process known as recrystallization, which also imposes freezing injury in plants. As freeze-dehydration proceeds, the cellular content becomes more and more concentrated. Although the mechanism of freezing tolerance is complex, an important plant defense mechanism involves the secretion of antifreeze proteins (AFPs) into the apoplast that represses the formation of ice crystals (Atıcı and Nalbantoğlu 2003; Griffith and Yaish 2004; Gupta and Deswal 2014). In plants, AFPs were first identified in winter rye (Griffith et al. 1992; Urrutia et al. 1992). Later on, AFPs were reported in more than 60 plant species, including ferns, gymnosperms, monocots, and dicots. These specialized proteins are thought to function by adsorbing onto the surface of ice crystals and slow down the rate of ice propagation together with recrystallization, thereby protecting cells from injury at subzero temperature (Raymond and DeVries 1977; Knight et al. 1991; Jia and Davies 2002). Although the specific role of AFPs has not been uncovered, it is known that they enhance freezing survival by lowering the freezing point of the tissues. Plant AFPs are different from animal AFPs, in that they do not act as cryoprotectants; rather, they act as an ice-interacting protein. Furthermore, plant AFPs have not been broadly categorized into four distinct groups (type I–IV) like animal AFPs due to their immense diversity (Bredow and Walker 2017). Surprisingly, a nucleotide sequencing study showed that plant AFPs are homologous to the pathogenesis-related proteins, which are vital for pathogen defense in plants (Hon et al. 1995). The regulation of AFPs involves tissue-specific and developmental responses. Previous reports have shown that transcripts and translation products of *AFP* genes are rapidly escalated in response to cold exposure (Yeh et al. 2000; Huang and Duman 2001).

8.2.2.1.3 Cold-Shock Domain Proteins

In addition to AFPs, cold-shock domain proteins (CSDPs) have been detected in a variety of plant species, including monocots and dicots. They are one of the most conserved nucleic acid-binding proteins, widely distributed from bacteria to mammals. CSDPs present in plants generally differ from CSPs found in prokaryotes (Sasaki and Imai 2012). They are typically characterized by two distinct nucleic acid-binding modules (a single N-terminus cold-shock domain and variable quantities of C-terminal retroviral-like CCHC zinc fingers), which are interspersed by glycine-rich regions (Kingsley and Palis 1994; Karlson and Imai 2003). Notably, certain CSDP family members also exert their roles by functioning as ribonucleic acid (RNA) chaperones (Jung et al. 2013). The first CSDP was functionally characterized in wheat. The wheat cold-shock protein 1 (WCSP1) accumulates in crown tissues during cold acclimation (Karlson et al. 2002). In a similar report, Kim et al. (2009) illustrated that overexpressing *AtCSDP3* in transgenic plants displayed induced freezing tolerance over wild-type plants in *Arabidopsis*. It was initially postulated that plant CSDPs might be involved in translation during cold acclimation. Later on, interactome analysis performed by Kim et al. (2013) suggested possible functions of plant CSDPs in RNA processing and stability control.

8.2.2.2 Signal Perception and Transduction Events during Temperature Extremes

8.2.2.2.1 High-Temperature Conditions

Plasma membranes act as primary heat sensors via alteration in membrane fluidity, resulting in the activation of CNGC (cyclic nucleotide-gated Ca^{2+} channel) heat sensors. CNGCs are cyclic nucleotide-gated ion channels that allow the transport of Ca^{2+}, K^{+}, and other cations (Véry and Sentenac 2002). They consist of transmembrane segments with a conserved calmodulin (CaM)-binding domain at the C-terminus that physically overlaps the nucleotide-binding domain (Clough et al. 2000). Heat exposure results in the activation of CNGCb and CNGC2/4/6 that leads to the transient entry of Ca^{2+} ions, promoting the expression of various HSPs (Saidi et al. 2009; Finka et al. 2012; Gao et al. 2012). In the recent past, Li et al. (2018) documented that *cngc16* mutants failed to induce the key thermotolerance genes (*HSFA2* and *HSFB1*). In addition, glutamate receptor-like (GLR) Ca^{2+}-channels and Ca^{2+}-dependent phospholipid-binding annexins, viz., ANN1 and ANN5, have been suggested to act as thermosensors and facilitate heat resistance (Qiao et al. 2015; Li et al. 2019). Several Ca^{2+}-binding proteins, such as calcium-dependent protein kinase (CDPKs) and calmodulin (CaM), play a key role in sensing stress cues and transduce them via second messengers to downstream intra- and intercellular signaling networks. The overexpression of AtCaM3 leads to the significant enhancement of HSP levels and confers protection against high-temperature stress, while mutants exhibited impaired heat resistance (Zhang et al. 2009). Previously, Liu et al. (2007, 2008) also revealed that CaM-binding protein kinase 3 (CBK3) and protein phosphatase 7 take part in thermotolerance, which is activated by AtCaM3. Moreover, CBK3 has also been suggested to activate HSFA1 by phosphorylation in *Arabidopsis*. It was also postulated that AtCBK3 is the key connection between CaM and heat stress transcription factor (HSF) in the Ca^{2+}/CaM pathway; thus, it acts as a positive regulator of AtHSFA1 function during heat exposure. In rice plants, OsCaM1-1 was shown to positively regulate Ca^{2+} signals, resulting in the acquisition of HSPs under thermal stress (Wu and Jinn 2012; Wu et al. 2012). Despite these, ROS accumulation during high-temperature conditions is an active process, which in turn activates downstream signaling components in the regulation of HSP- and HSF-dependent pathways. Respiratory burst oxidase homologs (RBOHs) are necessary for multiple signaling pathways under various stress conditions, including heat stress. Growing evidence established the critical role of RBOHD, ROS-generating NADPH oxidase during elevated temperature (Larkindale et al. 2005; Miller et al. 2007). The increased level of Ca^{2+}-ions in the cytosol mediated by Ca^{2+} channels also activate the RBOHD and further accumulates ROS, which in turn is necessary for the amplification of heat shock response by opening additional Ca^{2+} channels at the plasma membrane (Mittler et al. 2012; Yang et al. 2023). Therefore, ROS and Ca^{2+} signaling are highly interlinked in the activation of RBOHD, which can cause membrane depolarization to initiate heat shock responses. For example, Wang et al. (2014b) demonstrated the heat shock-induced impaired transcript levels of *HSP17.7* and *HSP21* in RBOH mutants (*rbohB*, *rbohD*, and *rbohB/D*). Heat stress-induced ROS overaccumulation has a direct effect on the activation and sequential phosphorylation of mitogen-activated protein kinase (MAPK) cascade mechanisms. The activation of MAPK is a prerequisite for signal transduction and amplification of environmental signals and activation of specific downstream targets and effector molecules to regulate their functions (Bharti et al. 2021). The expression of diverse *MAPK* genes such as *MAPK1* in rice and corn and *MPK1* in potato and tomato was reported to be induced under thermal exposure (Agarwal et al. 2003b; Nakagami et al. 2005; Blanco et al. 2006; Yu et al. 2019). It is suggested that the enhancement in Ca^{2+} concentration induced by heat exposure also needs the participation of MAPK activation. The activation of MPK6 specifically targets HSFA2 via phosphorylating T249 of HSFA2 and further regulates nuclear accumulation to combat heat stress (Evrard et al. 2013). Another report exhibited that heat induced the activation of MPK3 and MPK6 phosphorylate HSFA4A, which led to the upregulation of the transcript level of the *HSP17.6A* gene along with induced thermotolerance (Perez-Salamo Papdi et al. 2014).

8.2.2.2.2 Low-Temperature Conditions

Cold stress-mediated signals, perceived by the membrane and transduced via different transduction components, result in the stimulation of downstream transcription of several genes. Plants use diverse cytoplasmic membrane-localized receptor proteins/kinases like receptor-like proteins/kinases for signal perception. COLD1 (chilling-tolerance divergence 1), a plasma membrane and endoplasmic reticulum-localized specific transmembrane receptor protein, plays an important role in the recognition of cold stress signals in rice plants (Manishankar and Kudla 2015). Moreover, two Ca^{2+}-permeable mechanosensitive channels, MCA1 and MCA2, have been identified that mediate cold-induced cytosolic Ca^{2+} induction and cold tolerance in *Arabidopsis* (Mori et al. 2018). Transduction events are mediated by second messengers, such as calcium ions (Ca^{2+}) and ROS, which transmit the external cold signal to the intracellular system of the plants. Low temperature leads to an influx of Ca^{2+} in the cytosol, which is an early event of cold stress and is rapidly sensed by calcium-sensor proteins. In plants, CDPKs, CaM-like proteins (CML), calcineurin B-like (CBL) proteins, and CBL-interacting protein kinase (CIPK) act as Ca^{2+}-sensors. Following the binding of Ca^{2+}, CBL1 interacts with CIPK7 and binds with the dehydration-responsive element binding (DREB) core element of *COR* promoters to mediate cold stress tolerance in *Arabidopsis* (Huang et al. 2011). Recently, CML, viz., OsCML16 and its six putative targets, have been uncovered in rice during low-temperature stress (Yang et al. 2020). A rice OsCDPK7, a positive regulator of cold stress tolerance, operates predominantly through the putative target gene, *Rab16A* (Saijo et al. 2001a, b). OsCDPK17 also functions in the cold stress response in rice, via activating the membrane channels (aquaporin OsPIP2;1/OsPIP2;6) along with suppressing the activity of several components of sugar metabolism (Almadanim et al. 2017). The role of CaM-binding transcription activators was reported by Doherty et al. (2009) to regulate cold-mediated gene expression and freezing tolerance. The researchers showed that CAMTA3 functions as a positive regulator of *CBF2* expression via binding with the CM2 motif. In plants, CBL–CIPK interaction is observed in cold stress signaling. In *Arabidopsis*, CIPK7 after being activated by CBL1 may regulate other downstream components involved in cold stress responses (Huang et al. 2011). Similarly, CBL2 is escalated by cold stress along with CIPK7, suggesting their role in cold acclimation in cassava (Mo et al. 2018). In addition, ROS, produced during metabolism, is an integral part of cold signaling. Cold stress-induced ROS generation via RBOHs and plasma membrane-localized NADPH oxidase also result in a transient rise in Ca^{2+} in plant cells. In addition, a low temperature-inducible protein, AtSRC2 (soybean gene regulated by cold-2), was reported to interact with AtRbohF, a novel activator of Ca^{2+}-mediated NADPH oxidase activity during cold stress response (Kawarazaki et al. 2013). MAPK signaling also participates in cold stress signaling in plants. MAPK cascades are evolutionarily conserved signaling modules in eukaryotes, mainly composed of three consecutive serine/threonine protein kinases: MAPKK kinase (MAPKKK or MEKK), MAPK kinase (MAPKK, MKK, or MEK), and MAPK (Bogre et al. 2000; Taj et al. 2010). Studies indicated that ROS activates a MAP kinase cascade, which includes ANP1, AtMPK3, and AtMPK6, and its positive regulator, nucleoside diphosphate kinase 2 (NDPK2) in *Arabidopsis* (Kovtun et al. 2000; Moon et al. 2003). MKK2 is induced by low-temperature stress in association with MAPKKK kinase, MEKK1. MKK2 interacts with MPK4 (MAPK4) and MPK6 (MAPK6) to regulate their activity and triggers the transcript level of *COR* genes for the induction of freezing tolerance of plants (Teige et al. 2004).

8.2.2.3 Transcriptional Regulation during Temperature Extremes

8.2.2.3.1 High-Temperature Conditions

The induction of HSPs under the control of HSFs is believed to play a central role in response to heat stress and acquired thermotolerance in plants. HSFs are terminal components of the signal transduction chain, mediating the activation of genes responsive to high-temperature stress. Heat stress-induced protein denaturation response triggers transcriptional activation of HSP-encoding genes under the control of HSFs, which are crucial for the maintenance and restoration of protein homeostasis (Morimoto 2008; Pratt et al. 2010). In general, HSFs control the expression of genes during numerous other abiotic stress conditions (Mittler et al. 2009; Dossa et al. 2016; Guo et al. 2016; Chidambaranathan et al. 2018). Depending on the plant species, each HSF gene family has 15–50 members of HSFs (Nover et al. 2001). HSFs also possess modular structure, similar to various other proteins regulating gene structure. Each HSF shares a highly structured DNA-binding domain (DBD) close to the N-terminus, which is also the most conserved part of the HSFs. The oligomerization domain is connected to the DBD via a flexible linker of variable length (15–80 amino acid residues). HSFs can be categorized into three different types: class A, class B, and class C, according to the variable length of the insertion in the linker sequences (Baniwal et al. 2004). Among HSFs, HSF1s are master regulators of heat shock response, functioning upstream of numerous transcription factors and regulating the transcript levels of various downstream genes. Previously, Mishra et al. (2002) uncovered the unique role of HSFA1a, which acts as a master regulator for induced thermotolerance in tomato plants. It has been identified that HSFA2 plays a dominant role during repeated cycles of heat stress and recovery in both tomato and *Arabidopsis*, due to its high activating potential for transcription of *HSP* genes (Scharf et al. 1998; Schramm et al. 2006). The knockdown or constitutive expression study of *HSFA2* identified the major target genes encoding individual isoforms of HSP70 and HSP101 families as well as several small HSPs (Schramm et al. 2006; Nishizawa et al. 2006). Although the functions of class B and class C HSFs have not been identified, HSFB1 is reported to act synergistically with HSFA1a during heat stress (Bharti et al. 2004).

8.2.2.3.2 Low-Temperature Conditions

Cold acclimation induces profound alterations in the gene expression level. Until now, numerous cold-responsive genes have been identified and often they synergistically work to improve cold stress resistance. Among these, *inducer of CBF expression* (*ICE*), *CBF*, and *cold-regulated* genes (*CORs*) are the most widely reported and essential for cold stress response (Mehrotra and Verma 2020). These three are the key players in the *ICE–CBF–COR* cascade that are recruited to alleviate low-temperature stress in several plant species (Thomashow 1999; Badawi and Reddy 2008; Bremer and Kent 2017; Shu et al. 2017; Hao and Wang 2018; Guo et al. 2019; Li et al. 2020). The ICE is the pioneer for cold acclimation, which encodes a MYC-type basic helix-loop-helix (bHLH) transcription factor. In *Arabidopsis*, it is reported that ICE1 controls the expression of CBF3 by binding to MYC recognition elements to regulate cold stress and freezing tolerance (Chinnusamy et al. 2003; Fursova et al. 2009). However, *Ice1* mutant is defective in cold induction of *CBF3*, rendering chilling susceptibility and incapability of cold acclimation. During cold exposure, ICE1 is released from the JAZ repressor protein, bound by DELLA and consequently triggers the expression of CBF3. DELLA contributes to the cold induction of *CBF* genes by interacting with JAZs. CBFs are indispensable for activating the expression of *COR* genes via binding with the cis-element of promoter regions and positively control the expression during cold acclimation (Zhen and Ungerer 2008; Rihan et al. 2017). CBFs also belong to DREB proteins and are members of the AP2/ERF (APETALA2/ethylene-responsive factors) transcription factor superfamily. Three *CBF* (*CBF1*, *CBF2*, and *CBF3*) genes have been identified in *Arabidopsis*, among which *CBF2* was known to negatively regulate *CBF1* and *CBF3* expression during cold acclimation (Novillo et al. 2004).

8.3 CONCLUSION

Plants experience various kinds of environmental stress factors, of which temperature is of particular interest. The escalating threat of extreme temperature events (both high and low) imposes several negative effects on global plant life that are dependent on the rate of temperature change, intensity, and duration. Plants suffering from extreme temperature also face catastrophic losses in annual crop production and end up with massive food insecurity. In the present scenario, inducing plant resistance is critical for improvement in crop yield and environmental sustainability. Stress tolerance in plants is a convoluted network involving biochemical and molecular processes that are difficult to untangle. The development of plants that are able to maintain a greater yield under both high- and low-temperature conditions is of great importance in the face of climate change and global warming scenarios. A detailed understanding of the plant interactions during extreme temperature is crucial, which could serve as the most straightforward and perspective strategies for effective management. For this reason, plant responses during exposure to extreme temperature regimes are thoroughly discussed in this chapter. This will speed up the future research for achieving temperature stress resilience in plants, which will be ultimately helpful in crop improvement programs for generating high-yielding, stress-tolerant germplasm and climate-smart crop cultivars to meet the increasing global food demand for rapidly expanding populations in the next decade.

ACKNOWLEDGMENTS

Financial assistance from the Science and Engineering Research Board, Government of India, through the grant (EMR/2016/004799) and Department of Higher Education, Science and Technology and Biotechnology, Government of West Bengal, through the grant (264(Sanc.)/ST/P/S&T/1G-80/2017) to Prof. Aryadeep Roychoudhury is gratefully acknowledged.

REFERENCES

Agarwal, M., Sahi, C., Katiyar-Agarwal, S., Agarwal, S., Young, T., Gallie, D. R., Sharma, V. M., Ganesan, K., & Grover, A. (2003a). Molecular characterization of rice hsp101: Complementation of yeast hsp104 mutation by disaggregation of protein granules and differential expression in indica and japonica rice types. *Plant Molecular Biology*, *51*, 543–553.

Agrawal, G. K., Tamogami, S., Iwahashi, H., Agrawal, V. P., & Rakwal, R. (2003b). Transient regulation of jasmonic acid-inducible rice MAP kinase gene (OsBWMK1) by diverse biotic and abiotic stresses. *Plant Physiology and Biochemistry*, *41*, 355–361.

Ahn, S.-J., Im, Y.-J., Chung, G.-C., Seong, K.-Y., & Cho, B.-H. (2000). Sensitivity of plasma membrane H^+-ATPase of cucumber root system in response to low root temperature. *Plant Cell Reports*, *19*, 831–835.

Allen, D. J., & Ort, D. R. (2001). Impacts of chilling temperatures on photosynthesis in warm-climate plants. *Trends in Plant Science*, *6*, 36–42.

Almadanim, M. C., Alexandre, B. M., Rosa, M. T. G., Sapeta, H., Leitão, A. E., Ramalho, J. C., Lam, T. T., Negrão, S., Abreu, I. A., & Oliveira, M. M. (2017). Rice calcium-dependent protein kinase OsCPK17 targets plasma membrane intrinsic protein and sucrose-phosphate synthase and is required for a proper cold stress response. *Plant Cell & Environment*, *40*, 1197–1213.

Al-Niemi, T. S., & Stout, R. G. (2002). Heat-shock protein expression in a perennial grass commonly associated with active geothermal areas in western North America. *Journal of Thermal Biology*, *27*, 547–553.

Antikainen, M., & Griffith, M. (1997). Antifreeze protein accumulation in freezing-tolerant cereals. *Physiologia Plantarum*, *99*, 423–432.

Armstrong, A. F., Logan, D. C., & Atkin, O. K. (2006). On the developmental dependence of leaf respiration: Responses to short- and long-term changes in growth temperature. *American Journal of Botany*, *93*, 1633–1639.

Arun-Chinnappa, K. S., Ranawake, L., & Seneweera, S. (2017). Impacts and management of temperature and water stress in crop plants. In P. Minhas, J. Rane, & R. Pasala (Eds.), *Abiotic stress management for resilient agriculture* (pp. 221–233). Springer.

Atıcı, Ö., & Nalbantoğlu, B. (2003). Antifreeze proteins in higher plants. *Phytochemistry*, *64*, 1187–1196.

Badawi, M., & Reddy, Y. M. (2008). Structure and functional analysis of wheat *ICE* (inducer of CBF expression) genes. *Plant and Cell Physiology*, *49*, 1237–1249.

Bajguz, A. (2009). Isolation and characterization of brassinosteroids from algal cultures of *Chlorella vulgaris* Beijerinck (Trebouxiophyceae). *Journal of Plant Physiology*, *166*, 1946–1949.

Baker, N. R., & Rosenqvist, E. (2004). Applications of chlorophyll fluorescence can improve crop production strategies: An examination of future possibilities. *Journal of Experimental Botany*, *55*, 1607–1621.

Baniwal, S. K., Bharti, K., Chan, K. Y., Fauth, M., Ganguli, A., Kotak, S., Mishra, S. K., Nover, L., Port, M., Scharf, K.-D., Tripp, J., Weber, C., Zielinski, D., & von Koskull-Döring, P. (2004). Heat stress response in plants: A complex game with chaperones and more than twenty heat stress transcription factors. *Journal of Biosciences*, *29*, 471–487.

Bertrand, A., Bipfubusa, M., Claessens, A., Rocher, S., & Castonguay, Y. (2017). Effect of photoperiod prior to cold acclimation on freezing tolerance and carbohydrate metabolism in alfalfa (*Medicago sativa* L.). *Plant Science*, *264*, 122–128.

Bharti, J., Sahil, Mehta, S., Ahmad, S., Singh, B., Padhy, A. K., Srivastava, N., & Pandey, V. (2021). Mitogen-activated protein kinase, plants, and heat stress. In A. Husen (Ed.), *Harsh environment and plant resilience* (pp. 323–354). Springer.

Bharti, K., von Koskull-Döring, P., Bharti, S., Kumar, P., Tintschl-Körbitzer, A., Treuter, E., & Nover, L. (2004). Tomato heat stress transcription factor HsfB1 represents a novel type of general transcription coactivator with a histone-like motif interacting with the plant CREB binding protein ortholog HAC1. *Plant Cell*, *16*, 1521–1535.

Blanco, F. A., Zanetti, M. E., Casalongué, C. A., & Daleo, G. R. (2006). Molecular characterization of a potato MAP kinase transcriptionally regulated by multiple environmental stresses. *Plant Physiology and Biochemistry*, *44*, 315–322.

Bogre, L., Meskiene, I., Heberle-Bors, E., & Hirt, H. (2000). Stressing the role of the MAP kinases in mitogenic stimulation. *Plant Molecular Biology*, *43*, 705–718.

Bravo, L. A., Saavedra-Mella, F. A., Vera, F., Guerra, A., Cavieres, L. A., Ivanov, A. G., Huner, N. P. A., & Corcuera, L. J. (2007). Effect of cold acclimation on the photosynthetic performance of two ecotypes of *Colobanthus quitensis* (Kunth) Bartl. *Journal of Experimental Botany*, *58*, 3581–3590.

Bredow, M., & Walker, V. K. (2017). Ice-binding proteins in plants. *Frontiers Plant Science*, *8*, 2153.

Bremer, A., & Kent, B. (2017). Intrinsically disordered stress protein COR15A resides at the membrane surface during dehydration. *Biophysical Journal*, *113*, 572–579.

Brookes, P. S. (2005). Mitochondrial H^+ leak and ROS generation: An odd couple. *Free Radical Biology and Medicine*, *38*, 12–23.

Burke, J. J., & Wanjura, D. F. (2010). Plant responses to temperature extremes. In J. M. Stewart, D. M. Oosterhuis, J. J. Heitholt, & J. R. Mauney (Eds.), *Physiology of cotton* (pp. 123–128). Springer.

Campbell, C., Atkinson, L., Zaragoza-Castells, J., Lundmark, M., Atkin, O., & Hurry, V. (2007). Acclimation of photosynthesis and respiration is asynchronous in response to changes in temperature regardless of plant functional group. *New Phytologist*, *176*, 375–389.

Campbell, J. L., Klueva, N. Y., Zheng, H.-G., Nieto-Sotelo, J., Ho, T. D., & Nguyen, H. T. (2001). Cloning of new members of heat shock protein HSP101 gene family in wheat (*Triticum aestivum* (L.) *Moench*) inducible by heat, dehydration, and ABA. *Biochimica et Biophysica Acta (BBA) - Gene Structure and Expression*, *1517*, 270–277.

Chidambaranathan, P., Jagannadham, P. T. K., Satheesh, V., Kohli, D., Basavarajappa, S. H., Chellapilla, B., Kumar, J., Jain, P. K., & Srinivasan, R. (2018). Genome-wide analysis identifies chickpea (*Cicer arietinum*) heat stress transcription factors (*HSFs*) responsive to heat stress at the pod development stage. *Journal of Plant Research*, *131*, 525–542.

Chinnusamy, V., Ohta, M., Kanrar, S., Lee, B.-H., Hong, X., Agarwal, M., & Zhu, J.-K. (2003). ICE1: A regulator of cold-induced transcriptome and freezing tolerance in *Arabidopsis*. *Genes & Development*, *17*, 1043–1054.

Clough, S. J., Fengler, K. A., Yu, I. C., Lippok, B., Smith, R. K., & Bent, A. F. (2000). The Arabidopsis DND1 "defense, no death" gene encodes a mutated cyclic nucleotide-gated ion channel. *Proceedings of the National Academy of Sciences of the United States of America*, *9716*, 9323–9328.

Cvetkovic, J., Müller, K., & Baier, M. (2017). The effect of cold priming on the fitness of *Arabidopsis thaliana* accessions under natural and controlled conditions. *Scientific Reports*, *7*, 44055.

Doherty, C. J., Van Buskirk, H. A., Myers, S. J., & Thomashow, M. F. (2009). Roles for *Arabidopsis* CAMTA transcription factors in cold-regulated gene expression and freezing tolerance. *Plant Cell*, *21*, 972–984.

Dossa, K., Diouf, D., & Cisse, N. (2016). Genome-wide investigation of HSF genes in sesame reveals their segmental duplication expansion and their active role in drought stress response. *Frontiers in Plant Science*, *7*, 1522.

Durand, M., Mainson, D., Porcheron, B., Maurousset, L., Lemoine, R., & Pourtau, N. (2017). Carbon source–sink relationship in *Arabidopsis thaliana*: The role of sucrose transporters. *Planta*, *247*, 587–611.

Dutta, S., Mohanty, S., & Tripathy, B. C. (2009). Role of temperature stress on chloroplast biogenesis and protein import in pea. *Plant Physiology*, *150*, 1050–1061.

Evrard, A., Kumar, M., Lecourieux, D., Lucks, J., von Koskull-Döring, P., & Hirt, H. (2013). Regulation of the heat stress response in Arabidopsis by MPK6-targeted phosphorylation of the heat stress factor HSFA2. *Peer J*, *1*, e59.

Fan, J., Ren, J., Zhu, W., Amombo, E., Fu, J., & Chen, L. (2014). Antioxidant responses and gene expression in bermudagrass under cold stress. *Journal of the American Society for Horticultural Science*, *139*, 699–705.

Finka, A., Cuendet, A. F. H., Maathuis, F. J., Saidi, Y., & Goloubinof, P. (2012). Plasma membrane cyclic nucleotide gated calcium channels control land plant thermal sensing and acquired thermotolerance. *Plant Cell, 248*, 3333–3348.

Fursova, O. V., Pogorelko, G. V., & Tarasov, V. A. (2009). Identification of ICE2, a gene involved in cold acclimation which determines freezing tolerance in *Arabidopsis thaliana. Gene, 429*, 98–103.

Gao, F., Han, X., Wu, J., Zheng, S., Shang, Z., Sun, D., & Li, B. (2012). A heat-activated calcium-permeable channel – Arabidopsis cyclic nucleotide-gated ion channel 6 – is involved in heat shock responses. *The Plant Journal, 706*, 1056–1069.

Gomathi, R., Shiyamala, S., Vasantha, S., Johnson, D. E., & Janani, P. K. (2013). Kinetics of metabolism in sugarcane (*Saccharum officinarum* L.) under heat stress. *Indian Journal of Plant Physiology, 18*, 41–47.

Grabelnych, O. I., Sumina, O. N., Funderat, S. P., Pobezhimova, T. P., Voinikov, V. K., & Kolesnichenko, A. V. (2004). The distribution of electron transport between the main cytochrome and alternative pathways in plant mitochondria during short-term cold stress and cold hardening. *Journal of Thermal Biology, 29*, 165–175.

Griffith, M., Ala, P., Yang, D. S. C., Hon, W.-C., & Moffatt, B. A. (1992). Antifreeze protein produced endogenously in winter rye leaves. *Plant Physiology, 100*, 593–596.

Griffith, M., & Yaish, M. W. F. (2004). Antifreeze proteins in overwintering plants: A tale of two activities. *Trends in Plant Science, 9*, 399–405.

Guo, J., Ren, Y., & Tang, Z. (2019). Characterization and expression profiling of the *ICE-CBF-COR* genes in wheat. *Peer J, 7*, e8190.

Guo, M., Liu, J. H., Ma, X., Luo, D. X., Gong, Z. H., & Lu, M. H. (2016). The plant heat stress transcription factors (HSFs): Structure, regulation, and function in response to abiotic stresses. *Frontiers in Plant Science, 7*, 114.

Gupta, R., & Deswal, R. (2014). Antifreeze proteins enable plants to survive in freezing conditions. *Journal of Biosciences, 39*, 931–944.

Guy, C. L., Huber, J. L. A., & Huber, S. C. (1992). Sucrose phosphate synthase and sucrose accumulation at low temperature. *Plant Physiology, 100*, 502–508.

Hajihashemi, S., Noedoost, F., Geuns, J. M. C., Djalovic, I., & Siddique, K. H. M. (2018). Effect of cold stress on photosynthetic traits, carbohydrates, morphology, and anatomy in nine cultivars of *Stevia rebaudiana. Frontiers in Plant Science, 9*, 1430.

Hameed, A., Goher, M., & Iqbal, N. (2012). Heat stress-induced cell death, changes in antioxidants, lipid peroxidation, and protease activity in wheat leaves. *Journal of Plant Growth Regulation, 31*, 283–291.

Hannah, M. A., Wiese, D., Freund, S., Fiehn, O., Heyer, A. G., & Hincha, D. K. (2006). Natural genetic variation of freezing tolerance in Arabidopsis. *Plant Physiology, 142*, 98–112.

Hao, X., & Wang, L. (2018). Response and adaptation mechanisms of tea plant to low-temperature stress. In W. Y. Han, X. Li, & G, Ahammed (Eds.), *Stress physiology of tea in the face of climate change* (pp. 39–61). Springer.

Harndahl, U., Hall, R. B., Osteryoung, K. W., Vierling, E., Bornman, J. F., & Sundby, C. (1999). The chloroplast small heat shock protein undergoes oxidation-dependent conformational changes and may protect plants from oxidative stress. *Cell Stress and Chaperones, 4*, 129–138.

Hatfield, J. L., & Prueger, J. H. (2015). Temperature extremes: Effect on plant growth and development. *Weather and Climate Extremes, 10*, 4–10.

Havaux, M. (1993). Rapid photosynthetic adaptation to heat stress triggered in potato leaves by moderately elevated temperatures. *Plant Cell & Environment, 16*, 461–467.

Heidarvand, L., & Amiri, R. M. (2010). What happens in plant molecular responses to cold stress? *Acta Physiologiae Plantarum, 32*, 419–431.

Higashi, Y., Okazaki, Y., Myouga, F., Shinozaki, K., & Saito, K. (2015). Landscape of the lipidome and transcriptome under heat stress in *Arabidopsis thaliana. Scientific Reports, 5*, 10533.

Hon, W.-C., Griffith, M., Mlynarz, A., Kwok, Y. C., & Yang, D. S. C. (1995). Antifreeze proteins in winter rye are similar to pathogenesis-related proteins. *Plant Physiology, 109*, 879–889.

Hou, W., Sun, A. H., Chen, H. L., Yang, F. S., Pan, J. L., & Guan, M. Y. (2016). Effects of chilling and high temperatures on photosynthesis and chlorophyll fluorescence in leaves of watermelon seedlings. *Biologia Plantarum, 60*, 148–154.

Hu, Y., Jiang, Y., Han, X., Wang, H., Pan, J., & Yu, D. (2017). Jasmonate regulates leaf senescence and tolerance to cold stress: Crosstalk with other phytohormones. *Journal of Experimental Botany, 68*, 1361–1369.

Huang, C., Ding, S., Zhang, H., Du, H., & An, L. (2011). CIPK7 is involved in cold response by interacting with CBL1 in *Arabidopsis thaliana. Plant Science, 181*, 57–64.

Huang, T., & Duman, J. G. (2001). Cloning and characterization of a thermal hysteresis (antifreeze) protein with DNA-binding activity from winter bittersweet nightshade, *Solanum dulcamara*. *Plant Molecular Biology*, *48*, 339–350.

Huner, N. P. A., Öquist, G., & Sarhan, F. (1998). Energy balance and acclimation to light and cold. *Trends in Plant Science*, *3*, 224–230.

Janicka-Russak, M., Kabała, K., Wdowikowska, A., & Kłobus, G. (2012). Response of plasma membrane H^+-ATPase to low temperature in cucumber roots. *Journal of Plant Research*, *125*, 291–300.

Janka, E., Körner, O., Rosenqvist, E., & Ottosen, C. O. (2013). High temperature stress monitoring and detection using chlorophyll *a* fluorescence and infrared thermography in chrysanthemum *(Dendranthema grandiflora)*. *Plant Physiology and Biochemistry*, *67*, 87–94.

Jenner, C. F. (1991). Effects of exposure of wheat ears to high temperature on dry matter accumulation and carbohydrate metabolism in the grain of two cultivars. *Functional Plant Biology*, *18*, 165–177.

Jha, U. C., Nayyar, H., & Siddique, K. H. M. (2022). Role of phytohormones in regulating heat stress acclimation in agricultural crops. *Journal of Plant Growth Regulation*, *41*, 1041–1064.

Jia, Z., & Davies, P. L. (2002). Antifreeze proteins: An unusual receptor-ligand interaction. *Trends in Biochemical Sciences*, *27*, 101–106.

Jung, H. J., Park, S. J., & Kang, H. (2013). Regulation of RNA metabolism in plant development and stress responses. *Journal of Plant Biology*, *56*, 123–129.

Karlson, D., & Imai, R. (2003). Conservation of the cold shock domain protein family in plants. *Plant Physiology*, *131*, 12–15.

Karlson, D., Nakaminami, K., Toyomasu, T., & Imai, R. (2002). A cold-regulated nucleic acid-binding protein of winter wheat shares a domain with bacterial cold shock proteins. *Journal of Biological Chemistry*, *277*, 35248–35256.

Kaurin, A., Junttila, O., & Hansen, J. (1981). Seasonal changes in frost hardiness in cloudberry (*Rubus chamaemorus*) in relation to carbohydrate content with special reference to sucrose. *Plant Physiology*, *52*, 310–314.

Kawarazaki, T., Kimura, S., Iizuka, A., Hanamata, S., Nibori, H., Michikawa, M., Imai, A., Abe, M., Kaya, H., & Kuchitsu, K. (2013). A low temperature-inducible protein AtSRC2 enhances the ROS-producing activity of NADPH oxidase AtRbohF. *Biochimica et Biophysica Acta (BBA) - Molecular Cell Research*, *1833*, 2775–2780.

Kim, M.-H., Sasaki, K., & Imai, R. (2009). Cold shock domain protein 3 regulates freezing tolerance in *Arabidopsis thaliana*. *Journal of Biological Chemistry*, *284*, 23454–23460.

Kim, M.-H., Sonoda, Y., Sasaki, K., Kaminaka, H., & Imai, R. (2013). Interactome analysis reveals versatile functions of Arabidopsis COLD SHOCK DOMAIN PROTEIN 3 in RNA processing within the nucleus and cytoplasm. *Cell Stress and Chaperones*, *18*, 517–525.

Kingsley, P. D., & Palis, J. (1994). GRP2 proteins contain both CCHC zinc fingers and a cold shock domain. *Plant Cell*, *6*, 1522–1523.

Klotke, J., Kopka, J., Gatzke, N., & Heyer, A. G. (2004). Impact of soluble sugar concentrations on the acquisition of freezing tolerance in accessions of *Arabidopsis thaliana* with contrasting cold adaptation: Evidence for a role of raffinose in cold acclimation. *Plant Cell & Environment*, *27*, 1395–1404.

Knight, C. A., Cheng, C. C., & DeVries, A. L. (1991). Adsorption of alpha-helical antifreeze peptides on specific ice crystal surface planes. *Biophysics Journal*, *59*, 409–418.

Kosová, K., Prášil, I. T., Vítámvás, P., Dobrev, P., Motyka, V., Floková, K., Novák, O., Turečková, V., Rolčik, J., Pešek, B., Trávničková, A., Gaudinová, A., Galiba, G., Janda, T., Vlasáková, E., Prášilová, P., & Vanková, R. (2012). Complex phytohormone responses during the cold acclimation of two wheat cultivars differing in cold tolerance, winter Samanta and spring Sandra. *Journal of Plant Physiology*, *169*, 567–576.

Kovtun, Y., Chiu, W. L., Tena, G., & Sheen, J. (2000). Functional analysis of oxidative stress-activated mitogen-activated protein kinase cascade in plants. *Proceedings of the National Academy of Sciences of the United States of America*, *97*, 2940–2945.

Kuk, Y. I., Shin, J. S., Burgos, N. R., Hwang, T. E., Han, O., Cho, B. H., Jung, S., & Guh, J. O. (2003). Antioxidative enzymes offer protection from chilling damage in rice plants. *Crop Science*, *43*, 2109.

Lado, J., Rey, F., & Manzi, M. (2023). Phytohormones and cold stress tolerance. In G. J. Ahammed & J. Yu (Eds.), *Plant hormones and climate change* (pp. 207–226). Springer.

Larkindale, J., Hall, J. D., Knight, M. R., & Vierling, E. (2005). Heat stress phenotypes of Arabidopsis mutants implicate multiple signaling pathways in the acquisition of thermotolerance. *Plant Physiology*, *1382*, 882–897.

Leonardos, E. D., Savitch, L. V., Huner, N. P. A., Öquist, G., & Grodzinski, B. (2003). Daily photosynthetic and C-export patterns in winter wheat leaves during cold stress and acclimation. *Physiologia Plantarum*, *117*, 521–531.

Li, B., Gao, K., Ren, H., & Tang, W. (2018). Molecular mechanisms governing plant responses to high temperatures. *Journal of Integrative Plant Biology*, *609*, 757–779.

Li, X., Liu, C., & Zhao, Z. (2020). COR27 and COR28 are novel regulators of the COP1–HY5 regulatory hub and photomorphogenesis in Arabidopsis. *Plant Cell*, *32*, 1393154.

Li, Z. G., Ye, X. Y., & Qiu, X. M. (2019). Glutamate signaling enhances the heat tolerance of maize seedlings by plant glutamate receptor-like channels-mediated calcium signaling. *Protoplasma*, *256*, 1165–1169.

Liu, H.-T., Gao, F., Li, G.-L., Han, J.-L., Liu, D.-L., Sun, D.-Y., & Zhou, R.-G. (2008). The calmodulin-binding protein kinase 3 is part of heat-shock signal transduction in *Arabidopsis thaliana*. *The Plant Journal*, *55*, 760–773.

Liu, H.-T., Li, G.-L., Chang, H., Sun, D.-Y., Zhou, R.-G., & Li, B. (2007). Calmodulin-binding protein phosphatase PP7 is involved in thermotolerance in *Arabidopsis*. *Plant Cell & Environment*, *30*, 156–164.

Loka, D. A., & Oosterhuis, D. M. (2010). Effect of high night temperatures on cotton respiration, ATP levels and carbohydrate content. *Environmental and Experimental Botany*, *68*, 258–263.

Lokesha, A. N., Shivashankara, K. S., Laxman, R. H., Geetha, G. A., & Shankar, A. G. (2019). Effect of high temperature on fruit quality parameters of contrasting tomato genotypes. *International Journal of Current Microbiology and Applied Sciences*, *8*, 1019–1029.

Malik, M. K., Slovin, J. P., Hwang, C. H., & Zimmerman, J. L. (1999). Modified expression of a carrot small heat shock protein gene, Hsp17.7, results in increased or decreased thermotolerance. *The Plant Journal*, *20*, 89–99.

Manishankar, P., & Kudla, J. (2015). Cold tolerance encoded in one SNP. *Cell*, *160*, 1045–1046.

Mathur, S., Agrawal, D., & Jajoo, A. (2014). Photosynthesis: Response to high temperature stress. *Journal of Photochemistry and Photobiology B: Biology*, *137*, 116–126.

Meehl, G. A., Stocker, T. F., Collins, W. D., Gaye, A. J., Gregory, J. M., Kitoh, A., Knutti, R., Murphy, J. M., Noda, A., Raper, S. C. B., Watterson, J. G., Weaver, A. J., & Zhao, Z. (2007). Global climate projections. In S. Solomon, D. Qin, M. Manning, Z. Chen, M. Marquis, K. B. Averyt, M. Tignor, & H. L. Miller (Eds.), *Climate change 2007: The physical science basis. Contribution of working group i to the fourth assessment report of the intergovernmental panel on climate change* (pp. 747–846). Cambridge University Press.

Mehrotra, S., & Verma, S. (2020). Transcriptional regulation and signalling of cold stress response in plants: An overview of current understanding. *Environmental and Experimental Botany*, *180*, 104243.

Miittal, D., Chakrabarti, S., Sarkar, A., Singh, A., & Grover, A. (2009). Heat shock factor gene family in rice: Genomic organization and transcript expression profiling in response to high temperature, low temperature and oxidative stresses. *Plant Physiology and Biochemistry*, *47*, 785–795.

Miller, G., Suzuki, N., Rizhsky, L., Hegie, A., Koussevitzky, S., & Mittler, R. (2007). Double mutants deficient in cytosolic and thylakoid ascorbate peroxidase reveal a complex mode of interaction between reactive oxygen species, plant development, and response to abiotic stresses. *Plant Physiology*, *1444*, 1777–1785.

Miroslava, L., & Otília, G. (1999). The effect of low temperature on root respiration in maize. Biologia. Section A. Section B. Section C. Botany. Zoology. *Cellular and Molecular Biology*, *54*, 453–458.

Mishkind, M., Vermeer, J. E. M., Darwish, E., & Munnik, T. (2009). Heat stress activates phospholipase D and triggers PIP_2 accumulation at the plasma membrane and nucleus. *The Plant Journal*, *60*, 10–21.

Mishra, S. K., Tripp, J., Winkelhaus, S., Tschiersch, B., Theres, K., Nover, L., & Scharf, K.-D. (2002). In the complex family of heat stress transcription factors, HsfA1 has a unique role as master regulator of thermotolerance in tomato. *Genes & Development*, *16*, 1555–1567.

Mittler, R., Finka, A., & Goloubinoff, P. (2012). How do plants feel the heat? *Trends in Plant Science*, *37*, 118–125.

Mo, C., Wan, S., Xia, Y., Ren, N., Zhou, Y., & Jiang, X. (2018). Expression patterns and identified protein-protein interactions suggest that cassava CBL-CIPK signal networks function in responses to abiotic stresses. *Frontiers in Plant Science*, *9*, 269.

Mohanty, S., Baishna, B. G., & Tripathy, C. (2006). Light and dark modulation of chlorophyll biosynthetic genes in response to temperature. *Planta*, *224*, 692–699.

Moon, H., Lee, B., Choi, G., Shin, D., Theertha-Prasad, D., Lee, O., Kwak, S., Kim, D. H., Nam, J., Bahk, J., Hong, J. C., Lee, S. Y., Cho, M. J., Lim, C. O., & Yun, D. J. (2003). NDP kinase 2 interacts with two oxidative stress activated MAPKs to regulate cellular redox state and enhances multiple stress tolerance in transgenic plants. *Proceedings of the National Academy of Sciences of the United States of America*, *100*, 358–363.

Mori, K., Renhu, N., Naito, M., Nakamura, A., Shiba, H., Yamamoto, T., Suzaki, T., Iida, H., & Miura, K. (2018). Ca^{2+}-permeable mechanosensitive channels MCA1 and MCA2 mediate cold-induced cytosolic Ca^{2+} increase and cold tolerance in Arabidopsis. *Scientific Reports*, *8*, 550.

Morimoto, R. (2008). Proteotoxic stress and inducible chaperone networks in neurodegenerative disease and aging. *Genes & Development*, *22*, 1427–1438.

Morimoto, R. I., & Westerheide, S. D. (2010). The heat shock response and the stress of misfolded proteins. In R. A. Bradshaw & E. A. Dennis (Eds.), *Handbook of cell signaling* (Vol. II, pp. 2231–2239), Academic Press.

Mueller, S. P., Unger, M., Guender, L., Fekete, A., & Mueller, M. J. (2017). Diacylglycerol acyltransferase-mediated triacylglycerol synthesis augments basal thermotolerance. *Plant Physiology*, *175*, 486–497.

Munné-Bosch, S., & Alegre, L. (2002). Plant aging increases oxidative stress in chloroplasts. *Planta*, *214*, 608–615.

Muzi, C., Camoni, L., Visconti, S., & Aducci, P. (2016). Cold stress affects H^+-ATPase and phospholipase D activity in Arabidopsis. *Plant Physiology and Biochemistry*, *108*, 328–336.

Nakagami, H., Pitzschke, A., & Hirt, H. (2005). Emerging MAP kinase pathways in plant stress signalling. *Trends in Plant Science*, *10*, 339–346.

Nishizawa, A., Yabuta, Y., Yoshida, E., Maruta, T., Yoshimura, K., & Shigeoka, S. (2006). Arabidopsis heat shock transcription factor A2 as a key regulator in response to several types of environmental stress. *The Plant Journal*, *48*, 535–547.

Nover, L., Bharti, K., Doring, P., Mishra, S. K., Ganguli, A., & Scharf, K. D. (2001). Arabidopsis and the heat stress transcription factor world: How many heat stress transcription factors do we need? *Cell Stress and Chaperones*, *6*, 177–189.

Novillo, F., Alonso, J. M., Ecker, J. R., & Salinas, J. (2004). CBF2/DREB1C is a negative regulator of CBF1/DREB1B and CBF3/DREB1A expression and plays a central role in stress tolerance in *Arabidopsis*. *Proceedings of the National Academy of Sciences of the United States of America*, *101*, 3985–3990.

Parsell, D. A., & Lindquist, S. (1993). The function of heat-shock proteins in stress tolerance: Degradation and reactivation of damaged proteins. *Annual Review of Genetics*, *27*, 437–496.

Perez-Salamo Papdi, C., Rigo, G., Zsigmond, L., Vilela, B., Lumbreras, V., Nagy, I., Horvath, B., Domoki, M., Darula, Z., Medzihradszky, K., Bogre, L., Koncz, C., & Szabados, L. (2014). The heat shock factor A4A confers salt tolerance and is regulated by oxidative stress and the mitogen-activated protein kinases MPK3 and MPK6. *Plant Physiology*, *165*, 319–334.

Porter, J. R. (2005). Rising temperatures are likely to reduce crop yields. *Nature*, *436*, 174.

Pratt, W. B., Morishima, Y., Peng, H. M., & Osawa, Y. (2010). Proposal for a role of the Hsp90/Hsp70-based chaperone machinery in making triage decisions when proteins undergo oxidative and toxic damage. *Experimental Biology and Medicine*, *235*, 278–289.

Qiao, B., Zhang, Q., Liu, D., Wang, H., Yin, J., Wang, R., He, M., Cui, M., Shang, Z., Wang, D., & Zhu, Z. (2015). A calcium binding protein, rice annexin OsANN1, enhances heat stress tolerance by modulating the production of H_2O_2. *Journal of Experimental Botany*, *66*, 5853–5866.

Rashid, F. A. A., Crisp, P. A., Zhang, Y., Berkowitz, O., Pogson, B. J., Day, D. A., Masle, J., Dewar, R. C., Whelan, J., Atkin, O. K., & Scafaro, A. P. (2020). Molecular and physiological responses during thermal acclimation of leaf photosynthesis and respiration in rice. *Plant Cell & Environment*, *43*, 594–610.

Raymond, J. A., & DeVries, A. L. (1977). Adsorption inhibition as a mechanism of freezing resistance in polar fishes. *Proceedings of the National Academy of Sciences of the United States of America*, *74*, 2589–2593.

Rihan, H. Z., Al-Issawi, M., & Fuller, M. P. (2017). Upregulation of CBF/DREB1 and cold tolerance in artificial seeds of cauliflower (*Brassica oleracea* var. botrytis). *Scientia Horticulturae*, *225*, 299–309.

Saidi, Y., Finka, A., Muriset, M., Bromberg, Z., Weiss, Y. G., Maathuis, F. J. M., & Goloubinoff, P. (2009). The heat shock response in moss plants is regulated by specific calcium-permeable channels in the plasma membrane. *Plant Cell*, *21*, 2829–2843.

Saijo, Y., Hata, S., Kyozuka, J., Shimamoto, K., & Izui, K. (2001a). Over-expression of a single Ca^{2+}-dependent protein kinase confers both cold and salt/drought tolerance on rice plants. *The Plant Journal*, *23*, 319–327.

Saijo, Y., Kinoshita, N., Ishiyama, K., Hata, S., Kyozuka, J., Hayakawa, T., Nakamura, T., Shimamoto, K., Yamaya, T., & Izui, K. (2001b). A Ca^{2+}-dependent protein kinase that endows rice plants with cold- and salt-stress tolerance functions in vascular bundles. *Plant and Cell Physiology*, *42*, 1228–1233.

Sakata, T., Oshino, T., Miura, S., Tomabechi, M., Tsunaga, Y., Higashitani, N., Miyazawa, Y., Takahashi, H., Watanabe, M., & Higashitani, A. (2010). Auxins reverse plant male sterility caused by high temperatures. *Proceedings of the National Academy of Sciences of the United States of America*, *107*, 8569–8574.

Sami, F., Yusuf, M., Faizan, M., Faraz, A., & Hayat, S. (2016). Role of sugars under abiotic stress. *Plant Physiology and Biochemistry*, *109*, 54–61.

Sasaki, K., & Imai, R. (2012). Pleiotropic roles of cold shock domain proteins in plants. *Frontiers in Plant Science*, *2*, 116.

Scharf, K.-D., Heider, H., Höhfeld, I., Lyck, R., Schmidt, E., & Nover, L. (1998). The tomato Hsf system: HsfA2 needs interaction with HsfA1 for efficient nuclear import and may be localized in cytoplasmic heat stress granules. *Molecular and Cellular Biology*, *18*, 2240–2251.

Schramm, F., Ganguli, A., Kiehlmann, E., Englich, G., Walch, D., & von Koskull-Döring, P. (2006). The heat stress transcription factor HsfA2 serves as a regulatory amplifier of a subset of genes in the heat stress response in *Arabidopsis*. *Plant Molecular Biology*, *60*, 759–772.

Shi, Y., & Yang, S. (2014). ABA regulation of the cold stress response in plants. In D. P. Zhang (Ed.), *Abscisic acid: Metabolism, transport and signaling* (pp. 337–363). Springer.

Shu, Y., Li, W., Zhao, J., Zhang, S., Xu, H., Liu, Y., & Guo, C. (2017). Transcriptome sequencing analysis of alfalfa reveals *CBF* genes potentially playing important roles in response to freezing stress. *Genetics and Molecular Biology*, *40*, 824–833.

Silva, E. N., Silveira, J. A. G., Ribeiro, R. V., Oliveira, J. Ê., & Cardoso, R. A. (2017). Photosynthetic and antioxidant responses of *Jatropha curcas* plants to heat stress: On the relative sensitivity of shoots and roots. *Journal of Plant Growth Regulation*, *37*, 255–265.

Srivastava, A., Guissre, B., Greppin, H., & Strasser, R. J. (1997). Regulation of antenna structure and electron transport in Photosystem II of *Pisum sativum* under elevated temperature probed by the fast polyphasic chlorophyll *a* fluorescence transient: OKJIP. *Biochimica et Biophysica Acta*, *1320*, 95–106.

Steponkus, P. L., & Lanphear, F. O. (1968). The relationship of carbohydrates to cold acclimation of *Hedera helix* L. cv. Thorndale. *Physiologia Plantarum*, *22*, 777–791.

Stewart, S. B., Elith, J., Fedrigo, M., Kasel, S., Roxburgh, S. H., Bennett, L. T., Chick, M., Fairman, T., Leonard, S., Kohout, M., Cripps, J. K., Durkin, L., & Nitschke, C. R. (2021). Climate extreme variables generated using monthly time-series data improve predicted distributions of plant species. *Ecography*, *44*, 626–639.

Sun, W., Van Montagu, M., & Verbruggen, N. (2002). Small heat shock proteins and stress tolerance in plants. *Biochimica et Biophysica Acta (BBA) - Gene Structure and Expression*, *1577*, 1–9.

Taj, G., Agarwal, P., Grant, M., & Kumar, A. (2010). MAPK machinery in plants. *Plant Signaling & Behavior*, *5*, 1370–1378.

Teige, M., Scheikl, E., Eulgem, T., Dóczi, R., Ichimura, K., Shinozaki, K., Dangl, J. L., & Hirt, H. (2004). The MKK2 pathway mediates cold and salt stress signaling in *Arabidopsis*. *Molecular Cell*, *15*, 141–152.

Thomashow, M. F. (1999). Plant cold acclimation: Freezing tolerance genes and regulatory mechanisms. *Annual Review of Plant Physiology and Plant Molecular Biology*, *50*, 571–599.

Tiwari, B. S., & Tripathi, S. N. (1998). Effect of hydration and dehydration on initiation and dynamics of some physiological reactions in desiccation tolerant cyanobacterium *Scytonema geitleri*. *Indian Journal of Biochemistry and Biophysics*, *35*, 172–178.

Todorova, D., Genkov, T., Vaseva-Gemisheva, I., Alexieva, V., Karanov, E., Smith, A., & Hall, M. (2005). Effect of temperature stress on the endogenous cytokinin content in *Arabidopsis thaliana* (L.) Heynh plants. *Acta Physiologiae Plantarum*, *27*, 13–18.

Toh, S., Imamura, A., Watanabe, A., Nakabayashi, K., Okamoto, M., Jikumaru, Y., Hanada, A., Aso, Y., Ishiyama, K., Tamura, N., Iuchi, S., Kobayashi, M., Yamaguchi, S., Kamiya, Y., Nambara, E., & Kawakami, N. (2008). High temperature-induced abscisic acid biosynthesis and its role in the inhibition of gibberellin action in Arabidopsis seeds. *Plant Physiology*, *146*, 1368–1385.

Uemura, M., Tominaga, Y., Nakagawara, C., Shigematsu, S., Minami, A., & Kawamura, Y. (2006). Responses of the plasma membrane to low temperatures. *Physiologia Plantarum*, *126*, 81–89.

Urrutia, M. E., Duman, J. G., & Knight, C. A. (1992). Plant thermal hysteresis proteins. *Biochimica et Biophysica Acta*, *1121*, 199–206.

Vanlerberghe, G. C. (2013). Alternate oxidase: A mitochondrial respiratory pathway to maintain metabolic and signaling homeostasis during abiotic and biotic stress in plants. *International Journal of Molecular Sciences*, *14*, 6805–6847.

Venzhik, Y. V., Titov, A. F., Talanova, V. V., Frolova, S. A., Talanov, A. V., & Nazarkina, Y. A. (2011). Influence of lowered temperature on the resistance and functional activity of the photosynthetic apparatus of wheat plants. *Biology Bulletin*, *38*, 132–137.

Véry, A. A., & Sentenac, H. (2002). Cation channels in the *Arabidopsis* plasma membrane. *Trends in Plant Science*, *7*, 168–175.

Voshida, S., & Sakai, A. (1974). Phospholipid degradation in frozen cells associated with freezing injury. *Plant Physiology*, *53*, 509–511.

Wang, H., Gong, M., Xin, H., Tang, L., Dai, D., Gao, Y., & Liu, C. (2018). Effects of chilling stress on the accumulation of soluble sugars and their key enzymes in *Jatropha curcas* seedlings. *Physiology and Molecular Biology of Plants*, *24*, 857–865.

Wang, J., Fang, R., Yuan, L., Yuan, G., Zhao, M., Zhu, S., Hou, J., Chen, G., & Wang, C. (2020). Response of photosynthetic capacity and antioxidative system of chloroplast in two wucai (*Brassica campestris* L.) genotypes against chilling stress. *Physiology and Molecular Biology of Plants*, *26*, 219–232.

Wang, J., Rajakulendran, N., Amirsadeghi, S., & Vanlerberghe, G. C. (2011). Impact of mitochondrial alternative oxidase expression on the response of *Nicotiana tabacum* to cold temperature. *Physiologia Plantarum*, *142*, 339–351.

Wang, K., Zhang, X., Goatley, M., & Ervin, E. (2014a). Heat shock proteins in relation to heat stress tolerance of creeping bentgrass at different N levels. *PLoS One*, *9*, e102914.

Wang, L., Guo, Y., Jia, L., Chu, H., Zhou, S., Chen, K., Wu, D., & Zhao, L. (2014b). Hydrogen peroxide acts upstream of nitric oxide in the heat shock pathway in Arabidopsis seedlings. *Plant Physiology*, *164*, 2184–2196.

Wang, W.-X., Pelah, D., Alergand, T., Shoseyov, O., & Altman, A. (2002). Characterization of SP1, a stress-responsive, boiling-soluble, homo-oligomeric protein from aspen. *Plant Physiology*, *130*, 865–875.

Wise, R. R., Olson, A. J., Schrader, S. M., & Sharkey, T. D. (2004). Electron transport is the functional limitation of photosynthesis in field-grown Pima cotton plants at high temperature. *Plant Cell & Environment*, *27*, 717–724.

Wu, H. C., & Jinn, T. L. (2012). Oscillation regulation of Ca^{2+}/calmodulin and heat-stress related genes in response to heat stress in rice (*Oryza sativa* L.). *Plant Signaling & Behavior*, *7*, 1056–1057.

Wu, H. C., Luo, D. L., Vignols, F., & Jinn, T. L. (2012). Heat shock induced biphasic Ca^{2+} signature and OsCaM1-1 nuclear localization mediate downstream signalling in acquisition of thermotolerance in rice (*Oryza sativa* L.). *Plant Cell & Environment*, *35*, 1543–1557.

Yang, J., Liu, S., Ji, L., Tang, X., Zhu, Y., & Xie, G. (2020). Identification of novel OsCML16 target proteins and differential expression analysis under abiotic stresses in rice. *Journal of Plant Physiology*, *249*, 153165.

Yang, X., Guan, H., Yang, Y., Zhang, Y., Su, W., Song, S., Liu, H., Chen, R., & Hao, Y. (2023). Extra- and intranuclear heat perception and triggering mechanisms in plants. *Frontiers in Plant Science*, *14*, 1276649.

Yang, X., Pang, X., Xu, L., Fang, R., Huang, X., Guan, P., Lu, W., & Zhang, Z. (2009). Accumulation of soluble sugars in peel at high temperature leads to stay-green ripe banana fruit. *Journal of Experimental Botany*, *60*, 4051–4062.

Yang, X.-Y., Chen, Z.-W., Xu, T., Qu, Z., Pan, X.-D., Qin, X.-H., Ren, D.-T., & Liu, G.-Q. (2011). Arabidopsis kinesin KP1 specifically interacts with VDAC3, a mitochondrial protein, and regulates respiration during seed germination at low temperature. *Plant Cell*, *23*, 1093–1106.

Yeh, S., Moffatt, B. A., Griffith, M., Xiong, F., Yang, D. S. C., Wiseman, S. B., Sarhan, F., Danyluk, J., Xue, Y. Q., Hew, C. L., Doherty-Kirby, A., & Lajoie, G. (2000). Chitinase genes responsive to cold encode antifreeze proteins in winter cereals. *Plant Physiology*, *124*, 1251–1264.

Yin, H., Chen, Q., & Yi, M. (2008). Effects of short-term heat stress on oxidative damage and responses of antioxidant system in *Lilium longiflorum*. *Plant Growth Regulation*, *54*, 45–54.

Young, T. E., Ling, J., Geisler-Lee, C. J., Tanguay, R. L., Caldwell, C., & Gallie, D. R. (2001). Developmental and thermal regulation of the maize heat shock protein, HSP101. *Plant Physiology*, *127*, 777–791.

Yu, W., Wang, L., Zhao, R., Sheng, J., Zhang, S., Li, R., & Shen, L. (2019). Knockout of SlMAPK3 enhances tolerance to heat stress involving ROS homeostasis in tomato plants. *BMC Plant Biology*, *19*, 354.

Zandalinas, S. I., Fritschi, F. B., & Mittler, R. (2020). Signal transduction networks during stress combination. *Journal of Experimental Botany*, *71*, 1734–1741.

Zhang, C.-X., Feng, B.-H., Chen, T.-T., Fu, W.-M., Li, H.-B., Li, G.-Y., Jin, Q.-Y., Tao, L.-X., & Fu, G.-F. (2018). Heat stress-reduced kernel weight in rice at anthesis is associated with impaired source-sink relationship and sugars allocation. *Environmental and Experimental Botany*, *155*, 718–733.

Zhang, W., Zhou, R.-G., Gao, Y.-J., Zheng, S.-Z., Xu, P., Zhang, S.-Q., & Sun, D.-Y. (2009). Molecular and genetic evidence for the key role of AtCaM3 in heat-shock signal transduction in Arabidopsis. *Plant Physiology*, *149*, 1773–1784.

Zhao, F. C., Jing, L. Q., Yan, F. B., Lu, D. L., Wang, G. Y., & Lu, W. P. (2013). Effects of heat stress during grain filling on sugar accumulation and enzyme activity associated with sucrose metabolism in sweet corn. *Acta Agronomica Sinica*, *39*, 1644–1651.

Zhen, Y., & Ungerer, M. C. (2008). Relaxed selection on the CBF/DREB1 regulatory genes and reduced freezing tolerance in the southern range of *Arabidopsis thaliana*. *Molecular Biology and Evolution*, *25*, 2547–2555.

Zheng, Y., Yang, Z., Luo, J., Zhang, Y., Jiang, N., & Khattak, W. A. (2023). Transcriptome analysis of sugar and acid metabolism in young tomato fruits under high temperature and nitrogen fertilizer influence. *Frontiers in Plant Science*, *14*, 1197553.

9 Molecular Mechanisms and Breeding Strategies for Heat Tolerance in Plants

An Update

Deepti Srivastava, Sandeep Kumar Singh, Saba Siddiqui, Mohammed Haris Siddiqui, Deepak Kumar, Md. Shamim, and Rashmi Maurya

9.1 INTRODUCTION

According to FAO-STAT (http://faostat.fao.org) and Verisk Maplecroft (https://www.maplecroft.com) (Bita and Gerats, 2013), climate change, specifically an increase in ambient temperatures, is expected to have a major impact on plant growth and development. This will lead to a catastrophic reduction in crop productivity, causing severe famine and limiting global food security. The Intergovernmental Panel on Climate Change report states that the combustion of fossil fuels and the resulting greenhouse gas emissions are responsible for the buildup of atmospheric concentrations of greenhouse gases that absorb infrared radiation reflected from the earth's surface. Forecasts suggest that global air temperatures will rise by 0.2°C on average every 10 years, with a possible increase of 1.8–4.0°C by 2100. The rise in temperature above a threshold level for a length of time long enough to permanently harm plant growth and development is commonly referred to as heat stress. Heat shock or heat stress is often defined as a brief increase in temperature, typically 10–15°C above ambient temperature. The severity, duration, and rate of increase in temperature play intricate roles in the effect of heat stress on plants. The occurrence and length of high temperatures (HTs), whether during the day or at night, determine how heat stress manifests in different climate zones. The ability of a plant to flourish and produce commercially viable produce in HTs is known as heat tolerance. Heat stress profoundly affects essential biological processes, either directly or by modification of the surrounding ecological milieu. Of them, plants, being dependent on their environment, have significant difficulties in enduring HTs. HT stress thus seriously impairs the growth and developmental trajectories of plants, frequently with irreversible effects. Plant growth, development, physiological processes, and yield are all affected by heat stress in a variety of ways that are frequently detrimental (Figure 9.1 and Table 9.1).

The overproduction of reactive oxygen species (ROS), which causes oxidative stress, is one of the main effects of heat stress (Hasanuzzaman et al., 2012, 2013). Under a variety of environmental stress situations, including HTs, plants constantly fight for existence. A plant may withstand heat stress to some degree by internal physiological changes and typically by generating signals that alter metabolism. HTs lead to the inactivation of chloroplastic and mitochondrial enzymes including decline in protein synthesis, protein degradation, and a breakdown in membrane integrity (Howarth, 2005). Moreover, heat stress disrupts the arrangement of microtubules, leading to alterations such as splitting or elongation of spindles, formation of microtubule asters, and microtubule elongation (Smertenko et al., 1997). Starvation, growth inhibition, decreased ion flow, the generation of hazardous chemicals, and ROS are the final results of heat stress injuries (Schoffl et al., 1999; Howarth,

DOI: 10.1201/9781003457237-9

TABLE 9.1
Effects of High-Temperature Stress on Different Crop Species

S. No.	Crops	Heat Treatment	Growth Stage	Major Effects	References
1.	Chili pepper (*Capsicum annuum*)	38°C/30°C (day/night)	Reproductive, maturity, and harvesting stage	Fruit weight and width were decreased, and the percentage of aberrant seeds per fruit rose.	Cao et al. (2009)
2.	Rice (*Oryza sativa*)	Above 33°C, 10 days	Heading stage	Decreased the fertility of spikelets and pollen.	Hurkman et al. (2009)
3.	Wheat (*Triticum aestivum*)	37°C/28°C (day/night), 20 days	Grain filling and maturity stage	Reduced yield and kernel weight, shorter maturation and filling times.	Rahman et al. (2009)
4.	Wheat (*T. aestivum*)	30°C/25°C day/night	From 60 DAS to the maturity stage	Reduced leaf size, shorter timespan between heading, anthesis, and maturity; sharp decline in the number of grains and spikes; smaller grains and lower yield.	Djanaguiraman et al. (2010)
5.	Sorghum (*Hordeum vulgare*)	40°C/30°C (day/night)	65 DAS to maturity stage	Reduced yield due to thylakoid membrane damage, decreased chlorophyll (chl) concentration, decreased chl a fluorescence, decreased photochemistry of photosystem II (PSII), increased ROS content, Pn (net photosynthetic rate) and antioxidant enzyme activity.	Mohammed and Tarpley (2010)
6.	Rice (*O. sativa*)	32°C (night temperature)	Reproductive stage	Reduced grain length, width, and weight; increased spikelet sterility; and lowered yield.	Suwa et al. (2010)
7.	Maize (*Zea mays*)	35°C/27°C (day/night), 14 days	Reproductive stage	Decreased ear growth, specifically inhibiting the extensibility of the cob by impairing hemicellulose and cellulose synthesis by reducing the availability of photosynthate.	Yin et al. (2010)
8.	Rice (*O. sativa*)	25–42.5°C	Vegetative growth stage	Reduced the rate of CO_2 uptake.	Djanaguiraman et al. (2011)
9.	Soybean (*Glycine max*)	38°C/28°C (day/night), 14 days	Flowering stage	Reduced the stomatal conductance (gs) and Pn (net photosynthetic rate) of the leaf; thickened the spongy and palisade layers; thylakoid, chloroplast, and plasma membranes; and deformed the mitochondrial membranes, cristae, and matrix.	Tan et al. (2011)

(Continued)

TABLE 9.1 (*Continued*)
Effects of High-Temperature Stress on Different Crop Species

S. No.	Crops	Heat Treatment	Growth Stage	Major Effects	References
10.	Tobacco (*Nicotiana tabacum*)	43°C, 2 hours	Early growth stage	Reduced the apparent quantum yield, carboxylation efficiency, stomatal conductance, and net photosynthetic rate (Pn) of photosynthesis. Decreased antioxidant enzyme activity.	Gunawardhana and de Silva. (2011)
11.	Okra (*Abelmoschus esculentus*)	32°C and 34°C	Throughout the growing period	Lower yield, impacting the parameters of pod quality, including fiber content and Ca-pectate breakdown.	Edreira and Otegui (2012)
12.	Maize (*Z. mays*)	33–40°C, 15 days	During pre-anthesis and silking onward	Severe impact on the rates of plant and ear growth.	Zhang et al. (2013)
13.	Wheat (*T. aestivum*)	38°C, 24 and 48 hours	Seedling stage	Reduced antioxidative ability; decreased chloroplast concentration and relative water content.	Hasanuzzaman et al. (2013)
14.	Wheat (*T. aestivum*)	32°C/24°C (day/night), 24 hours	At the end of the spikelet initiation stage	Decreased grain yield and spikelet sterility.	Saitoh (2008)

DAS, days after sowing.

2005). Plants change their metabolism in many ways, most notably by generating compatible solutes that can reorganize proteins and cellular structures, adjust the osmotic pressure of cells to maintain turgor, and modify the antioxidant system to restore cellular homeostasis and redox balance (Valliyodan and Nguyen, 2006).

Plants confer heat resistance by heat avoidance, escape, and tolerance. In heat avoidance, plants reduce their leaf area and close their stomata so that high water status can be maintained, while in escape mechanism they complete their reproduction cycle before heat stress. In the heat tolerance mechanism, plants maintain their functions by efficiently scavenging free radicals, ROS. The key mechanism in maintaining the structure and function of protein is via heat shock protein (HSP) expression. The HSPs, which have molecular masses between 10 and 200 kDa, are involved in signal transduction during heat stress and have chaperone-like properties (Schoffl et al., 1999). The resilience provided by HSPs leads to enhancements in various physiological processes, including photosynthesis, distribution of nutrients, water utilization efficiency, and the stability of membranes (Camejo et al., 2005; Ahn and Zimmerman, 2006).

Heat stress also modifies the expression of genes that directly protect against heat stress at the molecular level (Chinnusamy et al., 2004). These comprise the genes that produce detoxification enzymes, regulatory proteins, transporters, and osmoprotectant expression (Krasensky and Jonak, 2012; Semenov and Halford, 2009). Under heat stress, alterations in gene expression that modify physiological and biochemical processes eventually result in the development of heat tolerance as either adaptation or acclimation (Moreno and Orellana, 2011; Hasanuzzaman et al., 2020). Exogenous protectant applications, such as osmoprotectants (proline, Pro; glycine betaine, GB; trehalose, Tre; etc.), phytohormones (abscisic acid, ABA; gibberellic acids, GAs; jasmonic acids, JAs;

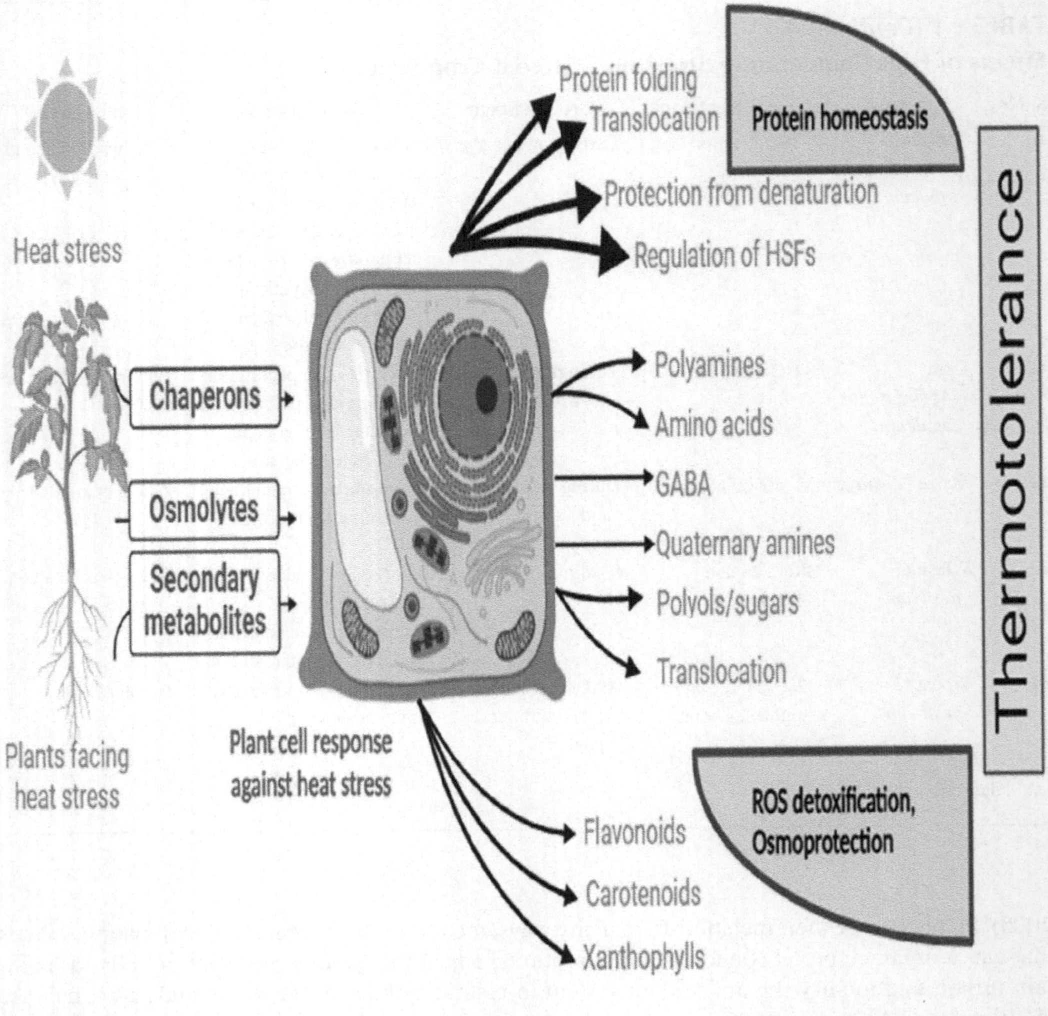

FIGURE 9.1 Effect of heat stress on plants.

brassinosteroids, BRs; salicylic acid, SA; etc.), signaling molecules (e.g., nitric oxide, NO), polyamines (putrescine, Put; spermidine, Spd; spermine, Spm), trace elements (selenium, Si; silicon, Si); and nutrients (nitrogen, N; phosphorus, P; potassium, K; calcium, Ca, etc.) are useful in reducing the damage caused by HT stress in plants (Waraich et al., 2012). Plant scientists have a significant problem in the development of novel agricultural cultivars tolerant to HT (Moreno and Orellana, 2011). The objective of plant scientists studying heat tolerance is to understand the mechanism of how plants respond and manage heat stress. Plant development and growth under heat stress are made possible by understanding the mechanism, and, subsequently, with the use of recent advancements heat-tolerant plants can be developed. Not every plant species, nor every genotype within a species, is equally able to withstand heat stress. Therefore, genetic variability both within and across species presents chances to enhance crop heat stress resistance. There have been some successful attempts to use traditional plant breeding techniques to create heat-tolerant genotypes (Camejo et al., 2005). But, more recently, breakthroughs in genetic engineering and molecular breeding have produced new tools that could be used to fight this ubiquitous environmental enemy and create crops that are more resilient to heat. However, coordinated efforts by plant physiologists, molecular biologists, and crop breeders are essential to ensure success. In the present chapter, we will describe how plants are affected by heat stress and different breeding methods by which heat-tolerant plants can be developed.

9.2 EFFECT OF HEAT STRESS ON PLANTS

Heat stress is a major limitation for plant productivity in most of the parts of the world. Heat stress causes morphological, physiological, and various biochemical changes in plants to withstand heat tolerance.

9.3 MORPHOLOGICAL EFFECT OF HEAT STRESS ON PLANTS

In the case of wheat, optimum temperatures for the anthesis and grain filling range from 12°C to 22°C. Exposure to HTs above this reduces the yield. HT causes both pre- and postharvest damage that includes scorching of leaves and twigs; sunburn of leaves, branches, and stems; senescence and abscission of leaves; inhibition of growth of root and shoot; discoloration of fruit; and reduced yield (Vollenweider and Günthardt-Goerg, 2005). Maize is also affected by heat stress (Giaveno and Ferrero, 2003). Prolonged exposure to heat stress causes delayed seed germination, loss of vigor, and reduced establishment of seed. A common issue associated with heat is reduced fertility, which has been linked to HTs during meiosis and fertilization in a variety of species, including rice, cowpea, barley, tomato, and *Arabidopsis* (Giorno et al., 2015; Jagadish et al., 2015; Ehlers and Hall, 1998; Sakata and Higashitani, 2008). The most heat-sensitive organs are the male reproductive system and, in particular, pollen formation. Reductions in viable and germinating pollen occur when plants are subjected to extreme temperature stress during flowering (Zinn et al., 2010). For instance, spikelet sterility in rice happens when temperatures rise beyond 35°C for even a single hour (Endo et al., 2009). According to Zhang and Yang (2014), tomatoes exhibit a significant reduction in fruit set when subjected to heat stress, particularly during the microsporogenesis stage. Male sterility occurs during this stage when exposure to temperatures merely a few degrees above ideal (32°C instead of 26°C during the day) or for an extended period above 40°C causes sterility (Sato et al., 2006; Giorno et al., 2013). Similarly, lack of tapetum differentiation and harm to the microsporogenesis process indicate that barley is susceptible to persistent mild heat stress (30°C/25°C day/night, for five days) (Sakata and Higashitani, 2008). It has long been believed that the maternal tissues of the female gametophyte and pistil are more thermotolerant. However, exposure to heat can cause abnormalities in the female tissues of certain species. There have been reports of embryosac abnormalities in peaches developed at 25°C, in wheat at 30°C, and in rapeseed at 32°C. In the latter two species, this has led to a decrease in seed set (Hedhly, 2011). Even a small elevation of 3°C over control circumstances during the final week of flower development in apricots produced aberrant ovaries and a shortening of the style (Rodrigo and Herrero, 2002). Heat shortens stigma receptivity in cherries and peaches as well as ovule lifetime in cherries and plums, in addition to its effects on male tissue (Endo et al., 2009; Hedhly, 2011). These changes cause the female and male reproductive tissues to get out of synchrony, which lowers the effectiveness of conception. However, timely pollination does not ensure fruit or seed set because heat sensitivity has also been demonstrated in post-pollination processes like pollen tube growth, fertilization, endosperm formation, and embryo development (Barnabás et al., 2008).

9.4 PHYSIOLOGICAL EFFECT OF HEAT STRESS ON PLANTS

HS has a negative impact on a number of physiological functions, including photosynthesis, respiration, transpiration, membrane thermostability, and osmotic control. HS frequently affects the physiological reactions, growth and development, and yield of plants. In general, HS shortens the plant life cycle and lowers output by reducing photosynthetic efficiency (Zahara et al., 2024).

9.5 EFFECT ON PHOTOSYNTHESIS

One of the physiological processes that is most sensitive to heat is photosynthesis. Carbon metabolism in the chloroplast stroma and photochemical processes in thylakoid lamellae are vulnerable to damage under HS (Zahara et al., 2024). Heat stress breaks down thylakoid membranes, which stops electron carriers and enzymes linked to membranes from working, slowing down photosynthesis. In particular, because the PSII complex is the most heat intolerant, photosystem II (PSII) activity is significantly decreased or completely ceases under HTs. Furthermore, HS decreases the quantities of photosynthetic pigments, the carbon fixation capability, and the activity of ribulose-1,5-bisphosphate carboxylase/oxygenase (Rubisco), as well as the structure of the chloroplast and the thermal stability of photosynthetic system components. Under HS, these variables greatly contribute to the decline in photosynthetic efficiency. Therefore, studying plant thermostability and the detrimental impacts of warming on crop productivity requires a basic understanding of photosynthetic physiology (Zahara et al., 2024). When temperatures fluctuate, plant water status is typically unpredictable. Dehydration brought on by heat stress hinders the growth and development of plants. When exposed to HS, water potential and relative water content significantly drop, which lowers photosynthetic production. On the contrary, plants control transpiration and rate of respiration under transitory or mild HS to maintain a balance between heat dissipation and water loss. To control the osmotic pressure inside the cell, soluble sugar and protein levels are also changed during HS.

9.6 EFFECT OF HEAT STRESS ON RESPIRATION AND TRANSPIRATION

Under HT, respiration might be the main cause of yield losses rather than photosynthesis. High crop yields are typically associated with low respiration rates. According to Walker et al. (2016), photorespiration reduced wheat and soybean yields in the United States by 20% and 36%, respectively. HTs were found to be the cause of a 10–12% and 17–35% drop in wheat and rice yields, respectively, in another study (Cai et al., 2016). Rice yield loss due to high night temperature (HNT) and wheat is mostly attributed to increased dark respiration, which reduces non-structural carbohydrates (NSC) in the stem and increases the level of photoassimilate consumption (Impa et al., 2021). Grapevines also experienced higher leaf respiratory carbon losses when night temperatures rose from 25°C to 35°C (Tombesi et al., 2019). This was shown by a drop in Non-structural carbohydrate (NSC s) of 0.025 and 0.041 mg/g dry weight, respectively. These losses are in line with research on the metabolite profiles of wheat and rice, which showed that leaves exposed to HNT had higher levels of TCA cycle intermediates, indicating enhanced respiration in the photosynthesizing tissue (Sharma et al., 2022). According to Xu et al. (2021), enhanced dark respiration restricts the availability of sources when high day and night-time temperatures are combined, resulting in a yield penalty from carbon loss that is noticeably more severe.

9.7 EFFECTS ON THE THERMOSTABILITY OF CELL MEMBRANE

Heat stress has a significant impact on cell membrane stability. Due to the heat stress conformation of the proteins and lipids of the membrane changes due to which functions of the biological membrane are also affected (Ibrahim and Quick, 2001). At very high pressure, chemical bonds unwind due to the breakdown of membrane lipids and an increase in membrane fluidity. This arises due to the biomolecules' higher kinetic energy and ability to move through membranes. Cells with HS are less able to function normally and have a decreased capacity to endure heat due to the increased permeability of their cellular membranes and electrolyte loss. Moreover, HS damages membranes by producing ROS, which lowers thermotolerance. Membrane thermostability primarily determines a plant's capacity to withstand HTs. Nijabat et al. (2020) investigated the stability of cell membranes and the degree of cell injury in various carrots (*Daucus carota* L.) germplasms during both early and late seedling stages under heat stress conditions. They identified that heat tolerance varied greatly between cultivated and wild accessions, and it was more prevalent in cultivated carrots than in wild ones.

9.8 OXIDATIVE DAMAGE

Like many other environmental stressors, elevated temperatures lead to heightened production of ROS, comprising hydroxyl radical (HO), superoxide anion radical (O2•–), hydrogen peroxide (H_2O_2), and singlet oxygen ($1O_2$). When biomolecules such as carbohydrates, proteins, lipids, pigments, and DNA are subjected to oxidative stress due to ROS accumulation, this can result in cellular damage and even death.

However, these molecules can act as molecular signals, triggering alterations in cellular metabolism that lead to adaptive responses or the initiation of programmed cell death; therefore, synthesis and scavenging are balanced carefully, hence regulating ROS levels (Mittler, 2017). The concentration of ROS in various organelles is solely dependent on the plant tissues. ROSs are primarily generated in mitochondria, peroxisomes, and the apoplast within organs characterized by limited photosynthetic metabolisms, such as fruits and flowers (Decros et al., 2019), whereas, in the case of photosynthetic tissue, chloroplasts are the main center of ROS generation. The primary environmental stress sensors, including heat, are chloroplasts (Hasanuzzaman et al., 2020). By disrupting PSII's reaction center, heat stress within chloroplasts diminishes photosynthetic efficiency, hindering electron transfer to the plastoquinone pool and the activation of non-photochemical quenching (Doğru, 2021). Damage to the photosynthetic components could lead to excess excitation energy, generating a chlorophyll triplet state, which ultimately forms a singlet oxygen state via interacting with molecular oxygen (Krieger-Liszkay et al., 2008). At elevated temperatures, the impairment of PSII and the occurrence of heat-induced lipid peroxidation in the thylakoid membrane are associated with the generation of singlet oxygen.

9.9 ROLE OF HSP IN HEAT STRESS TOLERANCE

Heat stress generally induces several heat-inducible genes that code for HSPs. Plant survival under HTs relies largely on their active products (Chang et al., 2007). The majority of these proteins work as chaperones because their constitutive production at HTs prevents denaturation and maintains the stability of intracellular proteins through protein folding (Baniwal et al., 2004). The HSP family protein has vast diversity and is constantly growing. Only specific plant developmental stages, such as seed germination, embryogenesis, microsporogenesis, and fruit maturation, are associated with HSP expression (Prasinos et al., 2005). Five distinct families of well-characterized HSPs found in plants include HSP 20 (small HSP, sHSP), HSP60 (GroE), HSP70 (DnaK), HSP90, and HSP100 (ClpB) (Swindell et al., 2007). The most conservation is found among families HSP60 and HSP70, which is in line with their essential function in the body's reaction to heat stress. Certain mutants with impaired thermotolerance exhibit typical HSP induction. With a molecular mass of only 12–40 kDa, plants also contain the greatest diversity of single-chain DNA polymorphisms (Morrow and Tanguay, 2012).

HSPs are induced under high heat conditions. HSF contains a heat shock element (HSE) which triggers the transcription of HSPs under heat treatment. HSE contains a palindromic nucleotide sequence (5-AGAANNTTCT-3), which is recognized and binds with heat shock factors (HSFs). Most plant species typically demonstrate constitutive expression of HSFs, proteins typically residing in the cytoplasm as monomers, associated with HSP70. Upon increase in temperature, HSP70 dissociates from cytoplasmic monomeric HSFs and translocates in the nucleus, forming a trimer that can attach to the HSEs. The expression of HSP happens within minutes of high heat induction via the interaction of HSFs with the transcription machinery. Since the HSE conserved region is present in all HSGs, overexpressing the HSF gene activates nearly all HSGs and hence protects against heat stress. Plants have a very complex fundamental system. It has been discovered that plants contain several copies of these genes, compared to animals and yeasts. For example, tomato and *Arabidopsis* contain at least 17 copies, distinct 21 distinct HSF genes, respectively. Based on the characteristics of their oligomerization domains and flexible linkers, these are classified into three classes (A, B, and C) (Nover et al., 2001). Plant thermotolerance can generally be increased by overexpressing HSFs; however, so far, screening with individual HSF gene knockouts has not shown significant effects on plant survival at HTs. Heat also induces several non-HSP genes in plants (Morrow and Tanguay, 2012). Interestingly, the *Arabidopsis* cytosolic ascorbate peroxidase gene (APX1) has been shown to have a functional HSE in its 5′-promoter region and induced under heat conditions.

While HSFs 1a and 1b in *Arabidopsis* seem to regulate many genes' early response to heat, additional HSFs – possibly including heat-inducible HSF – seem to be responsible for inducing the expression of genes that are expressed later. The 'master regulator' of the heat shock response in tomatoes may be one specific HSF (HSFA1). A plant that has this gene silenced is highly vulnerable to HTs and does not produce HSP normally (Mishra et al., 2002). There is evidence that various HSFs work in coordination. Thus, it seems that plants may precisely regulate how heat-induced genes are expressed using the HSF system. A positive correlation occurs between HSP level in the cell and related stress tolerance. Although the exact mechanism by which HSPs are yet to be identified, Miroshnichenko et al. (2005) examined the in vivo function of thermoprotection. The thermoprotection is provided by HSP assembly in stress granules. HSPs have been found alone or in the form of chaperones and regulate plant cell defense mechanisms under heat stress (Ul Haq et al., 2019). Because of the increased degradative molecular interactions caused by a smaller cellular volume, protein denaturation happens under HT. HSPs target improperly assembled and foreign proteins for breakdown and removal from cells while also maintaining and repairing the structure of partner proteins (Ul Haq et al., 2019). To determine the role of HSPs in plant stress response and tolerance, NtHSP70-1 was constitutively overexpressed in tobacco, which showed enhanced drought stress tolerance (Cho and Hong, 2006). The goal of modern genetic studies is to improve the capacity of crop plants to withstand HTs. The nature of heat stress tolerance is spatiotemporal as different components are required to regulate different phases of plant development or in different tissues (Ul Haq et al., 2019). Till today, promising methods to identify the genetic basis of plant thermotolerance include genetic stocks and breeding, correlation studies, co-segregation analysis, and using molecular markers to discover tolerance QTLs (Ul Haq et al., 2019). Further, the identified target could be utilized to validate via transgenics or breeding-based variety development.

9.10 MOLECULAR MECHANISM OF HEAT STRESS IN PLANTS

While small temperature increases may not result in significant cell damage, they can change immune response, biorhythms, and morphogenesis (Li et al., 2013). According to Zhang et al. (2019a), HS can denature the cytotoxic proteins, rupture ROS, damage the membrane, and eventually kill the plant. Different thermal responses have been created by plants to withstand heat stress. HTs alter the membrane's fluidity, which produces ROS that alters the lipid composition and membrane lipid saturation level. It affects the activation or repression of activities linked to the Ca2þ/CaM-related kinases, phosphatases, and transcription factors by directly or indirectly activating membrane-localized calcium channels, which raises the point of Ca in the cytoplasm (Kan and Lin, 2021). OsCNGC14 and OsCNGC16 react to heat stress by triggering the Ca signals (Cui et al., 2020). Leucine-repeat receptor-like kinases, ERECTA and Hwi1, control rice's temperature response via an unidentified transduction route (Kan and Lin, 2021). Under HS, membrane-localized OsNTL3 travels to the nucleus and controls the expression of OsZIP74, after which endoplasmic reticulum-localized OsbZIP74 travels to the nucleus and triggers the expression of OsNTL3 (Liu et al., 2020; Lu et al., 2012). De novo synthesis of the D1 subunit is necessary for the restoration of damaged photosystem II and PGL (pale green leaf, encoding chlorophyllide and oxygenase 1) in the chloroplast (Kan and Lin, 2021); the OsTRXz-OsFLN1/2 complex shields chloroplasts from heat-induced damage, while EG1 located in mitochondria preserves the organelle and maintains the robustness of the floral community (Zhang et al., 2016). OsANN1, SNAC3, OsHTAS, drought and salt tolerance (DST), and a few additional chloroplast genes work together to preserve ROS homeostasis (Kan and Lin, 2021). The calcium-binding annexin OsANN1 increases heat stress tolerance by regulating antioxidant accumulation (Qiao et al., 2015); the zinc finger transcription factor DST, C_2H_2, regulates the expression of genes that determine thermotolerance and regulate H_2O_2 homeostasis (Zhang et al., 2019b). OsHTAS, a ubiquitin E3 ligase, promotes thermotolerance by controlling hydrogen peroxide buildup to change the stomatal aperture in an ABA-dependent and DST-mediated process. It also helps with the clearance of unfolded proteins by interacting with ubiquitin/26S proteasome system components. SNAC3, a NAC transcription factor, maintains ROS homeostasis and provides thermotolerance to rice. OsHATS defend cells against harmful denatured proteins via the 26S proteasome system, and OsHSPs stabilize, renature, and aid in the degradation of unfolded proteins, TT1 (THERMOTOLERANCE1), and the a2 component of the 26S proteasome when protein homeostasis is disrupted (Li et al., 2013). While LS1 (local lesion 1) preserves genome stability against DNA damage for DNA homeostasis, TOGR1, OsNSUN2, SLG1, and AET1 sustain RNA processing, modification, and stability during HS to maintain RNA homeostasis (Dong et al., 2020). AET1, which codes for a tRNA His guanylyltransferase, is essential for the translation of mRNA. Rice is protected from heat damage by GSA1, which codes for a UDP-glucosyltransferase. This is achieved by preventing the buildup of flavonoid glycosides (Dong et al., 2020). Ultimately, three cycles of recurrent selection for heat tolerance in potato breeding produced a genetic gain that resulted in a significant increase in yield of up to 37.8% (Benites and Pinto, 2011). Heat-tolerant lines have been successfully developed through standard 'yield' breeding; nevertheless, the underlying genetic and physiological basis for these gains is still unknown.

This stops additional or molecular biomarkers from emerging, which would help with germplasm screening for enhanced heat tolerance and enable effective breeding of the complex trait. The fact that conventional breeding operations frequently include crossing rather advanced starting material that has already been employed in the specific breeding regions relevant to the targeted market segment is another disadvantage of the method.

9.11 BREEDING METHODS FOR HEAT STRESS

To lessen the detrimental impact of heat on agricultural productivity temperature tolerance should be increased through traditional breeding. Breeding programs are typically conducted in a climacteric zone that is similar to the final production location of the crop. Thus, heated circumstances are used to select breeding materials in comparatively hot places (Mickelbart et al., 2015). Therefore, breeders working locally should 'passively' select for thermotolerance features in hot climates. This approach is successful for cultivars of warm climates (Kugblenu et al., 2013). Several novel heat stress-tolerant varieties have been developed by conventional breeding approaches. For instance, an early maturing variety of broccoli produces a better-quality head because it keeps hot days later in the season from affecting the heat-sensitive flower initiation developmental stage (Farnham and Björkman, 2011). Furthermore, new cowpea varieties that were grown under hot, long days during reproduction have high grain yield (Ehlers and Hall, 1998). Similarly, wheat genotypes for the screening of various genetic resources of wheat are evaluated for the identification of heat-tolerant genotypes, and recurrent selection of the breeding method was used. Recurrent selection in BC1F6 generation resulted in an improved rate of grain filling and grain size (Gororo et al., 2002). In potato breeding, three cycles of recurrent selection for heat tolerance resulted in a genetic gain that resulted in a significant increase in yield of up to 37.8% (Benites and Pinto, 2011). However, conventional breeding is successful in the development of heat-tolerant lines, but the physiological and genetic basis of improvement is not clear, which prevents the development of molecular and biomarkers. Molecular markers facilitate the screening of germplasms for heat tolerance as well as allow the breeding of complex traits. Therefore, advanced breeding and molecular methods should be used for the development of heat-tolerant plants.

9.12 ADVANCED BREEDING AND MOLECULAR METHODS FOR HEAT TOLERANCE

9.12.1 Quantitative Trait Loci Mapping

To meet human needs, crops have undergone drastic morphological and physiological modifications through 'crop domestication', which is also the first step of plant breeding. These modifications include the improvement of seed and fruit size and number, seed shattering, dormancy of seed, photoperiod and flowering time, taste, nutrition, and overall plant architecture (Meyer and Purugganan, 2013). A major drawback of using selection and breeding of lines having beneficial traits is that it leads to the genetic bottleneck as cultivars have only a small subset of genes that are present in their wild ancestor's gene pool (Olsen and Wendel, 2013). Abiotic stress tolerance traits appear to have been lost as a result of this reduction in diversity, as many wild relatives and landraces exhibit greater stress tolerance than domesticated crops (Dolferus, 2014). Therefore, there is a need for the identification of superior wild genes which are not present in the cultivated germplasm (Feuillet et al., 2008). In only a few crops [(rice (Atwell et al., 2014), wheat (Pradhan et al., 2012), and maize (Prasanna, 2012)], crop improvement has been done by the introgression of wild alleles. A major reason for this is a lack of molecular markers for precision breeding and undesirable traits that are linked with wild genes. This hinders the use of wild relatives as the source of novel alleles. However, resequencing entire genomes has now become affordable because of the new sequencing techniques. Resequencing is done in wild tomatoes (Aflitos et al., 2014), cucumber (Qi et al., 2013), sorghum (Mace et al., 2013), grape (Lijavetzky et al., 2007), soybean (Li et al., 2013), rice (Xu et al., 2012), and maize (www.panzea.org). Heat tolerance is governed by multiple genes with cumulative effects on trait expression, also known as quantitative trait loci (QTL). The identification of genes or QTLs that provide increased resistance to HT stress has applications in the field of agriculture. Jha et al. (2021) discovered major QTLs and potential candidate genes for heat stress tolerance in chickpeas (*Cicer arietinum L.*). They developed a recombinant inbred line (RIL) population by crossing heat-sensitive DCP 92–3 and heat-tolerant ICCV 92944, and then genotyping was done using the genotyping-by-sequencing approach. Heat tolerance was evaluated for two successive years (2017 and 2018) under normal and late-sown or heat stress environments for the determination of the genomic regions responsible for heat stress tolerance. A high-density genomic map covering 1,125 cM was created using 788 single-nucleotide polymorphism markers. For 12 of the 13 characteristics, a total of 77 QTLs (37 major and 40 minor) were found using composite interval mapping. QTLs are present in a genomic region on CaLG07 that accounts for >30% of the phenotypic variance in days to pod initiation, 100 seed weight, and nitrogen balance index, which accounts for >10% of PVE. Thirty-two genes were also identified in the QTL, which regulates flowering time as well as encodes for heat shock transcription factors and HSPs. Identified QTLs will help in the development of superior lines that have heat-tolerant genes or heat tolerance.

Similarly, Wen et al. (2019) used traditional mapping of QTL, QTL-seq, and RNA-seq in tomatoes to identify heat-tolerant QTLs and HT stress-responsive genes. Five significant QTLs were found: qHII-1–2, qHII-1–3, qHII-2-1, qCC-1–5 (qREC-1–3), and HII-1-1. qHII-1-1, qHII-1–2, and qHII-1–3 were found on chromosome 1 in the intervals of 1.43, 1.17, and 1.19 Mb, respectively, whereas the interval of qHII-2-1 was found on chromosome 2 in the intervals of 1.87 Mb. By using traditional QTL mapping, the locations of qCC-1–5 and qREC-1–3 for CC and REC, respectively, were shown to be identical. Similarly, Chen et al. (2021) did QTL mapping and identified candidate genes for heat tolerance in rice at the flowering stage. Heat-tolerant QTLs mapped in the F2:3 population resulted from a hybrid between the heat-tolerant cultivar Huanghuazhan (HHZ) and the heat-sensitive variety 9311. On chromosome 8, a new significant QTL, qHTT8, regulating HT, was found by employing the bulked-segregant analysis (BSA)-seq technique. With a size of 0.965 Mb, the 3,555,000–4,520,000 bp region was assigned to the QTL qHTT8. Ten putative predicted genes were discovered to be connected to abiotic stress tolerance in the candidate region of qHTT8 on chromosome 8, which had 65 projected genes. Additionally, under HT settings, qRT-PCR was used to examine the differential expression of these ten genes between HHZ and 9311.

To decipher the complex genetic pathways behind heat tolerance in hot peppers, an F2 population was created by crossing two different genotypes that showed opposing traits for heat tolerance: DChBL-240 (heat susceptible) and DLS-161-1 (heat tolerant) (Aruna et al., 2023). The F2 population and the parental lines underwent extensive phenotyping in HT conditions (maximum temperature range: 31–46.5°C; minimum temperature range: 15.4–30.5°C) covering a wide range of morphological, physiological, and biochemical heat-related traits. Double digest restriction-site-associated DNA sequencing (ddRAD-seq) was utilized to produce 67.215 Gb of data by the Illumina Nova Seq-6000 platform. 218.93 million processed reads were then aligned to the *Capsicum annuum* reference genome. Twelve linkage groups containing 5,806 single-nucleotide polymorphism markers were formed. Subsequent QTL analysis revealed 64 QTLs that have phenotypic variance ranging from 4.05% to 19.39%, with a limit of detection values ranging from 2.517 to 11.170. Of them, 21 QTLs were revealed to be important QTLs governing nine morphological, three physiological, and two biochemical parameters, accounting for more than 10% of the phenotypic variance. Fascinatingly, several QTLs controlling different parameters were discovered to be co-localized, indicating either a strong genetic proximity or a strong link between the QTLs controlling various characteristics (Aruna et al., 2023). Apart from QTLs, they found 368380 simple sequence repeat (SSR) loci in the QTL regions, with dinucleotides being the most common form (211,381). These results shed important light on the genetics of hot peppers' heat tolerance. The discovered QTLs and SSR markers present chances to create heat-tolerant cultivars, guaranteeing improved agricultural performance in hot weather.

Heat stress during the development of cucumbers frequently results in sunburned leaves, stunted growth of stems and roots, deformed fruit, and even plant mortality, all of which have a significant effect on the output and quality of the fruit. Nevertheless, there hasn't been any published research on the genetic makeup or quantitative trait locus mapping of heat tolerance in mature cucumbers. To determine the heat tolerance QTL in the summers of 2018, 2019, and 2020, 86 RILs were produced from the heat-tolerant and heat-sensitive lines, '99281' and '931' (Liu et al., 2021). The heat damage index was utilized to assess the heat tolerance performance of 8-week-old plants that were placed in a naturally hot setting in the field. Genetic analysis showed that adult cucumber's ability to withstand heat is quantitatively inherited. On chromosome 1, a single QTL known as qHT1.1 was discovered. When 62 cucumber germplasm accessions were used to examine the efficacy of marker HT-1, which is closely linked to the locus, it was discovered that the marker's accuracy in heat-sensitive plants was 97.8%. Nine of the 98 genes in the 694.5-kb area that make up the defined qHT1.1 may be related to heat tolerance. Three single-base changes have been found in the coding sequences of Csa1G004990, according to further sequencing analysis. According to gene expression analyses, at 14d, 35d, 42d, and 49d after transplanting, '99281' had considerably higher Csa1G004990 expression than '931' (Liu et al., 2021).

9.13 ROLE OF GENETIC ENGINEERING IN DEVELOPMENT OF HEAT-TOLERANT PLANTS

The ineffectiveness of conventional breeding for HT stress tolerance can be attributed to several factors, including the complexity of the heat stress trait, inadequate knowledge of the genetic mechanisms underlying the HT tolerance response, and a lack of suitable gene sources in sexually compatible gene pools. The development of recombinant DNA (rDNA) technology helps to overcome the problem of complex genetic traits (Singh and Grover, 2008).

Transgenic breeding is an alternate method for enhancing heat tolerance. Beneficial genes from other plant species can be transferred to superior crop cultivars through transgenetic breeding without experiencing linkage drag. This makes it possible to use genes for purposes other than breeding through hybridization and to introduce novel features. Fifty-one countries have authorized the production of 540 distinct transgenic events in 32 crops, including maize, rice, and wheat (GM Approval Database, 2022). Recent instances include the overexpression of the transcription factor rive MYB in maize, which reduces the growth of plant and performance under heat and drought stress conditions, and the expression of a rice soluble starch synthase gene in transgenic wheat increases thousand kernel weight by up to 34% under heat stress conditions (Tian et al., 2018).

Gangadhar et al. (2021) enhanced thermotolerance in transgenic potato (*Solanum tuberosum* L.) overexpressing hydrogen peroxide-producing germin-like protein (GLP). Although it has been documented that germins and GLPs contribute to the formation of hydrogen peroxide (H_2O_2) in plants in response to biotic and abiotic challenges, little is known about how they help plants cope with heat stress. Using a yeast cDNA library created from heat-stressed potato plants, we examined the potential of a *Solanum tuberosum L. GLP (StGLP)* gene and described its function in conferring innate and/or learned thermotolerance to potatoes through genetic transformation. In comparison with severe heat shock, transgenic plants showed improved tolerance to gradual heat stress in terms of maximal cell viability, low ion leakage, and decreased chlorophyll degradation. Additionally, under both normal and heat stress conditions, three *StGLP* transgenic lines (G9, G12, and G15) showed an increase in H_2O_2 production, which was either decreased or prevented by H_2O_2 inhibitors. The upregulation of heat stress-responsive genes (*StHSP70, StHSP20, and StHSP90*) and antioxidant enzymes (catalase, ascorbate peroxidase, and glutathione reductase) in transgenic potato plants was the mechanism behind this tolerance. These findings show that the overexpression of StGLP in transgenic potato plants created H_2O_2, which in turn activated the signaling pathways that scavenge ROS and regulate antioxidant and heat stress-responsive genes, thereby providing the plants with a resistance to heat stress. Similarly, Liu et al. (2020) investigated *AtPLC9*'s ability to enhance heat tolerance in rice and ectopically expressed *AtPLC9*. Transgenic rice showed higher heat tolerance than the natural type; however, *AtPLC9* did not increase rice's resistance to cold, salt, or drought. Following heat stress, transgenic plants underwent substantial and dynamic transcriptome reprofiling, as shown by high-throughput RNA-seq. Furthermore, under heat stress, transgenic rice gene related to calcium ions and transcription factors activates that may induce increased heat tolerance.

Although transgenic crops are being farmed commercially on millions of hectares worldwide, there remain worries among the public over their safety and health. But with the help of alternative technologies like genome editing can provide alterations that are identical to those from conventional breeding, naturally occurring, or cause mutations, which may allay many societal concerns.

9.14 GENE EDITING

To create thermotolerant plants, breeders have employed several techniques, including transgenic approaches, molecular breeding approaches, mutation breeding, traditional breeding, and genome editing (Anwar and Kim, 2020). Genes associated with heat stress tolerance have been found through the functional genomics approaches (Anwar and Kim, 2020; Liu et al., 2015). Transgenic breeding has made it possible to create cultivars that express genes that provide resistance to temperature stress within the past 20 years (Liu et al., 2020). The adoption of transgenic crops has been hindered by public opinion and other concerns about market access (Lucht, 2015). Several site-directed nuclease approaches have been utilized in gene editing, such as zinc finger nucleases (ZFNs), meganucleases, transcription activator-like effector nucleases (TALENs), and the clustered regularly interspaced short palindromic repeats (CRISPR)/CRISPR-associated protein9 (Cas9) system (CRISPR/CAS9) (Sander and Joung, 2014). TALENs, ZFNs, and clustered regularly interspaced short palindromic repeats – Caspase 9 (CRISPR/Cas9) have become highly accurate gene-editing instruments that are particularly helpful in comprehending gene function. Transgenic plant species that can withstand HTs have already been created. The plant shows extremely complicated responses after facing HTs and shows different abnormalities related to their tissue, developmental stage, or even species-specific (Hedhly, 2011).

Major agricultural species have benefited from the effective application of gene editing, albeit many are still in the formative stage (Iqbal et al., 2020). Because new transgenes don't necessarily need to be inserted, customers may be more receptive to new commercial cultivars produced through gene editing. CRISPR was used to eliminate the seed-specific NAC transcription factors OsNAC127 and OsNAC129 (Ren et al., 2021). Under heat stress, the undeveloped grains and partial grain filling in the knockout lines became more pronounced. Thus, in the heat-stressed seed development stage, OsNAC127 and ONAC129 were both somewhat involved (Ren et al., 2021). Ca^{2+}-based sensor proteins called calcium-dependent protein kinases (CDPKs) are involved in several plants when exposed to abiotic stress (Hu et al., 2021). CRISPR-generated mutant lines of tomato plants cpk28 showed increased heat stress susceptibility as well as certain signs of damage caused by HTs in contrast to wild-type plants when the mutant plants faced HTs (Hu et al., 2021). Mutant lines of ZmCDPK7 (CDPKs) were more heat sensitive with lower antioxidant enzyme activity and reduced photosynthetic rate compared to wild-type plants after heat treatment at 42°C (day) and 35°C (night), according to a functional genomics study on heat stress in maize plants using CRISPR (Zhao et al., 2021).

9.15 CONCLUSION

Developing new varieties for heat tolerance involves a holistic approach that incorporates both selective breeding methods and molecular mechanisms in the plant system. This update highlights the critical need for a deeper comprehension of the complex biochemical processes underlying the HT response. By understanding the functions of different genes, proteins, and signaling pathways, scientists can identify important targets for breeding and genetic modification. Additionally, the development of HT-tolerant cultivars with improved resistance to heat stress is made possible by the extraordinary precision with which plant genomes can be edited, made possible by the application of cutting-edge molecular tools like CRISPR/Cas9. Nonetheless, it's critical to appropriately negotiate the legal and ethical issues raised by genetic engineering. To harness natural genetic variation for heat tolerance, conventional breeding techniques including hybridization, selection, and marker-assisted breeding are still very useful. With the help of new tools now breeders accelerate the breeding and development of robust crop varieties that are suited to endure increasingly erratic environmental circumstances by combining these traditional methods with state-of-the-art molecular procedures. Furthermore, cooperative efforts between scientists, breeders, legislators, and farmers are essential to guaranteeing the global sustainability of agricultural systems and the widespread adoption of heat-tolerant crops. In conclusion, novel breeding techniques and the ongoing development of our knowledge of molecular mechanisms and pathways present exciting opportunities for improving plant heat tolerance. To protect agricultural output and livelihoods in the face of a changing climate, it will be crucial to adopt interdisciplinary methods and innovative ideas with collaborative efforts that may be helpful for the fulfillment of sufficient food production.

ACKNOWLEDGMENT

The first author acknowledges the Seed Money Project 'IUL/IIRC/SMP/2022/010' given by the Integral University.

REFERENCES

Aflitos, S., Schijlen, E., de Jong, H., de Ridder, D., Smit, S., Finkers, R., Wang, J., Zhang, G., Li, N., Mao, L., Bakker, F., Dirks, R., Breit, T., Gravendeel, B., Huits, H., Struss, D., Swanson-Wagner, R., van Leeuwen, H., van Ham, R. C., … Peters, S. (2014). Exploring genetic variation in the tomato (*Solanum section* Lycopersicon) clade by whole-genome sequencing. *The Plant Journal*, 80(1), 136–148.

Ahn, Y. J., & Zimmerman, J. L. (2006). Introduction of the carrot HSP17. 7 into potato (Solanum tuberosum L.) enhances cellular membrane stability and tuberization in vitro. *Plant, Cell & Environment*, 29(1), 95–104.

Anwar, A., & Kim, J. K. (2020). Transgenic breeding approaches for improving abiotic stress tolerance: Recent progress and future perspectives. *International Journal of Molecular Science*, 21, 2695.

Aruna, T,S., Srivastava, A., Tomar, B. S., Behera, T. K., Krishna, H., Jain, P. K., Pandey, R., Singh, B., Gupta, R., & Mangal, M. (2023) Genetic analysis of heat tolerance in hot pepper: Insights from comprehensive phenotyping and QTL mapping. *Frontiers in Plant Science*, 14, 1232800. https://doi.org/10.3389/fpls.2023.1232800

Atwell, B. J., Wang, H., & Scafaro, A. P. (2014). Could abiotic stress tolerance in wild relatives of rice be used to improve Oryza sativa? *Plant Science*, 215, 48–58.

Baniwal, S. K., Bharti, K., Chan, K. Y., Fauth, M., Ganguli, A., Kotak, S., Mishra, S. K., Nover, L., Port, M., Scharf, K. D., Tripp, J., Weber, C., Zielinski, D., & von Koskull-Döring, P. (2004). Heat stress response in plants: A complex game with chaperones and more than twenty heat stress transcription factors. *Journal of Bioscience*, 29, 471–487.

Barnabás, B., Jäger, K., & Fehér, A. (2008). The effect of drought and heat stress on reproductive processes in cereals. *Plant, Cell & Environment*, 31(1), 11–38.

Benites, F. R.G., & Pinto, C. A. B. P. (2011). Genetic gains for heat tolerance in potato in three cycles of recurrent selection. *Crop Breeding and Applied Biotechnology*, 11, 133–140.

Bita, C. E., & Gerats, T. (2013). Plant tolerance to high temperature in a changing environment: Scientific fundamentals and production of heat stress-tolerant crops. *Frontiers in Plant Science*, 4, 273.

Cai, C., Yin, X., He, S., Jiang, W., Si, C., Struik, P. C., Luo, W., Li, G., Xie, Y., Xiong, Y., & Pan, G. (2016). Responses of wheat and rice to factorial combinations of ambient and elevated CO2 and temperature in FACE experiments. *Global Change Biology*, 22(2), 856–874.

Camejo, D., Rodríguez, P., Morales, M. A., Dell'Amico, J. M., Torrecillas, A., & Alarcón, J. J. (2005). High temperature effects on photosynthetic activity of two tomato cultivars with different heat susceptibility. *Journal of Plant Physiology*, 162(3), 281–289.

Cao, Y. Y., Duan, H., Yang, L. N., Wang, Z. Q., Liu, L. J., & Yang, J. C. (2009). Effect of high temperature during heading and early filling on grain yield and physiological characteristics in *indica* rice. *Acta Agronomica Sinica*, 35, 512–521.

Chang, H. C., Tang, Y. C., Hayer-Hartl, M., & Hartl, F. U. (2007). SnapShot: Molecular chaperones, part I. *Cell*, 128. https://doi.org/10.1016/j.cell.2007.01.001

Chen, L., Wang, Q., Tang, M., Zhang, X., Pan, Y., Yang, X., Gao, G., Lv, R., Tao, W., Jiang, L., & Liang, T. (2021). QTL mapping and identification of candidate genes for heat tolerance at the flowering stage in rice. *Frontiers Genetics*, 11, 621871. https://doi.org/10.3389/fgene.2020.621871

Chinnusamy, V., Schumaker, K., & Zhu, J. K. (2004). Molecular genetic perspectives on cross-talk and specificity in abiotic stress signalling in plants. *Journal of Experimental Botany*, 55(395), 225–236.

Cho, E. K., & Hong, C. B. (2006). Over-expression of tobacco NtHSP70-1 contributes to drought-stress tolerance in plants. *Plant Cell Reports*, 25, 349–358.

Cui, Y., Lu, S., Li, Z., Cheng, J., Hu, P., Zhu, T., Wang, X., Jin, M., Wang, X., Li, L., Huang, S., Zou, B.,& Hua, J. (2020). Cyclic nucleotide-gated ion channels 14 and 16 promote tolerance to heat and chilling in rice. *Plant Physiology*, 183(4), 1794–1808.

Decros, G., Baldet, P., Beauvoit, B., Stevens, R., Flandin, A., Colombié, S., Gibon, Y., & Pétriacq, P. (2019). Get the balance right: ROS homeostasis and redox signalling in fruit. *Frontiers Plant Science*, 10, 1091. doi: 10.3389/fpls.2019.01091

Djanaguiraman, M., Prasad, P. V. V., & Al-Khatib, K. (2011). Ethylene perception inhibitor 1-MCP decreases oxidative damage of leaves through enhanced antioxidant defense mechanisms in soybean plants grown under high temperature stress. *Environmental and Experimental Botany*, 71, 215–223.

Djanaguiraman, M., Prasad, P. V. V., & Seppanen, M. (2010). Selenium protects sorghum leaves from oxidative damage under high temperature stress by enhancing antioxidant defense system. *Plant Physiology and Biochemistry*, 48, 999–1007.

Doğru, A. (2021). Effects of heat stress on Photosystem II activity and antioxidant enzymes in two maize cultivars. *Planta*, 253, 85. https://doi.org/10.1007/s00425-021-03611-6

Dolferus, R. (2014). To grow or not to grow: A stressful decision for plants. *Plant Science*, 229, 247–261.

Dong, N. Q., Sun, Y., Guo, T., Shi, C. L., Zhang, Y. M., Kan, Y., Xiang, Y. H., Zhang, H., Yang, Y. B., Li, Y. C., Zhao, H. Y., Yu, H. X., Lu, Z. Q., Wang, Y., Ye, W. W., Shan, J. X., & Lin, H. X. (2020). UDP-glucosyltransferase regulates grain size and abiotic stress tolerance associated with metabolic flux redirection in rice. *Nature Communications*, 11(1), 1–16. 11:1.

Edreira, J. I. R., & Otegui, M. E. (2012). Heat stress in temperate and tropical maize hybrids: Differences in crop growth, biomass partitioning and reserves use. *Field Crops Research*, 130, 87–98.

Ehlers, J. D., & Hall, A. E. (1998). Heat tolerance of contrasting cowpea lines in short and long days. *Field Crops Research*, 55(1–2), 11–21.

Endo, M., Tsuchiya, T., Hamada, K., Kawamura, S., Yano, K., Ohshima, M., Higashitani, A., Watanabe, M., & Kawagishi-Kobayashi, M. (2009). High temperatures cause male sterility in rice plants with transcriptional alterations during pollen development. *Plant and Cell Physiology*, 50(11), 1911–1922.

Farnham, M. W., & Björkman, T. (2011). Evaluation of experimental broccoli hybrids developed for summer production in the Eastern United States. *HortScience*, 46(6), 858–863.

Feuillet, C., Langridge, P., & Waugh, R. (2008). Cereal breeding takes a walk on the wild side. *Trends in Genetics*, 24(1), 24–32.

Gangadhar, B. H., Mishra, R. K., Kappachery, S., Baskar, V., Venkatesh, J., Nookaraju, A., & Thiruvengadam, M. (2021). Enhanced thermo-tolerance in transgenic potato (Solanum tuberosum L.) overexpressing hydrogen peroxide-producing germin-like protein (GLP). *Genomics*, 113(5), 3224–3234.

Giaveno, C., & Ferrero, J. (2003). Introduction of tropical maize genotypes to increase silage production in the central area of Santa Fe, Argentina. *Crop Breeding and Applied Biotechnology*, 3(2), 89–94.

Giorno, F., Wolters-Arts, M., Grillo, S., Scharf, K. D., Vriezen, W. H., & Mariani, C. (2009). Developmental and heat stress-regulated expression of HsfA2 and small heat shock proteins in tomato anthers. *Journal of Experimental Botany*, *61*(2), 453–462.

Giorno, F., Wolters-Arts, M., Mariani, C., & Rieu, I. (2013). Ensuring reproduction at high temperatures: The heat stress response during anther and pollen development. *Plants*, 2, 489–506. https://doi.org/10.3390/plants2030489

GM Approval Database. (2022). https://www.isaaa.org/gmapprovaldatabase/

Gororo, N. N., Eagles, H. A., Eastwood, R. F., Nicolas, M. E., & Flood, R. G. (2002). Use of Triticum tauschii to improve yield of wheat in low-yielding environments. *Euphytica*, 123, 241–254.

Gunawardhana, M. D. M., & de Silva, C. S. (2011). Impact of temperature and water stress on growth yield and related biochemical parameters of okra. *Tropical Agricultural Research*, 23, 77–83.

Hasanuzzaman, M., Bhuyan, M. H. M. B., Zulfiqar, F., Raza, A., Mohsin, S. M., Al Mahmud, J., Fujita, M., & Fotopoulos, V. (2020). Reactive oxygen species and antioxidant defense in plants under abiotic stress: Revisiting the crucial role of a universal defense regulator. *Antioxidants*, 9, 681. https://doi.org/10.3390/antiox9080681

Hasanuzzaman, M., Hossain, M. A., da Silva, J. A. T., & Fujita, M. (2012). Plant response and tolerance to abiotic oxidative stress: Antioxidant defense is a key factor. In A. K. Madugund, M. Mandapaka, S. K. Yadav, A. K. Shanker, C. Shanker, & M. Maheswari (Eds.), *Crop stress and its management: Perspectives and strategies* (pp. 261–315). Springer Science+Business Media B.V.

Hasanuzzaman, M., Nahar, K., Alam, M. M., Roychowdhury, R., & Fujita, M. (2013). Physiological, biochemical, and molecular mechanisms of heat stress tolerance in plants. *International Journal of Molecular Sciences*, 14(5), 9643–9684.

Hedhly, A. (2011). Sensitivity of flowering plant gametophytes to temperature fluctuations. *Environmental and Experimental Botany*, 74, 9–16.

Howarth, C. J. (2005). Genetic improvements of tolerance to high temperature. In M. Ashraf, & P. J. C. Harris (Eds.), *Abiotic stresses: Plant resistance through breeding and molecular approaches*. Howarth Press Inc.

Heat Stress in Rice – Physiological Mechanisms and Adaptation Strategies. https://www.researchgate.net/publication/233841464_Heat_Stress_in_Rice_-_Physiological_Mechanismsand_Adaptation_Strategies

Hu, Z., Li, J., Ding, S., Cheng, F., Li, X., Jiang, Y., Yu, J., Foyer, C. H., & Shi, K. (2021). The protein kinase CPK28 phosphorylates ascorbate peroxidase and enhances thermotolerance in tomato. *Plant Physiology*, 186(2), 1302–1317.

Hurkman, W. J., Vensel, W. H., Tanaka, C. K., Whitehand, L., & Altenbach, S. B. (2009). Effect of high temperature on albumin and globulin accumulation in the endosperm proteome of the developing wheat grain. *Journal of Cereal Science*, 49, 12–23.

Ibrahim, A. M., & Quick, J. S. (2001). Heritability of heat tolerance in winter and spring wheat. *Crop Science*, *41*(5), 1446–1452.

Impa, S. M., Raju, B., Hein, N. T., Sandhu, J., Prasad, P. V., Walia, H., & Jagadish, S. K. (2021). High night temperature effects on wheat and rice: Current status and way forward. *Plant, Cell & Environment*, 44(7), 2049–2065.

Iqbal, Z., Iqbal, M. S., Ahmad, A., Memon, A. G., & Ansari, M. I. (2020). New prospects on the horizon: Genome editing to engineer plants for desirable traits. *Current Plant Biology*, 100171, https://doi.org/10.1016/j.cpb.2020.100171

Lucht, J. M. (2015). Public acceptance of plant biotechnology and GM crops. *Viruses*, 7, 4254–4281.

Jagadish, S. V. K., Murty, M. V. R., & Quick, W. P. (2015). Rice responses to rising temperatures–Challenges, perspectives and future directions. *Plant, Cell & Environment*, 38(9), 1686–1698.

Jagadish, S. V. K., Muthurajan, R., Oane, R., Wheeler, T. R., Heuer, S., Bennett, J., & Craufurd, P. (2010). Physiological and proteomic approaches to address heat tolerance during anthesis in rice (Oryza sativa L.). *Journal of Experimental Botany*, 61(1), 143–156.

Jha, U. C., Nayyar, H., Palakurthi, R., Jha, R., Valluri, V., Bajaj, P., Chitikineni, A., Singh, N. P., Varshney, R. K., & Thudi, M. (2021). Major QTLs and potential candidate genes for heat stress tolerance identified in chickpea (Cicer arietinum L.). *Frontiers in Plant Science*, 12, 655103.

Kan, Y., & Lin, H.X. (2021). Molecular regulation and genetic control of rice thermal response. *The Crop Journal*, 9(3), 497–505.

Krasensky, J., & Jonak, C. (2012). Drought, salt, and temperature stress-induced metabolic rearrangements and regulatory networks. *Journal of Experimental Botany*, 63(4), 1593–1608.

Krieger-Liszkay, A., Fufezan, C., & Trebst, A. (2008). Singlet oxygen production in Photosystem II and related protection mechanism. *Photosynthesis Research*, 98, 551–564. https://doi.org/10.1007/s11120-008-9349-3

Kugblenu, Y. O., Danso, E. O., Ofori, K., Andersen, M. N., Abenney-Mickson, S., Sabi, E. B., Plauborg, F., Abekoe, M. K., Ofosu-Anim, J., Ortiz, R., & Jørgensen, S. T. (2013). Screening tomato genotypes for adaptation to high temperature in West Africa. *Acta Agriculturae Scandinavica, Section B-Soil & Plant Science*, 63(6), 516–522.

Li, Y. H., Zhao, S. C., Ma, J. X., Li, D., Yan, L., Li, J., Qi, X. T., Guo, X. S., Zhang, L., He, W. M., Chang, R. Z., Liang, Q. S., Guo, Y., Ye, C., Wang, X. B., Tao, Y., Guan, R. X., Wang, J. Y., Liu, Y. L., … Qiu L. J. (2013). Molecular footprints of domestication and improvement in soybean revealed by whole genome re-sequencing. *BMC Genomics*, 14, 1–12.

Lijavetzky, D., Cabezas, J. A., Ibáñez, A., Rodríguez, V., & Martínez-Zapater, J. M. (2007). High throughput SNP discovery and genotyping in grapevine (Vitis vinifera L.) by combining a re-sequencing approach and SNPlex technology. *BMC Genomics*, 8, 1–11.

Liu, X. H., Lyu, Y. S., Yang, W., Yang, Z. T., Lu, S. J., & Liu, J. X. (2020). A membrane-associated NAC transcription factor OsNTL3 is involved in thermotolerance in rice. *Plant Biotechnology Journal*, 18(5), 1317–1329.

Liu, Y., Dong, S., Wei, S., Wang, W., Miao, H., Bo, K., Gu, X., & Zhang, S. (2021). QTL mapping of heat tolerance in cucumber (*Cucumis sativus* L.) at adult stage. *Plants*, 10(2), 324. https://doi.org/10.3390/plants10020324

Liu, Y., Liu, X., Wang, X., Gao, K., Qi, W., Ren, H., Hu, H., Sun, D., Bai, J., & Zheng, S. (2020). Heterologous expression of heat stress-responsive AtPLC9 confers heat tolerance in transgenic rice. *BMC Plant Biology*, 20, 1–11.

Lu, S. J., Yang, Z. T., Sun, L., Sun, L., Song, Z. T., & Liu, J. X. (2012). Conservation of IRE1-regulated bZIP74 mRNA unconventional splicing in rice (Oryza sativa L.) involved in ER stress responses. *Molecular Plant*, 5(2), 504–514.

Mace, E. S., Tai, S., Gilding, E. K., Li, Y., Prentis, P. J., Bian, L., Campbell, B. C., Hu, W., Innes, D. J., Han, X., Cruickshank, A., Dai, C., Frère, C., Zhang, H., Hunt, C. H., Wang, X., Shatte, T., Wang, M., Su, Z., … Wang, J. (2013). Whole-genome sequencing reveals untapped genetic potential in Africa's indigenous cereal crop sorghum. *Nature Communications*, 4(1), 2320.

Meyer, R. S., & Purugganan, M. D. (2013). Evolution of crop species: Genetics of domestication and diversification. *Nature Reviews Genetics*, 14(12), 840–852.

Mickelbart, M. V., Hasegawa, P. M., & Bailey-Serres, J. (2015). Genetic mechanisms of abiotic stress tolerance that translate to crop yield stability. *Nature Reviews Genetics*, 16(4), 237–251.

Miroshnichenko, S., Tripp, J., Nieden, U. Z., Neumann, D., Conrad, U., & Manteuffel, R. (2005). Immunomodulation of function of small heat shock proteins prevents their assembly into heat stress granules and results in cell death at sub-lethal temperatures. *Plant Journal*, 41, 269–281.

Mishra, S. K., Tripp, J., Winkelhaus, S., Tschiersch, B., Theres, K., Nover, L., & Scharf, K. D. (2002). In the complex family of heat stress transcription factors, *HsfA1* has a unique role as master regulator of thermotolerance in tomato. *Genes & Developement*, 16, 1555–1567.

Mittler, R. (2017). ROS are good. *Trends Plant Science*, 22, 11–19. https://doi.org/10.1016/j.tplants.2016.08.002

Mohammed, A. R., & Tarpley, L. (2010). Effects of high night temperature and spikelet position on yield-related parameters of rice (*Oryza sativa* L.) plants. *European Journal of Agronomy*, 33, 117–123.

Moreno, A. A., & Orellana, A. (2011). The physiological role of the unfolded protein response in plants. *Biological Research*, 44, 75–80.

Morrow, G., & Tanguay, R. M. (2012). Small heat shock protein expression and functions during development. *International Journal of Biochemistry & Cell Biology*, 44, 1613–1621.

Nijabat, A., Bolton, A., Mahmood-ur-Rehman, M., Shah, A. I., Hussain, R., Naveed, N. H., Ali, A., & Simon, P. (2020). Cell membrane stability and relative cell injury in response to heat stress during early and late seedling stages of diverse carrot (Daucus carota L.) germplasm. *Hort Science*, 55(9), 1446–1452.

Nover, L., Bharti, K., Döring, P., Mishra, S. K., Ganguli, A., & Scharf, K. D. (2001). Arabidopsis and the heat stress transcription factor world: how many heat stress transcription factors do we need? *Cell Stress Chaperones*, 6, 177–189

Olsen, K. M., & Wendel, J. F. (2013). A bountiful harvest: Genomic insights into crop domestication phenotypes. *Annual Review of Plant Biology*, 64, 47–70.

Hu, Zhangjian, Li, Jianxin, Ding, Shuting, Cheng, Fei, Li, Xin, Jiang, Yuping, Yu, Jingquan, Foyer, Christine H., & Shi, Kai. (2021). The protein kinase CPK28 phosphorylates ascorbate peroxidase and enhances thermotolerance in tomato. *Plant Physiology*, 186(2021), 1302–1317.

Pradhan, G. P., Prasad, P. V., Fritz, A. K., Kirkham, M. B., & Gill, B. S. (2012). Effects of drought and high temperature stress on synthetic hexaploid wheat. *Functional Plant Biology*, 39(3), 190–198.

Prasanna, B. M. (2012). Diversity in global maize germplasm: Characterization and utilization. *Journal of Bioscience*, 37, 843–855. https://doi.org/10.1007/s12038-012-9227-1

Prasinos, C., Krampis, K., Samakovli, D., & Hatzopoulos P. (2005). Tight regulation of expression of two *Arabidopsis* cytosolic *Hsp90* genes during embryo development. *Journal of Experimental Botany*, 56, 633–644.

Qi, J., Liu, X., Shen, D., Miao, H., Xie, B., Li, X., Zeng, P., Wang, S., Shang, Y., Gu, X., Du, Y., Li, Y., Lin, T., Yuan, J., Yang, X., Chen, J., Chen, H., Xiong, X., Huang, K., … Huang, S. (2013). A genomic variation map provides insights into the genetic basis of cucumber domestication and diversity. *Nature Genetics*, 45(12), 1510–1515.

Qiao, B., Zhang, Q., Liu, D., Wang, H., Yin, J., Wang, R., He, M., Cui, M., Shang, Z., Wang, D., & Zhu, Z. (2015). A calcium-binding protein, rice annexin OsANN1, enhances heat stress tolerance by modulating the production of H_2O_2. *Journal of Experimental Botany*, 66(19), 5853–5866.

Rahman, M. A., Chikushi, J., Yoshida, S., & Karim, A. J. M. S. (2009). Growth and yield components of wheat genotypes exposed to high temperature stress under control environment. *Bangladesh Journal of Agricultural Research*, 34, 361–372.

Ren, Y., Huang, Z., Jiang, H., Wang, Z., Wu, F., Xiong, Y., & Yao, J. (2021). A heat stress responsive NAC transcription factor heterodimer plays key roles in rice grain filling. *Journal of Experimental Botany*, 72, 2947–2964.

Rodrigo, J., & Herrero, M. (2002). Effects of pre-blossom temperatures on flower development and fruit set in apricot. *Scientia Horticulturae*, 92(2), 125–135.

Saitoh, H. (2008). *Ecological and physiology of vegetable*. Nousangyoson Bunka Kyoukai.

Sakata, T., & Higashitani, A. (2008). Male sterility accompanied with abnormal anther development in plants–genes and environmental stresses with special reference to high temperature injury. *International Journal of Plant Developmental Biology*, 2(4), 42–51.

Sander, J. D., & Joung, J. K. (2014). CRISPR-Cas systems for editing, regulating and targeting genomes. *Nature Biotechnology*, 32(4), 347–355.

Sato, S., Kamiyama, M., Iwata, T., Makita, N., Furukawa, H., & Ikeda, H. (2006). Moderate increase of mean daily temperature adversely affects fruit set of Lycopersicon esculentum by disrupting specific physiological processes in male reproductive development. *Annals of Botany*, 97(5), 731–738.

Schoffl, F., Prandl, R., & Reindl, A. (1999) Molecular responses to heat stress. In K. Shinozaki, & K. Yamaguchi-Shinozaki (Eds.), *Molecular responses to cold, drought, heat and salt stress in higher plants* (pp. 81–98). Austin, TX: R.G. Landes Co.

Semenov, M. A., & Halford, N. G. (2009). Identifying target traits and molecular mechanisms for wheat breeding under a changing climate. *Journal of Experimental Botany*, 60(10), 2791–2804.

Sharma, N., Thakur, M., Suryakumar, P., Mukherjee, P., Raza, A., Prakash, C. S., & Anand, A. (2022). 'Breathing Out' under heat stress – Respiratory control of crop yield under high temperature. *Agronomy*, 12(4), 806.

Singh, A., & Grover, A. (2008). Genetic engineering for heat tolerance in plants. *Physiology and Molecular Biology of Plants*, 14(1–2), 155–166. https://doi.org/10.1007/s12298-008-0014-2

Smertenko, A., Draber, P., Viklicky, V., & Opatrny, Z. (1997). Heat stress affects the organization of microtubules and cell division in Nicotiana tabacum cells. Plant Cell Environment, 20, 1534–1542. https://doi.org/10.1046/j.1365-3040.1997.d01-44.x

Suwa, R., Hakata, H., Hara, H., El-Shemy, H. A., Adu-Gyamfi, J. J., Nguyen, N. T., Kanai, S., Lightfoot, D. A., Mohapatra, P. K., & Fujita, K. (2010). High temperature effects on photosynthetic partitioning and sugar metabolism during ear expansion in maize (*Zea mays* L.) genotypes. *Plant Physiology and Biochemistry*, 48, 124–130.

Swindell, W. R., Huebner, M., & Weber, A. P. (2007). Transcriptional profiling of *Arabidopsis* heat shock proteins and transcription factors reveals extensive overlap between heat and non-heat stress response pathways. *BMC Genomics*, 8, 125.

Tan, W., Meng, Q. W., Brestic, M., Olsovska, K., & Yang, X. (2011). Photosynthesis is improved by exogenous calcium in heat-stressed tobacco plants. *Journal of Plant Physiology*, 168, 2063–2071.

Tian, B., Talukder, S. K., Fu, J., Fritz, A. K., & Trick, H. N. (2018). Expression of a rice soluble starch synthase gene in transgenic wheat improves the grain yield under heat stress conditions. *In Vitro Cellular & Developmental Biology-Plant*, 54, 216–227.

Tombesi, S., Cincera, I., Frioni, T., Ughini, V., Gatti, M., Palliotti, A., & Poni, S. (2019). Relationship among night temperature, carbohydrate translocation and inhibition of grapevine leaf photosynthesis. *Environmental and Experimental Botany*, 157, 293–298.

Ul Haq, S., Khan, A., Ali, M., Khattak, A. M., Gai, W. X., Zhang, H. X., Wei, A. M., & Gong, Z. H. (2019). Heat shock proteins: Dynamic biomolecules to counter plant biotic and abiotic stresses. *International Journal of Molecular Sciences*, 20(21), 5321. https://doi.org/10.3390/ijms20215321

Valliyodan, B., & Nguyen, H. T. (2006). Understanding regulatory networks and engineering for enhanced drought tolerance in plants. *Current Opinion in Plant Biology*, *9*, 189–195

Vollenweider, P., & Günthardt-Goerg, M. S. (2005). Diagnosis of abiotic and biotic stress factors using the visible symptoms in foliage. *Environmental Pollution*, 137(3), 455–465.

Walker, B. J., VanLoocke, A., Bernacchi, C. J., & Ort, D. R. (2016). The costs of photorespiration to food production now and in the future. *Annual Review of Plant Biology*, 67, 107–129.

Waraich, E. A., Ahmad, R., Halim, A., & Aziz, T. (2012). Alleviation of temperature stress by nutrient management in crop plants: A review. *Journal of Soil Science and Plant Nutrition*, 12(2), 221–244.

Wen, J., Jiang, F., Weng, Y., Sun, M., Shi, X., Zhou, Y., Yu, L., & Wu, Z. (2019). Identification of heat-tolerance QTLs and high-temperature stress-responsive genes through conventional QTL mapping, QTL-seq and RNA-seq in tomato. *BMC Plant Biology*, 2019(19), 398. https://doi.org/10.1186/s12870-019-2008-3

Xu, J., Misra, G., Sreenivasulu, N., & Henry, A. (2021). What happens at night? Physiological mechanisms related to maintaining grain yield under high night temperature in rice. *Plant, Cell & Environment*, 44(7), 2245–2261.

Xu, X., Liu, X., Ge, S., Jensen, J. D., Hu, F., Li, X., Dong, Y., Gutenkunst, R. N., Fang, L., Huang, L., Li, J., He, W., Zhang, G., Zheng, X., Zhang, F., Li, Y., Yu, C., Kristiansen, K., Zhang, X., Wang, J., … Wang, W. (2012). Resequencing 50 accessions of cultivated and wild rice yields markers for identifying agronomically important genes. *Nature Biotechnology*, 30(1), 105–111.

Yin, Y., Li, S., Liao, W., Lu, Q., Wen, X., & Lu, C. (2010). Photosystem II photochemistry, photoinhibition, and the xanthophyll cycle in heat-stressed rice leaves. *Journal of Plant Physiology*, 167, 959–966.

Zahara, M., Larassati, D., Rahmawati, L., Harahap, D., Hayatillah, R., & Kamaliah, K. (2024). The effect of palm oil (*Elaeis guineensis* Jacq) solid waste on the stomatal density and chlorophyll content of the pepper (Piper nigrum L.). *IOP Conference Series: Earth and Environmental Science*, 1297(1), 012013. IOP Publishing.

Zhang, B., Wu, S., Zhang, Y. E., Xu, T., Guo, F., Tang, H., Li, X., Wang, P., Qian, W., & Xue, Y. (2016). A high temperature-dependent mitochondrial lipase EXTRA GLUME1 promotes floral phenotypic robustness against temperature fluctuation in rice (Oryza sativa L.). *PLoS Genetics*, 12(7), e1006152.

Zhang, D., & Yang, L. (2014). Specification of tapetum and microsporocyte cells within the anther. *Current Opinion in Plant Biology*, 17, 49–55. https://doi.org/10.1016/j.pbi.2013.11.001

Zhang, H., Li, Y., Zhao, Y., Li, G., & Zhang, F. (2019b). Carbon black oxidized by air calcination for enhanced H2O2 generation and effective organics degradation. *ACS Applied Materials & Interfaces*, 11(31), 27846–27853.

Zhang, X., Cai, J., Wollenweber, B., Liu, F., Dai, T., Cao, W., & Jiang, D. (2013). Multiple heat and drought events affect grain yield and accumulations of high molecular weight glutenin subunits and glutenin macropolymers in wheat. *Journal of Cereal Science*, 57, 134–140.

Zhang, X., Xu, Z., Chen, L., & Ren, Z. (2019a). Comprehensive analysis of multiprotein bridging factor 1 family genes and SlMBF1c negatively regulate the resistance to Botrytis cinerea in tomato. *BMC Plant Biology*, 19, 1–13. https://doi.org/10.1186/s12870-019-2029-y

Zhao, Y., Du, H., Wang, Y., Wang, H., Yang, S., Li, C., Chen, N., Yang, H., Zhang, Y., Zhu, Y., Yang, L., & Hu, X. (2021). The calcium-dependent protein kinase ZmCDPK7 functions in heat stress tolerance in maize. *Journal of Integrative Plant Biology*, 63, 510–527.

Zinn, K. E., Tunc-Ozdemir, M., & Harper, J. F. (2010). Temperature stress and plant sexual reproduction: Uncovering the weakest links. *Journal of Experimental Botany*, 61(7), 1959–1968.

10 Plant Cold Stress Tolerance
Molecular Mechanisms and Breeding Strategies

Anupam Kumari, Shreni Agrawal, Richa Das, and Indrani Bhattacharya

10.1 INTRODUCTION

The maturation and evolution of plants are considerably impacted by abiotic stresses such as low temperatures, dryness, salt, and UV radiation (Mboup et al., 2012). Beyond drought-related stress, one of the most detrimental external stresses for higher plants is cold stress (Theocharis et al., 2012; Zhou et al., 2017). Depending on the ambient air temperature, cold stress is categorized as either freezing stress (<0°C) or chilling stress (<20°C). In addition to hindering the development of the crop and formation, cold stress significantly reduces the plant's geographic range (Mickelbart et al., 2015; Guo et al., 2017; Liu et al., 2018; Shi et al., 2018).

Tropical and subtropical plants are vulnerable to chilling effects because they cannot adapt to lower climates. However, after a period of time spent in non-freezing environments, tropical plants that have undergone cold adaptation may tolerate freezing temperatures (Chinnusamy et al., 2007). Temperate flora can tolerate periodic changes in temperature as well as the harshness of early spring and winter cold. Rice, corn, soybeans, potatoes, cotton, and tomatoes are just a few of the important crops that are chilling susceptible and cannot tolerate frigid temperatures. Certain crops, like oats, on the contrary, can withstand freezing but not chilling. All phases of plant development vegetative and reproductive are negatively impacted by low temperatures. Signs of damage such as chlorosis, wilting, and stunted growth can result in senescence or even death. Abiotic stresses have detrimental impacts on a plant's anatomy, physiology, biochemistry, and structural makeup in addition to stunting and decreasing yield. One major environmental element that restricts plant output in hilly areas is cold stress. In response to this stress, plants adjust and change at the molecular, cellular, physiological, and biochemical levels to live.

Cold stress induces a variety of physiological reactions, including membrane stiffness, lipid composition alterations, protoplasmic streaming, Ca^{2+} ion dispersion, and adjustments to the photosynthetic apparatus and electron transport in flora (Gusain et al., 2023). C-repeat binding factors (CBFs), inducers of CBF expression (ICE), and cold-regulated genes (COR genes) are the three primary genes in plants that respond to cold (Mehrotra and Verma, 2020). These three previously identified major participants, the ICE, CBF, and COR genes, model the ICE–CBF–COR cascade, a crucial signaling system that mitigates cold stress in plants (Shu et al., 2017; Thomashow et al., 1999; Guo et al., 2019; Li et al., 2020; Wang et al., 2017). Reduction in plant water loss, growth, photoperiod, and other physiological changes are often indicative of a plant's potential to withstand cold stress (Liu et al., 2019). Numerous kinds of plants have had their genomes described so far, and the ICE–CBF–COR cascade has been identified in tea (Hao and Wang, 2018), wheat (Guo et al., 2019; Badawi et al., 2008), and rice (Bremer et al., 2017).

Inducing and regulating the articulation of the C-repeat binding factor (CBF), the ICE operates upstream (Shu et al., 2017; Thomashow et al., 1999; Guo et al., 2019; Li et al., 2020). Therefore, the CBFs, also known as the DREB1 genes, regulate cold stress by attaching to the cold and

DOI: 10.1201/9781003457237-10

dehydration regulatory elements (CRT/DRE) in the promoter sequences of COR genes. This causes the expression of the COR genes to be activated in *Arabidopsis*, including RD29A (Cho et al., 2017) and COR15A (Chandler et al., 2018; Shi et al., 2018). CBF genes are likely to be more crucial for plants' ability to withstand cold temperatures among the three genes listed above. Several studies have reported on the various functions and reactions of the C-repeat binding factor/dehydration-responsive element binding 1 (CBF/DREB1) genes in various plant taxa (Guo et al., 2019; Wang et al., 2017). Despite having similar sequence structures and binding properties, CBFs (CBF1, CBF2, and CBF3) are known to perform distinctive tasks under severe cold conditions due to various changes in their distinct protein sequences (Ritonga et al., 2020; Guo et al., 2019; Jin et al., 2018). Numerous plant species have been shown to have homologs of the ICE gene (ICE1 and ICE2), and their functions in cold tolerance have been inferred (Verma et al., 2020). Research conducted recently (Vyse and Faivre, 2020; Hirsz et al., 2021; Yin et al., 2019) has demonstrated the crucial role of PTMs in modulating the ICE–CBF chain mechanism under extreme cold conditions. Numerous PTMs, such as SUMOylation, ubiquitination, and phosphorylation, have been shown to increase the reliability and efficacy of ICE genes' adhering to subsequent genes (Damaris et al., 2021). This is achieved by the transcription factors (TFs) OPEN STOMATA 1 (OST1) and others. ICEs and CBFs are modulated by the OST1 in distinct ways. Moreover, the phosphorylation of ICE genes is dependent on kinases in the MAPK pathway; ICE1 is phosphorylated at Ser403 to activate CBF modulation within the MAPK process and preserve stability (Praat et al., 2021). It has been shown that a constitutive kinase known as SaMMK2 favorably enhances the expression of SaCBF in potatoes under cold stress, which activates COR genes to produce cold stress resistance (Chen and Chen, 2022).

Conventional breeding strategies have only partially been accomplished in boosting the low-temperature resilience of main crops through interspecific or inter-generic hybridization. In vitro alterations have also been employed, albeit less successfully, to boost certain crops' resilience to abiotic stressors. The primary limitations of traditional breeding strategies include the complicated nature of stress sensitivity traits, the limited genetic diversity of produce elements under pressure, and a scarcity of useful chosen parameters. Thus, it's critical to explore alternate approaches to cultivate crops resistant to cold stress. Nowadays, molecular markers are seen to be superior to physiological and morphological characteristics since they are easily detectable without regard to growth or developmental phases, theoretically limitless, impervious to environmental influences, and stable. Also, they are perfect for positional cloning of valuable genes, evolutionary research, genetic diversity analysis, fingerprinting, tagging of useful genes, constructing genetic and physical maps, and marker-assisted selection (MAS; Nadeem et al., 2018; Gupta and Rustgi, 2004; Maqbool et al., 2016). It's becoming possible to use molecular markers in breeding programs at a reasonable cost. Breeding can be accelerated by making use of genome-level variation through QTL analysis, genotyping platforms, and genomics research (Roorkiwal et al., 2018).

10.2 PHYSIOLOGICAL RESPONSES TO COLD STRESS

Stress encompasses both biotic and abiotic stimuli that limit a plant's ability to grow, develop, and provide food (Lichtenthaler, 1998). Abiotic stressors that hinder plant growth and production include freezing (<0°C) and chilling (0°C–15°C), as reported by Guo et al. (2018) and Liu et al. (2018). Chilling temperatures have a variety of effects on the cells of plants, many of which restrict the plant's ability to grow and develop. According to Orvar et al. (2000), the first step that sets off a plant's subsequent cold stress reactions is the alteration in membrane rigidification in plant cells caused by chilling stress. Second, the strength of proteins or complicated proteins is affected by freezing stress, which also decreases the function of several enzymes, such as the ones that remove reactive oxygen species. Decreased light-dependent reactions, photodamage, and severe harm to membranes are the results of these processes (Siddiqui and Cavicchioli, 2006; Ruelland et al., 2009). Chilling temperature influences gene regulation as well as the production of proteins by encouraging the formation of secondary structures in RNA, according to Ruelland et al. (2009).

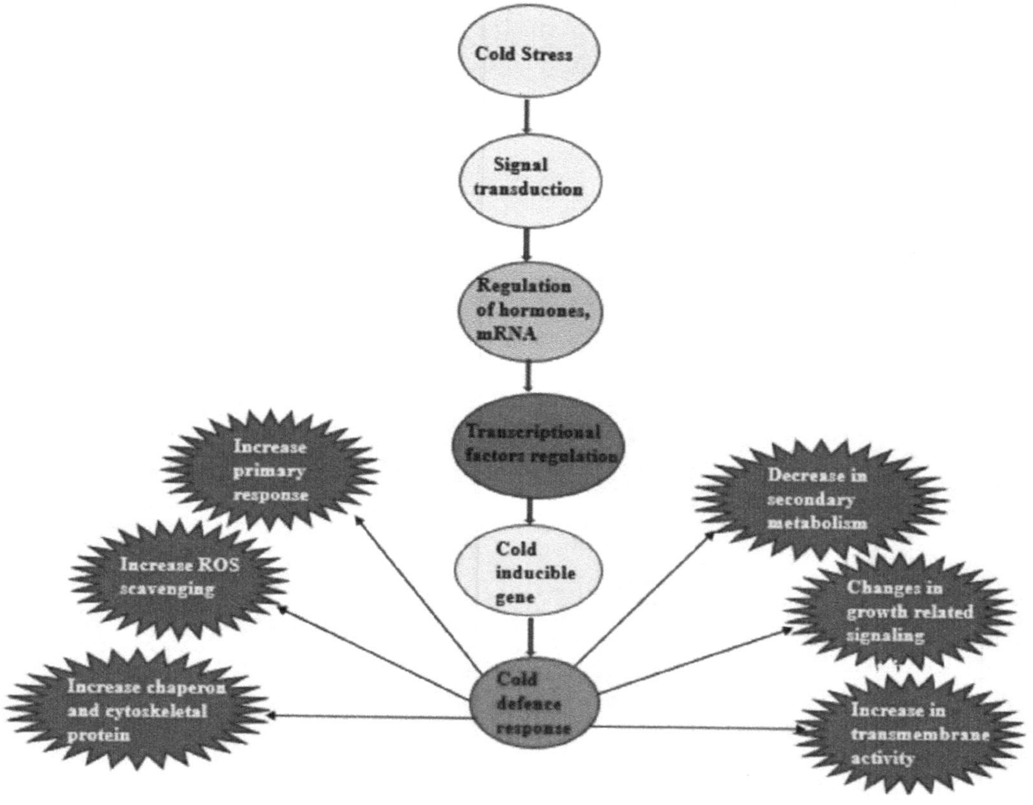

FIGURE 10.1 A schematic representation of plants' reaction to cold stress.

Freezing stress, which can potentially cause plant death, has a greater probability of affecting plants than chilling stress. Frosting harm in the environment originates from extracellular ice formation (Pearce, 2001). Once ice nuclei have formed, they expand into ice crystals that extend into the apoplast and cause water to be ejected, which dehydrates the cell. Ice crystals cause harm to cells that cannot be reversed (Dowgert and Steponkus, 1984; Pearce, 2001). Thus, for plants to withstand cold stress, it is crucial to stop the creation of intracellular ice crystals and ice crystal development. Plants most commonly respond to cold stress by accumulating cryoprotective polypeptides (like COR15a) and osmolytes (like proline and soluble carbohydrates) that enable the plant to withstand freezing. Cold stress has been shown to increase the sweetness level of *Euphorbia resinifera*, *Echinocactus grusonii*, *Aloe vera*, *Crassula lacteal*, *Bryophyllum pinnatum*, *Yucca aloifolia*, and *Sansevieria trifasciata* (Khan et al., 2015). Plants that have evolved to cold regularly retain higher levels of sugar in their subsurface tissues. This includes amylose, starch, maltose, and glucose (Londo et al., 2018; Janská et al., 2010) (Figure 10.1).

10.3 COLD STRESS RESILIENCE THROUGH MOLECULAR PROCESSES

Comprehending the molecular process beneath cold stress in plants requires an examination of the transcriptional modifications that transpire during their cold acclimation. Thus far, a considerable number of genes sensitive to cold have been found, and many gene regulatory networks have been documented. Cold stress initiates the ICE–CBF–COR process inside most plant species, which in turn causes downstream genes encoding osmoregulation-related compounds to express properly (Zhang et al., 2011). Inducer of CBF expression (ICE), an MYC-type basic helix-loop-helix family

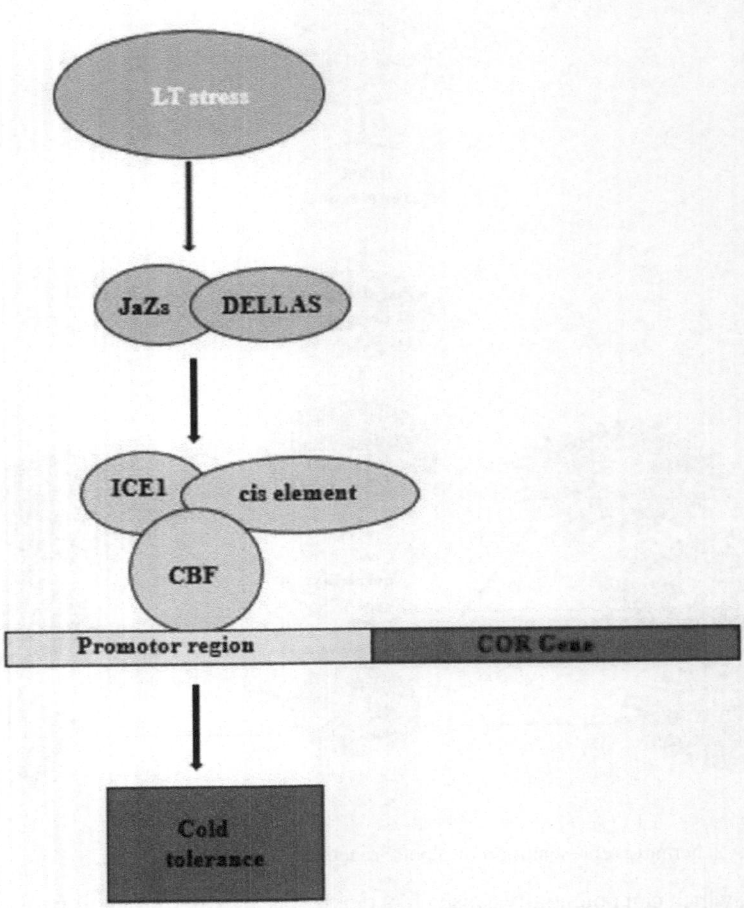

FIGURE 10.2 Plant resistance to cold stress via the ICE–CBF–COR pathways.

TF, is recognized as a trailblazer in cold adaptation (Chinnusamy et al., 2007; Zarka et al., 2003; Wang et al., 2018). ICE1, which is bound by DELLAs and released from JAZs when plants experience cold stress, can cause CBF3 expression. Gibberalic acid (GA) is a bioactive substance that raises the accumulation of DELLAs. CBF3 stimulates the production of GA2ox7 to lower the GA level. With ICE1 via JAZs, DELLAs may therefore control the cold induction of CBF3 (Figure 10.2).

10.3.1 Role of Cold-Responsive Genes and Proteins

10.3.1.1 Dehydration-Responsive Elements Binding 1 (CBF/DREB1) or C-Repeat Binding Factor

Plants' cold signaling system is mediated by CBF TFs (Yin et al., 2019; Jofuku et al., 1994). They were first discovered in *Arabidopsis thaliana* by Jofuku et al. (1994), which are activated by low temperatures and/or the ICE (Okamuro et al., 1997). Their three-dimensional (3D) structure is presented into a layer of three antiparallel β-sheets accompanied by a parallel α-helix sheet. They have about 60 amino acid residues (Nakano et al., 2006). The 3D structural protein prediction investigation identified the Arg- and Try-residues of the β-sheet, which link the nucleotides of the attachment locations in the critical groove of the DNA. Furthermore, the AP2/ERF family exhibits strong conservation of these important residues (Allen and Yamasaki, 1998). The CBF1/2/3 genes are located on chromosome IV, and the first sequencing of the CBF gene structure in *Arabidopsis* was done by Medina et al. (2011). CBF2 and CBF3 are located 3 and 7 kb downstream of CBF1, respectively. According to a gene ontology (GO) study of CBFs, the primary molecular function of these proteins is the binding of CRT/DRE binding domains to low temperature-responsive genes, hence promoting frigid temperature resilience (Shi et al., 2017). Recent research has demonstrated that some CBF sequences of amino acids from more comparable plant varieties have the same preserved patterns (Guo et al., 2019; Hu et al., 2020; Guan et al., 2021). Furthermore, Novillo et al. (2004) showed how the CBF/DREB1 TFs undergo an unfavorable feedback mechanism. In a similar vein, overexpression of CBF1/DREB1B prevents CBF3/DREB1A transcript collections. However, for cold stress-related genes to be expressed correctly, this detrimental feedback is required.

10.3.1.2 Inducer of CBF Expression (ICE)

The cold acclimatization process is facilitated by the ICE, which is a forerunner to the cold-response cascade (Chinnusamy et al., 2003; Xie et al., 2018). The bHLH binding domain is present in the ICE, and the basic section of its amino acid sequence bears a striking resemblance to other bHLH proteins. According to Sun et al. (2018) and Tang et al. (2020), the canonical MYC cis-elements (CANNTG) in the CBF3/DREB1A promoter are where the ICE proteins connect to trigger the CBF/DREB1 regulon. ICE2 is distinguished by an additional amino acid box toward the end of Box II that alters the conserved LPPT sequence, while the ICE1 genes lack Box I (Kurbidaeva et al., 2014). Moreover, ICE2's structure has a greater number of phosphorylation sites than ICE1's; other than this, their secondary structures – each including four exons and three introns – are similar (Kashyap et al., 2019). According to Tang et al. (2020), AtICE1's GO enrichment study, nucleic acid binding (GO: 0001071), binding sites for organic cyclic compounds (GO: 0097159), and binding sites for heterocyclic compounds (GO: 1901363) are among the groups in which the ICE1 binding sites are supplemented. Therefore, binding to CBF gene areas downstream is its main molecular function.

10.3.1.3 Cold-Regulated Genes

The CRT/DRE *Cis*-acting element, which has been shown in multiple reports to contain the gene, enhances freezing temperature sensitivity via the CBF-dependent process. These genes include COR genes, ABA-inducible protein-coding (KIN1 and KIN2), and reactive to dryness (Li et al., 2020). The transcription of COR genes is further activated by CBF binding to C-repeat (CRT/DRE) cis-elements found in their promoters, which are identified by a CCGAC sequence (Leuendorf et al., 2020). CBFs are thought to directly control between 10% and 20% of all COR genes in *Arabidopsis* (Kosová and Kilma, 2021). Exons (protein-coding sections) located in the 50UTR and 30UTR with a central intron surround the majority of investigated COR gene architectures. All COR gene families are related via a common CRT/DRE motif that attaches upfront of CBF genes to enable them to function, even if different patterns differentiate them (Li et al., 2008). As a result, crops express themselves to extreme cold conditions in a variety of manners, depending on the temperature range: pre-toughening up, toughening up, and crop recuperation (Welti and Li, 2002). To maintain ionic equilibrium and the elimination of reactive oxygen species, certain COR genes work to regulate cell membrane phosphoglyceride, proteins, and cytosolic proteins, depending on the temperature variation (Wang et al., 2003; Okawa et al., 2008).

10.3.1.4 Dehydrins

In angiosperms, the late-embryogenesis-abundant protein subgroup includes dehydrins (DHNs). They have a wide range of typical domain combinations and strong hydrophilicity. Plants respond to low-temperature distress by accumulating them, especially the SKn type, which acts as ion sequestration or molecular chaperones to prevent membrane lipids from dehydrating and freezing desiccation (Wan et al., 2014). As an example, the secondary structure of AtCOR15A serves as a chaperone to keep the membrane from freezing and desiccating and allows it to interact with other proteins. AtCOR15A's molecular binding ability to lipids, carbohydrates, heterocyclic compounds, as well as various tiny substances has been demonstrated by GO analysis of COR15A (Vazquez-Hernandez et al., 2017). Prior investigations have shown that the CBF process controls cold stress in rice through the action of dehydrins (OsDhn1, lip5, and lip9) and their analogs Wcor410 and AtCOR47. Apart from the dehydrins mentioned above, many DHN proteins, including Wcs120, COR47, and RD17, are being displayed to modulate frigid stress by means of the CBF process. According to recent studies, CBF1 can recognize the consensus sequence (CCGAC) of the CRT/DRE components from dehydrins in Triticeae and *Vitis vinera* species (Zhu et al., 2020).

10.3.1.5 Antifreeze Proteins

Another set of cold-responsive genes, the antifreeze proteins, were initially identified and studied in polar fishes. By inhibiting water molecules from sticking to the crystal's growing faces in a non-collaborative manner, AFPs can halt the growth of an ice crystal. They can be found in significant amounts. When AFPs cling to the surface of ice crystals, they positively alter their development and structure. This results in the development of several, smaller, slower-growing ice crystals rather than a single, larger one. AFPs may prevent bigger ice crystals from forming and from recrystallizing during thawing. The likelihood of physical harm occurring within frozen plant tissue is higher in larger ice crystals (Griffith et al., 1997). AFPs are commonly confused with thermal hysteresis proteins as they have the same characteristics (Guy, 1999; Houde et al., 2006). Following CA, freezing-tolerant mono-cotyledons and dicotyledons' apoplastic extracts are the only ones that exhibit AFP activity. According to research, antifreeze activity in response to cold is regulated by salicylic acid, drought, ABA, and the hormone ethylene (Atici and Nalbantoglu, 2003).

10.3.2 Signaling Pathways Involved in Cold Stress Tolerance

10.3.2.1 CBF Pathway

The most important findings in the fields of signal transduction and low-temperature adaptation are related to CBF genes. All significant field crops and several vegetable species have been found to have CBFs (Sanghera et al., 2011). It has been established that CBF genes are present in many herbaceous (Mizoi et al., 2012) and woody (Wisniewski et al., 2014) plants. The CBF/DREB1 TFs family was identified after a thorough investigation into *Arabidopsis* cold-response gene expression. According to Gilmour et al. (1998), these TFs are essential for plants' capacity to tolerate freezing stress and grow acclimated to low temperatures. The highly conserved core 5-bp region of CCGAC, which can regulate transcription under drought, salt, and low temperatures, is present in this pair of sequences (Baker et al., 1994; Yamaguchi-Shinozaki and Shinozaki, 1994; Gao et al., 2007). Mizoi et al. (2012) state that, as a result, CBF increases the articulation of COR genes or genes that include the COR sequence, which is essential for improving the cold-adapted response of plants. Membership in the ERF/AP2 transcription family includes the TF family CBF/DREB1 (Yamaguchi-Shinozaki and Shinozaki, 2005). Based on studies conducted by Yamasaki and Randall (2016), the first steps of the soybean CBF/DREB1 pathway have been confirmed. Thomashow (2001) found three different cold-inducible CBF/DREB1 genes in *Arabidopsis*: CBF/DREB1B, CBF2/DREB1C, and CBF3/DREB1A. Monroe et al. (2016) discovered a substantial correlation between the variance in CBF sequences and winter temperatures after screening 477 wild accession plants from *Arabidopsis* for CBF gene variants. Significant access to the CBF polymorphisms was demonstrated by the kinds that originate from milder winters. The consequences of reduced CBF articulation on resilience to frigid temperatures showed significant diversity between accessions; most of its consequences on development, however, were too minor to make definite results (Gery et al., 2011). While three closely associated CBF genes have been identified in *Arabidopsis*, this family has expanded to include wheat and a few other cereals (Pearce et al., 2013). According to Badawi et al. (2007), every one of the three hexaploid wheat genetic material has at least 15 CBF genes.

Nevertheless, Zeller et al. (2009) and Matsui et al. (2010) reported that no target was detected among the three CBF parameters. The activation of these genes in transgenic plants increased the plants' resistance to cold by triggering downstream COR genes, even at mild temperatures, according to Liu et al. (1998) and Kasuga et al. (1999). Numerous plant taxa have been shown to exhibit the CBF/DREB1 pathway, suggesting a high degree of conservation of the CBF transcriptional cascade under cold stress throughout the plant kingdom (Jaglo et al., 2001). There is evidence of a strong relationship between temperature changes and CBF and DREB expression, indicating that these genes express more as the temperature drops. According to Zarka et al. (2003), the expression may, nonetheless, desensitize at a certain low temperature.

Three CBFs (CBF1, 2, and 3) have been shown to have higher transcription levels after being exposed to cold temperatures for 15 minutes. Given the quick reaction to the cold effect, it is highly likely that warm temperatures are home to the low-temperature 'thermometer' and signal 'transducer' that controls CBF expressions. According to Gilmour et al. (1998), in this context, a TF that is present at warm temperatures is called an inducer of CBF expression (ICE). Additionally, it was noted that cold temperatures activate 'ICE' or the proteins it interacts with, and, as a result, it was proposed that the CBF gene promoter is responsive to cold temperatures and activated (Zarka et al., 2003). As per the findings of Chinnusamy et al. (2003), ICE1, a nuclear gene that encodes a bHLH protein like MYC, plays a vital function in controlling some CBF genes. Research has examined the function of the ICE1-CBF/DREB1 singling pathway in enhancing cold resilience in many plant species, including *A. thaliana* (Welling and Palva, 2008; Chen et al., 2009; Fursova et al., 2009; Yang et al., 2011), *Cucumis sativa* (Liu et al., 2010), *Triticum aestivum* (Miller et al., 2006; Morran et al., 2011), *Hordeum vulgare* L. (Stockinger et al., 2007; Campoli et al., 2009), *Solanum tuberosum* (Behnam et al., 2007), *Eucalyptus* (Navarro et al., 2009), *Nicotiana plumbaginfolia* (Gutha and Reddy, 2008), and apple (Artlip et al., 2014). TaICE141 and TaICE187, two ICE1 homologs, upregulate the expression of CBF group IV genes in wheat (Badawi et al., 2008). Numerous studies (Ito et al., 2006; Wang et al., 2008; Yang et al., 2009; Zhang et al., 2009; Cruz et al., 2013) have documented the presence of the CBF/DREB pathway and its role in improving rice plants' ability to withstand cold. Rice under cold stress expresses OsICE1 and OsICE2, which causes OsDREB1B, OSHsfA3, and OsTPP1 to be upregulated (Nakamura et al., 2011). Based on the induction of MaMYC2, Zhao et al. (2013) reported that jasmonate enhanced the freezing tolerance of the banana plant (*Musa acuminate*). However, the CBF-dependent pathway's cold signaling is activated by MaMYC2's interaction with MaICE1. Wheat's ability to withstand cold is mostly dependent on TaCBF14 and TaCBF15, according to Soltész et al. (2013). Additionally, they stated that transgenic plants with these genes had a slightly reduced ability to withstand cold compared to control plants. As a result of differentiating expression patterns in response to cold temperatures, at least five CBF genes have been found in both peaches (*Prunus persica*) and apples (*Malus domestica*) (Wisniewski et al., 2014, 2015).

10.3.2.2 MAPK Cascade

Cold stress is regulated by the MAPK pathway, which binds to calcium/calmodulin-regulated receptors Kinase-Like 1 (CRKL1) (Wei et al., 2021). According to studies on the MAPK cascade in *Arabidopsis*, CRLK1/2 interacts with MEKK1, a MAPK module that responds to decreasing temperatures (Lin et al., 2021). The expression of CBF genes is then enhanced via the MEKK1–MKK2–MPK4–MPK3/6 pathway, which is formed upstream (Yang et al., 2010), and results from the MKK2 being successively phosphorylated by MEKK1 (Teige et al., 2004). COR gene expression in *Arabidopsis* is induced by CAMTA3 binding to CBF2 promoters (Jiang et al., 2020). Initial investigation indicates that the synthesis of RD29A and COR6.6 is triggered by the overexpression of CAMTA3, which is regulated by CBF (Townley and Knight, 2002). Similarly, a wide range of studies demonstrating the freezing sensitivity of mpk3, mpk5, and camta3 mutants have demonstrated the significance of this route in the induction of the AtCBF2 (Zhao et al., 2017). By binding to the ICE1 promoters, MPK3 facilitates the degradation of ICE1, which in turn lowers CBF transcription (Li et al., 2017; Ding et al., 2018). However, it has been demonstrated that mpk3 and mpk6 mutants both upregulate CBF expression, which confers cold stress resilience to plants (Ding et al., 2019). All things considered, the expression of cold regulatory genes is controlled by a single adverse feedback system that comprises these two regulatory pathways. By triggering the negative regulator MYB15, phosphorylated MPK6 in *Arabidopsis* causes the negative expression of CBF3 (Li et al., 2017). On the contrary, by blocking the MKK4/5-MPK3/6 pathway, MPK4/6 increases the expression of CBF (Wang et al., 2021).

10.4 BREEDING APPROACHES FOR COLD STRESS TOLERANCE

10.4.1 Conventional Breeding Methods for Cold Tolerance

The foundation of conventional breeding is crossing; simple backcrosses to a recurrent parent are selected from landrace genotypes and have been frequently employed to bring novel features into breeding programs. Conventional breeding can be used to produce lines with variable maturities that are appropriate for production in various agroecological zones (Khoshro and Lotfi, 2021). The creation of cold-tolerant cultivars in a variety of crops has been made possible through conventional breeding. For example, the EP80 9 Puenteareas population in maize is an essential component of resilience to cold, according to findings from both field and controlled conditions (Rodriguez et al., 2007). Tiber (Kisha et al., 1992) wheat, which was descended from Redwin, has excellent resilience to cold. Conventional breeding produced OAC Elmira in barley (Falk et al., 1997). Through back-crossbreeding, efforts were taken to incorporate cold resistance traits from landraces and crop wild relatives into sensitive but highly productive rice cultivars like Plaisant in tomato (Vallejos and Tanksley, 1983), Guichao 2 (Liu et al., 2003), Towada (Xu et al., 2008; Zhou et al., 2012), Hitomebore (Shirasawa et al., 2012), HJX74 (Zhang et al., 2004), Norin-PL11 (Kuroki et al., 2007), Norin-PL8 (Saito et al., 1995; Dai et al., 2004), Milyang 23 (Oh et al., 2004), and Norin-PL11 (Kuroki et al., 2007). For creating cultivars with increased stress tolerance, conventional breeding is an incredibly sluggish procedure. Conventional breeding is also severely constrained by incompatibility in larger crossings and a scarcity of genetic resources for stress resistance. To enhance accuracy and speed in the creation of genotypes with low-temperature tolerance, it is thus necessary to integrate traditional breeding programs with contemporary suitable molecular tools and clarify the genetic processes of this intricate phenomenon (Najeeb et al., 2021). However, to improve the effectiveness of conventional breeding methods, genomics techniques have surfaced recently.

10.4.2 Molecular Breeding Methods for Cold Tolerance

The simplest type of MAS, known as marker-assisted backcrossing (MABC), aims to selectively transfer or incorporate genomic segments encoding QTLs or genes to vulnerable but elite genetic bases. This method is very effective for introgressing QTL, which significantly affects the target phenotype (Varshney et al., 2012). More crucially, according to Collard and Mackill (2008), MABC is still the most effective method for pyramiding multiple gene(s) or QTL(s) into a single genotype. Several LT-tolerance-related QTLs have been effectively introgressed and pyramided into sensitive genetic backgrounds using MABC. According to Yang et al. (2016), resistance was imparted at the bud-breaking and seedling phases by two QTLs (qCTBB-5 and qCTBB-6) and two QTLs (qCTS6 and qCTS-12) that were pyramided from the singular-segment substitution line (SSSL) SC1-1. The SSSL was created by hybridizing Nan-yang-zhan and Hua-jing-xian 74. Pyramiding of the QTLs (qCTF7, qCTF8, and qCTF12) by utilizing SSR markers was tried to increase rice's resilience to cold during fertilization (Shinada et al., 2014). High-yielding but cold-sensitive rice cultivars should benefit from an MABC approach that uses co-segregating markers like In1-c3 and In11-d1 to impart cold tolerance (Kim et al., 2014). qRC10–2 (Xiao et al., 2014) and qCTB3-Silewah (Mori et al., 2011) are two of these QTLs. Zhou et al. (2012) found that two near-isogenic lines bearing QTL-containing area (cold tolerance genes) from Kunmingxiaobaigu were recovered in the background of Towada. The documented successful cases in rice demonstrate that MABC can significantly increase the cold tolerance level of sensitive but highly productive cultivars. As a result, it presents itself as a feasible breeding technique for food sovereignty given the rise of global increases in cold stress.

10.4.2.1 Linkage-Map-Based QTL Analysis

The use of the genetic marker technique has proven to be efficient in mapping the QTLs linked to the characteristics that help plants withstand cold temperatures.

10.4.2.2 Maize

QTLs were identified in maize under cold stress conditions for a variety of phenotypes, including initial seedling vigor, root, leaf, and shoot attributes (Hund et al., 2004; Presterl et al., 2007). According to research, a significant QTL on chromosome 6 has been shown to contribute to light inhibition sensitivity at frigid temperatures (Fracheboud et al., 2004). Notably, Rodriguez et al. (2014) have identified three genomic areas that are situated on the 2, 4, and 8 chromosomal regions and contribute to the cold sensitivity of seedlings.

10.4.2.3 Sorghum

A mapping population produced from the cross Shan Qui Red 9 SRN39 was used to identify two QTLs that confer cold tolerance in sorghum at the germination stage on LGs SBI-03a and SBI07b, respectively (Knoll et al., 2008).

10.4.2.4 Pea

According to evidence gathered over several decades and places, three interesting frost-resilience-associated QTLs were discovered. These QTLs also showed co-localization with the Hr flowering locus (Lejeune-Henaut et al., 2008). Like this, Klein et al. (2014) found a QTL cluster LG IV that might account for 70% of PV in pea frost resistance damage. Regarding cold tolerance, two reliable QTLs were identified for LG V and LG VI. Interestingly, according to Dumont et al. (2009), these QTLs aligned with the chromosomal regions linked to raffinose and Rubisco activity.

10.4.2.5 Other Crops

Among approximately 600 double haploid lines, nine main-effect QTLs linked to cold tolerance and winter resilience in triticale were identified (Liu et al., 2014). Funatsuki et al. (2005) discovered that three QTLs (qCTTSW1, qCTTSW2, and qCTTSW3) control LT sensitivity in the reproductive stage of soybeans, whereas Ikeda et al. (2009) discovered a major QTL that conserved seed development under cold stress. QTLs that specify the orientation of the leaf shoot and root under cold extreme circumstances were found in *Medicago truncatula* (Avia et al., 2013). Six major QTLs granting cold tolerance in *Brassica napus* were shown to be present over many winter seasons (Kole et al., 2002). Wide cross using wild *Lycopersicon pimpinellifolium* accession LA722 and NC84173 (*Lycopersicon esculentum*) in tomatoes provided QTLs that confer cold stress resistance during the germination stage (Foolad et al., 1998). Likewise, under chilling stress, three QTLs originating from *L. esculentum* and nine from *Lycopersicon hirsutum* were identified by Truco et al. (2000) to regulate root ammonium absorption and wilting. After being introduced to *L. esculentum* from wild *L. hirsutum*, one QTL for shoot turgor maintenance (identified by stm9) gradually conditioned tomato tolerance to cold (Goodstal et al., 2005).

10.4.3 Transgenic Approaches

Finding cold-tolerant genes has allowed for the use of genetically modified crops that were previously limited by cold temperatures. According to Kodama et al. (1994), transgenic tobacco (*Nicotiana tabacum*) was able to withstand 7 days at 1°C due to the overproduction of a gene that expressed chloroplast w3 fatty acid desaturase. Furthermore, Ishizaki-Nishizawa et al. (1996)

reported that when a gene expressing an ambiguous cyanobacterial desaturase was utilized to confer tolerance at 1°C for 11 days, the genetically engineered tobacco cultivars demonstrated a decline in the quantity of saturated fatty acids in their membrane fats. In transgenic plants, the overexpression of glycerol-3-phosphate acyl transferase changed the fatty acid unsaturation and provided cold resilience (Ariizumi et al., 2002; Sakamoto et al., 2003; Sui et al., 2007). Therefore, alterations in the makeup of fatty acids that preserve cell membrane integrity and prevent cytoplasmic spillage are the cause of cold tolerance. The transgenic method offers a wealth of intriguing potential to improve plant cold resistance by adding or deleting a gene or genes that govern a certain trait (Kumar et al., 2006). Furthermore, it offers exceptional opportunities to enhance plant genetic ability by developing crop varieties with enhanced dietary value and increased resilience to biotic and cold stresses.

10.5 FUTURE DIRECTION AND CHALLENGES

Understanding molecular pathways and improving breeding tactics will be key components of improving plant resistance to cold stress in the future. The understanding of signal perception, transduction, and gene expression under cold stress has advanced, opening the door to more focused therapies (Adhikari et al., 2022).

10.5.1 Understanding Molecular Mechanisms

It is recommended that future studies expand on the complex molecular processes underlying plant responses to cold stress, with a particular emphasis on the ICE–CBF–COR cascade pathways (Younis et al., 2020).

10.5.2 Exploring Genetic Diversity

To find and use genes resistant to cold stress, efforts should be focused on investigating and using the genetic diversity found in plant species. Genetic diversity analysis and cutting-edge genomics research can do this (Forster et al., 2000).

10.5.3 Development of Cold-Tolerant Varieties

Breeding programs that use molecular markers can help produce crop types that can withstand cold stress. The limitations of conventional approaches can be overcome with the use of MAS for focused and efficient breeding (Younis et al., 2020).

10.5.4 Complexity of Stress Tolerance Traits

Stress tolerance qualities are complex and might be difficult to understand. To improve crops' resistance to cold stress, researchers must better understand the complexities of these features.

10.5.5 Limited Success with Conventional Breeding

The success of conventional breeding techniques has been patchy. Improving the effectiveness of traditional breeding methods requires overcoming the difficulties associated with limited genetic variation and the complexity of stress tolerance characteristics. Even though molecular markers have great potential, further study is needed to maximize their use in breeding programs. This entails improving QTL analysis methods, developing genotyping platforms, and guaranteeing an economical implementation (Younis et al., 2020).

10.6 CONCLUSION

One of the main environmental stressors that reduce agricultural crop production is cold stress, which has an impact on the quality and postharvest life of the crops. Through a process termed cold acclimation, most temperate plants gain resistance to freezing and chilling after having previously experienced sublethal cold stress. However, a large number of crops that are crucial to agronomy cannot acclimatize to the cold. Almost every part of a plant's cellular activity is impacted by cold stress. These alterations caused by cold stress have an adverse effect on plant development and proliferation. In times of frigid temperatures, ICE acts as a regulator of CBFs, which control the articulation of the COR gene. The crucial gene COR oversees plants' processes of cold acclimation and chilling tolerance. In a typical setting, the photoperiod, and the circadian clock control CBF levels. However, to control a plant's resistance to cold, CBFs activate several genes linked to cold stress. Significant functions in plant resistance to cold stress are played by the ICE–CBF–COR pathways, MAPK process, and DELLA signaling, which offer possible targets for molecular manipulation. Analysis of the articulation of genes impacted by cold temperatures indicates the existence of many channels of signaling between the starting stress signals and the regulation of genes. Breeding with the use of molecular markers may be a good way to focus on the targeted genes or QTLs. There is a lot of room to use transgenic technology to create genotypes that are resistant to cold and low temperatures. To guarantee sustainable crop production under cold stress, traditional breeding may be greatly aided by the latest genomics techniques.

REFERENCES

Adhikari, L., Baral, R., Paudel, D., Min, D., Makaju, S. O., Poudel, H. P., Acharya, J. P., and Missaoui, A. M. (2022). Cold stress in plants: Strategies to improve cold tolerance in forage species. *Plant Stress*, *4*, 100081.

Allen, M. D., & Yamasaki, K. (1998). A novel mode of DNA recognition by a beta-sheet revealed by the solution structure of the GCC-box binding domain in complex with DNA. *The EMBO Journal*, *17*, 5484–5496.

Ariizumi, T., Kishitani, S., Inatsugi, R., Nishida, I., Murata, N., & Toriyama, K. (2002). An increase in unsaturation of fatty acids in phosphatidylglycerol from leaves improves the rates of photosynthesis and growth at low temperatures in transgenic rice seedlings. *Plant and Cell Physiology*, *43*, 751–758.

Artlip, T. S., Wisniewski, M. E., & Norelli, J. L. (2014). Field evaluation of apple overexpressing a peach CBF gene confirms its effect on cold hardiness, dormancy, and growth. *Environmental and Experimental Botany*, *106*, 79–86.

Atici, O., & Nalbantoglu, B. (2003). Antifreeze proteins in higher plants. *Phytochemistry*, *64*, 1187–1196.

Avia, K., Pilet-Nayel, M. L., Bahrman, N., Baranger, A., Delbreil, B., Fontaine, V., Hamon, C., Hanocq, E., Niarquin, M., Sellier, H., Vuylsteker, C., Prosperi, J. M., & Lejeune-Hénaut, I. (2013). Genetic variability and QTL mapping of freezing tolerance and related traits in Medicago truncatula. *Theoretical and Applied Genetics*, *126*, 2353–2366.

Badawi, M., Danyluk, J., Boucho, B., Houde, M., & Sarhan, F. (2007). The CBF gene family in hexaploid wheat and its relationship to the phylogenetic complexity of cereal CBFs. *Molecular Genetics and Genomics*, *277*, 533–554.

Baker, S. S., Wilhelm, K. S., & Thomashow, M. F. (1994). The 5′-region of Arabidopsis thaliana cor15a has cis-acting elements that confer cold-, drought- and ABA-regulated gene expression. *Plant Molecular Biology*, *24*, 701–713.

Behnam, B., Kikuchi, A., Celebi-Toprak, F., Kasuga, M., YamaguchiShinozaki, K., & Watanabe, K. N. (2007). Arabidopsis rd29A: DREB1A enhances freezing tolerance in transgenic potato. *Plant Cell Reports*, *26*, 1275–1282.

Bremer, A., & Kent, B. (2017). Intrinsically disordered stress protein COR15A resides at the membrane surface during dehydration. *Biophysical Journal*, *113*, 572–579.

Campoli, C., Matus-Cádiz, M., Pozniak, C., Cattivelli, L., & Fowler, D. B. (2009). Comparative expression of CBF genes in the Triticeae under different acclimation induction temperatures. *Molecular Genetics and Genomics*, *282*, 141–152.

Chandler, J. W. (2018). Class VIIIb APETALA2 ethylene response factors in plant development. *Trends in Plant Science*, *23*, 151–162.

Chen, M., Xu, Z., Xia, L., Li, L., Cheng, X., Dong, J., Wang, Q., & Ma, Y. (2009). Coldinduced modulation and functional analyses of the DRE-binding transcription factor gene, GmDREB3, in soybean (Glycine max L.). *Journal of Experimental Botany*, *60*, 121–135.

Chen, Y., & Chen, L. (2022). The mitogen-activated protein kinase kinase MKK2 positively regulates constitutive cold resistance in the potato. *Environmental and Experimental Botany*, *19*, 104702.

Chinnusamy, V., Ohta, M., & Kanrar, S. (2003). ICE1: A regulator of cold-induced transcriptome and freezing tolerance in Arabidopsis. *Genes & Development, 17*, 1043–1054.

Chinnusamy, V., Zhu, J., & Zhu, J.-K. (2007). Cold stress regulation of gene expression in plants. *Trends in Plant Sciences*, *12*, 444–451.

Cho, S., Yu, S. L., & Park, J. (2017). Accession-dependent CBF gene deletion by CRISPR/Cas system in Arabidopsis. *Frontiers in Plant Science*, 8, 1910.

Collard, B. C. Y., & Mackill, D. J. (2008). Marker-assisted selection: An approach for precision plant breeding in the twenty-first century. *Philosophical Transactions of the Royal Society B*, *363*, 557–572.

Cruz, R. P. D., Sperotto, R. A., Cargnelutti, D., Adamski, J. M., Freitasterra, T., & Fett, J. P. (2013). Avoiding damage and achieving cold tolerance in rice plants. *Food and Energy Security*, *2*, 96–119.

Dai, L., Lin, X., Ye, C., Ise, K., Saito, K., Kato, A., Xu, F., Yu, T., & Zhang, D. (2004). Identification of quantitative trait loci controlling cold tolerance at the reproductive stage in Yunnan landrace of rice, Kunmingxiaobaigu. *Breeding Science*, *54*, 253–258.

Damaris, R. N., & Yang, P. (2021). Protein phosphorylation response to abiotic stress in plants. *Methods in Molecular Biology*, *2358*, 635–674.

Ding, Y., Jia, X., & Shi, Y. (2018). OST 1-mediated BTF 3L phosphorylation positively regulates CBF s during plant cold responses. *The EMBO Journal*, *37*, e98228.

Ding, Y., Shi, Y., & Yang, S. (2019). Advances and challenges in uncovering cold tolerance regulatory mechanisms in plants. *New Phytologist*, *222*, 1690–1704.

Dowgert, M. F., & Steponkus, P. L. (1984). Behavior of the plasma membrane of isolated protoplasts during a freeze-thaw cycle. *Plant Physiology*, *75*, 1139–1151.

Dumont, E., Fontaine, V., Vuylsteker, C., Sellier, H., Bodèle, S., Voedts, N., Devaux, R., Frise, M., Avia, K., Hilbert, J. L., Bahrman, N., Hanocq, E., Lejeune-Hénaut, I., & Delbreil, B. (2009). Association of sugar content QTL and PQL with physiological traits relevant to frost damage resistance in pea under field and controlled conditions. *Theoretical and Applied Genetics*, *118*, 1561–1571.

Falk, D. E., Reinbergs, E., & Meatherall, G. (1997). OAC Elmira winter barley. *Canadian Journal of Plant Science*, *77*, 639–640.

Foolad, M. R., Chen, F. Q., & Lin, G. Y. (1998). RFLP mapping of QTLs conferring cold tolerance during seed germination in an inter-specific cross of tomato. *Molecular Breeding*, *4*, 519–529.

Forster, B. P., Ellis, R. P., Thomas, W. T. B., Newton, A. C., Tuberosa, R., This, D., el-Enein, R. A., Bahri, M. H., & Ben Salem, M. (2000). The development and application of molecular markers for abiotic stress tolerance in barley. *Journal of Experimental Botany*, *51*(342), 19–27.

Fracheboud, Y., Jompuk, C., Ribaut, J.-M., Stamp, P., & Leipner, J. (2004). Genetic analysis of cold-tolerance of photosynthesis in maize. *Plant Molecular Biology*, *56*, 241–253.

Funatsuki, H., Kawaguchi, K., Matsuba, S., Sato, Y., & Ishimoto, M. (2005). Mapping of QTL associated with chilling tolerance during reproductive growth in soybean. *Theoretical and Applied Genetics*, *111*, 851–861.

Fursova, O. V., Pogorelko, G. V., & Tarasov, V. A. (2009). Identification of ICE2, gene involved in cold acclimation which determines freezing tolerance in Arabidopsis thaliana. *Gene*, *429*, 98–103.

Gao, J. P., Chao, D. Y., & Lin, H. X. (2007). Understanding abiotic stress tolerance mechanisms: Recent studies on stress response in rice. *Journal of Integrative Plant Biology*, *49*, 742–750.

Gery, C., Zuther, E., Schulz, E., Legoupi, J., Chauveau, A., Mckhann, H., Hincha, D. K., & Téoulé, E. (2011). Natural variation in the freezing tolerance of Arabidopsis thaliana: Effects of RNAi-induced CBF depletion and QTL localisation vary among accessions. *The Plant Science*, *180*, 12–23.

Gilmour, S. J., Zarka, D. G., Stockinger, E. J., Salazar, M. P., Houghton, J. M., & Thomashow, M. F. (1998). Low temperature regulation of the Arabidopsis CBF family of AP2 transcriptional activators as an early step in cold-induced COR gene expression. *The Plant Journal*, *16*, 433–442

Goodstal, F. J., Kohler, G. R., Randall, L. B., Bloom, A. J., & Clair, D. A. S. (2005). A major QTL introgressed from wild Lycopersicon hirsutum confers chilling tolerance to cultivated tomato (Lycopersicon esculentum). *Theoretical and Applied Genetics*, *111*, 898–905.

Griffith, M., Antikainen, M., Hon, W. C., Pihakaski-Maunsbach, K., Yu, X. M., Chun, J. U., & Yang, D. S. C. (1997). Antifreeze proteins in winter rye. *Physiologia Plantarum*, *100*, 327–332. https://doi.org/10.1111/j.1399-3054.1997.tb04790.x.

Guan, Y., Liu, S., & Wu, W. (2021). Genome-wide identification and cold stress-induced expression analysis of the CBF gene family in Liriodendron chinense. *Journal of Forestry Research*, *32*, 2531–2543.

Guo, J., Ren, Y., & Tang, Z. (2019). Characterization and expression profiling of the ICE-CBF-COR genes in wheat. *Peer J*, *7*, e8190.

Guo, X., Zhang, L., Zhu, J., Liu, H., & Wang, A. (2017). Cloning and characterization of SiDHN, a novel dehydrin gene from Saussurea involucrata Kar. et Kir. that enhances cold and drought tolerance in tobacco. *The Plant Science*, *256*, 160–169.

Guo, X. Y., Liu, D. F., & Chong, K. (2018). Cold signaling in plants: Insights into mechanisms and regulation. *Journal of Integrative Plant Biology*, *60*, 745–756.

Gupta, P. K., & Rustgi, S. (2004). Molecular markers from the transcribed/expressed region of the genome in higher plants. *Functional & Integrative Genomics*, *4*, 139–162.

Gusain, S., Joshi, S., & Joshi, R. (2023). Sensing, signalling, and regulatory mechanism of cold-stress tolerance in plants. *Plant Physiology and Biochemistry*, *197*, 107646.

Gutha, L. R., & Reddy, A. R. (2008). Rice DREB1B promoter shows distinct stress-specific responses, and the overexpression of cDNA in tobacco confers improved abiotic and biotic stress tolerance. *Plant Molecular Biology*, *68*, 533–555.

Guy, C. (1999). Molecular responses of plants to cold shock and cold acclimation. *Journal of Molecular Microbiology and Biotechnology*, *1*, 231–242.

Hao, X., & Wang, L. (2018). Response and adaptation mechanisms of tea plant to low-temperature stress. In W.-Y. Han, X. Li, & J. Ahammed (Eds.), *Stress physiology of tea in the face of climate change* (pp. 39–61). Springer.

Hirsz, D., & Dixon, L. E. (2021). The roles of temperature-related post-transcriptional regulation in cereal floral development. *Plants*, *10*, 2230.

Houde, M., Belcaid, M., Ouellet, F., Danyluk, J., Monroy, A. F., Dryanova, A., Gulick, P., Bergeron, A., Laroche, A., Links, M. G., MacCarthy, L., Crosby, W. L., & Sarhan, F. (2006). Wheat EST resources for functional genomics of abiotic stress. *BMC Genomics*, *7*, 149. https://doi.org/10.1186/1471-2164-7-149.

Hu, Z., Ban, Q., & Hao, J. (2020). Genome-wide characterization of the C-repeat binding factor (CBF) gene family involved in the response to abiotic stresses in tea plant (Camellia sinensis). *Frontiers in Plant Sciences*, *11*, 921.

Hund, A., Fracheboud, Y., Soldati, A., Frascaroli, E., Salvi, S., & Stamp, P. (2004). QTL controlling root and shoot traits of maize seedlings under cold stress. *Theoretical and Applied Genetics*, *109*, 618–629.

Ikeda, T., Ohnishi, S., Senda, M., Miyoshi, T., Ishimoto, M., Kitamura, K., & Funatsuki, H. (2009). A novel major quantitative trait locus controlling seed development at low temperature in soybean (Glycine max). *Theoretical and Applied Genetics*, *118*, 1477–1488.

Ishizaki-Nishizawa, O., Fuji, T., Azume, M., Sekiguchi, K., Murata, N., Ohtani, T., & Toguri, T. (1996). Low-temperature resistance of higher plants is significantly enhanced by a nonspecific cyanobacterial desaturase. *Nature Biotechnology*, *14*, 1003–1006.

Ito, Y., Katsura, K., Maruyama, K., Taji, T., Kobayashi, M., Seki, M., Shinozaki, K., & Yamaguchi-Shinozaki, K. (2006). Functional analysis of rice DREB1/CBF-type transcription factors involved in cold-responsive gene expression in transgenic rice. *Plant and Cell Physiology*, *47*, 141–153.

Jaglo, K. R., Kleff, S., Amundsen, K. L., Zhang, X., Haake, V., Zhang, J. Z., Deits, T., & Thomashow, M. F. (2001). Components of the Arabidopsis C-repeat/dehydration-responsive element binding factor cold-response pathway are conserved in Brassica napus and other plant species. *Plant Physiology*, *127*, 910–917.

Janská, A., Maršík, P., Zelenková, S., & Ovesná, J. (2010). Cold stress and acclimation – What is important for metabolic adjustment? *Plant Biology*, *12*, 395–405.

Jiang, X., Hoehenwarter, W., & Scheel, D. (2020). Phosphorylation of the CAMTA3 transcription factor triggers its destabilization and nuclear export. *Plant Physiology*, *184*, 1056–1071.

Jin, Y., Zhai, S., & Wang, W. (2018). Ideentification of genes from the ICE-CBF-COR pathway under cold stress in Aegilops-Triticum composite group and the evolution analysis with those from Triticeae. *Physiology and Molecular Biology of Plants*, *24*, 211–229.

Jofuku, K. D., den Boer, B. G., & Van Montagu, M. (1994). Control of Arabidopsis flower and seed development by the homeotic gene APETALA2. *Plant Cell*, *6*, 1211–1225.

Kashyap, P., & Deswal, R. (2019). Two ICE isoforms showing differential transcriptional regulation by cold and hormones participate in Brassica juncea cold stress signaling. *Gene*, *695*, 32–41.

Kasuga, M., Liu, Q., Miura, S., Yamaguchi-Shinozaki, K., & Shinozaki, K. (1999). Improving plant drought, salt, and freezing tolerance by gene transfer of a single stress-inducible transcription factor. *Nature Biotechnology*, *17*, 287–291.

Khan, H., Shah, S. H., Uddin, N., Azhar, N., Asim, M., Syed, S., Ullah, F., Tawab, F., & Inayat, J. (2015). Biochemical and physiological changes of different plants species in response to heat and cold stress. *ARPN Journal of Agricultural and Biological Science*, *10*, 213–216.

Khoshro, H. H., & Lotfi, R. (2021). Advanced breeding approaches for cold-tolerant chickpea and lentil in dryland areas. In J. C. Jimenez-Lopez & A. Clemente (Eds.), *Legumes research-volume 1*. IntechOpen.

Kim, S. M., Suh, J. P., Lee, C. K., Lee, J. H., Kim, Y. G., & Jena, K. K. (2014). QTL mapping and development of candidate gene-derived DNA markers associated with seedling cold tolerance in rice (Oryza sativa L.). *Molecular Genetics and Genomics*, *289*, 333–343.

Kisha, T. J., Taylor, G. A., Bowman, H. F., Wiesner, L. E., Jackson, G. D., Carlson, G. R., Bergman, J. W., Kushnak, J. D., Stallknecht, G. F., & Stewart, V. R. (1992). Registration of Tiber hard red winter wheat. *Crop Science*, *32*, 1292–1293.

Klein, A., Houtin, H., Rond, C., Marget, P., Jacquin, F., Boucherot, K., Huart, M., Riviere, N., Boutet, G., & Lejeune-Henaut, I. (2014). QTL analysis of frost damage in pea suggests different mechanisms involved in frost tolerance. *Theoretical and Applied Genetics*, 2014, 1319–1330.

Knoll, J., Gunaratna, N., & Ejeta, G. (2008). QTL analysis of early season cold tolerance in Sorghum. *Theoretical and Applied Genetics*, *116*, 577–587.

Kodama, H., Hamada, T., Horiguchi, G., Nishimura, M., & Iba, K. (1994). Genetic enhancement of cold tolerance by expression of a gene for chloroplast w-3 fatty acid desaturase in transgenic tobacco. *Plant Physiology*, *105*, 601–605.

Kole, C., Thormann, C. E., Karlsson, B. H., Palta, J. P., Gaffney, P., Yandell, B., & Osborn, T. C. (2002). Comparative mapping of loci controlling winter survival and related traits in oilseed Brassica rapa and B. napus. *Molecular Breeding*, *9*, 201–210.

Kosová, K., & Kilma, K. (2021). COR/LEA proteins as indicators of frost tolerance in Triticeae: A comparison of controlled versus field conditions. *Plants*, *10*, 789.

Kumar, N., & Bhatt, R. P. (2006). Transgenics: An emerging approach for cold tolerance to enhance vegetables production in high altitude areas. *Indian Journal of Crop Science*, *1*, 8–12.

Kurbidaeva, A., Ezhova, T., & Novokreshchenova, M. (2014). Arabidopsis thaliana ICE2 gene: Phylogeny, structural evolution and functional diversification from ICE1. *The Plant Science*, *229*, 10–22.

Kuroki, M., Saito, K., Matsuba, S., Yokogami, N., Shimizu, H., Ando, I., & Sato, Y. (2007). A quantitative trait locus for cold tolerance at the booting stage on rice chromosome 8. *Theoretical and Applied Genetics*, *115*, 593–600.

Lejeune-Henaut, I., Hanocq, E., Bethenourt, L., Fontaine, V., Delbreil, B., Morin, J., Petit, A., Devaux, R., Boilleau, M., & Stempniak, J. J. (2008). The flowering locus Hr colocalizes with a major QTL affecting winter frost tolerance in Pisum sativum L. *Theoretical and Applied Genetics*, *116*, 1105–1116.

Leuendorf, J. E., Frank, M., & Schmülling, T. (2020). Acclimation, priming and memory in the response of Arabidopsis thaliana seedlings to cold stress. *Scientific Reports*, *10*, 689.

Li, H., & Ding, Y. (2017). MPK3-and MPK6-mediated ICE1 phosphorylation negatively regulates ICE1 stability and freezing tolerance in Arabidopsis. *Developmental Cell*, *43*, 630–642.

Li, W., Wang, R., & Li, M. (2008). Differential degradation of extraplastidic and plastidic lipids during freezing and post-freezing recovery in Arabidopsis thaliana. *Journal of Biological Chemistry*, *283*, 461–468.

Li, X., Liu, C., & Zhao, Z. (2020). COR27 and COR28 are novel regulators of the COP1–HY5 regulatory hub and photomorphogenesis in Arabidopsis. *Plant Cell*, *32*, 1393154.

Lichtenthaler, H. K. (1998). The stress concept in plants: An introduction. *Annals of the New York Academy of Sciences*, *851*, 187–198.

Lin, L., Wu, J., & Wang, Y. (2021). Plant mitogen-activated protein kinase cascades in environmental stresses. *International Journal of Molecular Sciences*, *22*, 1543.

Liu, F., Sun, C., Tan, L., Fu, Y., Li, D., & Wang, X. (2003). Identification and mapping of quantitative trait loci controlling cold-tolerance of Chinese common wild rice (O. rufipogon Griff.) at booting to flowering stages. *Chinese Science Bulletin*, *48*, 2068–2071.

Liu, J., Shi, Y., & Yang, S. (2018). Insights into the regulation of CBF cold signaling in plants. *Journal of Integrative Plant Biology*, *9*, 780–795.

Liu, L., Duan, L., Zhang, J., Zhang, Z., Mi, G., & Ren, H. (2010). Cucumber Cucumis sativus L. over-expressing cold-induced transcriptome regulator ICE1 exhibits changed morphological characters and enhances chilling tolerance. *Scientia Horticulturae*, *124*, 29–33.

Liu, Q., Kasuga, M., Sakuma, Y., Abe, H., Miura, S., Yamaguchi-Shinozaki, K., & Shinozaki, K. (1998). Two transcription factors, DREB1 and DREB2, with an EREBP/AP2 DNA binding domain separate two cellular signal transduction pathways in drought- and low-temperature-responsive gene expression, respectively, in Arabidopsis. *Plant Cell*, *10*, 1391–1406.

Liu, W., Maurer, H. P., Li, G., Tucker, M. R., Gowda, M., Weissmann, E. A., Hahn, V., & Würschum, T. (2014). Genetic architecture of winter hardiness and frost tolerance in triticale. *PLoS One*, *9*, e99848.

Liu, Y., Dang, P., & He, C. (2019). Cold acclimation by the CBF–COR pathway in a changing climate: Lessons from Arabidopsis thaliana. *Plant Cell Reports*, *38*, 511–519.

Liu, Y., & Zhou, J. (2018). MAPping kinase regulation of ICE1 in freezing tolerance. *Trends in Plant Science*, *23*, 91–93.

Londo, J. P., Kovaleski, A. P., & Lillis, J. A. (2018). Divergence in the transcriptional landscape between low temperature and freeze shock in cultivated grapevine (Vitis vinifera). *Horticulture Research*, 5, 10.

Maqbool, M. A., Aslam, M., Ali, H., & Shah, T. M. (2016). Evaluation of advanced chickpea (Cicer arietinum L.) accessions based on drought tolerance indices and SSR markers against different water treatments. *Pakistan Journal of Botany*, *48*, 1421–1429.

Matsui, A., Ishida, J., Morosawa, T., Okamoto, M., Kim, J.-M., Kurihara, Y., Kawashima, M., Tanaka, M., To, T. K., & Nakaminami, K. (2010). Arabidopsis tiling array analysis to identify the stress-responsive genes. In R. Sunkar (Ed.), *Plant stress tolerance*. Springer, 141–155.

Mboup, M., Fischer, I., Lainer, H., & Stephan, W. (2012). Trans-species polymorphism and allele-specific expression in the CBF gene family of wild tomatoes. *Molecular Biology and Evolution*, *29*, 3641–3652.

Medina, J., Catalá, R., & Salinas, J. (2011). The CBFs: Three arabidopsis transcription factors to cold acclimate. *The Plant Science*, *180*, 3–11.

Mehrotra, S., & Verma, S. (2020). Transcriptional regulation and signalling of cold stress response in plants: An overview of current understanding. *Environmental and Experimental Botany*, *180*, 104243.

Mickelbart, M. V., Hasegawa, P. M., & Bailey-Serres, J. (2015). Genetic mechanisms of abiotic stress tolerance that translate to crop yield stability. *Nature Reviews Genetics*, *16*, 237–251.

Miller, A., Galiba, G., & Dubcovsky, J. (2006). A cluster of 11 CBF transcription factors is located at the frost tolerance locus Fr-A m 2 in Triticum monococcum. *Molecular Genetics and Genomics*, *275*, 193–203.

Mizoi, J., Shinozaki, K., & Yamaguchi-Shinozaki, K. (2012). AP2/ERF family transcription factors in plant abiotic stress responses. *Biochimica et Biophysica Acta*, *1819*, 86–96.

Monroe, J., Mcgovern, C., Lasky, J. R., Grogan, K., Beck, J., & Mckay, J. K. (2016). Adaptation to warmer climates by parallel functional evolution of CBF genes in Arabidopsis thaliana. *Molecular Ecology*, *25*, 3632–3644.

Mori, M., Onishi, K., Tokizono, Y., Shinada, H., Yoshimura, T., Numao, Y., Miura, H., & Sato, T. (2011). Detection of novel quantitative trait locus for cold tolerance at the booting stage derived from a tropical japonica rice variety Silewah. *Breeding Science*, *61*, 61–68.

Morran, S., Eini, O., Pyvovarenko, T., Parent, B., Singh, R., Ismagul, A., Eliby, S., Shirley, N., Langridge, P., & Lopato, S. (2011). Improvement of stress tolerance of wheat and barley by modulation of expression of DREB/CBF factors. *Plant Biotechnology Journal*, *9*, 230–249.

Nadeem, M. A., Nawaz, M. A., Shahid, M. Q., Doğan, Y., Comertpay, G., Yıldız, M., Hatipoğlu, R., Ahmad, F., Alsaleh, A., Labhane, N., Özkan, H., Chung, G., & Baloch, F. S. (2018). DNA molecular markers in plant breeding: Current status and recent advancements in genomic selection and genome editing. *Biotechnology & Biotechnological Equipment, 32*(2), 261–285.

Najeeb, S., Mahender, A., Anandan, A., Hussain, W., Li, Z., & Ali, J. (2021). Genetics and breeding of low-temperature stress tolerance in rice. In J. Ali & S. H. Wani (Eds.), *Rice improvement: Physiological, molecular breeding and genetic perspectives* (pp. 221–280). Springer.

Nakamura, J., Yuasa, T., Huong, T. T., Harano, K., Tanaka, S., Iwata, T., Phan, T., & Iwaya-Inoue, M. (2011). Rice homologs of inducer of CBF expression (OsICE) are involved in cold acclimation. *Plant Biotechnology*, *28*, 303–309.

Nakano, T., Suzuki, K., & Fujimura, T. (2006). Genome-wide analysis of the ERF gene family in Arabidopsis and rice. *Plant Physiology*, *140*, 411–432.

Navarro, M., Marque, G., Ayax, C., Keller, G., Borges, J. P., Marque, C., & Teulières, C. (2009). Complementary regulation of four Eucalyptus CBF genes under various cold conditions. *Journal of Experimental Botany*, *60*, 2713–2724.

Novillo, F., Alonso, J. M., & Ecker, J. R. (2004). CBF2/DREB1C is a negative regulator of CBF1/DREB1B and CBF3/DREB1A expression and plays a central role in stress tolerance in Arabidopsis. *Proceedings of the National Academy of Sciences of the United States of America*, *101*, 3985–3990.

Oh, C. S., Choi, Y. H., Lee, S. J., Yoon, D. B., Moon, H. P., & Ahn, S. N. (2004). Mapping of quantitative trait loci for cold tolerance in weedy rice. *Breeding Science*, *54*, 373–380.

Okamuro, J. K., Caster, B., & Villarroel, R. (1997). The AP2 domain of APETALA2 defines a large new family of DNA binding proteins in Arabidopsis. *Proceedings of the National Academy of Sciences of the United States of America*, *94*, 7076–7081.

Okawa, K., Nakayama, K., & Kakizaki, T. (2008). Identification and characterization of Cor413im proteins as novel components of the chloroplast inner envelope. *Plant, Cell & Environment*, *31*, 1470–1483.

Orvar, B. L., Sangwan, V., Omann, F., & Dhindsa, R. S. (2000). Early steps in cold sensing by plant cells: The role of actin cytoskeleton and membrane fluidity. *The Plant Journal*, *23*, 785–794.

Pearce, R. S. (2001). Plant freezing and damage. *Annals of Botany*, *87*, 417–424.

Pearce, S., Zhu, J., Boldizsár, Á., Vágújfalvi, A., Burke, A., Garland-Campbell, K., Galiba, G., & Dubcovsky, J. (2013). Large deletions in the CBF gene cluster at the Fr-B2 locus are associated with reduced frost tolerance in wheat. *Theoretical and Applied Genetics*, *126*, 2683–2697.

Praat, M., De Smet, I., & van Zanten, M. (2021). Protein kinase and phosphatase control of plant temperature responses. *Journal of Experimental Botany*, *72*, 7459–7473.

Presterl, T., Ouzunova, M., Schmidt, W., Moller, E. M., Rober, F. K., Knaak, C., Emst, K., Westhoff, P., & Geiger, H. H. (2007). Quantitative trait loci for early plant vigour of maize grown in chilly environments. *Theoretical and Applied Genetics*, *114*, 1059–1070.

Ritonga, F. N., & Chen, S. (2020). Physiological and molecular mechanism involved in cold stress tolerance in plants. *Plants*, *9*, 560.

Rodriguez, V. M., Burton, A., Rady, M. O. A., Soengas, P., & Revilla, P. (2014). Identification of quantitative trait loci involved in the response to cold stress in maize (Zea mays L). *Molecular Breeding*, *33*, 363–371.

Rodriguez, V. M., Malvar, R. A., Butron, A., Ordas, A., & Revilla, P. (2007). Maize populations as sources of favorable alleles to improve cold-tolerant hybrids. *Crop Science*, *47*, 1779–1786.

Roorkiwal, M., Jain, A., Kale, S. M., Doddamani, D., Chitikineni, A., Thudi, M., & Varshney, R. K. (2018). Development and evaluation of high-density Axiom® Cicer SNP Array for high-resolution genetic mapping and breeding applications in chickpea. *Plant Biotechnology Journal*, *16*(4), 890–901.

Ruelland, E., Vaultier, M. N., Zachowski, A., & Hurry, V. (2009). Cold signalling and cold acclimation in plants. *Advances in Botanical Research*, *49*, 35–150.

Saito, K., Miura, K., Nagano, K., Hayano-Saito, Y., Saito, A., Araki, H., & Kato, K. (1995). Chromosomal location of quantitative trait loci for cool tolerance at the booting stage in rice variety 'Norin-PL8'. *Breeding Science*, *45*, 337–340.

Sakamoto, A., Sulpice, R., Hou, C. X., Kinoshita, M., Higashi, S. I., Kanaseki, T., Nonaka, H., Moon, B. Y., & Murata, N. (2003). Genetic modification of the fatty acid unsaturation of phosphatidylglycerol in chloroplasts alters the sensitivity of tobacco plants to cold stress. *Plant, Cell & Environment*, 27, 99–105.

Sanghera, G. S., Wani, S. H., Hussain, W., & Singh, N. B. (2011). Engineering cold stress tolerance in crop plants. *Current Genomics*, *12*, 30.

Shi, Y., Ding, Y., & Yang, S. (2018). Molecular regulation of CBF signaling in cold acclimation. *Trends in Plant Science*, *23*, 623–637.

Shi, Y., Huang, J., & Sun, T. (2017). The precise regulation of different COR genes by individual CBF transcription factors in Arabidopsis thaliana. *Journal of Integrative Plant Biology*, *59*, 118–133.

Shinada, H., Iwata, N., Sato, T., & Fujino, K. (2014). QTL pyramiding for improving of cold tolerance at fertilization stage in rice. *Breeding Science*, *63*, 483–488.

Shirasawa, S., Endo, T., Nakagomi, K., Yamaguchi, M., & Nishio, T. (2012). Delimitation of a QTL region controlling cold tolerance at booting stage of a cultivar, 'Lijiangxintuanheigu', in rice, Oryza sativa L. *Theoretical and Applied Genetics*, *124*, 937–946.

Shu, Y., Li, W., Zhao, J., Zhang, S., Xu, H., Liu, Y., & Guo, C. (2017). Transcriptome sequencing analysis of alfalfa reveals CBF genes potentially playing important roles in response to freezing stress. *Genetics and Molecular Biology*, *40*, 824–833.

Siddiqui, K. S., & Cavicchioli, R. (2006). Cold-adapted enzymes. *Annual Review of Biochemistry*, *75*, 403–433.

Soltész, A., Smedley, M., Vashegyi, I., Galiba, G., Harwood, W., & Vágújfalvi, A. (2013). Transgenic barley lines prove the involvement of TaCBF14 and TaCBF15 in the cold acclimation process and in frost tolerance. *Journal of Experimental Botany*, *64*, 1849–1862.

Stockinger, E. J., Skinner, J. S., Gardner, K. G., Francia, E., & Pecchioni, N. (2007). Expression levels of barley CBF genes at the frost resistance-H2 locus are dependent upon alleles at Fr-H1 and Fr-H2. *The Plant Journal*, *51*, 308–321.

Sui, N., Li, M., Zhao, S. J., Li, F., Liang, H., & Meng, Q. W. (2007). Overexpression of glycerol-3-phosphate acyl transferase gene improves chilling tolerance in tomato. *Planta*, *226*, 1097–1108.

Sun, X., Wang, Y., & Sui, N. (2018). Transcriptional regulation of bHLH during plant response to stress. *Biochemical and Biophysical Research Communications*, *503*, 397–401.

Tang, K., Zhao, L., & Ren, Y. (2020). The transcription factor ICE1 functions in cold stress response by binding to the promoters of CBF and COR genes. *Journal of Integrative Plant Biology*, *62*, 258–263.

Teige, M., Scheikl, E., Euglem, T., Dóczi, R., Ichimura, K., Shinozaki, K., Dangl, J. L., & Hirt, H. (2004). The MKK2 pathway mediates cold and salt stress signaling in Arabidopsis. *Molecular Cell*, *15*, 141–152.

Theocharis, A., Clément, C., & Barka, E. A. (2012). Physiological and molecular changes in plants grown at low temperatures. *Planta*, *235*, 1091–1105.

Thomashow, M. F. (1999). Plant cold acclimation: Freezing tolerance genes and regulatory mechanisms. *Annual Review of Plant Physiology and Plant Molecular Biology*, *50*, 571–599.

Thomashow, M. F. (2001). So what's new in the field of plant cold acclimation? Lots! *Plant Physiology*, *125*, 89–93.

Townley, H. E., & Knight, M. R. (2002). Calmodulin as a potential negative regulator of Arabidopsis COR gene expression. *Plant Physiology*, *128*, 1169–1172.

Truco, M. J., Randall, L. B., Bloom, A. J., &Clair, D. A. S. (2000). Detection of QTLs associated with shoot wilting and root ammonium uptake under chilling temperatures in an interspecific backcross population from Lycopersicon esculentum 9L. hirsutum. *Theoretical and Applied Genetics*, *101*, 1082–1092.

Vallejos, C. E., & Tanksley, S. D. (1983). Segregation of isozyme markers and cold tolerance in an interspecific backcross of tomato. *Theoretical and Applied Genetics*, *66*, 241–247.

Varshney, R. K., Ribaut, J. M., Buckler, E. S., Tuberosa, R., Rafalski, J. A., & Langridge, P. (2012). Can genomics boost productivity of orphan crops? *Nature Biotechnology*, *30*, 1172–1176.

Vazquez-Hernandez, M., Romero, I., Escribano, I., Merodio, C., & Sanchez-Ballesta, M. T. (2017). Deciphering the role of CBF/DREB transcription factors and dehydrins in maintaining the quality of table grapes cv. autumn royal treated with high CO_2 levels and stored at 0°C. *Frontiers in Plant Science*, *8*, 1591.

Verma, R. K., & Kumar, V. S. (2020). Overxpression of Arabidopsis ICE1 enhances yield and multiple abiotic stress tolerance in indica rice. *Plant Signaling & Behavior*, *15*, 1814547.

Vyse, K., & Faivre, L. (2020). Transcriptional and post-transcriptional regulation and transcriptional memory of chromatin regulators in response to low temperature. *Frontiers in Plant Science*, *11*, 39.

Wan, F., Pan, Y., & Li, J. (2014). Heterologous expression of Arabidopsis C-repeat binding factor 3 (AtCBF3) and cold-regulated 15A (AtCOR15A) enhanced chilling tolerance in transgenic eggplant (Solanum melongena L.). *Plant Cell Reports*, *33*, 1951–1961.

Wang, D. Z., Jin, Y. N., & Ding, X. H. (2017). Gene regulation and signal transduction in the ICE-CBF-COR signaling pathway during cold stress in plants. *Biochemistry*, *82*, 1103–1117.

Wang, L., Ma, R., Yin, Y., & Jiao, Z. (2018). Role of carbon ion beams irradiation in mitigating cold stress in Arabidopsis thaliana. *Ecotoxicology and Environmental Safety*, *162*, 341–347.

Wang, Q., Guan, Y., Wu, Y., Chen, H., Chen, F., & Chu, C. (2008). Overexpression of a rice OsDREB1F gene increases salt, drought, and low temperature tolerance in both Arabidopsis and rice. *Plant Molecuar Biology*, 67, 589–602.

Wang, W., Vinocur, B., & Altman, A. (2003). Plant responses to drought, salinity and extreme temperatures: Towards genetic engineering for stress tolerance. *Planta*, *218*, 1–14.

Wang, Z., Wan, Y., & Meng, X. (2021). Genome-wide identification and analysis of MKK and MAPK gene families in Brassica species and response to stress in Brassica napus. *International Journal of Molecular Sciences*, *22*, 544.

Wei, X., Liu, S., & Sun, C. (2021). Convergence and divergence: Signal perception and transduction mechanisms of cold stress in Arabidopsis and rice. *Plants*, *10*, 1864.

Welling, A., & Palva, E. T. (2008). Involvement of CBF transcription factors in winter hardiness in birch. *Plant Physiology*, *147*, 1199–1211.

Welti, R., & Li, W. (2002). Profiling membrane lipids in plant stress responses. Role of phospholipase D alpha in freezing-induced lipid changes in Arabidopsis. *Journal of Biological Chemistry*, *277*, 31994–32002.

Wisniewski, M., Nassuth, A., Teulieres, C., Marque, C., Rowland, J., Cao, P. B., & Brown, A. (2014). Genomics of cold hardiness in woody plants. *Critical Reviews in Plant Sciences*, *33*, 92–124.

Wisniewski, M., Norelli, J., & Artlip, T. (2015). Overexpression of a peach CBF gene in apple: A model for understanding the integration of growth, dormancy, and cold hardiness in woody plants. *Frontiers in Plant Science*, *6*, 85.

Xiao, N., Huang, W. N., Zhang, X. X., Gao, Y., Li, A. H., Dai, Y., Yu, L., Liu, G. Q., Pan, C. H., Li, Y. H., Dai, Z. Y., & Chen, J. M. (2014). Fine mapping of qRC10-2, a quantitative trait locus for cold tolerance of rice roots at seedling and mature stages. *PLoS One*, *9*, e96046.

Xie, Y., Chen, P., & Yan, Y. (2018). An atypical R2R3 MYB transcription factor increases cold hardiness by CBF-dependent and CBFindependent pathways in apple. *New Phytologist*, *218*, 201–218.

Xu, L. M., Zhou, L., Zeng, W., Wang, F. M., Zhang, H. L., Shen, S. Q., & Li, Z. C. (2008). Identification and mapping of quantitative trait loci for cold tolerance at the booting stage in a japonica rice nearisogenic line. *Plant Science*, *174*, 340–347.

Yamaguchi-Shinozaki, K., & Shinozaki, K. (1994). A novel cis-acting element in an Arabidopsis gene is involved in responsiveness to drought, low-temperature, or high-salt stress. *Plant Cell*, *6*, 251–226.

Yamaguchi-Shinozaki, K., & Shinozaki, K. (2005). Organization of cis-acting regulatory elements in osmotic- and cold-stress-responsive promoters. *Trends in Plant Science*, *10*, 88–94.

Yamasaki, Y., & Randall, S. K. (2016). Functionality of soybean CBF/DREB1 transcription factors. *Plant Science*, *246*, 80–90

Yang, T., Chaudhuri, S., & Yang, L. (2010). A calcium/calmodulin-regulated member of the receptor-like kinase family confers cold tolerance in plants. *Journal of Biological Chemistry*, *285*, 7119–7126.

Yang, T., Zhang, S., Zhao, Z., Liu, Q., Huang, Z., Mao, X., Dong, J., Wang, X., Zhang, G., & Liu, B. (2016). Identification and pyramiding of QTLs for cold tolerance at the bud bursting and the seedling stages by use of single segment substitution lines in rice (Oryza sativa L.). *Molecular Breeding*, *36*, 96

Yang, W., Liu, X.-D., Chi, X.-J., Wu, C.-A., Li, Y.-Z., Song, L.-L., Liu, X.-M., Wang, Y.-F., Wang, F.-W., & Zhang, C. (2011). Dwarf apple MbDREB1 enhances plant tolerance to low temperature, drought, and salt stress via both ABA-dependent and ABA-independent pathways. *Planta*, *233*, 219–229.

Yang, Z., Chen, C., Xiaofen, J., Aisheng, X., Rihe, P., Yihuan, H., Quanhong, Y., & Jianmin, C. (2009). Expression of a rice DREB1 gene, OsDREB1D, enhances cold and high-salt tolerance in transgenic Arabidopsis. *BMB Reports (Biochemistry & Molecular Biology Reports)*, *42*, 486–492.

Yin, J., Yi, H., & Chen, X. (2019). Post-translational modifications of proteins have versatile roles in regulating plant immune responses. *International Journal of Molecular Sciences*, *20*, 2807.

Younis, A., Ramzan, F., Ramzan, Y., Zulfiqar, F., Ahsan, M., & Lim, K. B. (2020). Molecular markers improve abiotic stress tolerance in crops: A review. *Plants*, *9*(10), 1374.

Zarka, D. G., Vogel, J. T., Cook, D., & Thomashow, M. F. (2003). Cold induction of Arabidopsis CBF genes involves multiple ICE (inducer of CBF expression) promoter elements and a cold-regulatory circuit that is desensitized by low temperature. *Plant Physiology*, *133*, 910–918.

Zeller, G., Henz, S. R., Widmer, C. K., Sachsenberg, T., Rätsch, G., Weigel, D., & Laubinger, S. (2009). Stress-induced changes in the Arabidopsis thaliana transcriptome analyzed using whole-genome tiling arrays. *The Plant Journal*, *58*, 1068–1082.

Zhang, F., Jiang, Y., Bai, L.-P., Zhang, L., Chen, L.-J., Li, H. G., Yin, Y., Yan, W.-W., Yi, Y., & Guo, Z.-F. (2011). The ICE-CBF-COR pathway in cold acclimation and AFPs in plants. *Middle-East Journal of Scientific Research*, *8*, 493–498.

Zhang, G. Q., Zeng, R. Z., Zhang, Z. M., Ding, X. H., Li, W. T., Liu, G. F., He, F. H., Tulukdar, A., Huang, C. F., Xi, Z. Y., Qin, L. J., Shi, J. Q., Zhao, F. M., Feng, M. J., Shan, Z. L., Chen, L., Guo, X. Q., Zhu, H. T., & Lu, Y. G. The construction of a library of single segment substitution lines in rice (Oryza sativa L.). *Rice Genetics Newsletter*, *121*, 85–87.

Zhang, J., Xu, Y., Huan, Q., & Chong, K. (2009). Deep sequencing of Brachypodium small RNAs at the global genome level identifies microRNAs involved in cold stress response. *BMC Genomics*, *10*, 449.

Zhao, C., & Wang, P. (2017). MAP kinase cascades regulate the cold response by modulating ICE1 protein stability. *Developmental Cell*, *43*, 618–629.

Zhao, M. L., Wang, J. N., Shan, W., Fan, J. G., Kuang, J. F., Wu, K. Q., Li, X. P., Chen, W. X., He, F. Y., & Chen, J. Y. (2013). Induction of jasmonate signalling regulators MaMYC2s and their physical interactions with MaICE1 in methyl jasmonate-induced chilling tolerance in banana fruit. *Plant, Cell & Environment*, *36*, 30–51.

Zhou, L., Zeng, Y., Hu, G., Pan, Y., Yang, S., You, A., Zhang, H., Li, J., & Li, Z. (2012). Characterization and identification of cold tolerant nearisogenic lines in rice. *Breeding Science*, *62*(2), 196–201.

Zhou, M., Chen, H., Wei, D., Ma, H., & Lin, J. (2017). Arabidopsis I3 and DELLAs positively regulate each other in response to low temperature. *Scientific Reports*, *7*, 1–13.

Zhu, W., Zhou, H., & Lin, F. (2020). Cold-regulated gene27 integrates signals from light and the circadian clock to promote hypocotyl growth in Arabidopsis. *Plant Cell*, *32*, 3155–3169.

11 Challenges of Drought Stress
Unveiling Plant Responses and Approaches to Confer Tolerance

Asad Azeem, Sumbul Habib, Aqsa Parvaiz, Ghulam Mustafa, Muhammad Sarwar Khan, and Faiz Ahmad Joyia

11.1 INTRODUCTION

In the face of rapid climate change and its far-reaching impacts on ecosystems, agriculture, and food security, understanding the intricate relationship between plants and their environment has become a pressing necessity. An increasingly frequent and severe drought is one of the most significant problems brought about by climate change. Droughts can have disastrous consequences for crop harvests, plant growth, and the stability of entire ecosystems (Raihan, 2023). It is critical to understand the mechanisms underlying plant responses to drought stress and to create novel techniques to improve their tolerance, as the growing global population puts increasing pressure on food, fiber, and bioenergy supplies.

Drought stress is a serious hazard to agricultural and natural environments since it is defined by a lack of water to meet the needs of plant metabolism (Bali et al., 2023). Complex physiological, biochemical, and molecular systems have been evolved by plants to enable them to survive in the face of water constraints. However, the ability of many plant species to adapt is stressed as drought episodes become more frequent and extreme, which results in decreased crop yields, changed plant distribution patterns, and unstable ecosystems. Determining how plants respond to drought stress on a variety of levels is crucial to developing practical ways to lessen its negative consequences (Aslam et al., 2023).

Plants are able to detect, react, and adapt to drought stress through an amazing variety of mechanisms that are coordinated at different levels, ranging from the cellular to the whole-plant level. At the cellular level, plants experience complex alterations in gene expression that lead to the production of stress-responsive proteins and regulatory molecules. These molecular changes enable the restructuring of cellular processes, allowing the plant to maintain vital functions while reallocating resources to deal with water scarcity (Saharan et al., 2022). Probably the most obvious response to drought stress is the closure of stomata, which are tiny pores on the leaf surface that control both gas and water vapor exchange. To counteract the negative effects of reduced photosynthesis, plants employ a variety of strategies to optimize water use efficiency (WUE). For example, some plants roll or fold their leaves to minimize exposure to the atmosphere and thereby reduce water loss (Ali et al., 2022). Others may accumulate osmoprotectants – organic compounds that maintain cellular hydration and stabilize cellular structures – in their tissues, allowing them to withstand prolonged periods of drought without irreversible damage. Stomatal closure reduces water loss through transpiration, but it also limits the uptake of carbon dioxide necessary for photosynthesis. As a result, plants often experience decreased photosynthetic rates, making it more difficult for them to produce energy and synthesize essential compounds. Although plants have developed complex defense systems against drought stress, the rate of climate change has frequently accelerated beyond their ability to respond. This emphasizes the necessity of developing novel strategies to increase plant resistance to drought and raise the sustainability of agriculture. Genetics, physiology, biochemistry,

DOI: 10.1201/9781003457237-11

and agronomy must all be integrated in an interdisciplinary manner to tackle the challenging issue of developing crops with increased drought resilience (Langensiepen et al., 2020).

The selection and crossing of plants with desirable traits – the mainstay of conventional breeding methods – has helped plants become more drought-tolerant to some extent. Nonetheless, drought tolerance frequently has a polygenic genetic foundation, which means that several genes cooperate. It can be difficult and time-consuming to identify and modify these genes through traditional breeding. Recent developments in genomics and molecular biology have made it easier to identify the critical genes involved in drought response, opening the door for the development of genome editing and genetic engineering methods (Nuccio et al., 2018). Through the introduction of particular genes that improve osmotic control, stress-signaling pathways, and WUE, genetic engineering offers enormous promise for delivering drought tolerance. For example, introducing genes that code for osmoprotectants like proline or glycine betaine can help plants adapt more readily to water scarcity (Kido et al., 2019). Furthermore, by improving the plant's capacity to promptly activate genes linked to stress, altering the expression of transcription factors (TFs) implicated in stress responses can promote drought tolerance.

CRISPR/Cas9 and other modern genome-editing technologies provided a more focused and accurate method of modifying particular genes in plants (Razzaq et al., 2019). These approaches allow researchers to modify the genetic makeup of plants in a way that improves resistance to drought without introducing alien genes. The potential to customize genetic alterations based on the distinct traits of various plant species is a promising avenue for creating crop types resistant to drought that may sustain elevated yields in difficult circumstances. Finding out how plants react to water shortage is crucial since drought stress continues to present serious problems for ecosystems and agriculture around the world. The intricacy of this process is highlighted by the interaction of physiological, ecological, and molecular pathways that allow plants to endure drought stress. Scientists have the potential to enhance crop resistance against drought by utilizing genome editing and genetic engineering. This would help agricultural systems remain resilient and sustainable in the face of unpredictable climatic changes. Our ability to comprehend how plants react to drought stress and to find creative solutions will be crucial in guaranteeing future generations' access to food and environmental stability.

11.2 IMPACT OF DROUGHT STRESS ON PLANT ATTRIBUTES

Drought stress, which results from a shortage of water, has a significant impact on a number of plant characteristics, including growth, development, and general health (Figure 11.1). Stomatal closure is a physiological response to drought stress, which reduces a plant's capacity to absorb carbon dioxide and impedes photosynthesis (Wu et al., 2022). As a result, less energy is produced, which restricts the buildup of biomass and hinders growth. In addition, a lack of water causes imbalances in the plant's food supply and impairs the uptake and transport of nutrients, which are crucial for vital metabolic functions. The plant's reproductive systems are also impacted, which frequently results in poor flower and seed production and jeopardizes both yield and the success of reproduction (Bareke, 2018). Drought stress causes morphological changes, including reduced leaf expansion, changed leaf-to-shoot ratios, and root proliferation seeking water sources. Although the goal of these adaptations is to maintain water supplies, they may inadvertently obstruct capacity for general expansion and resource allocation.

Drought stress has genetic and molecular effects in addition to physiological and morphological ones. Protective proteins and metabolites are synthesized as a result of the activation of stress-responsive genes. This increases the plant's resistance, but it takes resources away from activities involved in growth. Additionally, as reactive oxygen species (ROS) build up, oxidative stress intensifies and causes cellular damage. To summarize, the consequences of drought stress lead to a series of changes in plant characteristics, including decreased growth, changed morphology, disturbed physiology, and disturbed molecular processes. Comprehending these complex effects is essential to developing methods to improve drought resistance and guarantee plant survival and yield in water-scarce settings.

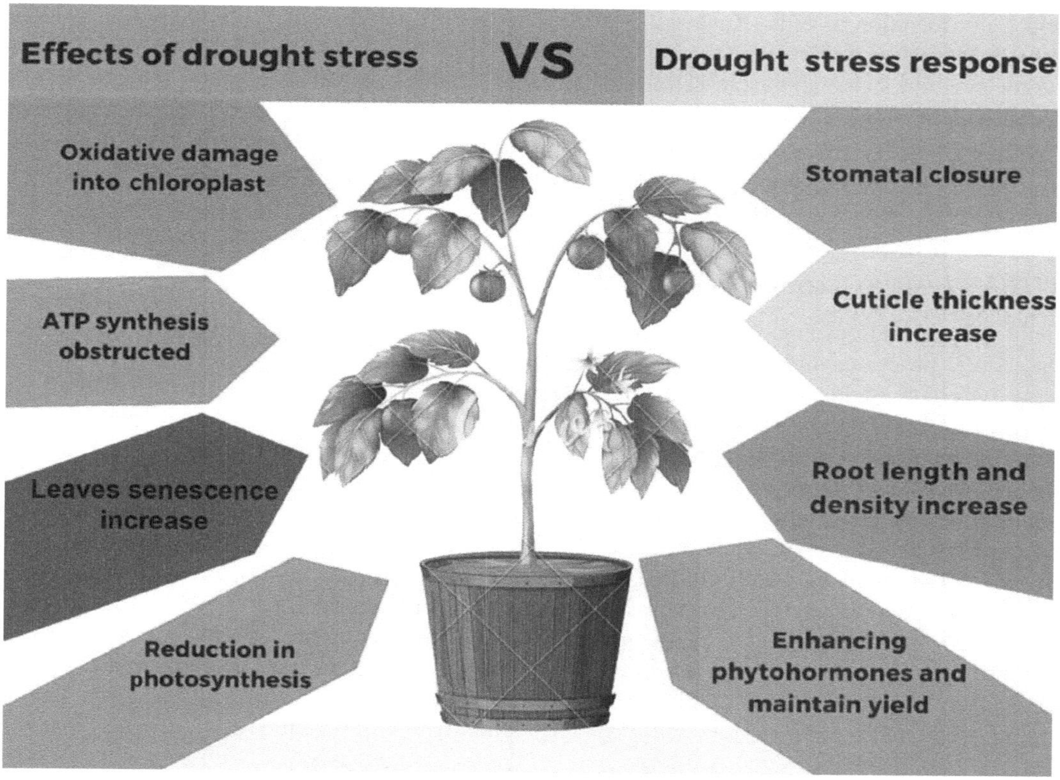

FIGURE 11.1 Adverse effects and plant response to drought stress.

11.2.1 Inhibition of Photosynthesis

Drought stress initiates a series of physiological reactions that culminate in the inhibition of photosynthesis, the essential mechanism that enables a plant to produce energy by converting light energy into carbohydrates (Pandey et al., 2022). The primary mechanism of this response is the closure of stomata, which are tiny pores on the surface of leaves that control gas exchange (Figure 11.1). Although this closure reduces transpiration, which is a method for conserving water, it also prevents the uptake of carbon dioxide (CO_2), which is an essential substrate for photosynthesis. As a result, the plant's capacity to convert CO_2 into sugars is hampered, which reduces the synthesis of energy and the accumulation of carbohydrates (Bhattacharya & Bhattacharya, 2021). Drought stress initiates a series of physiological reactions that culminate in the inhibition of photosynthesis, the essential mechanism that enables a plant to produce energy by converting light energy into carbohydrates. The primary mechanism of this response is the closure of stomata, which are tiny pores on the surface of leaves that control gas exchange. Although this closure reduces transpiration, which is a method for conserving water, it also prevents the uptake of CO_2, which is an essential substrate for photosynthesis. As a result, the plant's capacity to convert CO_2 into sugars is hampered, which reduces the synthesis of energy and the accumulation of carbohydrates (Figure 11.1).

11.2.2 Altered Plant Metabolism

Plants undergo a series of metabolic adjustments in response to drought stress to counteract water deprivation and preserve cellular equilibrium. One such adaptation is the buildup of organic substances called osmolytes, which are low molecular weight and aid in stabilizing cellular structures and water retention in cells (Kushwah et al., 2020). In times of drought, the body produces more osmolytes, such as proline, carbohydrates, and amino acids, which serve as osmoprotectants to keep turgor pressure stable and avoid cellular dehydration. Plants can withstand water scarcity thanks to these metabolic changes, but the main metabolism is altered in the process. The manufacture of osmolytes and stress-responsive proteins takes up resources, which leads to a decrease in the rate of photosynthesis and other growth-related processes, ultimately affecting the general productivity and fitness of the plant (Leisner et al., 2023).

11.2.3 Impaired Growth and Development

Plant growth and development are significantly impacted by drought stress due to its influence on cellular processes and the availability of resources. Cell turgor pressure decreases with water scarcity, which inhibits cell growth and leaf elongation (Coussement et al., 2021). This shows itself as thicker, smaller leaves with different morphologies, including folded or rolled leaves, which decrease the amount of surface area exposed to the air. Significant damage is also done to the root system, which is in charge of absorbing nutrients and water. Root elongation is restricted in drought conditions, whereas lateral root multiplication is encouraged to search a greater area of soil for accessible water. However, there's a chance that this adaptation won't be enough to offset the overall decline in root biomass, which would hamper the uptake of nutrients and water. Plant development is generally stunted, vegetative biomass is reduced, and the architecture is changed as a result of the combined impacts of poor leaf and root growth (Balliu et al., 2021). These modifications may result in fewer flowers and seeds being produced, which have an impact on yield potential and also have ramifications for reproductive architecture.

11.2.4 Membrane Integrity and Subsequent Cellular Damage

Plant cells accumulate ROS in response to drought stress. These extremely reactive ROS are the results of numerous metabolic processes and cause oxidative stress, which harms cellular components (Sachdev et al., 2021). Particularly vulnerable to oxidative damage are membranes because of their high unsaturated fatty acid content. Cell membrane integrity is compromised by a chain reaction called lipid peroxidation, which is started by ROS. Ions and other cellular components can leak out of permeable membranes when they lose their structural integrity (Tsubone et al., 2019), ultimately jeopardizing critical functions like signal transduction, energy production, and nutrient absorption. Furthermore, the ruptured membranes cause programmed cell death, which hastens the aging process of strained tissues and exacerbates the general deterioration in plant function.

11.2.5 Water and Nutrient Relationship

The delicate balance between the intake of water and the acquisition of nutrients is upset by drought stress. Under typical circumstances, a plant's intake of water produces a pressure gradient that enables nutrients to go from the soil to the root and subsequently to the shoot. Nevertheless, this hydraulic conductivity is hampered by a lack of water, which reduces nutrient uptake (Lambers et al., 2019). Moreover, water transport and nutrient mobility within the plant are strongly related. Unbalanced nutrient distribution is the result of a water deficit, which restricts the movement of nutrients to various plant sections (Yadav et al., 2020). Reduced intake of nitrogen, phosphorus, and potassium are examples of nutrient shortages that impede vital metabolic processes and further impair growth and production. On the contrary, nutritional imbalances may make drought stress worse. For instance, in drought-stricken situations, an imbalance between potassium and sodium ions can cause ion toxicity and cellular damage (Sharma & Bhateria, 2022). The complex connection between nutrient intake and water availability emphasizes how interrelated plant physiological responses are.

11.2.6 Assimilate Portioning

Plants under drought stress reassign assimilates (i.e. carbohydrates) generated by photosynthesis, to different parts of the plant. Assimilates are geared toward growth and reproduction in normal circumstances. The plant, however, reallocates these resources to support stress-reduction processes in response to drought. Much of the assimilates are directed toward the production of osmolytes, which are substances that control osmotic pressure and keep cells from drying out. Proline and soluble sugars are examples of osmolytes that need energy for synthesis and upkeep. As a result, these stress-related systems receive the resources that would have otherwise driven growth and reproduction. This change in resource distribution jeopardizes the plant's capacity for overall growth and output (Dietz et al., 2021).

11.2.7 Stomatal Movements

Gas exchange and water regulation are significantly influenced by stomata, which are small pores found on the surface of leaves and stems. A physiological reaction brought on by drought stress causes stomatal closure, which lowers water loss through transpiration. This adaptive system affects photosynthesis and leaf temperature regulation in addition to conserving water. The amount of CO_2 that is needed for photosynthesis is limited by stomatal closure (Harrison et al., 2020). Reduced photosynthetic rates result in decreased energy generation and carbohydrate buildup when CO_2 availability within the leaf decreases. This decrease in energy availability has an impact on the general vigor and growth of plants. Furthermore, the leaf's capacity to control its temperature through transpirational cooling is impacted by stomata closure. Lower transpiration raises the temperature of the leaves, which facilitates the production of ROS and oxidative stress (Qi et al., 2018). Plant performance under drought stress may be further hampered by the ensuing cellular damage.

11.2.8 Increased Susceptibility to Diseases and Pests

Examining how drought stress makes plants more susceptible to pests and diseases highlights the complex dance that occurs between plants and their surroundings. The complex effects of water shortage alter the biochemistry and defenses of plants, fostering an environment more favorable to the spread of diseases and pests. The defense mechanisms of the plant, which include the creation of specific chemicals that dissuade pathogens, are a key component in its resistance to diseases and pests. However, these fortifications face a major obstacle when there is a drought. The synthesis of chemicals linked to defense, such as phytoalexins, is frequently downregulated as a plant directs its resources toward survival tactics. Plants produce phytoalexins, which are chemical molecules that serve as their first line of defense against pathogen invasion. Because these substances are less readily available, the plant is less able to fight off invasive pathogens, which increases the plant's vulnerability to infections (Tiku, 2020). Plant tissues' metabolic makeup is significantly impacted by drought stress. This also applies to plant-emitted volatile organic compounds (VOCs), which function as cues for communication in a range of interactions, including interactions with herbivores. Herbivores and pests may be drawn to the changed VOC profile, which serves as a virtual dinner bell for would-be assailants (Murali-Baskaran et al., 2022). Herbivores may get a signal when a plant's VOC composition changes, indicating stress and possibly limited resources. Herbivore reactions to this may include increased feeding activity, faster rates of reproduction, and changes in their foraging habits. It is essential to comprehend how diseases, pests, and drought stress interact to develop efficient management plans. It is imperative to implement integrated pest management strategies that take into account both biotic and abiotic elements. Pesticides used selectively, the introduction of natural enemies of pests, and cultural practices like balanced irrigation and nutrient management that support plant health are examples of strategies. Finally, the complex web of relationships among diseases, pests, and drought stress draws attention to the weaknesses in stressed plants. Acknowledging these vulnerabilities and putting comprehensive strategies in place to manage them will be critical to protecting crop health, guaranteeing food security, and preserving the delicate balance of ecosystems as we cope with a changing climate and its accompanying problems.

11.2.9 Decreased Crop Productivity and Yield

Drought stress is an ecological problem that affects plant biology on all levels and has a severe negative impact on agricultural yield and productivity. The complex interactions between physiological, metabolic, and morphological changes brought about by water constraints result in a series of outcomes that have a significant influence on agricultural productivity (Kapoor et al., 2020). Photosynthesis, the process that converts light energy into chemical energy and ultimately propels growth and yield formation, is the foundation of a plant's productivity. However, this base of output is severely limited by drought stress. To preserve water, stomatal closure limits the input of CO_2, which is essential for photosynthesis. The plant's capacity to synthesize sugars is significantly reduced by this restriction in CO_2 availability, which has an immediate effect on biomass buildup and energy production (Gamage et al., 2018).

A degraded growth and development story also shapes the tale of yield decline under drought stress. Lack of water restricts cell growth, which results in smaller leaves and different leaf-to-shoot ratios (Albert et al., 2011). Furthermore, the plant's ability to obtain water and nutrients necessary for development and reproduction is hampered by the inhibition of root growth. Reduced yield potential is the direct result of stunted growth, decreased vegetative biomass, and lessened total plant stature. When a plant experiences drought stress, its metabolic dynamics are drastically changed. Instead of fueling development and output, resources are redirected into pathways that respond to stress. The delicate balance between growth and stress responses is upset by this adaptation, which is necessary for survival but ultimately affects crop yield. Resources available for yield production are further depleted by the synthesis of osmolytes, which are molecules that help the plant deal with water deprivation and require a large energy expenditure. Reduced agriculture output and productivity have wide-ranging effects that go beyond specific fields. A world that is battling rapidly increasing population and shifting climatic trends needs resilient and sustainable agriculture systems. Reductions in output brought on by drought put economies that rely heavily on agriculture under pressure and endanger food security (Ashraf et al., 2023). Consequently, the complex interplay between physiological, metabolic, and morphological reactions brought on by drought stress provided a clear picture of the difficulties that agricultural systems confront. Yield reduction, a direct consequence of these multifaceted effects, underscores the urgency of developing and implementing innovative strategies to enhance drought resilience in crops (Ali et al., 2023). By deciphering the underlying mechanisms and harnessing the power of advanced biotechnological tools, researchers strive to fortify global food systems against the mounting pressures of drought stress, ensuring a stable and secure future for generations to come.

11.3 GENOMIC APPROACHES FOR BREEDING DROUGHT-TOLERANT CROPS

Genomic techniques have emerged as effective tools for developing drought-tolerant crops that can flourish in water-stressed environments (Figure 11.2). These methods make use of the detailed insights provided by genetics to comprehend the molecular complexities underlying a plant's response to drought stress. Scientists have discovered a lot about individual genes, pathways, and regulatory processes that contribute to drought tolerance by deciphering the genetic code (Azeem et al., 2023).

Genome sequencing, a groundbreaking process that deciphers a plant's whole DNA sequence, is at the heart of these genomic tactics. This fundamental phase gives a detailed record of the plant's

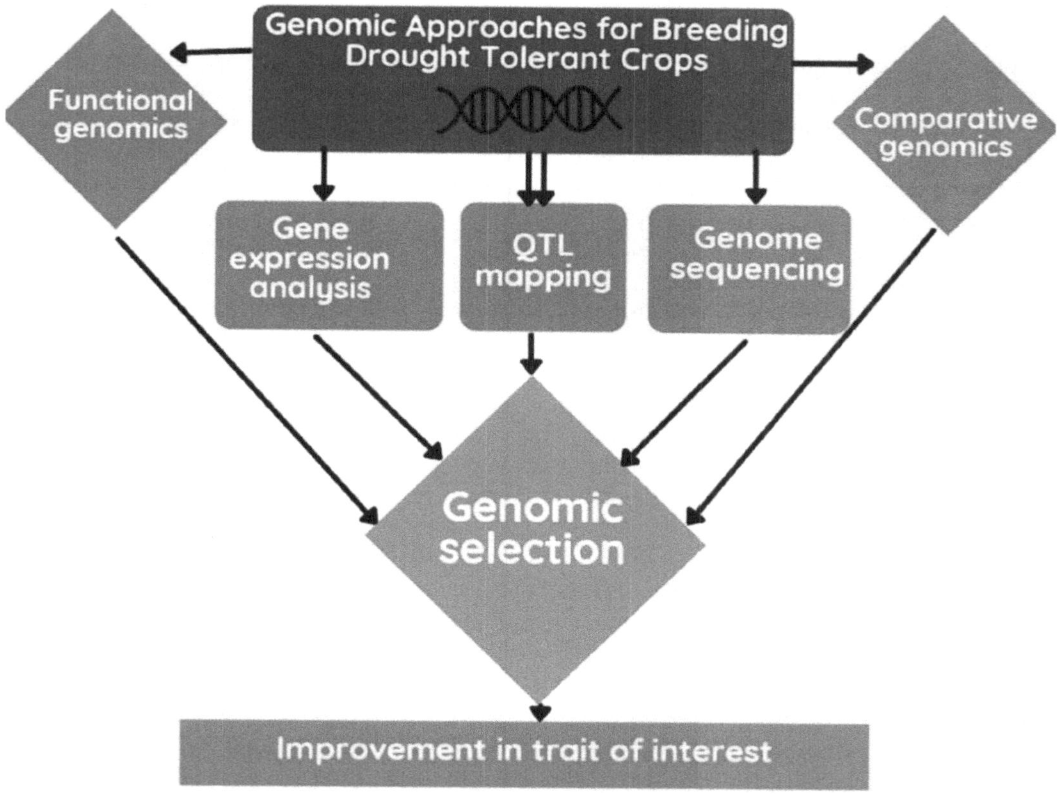

FIGURE 11.2 Genomic approaches for breeding drought-tolerant crops.

genetic makeup, providing vital insights into the genetic variations and genes related to drought resistance (Gupta et al., 2023). Researchers can find genetic markers linked to resilience by comparing the genomes of drought-tolerant and vulnerable cultivars, guiding breeding efforts.

Researchers established parallel pathways in genomics by using comparative genomics, an extension of genome sequencing. Breeders worked swiftly in the identification of potential genes for breeding programs by identifying conserved genes and areas among species with variable drought tolerance. The investigation of gene expression, particularly RNA sequencing (RNA-seq), digs into the dynamic changes in gene activity under drought stress. This method reveals the genes that govern the plant's reaction to water constraint, offering light on molecular pathways that can be altered to improve drought resistance (Nandy et al., 2023).

Quantitative trait locus (QTL) mapping refines genomic insights further by finding particular genomic areas associated with drought-related characteristics. Scientists validated gene functions by changing their expression and detecting the subsequent changes in plant behavior, thanks to functional genomics (Gojon et al., 2023). Genomic selection uses complex statistical models and genotyping to predict drought tolerance potential, allowing breeding efforts to be streamlined by finding promising individuals early in the process.

The once-difficult challenge of creating drought-tolerant crops is changing in the age of genomics. The convergence of cutting-edge technologies and the complicated genetic information contained within plants is paving the way for rapid breeding, targeted alterations, and the production of crops that can resist the difficulties of an unpredictable environment. As the globe grapples with the consequences of climate change, genomic techniques stand out as beacons of innovation, presenting hope for a more resilient and sustainable agricultural future.

11.3.1 Genome Sequencing

Genome sequencing emerged as a powerful and revolutionary ally to increase agricultural resilience in the face of climate change (Munaweera et al., 2022). This groundbreaking approach, which includes reading an organism's entire DNA sequence, has ushered in a new era of precision breeding for drought tolerance in crops. Scientists now have unprecedented access to the delicate genetic architecture that underpins a plant's ability to endure water scarcity, thanks to the introduction of high-throughput sequencing technology.

Genome sequencing, at its heart, gives a full map of a plant's genetic makeup. This molecular blueprint reveals how genes, regulatory elements, and genetic variants interact to shape a plant's features and responses. When used in drought tolerance breeding, genome sequencing allows researchers to find candidate genes that contribute to a plant's ability to withstand water-stressed conditions (van der Sanden et al., 2023). Researchers broadened their sights beyond specific species by using comparative genomics. They can identify shared genetic markers linked with resistance by comparing the genomic sequences of drought-tolerant and drought-susceptible plants. This method not only speeds up the finding of drought-related genes but also sheds light on the larger evolutionary background of these features, perhaps revealing common pathways across species (Sowmya et al., 2023).

Comprehensive gene expression study is also made possible by genome sequencing. Scientists can use techniques like RNA sequencing (RNA-seq) to determine which genes are active in response to drought stress. This dynamic view of gene activity highlights the complicated molecular reactions that allow plants to adapt and thrive in water-stressed conditions. Furthermore, genome sequencing lays the groundwork for QTL mapping. This method involves linking certain genetic areas to desirable features such as increased drought resistance. Researchers streamlined the breeding process by identifying these locations and prioritizing genes that contribute to drought resilience. Beyond identifying specific genes, genome sequencing has significant implications. It paves the way for the field of functional genomics, which manipulates genes to determine their specific functions in drought tolerance. Gene deletion or overexpression tests allow scientists to confirm the influence of particular genes on a plant's resistance to water constraints. Furthermore, the application of genomic selection is made possible by the abundance of genomic data. This method entails predicting a plant's capacity to withstand drought by linking genetic markers to characteristics and applying statistical models (Sinha et al., 2023). Breeders may choose the most promising candidates early on in the process thanks to genomic selection, which speeds up the breeding process A beacon of innovation in the face of shifting climate dynamics is genomic sequencing. It gives plant breeders a comprehensive grasp of the genetic underpinnings of drought tolerance, hastening the creation of crop varieties that can flourish in arid areas. Genome sequencing provides guidance for agriculture in navigating the uncertainties of the future and cultivating robust and sustainable crops that can protect world food security.

11.3.2 Comparative Genomics

In comparative genomics, genes that are conserved, regulatory components, and evolutionary links are found by comparing the genetic sequences of several species. Comparative genomics is a key tool in the breeding of drought-tolerant crops because it helps find genetic markers associated with resilience (Zoonomia Consortium, 2020). Breeders can use these markers to pinpoint the precise genes that have the potential to be drought-tolerant because they are frequently shared among species that have differing degrees of drought tolerance.

Comparative genomics also reveals the larger tapestry of genetic relationships that govern drought response, going beyond specific genes. Pathways that have endured evolutionarily and are shared by many species provide important insights into the fundamental mechanisms that enable plants to flourish in water-scarce situations (Zhang et al., 2021). Applying comparative genomics to species with different drought tolerance increases its power. Scientists can find genetic markers linked to resilience by examining species that have adapted to survive in arid environments. These signatures can take the form of gene networks, regulatory components, or genetic variants that all work together to aid in drought adaptation. This method aids in bridging the gap that exists between agricultural plants and model organisms. Although model plants offer an abundance of genetic data, it might be difficult to apply these discoveries to crops. By discovering genes with conserved activities across species, comparative genomics fills this gap and offers a path forward for converting basic research into useful breeding techniques.

Comparative genomics acts as a compass in the dynamic terrain of climate change, pointing breeders toward the genetic features that provide the best chance of survival in regions with limited water resources (Pourkheirandish et al., 2020). It presents a comprehensive view of the genetic fabric of drought tolerance, revealing the common tactics that have developed over millions of years across a wide range of species. Comparative genomics has revolutionized crop breeding as agricultural systems confront ever-greater problems, such as altered precipitation patterns and protracted droughts. Through the application of nature's genetic wisdom, scientists and breeders can pave the way for a future in which crops will flourish despite an increasingly unpredictable climate.

11.3.3 Gene Expression Analysis (RNA-seq)

Harvesting the genetic symphony underlying drought adaptation in crops has become possible with the advent of Gene Expression Analysis by RNA sequencing (RNA-seq) (Saidi & Hajibarat, 2020). This complex method, which is similar to listening in on a gene discussion, is the key to deciphering how plants react to water constraints and provides information that could completely change crop breeding. Fundamentally, RNA-seq provides a window into the genetic machinery of a plant. It gives an instantaneous picture of the genes that are actively generating RNA molecules, which act as bridges between the production of DNA and proteins. When used to study drought tolerance, RNA-seq revealed the genetic response of a plant to water stress and the molecular actors involved in the coordination of adaptive mechanisms (Liu et al., 2021). RNA-seq acts as a compass to help scientists navigate the genetic maze in the quest to produce drought-tolerant crops. Researchers can identify the genes that are activated or inactive in response to water shortage by comparing the gene expression patterns of plants that are prone to drought and those that are resistant. With the use of this dynamic understanding, important genetic mechanisms that help plants survive drought can be identified. RNA-seq explored the subtleties of gene regulation in addition to finding genes that are active. It reveals the complex dance of how genes are activated and inactive, frequently within complex networks that either enhance or attenuate particular reactions. With this level of knowledge, breeders can fine-tune gene expression to increase drought resistance through genetic manipulation (Mahdavi Mashaki et al., 2018).

A time dimension to genetic responses is another benefit of RNA-seq. Through examining the temporal variations in gene expression amid drought stress, researchers acquire a more profound comprehension of the plant's adaptation mechanism. The genes that respond quickly to stress and those that are later engaged to maintain resilience are both displayed in this dynamic picture. RNA-seq's observations offered concrete objectives for genetic advancement. Breeders can concentrate their breeding operations on genes that drought-tolerant plants routinely activate by identifying such genes. To learn more about these genes' precise functions in granting drought tolerance, functional investigations might be conducted to further validate them.

RNA-seq is a beacon of accuracy and possibilities in the era of climate uncertainty, as drought events become more frequent and intense. It provides a molecular toolkit for breeders to recognize, comprehend, and modify the genes underlying a plant's adaptation to water scarcity. RNA-seq provides a technique to develop crop types that can not only survive but thrive in water-limited environments, ensuring global food security as agriculture struggles to feed a growing population in the face of changing climate dynamics.

11.3.4 QTL Mapping

The method known as QTL mapping is used to pinpoint the areas of the genome linked to particular features, such as resistance to drought (Andleeb et al., 2020). Researchers identified genomic areas associated with drought-related features by examining genetic variants among populations of plants that display different levels of drought tolerance. Breeders can locate the genomic areas encoding genes responsible for drought tolerance by using these QTLs as signposts. The field of QTL mapping proves to be a valuable resource in the endeavor to grow crops that are resilient to the harsh conditions of water scarcity. This advanced method, which lies at the nexus of breeding and genetics, has the capacity to decipher the complex genetic elements that contribute to a plant's resistance to drought and pave the way for the creation of crops that can withstand it.

QTL mapping is fundamentally a genetic treasure hunt. It entails searching through plant genomes to find certain areas linked to desired characteristics, such as resistance to drought. These areas, referred to as quantitative trait loci, act as genetic markers that direct attention to the genes that bestow the desired trait. QTL mapping becomes an active tool in the breeding of drought-tolerant crops when it comes to identifying the genes that provide plants the capacity to survive and thrive in water-scarce situations. Through genetic analysis of plants with different levels of drought tolerance, scientists can find QTLs associated with the desired characteristic.

In addition, researchers located the genomic areas containing genes resistant to drought, and QTL mapping shaded light on the intricate genetic structure supporting drought resilience. Drought tolerance is one of those qualities that is impacted by several genes, each to differing degrees. This genetic complexity can be untangled by QTL mapping, which illuminates the relationships between the genes that together control how a plant reacts to water scarcity (Gupta et al., 2017). Moreover, QTL mapping fills in the gaps between basic genetic understanding and practical breeding techniques. Plant breeders can use the identified QTLs as signposts to direct their focus toward particular genomic areas that require more attention in breeding programs. Afterward, these areas can be included in breeding programs to create crops with increased resistance to drought. QTL mapping has the potential to completely change crop breeding techniques in a world where climate variability is posing an increasing challenge. It provided a road map for scientists and plant breeders to choose species that have the best chance of surviving drought. QTL mapping enabled breeders to develop novel cultivars that may flourish in water-scarce environments by leveraging the genetic diversity seen in crop populations, thereby enhancing global food security (Hafeez et al., 2023).

11.3.5 Functional Genomics

Functional genomics appeared to be a promising tool in the effort to strengthen crops against the growing threat of drought (Ashraf et al., 2018). This innovative strategy, which is based on deciphering the complex principles of gene activity, provided a revolutionary means of producing crops that are drought-tolerant and can flourish in the face of water constraints. Fundamentally, functional genomics explored the dynamic orchestration of genes in response to environmental cues such as drought stress, going beyond the static genetic blueprint. This knowledge depends on methods like gene deletion and overexpression, which let researchers change the way certain genes function and track how that changes the behavior of the plant. Functional genomics is like looking inside the gears that drive a plant's response to water shortage when it comes to the field of drought tolerance. Through targeted gene suppression or amplification, scientists can reveal the molecular mechanisms underlying a plant's adaptive mechanisms. This in-depth exploration of gene function reveals the complex networks that control osmotic balance, stress signaling, water retention, and other critical mechanisms supporting drought tolerance (Nounjan et al., 2018).

Functional genomics revealed the connectivity of genetic networks beyond the level of individual genes. Genes frequently cooperate to create intricate cascades that plan a plant's reaction to stress (Figure 11.2). Through the identification of key nodes in these networks, researchers might potentially unleash master regulators, which could improve a plant's resistance to water shortage and allow it to adapt more effectively to dry conditions. Functional genomics also helps to close the gap between research results in the lab and practical applications (Tiwari et al., 2017). It provided a means of verifying the significance of genes found using alternative genomic techniques, like comparative genomics and genome sequencing. Functional genomics offered a concrete means of converting scientific findings into useful breeding techniques by illustrating the precise influence of particular genes on a plant's ability to withstand drought. Functional genomics presented the potential of a more resilient and environmentally friendly agricultural landscape in the future, as drought episodes are more frequent and harsh. It represents the union of basic science and practical innovation, cooperating to give crops the genetic resources they require to face the difficulties posed by a changing climate.

11.3.6 Genomic Selection

Utilizing high-throughput genotyping and sophisticated statistical models, genomic selection makes use of genomic data to forecast plant trait performance. Breeders can forecast a plant's potential to withstand drought before it reaches maturity by linking particular genetic markers to desired characteristics. Enabling breeders to choose individuals with the highest chance of drought tolerance for additional breeding efforts, speeds up the breeding process. Genomic techniques are promising in the search for crops resistant to drought because they provide insights into the intricate genetic basis of plant tolerance to water constraints (Crossa et al., 2017). All of these techniques – from gene expression analysis and genomic selection to genome sequencing and comparative genomics – empower breeders to make well-informed decisions, increase breeding efficiency, and create crop varieties that can flourish in water-stressed situations (Figure 11.2). The combination of genomics and agriculture promises a sustainable future where robust crops protect food security in the face of climate change, even as the world's agricultural landscape faces hitherto unseen difficulties.

11.4 GENETIC MODIFICATION AND GENOME-EDITING APPROACHES FOR DEVELOPING DROUGHT-TOLERANT CROPS

Enhancing crop resilience against drought is becoming more and more important as the effects of climate change become more apparent on a worldwide scale. Technologies like genome editing and genetic modification (GM) have become effective techniques for engineering crops with increased drought tolerance. With the use of these techniques, scientists may accurately alter the genetic composition of the plant, opening up a world of possibilities for the conferment of traits that make crops resistant to water scarcity.

11.4.1 Overexpression or Downregulation of Drought-Responsive Genes

A crucial technique in creating crops resistant to drought is the exact alteration of genes that are essential for the plant's reaction to low water availability. This strategy aims to strengthen the plant's innate defenses against drought stress by overexpressing or downregulating genes that are responsive to the drought. This method involves inserting more copies of particular genes into the plant's genome, which raises the levels of protein expression that correspond to those genes. These proteins, which include those involved in osmotic control, water transport, antioxidant defense, and stress signaling, are frequently important actors in drought responses.

The gene encoding DREB1A, a TF that controls the expression of several stress-responsive genes, is a well-known example of overexpression (Singh & Chandra, 2021). This gene improves the plants' resistance to drought when it is included in crops like rice, wheat, or maize. By activating a group of drought-responsive genes, DREB1A helps plants retain more water, lessen oxidative stress, and be more resilient to stress in general (Singh & Chandra, 2021). On the contrary, downregulation is the process of inhibiting the expression of particular genes linked to adverse impacts of drought, like transpiration-induced water loss or ROS generation. Plants can better withstand drought conditions by conserving energy and water through the reduction of gene expression. One intriguing candidate for downregulation is the gene PIP1b, which codes for a protein involved in the transport of water across cell membranes. Reductions in PIP1b expression in some plants led to reduced transpirational water loss and enhanced water conservation (Liu et al., 2020). Under drought stress, this GM aids the plant in preserving its water balance. As a result, this strategy presents a viable route for creating crops that flourish in dry environments, eventually enhancing global food security in the face of climate change. But like with any genetic alteration approach, regulatory frameworks, ethical issues, and a thorough evaluation of safety are still crucial.

11.4.2 Exploitation of TFs

TFs regulate the expression of numerous genes involved in particular biological responses, much like directors in a genetic symphony. Using these master regulators to their full potential provides a powerful method for creating crops resistant to drought. Scientists can adjust the plant's response to drought stress and give it more resilience by modifying TFs. TFs are proteins that attach themselves to particular DNA sequences within the genome of plants, functioning as switches that regulate the activation or repression of genes downstream (Manna et al., 2021). Numerous TFs serve as key nodes in stress-signaling networks, controlling a range of processes including antioxidant defense, osmotic balance, and water retention. One well-known family of TFs that is used to improve drought resistance is the dehydration-responsive element binding (DREB) family. In particular, the DREB1/CBF subfamily regulates genes related to cold and drought responses. When these TFs are overexpressed in plants, they initiate a series of drought-responsive genes that in turn improve the plant's resistance to water scarcity by instigating mechanisms like osmotic adjustment and stomatal regulation (Singh & Chandra, 2021). Thus, the use of TFs is a novel approach to providing crops with increased resilience against drought stress. By gaining access to the complex gene regulatory networks governed by these master regulators, scientists are paving the way for a time when crops can flourish in water-scarce environments and contribute to global food security amid the challenges posed by climate change.

11.4.3 Activation of Stress-Signaling Pathways

Intricate signaling networks have evolved by plants to adapt to a variety of stressors, including drought. A viable approach to developing drought-tolerant crops is to take advantage of these stress-signaling pathways (Razi & Muneer, 2021). By initiating these pathways, researchers can endow plants with an enhanced capacity to perceive and adjust to water constraints, so augmenting their overall robustness. Within a plant, complex networks of biochemical signals known as stress-signaling pathways carry information about stress. Different chemicals known as signaling molecules are created when a plant experiences drought stress, initiating a series of processes that culminate in particular responses. Osmotic adjustment, stomatal closure, and the activation of stress-responsive genes are a few examples of these reactions. One well-known example of a stress-signaling pathway implicated in drought response is the abscisic acid (ABA) signaling pathway (Ali et al., 2020). One important hormone that builds up in response to a water shortage is ABA. It attaches itself to certain receptors on cell membranes, starting a chain of actions that shuts stomata, decreases transpirational water loss, and turns on genes related to stress tolerance. Scientists are trying to intentionally induce stress-signaling pathways to develop drought tolerance (Kuromori et al., 2018). This can be accomplished by either overexpressing the genes that produce signaling molecules or by modifying the pathway's constituent parts to increase their activity. Even in modest stress situations, engineering plants to overproduce ABA can activate the ABA signaling system, preparing the plant for a quick response to more severe drought. Through this designed reaction, the plant is able to better tolerate extended periods of water scarcity, activate genes that respond to stress, and anticipate changes in its water balance (Kuromori et al., 2018).

11.4.4 Modulation of Stomatal Behavior

One well-known example of a stress-signaling pathway implicated in drought response is the ABA signaling pathway. One important hormone that builds up in response to a water shortage is ABA (Kuromori et al., 2018). It attaches itself to certain receptors on cell membranes, starting a chain of actions that shuts stomata, decreases transpirational water loss, and turns on genes related to stress tolerance. Scientists are trying to intentionally induce stress-signaling pathways to develop drought tolerance. This can be accomplished by either overexpressing the genes that produce signaling molecules or modifying the pathway's constituent parts to increase their activity. But this closure may also restrict the intake of carbon dioxide necessary for photosynthesis. For instance, designing plants with modified sensitivity to environmental stimuli that cause stomatal closure can be used to manipulate stomatal behavior. To preserve water, plants can be genetically modified to display delayed or reduced stomatal closure in response to mild water stress (Sawicka et al., 2021). This enables the plants to continue photosynthesis for extended periods of time before shutting down. Modifying the expression of genes related to stomatal development, ion transport, and regulatory proteins is one method of engineering stomatal behavior. The timing and degree of stomatal closure during a drought may be impacted by these changes.

11.4.5 Improving Water Use Efficiency

Developing drought-tolerant crops requires improving a plant's WUE, or the amount of biomass produced per unit of water used. Through enhanced coordination of water supply and demand, genetic alterations can maximize a plant's WUE. Improved WUE can be attained, for example, by changing the root architecture to search deeper soil layers for water or by increasing the effectiveness of water transport inside the plant (Ullah et al., 2019). It's crucial to remember that, despite the enormous potential that GM and genome editing have for creating drought-tolerant crops, ethical and legal issues must be carefully considered. Concerns for society and the environment must continue to be addressed, as must the safety of the food supply. All things considered, genome editing and GM provided an amazing toolkit for giving crops the genetic characteristics they require to flourish in water-scarce environments. With the help of these strategies, agricultural sustainability may finally dawn, allowing drought-resistant crops to support world food security in the face of climate change.

11.5 CONVENTIONAL BREEDING FOR DROUGHT TOLERANCE

Conventional breeding is a fundamental technique in the field of elucidating plant responses to drought stress and devising strategies to augment tolerance. Using the inherent genetic diversity found in plant populations, conventional breeding enables breeders to identify and combine plants with desired characteristics to produce offspring that are more resistant to drought (Kumar et al., 2014). The first step in the conventional breeding process is to identify plants that show potential for tolerance to drought. Reduced water loss, effective water uptake, or modified root architecture that facilitates water extraction from deeper soil layers are a few examples of these qualities. These plants are discovered and then used as parental lines in carefully crafted crosses to produce offspring that carry the desired features. By carefully regulating pollination and selection over several generations, breeders hoped to enhance and maintain the desirable characteristics. To make sure that the progeny's performance continuously corresponds with drought tolerance, this iterative process entails assessing them under various drought situations.

Conventional breeding is in line with our knowledge of how plants react to water constraints in the wild. Conventional breeding produced crops more suited to survive in water-limited areas by looking for and cultivating features that mimic the plant's natural adaptive mechanisms (Bali et al., 2023). It worked in concert with the genetic composition of the plant, encouraging genetic diversity and shaping characteristics that correspond with the intended resistance to drought.

Nonetheless, traditional breeding encounters difficulties in tackling the intricacies of drought resilience. Conventional breeding techniques depend on the observable expression of traits, which are subject to the effect of numerous genetic factors. Many times, a trait like drought tolerance is polygenic, meaning it is regulated by several genes. It can take a lot of time and effort to identify and select these genes (Ul-Allah et al., 2023). Nevertheless, in the broader context of enhancing drought resistance, traditional breeding is still important and useful. It acts as a link between the practical objectives of agriculture and the innate genetic potential of plants. Conventional breeding is evolving as we learn more about how plants respond to drought stress. To hasten the generation of drought-tolerant cultivars, new techniques like genetic screening and marker-assisted selection are being included in the process (Dulakakharia et al., 2023). Conventional breeding continues to be a valuable and flexible strategy in the quest to understand plant responses and confer drought tolerance. This approach demonstrated the adaptability of conventional farming practices while incorporating scientific discoveries, providing a viable route forward for growing crops that can flourish in the face of water scarcity (Rasheed et al., 2023).

11.6 CHALLENGES AND FUTURE PERSPECTIVES

However, as we learn more about the difficulties caused by drought stress, it becomes clear that conventional breeding has drawbacks. Drought patterns are rapidly changing due to climate change, requiring quick solutions that typical breeding timeframes may not be able to provide. Furthermore, conventional breeding techniques that depend on a slow accumulation of genetic variants are challenged by the intricacy of drought tolerance, which is frequently regulated by several genes.

The combination of contemporary biotechnologies and conventional breeding shows promise in addressing these issues. Breeders can more effectively identify plants with desirable features by focusing on specific genetic markers associated with drought resistance through techniques such as marker-assisted selection. This hybrid strategy fills the gap between the need to address drought-related concerns urgently and the painstaking nature of conventional breeding.

11.7 CONCLUSION

Plant responses to drought stress are being uncovered, and one route to long-term solutions is the synergy between traditional breeding and state-of-the-art biotechnologies. The deeper our grasp of the genetic foundations of drought tolerance, the more equipped we are to use this knowledge to develop crops that can flourish in water-scarce conditions. Researchers, breeders, and legislators can work together strategically to overcome obstacles relating to time restrictions, genetic complexity, or ethical concerns. Combining genomics with traditional breeding provides a full toolkit to address the various aspects of drought stress. In summary, the search for crops resistant to drought is evidence of human resourcefulness and perseverance in the face of a changing environment. Through the discovery of plant responses and the application of genetic knowledge, we are at the cusp of a time when agricultural landscapes will be ornamented with drought-tolerant crops that can thrive in the face of water scarcity. Our shared endeavors hold the potential to safeguard our food systems and agricultural sustainability for future generations.

REFERENCES

Albert, K. R., Ro-Poulsen, H., Mikkelsen, T. N., Michelsen, A., Van Der Linden, L., & Beier, C. (2011). Effects of elevated CO_2, warming and drought episodes on plant carbon uptake in a temperate heath ecosystem are controlled by soil water status. *Plant, Cell & Environment*, *34*(7), 1207–1222.

Ali, N., Majeed, A., Saeed, S., Zulfiqar, N., & Ashraf, M. J. (2023). Climate change in Pakistan and its resilience efforts. *Al-Qanţara*, *9*(4).

Ali, S., Hayat, K., Iqbal, A., & Xie, L. (2020). Implications of abscisic acid in the drought stress tolerance of plants. *Agronomy*, *10*(9), 1323.

Ali, Z., Merrium, S., Habib-ur-Rahman, M., Hakeem, S., Saddique, M. A. B., & Sher, M. A. (2022). Wetting mechanism and morphological adaptation; leaf rolling enhancing atmospheric water acquisition in wheat crop – A review. *Environmental Science and Pollution Research*, *29*(21), 30967–30985.

Andleeb, T., Shah, T., Nawaz, R., Munir, I., Munsif, F., & Jalal, A. (2020). QTL mapping for drought stress tolerance in plants. In M. Hasanuzzaman & M. Tanveer (Eds.), *Salt and drought stress tolerance in plants: Signaling networks and adaptive mechanisms* (pp. 383–403). Springer.

Ashraf, J., Zuo, D., Wang, Q., Malik, W., Zhang, Y., Abid, M. A., Cheng, H., Yang, Q., & Song, G. (2018). Recent insights into cotton functional genomics: Progress and future perspectives. *Plant Biotechnology Journal*, *16*(3), 699–713.

Ashraf, M., Niaz, Y., Hafeez, M., Ahmad, M. U. D., & Shah, A. A. (2023). Water management strategies for agricultural disasters. In M. Ahmed & S. Ahmed (Eds.), *Disaster risk reduction in agriculture* (pp. 67–89). Springer.

Aslam, U., Azeem, A., Mumtaz, A., & Ul-Allah, S. (2023). Crop responses to climate change. In M. Hasanuzzaman (Ed.), *Climate-resilient agriculture, Vol 1: Crop responses and agroecological perspectives* (pp. 57–75). Springer.

Azeem, A., Abbas, N., Azeem, S., Iqbal, Z., & Ul-Allah, S. (2023). Physiological and molecular mechanism of nanoparticles induced tolerance in plants. In T. Aftab (Ed.), *Emerging contaminants and plants: Interactions, adaptations and remediation technologies* (pp. 233–248). Springer.

Bali, A., Rawal, S., & Singla, K. (2023). Challenges and strategies to improve drought tolerance in plants through agronomic managements. In A. Kumar, P. Dhansu, & A. Mann (Eds.), *Salinity and drought tolerance in plants: Physiological perspectives* (pp. 491–506). Springer.

Balliu, A., Zheng, Y., Sallaku, G., Fernández, J. A., Gruda, N. S., & Tuzel, Y. (2021). Environmental and cultivation factors affect the morphology, architecture and performance of root systems in soilless grown plants. *Horticulturae*, *7*(8), 243.

Bareke, T. (2018). Biology of seed development and germination physiology. *Advances in Plants & Agriculture Research*, *8*(4), 336–346.

Bhattacharya, A., & Bhattacharya, A. (2021). Effect of soil water deficit on growth and development of plants: A review. In A. Bhattacharya (Ed.), *Soil water deficit and physiological issues in plants* (pp. 393–488). Springer.

Coussement, J. R., Villers, S. L., Nelissen, H., Inzé, D., & Steppe, K. (2021). Turgor-time controls grass leaf elongation rate and duration under drought stress. *Plant, Cell & Environment*, *44*(5), 1361–1378.

Crossa, J., Pérez-Rodríguez, P., Cuevas, J., Montesinos-López, O., Jarquín, D., De Los Campos, G., Burgueño, J., González-Camacho, J. M., Pérez-Elizalde, S., & Beyene, Y. (2017). Genomic selection in plant breeding: Methods, models, and perspectives. *Trends in Plant Science*, *22*(11), 961–975.

Dietz, K. J., Zörb, C., & Geilfus, C. M. (2021). Drought and crop yield. *Plant Biology*, *23*(6), 881–893.

Dulakakharia, B., Longkho, K., Sharma, V., & Verma, R. K. (2023). Rice drought tolerance: Emerging molecular breeding strategies in the post-genomic era. In D. Sharma, S. Singh, S. K. Sharma, & R. Singh (Eds.), *Smart plant breeding for field crops in post-genomics era* (pp. 99–135). Springer.

Gamage, D., Thompson, M., Sutherland, M., Hirotsu, N., Makino, A., & Seneweera, S. (2018). New insights into the cellular mechanisms of plant growth at elevated atmospheric carbon dioxide concentrations. *Plant, Cell & Environment*, *41*(6), 1233–1246.

Gojon, A., Nussaume, L., Luu, D. T., Murchie, E. H., Baekelandt, A., Rodrigues Saltenis, V. L., Cohan, J. P., Desnos, T., Inzé, D., & Ferguson, J. N. (2023). Approaches and determinants to sustainably improve crop production. *Food and Energy Security*, *12*(1), e369.

Gupta, P. K., Balyan, H. S., & Gahlaut, V. (2017). QTL analysis for drought tolerance in wheat: Present status and future possibilities. *Agronomy*, *7*(1), 5.

Gupta, R., Verma, A., Yadav, H., Gupta, A. K., Srivastava, A., Singh, J., Kumar, P., & Maurya, A. K. (2023). Genome sequencing in field crops: Unlocking agricultural potential. *International Journal of Environment and Climate Change*, *13*(11), 3669–3677.

Hafeez, A., Ali, B., Javed, M. A., Saleem, A., Fatima, M., Fathi, A., Afridi, M. S., Aydin, V., Oral, M. A., & Soudy, F. A. (2023). Plant breeding for harmony between sustainable agriculture, the environment, and global food security: An era of genomics-assisted breeding. *Planta*, *258*(5), 97.

Harrison, E. L., Arce Cubas, L., Gray, J. E., & Hepworth, C. (2020). The influence of stomatal morphology and distribution on photosynthetic gas exchange. *The Plant Journal*, *101*(4), 768–779.

Kapoor, D., Bhardwaj, S., Landi, M., Sharma, A., Ramakrishnan, M., & Sharma, A. (2020). The impact of drought in plant metabolism: How to exploit tolerance mechanisms to increase crop production. *Applied Sciences*, *10*(16), 5692.

Kido, É. A., Ferreira-Neto, J. R. C., da Silva, M. D., Santos, V. E. P., da Silva Filho, J. L. B., & Benko-Iseppon, A. M. (2019). Osmoprotectant-related genes in plants under abiotic stress: Expression dynamics, in silico genome mapping, and biotechnology. In M. A. Hossain, V. Kumar, D. J. Burritt, M. Fujita, & P. S. A. Makela (Eds.), *Osmoprotectant-mediated abiotic stress tolerance in plants: Recent advances and future perspectives* (pp. 1–40). Springer.

Kumar, A., Dixit, S., Ram, T., Yadaw, R., Mishra, K., & Mandal, N. (2014). Breeding high-yielding drought-tolerant rice: Genetic variations and conventional and molecular approaches. *Journal of Experimental Botany*, *65*(21), 6265–6278.

Kuromori, T., Seo, M., & Shinozaki, K. (2018). ABA transport and plant water stress responses. *Trends in Plant Science*, *23*(6), 513–522.

Kushwah, N., Jain, V., & Yadav, D. (2020). Osmolytes: A possible therapeutic molecule for ameliorating the neurodegeneration caused by protein misfolding and aggregation. *Biomolecules*, *10*(1), 132.

Lambers, H., Oliveira, R. S., Lambers, H., & Oliveira, R. S. (2019). Plant water relations. In H. Lambers & R. S. Oliveira (Eds.), *Plant physiological ecology* (pp. 187–263). Springer.

Langensiepen, M., Jansen, M. A., Wingler, A., Demmig-Adams, B., Adams III, W. W., Dodd, I. C., Fotopoulos, V., Snowdon, R., Fenollosa, E., & De Tullio, M. C. (2020). Linking integrative plant physiology with agronomy to sustain future plant production. *Environmental and Experimental Botany*, *178*, 104125.

Leisner, C. P., Potnis, N., & Sanz-Saez, A. (2023). Crosstalk and trade-offs: Plant responses to climate change-associated abiotic and biotic stresses. *Plant, Cell & Environment*, *46*(10), 2946–2963.

Liu, S., Fukumoto, T., Gena, P., Feng, P., Sun, Q., Li, Q., Matsumoto, T., Kaneko, T., Zhang, H., & Zhang, Y. (2020). Ectopic expression of a rice plasma membrane intrinsic protein (OsPIP1; 3) promotes plant growth and water uptake. *The Plant Journal*, *102*(4), 779–796.

Liu, S., Zenda, T., Dong, A., Yang, Y., Wang, N., & Duan, H. (2021). Global transcriptome and weighted gene co-expression network analyses of growth-stage-specific drought stress responses in maize. *Frontiers in Genetics*, *12*, 645443.

Mahdavi Mashaki, K., Garg, V., Nasrollahnezhad Ghomi, A. A., Kudapa, H., Chitikineni, A., Zaynali Nezhad, K., Yamchi, A., Soltanloo, H., Varshney, R. K., & Thudi, M. (2018). RNA-Seq analysis revealed genes associated with drought stress response in kabuli chickpea (Cicer arietinum L.). *PLoS One*, *13*(6), e0199774.

Manna, M., Thakur, T., Chirom, O., Mandlik, R., Deshmukh, R., & Salvi, P. (2021). Transcription factors as key molecular target to strengthen the drought stress tolerance in plants. *Physiologia Plantarum*, *172*(2), 847–868.

Munaweera, T., Jayawardana, N., Rajaratnam, R., & Dissanayake, N. (2022). Modern plant biotechnology as a strategy in addressing climate change and attaining food security. *Agriculture & Food Security*, *11*(1), 1–28.

Murali-Baskaran, R. K., Mooventhan, P., Das, D., Dixit, A., Sharma, K. C., Senthil-Nathan, S., Kaushal, P., & Ghosh, P. K. (2022). The future of plant volatile organic compounds (pVOCs) research: Advances and applications for sustainable agriculture. *Environmental and Experimental Botany*, *200*, 104912.

Nandy, S., Mandal, S., Gupta, S. K., Anand, U., Ghorai, M., Mundhra, A., Rahman, M. H., Ray, P., Mitra, S., & Ray, D. (2023). Role of polyamines in molecular regulation and cross-talks against drought tolerance in plants. *Journal of Plant Growth Regulation*, *42*(8), 4901–4917.

Nounjan, N., Chansongkrow, P., Charoensawan, V., Siangliw, J. L., Toojinda, T., Chadchawan, S., & Theerakulpisut, P. (2018). High performance of photosynthesis and osmotic adjustment are associated with salt tolerance ability in rice carrying drought tolerance QTL: Physiological and co-expression network analysis. *Frontiers in Plant Science*, *9*, 1135.

Nuccio, M. L., Paul, M., Bate, N. J., Cohn, J., & Cutler, S. R. (2018). Where are the drought tolerant crops? An assessment of more than two decades of plant biotechnology effort in crop improvement. *Plant Science*, *273*, 110–119.

Pandey, K., Kumar, R. S., Prasad, P., Pande, V., Trivedi, P. K., & Shirke, P. A. (2022). Coordinated regulation of photosynthesis and sugar metabolism in guar increases tolerance to drought. *Environmental and Experimental Botany*, *194*, 104701.

Pourkheirandish, M., Golicz, A. A., Bhalla, P. L., & Singh, M. B. (2020). Global role of crop genomics in the face of climate change. *Frontiers in Plant Science*, *11*, 922.

Qi, J., Song, C. P., Wang, B., Zhou, J., Kangasjärvi, J., Zhu, J. K., & Gong, Z. (2018). Reactive oxygen species signaling and stomatal movement in plant responses to drought stress and pathogen attack. *Journal of Integrative Plant Biology*, *60*(9), 805–826.

Raihan, A. (2023). A review of the global climate change impacts, adaptation strategies, and mitigation options in the socio-economic and environmental sectors. *Journal of Environmental Science and Economics*, *2*(3), 36–58.

Rasheed , A., Zhao, L., Raza, A., Mahmood, A., Xing, H., Lv, X., Saeed, H., Alqahtani, F. M., Hashem, M., & Hassan, M. U. (2023). Role of molecular breeding tools in enhancing the breeding of drought-resilient cotton genotypes: An updated review. *Water, 15*(7), 1377.

Razi, K., & Muneer, S. (2021). Drought stress-induced physiological mechanisms, signaling pathways and molecular response of chloroplasts in common vegetable crops. *Critical Reviews in Biotechnology, 41*(5), 669–691.

Razzaq, A., Saleem, F., Kanwal, M., Mustafa, G., Yousaf, S., Imran Arshad, H. M., Hameed, M. K., Khan, M. S., & Joyia, F. A. (2019). Modern trends in plant genome editing: An inclusive review of the CRISPR/Cas9 toolbox. *International Journal of Molecular Sciences, 20*(16), 4045.

Sachdev, S., Ansari, S. A., Ansari, M. I., Fujita, M., & Hasanuzzaman, M. (2021). Abiotic stress and reactive oxygen species: Generation, signaling, and defense mechanisms. *Antioxidants, 10*(2), 277.

Saharan, B. S., Brar, B., Duhan, J. S., Kumar, R., Marwaha, S., Rajput, V. D., & Minkina, T. (2022). Molecular and physiological mechanisms to mitigate abiotic stress conditions in plants. *Life, 12*(10), 1634.

Saidi, A., & Hajibarat, Z. (2020). Application of next generation sequencing, GWAS, RNA seq, WGRS, for genetic improvement of potato (Solanum tuberosum L.) under drought stress. *Biocatalysis and Agricultural Biotechnology, 29*, 101801.

Sawicka, B., Skiba, D., Umachandran, K., & Dickson, A. (2021). Alternative and new plants. In C. Egbuna, A. P. Mishra, & M. R. Goyal (Eds.), *Preparation of phytopharmaceuticals for the management of disorders* (pp. 491–537). Elsevier.

Sharma, Y., & Bhateria, R. (2022). Alleviating abiotic stress in plants using nanoparticles. In A. Vaishnav, S. S. Arya, & D. K. Choudhary (Eds.), *Plant stress mitigators: Action and application* (pp. 541–558). Springer.

Singh, K., & Chandra, A. (2021). DREBs-potential transcription factors involve in combating abiotic stress tolerance in plants. *Biologia, 76*(10), 3043–3055.

Sinha, D., Maurya, A. K., Abdi, G., Majeed, M., Agarwal, R., Mukherjee, R., Ganguly, S., Aziz, R., Bhatia, M., & Majgaonkar, A. (2023). Integrated genomic selection for accelerating breeding programs of climate-smart cereals. *Genes, 14*(7), 1484.

Sowmya, M., Nadaf, H., Naidu, G., Chimmad, V., Mirajkar, K. K., Shirasawa, K., Gangurde, S., & Bhat, R. (2023). Identification of genomic differences and the candidate genes for drought tolerance in peanut. *Euphytica, 219*(10), 1–10.

Tiku, A. R. (2020). Antimicrobial compounds (phytoanticipins and phytoalexins) and their role in plant defense. In J.-M. Merillon & K. G. Ramawat (Eds.), *Co-evolution of secondary metabolites* (pp. 845–868). Springer.

Tiwari, S., Lata, C., Singh Chauhan, P., Prasad, V., & Prasad, M. (2017). A functional genomic perspective on drought signalling and its crosstalk with phytohormone-mediated signalling pathways in plants. *Current Genomics, 18*(6), 469–482.

Tsubone, T. M., Baptista, M. S., & Itri, R. (2019). Understanding membrane remodelling initiated by photosensitized lipid oxidation. *Biophysical Chemistry, 254*, 106263.

Ul-Allah, S., Azeem, A., Sher, A., Ijaz, M., Sattar, A., Naeem, M., & Ullah, I. (2023). Plant breeding approaches in developing stress tolerance. In A. Roychoudhury (Ed.), *Biology and biotechnology of environmental stress tolerance in plants* (pp. 167–186). United States: Apple Academic Press.

Ullah, H., Santiago-Arenas, R., Ferdous, Z., Attia, A., & Datta, A. (2019). Improving water use efficiency, nitrogen use efficiency, and radiation use efficiency in field crops under drought stress: A review. *Advances in Agronomy, 156*, 109–157.

van der Sanden, B. P., Schobers, G., Corominas Galbany, J., Koolen, D. A., Sinnema, M., van Reeuwijk, J., Stumpel, C. T., Kleefstra, T., de Vries, B. B., & Ruiterkamp-Versteeg, M. (2023). The performance of genome sequencing as a first-tier test for neurodevelopmental disorders. *European Journal of Human Genetics, 31*(1), 81–88.

Wu, J., Wang, J., Hui, W., Zhao, F., Wang, P., Su, C., & Gong, W. (2022). Physiology of plant responses to water stress and related genes: A review. *Forests, 13*(2), 324.

Yadav, S., Modi, P., Dave, A., Vijapura, A., Patel, D., & Patel, M. (2020). Effect of abiotic stress on crops. *Sustainable Crop Production, 3*(17), 5–16.

Zhang, T., Xiao, J., Zhao, Y., Zhang, Y., Jie, Y., Shen, D., Yue, C., Huang, J., Hua, Y., & Zhou, T. (2021). Comparative physiological and transcriptomic analyses reveal ascorbate and glutathione coregulation of cadmium toxicity resistance in wheat genotypes. *BMC Plant Biology, 21*, 1–15.

Zoonomia Consortium. (2020). A comparative genomics multitool for scientific discovery and conservation. *Nature, 587*(7833), 240–245.

12 Harnessing CRISPR/Cas for Enhanced Drought Tolerance and Sustainable Crop Management

Rajaswini Rupali, Nandakumar Vidya, Krishnagowdu Saravanan, Muthukrishnan Anand, Arumugam Vijaya Anand, and Muthukrishnan Arun

12.1 INTRODUCTION

Climate change creates a considerable challenge to international agriculture and food security. Worldwide crop yields, rural communities, and agricultural livelihoods are seriously threatened by the rising frequency of severe weather events like drought (Alotaibi et al., 2021). This condition highlights the critical need for creative approaches to improve crop resilience in changing climate circumstances. Drought stress, soil erosion, and water scarcity are some major problems made worse by climate change, resulting from an interaction between natural and human forces. It also impacts temperature and precipitation patterns (Li et al., 2022). To mitigate the negative consequences on agricultural productivity (Tsarouchi and Buytaert, 2018) and food safety, it is important to recognize their causes and implications.

Drought stress is characterized by low soil moisture content, which inhibits physiological processes in plants and reduces photosynthesis, nutrient uptake, and water consumption efficiency. Drought stress is caused due to climate alteration in several ways, such as altered precipitation patterns, increased evapotranspiration from rising temperatures, melting of glaciers, altered atmospheric circulation patterns, rise in the frequency of severe weather events, changes in land use such as deforestation, and changes in the ocean (Tsarouchi and Buytaert, 2018). Together, these variables cause regional climates to change, water availability to decrease, and drought conditions to worsen, creating severe problems for water resources and agriculture around the globe (Khan et al., 2018). Drought stress influences crop growth, production, and development. Long-lasting drought stress endangers food security and livelihoods in places vulnerable to drought (Hura et al., 2022). Every year, millions of hectares of land are affected by drought, which results in substantial crop losses, water shortages, and social disturbance. Asia, Australia, North and South America, and Africa are the most affected areas by drought stress, representing the urgent need for adaptive solutions to mitigate its consequences (Deng et al., 2022). Drought stress affects many crops and causes disruptions to physiological processes, which are necessary for the growth and development of several crops. Some specific crops vulnerable to drought are maize, wheat, rice, and sorghum, which can have reduced grain filling, early flowering, and restricted plant growth, leading to yield reduction (Shanker et al., 2014).

Many approaches have been designed to mitigate various negative impacts of drought stress on crops. Traditional methods in agronomy include conservation tillage, mulching, and irrigation scheduling (Kim et al., 2021). The breeding technique is the important key to developing different crop varieties that can resist drought and water scarcity (Raza et al., 2023).

DOI: 10.1201/9781003457237-12

The traditional breeding method involves identifying plants with desirable traits, cross-breeding them, and repeating the selection process in subsequent generations. Genomics and marker-assisted selection (MAS) are two genetic engineering techniques that can be exploited to create crops that are more drought-resistant (https://washingtonindependent.com/w/plant-development). Through MAS and other molecular breeding techniques, it is possible to expedite the breeding process by identifying and incorporating drought-resistant genes into high-quality crop varieties (Zia et al., 2021). Genetic engineering is a highly effective technology to enhance crop drought tolerance by introducing new genes or modifying existing genes to promote resilience to water constraints.

Gene editing allows for precise genome modification of crops to integrate drought tolerance traits (Zia et al., 2021). CRISPR/Cas9 technology has revolutionized genetic engineering with its correct accuracy and efficiency in genome editing. This versatile tool makes it possible to effectively and precisely modify the DNA sequence of organisms, including crops. The CRISPR/Cas9 technology shows much potential for accelerating the development of crop varieties resistant to drought (Xu and Zhang, 2009) and enhancing sustainable farming methods in the face of climate volatility (Kumar et al., 2023). For addressing drought stress and advancing sustainable crop management, employing CRISPR/Cas techniques is a favorable approach. Sustaining global food production and food security in the context of climate change can be addressed through this technique (Sami et al., 2021). Crop drought tolerance can be increased using CRISPR/Cas technology, which accurately identifies and alters genes connected to stress response pathways. Researchers can increase crops' resistance to water scarcity through targeted genome editing by adding beneficial genetic variations or deleting genes that indicate sensitivity to drought stress. Resilient crop varieties that can provide global food security in water-scarce environments may be developed by applying CRISPR/Cas to control drought stress (Kumar et al., 2023; Zia et al., 2021). This chapter delves into the principles, mechanism applications, and advancements of CRISPR/Cas technology in plant biotechnology, focusing on enhancing drought resistance and sustainable agricultural practices.

12.2 DROUGHT STRESS IN AGRICULTURE

Drought stress is an ongoing threat to agricultural production since it impacts yield, growth, and development of crops and sustainability. Drought can damage agricultural systems since they rely significantly on consistent water sources to grow crops. One of the signs of drought stress, which affects plants physiologically and reduces their ability to function at their peak, is a drying out of the soil (Hura et al., 2022). Reduction in water availability causes photosynthesis, nitrogen uptake, and metabolic processes to be disrupted, leading to slower growth, lower yield, and low-quality crops (Khan et al., 2018). Mitigating the effects of drought is a formidable task for the agricultural sector, made worse by changes in patterns of precipitation, brought about by climate change (Table 12.1) (https://www.huffpost.com/entry/drowning-nations-sea-level-rise_n_1783931).

12.3 IMPACT OF DROUGHT STRESS ON CROP PRODUCTION

Drought stress reduces crop productivity significantly, endangering food security by lowering yields. Lack of water disrupts physiological processes essential for crop growth and development, leading to decreased photosynthetic activity, reduced biomass building, and poorer nutrient uptake (Figure 12.1). The physiological changes causing the leaves' wilting, rolling, and senescence are observable symptoms of drought stress and reduced strength in plants (Hura et al., 2022). Plants that are stressed by drought also exhibit some other physiological responses such as altered hormone levels (Sami et al., 2021), reduced leaf development (Sami et al., 2021), and stomatal closure (Li et al., 2022). These adaptive mechanisms maintain cellular homeostasis and conserve water during scarcity (You and Chan, 2015). Drought-related agriculture productivity losses vary depending on crop type and degree of water stress. Main crops, such as maize, rice, and wheat (Melandri et al.,

TABLE 12.1
Application of CRISPR/Cas9 Genome Editing in Various Crops in Response to Drought Stress

Crop	Target Phenotypic Trait	Gene Responsible for That Trait	Knock-In or Knockdown of Gene	Output/Changes in Phenotypic Trait	References
Rice	Yield	*DEP1*	Knockdown of DEP1	Increased yield in rice by promoting branching and tillering	Huang et al. (2022)
Arabidopsis thalliana	Flowering	FT gene	Upregulation of FT gene, by fusing TA VP64 with dcas9, targeted FT gene	Early flowering	Lee et al. (2019a)
Cotton	Root development, growth, and development of whole crop	*ARG* gene	Knockout	Increase in the number of lateral roots and root surface area, enhance growth and development	Zhang et al. (2021); Wang et al. (2017)
Rice	Endosperm cellularization and seed development	*MADS78* *MADS79*	Knockout	Developed endosperm cellularization and early seed development	Paul et al. (2020); Zhang et al. (2021)
Rice	Reproductivity	*Hxk5*	Knockout	Male sterility	Lee et al. (2020)
Arabidopsis	Male sterility, pollen development	*COPII*	Knockout	Unable to produce fertile pollen and caused male sterility	Liang et al. (2020)
Brassica napus	Developmental delay phenotype	*Bnspl3*	Bnspl3 mutant generated by CRISPR/Cas9	Rapid generation of simultaneous mutagenesis of multiple gene homologs in *B. napus*	Li et al. (2018a)
B. napus	Flowering timing	*Tfl1*	Knockout	Early flowering but reduced plant height and branching	Sriboon et al. (2020)
Rice	Growth and yield	*gn1a*, *dep1*, or *gs3* gene	Knockout	Increased grain size, enhanced grain weight, and greater grain quantity	Li et al. (2016)
Rice	Grain weight and size	*gw2*, *gw5*, and *tgw 6*	Knockout	Increased grain mass and dimensions	Xu et al. (2016)
Rice	Seed weight and size	*Vin2*	Knockout	Diminished seed size and grain weight	Lee et al. (2019a)
Wheat	Grain weight	*GASR7*	*GASR7* mutant generated by CRISPR/cas9	Improved weight of the grain	Zhang et al. (2016)
Maize	Grain yield	*AGROS8*	Overexpression	Enhanced tolerance to ethylene, increased crop productivity	Shi et al. (2017)
Rice	Growth of agronomic trait	*gs3*, *dep1*	Knockout	Increased grain size and no per panicle	Li et al. (2016)
B. napus	Starch accumulation	*Pat2/5*	Knockout	Increased accumulation of starch in mature seed	Zhang et al. (2019b)
Rice	Starch content	*SBEIIB*	Knockout	High content of amylose	Sun et al. (2017)
Sweet potato	Starch content	*SBEII*	Knockout	Increased amylose content	Wang et al. (2019)
Potato	Starch quality	*GBSS*	Knockout	Altered starch quality	Andersson et al. (2017)
Barley	Size of starch granules and the content of prolamines and glutenins	*d-hordein*	Knockout	The starch granules have become smaller, the prolamines have decreased, and the glutenins have increased	Yang et al. (2020)

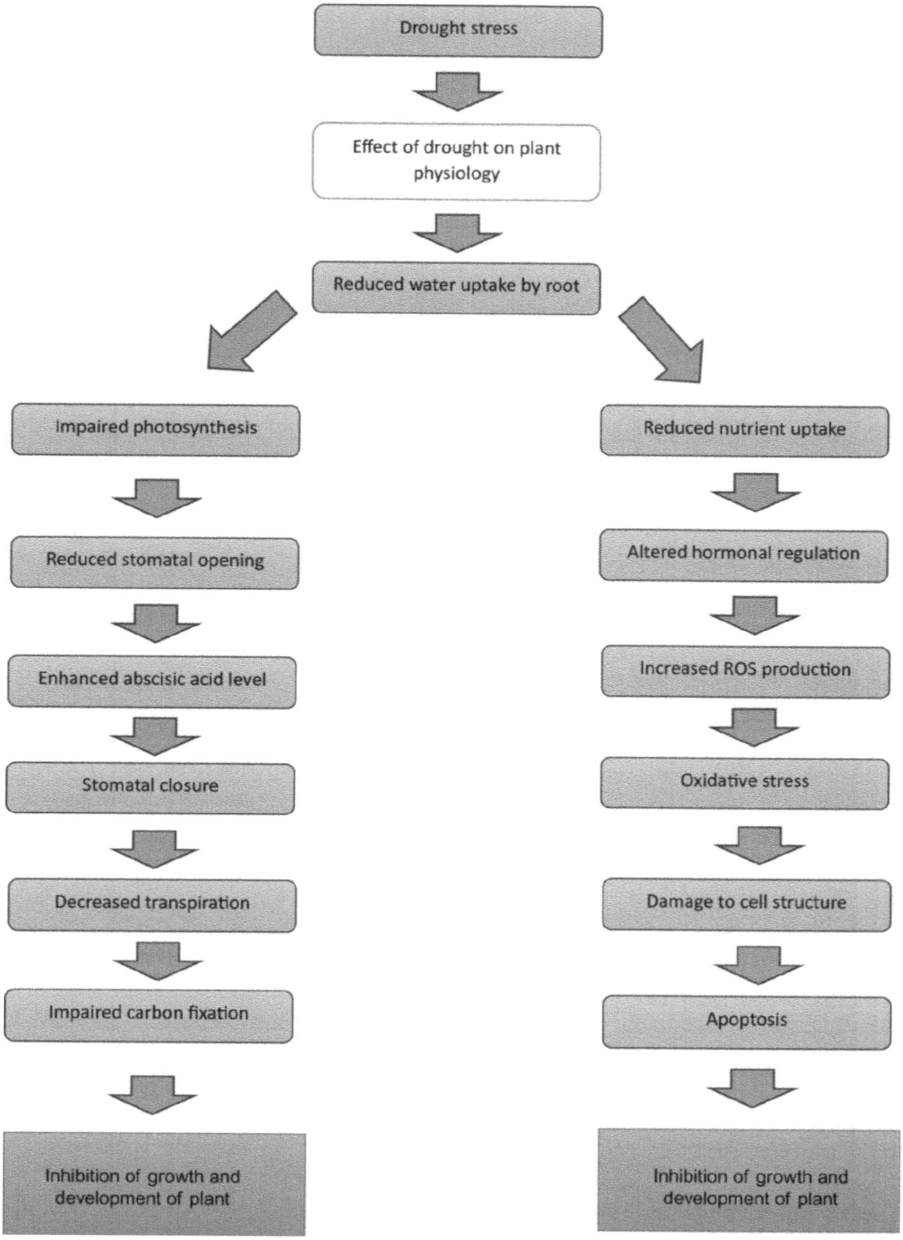

FIGURE 12.1 Suppression of plant growth and development by drought stress.

2020; Sami et al., 2021), are particularly vulnerable to drought, with yield losses ranging from 10% to 50% or more during extreme drought conditions. Several factors influence crop sensitivity to drought stress, including genetic diversity (Khan et al., 2018), root structure (Li et al., 2022), and water utilization efficiency (Khan et al., 2018). Drought-sensitive crops may experience permanent damage to reproductive structures (Yu et al., 2019), leading to poor seed and diminished grain filling, in contrast to drought-tolerant crops, which are resilient to water constraints due to adaptive traits such as deep root systems and osmotic adjustment (Dietz et al., 2021).

12.4 MOLECULAR PATHWAYS UNDERLYING PLANT DROUGHT TOLERANCE

During drought stress, the sustainability of crop production is a major issue in many countries. Drought stress intensity on plants varies both temporally and spatially. However, with a broad spectrum of physiological and morphological defense mechanisms against environmental stressors, plants have experienced extraordinary evolutionary advances. They exhibit a variety of adaptive, preventative, and evasive tactics, particularly in their reaction to the drought (Ray et al., 2018; Sourour et al., 2017).

To deal with the stress of drought, plants utilize a range of physiological strategies, such as antioxidant defense systems, root structure modification, osmotic adjustment, and stomatal closure (Biju et al., 2023). They activate signaling pathways, including abscisic acid (*ABA*) signaling, control closure of stomatal, and the expression of the stress-responsive gene (Yoshida et al., 2014). Similarly, some hormonal signaling pathways in plants, like ethylene and *ABA* interact to modulate stress tolerance followed by growth and development (Waadt et al., 2022). In response to drought stress, various transcription factors such as *MYB*, *NAC*, and *bZIP* (Golldack et al., 2014; Sharma et al., 2018) influence osmotic adjustment, antioxidant defense, and stomatal control (Baldoni et al., 2015; Wang et al., 2021). Plants make osmotic modifications in times of water constraint to maintain cellular turgor pressure (Turner, 2018). Compatible solutes like proline and sugars accumulate in this process to preserve cell integrity and stabilize cellular structures (Augé et al., 1986). Moreover, stomata have microscopic openings on plant surfaces close to store water in reaction to drought stress. So photosynthesis requires less carbon dioxide, yet less excessive water loss exists (Daems et al., 2022). In reaction to drought stress, plants change the structure of their roots. Improving water absorption from soil may require expanded root branching, deeper root penetration, or modifications to root exudation patterns (Valarmathi et al., 2023). Furthermore, lipid peroxidation caused by drought might result in membrane damage (Golldack et al., 2014). To preserve membrane stability, plants react by producing defense compounds such as HSP/heat shock proteins (Singh et al., 2016) and osmoprotectants (Ozturk et al., 2021). Photosynthetic activity is frequently decreased during drought stress. Certain plants have developed adaptive mechanisms, such as *CAM* (crassulacean-acid metabolism), to maximize photosynthesis in the face of low water availability (Wai et al., 2019).

The transcription factor of AP2/ERF orchestrates plant responses to osmotic stress, antioxidant defense, and stomatal control through gene regulation. Under osmotic stress, *AP2/ERF* factors activate genes involved in osmotic adjustment by promoting the synthesis of compatible solutes including sugar and proline. They also enhance antioxidant defense mechanisms by upregulating genes encoding enzymes like SOD and CAT, mitigating oxidative damage. Furthermore, these factors regulate stomatal closure by modulating the expression of ion channels and transporters, reducing water loss, and maintaining cellular hydration (Golldack et al., 2014); *MYB* transcription factor is crucial for plants to adapt to osmotic stress, regulate antioxidant defense, and control stomata movement. In response to osmotic stress, the expression of some genes (Gowrishankar, 1988) concerned with the synthesis of osmolytes like flavonoids and phenylpropanoids are regulated by the *MYB* factor, aiding in osmotic adjustment. They also modulate the expression of antioxidant genes, bolstering the plant's defense against oxidative stress. Additionally, *MYB* factors regulate stomatal aperture by fine-tuning the gene expression implicated in stomatal movement (Negi et al., 2013), thereby optimizing water use efficiency under drought stress (Sharma et al., 2018).

ABA is vital in regulating a plant's reaction to drought because it limits water loss through transpiration, controls stomatal closure, and regulates gene expression (Osakabe et al., 2014). ABRE-binding proteins, also known as *bZIP* or basic leucine zipper transcription factors, are essential for *ABA* signaling (Nakashima et al., 2014). Overexpressing *AREB1* (*ABF*) in rice, *Arabidopsis*, and soybean increased their resistance to drought while reducing the function of *AREB1* made them more susceptible to drought (Sami et al., 2021). *ABF*, also known as *AREB*

TF, regulates the expression of several genes which can respond to drought. When plants are under drought stress, the hormone ABA controls the water content and facilitates its release more effectively (Yoshida et al., 2010). Under drought stress, various genes downstream of the *ABA* signaling pathway are regulated by *AREB1* (Yoshida et al., 2010; Sami et al., 2021). The ABA response element in the promoter region of target genes is bound by *the ABF* transcription factor, which then stimulates the production of genes such as dehydrins and osmoprotectants. Proteins that aid the plant in surviving drought stress are produced due to this activation. Dehydrins, osmoprotectants, and numerous proteins from late embryogenesis are some of these proteins (Joshi et al., 2016).

The protein that binds to the dehydration-responsive element (*DREB*) is another significant player in the molecular mechanism that underlies plant resistance to drought. Plants that are stressed by drought have higher *DREB* transcription factor levels. Target genes' promoter regions contain a dehydration-responsive element to which, *DREB* binds to activate the target gene's expression. This activation triggers the synthesis of essential proteins for improving the plant's tolerance to drought (Lata and Prasad, 2011). The aquaporin gene is one of the significant genes that the *DREB* transcription factor activates. In drought stress, these proteins regulate water flow into and out of plant cells, enabling the plant to maintain its water balance (Lata and Prasad, 2011). *NAC* is another transcription factor that controls the molecular pathway that underlies plant drought tolerance. The NAC protein plays a vital role in regulating plant responses to drought (Lata and Prasad, 2011).

Under drought stress, various molecular markers also play an essential part in mitigating drought stress. Numerous crops possess molecular markers like QTL that give drought tolerance (Khan et al., 2018). QTL identification is still in need of improvement in terms of accuracy and reliability. Crops have been made more resistant to abiotic stress through genetic engineering (Shinwari et al., 2020; Sami et al., 2021).

12.5 TECHNIQUES USED TO MITIGATE DROUGHT

Crop resilience to drought stress must be increased using creative methods that strengthen plant resistance to water constraints. Traditional breeding methods and genetic engineering provide complementary techniques to create drought-resistant crops. Conventional breeding involves selecting different genotypes for characteristics related to drought tolerance and combining them through carefully regulated crossings to create cultivars that are resistant to drought (Raza et al., 2023; Kumar et al., 2023). MAS is a breeding technique that commonly uses various molecular markers for speed selection (Ahmed et al., 2016) of genotypes with desired features, such as drought resistance. Genomic selection forecasts drought tolerance by statistical models and genome-wide molecular markers to maximize breeding efficiency (Raza et al., 2023; Kumar et al., 2023). Conventional breeding has helped produce crop types with improved traits, but it could be more effective in addressing the complex issues that drought stress poses. Limitations of traditional breeding methods include their dependence on environmental conditions, long breeding cycles, and limited genetic diversity. Understanding those limitations is critical to identifying opportunities to influence creative strategies such as genetic engineering to improve traditional breeding and developing varieties of drought-tolerant crops for sustainable agriculture.

12.6 GENOME EDITING

Genome-editing techniques like CRISPR/Cas systems have revolutionized plant biotechnology by enabling precise modifications to the genetic composition of agricultural plants. Scientists can quickly create crop types more resistant to environmental stressors and water scarcity by precisely targeting genes relevant to drought response pathways through CRISPR/Cas-mediated genome editing.

12.6.1 Techniques of Genome Editing

Base editing is a precise technique of genome editing, which allows to conversion of one DNA base pair to another without driving double-strand breaks (DSBs). Another gene-editing technique i.e. transcription activator-like effector nucleases/TALEN are programmable nucleases that generate DSB at a specific sequence of DNA to facilitate targeted gene editing. Zinc finger nucleases/ZFN are modified DNA-binding proteins that fused with FokI endonuclease, enabling site-specific genome editing via DSB induction (Zhang et al., 2019a). Genome-editing methods hold the possibility to enhance crops despite their inherent drawbacks. We discuss issues with base editing, zinc finger nucleases, TALENs, and other methods. Understanding these restrictions is crucial to removing barriers and maximizing the potential of CRISPR/Cas genome editing for enhancing drought tolerance and sustainable farming methods.

12.6.2 CRISPR/Cas Technology

CRISPR/Cas techniques have revolutionized genome editing by enabling unprecedented precision and versatility in genetic material change. Initially, CRISPR/Cas9 was recognized as a 'bacterial defense system' against viral infections (Figure 12.2). CRISPR/Cas systems work by viral DNA fragments or spacers into the bacterial genome. When the virus re-enters, these spacers are a memory of previous infections, directing Cas nucleases to the appropriate location to cleave the viral DNA (Jiang and Doudna, 2017).

Based on their molecular mechanisms and protein constituents, there are different types and subtypes of CRISPR/Cas systems, which can be categorized into two classes, namely class 1 and class 2. Every CRISPR/Cas system demonstrates distinct features and functional qualities (Makarova et al., 2011). The CRISPR array, guide RNAs (gRNAs), and Cas proteins are the functional elements of the CRISPR system. Single-guide RNAs direct Cas nuclease to precisely cleave particular target DNA sequences complementary to the gRNA (Figure 12.3), (Horodecka and Düchler, 2021). Base pairing between guide RNA and target DNA produces an RNA–DNA duplex, which

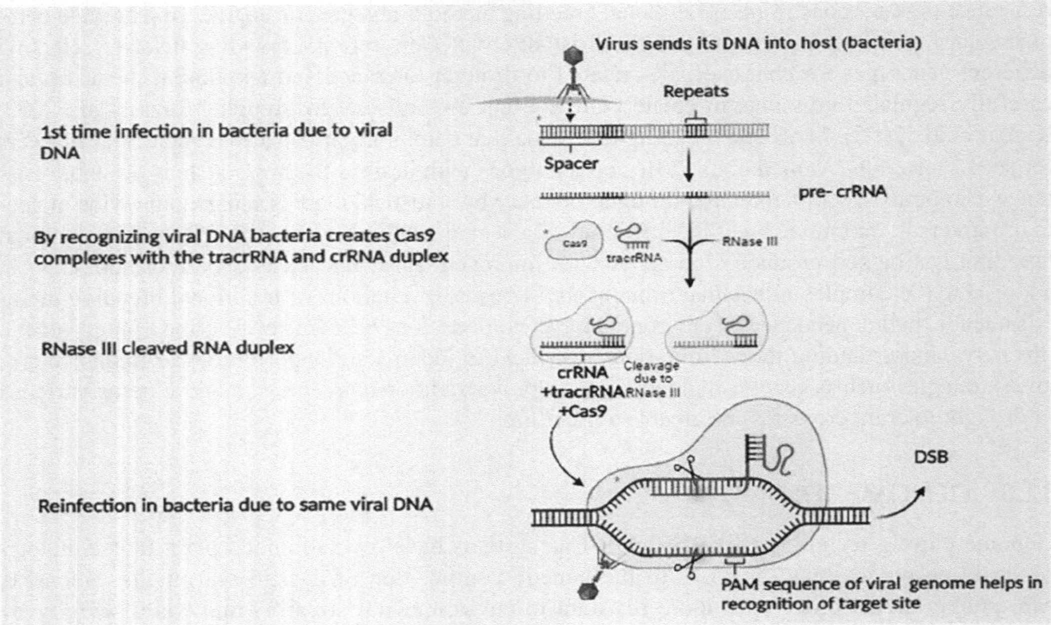

FIGURE 12.2 Bacterial defense mechanism against virus.

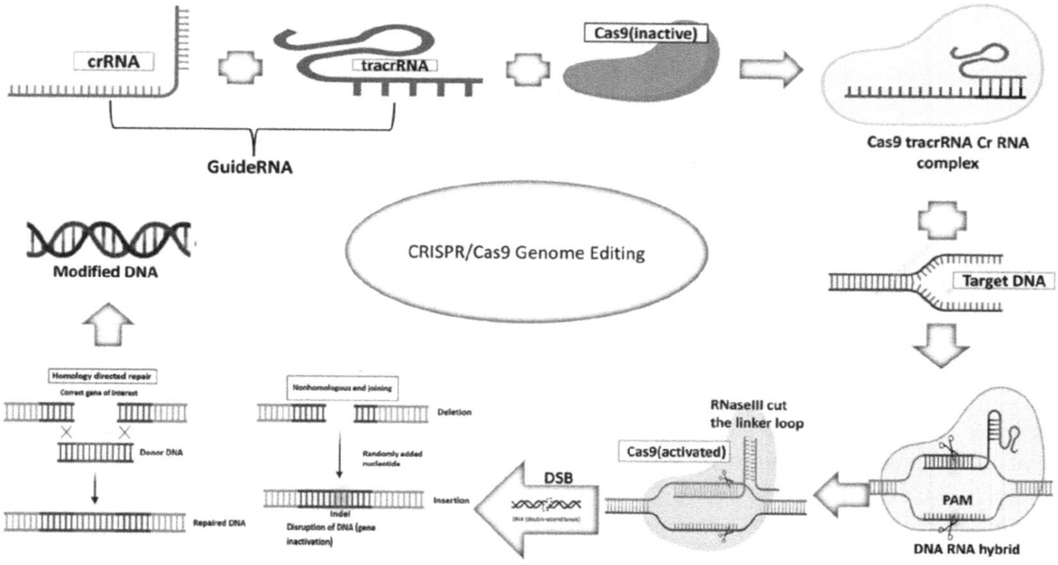

FIGURE 12.3 Mechanism of CRISPR/Cas9 genome editing.

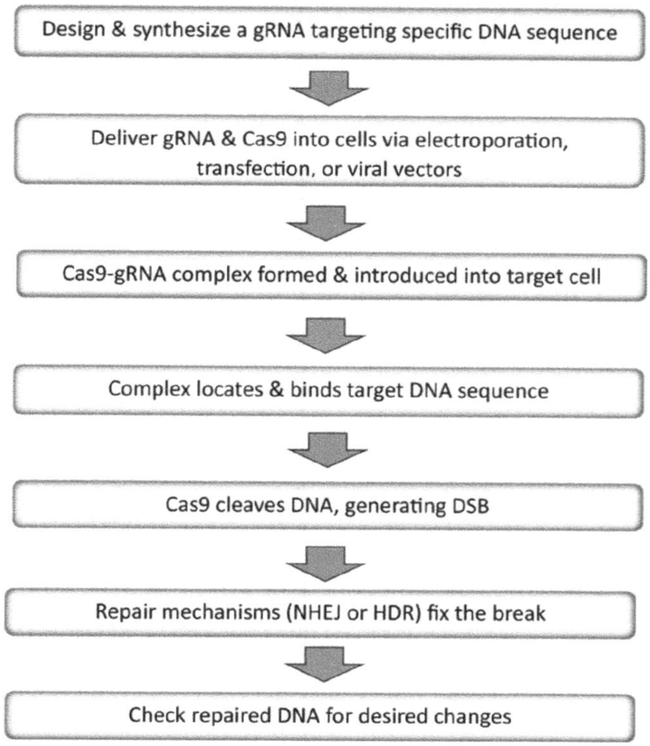

FIGURE 12.4 Steps involved in the CRISPR/Cas9 system.

can recognize target DNA sequences. Then the target DNA is cleaved by the Cas endonuclease at a location next to the protospacer-adjacent motif (PAM), a brief DNA sequence (Figure 12.4) (Jiang and Doudna, 2017).

CRISPR/Cas systems come in various forms. Class 2 CRISPR system has a single-effector protein, whereas multi-subunit effector complexes typically characterize class 1 systems. Class 2 CRISPR systems, particularly type II and Type V like Cas9 and Cas12, have been extensively employed for genome-editing applications due to their significance and simplicity (Makarova et al., 2020). Due to the extensive evolutionary adaptations that CRISPR/Cas systems have observed across various organisms, CRISPR arrays, spacer sequences, and Cas protein activities have all become increasingly diversified. Comprehending the evolutionary dynamics of CRISPR/Cas systems can facilitate understanding their origins, functional variety, and potential applications in crop development (Makarova et al., 2015).

12.7 CRISPR/Cas9 GENOME EDITING FOR MITIGATING DROUGHT TOLERANCE

Genomic adaptation is the mechanism for achieving drought tolerance since drought stress is a dynamic phenomenon. This is because several genes as well as transcription factors involved in drought signaling are upregulated (Fang and Xiong, 2015). Plant drought susceptibility can be increased by sensitive gene expressions, which can cause an imbalance in hormones, lower the antioxidative enzyme activity, and the generation of ROS (Yoshikawa et al., 2022). Stress-related ring finger protein 1 (*OsSRFP1*), dry- and salt-tolerant protein 1 (*OsD*ST), and drought-induced SINA protein 1 (*OsD*IS1) are all known to inhibit drought tolerance (Santosh Kumar et al., 2020). Silencing the gene has been demonstrated to increase drought tolerance in rice while also decreasing H_2O_2 concentrations and increasing antioxidative enzyme levels (Santosh Kumar et al., 2020).

Gene-editing technology is utilized to target genes that are sensitive to drought or to regulate the abiotic stress response negatively. The *Arabidopsis OST2* encoding gene, a vital plasma membrane H^+ ATPase required for the function of stomata, has new alleles that were introduced using CRISPR technology (Fahad et al., 2015). The proton gradient, produced by plasma membrane H^+ ATPases triggers the initiation of stomatal conductance. When dehydrated, the C-terminal of the proton pump binds with ABA, leading to stomatal closure by suppressing the H^+ ATPase's function (Merlot et al., 2007). The *OsT2* locus contains two mutations that prevent stomata from responding to ABA, leading to continuous proton pump activity and necrotic induction. The mutant created by *OST2*-CRISPR demonstrated an increased rate of stomatal closing, along with a decreased rate of transcriptional water loss (Osakabe et al., 2014). Under conditions mediated by ABA, the stomatal response was assessed and corresponded to the wild type. Thus, in response to drought stress, CRISPR/Cas9-mediated mutations at the locus of *OST2* improved drought resistance by enhancing stomatal response against drought. Mutations with greater than 32% efficacy and no off-target alterations were identified in transgenic crops using a truncated sgRNA and CRISPR/Cas9 (Sami et al., 2021). Through CRISPR/Cas9, simultaneous knockout of two genes, i.e. *ABF1* and bZIP40 (homolog of *ABF*1) resulted in an increased rate of flowering, while simultaneous overexpression of *ABF1* and *SAPK8* resulted in delayed flowering (Tang et al., 2024). In the *Arabidopsis* plant, overexpression of *AtbZIP62* and silencing of *AtbZIP62* by CRISPR/Cas9 genome editing resulted in hypersensitivity and hyposensitivity to ABA respectively (Hussain et al., 2022). In *Arabidopsis* plants, silencing of the gene *FDP* by CRISPR/Cas9 shows late flowering than wild-type nonmutated plants (Romera-Branchat et al., 2020).

Furthermore, another study revealed that CRISPR/Cas9 technique was used to create mutant lines of non-expresser tomato of pathogenesis-related 1 (*NPR*1) to confirm its activity in drought resistance (Li et al., 2019). The defense system of plants heavily relies on *NPR1* as a crucial activator, but *NPR1* plays a minor role during abiotic stress conditions (Li et al., 2019). Downregulation of *MdNPR*1 has been observed in apple plants that respond to drought stress (Bhat et al., 2016). Moreover, overexpression of *AtNPR1* in rice led to hypersensitivity in response to drought stress (Deng et al., 2018). Mutants of wild tomatoes were created through CRISPR/Cas9 with a loss of function of *SlNPR1*. This mutant plant showed weaker tolerance to drought, a higher rate of stomatal aperture, a higher content of *MDA* (malondialdehyde), more significant leakage of electrolytes, higher levels of H_2O_2, and lower levels of antioxidants production compared to wild-type tomato plant (Li et al., 2019).

The study of ABA and CRISPR/Cas9 molecular research has shown that, in plants, *ABA* serves as a significant drought response element. It does this by controlling the expression of some stress-related genes and adjusting the closure of stomata to control water loss (Osakabe et al., 2014). The transmission of ABA signals is facilitated by specific transcription factors known as ABA-sensitive factors. These factors consist of a binding domain (*ABRE)* and a *bZIP* unit (*AREBs*) (Nakashima et al., 2014). Drought vulnerability increased with AREB1 knockdown, but drought tolerance was enhanced by *AREB1* overexpression. As a critical component of the *ABA* manufacturing process, antioxidant signaling, and osmotic stress response, *AREB1* controls the expression level of several genes across the ABA signaling pathway. To increase plant resistance to drought, *AREB1* may be a key target. (Sami et al., 2021). *AREB1*'s promoter region was accessed in *Arabidopsis* using a modified CRISPR system that integrated sgRNA, the catalytic domain of *HAT*/histone acetyltransferase, and dCas9 (Deng et al., 2012). The acetylation of central histone protein resulting from the interaction of the *Arabidopsis* HAT catalytic domain enhanced the sensitivity of the *AREB1* promoter region to the transcriptional region. Under drought, mutants' physiologies and molecular investigations revealed the enhanced expression level of *AREB1*, stomatal opening rate, and content of chlorophyll. In addition, in water scarcity conditions, *AREB1* induced the expression of *RD29A*. During drought conditions, the transgenic CRISPR-edited lines revealed an increased rate of survival (Sami et al., 2021).

ABA-dependent hyperosmotic stress response is initiated by a plant-specific protein kinase called *SNF1*-associated protein kinase 2 (*SnRK2*). Members of the *SnRK2* family often play an important role in enhancing drought resistance, facilitating stomatal closure through ABA signaling, responding to hyperosmotic stress, promoting seed germination, and aiding in seedling development (Kulik et al., 2011). Studies in *Arabidopsis* have revealed that *AtSnRK* activates genes that respond to stress, which in turn creates a regulatory network that enhances drought tolerance. This network is active every day and helps to regulate stress levels positively (Xu et al., 2013). According to a microarray study, *SnRK2* is involved in regulating the *AREB-ABF* pathway and its genes of target. It is worth noting that both subclass 1 and 3 members of *SnRK2* have been found to enhance plant growth and improve plant response against abiotic stress in rice, despite no major differences in stomatal damage or rate of survival between wild type and CRISPR-edited *SnRK2* mutant (Ruan et al., 2003; Sami et al., 2021).

12.8 CURRENT PROGRESS AND CHALLENGES

Using CRISPR/Cas has yielded notable advancements in improving crop resilience to drought, particularly in model species and functional crops. However, still there are many obstacles to overcome. Off-target effects, delivery efficiency, and genetic and regulatory complexity are among the ongoing difficulties. More research is required to overcome these obstacles for CRISPR/Cas to realize its potential for sustainable crop management fully.

12.9 THE PROSPECTS FOR CRISPR/Cas IN SUSTAINABLE CROP MANAGEMENT

CRISPR/Cas has a bright future in sustainable crop management. CRISPR/Cas has the potential to completely transform agriculture because of continuing advances in genome-editing technology and a thorough understanding of how plants react during drought stress. Customized crop types with improved resistance to drought might be used in the future, enhancing environmental sustainability and promoting global food security.

12.10 CONCLUSION

Modern agriculture has utilized various gene-editing methods like TALEN, ZFN, and CRISPR systems to develop plants with improved tolerance to abiotic stresses like temperature, salinity, and drought. The CRISPR/Cas system has significantly enhanced the speed and precision of genome editing. Over time, the field of biotechnology has made remarkable progress in modifying known structures and discovering new enzymes or methodologies to enhance genome-editing tools for plants. These advances in genome-editing technologies have led to surprising progress and overcome many barriers in the field of biotechnology.

REFERENCES

Ahmed, F., Rafii, M. Y., Ismail, M. R., Juraimi, A. S., Rahim, H. A., Tanweer, F. A., & Latif, M. A. (2016). Recurrent parent genome recovery in different populations with the introgression of Sub1 gene from a cross between MR219 and Swarna-Sub1. *Euphytica*, *207*, 605–618.

Alotaibi, B. A., Abbas, A., Ullah, R., Nayak, R. K., Azeem, M. I., & Kassem, H. S. (2021). Climate change concerns of Saudi Arabian farmers: The drivers and their role in perceived capacity building needs for adaptation. *Sustainability*, *13*(22), 12677.

Andersson, M., Turesson, H., Nicolia, A., Fält, A. S., Samuelsson, M., & Hofvander, P. (2017). Efficient targeted multiallelic mutagenesis in tetraploid potato (Solanum tuberosum) by transient CRISPR-Cas9 expression in protoplasts. *Plant Cell Reports*, *36*, 117–128.

Augé, R. M., Schekel, K. A., & Wample, R. L. (1986). Osmotic adjustment in leaves of VA mycorrhizal and nonmycorrhizal rose plants in response to drought stress. Plant Physiology, *82*(3), 765–770.

Baldoni, E., Genga, A., & Cominelli, E. (2015). Plant MYB transcription factors: Their role in drought response mechanisms. *International Journal of Molecular Sciences*, *16*(7), 15811–15851.

Bhat, J. A., Ali, S., Salgotra, R. K., Mir, Z. A., Dutta, S., Jadon, V., Tyagi, A., Mushtaq, M., Jain, N., Singh, P. K., Singh, G. P., & Prabhu, K. (2016). Genomic selection in the era of next generation sequencing for complex traits in plant breeding. Frontiers in Genetics, *7*, 221.

Biju, S., Fuentes, S., & Gupta, D. (2023). Novel insights into the mechanism(s) of silicon-induced drought stress tolerance in lentil plants revealed by RNA sequencing analysis. BMC Plant Biology, *23*(1), 498.

Daems, S., Ceusters, N., Valcke, R., & Ceusters, J. (2022). Effects of chilling on the photosynthetic performance of the CAM orchid Phalaenopsis. *Frontiers in Plant Science*, *13*, 981581.

Deng, F., Tu, L., Tan, J., Li, Y., Nie, Y., & Zhang, X. (2012). GbPDF1 is involved in cotton fiber initiation via the core cis-element HDZIP2ATATHB2. Plant Physiology, *158*(2), 890–904.

Deng, L., Wang, H., Sun, C., Li, Q., Jiang, H., Du, M., Li, C.-B., & Li, C. (2018). Efficient generation of pink-fruited tomatoes using CRISPR/Cas9 system. *Journal of Genetics and Genomics = Yi chuan xue bao*, *45*(1), 51–54.

Deng, S., Tan, X., Liu, B., Yang, F., & Yan, T. (2022). A reversal in global occurrences of flash drought around 2000 identified by rapid changes in the standardized evaporative stress ratio. *Science of the Total Environment*, *848*, 157427.

Dietz, K. J., Zörb, C., & Geilfus, C. M. (2021). Drought and crop yield. *Plant Biology*, *23*(6), 881–893.

Fahad, S., Nie, L., Chen, Y., Wu, C., Xiong, D., Saud, S., & Huang, J. (2015). Crop plant hormones and environmental stress. *Sustainable Agriculture Reviews*, *15*, 371–400.

Fang, Y., & Xiong, L. (2015). General mechanisms of drought response and their application in drought resistance improvement in plants. *Cellular and Molecular Life Sciences*, *72*, 673–689.

Golldack, D., Li, C., Mohan, H., & Probst, N. (2014). Tolerance to drought and salt stress in plants: Unraveling the signaling networks. *Frontiers in Plant Science*, *5*, 151.

Gowrishankar, J. (1988). Osmoregulation in Enterobacteriaceae: Role of proline/betaine transport systems. *Current Science*, *57*, 225–234.

Horodecka, K., & Düchler, M. (2021). CRISPR/Cas9: Principle, applications, and delivery through extracellular vesicles. *International Journal of Molecular Sciences*, *22*(11), 6072.

Huang, H., Ye, Y., Song, W., Li, Q., Han, R., Wu, C., Wang, S., Yu, J., Liu, X., Fu, X., Liu, Q., & Wu, K. (2022). Modulating the C-terminus of DEP1 synergistically enhances grain quality and yield in rice. *Journal of Genetics and Genomics = Yi chuan xue bao*, *49*(5), 506–509.

Hura, T., Hura, K., & Ostrowska, A. (2022). Drought-stress induced physiological and molecular changes in plants. *International Journal of Molecular Sciences*, *23*(9), 4698.

Hussain, S., Cheng, Y., Li, Y., Wang, W., Tian, H., Zhang, N., Wang, Y., Yuan, Y., Hussain, H., Lin, R., Wang, C., Wang, T., & Wang, S. (2022). AtbZIP62 acts as a transcription repressor to positively regulate ABA responses in Arabidopsis. *Plants*, *11*(22), 3037.

Jiang, F., & Doudna, J. A. (2017). CRISPR–Cas9 structures and mechanisms. *Annual Review of Biophysics*, *46*, 505–529.

Joshi, R., Wani, S. H., Singh, B., Bohra, A., Dar, Z. A., Lone, A. A., ... & Singla-Pareek, S. L. (2016). Transcription factors and plants response to drought stress: current understanding and future directions. *Frontiers in Plant Science*, *7*, 1029.

Khan, A., Pan, X., Najeeb, U., Tan, D. K. Y., Fahad, S., Zahoor, R., & Luo, H. (2018). Coping with drought: Stress and adaptive mechanisms, and management through cultural and molecular alternatives in cotton as vital constituents for plant stress resilience and fitness. *Biological Research*, 51, 47.

Kim, M., Lee, C., Hong, S., Kim, S. L., Baek, J. H., & Kim, K. H. (2021). High-throughput phenotyping methods for breeding drought-tolerant crops. *International Journal of Molecular Sciences*, *22*(15), 8266.

Kulik, A., Wawer, I., Krzywińska, E., Bucholc, M., & Dobrowolska, G. (2011). SnRK2 protein kinases – Key regulators of plant response to abiotic stresses. *Omics: A Journal of Integrative Biology*, *15*(12), 859–872.

Kumar, M., Prusty, M. R., Pandey, M. K., Singh, P. K., Bohra, A., Guo, B., & Varshney, R. K. (2023). Application of CRISPR/Cas9-mediated gene editing for abiotic stress management in crop plants. *Frontiers in Plant Science*, *14*, 1157678.

Lata, C., & Prasad, M. (2011). Role of DREBs in regulation of abiotic stress responses in plants. *Journal of experimental botany*, *62*(14), 4731–4748.

Lee, D. W., Lee, S. K., Rahman, M. M., Kim, Y. J., Zhang, D., & Jeon, J. S. (2019a). The role of rice vacuolar invertase2 in seed size control. *Molecules and Cells*, *42*(10), 711.

Lee, S. K., Kim, H., Cho, J. I., Nguyen, C. D., Moon, S., Park, J. E., Park, H. R., Huh, J. H., Jung, K.-H., Guiderdoni, E., & Jeon, J. S. (2020). Deficiency of rice hexokinase HXK5 impairs synthesis and utilization of starch in pollen grains and causes male sterility. *Journal of Experimental Botany*, *71*(1), 116–125.

Li, C., Hao, M., Wang, W., Wang, H., Chen, F., Chu, W., Zhang, B., Mei, D., Cheng, H., & Hu, Q. (2018a). An efficient CRISPR/Cas9 platform for rapidly generating simultaneous mutagenesis of multiple gene homoeologs in allotetraploid oilseed rape. *Frontiers in Plant Science*, *9*, 442.

Li, M., Li, X., Zhou, Z., Wu, P., Fang, M., Pan, X., Lin, Q, Luo, W., Wu, G., Li, H., & Li, H. (2016). Reassessment of the four yield-related genes Gn1a, DEP1, GS3, and IPA1 in rice using a CRISPR/Cas9 system. *Frontiers in Plant Science*, *7*, 377.

Li, R., Liu, C., Zhao, R., Wang, L., Chen, L., Yu, W., Zhang, S., Sheng, J. & Shen, L. (2019). CRISPR/Cas9-Mediated SlNPR1 mutagenesis reduces tomato plant drought tolerance. *BMC Plant Biology*, *19*(1), 1–13.

Li, X., Xu, S., Fuhrmann-Aoyagi, M. B., Yuan, S., Iwama, T., Kobayashi, M., & Miura, K. (2022). CRISPR/Cas9 technique for temperature, drought, and salinity stress responses. *Current Issues in Molecular Biology*, *44*(6), 2664–2682.

Liang, X., Li, S. W., Gong, L. M., Li, S., & Zhang, Y. (2020). COPII components Sar1b and Sar1c play distinct yet interchangeable roles in pollen development. *Plant Physiology*, *183*(3), 974–985.

Makarova, K. S., Haft, D. H., Barrangou, R., Brouns, S. J., Charpentier, E., Horvath, P., Moineau, S., Mojica, F. J. M., Wolf, Y. I., Yakunin, A. F., van der Oost, J., & Koonin, E. V. (2011). Evolution and classification of the CRISPR–Cas systems. *Nature Reviews Microbiology*, *9*(6), 467–477.

Makarova, K. S., Wolf, Y. I., Alkhnbashi, O. S., Costa, F., Shah, S. A., Saunders, S. J., Barrangou, R., Brouns, S. J. J., Charpentier, E., Haft, D. H., Horvath, P., Moineau, S., Mojica, F. J. M., Terns, R. M., Terns, M. P., White, M. F., Yakunin, A. F., Garrett, R. A., van der Oost, J., … Koonin, E. V. (2015). An updated evolutionary classification of CRISPR–Cas systems. *Nature Reviews Microbiology*, *13*(11), 722–736.

Makarova, K. S., Wolf, Y. I., Iranzo, J., Shmakov, S. A., Alkhnbashi, O. S., Brouns, S. J., Charpentier, E., Cheng, D., Haft, D. H., Horvath, P., Moineau, S., Mojica, F. J. M., Scott, D., Shah, S. A., Siksnys, V., Terns, M. P., Venclovas, Č., White, M. F., Yakunin, A. F., … Koonin, E. V. (2020). Evolutionary classification of CRISPR–Cas systems: A burst of class 2 and derived variants. *Nature Reviews Microbiology*, *18*(2), 67–83.

Melandri, G., AbdElgawad, H., Riewe, D., Hageman, J. A., Asard, H., Beemster, G. T., Kadam, N., Jagadish, K., Altmann, T., Ruyter-Spira, C., & Bouwmeester, H. (2020). Biomarkers for grain yield stability in rice under drought stress. *Journal of Experimental Botany*, *71*(2), 669–683.

Merlot, S., Leonhardt, N., Fenzi, F., Valon, C., Costa, M., Piette, L., Vavasseur, A., Genty, B., Boivin, K., Müller, A., Giraudat, J., & Leung, J. (2007). Constitutive activation of a plasma membrane H^+-ATPase prevents abscisic acid-mediated stomatal closure. *The EMBO Journal*, *26*(13), 3216–3226.

Mugo, G. (2023). *Kiambu county distributes Sh103m farm inputs to farmers*. STAR. https://www.the-star.co.ke/counties/central/2023-03-30-kiambu-county-distributes-sh103m-farm-inputs-to-farmers/

Nakashima, K., Yamaguchi-Shinozaki, K., & Shinozaki, K. (2014). The transcriptional regulatory network in the drought response and its crosstalk in abiotic stress responses including drought, cold, and heat. *Frontiers in Plant Science*, *5*, 170.

Negi, J., Moriwaki, K., Konishi, M., Yokoyama, R., Nakano, T., Kusumi, K., Hashimoto-Sugimoto, M., Schroeder, J. I., Nishitani, K., Yanagisawa, S., & Iba, K. (2013). A Dof transcription factor, SCAP1, is essential for the development of functional stomata in Arabidopsis. *Current Biology*, *23*(6), 479–484.

Osakabe, Y., Osakabe, K., Shinozaki, K., & Tran, L. S. P. (2014). Response of plants to water stress. *Frontiers in Plant Science*, *5*, 86.

Ozturk, M., Turkyilmaz Unal, B., García-Caparrós, P., Khursheed, A., Gul, A., & Hasanuzzaman, M. (2021). Osmoregulation and its actions during the drought stress in plants. *Physiologia Plantarum*, *172*(2), 1321–1335.

Paul, P., Dhatt, B. K., Miller, M., Folsom, J. J., Wang, Z., Krassovskaya, I., Liu, K., Sandhu, J., Yu, H., Zhang, C., Obata, T., Staswick, P., & Walia, H. (2020). MADS78 and MADS79 are essential regulators of early seed development in rice. *Plant Physiology*, *182*(2), 933–948.

Plant Development. (2023). *What are the effects of high temperatures on plant development*? https://washingtonindependent.com/w/plant-development/

Ray, R. L., Fares, A., & Risch, E. (2018). Effects of drought on crop production and cropping areas in Texas. *Agricultural & Environmental Letters*, *3*(1), 170037.

Raza, A., Mubarik, M. S., Sharif, R., Habib, M., Jabeen, W., Zhang, C., Chen, H., Chen, Z-H., Siddique, K. H. M., Zhuang, W., & Varshney, R. K. (2023). Developing drought-smart, ready-to-grow future crops. *The Plant Genome*, *16*(1), e20279.

Romera-Branchat, M., Severing, E., Pocard, C., Ohr, H., Vincent, C., Née, G., Martinez-Gallegos R, Jang S, Andrés F, Madrigal P, & Coupland, G. (2020). Functional divergence of the Arabidopsis florigen-interacting bZIP transcription factors FD and FDP. *Cell Reports*, 31(9), 107717.

Ruan, Y. L., Llewellyn, D. J., & Furbank, R. T. (2003). Suppression of sucrose synthase gene expression represses cotton fiber cell initiation, elongation, and seed development. *The Plant Cell*, *15*(4), 952–964.

Sami, A., Xue, Z., Tazein, S., Arshad, A., He Zhu, Z., Ping Chen, Y. , Hong, Y., Tian Zhu, X., & Jin Zhou, K. (2021). CRISPR–Cas9-based genetic engineering for crop improvement under drought stress. *Bioengineered*, *12*(1), 5814–5829.

Santosh Kumar, V. V., Verma, R. K., Yadav, S. K., Yadav, P., Watts, A., Rao, M. V., & Chinnusamy, V. (2020). CRISPR-Cas9 mediated genome editing of drought and salt tolerance (OsDST) gene in indica mega rice cultivar MTU1010. *Physiology and Molecular Biology of Plants*, *26*, 1099–1110.

Shanker, A. K., Maheswari, M., Yadav, S. K., Desai, S., Bhanu, D., Attal, N. B., & Venkateswarlu, B. (2014). Drought stress responses in crops. *Functional & Integrative Genomics*, *14*, 11–22.

Sharma, R., Singh, G., Bhattacharya, S., & Singh, A. (2018). Comparative transcriptome meta-analysis of Arabidopsis thaliana under drought and cold stress. *PloS One*, 13(9), e0203266.

Shi, J., Gao, H., Wang, H., Lafitte, H. R., Archibald, R. L., Yang, M., Hakimi, S. M., Mo, H., & Habben, J. E. (2017). ARGOS 8 variants generated by CRISPR-Cas9 improve maize grain yield under field drought stress conditions. *Plant Biotechnology Journal*, *15*(2), 207–216.

Shinwari, Z. K., Jan, S. A., Nakashima, K., & Yamaguchi-Shinozaki, K. (2020). Genetic engineering approaches to understanding drought tolerance in plants. *Plant Biotechnology Reports*, *14*, 151–162.

Singh, R., Pandey, N., Kumar, A., & Shirke, P. A. (2016). Physiological performance and differential expression profiling of genes associated with drought tolerance in root tissue of four contrasting varieties of two Gossypium species. *Protoplasma*, *253*, 163–174.

Sourour, A., Afef, O., Mounir, R., & Mongi, B. Y. (2017). A review: Morphological, physiological, biochemical and molecular plant responses to water deficit stress. *The International Journal of Engineering and Science*, *6*(1), 2319–1805.

Sriboon, S., Li, H., Guo, C., Senkhamwong, T., Dai, C., & Liu, K. (2020). Knock-out of TERMINAL FLOWER 1 genes altered flowering time and plant architecture in Brassica napus. *BMC Genetics*, *21*(1), 1–13.

Sun, Y., Jiao, G., Liu, Z., Zhang, X., Li, J., Guo, X., Du, W., Du, J., Francis, F., Zhao, Y., & Xia, L. (2017). Generation of high-amylose rice through CRISPR/Cas9-mediated targeted mutagenesis of starch branching enzymes. *Frontiers in Plant Science*, *8*, 298.

Tang, L., Li, G., Wang, H., Zhao, J., Li, Z., Liu, X., Shu, Y., Liu, W., Wang, S., Huang, J., Ying, J., Tong, X., Yuan, W., Wei, X., Tang, S., Wang, Y., Bu, Q., & Zhang, J. (2024). Exogenous abscisic acid represses rice flowering via SAPK8-ABF1-Ehd1/Ehd2 pathway. *Journal of Advanced Research*, 59, 35–47.

Tsarouchi, G., & Buytaert, W. (2018). Land-use change may exacerbate climate change impacts on water resources in the Ganges basin. *Hydrology and Earth System Sciences*, *22*(2), 1411–1435.

Turner, N. C. (2018). Turgor maintenance by osmotic adjustment: 40 years of progress. *Journal of Experimental Botany*, *69*(13), 3223–3233.

Valarmathi, R., Mahadeva Swamy, H. K., Appunu, C., Suresha, G. S., Mohanraj, K., Hemaprabha, G., Mahadevaiah, C., & Ulaganathan, V. (2023). Comparative transcriptome profiling to unravel the key molecular signalling pathways and drought adaptive plasticity in shoot borne root system of sugarcane. *Scientific Reports*, *13*(1), 12853.

Waadt, R., Seller, C. A., Hsu, P. K., Takahashi, Y., Munemasa, S., & Schroeder, J. I. (2022). Plant hormone regulation of abiotic stress responses. *Nature Reviews Molecular Cell Biology*, *23*(10), 680–694.

Wai, C. M., Weise, S. E., Ozersky, P., Mockler, T. C., Michael, T. P., & VanBuren, R. (2019). Time of day and network reprogramming during drought induced CAM photosynthesis in Sedum album. *PLoS Genetics*, 15(6), e1008209.

Wang, H., Wu, Y., Zhang, Y., Yang, J., Fan, W., Zhang, H., Zhao, S., Yuan, L., & Zhang, P. (2019). CRISPR/Cas9-based mutagenesis of starch biosynthetic genes in sweet potato (Ipomoea Batatas) for the improvement of starch quality. *International Journal of Molecular Sciences*, *20*(19), 4702.

Wang, Y., Meng, Z., Liang, C., Meng, Z., Wang, Y., Sun, G., Zhu, T., Cai, Y., Guo, S., Zhang, R., & Lin, Y. (2017). Increased lateral root formation by CRISPR/Cas9-mediated editing of arginase genes in cotton. *Science China. Life Sciences*, *60*(5), 524.

Wang, X., Niu, Y., & Zheng, Y. (2021). Multiple functions of MYB transcription factors in abiotic stress responses. *International Journal of Molecular Sciences*, *22*(11), 6125.

Xu, Q., & Zhang, L. (2009, August). Subcellular localization of salt stress-induced reactive oxygen species production in Arabidopsis. In *Conference on Lasers and Electro-Optics/Pacific Rim* (p. ThF1_2). Optica Publishing Group.

Xu, R., Yang, Y., Qin, R., Li, H., Qiu, C., Li, L., Wei, P., & Yang, J. (2016). Rapid improvement of grain weight via highly efficient CRISPR/Cas9-mediated multiplex genome editing in rice. *Journal of Genetics and Genomics = Yi chuan xue bao*, *43*(8), 529–532.

Xu, W. L., Zhang, D. J., Wu, Y. F., Qin, L. X., Huang, G. Q., Li, J., Li, L., & Li, X. B. (2013). Cotton PRP5 gene encoding a proline-rich protein is involved in fiber development. *Plant Molecular Biology*, *82*, 353–365.

Yang, Q., Zhong, X., Li, Q., Lan, J., Tang, H., Qi, P., Ma, J., Wang, J., Chen, G., Pu, Z., Li, W., Lan, X., Deng, M., Harwood, W., Li, Z., Wei, Y., Zheng, Y., & Jiang, Q. (2020). Mutation of the D-hordein gene by RNA-guided Cas9 targeted editing reducing the grain size and changing grain compositions in barley. *Food Chemistry*, 311, 125892.

Yoshida, T., Fujita, Y., Sayama, H., Kidokoro, S., Maruyama, K., Mizoi, J., Shinozaki, K., & Yamaguchi-Shinozaki, K. (2010). AREB1, AREB2, and ABF3 are master transcription factors that cooperatively regulate ABRE-dependent ABA signaling involved in drought stress tolerance and require ABA for full activation. *The Plant Journal*, *61*(4), 672–685.

Yoshida, T., Mogami, J., & Yamaguchi-Shinozaki, K. (2014). ABA-dependent and ABA-independent signaling in response to osmotic stress in plants. *Current Opinion in Plant Biology*, *21*, 133–139.

Yoshikawa, S., Taniguchi, K., Sawamura, H., Ikeda, Y., Tsuji, A., & Matsuda, S. (2022). Potential diets to improve mitochondrial activity in amyotrophic lateral sclerosis. *Diseases*, *10*(4), 117.

You, J., & Chan, Z. (2015). ROS regulation during abiotic stress responses in crop plants. *Frontiers in Plant Science*, 6, 1092.

Yu, J., Jiang, M., & Guo, C. (2019). Crop pollen development under drought: From the phenotype to the mechanism. *International Journal of Molecular Sciences*, *20*(7), 1550.

Zhang, D., Zhang, Z., Unver, T., & Zhang, B. (2021). CRISPR/Cas: A powerful tool for gene function study and crop improvement. *Journal of Advanced Research*, *29*, 207–221.

Zhang, H. X., Zhang, Y., & Yin, H. (2019a). Genome editing with mRNA encoding ZFN, TALEN, and Cas9. *Molecular Therapy*, *27*(4), 735–746.

Zhang, K., Nie, L., Cheng, Q., Yin, Y., Chen, K., Qi, F., Zou, D., Liu, H., Zhao, W., Wang, B., & Li, M. (2019b). Effective editing for lysophosphatidic acid acyltransferase 2/5 in allotetraploid rapeseed (Brassica napus L.) using CRISPR-Cas9 system. *Biotechnology for Biofuels*, *12*(1), 1–18.

Zhang, Y., Liang, Z., Zong, Y., Wang, Y., Liu, J., Chen, K., Qiu, J.-L., & Gao, C. (2016). Efficient and transgene-free genome editing in wheat through transient expression of CRISPR/Cas9 DNA or RNA. *Nature Communications*, *7*(1), 12617.

Zia, R., Nawaz, M. S., Siddique, M. J., Hakim, S., & Imran, A. (2021). Plant survival under drought stress: Implications, adaptive responses, and integrated rhizosphere management strategy for stress mitigation. *Microbiological Research*, *242*, 126626.

13 Adaptation to Drought Stress in Cereals
A Molecular Perspective

Chayma Ikan, Fatima Ezzahra Soussani, Salma Toubali, Nizar El Mazouni, Rachid Lahlali, and Abdelilah Meddich

13.1 INTRODUCTION

Water shortage is a persistent natural stressor that has an enormous effect on plant development and production. It is considered that one is responsible for more than 70% of potential crop yields and damages in agriculture globally (Dolan et al., 2021; Seleiman et al., 2021). Drought is one of the many stresses that plants encounter as they develop and thrive (Adel & Carels, 2023) but its effects on plant morphology are noticeable at every stage of development and have particular consequences for plant metabolism, which might result in a predicted 30% decrease in global productivity by 2050. Persistent droughts spurred on by climate change represent a severe danger to agriculture since forecasts indicate a rise in the frequency of water scarcity (Malhi et al., 2021). Crop plants use acclimatization and particular tolerance mechanisms to deal with drought, providing significant evolutionary insights that researchers may use to reduce production losses and satisfy an expanding population's increasing food demands. In response to water scarcity, various biochemical, physiological, and molecular functions affect individual cells or the entire plant, which is crucial for reducing stress (Hirwa et al., 2022). Plants can respond to water scarcity in various ways, including drought escape (such as early flowering in especially dry environments), avoidance (such as regulating transpiration through stomata and developing deep root systems), trait adaptability, maintaining water management in tissues, antioxidant scavenging systems, regulating plant growth through plant growth substances, and osmotic enforcement (Ozturk et al., 2021). In agricultural plants, drought stress (DS) significantly impacts stomatal conductance, photosynthetic activity, cell turgidity, and altered oxidative metabolism, leading to membrane instability (Kapoor et al., 2020), a reduction in leaf size, stem elongation, and efficiency of water use (Zhang et al., 2020; Ahmad et al., 2023). Fortunately, the principal plant biological and molecular reactions of plants to water scarcity are the buildup of proteins and osmolytes (Negi et al., 2023) or the expression of genes sensitive to drought. At the leaf or root system level, plants that have experienced drought increase their production of abscisic acid (ABA). According to Muhammad Aslam et al. (2022), this raises transpiration costs and stomatal closure, which then sets off the expression of several genes linked to the drought, including the upward and downregulation of several gene transcripts. When plants are stressed by drought, they activate a unique sequence-specific cascade of signal molecules, which involves signal transduction, receipt, and responsiveness. This stimulation of particular genes triggers the activation of critical physiological and biochemical responses (Joshi et al., 2016a; Askari-Khorasgani et al., 2021; Hussain et al., 2021). Plants have been shown to have a high number of differentially expressed genes that participate in cellular responses such as transcriptional regulation and signaling cascades (Meraj et al., 2020). Typically, proteins with a role in regulation or metabolism, such as cellular membranes, late-embryogenesis-abundant (LEA) proteins, and osmoprotectants, are encoded by genes sensitive to drought (Manna et al., 2021). Moreover, transcription factors (AP2/ERF, bZIP, AREB, NAC, MYC,

DOI: 10.1201/9781003457237-13

and MYB), protein kinases (such as mitogen-activated protein kinases [MAPK], CDPK, receptor protein kinases, ribosomal protein kinases, and transcription regulation protein kinases), and protein phosphatases (including phosphoesterases and phospholipases) are classified as regulatory substances (Khan et al., 2020a). Additionally, phytohormones play a role in regulating signal transduction and imparting tolerance to water scarcity (Khan et al., 2020b). This chapter's goal is to provide a complete summary of current research on inducing drought tolerance in cereal crops, with a focus on conventional breeding, marker-assisted selection (MAS), transgenic approaches, and new genomic technologies. Research in this field is crucial, particularly considering how climate change is affecting agriculture. To deal with the negative impact of drought on agricultural plants, the plants use specialized processes that improve drought resistance. Plants use both gene expression modification and gene modulation as essential defense strategies against DS (Raza et al., 2023). Plants can be made more drought-tolerant by combining traditional breeding techniques with biotechnology treatments. Hence, it is essential to explore the uncharted adaptive traits present in different plant species and understand how these traits translate into genotypes capable of enduring the detrimental impacts of water scarcity while maintaining productivity (Bapela et al., 2022). Furthermore, Omics applications, including genomes, transcriptomics, proteomics, and metabolomics, provide a vital part in developing stress-resistant crops and modern advances in the field of molecular markers have shown that molecular tags linked to the desired trait/QTLs may be used in crop development efforts via genome-assisted breeding programs (Joshi et al., 2016a; Bhardwaj et al., 2021; Roychowdhury et al., 2023). Although there are a number of marker-assisted breeding techniques for introducing or transferring the desired trait, such as genomic selection (GS), which enhances the genetics of DS traits in crops, particularly cereals, MAS, marker-assisted backcrossing (MABC), and marker-aided gene pyramiding (MAGP). These innovative techniques pose a potential risk to global food security (Tiwari et al., 2023).

13.2 DROUGHT-RELATED PHYSIOLOGICAL AND BIOCHEMICAL CHANGES IN CEREALS

13.2.1 Impact of DS on Physiological Traits

Plant reactions to water deficit encompass a complex interplay of factors, and unraveling the mechanisms behind this coordination stands as a significant challenge in forecasting the consequences of drought. In this context, we introduce a straightforward novel framework for categorizing these elements, encompassing both the exposure to drought and the physiological responses over different time intervals (Kannenberg et al., 2022). The vital molecule of water modulates root water uptake, and transpiration and contributes to carbohydrate transport. Mortality of the plant becomes probable due to perturbed photosynthesis when water becomes scarce. This is because interactions within the system involving soil, plants, and the atmosphere under limited water are likely to lead to water content losses (Feldman et al., 2020). Moreover, mortality resulting from drought typically originates from the decline in the plant xylem's ability to transport water, a phenomenon referred to as hydraulic failure (Li et al., 2022). Drought increases xylem tensions, which heightens the possibility that air bubbles may form inside the conduits and eventually obstruct water flow. To minimize water loss and xylem tensions, stomata often shut before significant hydraulic losses are seen in the xylem (Anderegg et al., 2016). The slow increase in xylem tension even after the stomata close is caused by water loss via cuticle conductivity and stomatal leakage (Duursma et al., 2019). In the prolonged absence of adequate moisture, xylem tension eventually reaches a crucial threshold, initiating the spread of emboli throughout the xylem. This process induces hydraulic dysfunction, and under sustained drought conditions eventually, it may result in a systemic vascular failure that destroys the plant by drying out its tissues (McDowell et al., 2022).

13.2.2 DS Brings an Imbalance between Reactive Oxygen Species Production and Scavenging

Plants have the capability to enhance their scavenging abilities for reactive oxygen species (ROS) and modulate the expression levels of stress-responsive genes (Nadarajah, 2020). Within living cells, harmful ROS such as superoxide, hydrogen peroxide (H_2O_2), and hydroxyl radicals are produced through energy-based metabolism. According to Verma et al. (2019), significant ROS buildup occurs during droughts, endangering proteins, carbohydrates, and DNA. Metabolic abnormalities, aberrant cell death, and disturbances in cellular homeostasis can all be caused by this buildup. High ROS concentrations are neutralized, scavenged, or detoxified by plants utilizing a combination of enzymatic and non-enzymatic defensive mechanisms in response to oxidative stress. As NADPH oxidases are located in the plasma membrane, the respiratory burst oxidase (RBOH) homologs are the primary source of ROS (Wang et al., 2018). The activity of these enzymes is controlled by post-translational modification of different proteins (Filtz et al., 2014). By reducing oxygen, activated RBOHs produce apoplastic superoxide anions (O_2^-), which are then transformed into hydrogen peroxide. After entering guard cells, these molecules control stomatal motility (Rodrigues et al., 2017).

13.2.3 The Detrimental Effect of Drought on Seed Quality

The phenological, morphological, physiological, and agronomic characteristics of cereals are impacted by water shortage (Choukri et al., 2020; Valdivia-Cea et al., 2021). The reproductive development of a variety of crops is impacted by DS, which lowers eventual economic output. The complicated processes of gametogenesis, fertilization, endosperm, gametogenesis, mobilization of nutrients to various reproductive organs, and accumulation of reserves in stems are all involved in how drought affects grain yield. Any stage of crop growth can be affected by stress, but the impact is greatest during the seed-filling stage because it influences the average seed weight, seed composition, and ultimately the amount and quality of the final harvest. Seed filling is an important developmental step for all crops, involving the mobilization and transport mechanisms needed to import various components. Additionally, when seeds develop, a range of metabolic processes help to produce lipids, proteins, and carbs (El-Metwally et al., 2021).

13.2.4 DS Effect on Soil Properties and Fertility

Drought modifies the soil's biological and physicochemical properties, which directly affects the soil ecology. As a result, the physical qualities of the soil deteriorate and soil organic matter (SOM) and biodiversity decline. The depletion of SOM stands out as a primary contributor to ecosystem degradation and a loss of resilience (Soudek et al., 2024). Studies have shown that soil microbial activity can be severely damaged, and community structure can be upset by DS. Under adverse conditions, microorganisms may disappear as they struggle to adapt to severe stress (Figure 13.1) (Bogati & Walczak, 2022).

13.3 EFFECTIVE DROUGHT RESISTANCE MECHANISMS IN CEREALS

13.3.1 Morphophysiological Responses

To alleviate DS, morphophysiological responses in cereals become crucial, including changes in root architecture, leaf morphology, and stomatal characteristics offer tangible ways for cereals to optimize water uptake and minimize transpiration losses and osmotic adjustment, photosynthetic adaptations, and hormonal regulation reveal intricate biochemical strategies employed by cereals to withstand water scarcity (Rijal et al., 2021). Cereals adapt to drought by increasing root depth to

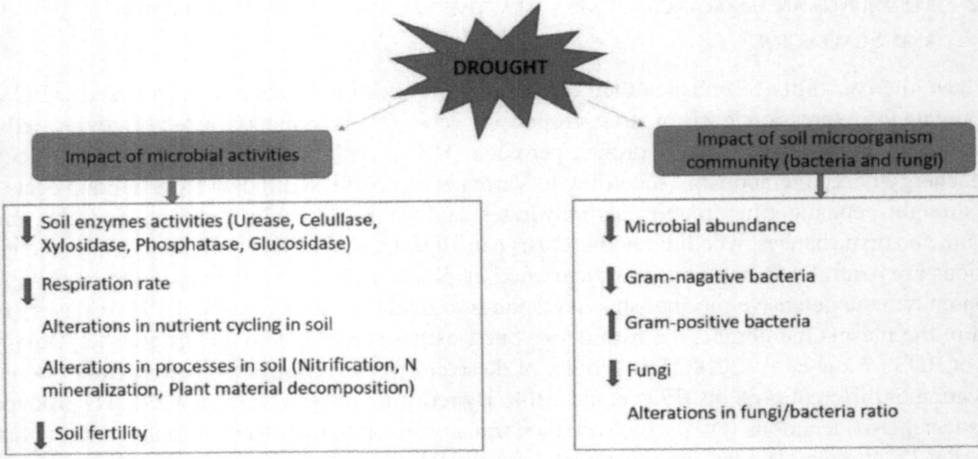

FIGURE 13.1 Implications of soil composition during drought stress.

access less evaporative water sources, providing a crucial advantage during water scarcity. While genotypes with deeper roots exhibit superior performance under DS. The development of taproots, exemplified by wheat, acts as efficient conduits for water uptake from deeper tables, strengthening the crop's sensitivity to extended periods of drought (Ouyang et al., 2020; Dixon et al., 2022). Additionally, cereals exhibit remarkable coordination between root and shoot systems, strategically allocating resources based on prevailing water conditions. This dynamic coordination optimizes water uptake while maintaining essential aerial functions, instrumental in overall drought resistance (Wahab et al., 2023). The balance achieved between root and shoot growth enhances the plant's ability to adapt to drought conditions through a boost in the root-to-leaf surface ratio, facilitating the acquisition of water to sustain existing shoots (Comas et al., 2013). Leaf morphology adaptations in cereals under DS include the reduction in leaf size, attributed to a reduction in both cell division and elongation (Liu et al., 2022). This strategic response minimizes reducing water loss via transpiration by decreasing the surface area of leaves (Chaturvedi et al., 2012). Thickening of leaves, associated with alterations in cell structure, enhances water retention and reduces transpiration rates (Jogawat et al., 2021). However, stomatal characteristics, including closure and alterations in stomatal density, further regulate water loss during drought (Talukder et al., 2021). Cereals orchestrate a complex physiological response to combat DS by photosynthetic adjustments, including temporary reduction and enhancement of water use efficiency, balance carbon assimilation with water conservation, and osmotic adjustments (Robertson et al., 2021). In cereals, ABA mediates physiological responses to DS by inducing the closing of stomatal pores to minimize transpiration-related water loss. It also influences gene expression related to stress tolerance, enhancing adaptive responses. This delicate hormonal balance allows cereals to adjust their biological systems and adapt dynamically to the changing circumstances of water exposure (Toulotte et al., 2022).

13.3.2 Biochemical Responses

Cereals exploit a wide range of molecular processes to protect their survival by coordinating sophisticated biochemical responses to DS. These adaptive processes are functioning cereals' resilience at the microscopic and molecular scales to address the limitations provided by reduced water supply. DS induces the generation of ROS within plant cells, resulting in oxidative stress (Chandrasekaran, 2022). In response, cereals activate antioxidant defense systems to mitigate the detrimental consequences of ROS (Tunc-Ozdemir et al., 2009). Enzymes such as catalase (CAT), superoxide dismutase (SOD), and peroxidase (POX) are upregulated to neutralize excess ROS, preventing cellular damage and maintaining cellular balance (Bratovcic et al., 2020). DS stimulates cereals to produce stress-responsive proteins, like LEA proteins and dehydrins (Joshi et al., 2016b). These proteins contribute to cellular protection by stabilizing membranes, preserving protein structures, and preventing aggregation under challenging or adverse conditions (López et al., 2004). The induction of these proteins reflects a molecular strategy employed by cereals to enhance their resilience in the face of water scarcity (Minh et al., 2019). However, at the hormonal level, cereals experiencing DS undergo significant metabolic reprogramming to optimize resource utilization. This includes redirecting metabolic pathways toward the synthesis of protective compounds (e.g. flavonoids and phenolics) which act as antioxidants (Sarker, 2018). These metabolic shifts ensure that cereals prioritize the production of molecules essential for stress tolerance, enhancing their ability to withstand water scarcity. For example, modification of brassinosteroid (BR) signaling might be optimized for enhancing tolerance to stress and productivity in cereal crops (Gruszka, 2020).

13.4 MOLECULAR ADAPTATION IN CEREALS ASSOCIATED WITH DROUGHT TOLERANCE

13.4.1 Molecular-Based Strategies as an Adaptation for Protecting Plants against DS

13.4.1.1 Conventional Breeding

Conventional breeding is an essential method for developing new genetic variation, preserving natural germplasm, performing sexual hybridization between lines of opposite parentage, and mutating. Inter-generic and interspecific hybridization, as well as induced modification, are examples of conventional breeding techniques (Ahmar et al., 2020). Recent developments in plant physiology have resulted in more effective drought resistance strategies. Due to the substantial impact of genotype–environment interaction and low genetic inheritance, the grain production rate has been utilized to choose superior cultivars during droughts; however, this strategy has not proven to be effective (Schultz et al., 2020). Because it is less laborious and depends on genetic variety, traditional breeding has evolved over time to become a selection strategy focused on physiological features (Gaikwad et al., 2020). The greater complexity of abiotic stress sensitivity is caused by several factors, including the absence of useful selection criteria, the low genetic heterogeneity for the elements that influence yield in drought-prone environments, and the intricate relationships between stress parameters and different physiological, biochemical, and molecular reactions that alter the growth and development of the plant pathway (Oladosu et al., 2019). Ultimately, CB practices are essential for the conservation of genetic material, for hybridization between sexually distinctive parents and for the establishment of novel genetic traits (Miah et al., 2013). More recently, hereditary selection, recurring selection, crossing back, and induced mutagenesis have emerged as the primary approaches employed in traditional breeding.

13.4.1.1.1 The Selection of Pedigrees

The selection of pedigrees represents a breeding approach for improving the genetics of self-fertile crops/species in which superior genotypes are picked from the separating generations and for keeping a correct record of the ancestry of the individually selected plant in each generation. Pedigree breeding involves selecting and mating specific plants to produce offspring with desired traits (Bakhtiar et al., 2020). This process aims to maintain or enhance certain characteristics within a population. The main disadvantage of lineage selection is the time it takes to maintain evidence of a decision that is made at maturity and to evaluate lines on a regular basis during the growing season (Ankamah-Yeboah et al., 2020). The pedigree approach places the highest demands on knowledge of the subject matter and the relative contributions of environment and genetics to trait expression. The pedigree selection was done on two bread wheat populations in DS conditions, Debeira × Sahel

1 and Sids 6×Sahel 1, as the first and second populations, respectively, to improve specific morphological traits (Abd-El-Haleem et al., 2012). Under severe climatic conditions, the Cotton Research Station, in Bahawalpur evolved five cotton varieties namely Rohi (desi), Gohar-87, BH-118 (Minhas et al., 2018), BH-136, and BH-160 (Khan et al., 2006), and, now, a new variety 'BH-167' has been released based on the pedigree method. The new variety is a drought-tolerant one, requires less number of irrigations, and possesses the required traits to survive in stressful conditions. Another study on rice aimed to develop a new Japanese cultivar, HuHan2B. The researchers employed genome resequencing to analyze the genetic makeup of HuHan2B, utilizing five pedigree-breeding variants. Additionally, they investigated genes with differential expressions (DEGs) in the leaves by conducting transcriptome sequencing of HuHan2B and its recurrent parent gene, HanFengB (Wei et al., 2016). The knowledge gained from this investigation concerns genetic variation in drought resistance qualities that were transmitted down from parents to children after artificial selection through pedigree breeding is made possible.

13.4.1.1.2 Mutation Breeding

Mutation breeding is an advanced technique in plant breeding utilized to enhance crop adaptability, develop improved cultivars, induce haploids, induce male sterility, and encourage genetic diversity (Yali & Mitiku, 2022). Genetic mutations can play an important role by introducing new traits into a population. Improving crops through plant breeding relies on the presence of genetic diversity in advantageous traits (Swarup et al., 2021). Mutagenic agents, such as radiation and specific chemicals, can induce desired alterations. This approach serves in addition to traditional selection as it is highly successful in enhancing important qualities such as agronomic features, the physical attributes of grains, and the quality of food (Mba et al., 2013). The purpose of mutations created in conventional mutation breeding is to produce a novel variety, although it is exceedingly challenging to identify the affected genes in later selections. However, mutated genes can now be marked, assembled into an entire line of the highest selection, and tracked in later breeding programs (Bado et al., 2015). The benefits of selection by mutation comprise the making of novel genetic alleles not found in germplasm banks at this time, as well as the invention of new genetic ones in new varieties. The downside of selection by mutation is the limited capacity to generate the necessary dominant alleles; it is also not as efficient as interbreeding for features, which requires the integration of multiple genotypes. Cereal crops have consistently been the primary focus for cultivators. Regarding this, mutations offer the opportunity to create previously unknown alleles, allowing plant breeders to overcome limitations in allelic variation at specific gene loci of interest (Monroe et al., 2022). Globally, it appears that there have been notable advancements in the diversity of mutant plant species. The top five crops and 3,222 items in the mutation breed database are maize, rice, soybeans, wheat, and barley. It's interesting to note that more than half of the mutant varieties included in the database have improved botanic and agronomic traits. This might be owing to the simplicity with which these traits can be observed, as they rarely need specialist screening equipment (Tiwari et al., 2021).

13.4.1.1.3 Single Backcrossing

The process of backcrossing is a widespread breeding technique for plants. In many instances, it involves transferring only a handful of donor parental genes to a line that has been modified; its main aim is to distribute the material of the parental genome to the descendants (Lamichhane & Thapa, 2022). The single backcrossing method has proved highly effective in CIMMYT's wheat breeding system, enabling the transfer of a durable resistance. In-depth similarities carried out as part of a research study demonstrated the strengths of the strategy in comparison with single-crossing and repeated backcrossing methods. In addition, this approach favored the transfer of resistance genes and other low-impact secondary genes that may boost an adapted cultivar's grain performance or raise its yield potential (Wang et al., 2009). In rice breeding, backcrossing is an effective method, which entails introducing, or substituting an inherited gene transferred to the recipient parents from the original generation. This offers an accurate method for enhancing cultivars that differ in a multitude of aspects (Zhang et al., 2020). There is a history of crossbreeding throughout South and Southeast Asia, as a reproductive strategy to enhance reach their highest cultivars including KDML105, Basmati, and Manawthukha (Khan et al., 2020a).

13.4.1.1.4 Recurrent Selection

Recurrent selection is applied as a substitute for traditional breeding. It offers well-defined, shorter selection cycles, more detailed tracking of genetic benefits, and the ability to develop breeding lines with substantial genetic diversity (Bijma et al., 2020). This method was employed in the selection of the rice-resistant cultivar for drought (Khanna et al., 2022)17 years of historical data from IRRI's rice drought breeding program was used to estimate the genetic trends and assess the breeding program's success. We also identified top-performing lines based on grain yield breeding values as an elite panel for implementing future population improvement-based breeding schemes. A two-stage approach of pedigree-based mixed model analysis was used to analyze the data and extract the breeding values and estimate the genetic trends for grain yield under non-stress, drought, and in combined data of non-stress and drought. Lower grain yield values were observed in all the drought trials. Heritability for grain yield estimates ranged between 0.20 and 0.94 under the drought trials and 0.43–0.83 under non-stress trials. Under non-stress conditions, the genetic gain of 0.21% (10.22 kg/ha/year. Drought resilience through recurrent selection was once recorded in maize, and it has also been applied in wheat to get lines with higher early vigor that boost water and nutrient efficiency (Ramya et al., 2016). Recurrent selection has been chosen by breeders as a substitute for pedigree selection in practically all self-pollinating crops, including rice. Furthermore, breeding which involves the use of both genetic and phenotypic tags speeds up and boost's reliability (Spindel & Iwata, 2018).

Conventional breeding, which relies on gene expressions, has limitations, such as the epistatic effect and it can be a long-term and highly intensive task to produce a resistant variety. In contrast, biotechnological approaches such as MAS, marker-assisted backcrossing, mapping quantitative trait loci, marker-assisted gene pyramiding, and omics approaches, which necessitate the use of molecular indicators; the most often used DNA markers are restriction fragment length polymorphism (RFLP), simple sequence repeat (SSR), amplified fragment length polymorphism (AFLP), randomly amplified polymorphic DNA (RAPD), or single-nucleotide polymorphism (SNP). These modern approaches are crucial for introducing genes that provide tolerance from abiotic triggers (Shabbir et al., 2021).

13.4.1.2 Marker-Assisted Selection

Genetic markers help in selecting parents with different features as a basis for the emergence of newly appealing traits (Hasan et al., 2021). Plant breeders may use molecular markers to quickly and precisely improve elite breeds, granting resistance to abiotic stressors including drought (Shokat et al., 2020). Due to large variations between hybridized and native populations, it is critical to define them to select genotypes appropriate for farming and conserve them in a gene bank for future use. However, there are difficulties in mapping morphological markers, which are employed to measure genetic diversity. When variability in genetics is undetermined, phenotypic are expensive or difficult to evaluate, or observable variance results from intricate interactions between a few genes and/or DNA products, MAS gets adopted more and more (Visscher et al., 2021). The most important accomplishment of MAS aims to expedite the backcross gene selection procedure by incorporating a couple of donor genes into a recipient genotype. MAS is indeed straightforward, sturdy, resilient, and precise compared to standard breeding methods (Jamaloddin et al., 2020). It makes it possible to choose desirable qualities more quickly and precisely, which promotes the creation of superior crop types (Shrestha et al., 2020). For a situation of sorghum, natural variables can impact morphological characteristics. Although isoenzymes have been utilized, they do not provide sufficient data about sorghum varieties (Mwamahonje et al., 2021). The viability of atomic markers in deciding the connections between promotions fluctuates, for example, RFLP, AFLP, and SSR markers display high polymorphic information content. Moreover, RAPD indicators require modest quantities of DNA and are straightforward and efficient (Serrote et al., 2020). However, the benefits of MAS for molecular biotechnology studies include codominance, polymorphism, high informativeness, simplicity of application with relatively high resolution, and lower cost (Lande & Thompson, 1990). Currently, the most widely used SNP markers ordinarily involve sub-atomic markers for sorghum genotyping because of their adaptability and overflow (Enyew et al., 2022). In any case, ten biparental tropical maize populaces in sub-Saharan Africa led to a marker-helped choice (MAS) way to deal with increment grain creation under dry spell pressure (Beyene et al., 2016). The reason for the study was to assess the exhibition of half breeds created from C1S2 after three MARS cycles and research yield security in both very much watered (WW) and dry season pushed (DS) conditions. It's captivating to perceive how biotechnology has added to speed up how we might interpret the atomic elements involved in agricultural plants' responses to abiotic stress, especially with regard to dry season pressure in rice (Villalobos-López et al., 2022). The distinguishing proof of qualities like DRO18, OsNAC59, and OsGL1–6, and the investigation of their administrative components, are significant stages in improving dry spell obstruction. The utilization of quality organizations and profound sequencing advances have assisted how we might interpret pressure-inducible qualities at the entire genome level in rice (McNally et al., 2006). In any case, translating the complex systems' fundamental pressure resistance remains a provoking undertaking because of the contribution of numerous metabolic and morphologically versatile pathways (Sreenivasulu et al., 2007). MAS has brought about the advancement of various assortments in the field of cereals, such as HHB67-Further invented for pearl millet, HUW510 in wheat-based (Vasistha et al., 2017), Pusa Basmati rice1 upgraded (Gopalakrishnan et al., 2008), Pusa Basmati 1728, Pusa Basmati 1637, Pusa Samba 1850 (Singh et al., 2019), and afterward proposed Samba Mahsuri (Madhavi et al., 2016). Through MAS, the C214 genome was gained in chickpeas (Varshney et al., 2014). Fortunately, MAS is only attainable if the feature in the issue is associated with one or a few main genes, and it is neither practical nor relevant for quantitative traits, which reinforce the majority of stress tolerance traits (Bernardo, 2008).

There are several approaches through which molecular markers might facilitate the spreading of improved germplasm, such as marker-aided backcrossing/ pyramiding of genes (MABC, MAGP), and GS. MABC is the most basic sort of marker-assisted choice used to transfer one or few genes or main impact QTL that have been fine-mapped and confirmed over diverse ancestral backgrounds (Cobb et al., 2019).

13.4.1.3 Marker-Assisted Backcrossing

MABC is a raising action that uses molecular indicators to deal with the exchange of a trait or qualities of interest from a benefactor parent to a beneficiary parent (Shasidhar et al., 2020). This method allows for the rapid and precise introgression of specific features into an entirely novel gene pool, while minimizing the linkage drag of undesirable genes (Batieno et al., 2016). Crop modification programs have made extensive use of MABC in generating interesting varietals with enhanced properties and characteristics including tolerance to abiotic stress. This method has demonstrated efficacy and efficiency in accelerating the breeding process while preserving the recipient parent's genetic integrity (Kadirvel et al., 2015). MABC has been applied with success to multiple crops, particularly maize, rice, and wheat, to enhance both qualitative and quantitative traits without significant alterations to the genetic background (Benavente & Giménez, 2021). However, an MABC study was undertaken on wheat cultivar GW322 to explore the incorporation of QTLs that were for tolerance to drought attributes (Wani et al., 2020). Located in central India and peninsular regions, GW322 is commonly grown. The giver parent, HI1500, is drought-resistant and exhibits constant output under rainfed circumstances (Manu et al., 2020). In wheat, marker-assisted backcrossing was performed using the F1 parent line from the HI1500 donor parent and the recurrent parent GW322 to produce BC1F1 seeds. DNA markers linked to the targeted QTL region were used in MABC: Xgdm93 upon

2D/2A for NDVI (Olivares-Villegas et al., 2007); Xbarc68-Xbarc101 on the 3B for chlorophyll content/chlorophyll fluorescence (Kumar et al., 2012); Xwmc89 on 4A for yield (Kirigwi et al., 2007) under stress; and Xgwm111 on 7D for income (Kumar et al., 2010). A collection of 590 SSR markers across 21 chromosomes was used for background selection with at least ten loci per chromosomal arm chosen (Todkar et al., 2020). Kumar et al. (2023) demonstrate that using integration markers and phenotypic selection can be very beneficial for generating novel cultivars with desired characteristics, like drought tolerance. These methods enable more accurate and efficient plant selection, which can accelerate the creation of new agricultural breeds that are ideally suited to tough environments. Furthermore, the study clearly shows the advantages of employing marker-assisted backcrossing to precisely integrate desired traits into the wheat genomes to strengthen production, quality, and endurance to stress. A hybrid among HD2733 and C306 was carried out to incorporate DS tolerance QTLs into HD2733. Foreground SSR markers were used to identify the actual F1 generation, which was subsequently backcrossed with the recurrent parent. First, a set of 64 polymorphic loci was utilized to select both the foreground and the background for the BC1F1 progeny. A total of 25 lines were picked based on favorable QTLs, the most substantial recovery of the repeated parent's DNA, and similarity in phenotype to the receiving generation. To get BC2F1 and BC1F2 seeds, after, the repeating offspring crossed and backcrossed 21 chosen lines. Following that, the BC2F1 and BC1F2 lines were selected using 120 polymorphic SSR markers for both foreground and background traits. The markers associated with QTL characteristics were (Xbarc68-Xbarc101, Xgdm93, Xgwm165, and Xgwm301). These lines showed greater mean phenotype values for the physiological features and stabilizing attributes of HD2733, as well as greater production under water-stressed circumstances as compared to the receiver parent. MABC was used in wheat to transfer QTLs linked with drought tolerance, which influence characteristics including chlorophyll content, yield of grains, and 1,000 kernel weight, into outstanding Indian hybrids HD2733 and GW322 (Jain et al., 2014). Qyld.csdh.7AL, an important adaptable to drought QTL for yield, has been recently incorporated into the key wheat cultivars in India HUW234, HUW468, K307, and DBW17 by the use of MABC (Gupta et al., 2017). Ribaut and Ragot (2007) underlined the high value of choosing 10–20 genotypes using the MABC cycle to maximize genetic benefits in maize plants.

13.4.1.4 Mapping QTLs

A recent advance in the cultivation of plants called quantitative trait loci mapping renders genetically complex systems for abiotic stress tolerance appropriate for investigation, controlling particular agronomic features and physiological mechanisms through QTL manipulation, crop yield has been improved under environmental and abiotic stress circumstances (Paudel et al., 2020). Quantitative characteristics are those that demonstrate continuous variation (polygenic), and the genes that cause them are commonly referred to as QTLs (Beavis et al., 2021). According to Hussain et al. (2017), mapping entails placing the genes or QTL in a certain order, identifying their relative distances, and allocating them to linkage groups according with relation to their hybridization values. The assignment population frequently results from crosses between identical species with distinct traits. Several characteristics crucial for drought tolerance, including the ability to stay green, chlorophyll levels, leaf quantity, and leaf dimensions, like root traits, are governed by numerous genes situated in genomic regions known as QTLs (Anwar & Kim, 2020). QTL mapping is proposed as a promising method for studying the genetic basis of drought resistance (Price et al., 2002). It is possible to effectively boost resistance to drought in elite varieties or popular cultivars that exhibit good qualitative traits but are sensitive to DS by mapping loci closely linked to the genes that govern water stress traits (Swamy et al., 2013; Sandhu et al., 2018). To develop drought-resistant sorghum cultivars, it is critical to recognize and understand the QTLs associated with this trait. Particularly during the grain blossoming stage, the characteristic of staying green that is linked to chlorophyll content is crucial for sustained grain production under DS. Focusing on Genome-wide mapping of connections with high-density SNP marker arrays is crucial to properly identify QTLs linked with this trait and other features linked to drought tolerance (Harris et al., 2007; Abreha et al.,

2022). The steps involved in QTL mapping include the establishment of the mapping population, DNA isolation, PCR, urea page, and linkage analysis. In traditional breeding of drought-affected characteristics, Numerous QTLs have been discovered and mapped; for example, Diab et al. (2013) recently discovered QTLs for canopy heat, traits connected to transpiration, and the water status index. Additionally, the analysis emphasized potential genes and sequences with differential expression linked to QTLs for attributes related to drought tolerance. The agronomic value of the highest-performing cultivar frequently decreases when it incorporates the QTL allele from the most advantageous parental line (Tuberosa & Salvi, 2006). Numerous genes have been found and linked to molecular markers (Table 13.1); however, some of the aforementioned markers depend on RFLPs, which are too expensive and have a limited ability to detect polymorphisms for widespread MAS (Idrees & Irshad, 2014). Recently, in *Medicago sativa* L., There were 25 QTL for blooming time, 17 QTL for springtime production, and 6 QTL for cumulative summer biomass (Adhikari et al., 2019). Wang et al. (2014) found that a maximum of five qualitative trait loci (QSnp1b, QGyp2a, QSnp3a, QSnp11, and QSf8) remained frequently expressed among 20–30 QTLs linked with drought resistance characteristics. These QTLs were found in a hybrid between the somewhat water-tolerant Teqing and the drought-sensitive Lemont. It was found that qDTY1.1 (Vikram et al., 2011) and qDTY2.1 (Venuprasad et al., 2009) were connected with QSnp1b and QGyp2a, correspondingly (Wang et al., 2014). In light of chromosome 12, the qDTY12.1 was quick to be recognized for rice seed creation under raised regenerative stage DS (Bernier et al., 2007). As mentioned by Bernier et al. (2007, 2009), there is a significant improvement in grain yield under drought tolerance conditions, which is primarily due to the explained genetic variation (ranging from 43% to 51%) and the consistent effects observed across various environments. qDTY12.1 is the main locus used to make dry spell-safe lines with further developed grain creation across exceptionally shifted rainfed and marsh rice conditions (Bernier et al., 2007; Mishra et al., 2013). Among the foremost recent cases of crucial QTLs for yields of grains discovered under dry season pressure is the QTL qDTY1.1 on chromosome 1, which has added substance influences going from 8.55% to 30%. The favorable allele from dry spell-safe assortments has been detected in several hereditary foundations and contributors, including N22, according to research by Vikram et al. (2011) and Ghimire et al. (2012).

TABLE 13.1
Major Cereal Genes Related to Drought Tolerance

Crops	Gene Name	Category Function	Abiotic Stress	References
Wheat (*Triticum aestivum*)	*TaPYL4*	ABA receptor	Drought stress	Mega et al. (2019); Feng et al. (2022); Rauf et al. (2024); Wang et al. (2023)
	TaSnRK2.9	Protein kinase and protein phosphatase		
	TaAREB3	Transcription factor and cofactor		
	TaSDIRl	Protein turnover		
Rice (*Orysa sativa*)	*OsPYL9*	ABA receptor		Hou et al. (2009); Jiang et al. (2012); Zhang et al. (2012); You et al. (2013)
	OsCIPK, OsMPKKl0.2, OsPP2C09	Protein kinase and protein phosphatase		
	OsMYB102, OsDST, OsSKIPa, OsRIP18	Transcription factor and cofactor		
	OsLEA3-2	Osmotic regulation		
	OsSIZl	Protein modification		
Maize (*Zea mays*)	*ZmGLOSSY13*	Cuticle wax		Wei et al. (2011); Li et al. (2013); Wang et al. (2016); Nuccio et al. (2015)
	ZmVPPl, ZmbetA, ZmTsVP	Transporter		

In tough highland and marsh dry season circumstances, two additional QTLs from the rice cultivar – on chromosome 2, qDTY2.3, and on chromosome 3, qDTY3.2, when combined with qDTY12.1 – increase yield and record of harvest in qDTY12.1-positive lines (Dixit et al., 2012; Tang et al., 2013). Under outrageous drought pressure, qDTY2.3 had a huge effect, adding 18–31% expanded grain creation (Yadaw et al., 2013; Saikumar et al., 2014). In addition, significant QTLs with pleiotropic consequences for yield under dry spell pressure were found on chromosome 4. Along these lines, Bernier et al. (2009) revealed a QTL, qtl12.1, that significantly impacted the yield under DS. An examination of twofold haploids obtained from rice lines, exposed to water pressure, distinguished 47 QTLs connected with credits connected with regenerative development and yield with a scope of the complete phenotypic impacts fluctuated from 9.5% to 35.6% (Sellamuthu et al., 2011), likewise, Uga et al. (2011) recognized an important QTL (Dro 1) on a ninth chromosomal that is necessary to deep roots, and Obara et al. (2010) discovered a crucial gene locus (qRL6.1) for determining the root length on the sixth chromosomal which is necessary for the development of a root system that can withstand severe drought by absorbing water from the soil's deep layers.

13.4.1.5 Marker-Assisted Gene Pyramiding

The technique known as 'gene pyramiding' involves combining a number of desirable genes from several excellent parent plants into one offspring to increase the plant's capacity to flourish under harsh conditions. Identifying and selecting the original parent plants possessing the desired genes, repeatedly crossing these plants to produce a genotype harboring all target genes, and ultimately stabilizing the selected genes for various desirable traits into a recessive condition to prevent segregation in subsequent generations are the three primary steps involved in successful gene pyramiding (Wani et al., 2020; Haque et al., 2021). For improved selection, appropriate surround markers are frequently connected to the genes of interest rather than functional ones. These days, one of the most often used molecular markers is single-nucleotide polymorphisms. Three common approaches are frequently used to manage MAGP; stepwise move, simultaneous exchange, and simultaneous and stepwise movement (Mekonnen et al., 2017). In the stepwise transfer process, Repeated Parent 1 (RP1) and Donor Parent 1 (DP1) merge to produce an F1 hybrid. To produce the improved recurrent parent (IRP1), this hybrid is then self-crossed and backcrossed until it reaches the third generation (BC3). The improved repeated parent is then obtained by backcrossing these enhanced individuals with the recurrent parent. The simultaneous and stepwise transfer approach, which combines the preceding two methods, is the most practical. In this method, RP1 is crossed with numerous donor parents at the same time and then backcrossed till the BC3 generation. After that, the backcross hybrids are intercrossed to create IF1, which is then self-crossed once again to create the pyramided lines. Targeted genes for drought tolerance in key crops include OsCDPK7/calcium-dependent protein kinase in rice, HVA1/LEA proteins in barley and wheat, and MAPKKK/NPK1 in maize (Nishihama & Machida, 2000; Hussain et al., 2021; Samtani et al., 2022). Similar to this, MAGP has been employed to generate drought-tolerant pyramided lines (MR219) in rice, which have the ability to produce more than 1,500 kg/ha in areas with a shortage of water (Selamat & Nadarajah, 2021).

13.4.2 Omics-Based Strategies

The most recent breakthroughs in omics methods, encompassing proteomics, genomics, and metabolomics, have created novel prospects for understanding plant adaptations to abiotic stress at several scales, spanning the entire plant to the molecular level. Deep sequencing technologies and other omics methods, especially metabolomics, have the potential to fast-track the creation of resilient to stress crops by providing valuable insights into the processes of cultivar stress tolerance. Furthermore, a study on epigenetics enriches our understanding of plant stress memory mechanisms. Moreover, functional molecular markers may be used in MAS more quickly because of developments in molecular breeding techniques. Genomic technologies can also help interpret the gene networks linked to plant stress as mentioned in Figure 13.2 (Zargar et al., 2022).

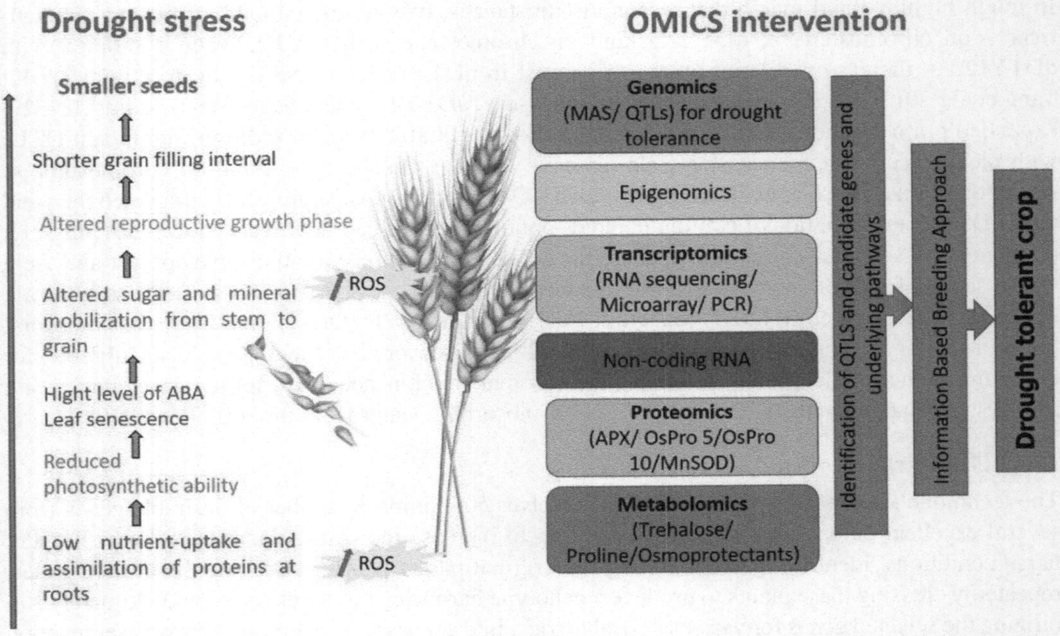

FIGURE 13.2 Representation of the OMICS strategies used to discover genes needed for breeding drought-resistant plants.

13.4.2.1 Transcriptomic

To comprehend the physiological and molecular functions behind grain drought resistance, it is essential to identify putative genes. This will make it possible to use transgenic methods in breeding to strengthen defenses against abiotic stress (Onyemaobi et al., 2022). A transgenic strategy is one in which genes are transferred from one species to another to modify some structural aspects of a characteristic (Ashraf & Ahmad, 2023). Since the abiotic stress reactions' fundamental regulatory structures are thoroughly understood, it will become evident that a growing number of suspected genes are currently applied in the production of transgenic plants (Barnabás et al., 2008). Numerous genes linked to resistance to drought have been revealed. Transcription factors influence gene expression, both upward and downward (Manna et al., 2021).

The complex stress response systems that plants have developed include transcription factor-encoding genes (TFs), which regulate the expression of transcripts that are responsive to stress. Adaptation to several environmental stressors can be strengthened by TFs. Although TFs have been overexpressed and modulated in crop plants (Hrmova & Hussain, 2021). The functions of TF relatives, comprising NAC, MYB, WRKY, bZIP, and ERF/drought-responsive element binding (DREB) in initiating reactions to abiotic stress have been well investigated (Santos et al., 2011). TFs are encoded by around 10% of genes and perform important functions throughout the plant's life process. Understanding the mechanisms behind TF activities is crucial for studying how plants respond to various abiotic stresses (Joshi et al., 2016b). Next-generation sequencing and chromatin immunoprecipitation with massively concurrent sequencing (ChIP-Seq) are instances of novel methods that may help discover genomic areas that further clarify the role of TFs (Werner, 2010). Thus, the CRISPR/Cas gene-editing method may be used to modify TFs; it is crucial to address the functional redundancy of TFs (Wolter & Puchta, 2018). Although previous research on TF overexpression in response to specific stresses has been beneficial, more investigation is required to determine whether overexpressing TFs linked to stress in transgenic plants promotes growth and stress tolerance without negatively impacting yield (Bhatnagar-Mathur et al., 2008). To fully understand the modes of action of different TFs and how they contribute to enhancing the resilience of food crop grains under stress, more study is needed. Further stress-responsive genes found include functional genes that code for metabolic components such as abundant late embryogenesis proteins (LEAs) and osmo-protectant-synthesizing enzymes (Abdeeva et al., 2012). DREB factors are the greatest and most widely researched family of transcription factors. DREB1A and DREB2A have been identified in cereal crops and *Arabidopsis* (Hussain et al., 2021). Moreover, research on DREBs was initially conducted on *Arabidopsis*, according to Akhtar et al. (2012), overexpression of DREB1/CBF in *Arabidopsis* activated the expression of many genes involved in stress tolerance, greatly increasing the plant's resistance to abiotic stressors. Generally, one may use either the constitutive promoter (CaMV 35S) or the dehydration-inducible promoter (rd29A) to induce DREB1A overexpression (Conforte et al., 2017). Kasuga et al. (1999) discovered that constitutive CaMV 35S promoter-induced overexpression of CBF3/DREB1A significantly enhances the plants' resilience to abiotic stresses like water shortages, in transgenic *Arabidopsis* plants. Parallel to this, wheat showed better drought tolerance when the stress-inducible promoter (rd29A) was used in combination with DREB1 (Pellegrineschi et al., 2004). Plants can withstand desiccation because these conditions activate the RD29 genes in them (Jia et al., 2012).

13.4.2.2 Proteomic

To comprehend the molecular mechanisms behind water stress in cereal crops, proteomic analysis has been employed (Ghatak et al., 2017). Recognizing the mechanisms underlying drought resistance in wheat requires identifying several proteins to better understand their function. In this context, several instruments are employed comprising 2-DE, DIGE, iTRAQ, and LC–MS/MS (Singh et al., 2022).

13.4.2.2.1 Proteins Involved in Biogenesis Adjust to DS

Plants adjust their system of roots to adapt to severe drought. According to many studies, the presence of drought has been found to impact the production of lignin and cellulose, and variations in the form of the roots or shoots influence the degree of biogenesis-related proteins (Niu et al., 2019; Ganie & Ahammed, 2021; Huang et al., 2024). According to Guo et al. (2014), actin and *S*-adenosylmethionine synthetase take part in the methylation of lignin monomers, thereby helping plants endure DS. Overexpressing *S*-adenosylmethionine synthetase during DS enhances the capacity of the plant to endure the stress. When actin isoforms are present, thicker filaments are produced more often, which, in reaction to stress, raises the density of filamentous generated by actin and improves the rigidity of the cell structure (Kijima et al., 2018). According to Agrawal et al. (2016), a number of rice proteins, among them α-tubulin, putative elongation factor 2, cysteine synthase, putative beta-alanine synthase, and putative thiamine biosynthesis protein, were recognized as being prevalent under DS. Nevertheless, other proteins such as triose phosphate isomerase, tricin synthase 1, proteasome subunit alpha type 1, proteasome subunit beta type 3, homocysteine methyltransferase, and ubiquitin-activating enzyme E1 and E2 were also detected. Peptidyl-prolyl cis/trans isomerases were shown to be prevalent in wheat under drought (Ford et al., 2011).

13.4.2.2.2 Modulation of Antioxidants Defense-Related Proteins during Extreme Drought

During water stress, plants stimulate their enzymatic machinery, boosting the amount of guaiacol peroxidase (GPOX) and other antioxidant proteins, monodehydroascorbate reductase, and CAT. To combat the elevated levels of ROS brought on by DS, antioxidant enzyme synthesis is increased (Ilyas et al., 2021). Several antioxidant proteins linked to cell defense and recovery under DS were found in proteomic research on rice (Wang et al., 2017; Auler et al., 2021). Paradoxically, amino acids such as HSPs, trigger factor-like proteins, ER-luminal binding proteins, and other molecular chaperones were discovered in massive quantities. These polypeptides have a variety of functions that include preserving protein structure and renaturing damaged proteins caused by DS (Singh et al., 2022). However, in attempts to mitigate against the emergence of ROS, chickpea cultivars increase the oxidative stress defense response (Gupta et al., 2020).

13.4.2.2.3 Adjustment of Proteins Related to Photosynthesis under Drought Conditions

Respiration, metabolic activities, signal transduction, phosphorus transmission, photosynthesis, protection against oxidative stress, and methyl transduction are all examples of important protein expressions in cellular metabolism (Muhammad et al., 2021). The plant can adapt to severe circumstances thanks to these expressions. Furthermore, DS can disrupt photosynthesis and trigger the immune system's defensive mechanisms against free radicals. Most of the proteins implicated in photosynthesis include the ferredoxin-NDP reductase, the Rubisco activase, the chlorophyll a-b-binding protein LHCII, and the short-chain ribulose bisphosphate carboxylase (Figure 13.3) (Das et al., 2016). As Yamori and Shikanai (2016) point out, these properties are required for the electron transport chain, the Calvin cycle, and light harvesting, the three phases of photosynthesis. In a case study by Yu et al. (2016), the abundance of the enzyme Rubisco activase decreased 472-fold in the stressed genotype under severe DS conditions, while it increased 67-fold in the drought-tolerant genotype of soybean cultivars, including the drought-tolerant genotype (GN-2032) and the drought-sensitive one (GN-3074). Overall, the results demonstrate that the impact of DS on light response is significant, with the amount of light response-related proteins increasing relative to normal conditions. However, under DS, the number of photosynthetic proteins fell in the GN-2032 genotype. One Rubisco-linked protein that is crucial for regulating photosynthesis is Rubisco activase (Galmés et al., 2013). Under DS, the expression of this enzyme did not alter in the drought-resistant genotype GN-3074, while it declined in the hypersensitive genotype GN-2032. In a similar vein, under DS circumstances, Rubisco's expression in GN-3074 did not alter, while it significantly decreased in GN-2032 (Yahoueian et al., 2021).

Stay green, a mechanism in plants that allows them to withstand dryness, is a valuable crop species agronomic property (Kamal et al., 2019). Under extreme drought circumstances, it permits the preservation of photosynthetic activity and chlorophyll concentration in leaves for an extended period of time (Kumar et al., 2022). When Galyuon et al. (2019) looked into the molecular processes of remaining green in drought-stressed Sorghum, they discovered that genotypes with high amounts of photosynthetic proteins stayed green. This suggests that in stressful circumstances, protecting and maintaining the ability to photosynthesis is an essential component of the stay-green phenotypes. Bazargani et al. (2011) observed reduced photosynthetic protein content in aging leaves compared

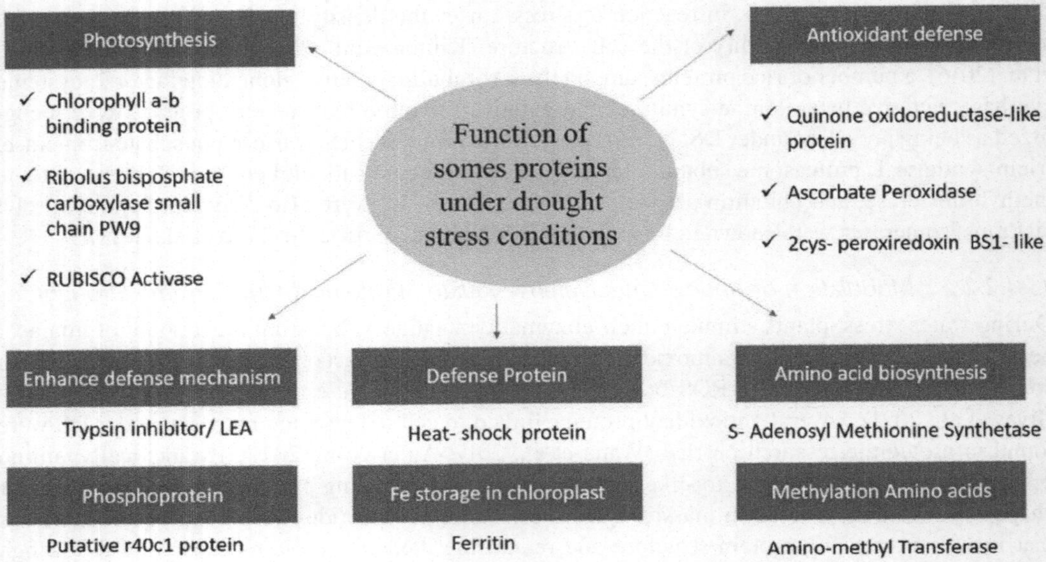

FIGURE 13.3 Some fundamental proteins involved in plants survived drought stress.

to previous wheat studies. Iqbal et al. (2020) revealed that when there is a rise in stomatal conductivity, a drought-tolerant genotype seeks to maintain sugar production under stress by stabilizing or enhancing photosynthetic enzymes (stay green). The preservation of stomatal conductivity may have enabled the tolerant genotype to achieve greater gas exchange, hence, there was no decrease in enzyme expression, resulting in increased sugar synthesis. The findings provide a new understanding of the stay-green mechanism for soybean tolerance to DS by controlling proteins linked to the photosynthetic apparatus to assist the plant survive under water deficit (Sabagh et al., 2020). Essentially, proteins are markers that may be applied to breeding initiatives (Jonah et al., 2011).

13.4.2.3 Metabolomic Responses

Molecular breeding and molecular genetics are critical in raising the genetic resistance of plants to abiotic stress, leading to stable yields in unfavorable environmental conditions (De la Rosa et al., 2020). Under stress, plants usually slow down their vegetative growth and speed up their reproductive processes. We still don't completely understand the precise genetic variants and processes that drought-tolerant plants use to save soil moisture and postpone biomass increase until grain filling. Metabolomics encompasses a wide variety of analytical techniques used to recognize the components of organic molecular metabolomics. These methods include metabolic fingerprinting, metabolite profiling, gas chromatography–mass spectrometry, liquid chromatography–mass spectrometry, nuclear magnetic resonance, and targeted analysis. These tools may be used to precisely identify metabolomic components (Guo et al., 2020). Additionally, distinct metabolic modifications in sugar accumulation, amino acids, and proline content were noted with extreme dehydration stress, Maize yields under well-watered regimes and mild drought during blooming have been demonstrated to rise with the production of a biochemical enzyme that transforms trehalose-6-phosphate (T6P) to trehalose in the phloem companion cells at the base of the ear and developing flowers (Zhou et al., 2014). T6P modulation enhances photosynthesis during grain filling, improves the activity of leaves during transpiration, and facilitates photosynthates' access to unfertilized heads (Bailey-Serres et al., 2019). Thus, to enhance agricultural output stability under unfavorable climatic circumstances, unique genetic variants and enhancements can integrate metabolism and stress tolerance (Mickelbart et al., 2015).

13.4.2.4 Genomic Approaches

Drought is a complicated characteristic since it is polygenic and has poor heritability (Cattivelli et al., 2008). Knowing the process behind gene action for different agronomic characteristics under DS and ideal circumstances is accomplished via research on gene expression and trait combination. Moreover, GS and high-throughput phenotyping (HTP) are potential techniques for accelerating breeding progress for cereal varieties that are tolerant to climate change and produce massive yields. Plant breeding communities worldwide are showing interest in GS and HTP technologies (Cabrera et al., 2012). These methods have the potential to completely change the complexity of qualities like growth, yield, and stress tolerance. Although HTP has the potential to enhance comprehension of agricultural physiology, its most effective uses are empirical, as opposed to analytical, and are similar to GS (Gehan & Kellogg, 2017).

13.4.2.4.1 Genomic Selection

Precision and repeatability are two issues with traditional abiotic stress selection techniques. Even though yield QTLs under abiotic stress circumstances have been found and transmitted via DNA markers (Ribaut & Ragot, 2007), These strategies might not work because genetic history has a substantial influence on QTLs with limited genetic resources, which only partially explain variance in grain production under stress, the environment, and its interactions (Semagn et al., 2013). According to Cerrudo et al. (2018), GS is better than MAS and has a greater prediction effectiveness for abiotic stress tolerance. For drought resistance, GS has been shown to be useful in rice, wheat, and maize. GS, which integrates genome-wide indicator information to estimate genomic breeding values for line selection, has been proposed as a method to exploit data from molecular markers for complicated characteristics (Desta & Ortiz, 2014). However, in GS, marker effects are assessed in a 'training population' of phenotypes and extracted lines, which are then used to forecast the potential for breeding of genotyped lines in the 'validation population' (Lorenz et al., 2011). It has been shown how applying GS to wheat might improve quality features such as flour yield, flour protein, lactic acid, water absorption, softness equivalence, and stem rust in addition to other significant quantitative qualities like grain yield (Heffner et al., 2011; Poland et al., 2012). Depending on the GS models, marker imputed strategies, population size, and number of markers hired, the precision of GS in predicting these elements varies (Rutkoski et al., 2015). In breeding tasks, GS is a potentially helpful method to enhance individuals' genetic gains (Yuan et al., 2019). Genotypic and genotypic knowledge people comprise the training population, which is the first step in the basic process of GS. Next, a model is constructed using this population, with the phenotype serving as the responder and the genotype serving as the predictor. The breeding population, which is made up of individuals with just genotypic information, has its genomic estimated breeding value determined using the information derived from the developed model. Adopting GS has the primary advantage of allowing for a significant reduction in the breeding cycle's duration compared to standard breeding and minimizing the expense of comprehensive phenotyping (Xu et al., 2020). This expedites reproductive gains and ensures the security of food and nutrition (Sinha et al., 2021). Beyene et al. (2015) indicated that using the quick cycle GS technique resulted in a grain production increase of 0.086 t/ha in eight biparental populations of drought-stricken maize. The final gain, following three selection cycles, was 0.176 t/ha. In an analogous line, Das et al. (2020) indicated that fast cycle selection for genomes allowed for a genetic boost of 0.110 and 0.135 t/ha/year for grain output during drought and 0.038 and 0.113 t/ha/year for grain production during floods in two maize individuals, Maize Yellow Synthetic 1 and Maize Yellow Synthetic 2.

13.4.2.4.2 High-Throughput Phenotyping

Phenomics is the investigation of high-throughput phenotypic assessment. Breeders are able to examine large breeding populations for grain output under challenged conditions now that accurate, quick, and cost-effective phenotyping methods are available (Yang et al., 2020). Because of its correlation with grain output, canopy heat is a physiological feature that is frequently used to screen for adaptation to drought-stressed conditions (Blum et al., 1989; Reynolds et al., 1999; Mason & Singh, 2014). Moreover, spectral reflectance matching to various radiation frequency ranges, which can be remotely observed without causing destruction, incorporates the crop's productivity throughout time (Thenkabail et al., 2018). This trait helps select physiologically superior plants, is repeatable, and is highly heritable. The advent of improved HTP platforms, including piloted airplanes and drones, has made it possible to automate accurate phenotypes of huge breeding populations based on spectral reflectance (Smith et al., 2021). According to Haghighattalab et al. (2016), these platforms have the ability to completely transform selection processes, help with physiological screening, and improve our knowledge of the genetic regulation of characteristics. Vegetable indices, which provide information on the amount of pigment and green biomass of the plant, level of hydration, and more, are able to be made using the spectrum reflectance data gathered from HTP platforms (Polivova & Brook, 2021; Tayade et al., 2022). HTP methods, notably research on root characteristics, estimation of water usage efficiency, and studies on evapotranspiration, have provided a fresh boost to accuracy phenotyping, which is necessary for genomics-assisted selection (Tuberosa, 2012).

13.5 DIFFICULTIES RELATING TO THE EFFECTIVENESS OF USING MOLECULAR TOOLS AGAINST DS

The epistatic effect is one of the limitations of conventional breeding, which depends on gene expressions (Xu & Crouch, 2008). Using traditional breeding techniques to create a resistant variety can be an extensive and highly challenging process. Furthermore, to effectively use omics technology in enhancing agricultural plant resistance to abiotic stress, present knowledge gaps must be filled using a systems biology approach (Jauhar, 2006). Optimizing HTP instruments and techniques is necessary to find pertinent genes for crop breeding. These gaps can be filled with new machine learning algorithm-based techniques for sophisticated data analysis. Furthermore, integrated remote sensing technologies are increasingly used in addition to traditional agricultural production techniques to assess crop stress (Kamarudin et al., 2021). Improvements in high-throughput capabilities phenotyping, remote sensing methods, and imaging and sensing technologies enable prompt assessments of plant health and stress sensitivity or tolerance at the field scale. These assessments inform management decisions for cereal crops under stressful conditions. To find tolerant genes, it is crucial to methodically assess large germplasm collections and genotypes and to create focused breeding programs that use both old and innovative approaches to reducing environmental stress to boost yield and cereals endurance (Nkomo et al., 2021; Adhikari et al., 2022). While a number of studies (Crossa et al. 2014; Hayes et al. 2017; Juliana et al. 2017) have shown the value of GS for intricate characteristics in wheat, evaluating forward and backward genetic predictions was one of its primary goals for grain yield across various preschools and years, as well as to look into the possibility of utilizing GS in breeding programs. MAGP needs more efficient transformation approaches for the establishment of multiple stress tolerance with all other management strategies on a large scale (Dormatey et al., 2020).

13.6 CONCLUSION

The creation of climate-tolerant agricultural cultivars is crucial for ensuring a stable supply of food and energy in a changing environment. Drought poses an immediate threat to worldwide food supply as it drastically reduces plant growth and productivity. Growth and metabolism are severely impacted by DS, which has a profound impact on morphophysiological and biochemical traits. Crop plants use specialized mechanisms to make them more resilient to drought to be able to counteract the detrimental effects of water deficit. Certain crops exhibit avoidance tactics or acclimate to DS, while some plants tolerate it. Gene expression modulation and regulation are crucial plant defense strategies against DS. The combination of conventional breeding strategies and biotechnology (MAS, MABC, QTLs mapping, MAGP) can increase drought tolerance in plants and present another strategy to manage DS to generate drought-resistant alleles. Consequently, it is critical to look at unidentified adaptive traits in a variety of plant species and combine them into genotypes that are resistant to the detrimental effects of DS without compromising productivity. In this regard, omics applications utilizing specialized methodologies including transcriptomics, proteomics, metabolomics, genomics, and phenomics are crucial.

ACKNOWLEDGMENTS

Financial support for this study was provided by Cadi Ayyad University and the Ecole Nationale d'Agriculture de Meknès.

Funding: The authors gratefully acknowledge the FOSC project (Sus-Agri-CC) from the European Union's Horizon 2020 research and innovation program under grant agreement N 862555.

REFERENCES

Abdeeva, I., Abdeev, R., Bruskin, S., & Piruzian, E. (2012). Transgenic plants as a tool for plant functional genomics. In Y. Ö. Çiftçi (Ed.), *Transgenic plants-advances and limitations* (pp. 259–284). London: IntechOpen.

Abd-El-Haleem, S. H. M., Ehab, M. R., & SMS, M. (2012). Efficiency of pedigree selection in bread wheat under drought stress conditions. I. Morphological traits. *Life Science Journal*, *9*(4), 3423.

Abreha, K. B., Enyew, M., Carlsson, A. S., Vetukuri, R. R., Feyissa, T., Motlhaodi, T., Ng'uni, D., & Geleta, M. (2022). Sorghum in dryland: Morphological, physiological, and molecular responses of sorghum under drought stress. *Planta*, *255*, 1–23.

Adel, S., & Carels, N. (2023). Plant tolerance to drought stress with emphasis on wheat. *Plants*, *12*(11).

Adhikari, L., Makaju, S. O., & Missaoui, A. M. (2019). QTL mapping of flowering time and biomass yield in tetraploid alfalfa (Medicago sativa L.). BMC Plant Biology, *19*(1), 1–15.

Adhikari, S., Kumari, J., Jacob, S. R., Prasad, P., Gangwar, O. P., Lata, C., Thakur, R., Singh, A. K., Bansal, R., Kumar, S., Bhardwaj, S. C., & Kumar, S. (2022). Landraces-potential treasure for sustainable wheat improvement. *Genetic Resources and Crop Evolution*, *69*(2), 499–523.

Agrawal, L., Gupta, S., Mishra, S. K., Pandey, G., Kumar, S., Chauhan, P. S., Chakrabarty, D., & Nautiyal, C. S. (2016). Elucidation of complex nature of PEG induced drought-stress response in rice root using comparative proteomics approach. *Frontiers in Plant Science*, *7*, 1466.

Ahmar, S., Gill, R. A., Jung, K. H., Faheem, A., Qasim, M. U., Mubeen, M., & Zhou, W. (2020). Conventional and molecular techniques from simple breeding to speed breeding in crop plants: Recent advances and future outlook. *International Journal of Molecular Sciences*, *21*(7), 2590.

Ahmad, M., Waraich, E. A., Shahid, H., Ahmad, Z., Zulfiqar, U., Mahmood, N., Al-Ashkar, I., Ditta, A., & Sabagh, A. El. (2023). Exogenously applied potassium enhanced morpho-physiological growth and drought tolerance of wheat by alleviating osmotic imbalance and oxidative damage. *Polish Journal of Environmental Studies*, *32*(5), 4447–4459.

Akhtar, M., Jaiswal, A., Taj, G., Jaiswal, J. P., Qureshi, M. I., & Singh, N. K. (2012). DREB1/CBF transcription factors: Their structure, function and role in abiotic stress tolerance in plants. *Journal of Genetics*, *91*, 385–395.

Anderegg, W. R., Klein, T., Bartlett, M., Sack, L., Pellegrini, A. F., Choat, B., & Jansen, S. (2016). Meta-analysis reveals that hydraulic traits explain cross-species patterns of drought-induced tree mortality across the globe. *Proceedings of the National Academy of Sciences*, *113*(18), 5024–5029.

Ankamah-Yeboah, T., Janss, L. L., Jensen, J. D., Hjortshøj, R. L., & Rasmussen, S. K. (2020). Genomic selection using pedigree and marker-by-environment interaction for barley seed quality traits from two commercial breeding programs. *Frontiers in Plant Science*, *11*, 539.

Anwar, A., & Kim, J. K. (2020). Transgenic breeding approaches for improving abiotic stress tolerance: Recent progress and future perspectives. *International Journal of Molecular Sciences*, *21*(8), 2695.

Ashraf, M. S., & Ahmad, A. (Eds.). (2023). Current understanding of genomics, transcriptomics, proteomics, and metabolomics of crop plants under low nutrient stress. In *Genomics, transcriptomics, proteomics and metabolomics of crop plants* (pp. 89–130). Academic Press.

Askari-Khorasgani, O., Pardo, F. B. F., & Pessarakli, M. (2021). Drought stress sensing-signaling in plants. In M. Pessarakli (Ed.), *Handbook of plant and crop physiology* (pp. 465–481). CRC Press.

Auler, P. A., do Amaral, M. N., Braga, E. J. B., & Maserti, B. (2021). Drought stress memory in rice guard cells: Proteome changes and genomic stability of DNA. *Plant Physiology and Biochemistry*, *169*, 49–62.

Bado, S., Forster, B. P., Nielen, S., Ali, A. M., Lagoda, P. J., Till, B. J., & Laimer, M. (2015). Plant mutation breeding: Current progress and future assessment. *Plant Breeding Reviews*, *39*, 23–88.

Bailey-Serres, J., Parker, J. E., Ainsworth, E. A., Oldroyd, G. E., & Schroeder, J. I. (2019). Genetic strategies for improving crop yields. *Nature*, *575*(7781), 109–118.

Bakhtiar, F., Najafian, G., & Sharsoda, N. (2020). Selection of drought tolerant bread wheat lines based on tolerance indexes and pedigree analysis. *Applied Field Crops Research*, *33*(02), 114–134.

Bapela, T., Shimelis, H., Tsilo, T. J., & Mathew, I. (2022). Genetic improvement of wheat for drought tolerance: Progress, challenges and opportunities. *Plants*, *11*(10), 1331.

Barnabás, B., Jäger, K., & Fehér, A. (2008). The effect of drought and heat stress on reproductive processes in cereals. *Plant, Cell & Environment*, *31*(1), 11–38.

Batieno, B. J., Danquah, E., Tignegre, J. B., Huynh, B. L., Drabo, I., Close, T. J., Ofori, K., & Ouedraogo, T. J. (2016). Application of marker-assisted backcrossing to improve cowpea (Vignaunguiculata L. Walp) for drought tolerance. *Journal of Plant Breeding and Crop Science*, *8*(12), 273–286.

Bazargani, M. M., Sarhadi, E., Bushehri, A. A. S., Matros, A., Mock, H. P., Naghavi, M. R., Hajihoseini, V., Mardi, M., Hajirezaei, M.-R., Moradi, F., Ehdaie, B., & Salekdeh, G. H. (2011). A proteomics view on the role of drought-induced senescence and oxidative stress defense in enhanced stem reserves remobilization in wheat. *Journal of Proteomics*, *74*(10), 1959–1973.

Beavis, W., Merrick, L., Meade, K., Campbell, A., Muenchrath, D., & Fei, S. Z. (2021). *Inheritance of quantitative traits* (pp. 1–64). Ames, IW: Plant Breeding E-Learning in Africa.

Benavente, E., & Giménez, E. (2021). Modern approaches for the genetic improvement of rice, wheat and maize for abiotic constraints-related traits: A comparative overview. *Agronomy*, *11*(2), 376.

Bernardo, R. (2008). Molecular markers and selection for complex traits in plants: Learning from the last 20 years. *Crop Science*, *48*(5), 1649–1664.

Bernier, J., Kumar, A., Ramaiah, V., Spaner, D., & Atlin, G. (2007). A large-effect QTL for grain yield under reproductive-stage drought stress in upland rice. *Crop Science*, *47*(2), 507–516.

Bernier, J., Kumar, A., Venuprasad, R., Spaner, D., Verulkar, S., Mandal, N. P., Sinha, P. K., Peeraju, P., Dongre, P. R., Mahto, N., & Atlin, G. (2009). Characterization of the effect of a QTL for drought resistance in rice, qtl12.1, over a range of environments in the Philippines and eastern India. *Euphytica*, *166*, 207–217.

Beyene, Y., Semagn, K., Crossa, J., Mugo, S., Atlin, G. N., Tarekegne, A., Meisel, B., Sehabiague, P., Vivek, B. S., Oikeh, S., Alvarado, G., Machida, L., Olsen, M., Prasanna, B. M., & Bänziger, M. (2016). Improving maize grain yield under drought stress and non-stress environments in sub-Saharan Africa using marker-assisted recurrent selection. *Crop Science*, *56*(1), 344–353.

Beyene, Y., Semagn, K., Mugo, S., Tarekegne, A., Babu, R., Meisel, B., Sehabiague, P., Makumbi, D., Magorokosho, C., Oikeh, S., Gakunga, J., Vargas, M., Olsen, M., Prasanna, B., Bänziger, M., & Crossa, J. (2015). Genetic gains in grain yield through genomic selection in eight bi-parental maize populations under drought stress. *Crop Science*, *55*(1), 154–163.

Bhardwaj, A., Devi, P., Chaudhary, S., Rani, A., Jha, U. C., Kumar, S., Bindumadhava, H., Vara Prasad, P. V., Sharma, K. D., Siddique, K. H. M., & Nayyar, H. (2021). 'Omics' approaches in developing combined drought and heat tolerance in food crops. Plant Cell Reports, *41*, 1–41.

Bhatnagar-Mathur, P., Vadez, V., & Sharma, K. K. (2008). Transgenic approaches for abiotic stress tolerance in plants: Retrospect and prospects. *Plant Cell Reports*, *27*, 411–424.

Bijma, P., Wientjes, Y. C., & Calus, M. P. (2020). Breeding top genotypes and accelerating response to recurrent selection by selecting parents with greater gametic variance. *Genetics*, *214*(1), 91–107.

Blum, A., Shpiler, L., Golan, G., & Mayer, J. (1989). Yield stability and canopy temperature of wheat genotypes under drought-stress. *Field Crops Research*, *22*(4), 289–296.

Bogati, K., & Walczak, M. (2022). The impact of drought stress on soil microbial community, enzyme activities and plants. *Agronomy*, *12*(1), 189.

Bratovcic, A. (2020). Antioxidant enzymes and their role in preventing cell damage. *Acta Scientific Nutritional Health*, *4*, 1–7.

Cabrera-Bosquet, L., Crossa, J., von Zitzewitz, J., Serret, M. D., & Luis Araus, J. (2012). High-throughput phenotyping and genomic selection: The frontiers of crop breeding converge F. *Journal of Integrative Plant Biology*, *54*(5), 312–320.

Cattivelli, L., Rizza, F., Badeck, F. W., Mazzucotelli, E., Mastrangelo, A. M., Francia, E., Marè, C., Tondelli, A., & Stanca, A. M. (2008). Drought tolerance improvement in crop plants: An integrated view from breeding to genomics. *Field Crops Research*, *105*(1–2), 1–14.

Cerrudo, D., Cao, S., Yuan, Y., Martinez, C., Suarez, E. A., Babu, R., Zhang, X., & Trachsel, S. (2018). Genomic selection outperforms marker assisted selection for grain yield and physiological traits in a maize doubled haploid population across water treatments. *Frontiers in Plant Science*, *9*, 366.

Chandrasekaran, M. (2022). Arbuscular mycorrhizal fungi mediated alleviation of drought stress via non-enzymatic antioxidants: A meta-analysis. *Plants*, *11*(19), 2448.

Chaturvedi, G. S., Singh, A., & Bahadur, R. (2012). Screening techniques for evaluating crop germplasm for drought tolerance. *Plant Archives*, *12*(1), 11–18.

Choukri, H., Hejjaoui, K., El-Baouchi, A., El Haddad, N., Smouni, A., Maalouf, F., Thavarajah, D., & Kumar, S. (2020). Heat and drought stress impact on phenology, grain yield, and nutritional quality of lentil (Lens culinaris Medikus). *Frontiers in Nutrition*, *7*, 596307.

Cobb, J. N., Biswas, P. S., & Platten, J. D. (2019). Back to the future: Revisiting MAS as a tool for modern plant breeding. *Theoretical and Applied Genetics*, *132*, 647–667.

Comas, L. H., Becker, S. R., Cruz, V. M. V., Byrne, P. F., & Dierig, D. A. (2013). Root traits contributing to plant productivity under drought. *Frontiers in Plant Science*, *4*, 442.

Conforte, A. J., Guimarães-Dias, F., Neves-Borges, A. C., Bencke-Malato, M., Felix-Whipps, D., & Alves-Ferreira, M. (2017). Isolation and characterization of a promoter responsive to salt, osmotic and dehydration stresses in soybean. *Genetics and Molecular Biology*, *40*, 226–237.

Crossa, J., Perez, P., Hickey, J., Burgueno, J., Ornella, L., Cerón-Rojas, J., Zhang, X., Dreisigacker, S., Babu, R., Li, Y., Bonnett, D., & Mathews, K. (2014). Genomic prediction in CIMMYT maize and wheat breeding programs. *Heredity*, *112*(1), 48–60.

Das, A., Eldakak, M., Paudel, B., Kim, D. W., Hemmati, H., Basu, C., & Rohila, J. S. (2016). Leaf proteome analysis reveals prospective drought and heat stress response mechanisms in soybean. *BioMed Research International*, 2016, 6021047.

Das, R. R., Vinayan, M. T., Patel, M. B., Phagna, R. K., Singh, S. B., Shahi, J. P., Sarma, A., Barua, N. S., Babu, R., Seetharam, K., Burgueño, J. A., & Zaidi, P. H. (2020). Genetic gains with rapid-cycle genomic selection for combined drought and waterlogging tolerance in tropical maize (Zea mays L.). *The Plant Genome*, *13*(3), e20035.

De la Rosa, L., Zambrana, E., & Ramirez-Parra, E. (2020). Molecular bases for drought tolerance in common vetch: Designing new molecular breeding tools. *BMC Plant Biology*, *20*(1), 1–18.

Desta, Z. A., & Ortiz, R. (2014). Genomic selection: Genome-wide prediction in plant improvement. *Trends in Plant Science*, *19*(9), 592–601.

Diab, A. A., Atia, M. A., Hussein, E. H., Hussein, H. A., & Adawy, S. S. (2013). A multidisciplinary approach for dissecting QTL controlling high-yield and drought tolerance-related traits in durum wheat. *International Journal of Agricultural Science and Research (IJASR)*, *3*(3), 99–116.

Dixit, S., Swamy, B. M., Vikram, P., Ahmed, H. U., Sta Cruz, M. T., Amante, M., Sta Cruz, M. T., Amante, M., Atri, D., Leung, H., & Kumar, A. (2012). Fine mapping of QTLs for rice grain yield under drought reveals sub-QTLs conferring a response to variable drought severities. *Theoretical and Applied Genetics*, *125*, 155–169.

Dixon, L., Esse, G., Hirsz, D., Willemsen, V., & McKim, S. (2022). Cereal architecture and its manipulation. *Annual Plant Reviews*, *5*, 1–54.

Dolan, F., Lamontagne, J., Link, R., Hejazi, M., Reed, P., & Edmonds, J. (2021). Evaluating the economic impact of water scarcity in a changing world. *Nature Communications*, *12*(1), 1915.

Dormatey, R., Sun, C., Ali, K., Coulter, J. A., Bi, Z., & Bai, J. (2020). Gene pyramiding for sustainable crop improvement against biotic and abiotic stresses. *Agronomy*, *10*(9), 1255.

Duursma, R. A., Blackman, C. J., Lopéz, R., Martin-StPaul, N. K., Cochard, H., & Medlyn, B. E. (2019). On the minimum leaf conductance: Its role in models of plant water use, and ecological and environmental controls. *New Phytologist*, *221*(2), 693–705.

El-Metwally, I. M., & Saudy, H. S. (2021). Interactional impacts of drought and weed stresses on nutritional status of seeds and water use efficiency of peanut plants grown in arid conditions. *Gesunde Pflanzen*, *73*(4), 407–416.

Enyew, M., Feyissa, T., Carlsson, A. S., Tesfaye, K., Hammenhag, C., & Geleta, M. (2022). Genetic diversity and population structure of sorghum [*Sorghum bicolor* (L.) moench] accessions as revealed by single nucleotide polymorphism markers. *Frontiers in Plant Science*, *12*, 799482.

Feldman, A. F., Short Gianotti, D. J., Trigo, I. F., Salvucci, G. D., & Entekhabi, D. (2020). Land-atmosphere drivers of landscape-scale plant water content loss. *Geophysical Research Letters*, *47*(22), e2020GL090331.

Feng, X., Yu, Q., Zeng, J., He, X., & Liu, W. (2022). Genome-wide identification and characterization of GATA family genes in wheat. *BMC Plant Biology*, *22*(1), 1–15.

Filtz, T. M., Vogel, W. K., & Leid, M. (2014). Regulation of transcription factor activity by interconnected post-translational modifications. *Trends in Pharmacological Sciences*, *35*(2), 76–85.

Ford, K. L., Cassin, A., & Bacic, A. (2011). Quantitative proteomic analysis of wheat cultivars with differing drought stress tolerance. *Frontiers in Plant Science*, *2*, 44.

Gaikwad, K. B., Rani, S., Kumar, M., Gupta, V., Babu, P. H., Bainsla, N. K., & Yadav, R. (2020). Enhancing the nutritional quality of major food crops through conventional and genomics-assisted breeding. *Frontiers in Nutrition*, *7*, 533453.

Galmés, J., Aranjuelo, I., Medrano, H., & Flexas, J. (2013). Variation in Rubisco content and activity under variable climatic factors. *Photosynthesis Research*, *117*, 73–90.

Galyuon, I. K. A., Gay, A., Borrell, A. K., & Howarth, C. J. (2019). Physiological and biochemical basis for stay-green trait in sorghum. *African Crop Science Journal*, *27*(4), 653–677.

Ganie, S. A., & Ahammed, G. J. (2021). Dynamics of cell wall structure and related genomic resources for drought tolerance in rice. *Plant Cell Reports*, *40*, 437–459.

Gehan, M. A., & Kellogg, E. A. (2017). High-throughput phenotyping. *American Journal of Botany*, *104*(4), 505–508.

Ghatak, A., Chaturvedi, P., & Weckwerth, W. (2017). Cereal crop proteomics: Systemic analysis of crop drought stress responses towards marker-assisted selection breeding. *Frontiers in Plant Science*, *8*, 757.

Ghimire, K. H., Quiatchon, L. A., Vikram, P., Swamy, B. M., Dixit, S., Ahmed, H., Hernandez, J. E., Borromeo, T. H., & Kumar, A. (2012). Identification and mapping of a QTL (qDTY1.1) with a consistent effect on grain yield under drought. *Field Crops Research*, *131*, 88–96.

Gopalakrishnan, S., Sharma, R. K., Anand Rajkumar, K., Joseph, M., Singh, V. P., Singh, A. K., Bhat, K. V., Singh, N. K., & Mohapatra, T. (2008). Integrating marker assisted background analysis with foreground selection for identification of superior bacterial blight resistant recombinants in Basmati rice. *Plant Breeding*, *127*(2), 131–139.

Gruszka, D. (2020). Exploring the brassinosteroid signaling in monocots reveals novel components of the pathway and implications for plant breeding. *International Journal of Molecular Sciences*, *21*(1), 354.

Guo, M., Wang, R., Wang, J., Hua, K., Wang, Y., Liu, X., & Yao, S. (2014). ALT1, a Snf2 family chromatin remodeling ATPase, negatively regulates alkaline tolerance through enhanced defense against oxidative stress in rice. *PLoS One*, *9*(12), e112515.

Guo, X., Xin, Z., Yang, T., Ma, X., Zhang, Y., Wang, Z., Ren, Y., & Lin, T. (2020). Metabolomics response for drought stress tolerance in Chinese wheat genotypes (*Triticum aestivum*). *Plants*, *9*(4), 520.

Gupta, P. K., Balyan, H. S., & Gahlaut, V. (2017). QTL analysis for drought tolerance in wheat: Present status and future possibilities. *Agronomy*, *7*(1), 5.

Gupta, S., Mishra, S. K., Misra, S., Pandey, V., Agrawal, L., Nautiyal, C. S., & Chauhan, P. S. (2020). Revealing the complexity of protein abundance in chickpea root under drought-stress using a comparative proteomics approach. *Plant Physiology and Biochemistry*, *151*, 88–102.

Haghighattalab, A., González Pérez, L., Mondal, S., Singh, D., Schinstock, D., Rutkoski, J., Ortiz-Monasterio, I., Singh, R. S., Goodin, D., & Poland, J. (2016). Application of unmanned aerial systems for high throughput phenotyping of large wheat breeding nurseries. *Plant Methods*, *12*, 1–15.

Haque, M. A., Rafii, M. Y., Yusoff, M. M., Ali, N. S., Yusuff, O., Datta, D. R., Anisuzzaman, M., & Ikbal, M. F. (2021). Recent advances in rice varietal development for durable resistance to biotic and abiotic stresses through marker-assisted gene pyramiding. *Sustainability*, *13*(19), 10806.

Harris, K., Subudhi, P. K., Borrell, A., Jordan, D., Rosenow, D., Nguyen, H., Klein, P., Klein, R., & Mullet, J. (2007). Sorghum stay-green QTL individually reduce post-flowering drought-induced leaf senescence. *Journal of Experimental Botany*, *58*(2), 327–338.

Hasan, N., Choudhary, S., Naaz, N., Sharma, N., & Laskar, R. A. (2021). Recent advancements in molecular marker-assisted selection and applications in plant breeding programmes. *Journal of Genetic Engineering and Biotechnology*, *19*(1), 1–26.

Hayes, B. J., Panozzo, J., Walker, C. K., Choy, A. L., Kant, S., Wong, D., Tibbits, J., Daetwyler, H. D., Rochfort, S., Hayden, M. J., & Spangenberg, G. C. (2017). Accelerating wheat breeding for end-use quality with multi-trait genomic predictions incorporating near infrared and nuclear magnetic resonance-derived phenotypes. *Theoretical and Applied Genetics*, *130*, 2505–2519.

Heffner, E. L., Jannink, J. L., & Sorrells, M. E. (2011). Genomic selection accuracy using multifamily prediction models in a wheat breeding program. *The Plant Genome*, *4*(1), 65.

Hirwa, H., Peng, Y., Zhang, Q., Qiao, Y., Leng, P., Tian, C., Yang, G., Muhirwa, F., Diop, S., Kayiranga, A., Li, F., & Chen, G. (2022). Virtual water transfers in Africa: Assessing topical condition of water scarcity, water savings, and policy implications. *Science of the Total Environment*, *835*, 155343.

Hou, X., Xie, K., Yao, J., Qi, Z., & Xiong, L. (2009). A homolog of human ski-interacting protein in rice positively regulates cell viability and stress tolerance. *Proceedings of the National Academy of Sciences*, *106*(15), 6410–6415.

Hrmova, M., & Hussain, S. S. (2021). Plant transcription factors involved in drought and associated stresses. *International Journal of Molecular Sciences*, *22*(11), 5662.

Huang, Y., Yi, J., Li, X., & Li, F. (2024). Transcriptomics and physiological analyses reveal that sulfur alleviates mercury toxicity in rice (*Oryza sativa* L.). *Journal of Environmental Sciences*, *135*, 10–25.

Hussain, Q., Asim, M., Zhang, R., Khan, R., Farooq, S., & Wu, J. (2021). Transcription factors interact with ABA through gene expression and signaling pathways to mitigate drought and salinity stress. *Biomolecules*, *11*(8), 1159.

Hussain, W., Baenziger, P. S., Belamkar, V., Guttieri, M. J., Venegas, J. P., Easterly, A., Sallam, A., & Poland, J. (2017). Genotyping-by-sequencing derived high-density linkage map and its application to QTL mapping of flag leaf traits in bread wheat. *Scientific Reports*, *7*(1), 16394.

Idrees, M., & Irshad, M. (2014). Molecular markers in plants for analysis of genetic diversity: A review. *European Academic Research*, *2*(1), 1513–1540.

Ilyas, M., Nisar, M., Khan, N., Hazrat, A., Khan, A. H., Hayat, K., Fahad, S., Khan, A., & Ullah, A. (2021). Drought tolerance strategies in plants: A mechanistic approach. *Journal of Plant Growth Regulation*, *40*, 926–944.

Iqbal, A., Fahad, S., Iqbal, M., Alamzeb, M., Ahmad, A., Anwar, S., Khan, A. A., Amanullah, Arif, M., Inamullah, Shaheenshah, Saeed, M., & Song, M. (2020). Special adaptive features of plant species in response to drought. In M. Hasanuzzaman & M. Tanveer (Eds.), *Salt and drought stress tolerance in plants: Signaling networks and adaptive mechanisms* (pp. 77–118). Springer.

Jain, N., Singh, G. P., Singh, P. K., Ramya, P., Krishna, H., Ramya, K. T., Todkar, L., Amasiddha, B., Prashant Kumar, K. C., Vijay, P., Jadon, V., Dutta, S., Rai, N., Sinha, N., & Prabhu, K. V. (2014). Molecular approaches for wheat improvement under drought and heat stress. Indian Society of Genetics and Plant, *74*, 578.

Jamaloddin, M., Durga Rani, C. V., Swathi, G., Anuradha, C., Vanisri, S., Rajan, C. P. D., Krishnam Raju, S., Bhuvaneshwari, V., Jagadeeswar, R., Laha, G. S., Prasad, M. S., Satyanarayana, P. V., Cheralu, C., Rajani, G., Ramprasad, E., Sravanthi, P., Arun Prem Kumar, N., Aruna Kumari, K., Yamini, K. N., & Madhav, M. S. (2020). Marker Assisted Gene Pyramiding (MAGP) for bacterial blight and blast resistance into mega rice variety "Tellahamsa". *PLoS One*, *15*(6), e0234088.

Jauhar, P. P. (2006). Modern biotechnology as an integral supplement to conventional plant breeding: The prospects and challenges. *Crop Science*, *46*(5), 1841–1859.

Jia, H., Zhang, S., Ruan, M., Wang, Y., & Wang, C. (2012). Analysis and application of RD29 genes in abiotic stress response. *Acta Physiologiae Plantarum*, *34*, 1239–1250.

Jiang, H., Feng, Y., Bao, L., Li, X., Gao, G., Zhang, Q., ... & He, Y. (2012). Improving blast resistance of Jin 23B and its hybrid rice by marker-assisted gene pyramiding. *Molecular Breeding*, *30*, 1679–1688.

Jogawat, A., Yadav, B., Lakra, N., Singh, A., & Narayan, O. (2021). Crosstalk between phytohormones and secondary metabolites in the drought stress tolerance of crop plants: A review. *Physiologia Plantarum*, *172*(2), 1106–1132.

Jonah, P. M., Bello, L. L., Lucky, O., Midau, A., & Moruppa, S. M. (2011). The importance of molecular markers in plant breeding programmes. *Global Journal of Science Frontier Research*, *11*(5), 5–12.

Joshi, R., Wani, S. H., Singh, B., Bohra, A., Dar, Z. A., Lone, A. A., Pareek, A., & Singla-Pareek, S. L. (2016b). Transcription factors and plants response to drought stress: Current understanding and future directions. *Frontiers in Plant Science*, *7*, 1029.

Juliana, P., Singh, R. P., Singh, P. K., Crossa, J., Huerta-Espino, J., Lan, C., Bhavani, S., Rutkoski, J. E., Poland, J. A., Bergstrom, G. C., & Sorrells, M. E. (2017). Genomic and pedigree-based prediction for leaf, stem, and stripe rust resistance in wheat. *Theoretical and Applied Genetics*, *130*, 1415–1430.

Kadirvel, P., Senthilvel, S., Geethanjali, S., Sujatha, M., & Varaprasad, K. S. (2015). Genetic markers, trait mapping and marker-assisted selection in plant breeding. In P. Kadirvel, S. Senthilvel, S. Geethanjali, M. Sujatha, & K. S. Varaprasad (Eds.), *Plant biology and biotechnology: Volume II: Plant genomics and biotechnology* (pp. 65–88). Springer.

Kamal, N. M., Gorafi, Y. S. A., Abdelrahman, M., Abdellatef, E., & Tsujimoto, H. (2019). Stay-green trait: A prospective approach for yield potential, and drought and heat stress adaptation in globally important cereals. *International Journal of Molecular Sciences*, *20*(23), 5837.

Kamarudin, M. H., Ismail, Z. H., & Saidi, N. B. (2021). Deep learning sensor fusion in plant water stress assessment: A comprehensive review. *Applied Sciences*, 11(4), 1403.

Kannenberg, S. A., Cabon, A., Babst, F., Belmecheri, S., Delpierre, N., Guerrieri, R., Maxwell, J. T., Meinzer, F. C., Moore, D. J. P., Pappas, C., Ueyama, M., Ulrich, D. E. M., Voelker, S. L., Woodruff, D. R., & Anderegg, W. R. (2022). Drought-induced decoupling between carbon uptake and tree growth impacts forest carbon turnover time. *Agricultural and Forest Meteorology*, *322*, 108996.

Kapoor, D., Bhardwaj, S., Landi, M., Sharma, A., Ramakrishnan, M., & Sharma, A. (2020). The impact of drought in plant metabolism: How to exploit tolerance mechanisms to increase crop production. *Applied Sciences*, *10*(16), 5692.

Kasuga, M., Liu, Q., Miura, S., Yamaguchi-Shinozaki, K., & Shinozaki, K. (1999). Improving plant drought, salt, and freezing tolerance by gene transfer of a single stress-inducible transcription factor. *Nature Biotechnology*, *17*(3), 287–291.

Khan, N. I., Shah, R. A., Asad, M. N., Shah, S. M., Ali, S., & Zafar, M. (2006). BH-160: A high yielding and early maturing variety of cotton (*Gossypium hirsutum* L.). *Science, Technology and Development*, *25*, 13–18.

Khan, A. H., Hassan, M., & Khan, M. N. (2020a). Conventional plant breeding program for disease resistance. In I. U. Haq & S. Ijaz (Eds.), *Plant disease management strategies for sustainable agriculture through traditional and modern approaches* (pp. 27–51). Springer.

Khan, M., Jannat, A., Munir, F., Fatima, N., & Amir, R. (2020b). Biochemical and molecular mechanisms of abiotic stress tolerance. In M. Hasanuzzaman (Eds.), *Plant ecophysiology and adaptation under climate change: Mechanisms and perspectives II: Mechanisms of adaptation and stress amelioration* (pp. 187–230). Springer.

Khanna, A., Anumalla, M., Catolos, M., Bartholomé, J., Fritsche-Neto, R., Platten, J. D., Pisano, D. J., Gulles, A., Ma, M. T., Ramos, J., Faustino, G., Bhosale, S., & Hussain, W. (2022). Genetic trends estimation in IRRIs rice drought breeding program and identification of high yielding drought-tolerant lines. *Rice*, *15*(1).

Kijima, S. T., Staiger, C. J., Katoh, K., Nagasaki, A., Ito, K., & Uyeda, T. Q. (2018). Arabidopsis vegetative actin isoforms, AtACT2 and AtACT7, generate distinct filament arrays in living plant cells. *Scientific Reports*, *8*(1), 4381.

Kirigwi, F. M., van Ginkel, M., Brown-Guedira, G., Gill, B. S., Paulsen, G. M., & Fritz, A. K. (2007). Markers associated with a QTL for grain yield in wheat under drought. *Molecular Breeding: New Strategies in Plant Improvement*, *20*, 401–413

Kumar, P. K., Bellundagi, A., Krishna, H., Mallikarjuna, M. G., Thimmappa, R. K., Rai, N., Shashikumara, P., Sinha, N., Jain, N., Singh, P. K., Singh, G. P., & Prabhu, K. V. (2023). Development of bread wheat (Triticum aestivum L) variety HD3411 following marker-assisted backcross breeding for drought tolerance. *Frontiers in Genetics*, *14*, 1046624.

Kumar, R., Harikrishna, Barman, D., Ghimire, O. P., Gurumurthy, S., Singh, P. K., Chinnusamy, V., Padaria, J. C., & Arora, A. (2022). Stay-green trait serves as yield stability attribute under combined heat and drought stress in wheat (Triticum aestivum L.). *Plant Growth Regulation*, *96*, 1–12.

Kumar, S., Sehgal, S. K., Kumar, U., Prasad, P. V., Joshi, A. K., & Gill, B. S. (2012). Genomic characterization of drought tolerance-related traits in spring wheat. *Euphytica*, *186*, 265–276.

Kumar, U., Joshi, A. K., Kumari, M., Paliwal, R., Kumar, S., & Röder, M. S. (2010). Identification of QTLs for stay green trait in wheat (*Triticum aestivum* L.) in the 'Chirya 3'×'Sonalika'population. *Euphytica*, *174*, 437–445.

Lamichhane, S., & Thapa, S. (2022). Advances from conventional to modern plant breeding methodologies. *Plant Breeding and Biotechnology*, *10*(1), 1–14.

Lande, R., & Thompson, R. (1990). Efficiency of marker-assisted selection in the improvement of quantitative traits. *Genetics*, *124*(3), 743–756.

Li, H., Peng, Z., Yang, X., Wang, W., Fu, J., Wang, J., ... & Yan, J. (2013). Genome-wide association study dissects the genetic architecture of oil biosynthesis in maize kernels. *Nature Genetics*, 45(1), 43–50.

Li, X., Xi, B., Wu, X., Choat, B., Feng, J., Jiang, M., & Tissue, D. (2022). Unlocking drought-induced tree mortality: physiological mechanisms to modeling. *Frontiers in Plant Science*, 13, 835921.

Liu, Z., Guo, C., Wu, R., Wang, J., Zhou, Y., Yu, X., Zhang, Y., Zhao, Z., Liu, H., Sun, S., Hu, M., Qin, A., Liu, Y., Yang, J., Bawa, G., & Sun, X. (2022). Identification of the regulators of epidermis development under drought- and salt-stressed conditions by single-cell RNA-seq. *International Journal of Molecular Sciences*, *23*(5), 2759.

López, C., Banowetz, G. M., Peterson, C. J., & Kronstad, W. E. (2004). Differential accumulation of a 24-kd dehydrin protein in wheat seedlings correlates with drought stress tolerance at grain filling. *Hereditas*, *135*(2–3), 175–181.

Lorenz, A. J., Chao, S., Asoro, F. G., Heffner, E. L., Hayashi, T., Iwata, H., Smith, K. P., Sorrells, M. E., & Jannink, J. L. (2011). Genomic selection in plant breeding: Knowledge and prospects. *Advances in Agronomy*, *110*, 77–123.

Madhavi, K., Rambabu, R., Abhilash Kumar, V., Vijay Kumar, S., Aruna, J., Ramesh, S., Sundaram, R., Laha, G. S., Madhav, M., Babu, V. R., & Prasad, M. S. (2016). Marker assisted introgression of blast (Pi-2 and Pi-54) genes in to the genetic background of elite, bacterial blight resistant indica rice variety, Improved Samba Mahsuri. *Euphytica*, *212*, 331–342.

Malhi, G. S., Kaur, M., & Kaushik, P. (2021). Impact of climate change on agriculture and its mitigation strategies: A review. *Sustainability*, *13*(3), 1318.

Manna, M., Thakur, T., Chirom, O., Mandlik, R., Deshmukh, R., & Salvi, P. (2021). Transcription factors as key molecular target to strengthen the drought stress tolerance in plants. *Physiologia Plantarum*, *172*(2), 847–868.

Manu, B., Kumara, P. S., Biradar, S., Phuke, R., Ambati, D., Prasad, S. S., Mishra, P., Mishra, K. K., Harikrishna, Jain, N., Singh. P., Singh, G., Prabhu, K. V., & Prabhu, K. V. (2020). Genetic gain and morpho-physiological characterisation of BILs (backcross inbred lines) under different moisture regimes in wheat (Triticum aestivum. L). *Indian Journal of Genetics and Plant Breeding*, *80*(01), 84–93.

Mason, R. E., & Singh, R. P. (2014). Considerations when deploying canopy temperature to select high yielding wheat breeding lines under drought and heat stress. *Agronomy*, *4*(2), 191–201.

Mba, C. (2013). Induced mutations unleash the potentials of plant genetic resources for food and agriculture. *Agronomy*, *3*(1), 200–231.

McDowell, N. G., Sapes, G., Pivovaroff, A., Adams, H. D., Allen, C. D., Anderegg, W. R., Arend, M., Breshears, D. D., Brodribb, T., Choat, B., Cochard, H., De Cáceres, M., De Kauwe, M. G., Grossiord, C., Hammond, W. M., Hartmann, H., Hoch, G., Kahmen, A., Klein, T., … Xu, C. (2022). Mechanisms of woody-plant mortality under rising drought, CO_2 and vapour pressure deficit. *Nature Reviews Earth & Environment*, *3*(5), 294–308.

McNally, K. L., Bruskiewich, R., Mackill, D., Buell, C. R., Leach, J. E., & Leung, H. (2006). Sequencing multiple and diverse rice varieties. Connecting whole-genome variation with phenotypes. *Plant Physiology*, *141*(1), 26–31.

Mega, R., Abe, F., Kim, J. S., Tsuboi, Y., Tanaka, K., Kobayashi, H., ... & Okamoto, M. (2019). Tuning water-use efficiency and drought tolerance in wheat using abscisic acid receptors. *Nature Plants*, *5*(2), 153–159.

Mekonnen, T., Haileselassie, T., & Tesfaye, K. (2017). Identification, mapping and pyramiding of genes/quantitative trait loci (QTLs) for durable resistance of crops to biotic stresses. *Journal of Plant Pathology and Microbiology*, *8*(6), 412.

Meraj, T. A., Fu, J., Raza, M. A., Zhu, C., Shen, Q., Xu, D., & Wang, Q. (2020). Transcriptional factors regulate plant stress responses through mediating secondary metabolism. *Genes*, *11*(4), 346.

Miah, G., Rafii, M. Y., Ismail, M. R., Puteh, A. B., Rahim, H. A., Asfaliza, R., & Latif, M. A. (2013). Blast resistance in rice: A review of conventional breeding to molecular approaches. *Molecular Biology Reports*, *40*, 2369–2388.

Mickelbart, M. V., Hasegawa, P. M., & Bailey-Serres, J. (2015). Genetic mechanisms of abiotic stress tolerance that translate to crop yield stability. *Nature Reviews Genetics*, *16*(4), 237–251.

Minh, B., Linh, N., Hanh, H., Hien, L., Thang, N., Hai, N., & Hue, H. (2019). A lea gene from a Vietnamese maize landrace can enhance the drought tolerance of transgenic maize and tobacco. *Agronomy*, *9*(2), 62.

Minhas, R., Shah, S. M., Akhtar, L. H., Awais, S., & Shah, S. (2018). Development of a new drought tolerant cotton variety "BH-167" by using pedigree method. *Journal of Environmental and Agricultural Sciences*, *14*, 54–62.

Mishra, K. K., Vikram, P., Yadaw, R. B., Swamy, B. M., Dixit, S., Cruz, M. T. S., Maturan, P., Marker, S., & Kumar, A. (2013). qDTY12.1: A locus with a consistent effect on grain yield under drought in rice. *BMC Genetics*, *14*, 1–10.

Monroe, J. G., Srikant, T., Carbonell-Bejerano, P., Becker, C., Lensink, M., Exposito-Alonso, M., Klein, M., Hildebrandt, J., Neumann, M., Kliebenstein, D., Weng, M.-L., Imbert, E., Ågren, J., Rutter, M. T., Fenster, C. B., & Weigel, D. (2022). Mutation bias reflects natural selection in *Arabidopsis thaliana*. *Nature*, *602*(7895), 101–105.

Muhammad, I., Shalmani, A., Ali, M., Yang, Q. H., Ahmad, H., & Li, F. B. (2021). Mechanisms regulating the dynamics of photosynthesis under abiotic stresses. *Frontiers in Plant Science*, *11*, 615942.

Muhammad Aslam, M., Waseem, M., Jakada, B. H., Okal, E. J., Lei, Z., Saqib, H. S. A., Yuan, W., Xu, W., & Zhang, Q. (2022). Mechanisms of abscisic acid-mediated drought stress responses in plants. *International Journal of Molecular Sciences*, *23*(3), 1084.

Mwamahonje, A., Eleblu, J. S. Y., Ofori, K., Deshpande, S., Feyissa, T., & Tongoona, P. (2021). Drought tolerance and application of marker-assisted selection in sorghum. *Biology*, *10*(12), 1249.

Nadarajah, K. K. (2020). ROS homeostasis in abiotic stress tolerance in plants. *International Journal of Molecular Sciences*, *21*(15), 5208.

Negi, K. S., Das, N., Khan, T., & Sen, P. (2023). Osmolyte induced protein stabilization: Modulation of associated water dynamics might be a key factor. *Physical Chemistry Chemical Physics*, *25*(47), 32602–32612.

Nishihama, R., & Machida, Y. (2000). The MAP kinase cascade that includes MAPKKK-related protein kinase NPK1 controls a mitotic process in plant cells. In H. Hirt (Eds.), *MAP kinases in plant signal transduction* (pp. 119–130). Springer.

Niu, L., Wu, Z., Liu, H., Wu, X., & Wang, W. (2019). 2-DE-based proteomic analysis of protein changes associated with etiolated mesocotyl growth in Zea mays. *BMC Genomics*, *20*(1), 1–17.

Nkomo, G. V., Sedibe, M. M., & Mofokeng, M. A. (2021). Production constraints and improvement strategies of cowpea (*Vigna unguiculata* L. Walp.) genotypes for drought tolerance. *International Journal of Agronomy*, 2021, 1–9.

Nuccio, M., Chen, X., Conville, J., Zhou, A., & Liu, X. (2015). Plant trait gene expression cassette design. In *Recent advancements in gene expression and enabling technologies in crop plants* (pp. 41–77).

Obara, M., Tamura, W., Ebitani, T., Yano, M., Sato, T., & Yamaya, T. (2010). Fine-mapping of qRL6.1, a major QTL for root length of rice seedlings grown under a wide range of NH_4^+ concentrations in hydroponic conditions. *Theoretical and Applied Genetics*, *121*, 535–547.

Oladosu, Y., Rafii, M. Y., Samuel, C., Fatai, A., Magaji, U., Kareem, I., Kamarudin, Z. S., Muhammad, I., & Kolapo, K. (2019). Drought resistance in rice from conventional to molecular breeding: A review. *International Journal of Molecular Sciences*, *20*(14), 3519.

Olivares-Villegas, J. J., Reynolds, M. P., & McDonald, G. K. (2007). Drought-adaptive attributes in the Seri/Babax hexaploid wheat population. *Functional Plant Biology*, *34*(3), 189–203.

Onyemaobi, O., Sangma, H., Garg, G., Wallace, X., Kleven, S., & Dolferus, R. (2022). Transcriptome profiling of the chilling response in wheat spikes: I, acclimation response to long-term chilling treatment. *Current Plant Biology*, *31*, 100255.

Ouyang, W., Yin, X., & Struik, P. (2020). Comparisons with wheat reveal root anatomical and histochemical constraints of rice under water-deficit stress. *Plant and Soil*, *452*(1–2), 547–568.

Ozturk, M., Turkyilmaz Unal, B., García-Caparrós, P., Khursheed, A., Gul, A., & Hasanuzzaman, M. (2021). Osmoregulation and its actions during the drought stress in plants. *Physiologia Plantarum*, *172*(2), 1321–1335.

Paudel, D., Dhakal, S., Parajuli, S., Adhikari, L., Peng, Z., Qian, Y., Shahi, D., Avci, M., Makaju, S. O., & Kannan, B. (2020). Use of quantitative trait loci to develop stress tolerance in plants. In D. K. Tripathi (Eds.), *Plant life under changing environment* (pp. 917–965). Academic Press.

Pellegrineschi, A., Reynolds, M., Pacheco, M., Brito, R. M., Almeraya, R., Yamaguchi-Shinozaki, K., & Hoisington, D. (2004). Stress-induced expression in wheat of the *Arabidopsis thaliana* DREB1A gene delays water stress symptoms under greenhouse conditions. *Genome*, *47*(3), 493–500.

Poland, J., Endelman, J., Dawson, J., Rutkoski, J., Wu, S., Manes, Y., Dreisigacker, S., Crossa, J., Sánchez-Villeda, H., Sorrells, M., & Jannink, J. L. (2012). Genomic selection in wheat breeding using genotyping-by-sequencing. *The Plant Genome*, *5*(3), 103.

Polivova, M., & Brook, A. (2021). Detailed investigation of spectral vegetation indices for fine field-scale phenotyping. In E. C. Carmona, A. C. Ortiz, R. Q. Canas, & C. M. Musarella (Eds.), *Vegetation index and dynamics* (pp. 103, 338, 103–139). London: Intechopen.

Price, A. H., Cairns, J. E., Horton, P., Jones, H. G., & Griffiths, H. (2002). Linking drought-resistance mechanisms to drought avoidance in upland rice using a QTL approach: Progress and new opportunities to integrate stomatal and mesophyll responses. *Journal of Experimental Botany*, *53*(371), 989–1004.

Ramya, P., Singh, G. P., Jain, N., Singh, P. K., Pandey, M. K., Sharma, K., Kumar, A., Harikrishna, & Prabhu, K. V. (2016). Effect of recurrent selection on drought tolerance and related morpho-physiological traits in bread wheat. *PLoS One*, *11*(6), e0156869.

Rauf, A., Sher, M. A., Farooq, U., Rasheed, A., Sajjad, M., Jing, R., Khan, Z., Attia, K. O. T. B. A., Mohammed, A. A., Fiaz, S., Chen, J., & Rehman, S. U. (2024). An SNP based genotyping assay for genes associated with drought tolerance in bread wheat. *Molecular Biology Reports*, *51*(1).

Raza, A., Mubarik, M. S., Sharif, R., Habib, M., Jabeen, W., Zhang, C., Chen, H., Chen, Z.-H., Siddique, K. H. M., Zhuang, W., & Varshney, R. K. (2023). Developing drought-smart, ready-to-grow future crops. *The Plant Genome*, *16*(1), e20279.

Reynolds, M. P., Rajaram, S., & Sayre, K. D. (1999). Physiological and genetic changes of irrigated wheat in the post–green revolution period and approaches for meeting projected global demand. *Crop Science*, *39*(6), 1611–1621.

Ribaut, J. M., & Ragot, M. (2007). Marker-assisted selection to improve drought adaptation in maize: The backcross approach, perspectives, limitations, and alternatives. *Journal of Experimental Botany*, *58*(2), 351–360.

Rijal, B., Baduwal, P., Chaudhary, M., Chapagain, S., Khanal, S., Khanal, S., & Poudel, P. B. (2021). Drought stress impacts on wheat and its resistance mechanisms. *Malaysian Journal of Sustainable Agriculture*, *5*, 67–76.

Robertson, B., He, T., & Li, C. (2021). The genetic control of stomatal development in barley: New solutions for enhanced water-use efficiency in drought-prone environments. *Agronomy*, *11*(8), 1670.

Rodrigues, O., Reshetnyak, G., Grondin, A., Saijo, Y., Leonhardt, N., Maurel, C., & Verdoucq, L. (2017). Aquaporins facilitate hydrogen peroxide entry into guard cells to mediate ABA-and pathogen-triggered stomatal closure. *Proceedings of the National Academy of Sciences*, *114*(34), 9200–9205.

Roychowdhury, R., Das, S. P., Gupta, A., Parihar, P., Chandrasekhar, K., Sarker, U., Kumar, A., Ramrao, D. P., & Sudhakar, C. (2023). Multi-omics pipeline and omics-integration approach to decipher plant's abiotic stress tolerance responses. *Genes*, *14*(6), 1281.

Rutkoski, J., Singh, R. P., Huerta-Espino, J., Bhavani, S., Poland, J., Jannink, J. L., & Sorrells, M. E. (2015). Efficient use of historical data for genomic selection: A case study of stem rust resistance in wheat. *The Plant Genome*, *8*(1), 0046.

Sabagh, A. E., Hossain, A., Islam, M. S., Iqbal, M. A., Fahad, S., Ratnasekera, D., & Llanes, A. (2020). Consequences and mitigation strategies of heat stress for sustainability of soybean (Glycine max L. Merr.) production under the changing climate. In E. Rami (Eds.), *Plant stress physiology* (pp. 119–140). Delve Publishing.

Saikumar, S., Gouda, P. K., Saiharini, A., Varma, C. M. K., Vineesha, O., Padmavathi, G., & Shenoy, V. V. (2014). Major QTL for enhancing rice grain yield under lowland reproductive drought stress identified using an O. sativa/O. glaberrima introgression line. *Field Crops Research*, *163*, 119–131.

Samtani, H., Sharma, A., & Khurana, P. (2022). Overexpression of HVA1 enhances drought and heat stress tolerance in Triticum aestivum doubled haploid plants. *Cells*, *11*(5), 912.

Sandhu, N., Dixit, S., Swamy, B. M., Vikram, P., Venkateshwarlu, C., Catolos, M., & Kumar, A. (2018). Positive interactions of major-effect QTLs with genetic background that enhances rice yield under drought. *Scientific Reports*, *8*(1), 1626.

Santos, A. P., Serra, T., Figueiredo, D. D., Barros, P., Lourenço, T., Chander, S., Oliveira, M. M., & Saibo, N. J. (2011). Transcription regulation of abiotic stress responses in rice: A combined action of transcription factors and epigenetic mechanisms. *Omics: A Journal of Integrative Biology*, *15*(12), 839–857.

Sarker, U. (2018). Drought stress enhances nutritional and bioactive compounds, phenolic acids and antioxidant capacity of amaranthus leafy vegetable. *BMC Plant Biology*, *18*(1), 258.

Schultz, B., Serão, N., & Ross, J. W. (Eds.). (2020). Genetic improvement of livestock, from conventional breeding to biotechnological approaches. In *Animal agriculture* (pp. 393–405). Amsterdam: Academic Press.

Selamat, N., & Nadarajah, K. K. (2021). Meta-analysis of quantitative traits loci (Qtl) identified in drought response in rice (oryza sativa l.). *Plants*, *10*(4).

Seleiman, M. F., Al-Suhaibani, N., Ali, N., Akmal, M., Alotaibi, M., Refay, Y., Dindaroglu, T., Abdul-Wajid, H. H., & Battaglia, M. L. (2021). Drought stress impacts on plants and different approaches to alleviate its adverse effects. *Plants*, *10*(2), 259.

Sellamuthu, R., Liu, G. F., Ranganathan, C. B., & Serraj, R. (2011). Genetic analysis and validation of quantitative trait loci associated with reproductive-growth traits and grain yield under drought stress in a doubled haploid line population of rice (*Oryza sativa* L.). *Field Crops Research*, *124*(1), 46–58.

Semagn, K., Beyene, Y., Warburton, M. L., Tarekegne, A., Mugo, S., Meisel, B., Sehabiague, P., & Prasanna, B. M. (2013). Meta-analyses of QTL for grain yield and anthesis silking interval in 18 maize populations evaluated under water-stressed and well-watered environments. BMC Genomics, *14*(1), 1–16.

Serrote, C. M. L., Reiniger, L. R. S., Silva, K. B., dos Santos Rabaiolli, S. M., & Stefanel, C. M. (2020). Determining the polymorphism information content of a molecular marker. *Gene*, *726*, 144175.

Shabbir, R., Javed, T., Afzal, I., Sabagh, A. E., Ali, A., Vicente, O., & Chen, P. (2021). Modern biotechnologies: Innovative and sustainable approaches for the improvement of sugarcane tolerance to environmental stresses. *Agronomy*, *11*(6), 1042.

Shasidhar, Y., Variath, M. T., Vishwakarma, M. K., Manohar, S. S., Gangurde, S. S., Sriswathi, M., Sudini, H. K., Dobariya, K. L., Bera, S. K., Radhakrishnan, T., Pandey, M. K., Janila, P., & Varshney, R. K. (2020). Improvement of three popular Indian groundnut varieties for foliar disease resistance and high oleic acid using SSR markers and SNP array in marker-assisted backcrossing. *The Crop Journal*, *8*(1), 1–15.

Shokat, S., Sehgal, D., Vikram, P., Liu, F., & Singh, S. (2020). Molecular markers associated with agro-physiological traits under terminal drought conditions in bread wheat. *International Journal of Molecular Sciences*, *21*(9), 3156.

Shrestha, S., Subedi, S., & Shrestha, J. (2020). Marker-assisted selection: A smart biotechnological strategy for modern plant breeding. *Peruvian Journal of Agronomy*, *4*(3), 104–120.

Singh, A. K., Krishnan, S. G., Vinod, K. K., Ellur, R. K., Bollinedi, H., Bhowmick, P. K., & Nagarajan, M. (2019). Precision breeding with genomic tools: A decade long journey of molecular breeding in rice. *Indian Journal of Genetics and Plant Breeding*, *79*(Sup-01), 181–191.

Singh, P. K., Indoliya, Y., Agrawal, L., Awasthi, S., Deeba, F., Dwivedi, S., Chakrabarty, D., Shirke, P. A., Pandey, V., Singh, N., Dhankher, O. P., Barik, S. K., & Tripathi, R. D. (2022). Genomic and proteomic responses to drought stress and biotechnological interventions for enhanced drought tolerance in plants. *Current Plant Biology*, *29*, 100239.

Sinha, P., Singh, V. K., Bohra, A., Kumar, A., Reif, J. C., & Varshney, R. K. (2021). Genomics and breeding innovations for enhancing genetic gain for climate resilience and nutrition traits. *Theoretical and Applied Genetics*, *134*(6), 1829–1843.

Smith, D. T., Potgieter, A. B., & Chapman, S. C. (2021). Scaling up high-throughput phenotyping for abiotic stress selection in the field. *Theoretical and Applied Genetics*, *134*(6), 1845–1866.

Soudek, P., Langhansová, L., Dvořáková, M., Revutska, A., Petrová, Š., Hirnerová, A., Hirnerová, A., Bouček, J., Trakal, L., Hošek, P., & Soukupová, M. (2024). The impact of the application of compochar on soil moisture, stress, yield and nutritional properties of legumes under drought stress. *Science of the Total Environment*, *914*, 169914.

Spindel, J., & Iwata, H. (2018). Genomic selection in rice breeding. In T. Sasaki, & M. Ashikari (Eds.), *Rice genomics, genetics and breeding* (pp. 473–496). Springer.

Sreenivasulu, N., Sopory, S. K., & Kishor, P. K. (2007). Deciphering the regulatory mechanisms of abiotic stress tolerance in plants by genomic approaches. *Gene*, *388*(1–2), 1–13.

Swamy, B. P. M., Ahmed, H. U., Henry, A., Mauleon, R., Dixit, S., Vikram, P., Tilatto, R., Verulkar, S. B., Perraju, P., Mandal, N. P., Variar, M., Robin, S., Chandrababu, R., Singh, O. N., Dwivedi, J. L., Das, S. P., Mishra, K. K., Yadaw, R. B., Aditya, T. L., … Kumar, A. (2013). Genetic, physiological, and gene expression analyses reveal that multiple QTL enhance yield of rice mega-variety IR64 under drought. *PLoS One*, *8*(5), e62795.

Swarup, S., Cargill, E. J., Crosby, K., Flagel, L., Kniskern, J., & Glenn, K. C. (2021). Genetic diversity is indispensable for plant breeding to improve crops. *Crop Science*, *61*(2), 839–852.

Talukder, M. M. U., Debnath, P., Nasrin, S., Akter, S., Ali, M. R., Islam, M. R., & Abdul-Awal, S. M. (2021). Stomatal responses at different vegetative stages of selected maize varieties of Bangladesh under water deficit condition. *bioRxiv*, 2021–09.

Tang, S. Q., Shao, G. N., Wei, X. J., Chen, M. L., Sheng, Z. H., Luo, J., Gui-Ai, J., Li-Hong, X. & Hu, P. S. (2013). QTL mapping of grain weight in rice and the validation of the QTL qTGW3.2. *Gene*, *527*(1), 201–206.

Tayade, R., Yoon, J., Lay, L., Khan, A. L., Yoon, Y., & Kim, Y. (2022). Utilization of spectral indices for high-throughput phenotyping. *Plants*, *11*(13), 1712.

Thenkabail, P. S., Lyon, J. G., & Huete, A. (2018). Advances in hyperspectral remote sensing of vegetation and agricultural crops. In P. S. Thenkabail, J. G. Lyon, & A. Huete (Eds.), *Fundamentals, sensor systems, spectral libraries, and data mining for vegetation* (pp. 3–37). CRC Press.

Tiwari, A., Tikoo, S. K., Angadi, S. P., Kadaru, S. B., Ajanahalli, S. R., & Vasudeva Rao, M. J. (2023). Use of molecular technologies in plant breeding. In A. Tiwari, S. K. Tikoo, S. P. Angadi, S. B. Kadaru, S. R. Ajanahalli, & M. J. V. Rao (Eds.), *Market-driven plant breeding for practicing breeders* (pp. 157–203). Springer Nature.

Tiwari, M., Kumar Trivedi, P., & Pandey, A. (2021). Emerging tools and paradigm shift of gene editing in cereals, fruits, and horticultural crops for enhancing nutritional value and food security. *Food and Energy Security*, *10*(1), e258.

Todkar, L., Singh, G. P., Jain, N., Singh, P. K., & Prabhu, K. V. (2020). Introgression of drought tolerance QTLs through marker assisted backcross breeding in wheat (*Triticum aestivum* L.). Indian *Journal of Genetics and Plant Breeding, 80*(02), 209–212.

Toulotte, J. M., Pantazopoulou, C. K., Sanclemente, M. A., Voesenek, L. A., & Sasidharan, R. (2022). Water stress resilient cereal crops: Lessons from wild relatives. *Journal of Integrative Plant Biology*, *64*(2), 412–430.

Tuberosa, R. (2012). Phenotyping for drought tolerance of crops in the genomics era. *Frontiers in Physiology*, *3*, 347.

Tuberosa, R., & Salvi, S. (2006). Genomics-based approaches to improve drought tolerance of crops. *Trends in Plant Science*, *11*(8), 405–412.

Tunc-Ozdemir, M., Miller, G., Song, L., Kim, J., Sodek, A., Koussevitzky, S., Misra, A. N., Mittler, R., & Shintani, D. K. (2009). Thiamin confers enhanced tolerance to oxidative stress in *Arabidopsis*. *Plant Physiology*, *151*(1), 421–432.

Uga, Y., Okuno, K., & Yano, M. (2011). Dro1, a major QTL involved in deep rooting of rice under upland field conditions. *Journal of Experimental Botany*, *62*(8), 2485–2494.

Valdivia-Cea, W., Bustamante, L., Jara, J., Fischer, S., Holzapfel, E., & Wilckens, R. (2021). Effect of soil water availability on physiological parameters, yield, and seed quality in four quinoa genotypes (*Chenopodium quinoa* Willd.). *Agronomy*, *11*(5), 1012.

Varshney, R. K., Mohan, S. M., Gaur, P. M., Chamarthi, S. K., Singh, V. K., Srinivasan, S., Swapna, N., Sharma, M., Pande, S., Singh, S., & Kaur, L. (2014). Marker-assisted backcrossing to introgress resistance to Fusarium wilt race 1 and Ascochyta blight in C214, an elite cultivar of chickpea. *The Plant Genome*, *7*(1), 2013–10.

Vasistha, N. K., Balasubramaniam, A., Mishra, V. K., Srinivasa, J., Chand, R., & Joshi, A. K. (2017). Molecular introgression of leaf rust resistance gene Lr34 validates enhanced effect on resistance to spot blotch in spring wheat. *Euphytica*, *213*, 1–10.

Venuprasad, R., Dalid, C. O., Del Valle, M., Zhao, D., Espiritu, M., Sta Cruz, M. T., Amante, M., Kumar, A., & Atlin, G. N. (2009). Identification and characterization of large-effect quantitative trait loci for grain yield under lowland drought stress in rice using bulk-segregant analysis. *Theoretical and Applied Genetics*, *120*, 177–190.

Verma, G., Srivastava, D., Tiwari, P., & Chakrabarty, D. (2019). ROS modulation in crop plants under drought stress. In M. Hasanuzzaman, V. Fotopoulos, K. Nahar, & M. Fujita (Eds.), *Reactive oxygen, nitrogen and sulfur species in plants: Production, metabolism, signaling and defense mechanisms* (pp. 311–336). John Wiley & Sons.

Vikram, P., Swamy, B. M., Dixit, S., Ahmed, H. U., Teresa Sta Cruz, M., Singh, A. K., & Kumar, A. (2011). qDTY1.1, a major QTL for rice grain yield under reproductive-stage drought stress with a consistent effect in multiple elite genetic backgrounds. *BMC Genetics*, *12*, 1–15.

Villalobos-López, M. A., Arroyo-Becerra, A., Quintero-Jiménez, A., & Iturriaga, G. (2022). Biotechnological advances to improve abiotic stress tolerance in crops. *International Journal of Molecular Sciences*, *23*(19), 12053.

Visscher, P. M., Van der Beek, S., & Haley, C. S. (2021). Marker assisted selection. In A. J. Clark (Ed.), *Animal breeding* (pp. 119–136). London: Routledge.

Wahab, A., Batool, F., Muhammad, M., Zaman, W., Mikhlef, R. M., Qaddoori, S. M., Ullah, S., Abdi, G., & Saqib, S. (2023). Unveiling the complex molecular dynamics of arbuscular mycorrhizae: A comprehensive exploration and future perspectives in harnessing phosphate-solubilizing microorganisms for sustainable progress. *Environmental and Experimental Botany*, 219, 105633.

Wang, J., Li, C., Li, L., Gao, L., Hu, G., Zhang, Y., Reynolds, M. P., Zhang, X., Jia, J., Mao, X., & Jing, R. (2023). DIW1 encoding a clade I PP2C phosphatase negatively regulates drought tolerance by de-phosphorylating TaSnRK1.1 in wheat. *Journal of Integrative Plant Biology*, *65*(8), 1918–1936.

Wang, J., Singh, R. P., Braun, H. J., & Pfeiffer, W. H. (2009). Investigating the efficiency of the single backcrossing breeding strategy through computer simulation. *Theoretical and Applied Genetics*, *118*, 683–694.

Wang, R., Yang, X., Wang, N., Liu, X., Nelson, R. S., Li, W., ... & Zhou, T. (2016). An efficient virus-induced gene silencing vector for maize functional genomics research. *The Plant Journal*, *86*(1), 102–115.

Wang, W., Chen, D., Zhang, X., Liu, D., Cheng, Y., & Shen, F. (2018). Role of plant respiratory burst oxidase homologs in stress responses. *Free Radical Research*, *52*(8), 826–839.

Wang, Y., Xu, C., Zhang, B., Wu, M., & Chen, G. (2017). Physiological and proteomic analysis of rice (Oryza sativa L.) in flag leaf during flowering stage and milk stage under drought stress. *Plant Growth Regulation*, *82*, 201–218.

Wang, Y., Zhang, Q., Zheng, T., Cui, Y., Zhang, W., Xu, J., & Li, Z. (2014). Drought-tolerance QTLs commonly detected in two sets of reciprocal introgression lines in rice. *Crop and Pasture Science*, *65*(2), 171–184.

Wani, S. H., Choudhary, J. R., Choudhary, M., Rana, M., & Gosal, S. S. (2020). Recent advances in genomics assisted breeding for drought stress tolerance in major cereals. *Journal of Cereal Research*, *12*(1), 1–12.

Wei, H., Feng, F., Lou, Q., Xia, H., Ma, X., Liu, Y., Xu, K., Yu, X., Mei, H., & Luo, L. (2016). Genetic determination of the enhanced drought resistance of rice maintainer HuHan2B by pedigree breeding. *Scientific Reports*, *6*(1), 37302.

Wei, A., He, C., Li, B., Li, N., & Zhang, J. (2011). The pyramid of transgenes TsVP and BetA effectively enhances the drought tolerance of maize plants. *Plant Biotechnology Journal*, *9*(2), 216–229.

Werner, T. (2010). Next generation sequencing in functional genomics. *Briefings in Bioinformatics*, *11*(5), 499–511.

Wolter, F., & Puchta, H. (2018). Application of CRISPR/Cas to understand cis-and trans-regulatory elements in plants. In N. Yamaguchi (Ed.), *Plant transcription factors: Methods and protocols* (pp. 23–40). Humana Press.

Xu, Y., & Crouch, J. H. (2008). Marker-assisted selection in plant breeding: From publications to practice. *Crop Science*, *48*(2), 391–407.

Xu, Y., Liu, X., Fu, J., Wang, H., Wang, J., Huang, C., Prasanna, B. M., Olsen, M. S., Wang, G., & Zhang, A. (2020). Enhancing genetic gain through genomic selection: From livestock to plants. *Plant Communications*, *1*(1), 100005.

Yadaw, R. B., Dixit, S., Raman, A., Mishra, K. K., Vikram, P., Swamy, B. M., Cruz, M. T. S., Maturan, P. T., Pandey, M. & Kumar, A. (2013). A QTL for high grain yield under lowland drought in the background of popular rice variety Sabitri from Nepal. *Field Crops Research*, *144*, 281–287.

Yahoueian, S. H., Bihamta, M. R., Babaei, H. R., & Bazargani, M. M. (2021). Proteomic analysis of drought stress response mechanism in soybean (*Glycine max* L.) leaves. *Food Science & Nutrition*, *9*(4), 2010–2020.

Yali, W., & Mitiku, T. (2022). Mutation breeding and its importance in modern plant breeding. *Journal of Plant Sciences*, *10*(2), 64–70.

Yamori, W., & Shikanai, T. (2016). Physiological functions of cyclic electron transport around photosystem I in sustaining photosynthesis and plant growth. *Annual Review of Plant Biology*, *67*, 81–106.

Yang, W., Feng, H., Zhang, X., Zhang, J., Doonan, J. H., Batchelor, W. D., Xiong, L., & Yan, J. (2020). Crop phenomics and high-throughput phenotyping: Past decades, current challenges, and future perspectives. *Molecular Plant*, *13*(2), 187–214.

You, J., Zong, W., Li, X., Ning, J., Hu, H., Li, X., … & Xiong, L. (2013). The SNAC1-targeted gene OsSRO1c modulates stomatal closure and oxidative stress tolerance by regulating hydrogen peroxide in rice. *Journal of Experimental Botany*, *64*(2), 569–583.

Yu, X., James, A. T., Yang, A., Jones, A., Mendoza-Porras, O., Bétrix, C. A., Ma, H., & Colgrave, M. L. (2016). A comparative proteomic study of drought-tolerant and drought-sensitive soybean seedlings under drought stress. *Crop and Pasture Science*, *67*(5), 528–540.

Yuan, Y., Cairns, J. E., Babu, R., Gowda, M., Makumbi, D., Magorokosho, C., Zhang, A., Liu, Y., Wang, N., Hao, Z., Vicente, F. S., Olsen, M. S., Prasanna, B. M., Lu, Y., & Zhang, X. (2019). Genome-wide association mapping and genomic prediction analyses reveal the genetic architecture of grain yield and flowering time under drought and heat stress conditions in maize. *Frontiers in Plant Science*, *9*, 1919.

Zargar, S. M., Mir, R. A., Ebinezer, L. B., Masi, A., Hami, A., Manzoor, M., Salgotra, R. K., Sofi, N. R., Mushtaq, R., Rohila, J. S., & Rakwal, R. (2022). Physiological and multi-omics approaches for explaining drought stress tolerance and supporting sustainable production of rice. *Frontiers in Plant Science*, *12*, 803603.

Zhang, W., Wu, Y., Schnable, J. C., Zeng, Z., Freeling, M., Crawford, G. E., & Jiang, J. (2012). High-resolution mapping of open chromatin in the rice genome. *Genome Research*, *22*(1), 151–162.

Zhang, Y. B., Yang, S. L., Dao, J. M., Deng, J., Shahzad, A. N., Fan, X., Li, R.-D., Quan, Y.-J., Bukhari, S. A. H., & Zeng, Z. H. (2020). Drought-induced alterations in photosynthetic, ultrastructural and biochemical traits of contrasting sugarcane genotypes. *PLoS One*, *15*(7), e0235845.

Zhou, M. L., Zhang, Q., Sun, Z. M., Chen, L. H., Liu, B. X., Zhang, K. X., Zhu, X.-M., Shao, J.-R., Tang, Y.-X., & Wu, Y. M. (2014). Trehalose metabolism-related genes in maize. *Journal of Plant Growth Regulation*, *33*, 256–271.

14 Phytohormones Increase Plant Adaptation under Soil and Water Stresses

Augustine Antwi-Boasiako, Tuanjie Zhao, Padmore Adu-Antwi, Chunting Zhang, Priscilla Amponsah, Zhang NanNan, Naheeda Begum, and Muhammad Khuram Razzaq

14.1 INTRODUCTION

Stresses in crop plants caused by abiotic factors are constantly causing a reduction in crop yields globally. The abiotic stresses imposed on plants by elements such as freezing and extreme temperatures, dry period, waterlogging, brininess, heavy metals, and nutrient stress impact the development, productivity, as well as yields of crops (Chaudhry & Sidhu, 2022; Pandey et al., 2021; Sharma et al., 2020; Vaughan et al., 2018). Abiotic stressors damage agriculture and ecosystems. Because plants are sessile by nature, they severely strain them, resulting in disproportionate economic losses. To preserve sustainable agriculture and to satisfy the combined demands on food from a world population that is growing at an accelerating pace, crop plants must be protected from climate change (Praveen, 2022). Abiotic stresses have a significant impact on crop performance since they are influenced by several stress combinations. These stresses are linked to a major impact on the development and growth of plants, which ultimately results in a decrease in agricultural output (Chaudhry & Sidhu, 2022; Zörb et al., 2019). Specifically, drought and other abiotic stresses have decreased and continued to impair agricultural output globally, particularly for grains like wheat and maize (Matiu et al., 2017). According to estimations by Hendrawan et al. (2023), between one-third and two-thirds of the globe's harvested lands of wheat, rice, soybeans, and corn were negatively impacted by various drought spells. In contrast to rice and maize, which may respond to droughts that last for a brief to medium amount of time, soybeans and wheat may respond to longer-lasting droughts. The various soil elements present in a soil contribute to the performance of a crop. However, it noted that nutritional levels in soils are inequitably distributed, resulting in abiotic stress impacting crop yield (Sharma et al., 2020) and directly affecting human dietary nutrition (Copeland, 2016; Day, 2013).

Since most plant species are stationary and impacted by various environmental pressures, plants are frequently vulnerable to a wide range of abiotic stresses. Plants respond to these stresses by developing strong adaptation responses to mitigate or tolerate them. These complex mechanisms consist of a network of signaling molecules such as phytohormones, polyamines (PAs), gasotransmitters, phytochromes, and calcium, which are involved in regulatory functions (He et al., 2018). Moreover, PAs, which are tiny amine molecules, are essential for crops to have the ability to withstand a range of abiotic stresses (Napieraj et al., 2023). Among them, phytohormones are crucial in the adaptive mechanisms of crops for stress tolerance. Phytohormones and growth regulators such as abscisic acid (ABA), auxin/indole acetic acid (IAA), gibberellic acid (GA), jasmonic acid (JA), salicylic acid (SA), brassinosteroids (BR), cytokinin (CK), ethylene (ET), strigolactones (SLs), and PA have been used to treat abiotic stresses to decrease the harmful effects of ions, ameliorate oxidative stress, maintain water-related balance, as well as modify gaseous exchange. They improve

DOI: 10.1201/9781003457237-14

plant tolerance by eliminating reactive oxygen species (ROS) and raising the function of antioxidant enzymes. These hormones also activate the stress-signaling processes regulated by ET, salicylic, abscisic, and salicylate (Swain et al., 2023).

Previous research findings have concentrated on the roles of endogenous phytohormones in crops concerning stress tolerance. Phytohormones are chemical messengers of minute amounts that significantly function in controlling the growth, development, and metabolism of plants in response to a variety of biotic and abiotic stresses. They are recyclable, greatly improve plant growth and development, and raise agricultural output in both ideal and stressful environments. The focus of this chapter is to comprehend the key role of the external use of phytohormones on the growth and development of crops in terms of their morphological changes, physiological and biochemical response, and gene network related to phytohormone regulation in plants under abiotic stresses. In this chapter, we highlight the significant roles of phytohormones, their crosstalk, as well as their stress mitigative roles in plants that are subject to abiotic stressors. This insight into exogenous phytohormones will contribute to the forthcoming crop development programs.

14.2 EFFECTS OF ABIOTIC STRESS ON PLANT GROWTH AND DEVELOPMENT

Amid the growing demand for food production owing to the expanding global population, crop plants are seriously threatened by climate change (Malhi et al., 2021). For instance, agricultural productivity decreases as water resources drop by 5% due to the temperature rise globally posing a threat to the security of food (Naumann et al., 2018; Neupane et al., 2021). Abiotic stressors also have an impact on the growth and development of plants. During such conditions, plants undergo morphophysiological changes in adapting to the pressure resulting in biochemical responses and activation of gene networks. Condition, plants undergo morphophysiological changes in adapting to the pressure, resulting in biochemical responses and activation of gene.

14.2.1 Growth and Morphological Changes

During critical phases of their life cycle, plants are severely susceptible to stresses caused by temperature extremes like heat, cold, and drought, which can negatively impact their growth and development (Zhou et al., 2018). Heat stress refers to a rise in ambient temperature (beyond a particular limit) that lasts long enough to permanently impede plant growth and development. The duration, intensity, and rate of rising temperatures, as well as the plant's growth phase, all influence how much the plant is impacted. For instance, a sudden temperature rise (5–15°C) is more injurious to plants (Muhlemann et al., 2018).

Water is still indispensable for plant growth and development. The adverse effects of water constraints on losses in the manufacturing and distribution plants on land are significant. Water remarkably affects plants because it is physiologically significant and necessary for healthy plant growth. Plants primarily absorb water from the soil through the root system, which distributes it to the plant's growing points through the anatomical system. Water reserves, their availability to plants, and, subsequently, plant uptake of water are all impacted by the water potential of the soil. The occurrences and severity of water stress and severe temperatures are increasing with climate change, which poses a hazard to several natural environments (Čanak et al., 2020; Cardinale et al., 2012) because of the implication for plant growth and development (Du et al., 2019). Water is still among the most critical ingredients for plant growth and development (Figure 14.1). However, paucity or excess presents a severe threat to the survival of plants as they serve as a constraint regulating primary production in terrestrial ecosystems. Plant morphology is impacted by water stress. When a plant's transpiration rate from its leaf surfaces exceeds its root water uptake, it has a water deficit. Plants shut their stomata as a result, which lowers stomatal conductance and increases water usage efficiency.

Different parts of plant growth and development are impacted by drought, and the extent of the impact on plants varies depending on the duration, timing, and severity of the stress.

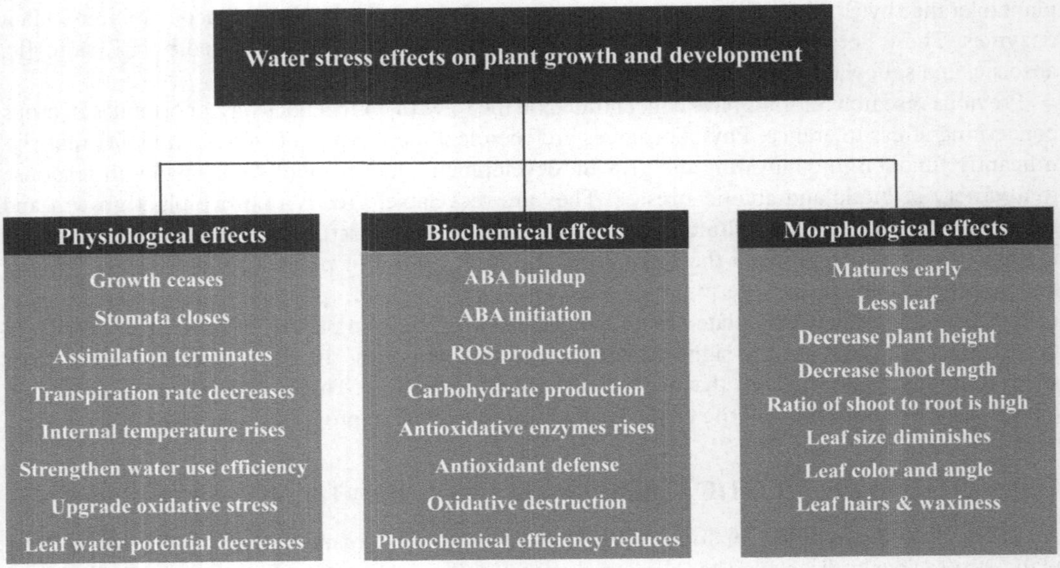

FIGURE 14.1 Water stress effects on plant growth and development.

Three essential processes that affect plant growth and development are the division of cells, elongation, and differentiation. Due to turgor loss, disordered enzyme activity, and reduced photosynthesis-derived energy, drought circumstances harm each phase (Osakabe et al., 2014; Thomas et al., 2022). Drought reduces plant water capacity and strength; thus, plant cells cannot function normally (Figure 14.1). Reduction in the turgor of plants leads to suppressed cell expansion and growth. Cell proliferation and expansion are essential phenomena in the early stages of plant growth and establishment. Drought modifies leaves' morphology and ultrastructure (Khan et al., 2022; Rao et al., 2000). Reduction in leaf size, decrease in stomata number and opening, thickening of the cell wall, development of cuticles on the leaf surface and conductive system (an upsurge in large vessels), stomata immersion in xerophyte and succulent plants, and the development of tube leaves in cereal plants are known examples of these changes (Jaleel et al., 2009; Khan et al., 2022). Due to the environment's decreased water potential, high temperatures, droughts, or elevated salinities, plant water relations are disturbed. A lack of water generally results in decreased stomatal and root hydraulic conductivity, which jeopardizes the plant's water status (Li et al., 2021; Siemens & Zwiazek, 2004).

Waterlogging and flooding stress impair plants' growth, subsequently influencing their morphology. In cases where high rainfall, coupled with poor soil drainages, results in waterlogging, it becomes a severe issue of crop production, especially in cereals, altering their development (Lone et al., 2018; Manik et al., 2019). This happens because waterlogging hinders the soil's ability to provide the required biological, chemical, and physical support essential for the growth of plants (Ferronato et al., 2019; Glinski, 2018). Plant growth under waterlogging stress is hindered as the spread of oxygen to plants is deficient since the soil pores are stuffed mainly with water, affecting the nitrogen (N) availability in the soil and thereby affecting N-loving plants (Nguyen et al., 2018). Also, the plant nutrient required for the shoot and the soil oxygen concentration are not met (da Ponte et al., 2019; Morales-Olmedo et al., 2015).

Flooding causes whole or partial submergence stress, which adversely affects plant reproductive development, vegetative composition, and the sprouting of seeds (Chattopadhyay et al., 2021). Flooding creates oxygen-deficient situations, which negatively impact the phases of development of a plant's life cycle as a result of switching from respiration, that is aerobic to anaerobic fermentation, as well as limit root growth, leading to its death (Christianson et al., 2009; Xuewen et al., 2014).

Salinity stress is well noted to be the primary factor of dehydration in crop plants (Adnan et al., 2020). One of the significant issues facing many parts of the world is salinity stress, particularly in arid and semi-arid areas where irrigation uses saline-sodic water (Qadir et al., 2021; Zaman et al., 2018). A third of the world's food is said to be produced on irrigated land; however, salt stress affects 20% of this land, or roughly 45 million hectares, reducing the potential yield of crops every year by 50% (Gupta et al., 2018). Stress imposed by saline soils impacts the growth and development systems of plants. It is usually brought about due to the presence of components such as Na^+, Mg^{2+}, K^+, Ca^{2+}, Cl^-, and SO_4^{2-} in soils, contributing to high soil electrical conductivity (Yan et al., 2023; Zhang et al., 2006). Similarly, it affects the plant's morphology, biomass, and life processes (Rahneshan et al., 2018). Also, salinity stress is noted for retarding seed sprouting rate, development, flowering, and fruiting (Park et al., 2013; Taghvaei et al., 2022), which affects plant production and quality. Plants exposed to saline stress have growth retard (Parida et al., 2016), and morphological characteristics such as the length of the petiole, internode, leaf, and its width are reduced (Avestan et al., 2021; Kumar et al., 2021).

Plants experiencing nutrient stress undergo morphological modification to cope with the changing environment, usually at the stem, root, leaf, and branch levels. For instance, the plant root system alters its structure by suppressing the growth of primary roots while stimulating the growth of lateral roots and root hairs for exploring more soil zones. For instance, in contrast to plants with high P availability, *Arabidopsis* subjected to low P had significantly greater root hair density (Francis et al., 2023). The root structure exhibits remarkable plasticity in varying conditions with changing nitrogen concentrations and leaf features, influencing plant biomass (Luo & Luo, 2017; Mao et al., 2018).

14.2.2 Physiological and Biochemical Response

From an agricultural and physiological perspective, aridity/heat stress happens when low soil moisture at a specific period reduces the amount of water available in the soil for plants. For plants to engage in photosynthesis at its best, stomatal opening and leaf area development must be at their ideal levels. Drought significantly affects net photosynthesis in plants by reducing chlorophyll content and leaf area, increasing premature leaf senescence and resistance to gas exchange in stomata (Canavar, 2015; Shao et al., 2008; Ullah et al., 2022). These result in a lowered photosynthetic rate, causing a substantial decrease in plant size and biomass output (Figure 14.1).

Heat stress causes severe changes in the physiological and biochemical alterations like increased respiration, increased membrane permeability, decreased photosynthesis, and production of ROS, and biochemical processes like the synthesis of energetic composites and new tissues (Hasanuzzaman et al., 2013; Huang et al., 2022). Heat stress leads to high transpiration in plant leaves, leading to a substantial decrease in water potential in addition to relative humidity, thus reducing photosynthetic productivity (Liu et al., 2020). Photosynthesis is an integral phenomenon for plant growth. It has been demonstrated that heat stress disrupts the photosynthesis process, thereby severely impacting plant productivity (Netshimbupfe et al., 2022). Additionally, heat stress has an impact on chlorophyll production, photochemical reactions, the turnover of D1 and D2 proteins, and CO_2 assimilation. One of the heat-sensitive proteins, Rubisco activase, is deactivated owing to chloroplast damage caused by heat stress. A redox imbalance is produced due to this, which reduces photosynthetic performance and can potentially cause cell death (Zahra et al., 2023). Furthermore, the adverse effects on plant growth are manifested in rolling and aging of leaves, restriction of root and shoot growth along with a decline in seed germination, discoloration of fruit, and decrease in pollen viability and output (Hasanuzzaman et al., 2013; Kaushal et al., 2016).

The primary pigment used in photosynthetic processes in higher plants is chlorophyll, and plants exposed to high temperatures lose some of their ability to synthesize chlorophyll (Hu et al., 2020). Similarly, high temperatures (28–38°C) reduced the amount of total chlorophyll (18%), chlorophyll 'a' content (7%), chlorophyll a/b ratio (3%), and sucrose content (9%) and increased the amount of reducing sugar (47%) and leaf-soluble sugars (36%) in soybeans. Chlorophyll is the primary pigment utilized in photosynthesis by higher plants. It has been found that exposure to high temperatures can hamper the plant's ability to synthesize chlorophyll, according to research conducted by Hu et al. (2020). Additionally, a study by Hasanuzzaman et al. (2013) revealed that soybeans, when exposed to temperatures ranging from 28°C to 38°C, demonstrated a decrease in total chlorophyll (18%), chlorophyll 'a' content (7%), and chlorophyll a/b ratio (3%). The sucrose content also decreased by 9%. Conversely, the amount of reduced sugar increased by 47%, and leaf-soluble sugars increased by 36%. High-temperature stress can severely impact photosystems I and II, electron transfer, carbon dioxide fixation, and stomal activity, thereby hindering plant growth and development (Hasanuzzaman et al., 2013). Furthermore, high-temperature stress has been indicated to drastically decrease the photosynthetic process by severely affecting photosystems I and II, electron transfer, carbon dioxide fixation, and stomal activity in several ways, consequently hampering plants' growth and development.

Salinity stress brings about physiological, biochemical, and molecular changes, which primarily inhibit photosynthesis influencing cell division and growth (Van Zelm et al., 2020). Moreover, sugar signaling and production levels are affected when plants experience salt stress (Shumilina et al., 2019). Some enzymes and proteins like ribulose-1,5-bisphosphate carboxylase/oxygenase (Rubisco) play an important role during photosynthesis; however, the actions are limited under a stressful environment (Chen et al., 2015). When roots are subjected to salt stress, they typically absorb less water. Depending on the concentration of salt, the decline may result from either harmful or osmotic effects. For instance, the rate of water uptake was reduced when pepper plants were supplied with a small concentration (30 mM) of NaCl or with a nutrient solution having a similar osmotic value (Gupta et al., 2018). However, a further increase in the NaCl concentration to 60 mM and the osmotic pressure of the nutrient solution to the same value (−0.290 MPa) caused a reduction in root water uptake only in plants administered with NaCl. Therefore, either an imbalance in the acquisition of other nutrients or particular toxic effects resulting from the buildup of Na^+ and Cl^- ions in root tissues caused the drop in the root water uptake rate under higher concentrations of NaCl (Gupta et al., 2018). Salt stress results in ionic toxicity and osmotic stress in crop plants which adversely affects cellular growth, membrane functions, and cytosolic metabolism, ultimately impeding the growth and development of plants (Osakabe et al., 2014). Higher plants can endure salinity stress with the aid of nitric oxide, PAs, and osmolytes (Choudhary et al., 2023). Osmolytes quench free radicals while also activating antioxidant mechanisms and managing osmotic balance. PAs stabilize membranes, neutralize acids, and stop the formation of ROS. Increased plant stress tolerance is predicted by the link between PA levels and nitric oxide synthesis. These methods may be the ones that researchers concentrate on to assist plants to cope with salt stress.

Under flooding stress, the plant leaf is significantly affected. For example, its water and nitrogen content are reduced, thereby accelerating leaf chlorosis and senescence as well as CO_2 assimilation (Zheng et al., 2009). Moreover, plants under flooding stress conditions trigger ROS production, more frequently resulting in oxidative harm and the deterioration of cellular membranes, proteins, and lipids (Lal et al., 2019). Likewise, ROS production is instigated by plants under waterlogging, thereby producing lipid peroxidation leading to damage in membranes, enzyme deactivation, and eventually killing of cells (Hasanuzzaman et al., 2017; Sharma et al., 2012). Plants deal with flooding stress by turning on their antioxidant superoxide dismutase (SOD), catalase (CAT), glutathione (GSH) reductase, ascorbate peroxidase, and non-antioxidant (proline, and sugars) defense mechanisms to avert the flooding-induced ROS production (Alizadeh-Vaskasi et al., 2018; Sharma et al., 2018).

Cold stress impedes plant growth and development and has numerous vital outcomes on plant cells by negatively impacting the biochemical and physiological activities in plants. Low temperature interferes with membrane structure and protoplasmic streaming, causing electrolyte leakage or plasmolysis (Yadav et al., 2020). It impacts plant cell membrane stiffness, which is thought to be the fundamental mechanism underlying downstream cold stress responses in plants (Örvar et al., 2000). Furthermore, cooling stress diminishes enzymatic activity like ROS-scavenging enzymes and tampers with protein stability or protein complexes. These methods lead to photoinhibition and photosynthesis impairment in addition to membrane damage (Foyer, 2018; Lima-Melo et al., 2019; Ruelland et al., 2009). Again, freezing stress inhibits the synthesis of proteins and the expression of genes that support the development of secondary structures in RNA (Ruelland et al., 2009; Tang et al., 2022). Then again, cold stress inhibits plant growth by influencing the synthesis of proteins and secondary metabolites, antioxidant defense enzyme activity, membrane integrity, and phytohormonal signaling. The effects of cold stress vary according to its interval, plant species, source, growth stage, and tissues/organs of the plant. Plant stress modifies the structure of proteins, the fluidity of cell membranes, the activity of antioxidant enzymes, and the buildup of ROS in plants. Moreover, ion leakage, membrane deterioration, and photosynthesis are all impacted by stress (Raza et al., 2023). Freezing stress is identified as more destructive to plants and may be fatal to plants. Extracellular ice nucleation is the first step in freezing damage under natural circumstances (Pearce, 2001; Stegner et al., 2022). Following the formation of ice nuclei, these crystals expand and disseminate into the apoplast, where they cause water outflow and cause cell dehydration. Usually, when ice crystals penetrate cells, permanent harm results.

14.2.3 Gene Network Related to Phytohormone Regulation

Phytohormones are minute molecules that govern the formation and maturation of plants while their survival and adaptation to different environmental stresses. A good document is kept on the induction, production, distribution, and role of phytohormones (Amrine et al., 2015; Li et al., 2017) and their function in adverting abiotic stressors in plants for the enhancement of their growth and productivity (Asami & Nakagawa, 2018; Oguz et al., 2022; Shah et al., 2021; Wang & Komatsu, 2022). Plant hormones are essential in their roles in growth-related processes ranging from organ induction to maturation. The application of omics techniques supports the discovery of novel genes, metabolites, and metabolic pathways induced by phytohormones. Recently, it has been discovered that plant growth regulators and phytohormones are beneficial resources to minimize plant stresses. These regulators target transcription factors (TFs) and genes linked to stress to boost agricultural output (Altaf et al., 2023). Hence, phytohormones regulate the critical stages in the life phases of plants. A way for plants to withstand abiotic stress requires activation of sensing of the stress followed by signal transduction, which then results in gene induction and its related products to provide protection and necessary repair and tolerance mechanisms (Rao et al., 2006). It is recognized that phytohormone signaling and TFs govern plant resistance to abiotic influences (Waadt et al., 2022; Yoon et al., 2020).

Phytohormones like ABA and JA modify how plants cope with abiotic stress and are vital in activating TFs (Verma et al., 2016; Waadt et al., 2022). For instance, TFs like ABFs, MYCs, NAC, AP2/ERF, MYB, and WRKY families determine how plants adapt and endure abiotic stress (Yoon et al., 2020). Especially in soybeans, *GmNAC12* improves drought tolerance with a survival rate of 57% in comparison to the wild plant through overexpression (Yang et al., 2022). Similarly, previous research on JA demonstrated how exogenous JA application quickly and dynamically regulates how genes associated with plant defenses reveal the role of JA in signaling pathways and governing stress connected to transcription networks (Genva et al., 2019). For example, under circumstances involving drought stress, plants utilize signaling mechanisms that are both ABA-dependent and ABA-independent to identify and deal with it (Kim, 2014). ABA-independent TFs operate as protein switches upon signaling, thereby regulating the expression of corresponding genes by interacting with cis-elements in the protein-coding region of genes (Franco-Zorrilla et al., 2014; Villano et al., 2020). ROS generation peaks with stress signaling, and it is associated with ABA, activating gene expression due to the synthesis of protein (Hasanuzzaman, 2020; Kaur & Asthir, 2017). Modifications in the transcription of proteins mediate numerous impacts of phytohormones. ABA increases in plants under abiotic stresses are associated with changing multiples of genes at the mRNA levels, with evidence in mutant plants that are ABA-insensitive (Finkelstein, 2013; Goda et al., 2008).

14.3 ROLES OF PHYTOHORMONES IN REGULATING ABIOTIC STRESS RESPONSE IN PLANTS

Low amounts of phytohormones exhibit significant physiological consequences which impact a variety of plant processes, including sex determination; dormancy; organ shedding; germination and rooting; blooming and fruiting; cell growth, multiplication, and differentiation; as well as the development of plants and resilience to stress (Jiang & Asami, 2018). The majority of the phytohormones, including ABA, auxin/IAA, BR, ET, JA, GA, SA, and CK, have been studied extensively. Others, such as SLs and PA, are generating an increasing investigation. Additional plant growth regulators have not been fully explained, including karrikins and PAs.

14.3.1 Seed Germination and Seedling Growth

ABA is one essential phytohormone that broadly impacts seed germination, dormancy, and development. It is essential to seed dormancy because it prevents the growth of embryos till germination (Jhanji et al., 2024). The equilibrium between dormancy and germination is controlled by the ratio of ABA synthesis to catabolism, with endogenous ABA levels in ingested seeds being connected with seed dormancy. ABA also controls critical processes in seed production, like the deposition of store reserves and the inhibition of early germination (Kermode, 2005). Also, PAs support the germination and growth of seeds by promoting seed priming and germination rate.

Cytokinins are crucial for early development plants as they control stress tolerance, encourage cell division, and shoot growth in seedlings. To shape the overall architecture of the seedling, they are essential in preserving the activity of the shoot's uppermost meristem and encouraging the growth of fresh foliage and branches (Sharma et al., 2022). Additionally, cytokinin controls stress tolerance, affecting numerous plant growth, development, and physiology characteristics. These characteristics include apical dominance, flower and fruit development, delayed senescence, seed germination, and stress response. The interactions between cytokinins and other signaling molecules ultimately influence plant resilience and stress tolerance, enabling plants to alter their stress reactions more accurately (Kissoudis et al., 2014). For plants to establish the growth of seedlings, they must intricately coordinate their cytokinin signaling pathways with other hormonal and environmental cues. Due to this, plants can adapt to the conditions of their surroundings. With their diverse effects on stress responses and seedling growth, cytokinins provide essential information for the creation of stress-tolerant crops and environmentally friendly farming methods (Mandal et al., 2022). Comprehending the molecular mechanisms that underpin cytokinin function and its interaction with environmental stimuli is imperative in deciphering the multifaceted systems of regulation that oversee plant reactions to stress and seedling growth.

The gaseous plant growth hormone ET influences root development and stress responses in addition to controlling seedling growth and either boosting or suppressing growth under stressful circumstances. Plant life and environment adaptation depend on ET's participation in these processes. Because it may both restrict and encourage growth under stress, ET has various applications in the growth of seedlings. In adverse conditions, such as elevated temperature, salt, inadequate nutrient availability, drought, and flooding, plants produce ET to initiate a stress reaction that may impede the growth of seedlings (Bharadwaj et al., 2022). Despite this, ET promotes seedling growth by interrelating alongside nutrients and other phytohormones to enhance plant development and resilience to stress (Fatma et al., 2022).

Plant defense responses are mediated under an abiotic stressor by jasmonates, especially JA. Plants create jasmonates as a reaction to environmental stressors such as heat, drought, and heavy metals, to be able to initiate reactions to stress and improve tolerance levels. As a means of adjusting to unfavorable environmental circumstances, jasmonates are engaged in the control of several reactions arising from physiological, biochemical, and molecular pathways to the stressors, such as the initiation of defense mechanisms and the modification of growth and development (Wang et al., 2020). Jasmonates synchronize plant growth and its response to stressors, which is essential in managing seedlings under stressful situations. The plants can fine-tune their growth and adapt to the stress conditions by establishing relationships with additional signaling molecules, which ultimately impacts the resilience of seedlings (Kissoudis et al., 2014). Thus, jasmonates control seedling growth and mediate defense mechanisms in stressed plants. Deciphering the intricate regulatory networks governing the response of plants to stress from abiotic factors and devising methods to improve stress tolerance in crops requires a comprehension of the molecular processes underpinning jasmonate action and its interaction with environmental cues (Wang et al., 2021).

14.3.2 Root Development

An increase in length, volume, and area characterizes root growth. Ethylene plays a crucial part in root formation and reactions to stress. For instance, by altering plants physiological and developmental procedures of plants, ET enhances plants' ability to withstand salt stress (Chen et al., 2021). Additionally, it increases the liverwort *Marchantia polymorpha*'s ability to withstand abiotic stress, demonstrating the stress hormone's universal function in plant types (Bharadwaj et al., 2022). The dual nature of ET in seedling growth is seen in its ability to promote or restrict growth under stressful environments. Also, PAs support root development as they increase the root's quantity and length (Tiburcio & Alcázar, 2018). These phytohormones' role in stress responses and root growth is crucial for plant life and environment adaptation.

14.3.3 Leaf Growth and Development

Auxin's involvement in the leaf life cycle cannot be underestimated. Auxin manifests to improve the shoot apical meristem during leaf inception via its enhanced biosynthesis (Pinon et al., 2013). Specifically, auxin modulates the emergence of fresh leaves and vascular leaves in tomatoes (Mattsson et al., 2003). *Arabidopsis* shoot, radial location, and leaf size are moderated by IAA (Reinhardt et al., 2000). The occurrences of leaf epinasty are attributed to auxin or auxin-stimulated ET (Hansen & Grossmann, 2000; Roe et al., 1997). Similarly, GAs facilitate the maturation and development of plants by fostering seed germination, emergence, establishment, and leaf enlargement. Thus, supporting the plant's total life phases. Several works prove that applying GA to plants results in larger root and leaf development (Bagautdinova et al., 2022; Junttila et al., 1997). The SA mainly manages plant growth by regulating processes involved in flowering and stomata (Miura & Tada, 2014). Other researchers confirm that SA considerably impacts the growth and development of maize and rice (Bagautdinova et al., 2022; Ruan et al., 2002; Szalai et al., 2016). Plants' reaction to SA depends on its concentration and varies with the plant species (Li et al., 2022). For example, 0.1 mm SA induced flowering and grain development in finger millet (Appu & Muthukrishnan, 2014), while the strawberry leaf area was increased by 0.25 mm (Kazemi, 2013). Also, in *Arabidopsis*, SA of less than 0.05 mm encouraged adventitious root growth, but SA higher than 0.05 mm limited root growth (Pasternak et al., 2019). The differences in the observed effects on the plants demonstrate that the functions of exogenous SA are complex in plant development.

14.3.4 Flower Development

Flowers are the crucial reproductive organ that contributes to ensuring the survival of significant plants. Flower development is a multifaceted process comprising numerous regulatory networks, including plant hormone signals. The period when plants flower is of concern as it directly affects the stage of plant reproduction. Plant flowering is the basis for grouping plants into short, long, and neutral days. BR, ABA, and auxin regulate the photoperiod of floral growth (Pearce et al., 2016). Previous studies show the applicability of interaction among phytohormones (auxin, GAGA, and ABA) in crops like *Arabidopsis* (Matsoukas, 2014) and apples (Frankowski et al., 2014). DELLA protein found in the GA pathway supports the governing of various signaling and regulation of flowering development. When GA20OX downregulates GA biosynthesis, it leads to late flowering in FR-enriched conditions, showing its function in regulating the flowering period (Hisamatsu & King, 2008). Similarly, varied amounts of GA demonstrate differences in flowering activity in many plants (Fukazawa et al., 2021; Zhang et al., 2022). Also, GAs (GA3–GA7) trigger inconsistent flowering induction processes in plants of the species (He et al., 2020). For instance, in *Malu domestica*, flowering is positively regulated, while GA3 functions inhibit flowering (Song et al., 2014).

14.3.5 Plant Photosynthesis

Photosynthesis is crucial for plants and also maintains life on the earth. The function of phytohormones including SA, BR, GA, and auxin has demonstrated an involvement in photosynthesis. Phytohormones such as SA regulate the photosynthetic activities of plants involving parameters like stomatal closure by manipulating the assembly of chloroplasts and enzyme actions (Kohli et al., 2017). Applying SA for seed priming results in high photosynthetic efficiency and grain production in corn (Szalai et al., 2016). BRs are critical components in regulating photosynthesis in plants (Siddiqui et al., 2018). Phytohormones are used by plants experiencing salt stress to lessen its effects by enhancing their capacity for photosynthesis. For instance, BRs applied to *Brassica juncea* under salt conditions demonstrated a high photosynthetic rate (Hayat et al., 2012). BRs reduce damage caused by sodium chloride in the chlorophyll content, restoring the stomatal pores, photorespiration rate, and photosynthetic rate (Ekinci et al., 2012; Rady, 2011). Furthermore, BR signaling through the TF BRASSINAZOLE-RESISTANT1 (BZR1) controls light signaling mechanisms by involving PIF4 to enhance etiolation (Oh et al., 2012). GAs stimulate the advance of significant structures via cell elongation and the promotion of mitosis and meiosis (Colebrook et al., 2014). Similarly, auxins control the plants' various sections for their growth (Salazar-Iribe & De-la-Pena, 2020). In addition, auxin influences the photorespiration rate by regulating leaf venation and stomata formation. Extreme quantities of exogenous auxin application or loss of function through mutation led to the stomatal cluster, but auxin overexpression limited the number of stomata (Balcerowicz et al., 2014; Zhang et al., 2014). Auxin supports the regulatory role of TF MONOPTEROS (MP) and constrains the pathway of the stomata by downregulating the STOMAGEN (Qi & Torii, 2018).

14.4 EXOGENOUS APPLICATION OF PHYTOHORMONES AND PLANT GROWTH REGULATORS TO IMPROVE PLANT TOLERANCE TO ABIOTIC STRESS

Plants often undergo stress by interacting with their environment. Hence, plants should be able to tolerate abiotic stress to be productive. Most of the time, farmers link abiotic stress to economic losses. Bestowing environmental stress on plants is a reasonably multifaceted procedure. Unfortunately, the distribution of plants and their growth is impeded by abiotic stressors ranging from drought, heat, nutrient deficiency, cold, and heavy metals (Zhu et al., 2016). Phytohormones can govern, integrate, and assist plants to overcome stresses (drought, heavy metals, salinity, nutrient stress, and cold) as they interact with their surroundings to sustain their development (Bücker-Neto et al., 2017).

Due to their vulnerability to numerous abiotic threats, climate change significantly threatens horticulture crops. Defense systems have evolved for plants to sense stress signals, allowing for the most significant development responses. Phytohormones can increase a plant's resistance to abiotic stressors by boosting the germination of seeds, root and seedling development, and reducing ROS, malonaldehyde, and electrolyte leakage. According to recent studies, numerous phytohormones may increase horticulture plants' capacity to endure abiotic stress (Zheng et al., 2023). Hormones produced by plants have historically been classified as either growth-inhibiting or growth-promoting, and their critical roles in supporting plant functions under varied abiotic conditions are revealed.

14.4.1 Abscisic Acid

ABA is an isoprenoid plant hormone previously identified to overcome seed dormancy (Cornforth et al., 1965). Now, it is known to aid in plant adjustment to abiotic stressors (Arkhipova et al., 2020). ABA regulates plant stresses and may involve physiological processes adapting to cold, drought, and salt stresses. ABA influences the plants' survival by serving as an intermediary of stresses resulting from the environment, regulating abiotic stressors, and coordinating a variety of tasks (Finkelstein, 2013; Sreenivasulu et al., 2012; Wani & Kumar, 2015), thereby increasing the plant adaptation to stress.

Most often, ABA inhibits plant development to coordinate their acclimatization to ecological stresses (Suzuki et al., 2016). Several species of plant have witnessed an increased level of endogenous ABA when subject to stress situations. For example, endogenous ABA has been recorded under drought conditions in plants such as soybean (Arya et al., 2021), maize (Cao et al., 2021; Lei et al., 2023), and barley (Collin et al., 2020) with evidence of increase plants resilience level to drought (Collin et al., 2020; Wei et al., 2015). Furthermore, in heat stress situations, there is an indication of a swift rise in the quantity of endogenous ABA produced in the *Arabidopsis* (Larkindale et al., 2005). Similarly, ABA's role in plant–water relations results in improving plant resistance to water scarcity and their acclimation (Arkhipova et al., 2020). ABA boosts drought via drought priming treatment in *Arabidopsis* and tall fescue (Zhang et al., 2019b). Biosynthesis of ABA in the root is triggered under water shortage conditions. The ABA increases its concentration in the leaves to control the cell turgor (Awan et al., 2017). The regulation of the ABA leads to limiting water loss by stomata closure, a decreased in transpiration rate, and restriction of cell development (Peleg & Blumwald, 2011).

ABA interacts with other phytohormones of plants subjected to abiotic conditions to intervene in physiological responses (He et al., 2018; Zhang et al., 2021). To illustrate, the physiological responses of ABA, including morphological, cytological, and anatomical changes induced by interactions with phytohormones in extreme conditions such as drought, cold, salinity, and toxicity of heavy metals are examined. ABA controls physiological processes by highlighting its favorable or detrimental interactions with plant hormones such as GA, melatonin, cytokinin, auxin, SA, JA, ET, BR, and SL (Singh & Roychoudhury, 2023). For ABA to lessen the consequences of non-living stressors to a minimal level, it uses an array of biochemical defense which may include reactive ROS purifying enzymes, antioxidants, heat shock proteins, proline biosynthesis, and unsaturated fatty acids that strengthen the plants' cuticular waxes (Guan et al., 2020; He et al., 2020; Liu et al., 2018). Similarly, ABA participates in the governing of hormone-responsive TFs to facilitate phosphatases and kinase roles in managing abiotic stressors (Dar et al., 2017). For example, TFs in the families of NAC and MYB are supported by ABA to regulate abiotic stressors including salt and drought in plant fields (Hossain et al., 2021). Specifically, the key *OsNAC3* from rice is a key component governing ABA signal response and tolerance to salinity (Zhang et al., 2021).

14.4.2 Auxin/IAA

Auxin is a natural compound existing as IAA (Tromas & Perrot-Rechenmann, 2010) and is present in all plants. They are small and have small molecular weights. Auxin supports the development of shoot, root, and vascular tissues (Woodward & Bartel, 2005) and mitigates against stresses (Kazan, 2013). Several findings support that auxin is essential when plant adaptability in salt conditions is discussed (Fahad et al., 2015; Iqbal et al., 2014). Plants under salt and heavy metal stress lead to growth in their root and shoot fields (Egamberdieva, 2009; Sheng & Xia, 2006). The anomalous levels of metals in plant-growing mediums are deadly to them and inhibit their maturation due to their buildup in their roots and shoots, influencing plant adaptation. This condition is usually managed by applying auxin, which improves biomass production (Zhang et al., 2020).

Auxin manufactures IAA when plants are under mental stress due to the generation of ROS in plant parts (cells, cytoplasm, mitochondria, and plastids) (Singh et al., 2021). At times, the amount of IAA produced is compromised since the plant organs contain heavy metals. Plants' reactions to heavy metals result in controlling genes linked to IAA synthesis, which possibly may direct the locality and accumulation of IAA (Bruno et al., 2017). A number of genes are engaged through the interaction between IAA and ROS to maintain homeostasis by controlling the amount of H_2O_2 via governing and energizing the antioxidant enzymes and the chlorophyll present in the plant (Fahad et al., 2015). IAA is complicated in improving acceptance of drought by regulating genes linked to leaf senescence, and drought stress reaction (Zhang et al., 2020).

14.4.3 Gibberellic Acid

The application of exogenous GA controls the processes involved in oxidative stresses and the different roles that antioxidant defense enzymes play. The mechanism suppresses the negative effects on plants of environmental stress (Khan et al., 2012). GAs play a role in plant systems related to coping with environmental stress (Colebrook et al., 2014; Hamayun et al., 2017). Under extreme metal stress conditions such as cadmium (Cd) and lead (Pb) active gibberellin (GA_3) promotes green algae acclimatization (Falkowska et al., 2011), by initiating defense genes, synthesis of CAT, and accretion of photoheating (Grill et al., 1985). According to study findings, GA may strengthen the soybean's mechanism for resisting Nickel (Ni) toxicity under stress. GA enhanced soybean seed germination, plant growth, biomass indices, photosynthetic apparatus, and relative water levels during Ni-induced stress. It was found that by reducing the quantity of hemicellulose, GA lowered Ni uptake, diffusion, and fixing in the root cell wall. In soybean plants, GA also reduced Ni uptake and dispersion (Bhat et al., 2023).

Similarly, maize plants grown under a metal stress environment showed high productivity for membrane permeability and growth in nutrient status under the application of GA_3 (Tuna et al., 2008). Moreover, adding GA_3 to soybean plants stressed by salinity conditions improved their growth (Hamayun et al., 2010). Seeds soaked with appropriate concentrations of GA prevent nutrient disparity, ion harmfulness, and osmotic stress, thereby overcoming the consequences of salinity. GA improves plant's tolerance to salt stress when used in seed treatment (Ashraf et al., 2004). Stresses imposed by high temperatures and drought are critical factors that impede plant yields. These result in the readiness of spontaneous electrons in mitochondria and chloroplast, controlling the generation of ROS (Caverzan et al., 2016). ROS presence in the plant is detrimental to the function of the plants as it impairs the protein, lipids, and nucleic acids. Wheat plants growing under heat stress exhibited tolerance via an exogenous spray of GA_3, which augmented its antioxidant enzyme actions and photosynthesis rate and limited lipid peroxidase activities (Nagar et al., 2021). The glyoxalase system is upregulated to support wheat's seedling rate during water shortage (Al Mahmud et al., 2019). Previous research confirms that GA enables plants such as tomatoes and wheat to cope with drought (Al Mahmud et al., 2019; Colebrook et al., 2014; Li et al., 2012).

14.4.4 Jasmonic Acid

The application of exogenous JA has direct consequences on plants. JA is a plant-signaling molecule related to abiotic stressor tolerance by controlling the molecular and physiological responses. Exogenous JA initiates antioxidant structure and protein buildup and modulates stomata (Acharya & Assmann, 2009; Karpets et al., 2014; Wasternack, 2014). Molecular response leads to the expression of genes linked with JA (Hu et al., 2017) and engagement with TFs and other plant hormones (Yang et al., 2019; Zhao et al., 2013). These responses permit JA to defend plants from the poisonous outcomes of abiotic stressors arising from light, freeze, heavy metals, drought, and nutrient deficiencies. For instance, in the apricot (Ge et al., 2010) and soybean (Mohamed & Latif, 2017), the exogenous application of JA improved their drought tolerances. Similarly, salt stress was improved in maize and common wheat (Qiu et al., 2014; Shahzad et al., 2015). Stresses arising from heavy metals (nickels) were tolerated in soybean and maize, improving the activities of antioxidant enzymes (Azeem, 2018; Sirhindi et al., 2015). Also, under the influence of cadmium stress, soybean and faba bean inhibited its accumulation via the actions of antioxidants (Ahmad et al., 2017; Noriega et al., 2012).

14.4.5 Salicylic Acid

SA is useful in plants by serving as a signal molecule when plants encounter stresses resulting from their environment (Janda & Ruelland, 2015). SA increases PA content, which promotes plants' ability to survive under stress (Szalai et al., 2016). SA regulates ion channels, photosynthetic rates, antioxidant enzymes, and defense mechanisms in plants (Pál et al., 2015). The usage of SA on plants during stressful conditions improves their antioxidant activities (Khandaker et al., 2011). SA functions effectively under high metal stress in plants among several plant species. SA can lessen the toxic effects of heavy metal stress caused by lead (Pb) in Indian mustard, pea, and common bean (Boroumand et al., 2011; Ghani et al., 2015; Khalil et al., 2021); cadmium (Cd) in maize, *Thlaspi*, wheat, milk thistle, common duckweed, and rice (Gondor et al., 2016; Kovács et al., 2014; Llugany et al., 2013; Lu et al., 2018; Mostofa et al., 2019); and mercury (Hg) in alfalfa (Zhou et al., 2009). This is achieved as SA reduces the uptake of heavy metals and their distribution within the plants while promoting the actions of antioxidants enzymes, CAT, SOD, and POD.

14.4.6 Brassinosteroids

BRs are present in organs and parts of plants, with significant concentrations in fruits, seeds, and pollen (Li & He, 2020; Ohnishi, 2018). The presence of BRs in plants supports their adaptation to environmental stresses by acting as biostimulants and contributing to their performance (Trevisan et al., 2020). BRs regulate several metabolic pathways and interact with other phytohormones by enhancing antioxidant enzymes, detoxification, and photosynthetic rate and activating the expression of associated genes, thus assisting plants' adaptation to changes in their environment (Rajewska et al., 2016). BRs engage in oxidative processes and abiotic stressors associated with heavy metals (Hao et al., 2013; Rajewska et al., 2016) while relieving plants from stress arising from cold, heat, drought, and salinity (Fàbregas et al., 2018; Ibañez et al., 2018; Zhu et al., 2016).

When plants are stressed by conditions like low light and temperature or calcium nitrate, as evidenced by numerous prior research studies, their photosynthetic activities are boosted with BR (Li, 2016; Niu et al., 2016). In addition, BR maintains the rigidity and upkeep of the thylakoid membrane and the regulation of chlorophyll molecules, making the activities of the chlorophyllase easier. In wheat plants, BR initiates the processes involved in photosynthesis and antioxidant enzymes in heat and drought stress (Zhao et al., 2017). BR regulates abiotic stresses in plants by controlling defense-related genes, thus reducing the negative consequences in cucumber plants (Li et al., 2013).

14.4.7 Cytokinin

The usual description of cytokinin is a growth-promoting hormone. However, several substances possessing cytokinin activity have been discovered to govern plants' various physiological and developmental functions. For instance, CK functions in modulating cell disparity, delaying leaf senescence, and regulating and establishing the relationship between source and sinks (Kuiper, 1993; Sakakibara et al., 2006). Cytokinin signaling cascades are triggered by alterations in the plant environment, which have an evolutionary connection to the signal transduction schemes consisting of components (Hwang et al., 2002). Plants responding to these stresses may increase CK levels for short or long periods (Alvarez et al., 2008; Dobra et al., 2010). CK effects or its signaling under abiotic stresses are established in circumstances like salt (Joshi et al., 2018; Mason et al., 2010), drought (Nguyen et al., 2016), and temperature (Wu et al., 2017). In *Arabidopsis*, CK reduces the expression of high-affinity K^+ transporter 1; 1 (AtHKT1; 1) in reacting to salt stress by taking sodium ions out of the root xylem via ARR1 and ARR12 (Mason et al., 2010). Similarly, under drought stress, exogenous CK enhances plant adaptation by its spatiotemporal expression, quick reaction, and extensively linked pathways, which are essential components of transduction and cytokinin biosynthesis participating as constitutive signals that govern drought adaptation and specify the plant reaction when subjected to drought stress. Due to their extensively connected pathways, quick reactions, and spatiotemporal activation, they facilitate the governing of morphogenetic process mechanisms during water stress (Mandal et al., 2022).

14.4.8 Ethylene

Mankind is aware of ET's fundamental function as the hormone that causes fruit to mature. Still being researched, yet, is ET's ability to carry out multiple tasks. It is among the most varied signaling molecules that help plants adapt to harsh environments. Plants, for instance, produce ET in their interaction with abiotic stressors including nutrient deficiency (Romera et al., 1999), salinity (Riyazuddin et al., 2020), and waterflooding (Fukao & Bailey-Serres, 2008). Several plant species such as orange, French bean, and faba witness high levels of ET when they are subjected to drought stress (Murti & Upreti, 2007). *M. polymorpha* mutants of Mpein3 and Mpctrl are involved in ET-mediated transduction pathways and are tolerant and sensitive to ET responses, respectively. Mpctrl mutants showed resilience to stress on par with or probably greater than that of the wild kind, while resilience was dropped in Mpein3 mutants compared to the wild type under abiotic stress conditions such as far-red light, heat, salinity, and nutritional shortage (Bharadwaj et al., 2022). Also, in mustard plants that undergo stress by salt and *Arabidopsis* under ozone stress, ET modulates the synthesis of GSH (Fatma et al., 2021; Yoshida et al., 2009). It is suspected that cadmium is the most toxic inorganic ion that causes plants to release excess ET. However, the repercussions of cadmium concentration on mustard crops demonstrated the critical function ET plays in metal resistant to stress (Masood et al., 2012).

14.4.9 Strigolactones

SLs are unique phytohormones that regulate plant growth, metabolism, and development. It is a naturally occurring carotenoid derivative in plants, with production primarily occurring in the root system and spreading to the plant's uppermost sections. SLs usually function in plants by promoting resistance to stressors caused by drought and nutrient deficiency stressors by triggering physiological reactions while mediating the relationship with soil microbes such as arbuscular mycorrhiza fungi (AMF). Low soil phosphorus (P) levels stimulate the release and production of SLs, consistent with their function in the recruitment of AMF (Jamil et al., 2014). Application of orobanchol from red clover under nutrient-stress soil was recorded to be 20-fold higher under phosphorus starvation (Yoneyama et al., 2007). This is demonstrated by applying exogenous SL results in facilitating wild-type plants' resilience to drought stress in maize (Yu et al., 2016), tomato (Ali et al., 2020), and wheat (Yoon et al., 2020). Moreover, when tomato was grafted onto SL-depleted rootstock, wild-type tomato scions showed reduced transpiration characteristics under drought conditions and more transcripts for SL biosynthesis genes than control grafted plants (Visentin et al., 2016).

14.4.10 Polyamines

PAs are aliphatic, polycationic, and widely distributed amines that participate in several growth-oriented and physiological events comprising root growth, leaf senescence, fruit maturity, organogenesis, embryogenesis, and cell division (Gill & Tuteja, 2010; Lenis et al., 2017). The enhanced buildup of PA is among the most notable metabolism indicators in abiotically stressed plants (Alcázar et al., 2020). Additionally, PAs function in how plants respond to abiotic stresses (salinity, drought, and fluctuations in temperature) (Cuevas et al., 2008; Jinlong Liu et al., 2020; Yamaguchi et al., 2007). Recent studies have established the existence of exogenous PAs. Exogenous PAs have been proven in recent research to have a strong correlation with plant growth, resilience to stresses, and the continued existence of plants (Agudelo-Romero et al., 2013; Pál et al., 2015). For instance, the application of endogenous PA on olive (*Olea europae*) promoted flower development (Alagna et al., 2016).

14.5 CROSSTALK AMONG PHYTOHORMONES TO MITIGATE STRESS-INDUCED DAMAGES BY ABIOTIC FACTORS

Studies on the biological function of just one phytohormone in isolation are challenging because of the intricate relationships among phytohormones throughout plant growth and stress responses. The locations where phytohormones converge signal transduction cascades are noted as crosstalk essential for plant growth and adapting to stresses arising from biotic and abiotic sources. During abiotic stress conditions, these routes interact for synergies and signaling (Knight & Knight, 2001). These interactions may activate another messenger via a phosphorylation cascade. The transduction pathways lead to the regulation of gene expression that may influence other hormones, biosynthesis, or developmental processes. Crosstalk exists among phytohormone to collectively activate and organize a variety of genes for protection from the danger of the stressors (He et al., 2017; Nishiyama et al., 2013; Pieterse et al., 2009; Pieterse et al., 2012). Hence, hormone production, signal transduction, perception, and crosstalk develop a multifaceted network. Consequently, it is essential to appreciate the complex relationship between phytohormone crosstalk, as discussed below.

14.5.1 JA and GA

Many investigations demonstrate that the interaction between growth-related and stress-related hormones determines how plant development responds to stress. Data support how JA and GA engage to increase and advance the plant organs and their existence (Hou et al., 2013; Hu et al., 2017; Jang et al., 2017). In this regard, the male-sterile phenotype is shown in both ga1–3 and opr3, mutants lacking in GA and JA. The results are poor filament elongation, lowered pollen viability, and deficient anther dehiscence (Li et al., 2021; Stintzi & Browse, 2000). Also, the JA biosynthesis genes DEFECTIVE IN ANTHER DEHISCENCE 1 (DAD1) and LIPOXYNENASE 1 (LOX1) are upregulated in reaction to GA's suppression of DELLAs. This stimulates JA synthesis and upregulates MYB21, MYB24, and MYB57 expression, which in turn activates the JASMONATE-ZIM DOMAIN (JAZ) protein to regulate the JA-mediated for stamen formation (Chen et al., 2009; Li et al., 2021). Plants utilize the DELLA and JAZs to act as the regulatory regions between various signaling pathways of JA, GA, and others and alter their environment for their development and defense against stresses.

14.5.2 JA and Auxin

For the development of plants, the crosstalk among auxin signaling and JA is also crucial as both hormones are engaged in the growth of roots. It has been established that JA promotes lateral root growth by encouraging auxin production (Sun et al., 2009). Subsequent research showed that by eliminating the JAZ suppression, the wound-induced JA-activated ERF109. Following that, studies demonstrated that wound-induced JA activates ERF109 by getting rid of JAZ suppression. The promoters' GCC box component of YUCCA2 (YUC2) and ANTHRANILATE SYNTHASE α1 (ASA1) is bound by ERF109, which enhances auxin biosynthesis and production, which are key components necessary for lateral root development (Cai et al., 2014; Sun et al., 2009; Zhang et al., 2019a). The auxin signaling suppressor IAA29 acts as a positive regulator in leaf senescence reliant on JA, while the JA signaling suppressors JAZ4 and JAZ8 act as negative controllers. Also, WRKY57 suppresses the expression of genes linked to senescence, revealing their rivalry existing among WRKY57–JAZ4/8 and WRKY57–IAA29 exchanges and the function JA and auxin in the senescence of leaves (Jiang et al., 2014).

14.5.3 JA and ET

Through collaboration and antagonistic interactions, JA and ET control how plants react when faced with abiotic stresses, including salinity, drought, and cold (van der Fits & Memelink, 2000; Zhai et al., 2013). For instance, ET response factors (ERFs) are triggered by both ET and JA signaling pathways. Hence, they might promote communication between different abiotic stress response pathways (Ramegowda & Senthil-Kumar, 2015).

14.5.4 SA and Auxin

In signal perception, mediation, and transduction, stresses are mainly affected by the interaction among phytohormones. For instance, in maize plants, salinity limits IAA quantities in the plant, whereas SA application increases IAA, which shows hormonal balance and crosstalk (Fahad et al., 2015;). Also, auxin's roles in defense response are key by regulating several genes and balancing the interplay between stress reactions.

14.5.5 SA and GA

The interaction between SA and GA and the exogenous GA_3 facilitates the process of biosynthesis of SA, leading to its increase, thereby promoting the plant's ability to withstand metal stress (Emamverdian et al., 2020). Similarly, during stress from abiotic factors circumstances, the crosstalk between Jasmonates and SA is crucial for plant development (Per et al., 2018); however, the signaling pathway functions in an adversary form (Khan & Khan, 2013). The antagonistic relationship between JA and SA signaling in cells is mediated by the MAPK signaling pathway (Petersen et al., 2000). On the contrary, nonhostile interaction is observed among JA and SA (Per et al., 2018). A clear example is demonstrated in the maize crop when copper stress activated the process via SA is biosynthesized, which in turn induces JA-activated volatile organic compounds and JA priming (Engelberth et al., 2011; Rostás et al., 2013).

14.5.6 ABA and SA

Interaction between ABA and SA aids in conserving water by balancing transpiration, leaf senescence, stomata openings, and osmotic adaptation. Equally, ET interacts with ABA to regulate stomata openings and shoot and root growth during drought situations (Gazzarrini & McCourt, 2003). During an abundance of water, ABA works against ET in the adventitious root's development. Also, GA and JA collaborate and promote metabolic processes in each other (Tanaka et al., 2006). The crosstalk of SA-ABA involves PAMP reactions to *Pseudomonas syringae* by restricting the pathogen entry via the stomata (Schulze-Lefert & Robatzek, 2006). A number of studies confirm that JA-CK plays a similar function in activating plant defense machinery in *Arabidopsis*. They are needed to overcome *Alternaria brassicicola* by reducing the JA gene regulation level among CK mutants (Argueso et al., 2012).

14.5.7 ABA and GA

To regulate seed dormancy and germination, understanding the ABA's molecular processes and how they engage with GAs is crucial. This interaction involves the coordination of ABA biosynthesis, catabolism, and its modulation through the action of key regulatory proteins such as SNF1-related protein kinase2, ABI3, ABI4, ABI5, VP1, and ABF (ABRE-binding factor) (Parwez et al., 2022). By coordinating ABA production and catabolism and modifying ABA sensitivity through important regulatory proteins, ABA interacts antagonistically with GAs to regulate the balance involving seed dormancy and germination-induced germination (Schopfer & Plachy, 1984). The process via which seeds go from dormancy to germination depends on the interaction of ABA and GAs. ABA prevents water from being absorbed by embryos whereas GAs increase seed germination by offsetting the restricting impacts of ABA. The survival of crop plants and the dispersal of species are ultimately impacted by the delicate balance between ABA and GAs, which govern seed dormancy and germination. ROS, ascorbic acid, and their crosstalk all have a part in facilitating the communication between GA and ABA (Considine & Foyer, 2021). Additionally, GAs influence seed germination by interacting with environmental stressors including light. The role that GAs play in regulating the germination of seeds under varying lighting circumstances and modulating plant responses to environmental stimuli is highlighted by this interaction. Improved crop yields depend on our ability to comprehend the molecular underpinnings of ABA signaling and how it interacts with other hormones (Gupta et al., 2022).

14.5.8 CK and ABA

ABA and cytokinin antagonistically determine numerous plant developmental processes and reactions to environmental stress (Huang et al., 2018). This antagonistic activity is essential for managing the trade-offs amid resilience to stress and plant growth throughout the plant's existence. According to research, ABA phosphorylates type-A protein ARR5, an adverse modulator of cytokinin signaling, activating ABA-responsive genes. This eliminates the antagonistic relationship between cytokinin signaling and ABA (Huang et al., 2018). Furthermore, ABA and cytokinin have antagonistic functions in plants' response to environmental stressors like salt and drought (Lu et al., 2014). ABA is essential for controlling root growth and responses to environmental stressors, such as the prevention of germination, seed dormancy, dehydration, and root elongation during drought stress. Conversely, it has been demonstrated that cytokinin improves plant roots' growth, drought resistance, and mineral richness in leaves (Avni et al., 2020; Wani et al., 2016). The complex interactions between these phytohormones in regulating root responses to environmental cues are further highlighted by the antagonistic activity of auxin and cytokinin in controlling the importance of the shoot/root growth ratio for coping with nutrient shortage and drought-related stress (Kurepa & Smalle, 2022). An essential component of plant physiology is the antagonistic interaction between ABA and cytokinin in controlling root development in the face of abiotic stress. These phytohormones' complex interactions are essential for moderating the balance between tolerance to stress and growth, affecting how resilient plants are to environmental obstacles.

The equilibrium between CK and ABA is fundamental. Plant resilience to environmental adversities is ultimately determined by the antagonistic yet complementary functions that ABA and CK have a role in controlling adaptation to stress and root growth (Bielach et al., 2017). Elevated ABA levels in response to abiotic stress circumstances cause stomatal closure and change gene expression, which helps the body respond adaptively. In reaction to water constraints, ABA is also essential for changing the root architecture and growth patterns, which affects the plant's stress tolerance. Conversely, more drought-tolerant plants with more extensive roots produce cytokinin. Their respective regulation systems make clear the antagonistic link between ABA and CK. Under stressful circumstances, ABA promotes root growth by downregulating the production of cytokinin oxidase genes, which accumulate cytokinin. On the contrary, cytokinin decreases ABA levels, which affects how the body responds to stress and controls root growth. This complex equilibrium enables plants to modify their physiological reactions to external stimuli, augmenting their resistance to abiotic stress (Table 14.1).

TABLE 14.1
Effects of Exogenous Phytohormones on Plant Growth and Development from 2000 to 2022

Hormone	Species	Stress	Plant Material	Effects	References
ABA	*Arabidopsis thaliana*	Light	4-day-old wild type and mutants	Regulate light response; upregulate ABA biosynthetic genes	Huang et al. (2019)
ABA	Spring barley	Drought	4-day-old seedlings	Impact the electron destiny associated with the PSII side's QA and photosynthetic antenna absorption	Daszkowska-Golec et al. (2019)
ABA	*A. thaliana*	Light	30–45-day-old wild type and mutants	Regulate stomatal closure	Devireddy et al. (2018)
ABA	*A. thaliana*	Light	8–21-day-old wild type and mutants	Accumulate of 20% transcript reacts to ABA within seconds	Suzuki et al. (2015)
ABA	*Lycium chinense*	Drought	15-week-old plants with 35–40 leaves	Enhance non-photochemical quenching and a delayed PSII efficiency reduction	Guan et al. (2015)
ABA	*Oryza sativa*	Drought	Mutant seedlings at the four-leaf stage	Enhance quenching without photochemistry as well as PSII effectiveness	Du et al. (2010)
ABA	Common bean, tobacco, beetroot, *Zea mays*	Drought	Seedlings stage	Enhance PSII performance	Haisel et al. (2006)
ABA	*Typha latifolia L. Phragmites australis (Cav.) Trin. ex Steud.*	Cadmium	0–5 hours and 7 days	Application of heavy metals boosts the endogenous levels of ABA	Fediuc et al. (2005)
ABA	*O. sativa*	Arsenic	7-day wild type and mutant	Enhance ABA biosynthesis and its signaling gene expression	Xin et al. (2005)
Auxin	*Nicotiana tabacum*	Potassium deficiency	Seedlings	Increase the length and growth of lateral roots	Song et al. (2015)
Auxin	*A. thaliana*	Salinity	Seedling stage	Maintain ion homeostasis; reduce auxin concentration and signaling to cope with salinity stress	Iglesias et al. (2014)
Auxin	*A. thaliana*	Cadmium		Lower the amount of cadmium in the shoots	Zhu et al. (2013)
Auxin	*A. thaliana, O. sativa*	Arsenic toxicity		Improve arsenic tolerance in plants	Krishnamurthy and Rathinasabapathi (2013); Pandey and Gupta (2015)
Auxin	*O. sativa*	Drought	Seedling stage	Increase compatible solutes, antioxidants, and root branching	Du et al. (2013)
Auxin	*Hordeum vulgare; O. sativa*	Temperature	Seedling stage	Reduce stress-induced plant injury	Du et al. (2013); Oshino et al. (2011)

(Continued)

TABLE 14.1 (*Continued*)
Effects of Exogenous Phytohormones on Plant Growth and Development from 2000 to 2022

Hormone	Species	Stress	Plant Material	Effects	References
Auxin	*A. thaliana*	Drought and salt	Middle-aged leaves and 1-month-old seedlings	Improve electron transfer rate, photochemical quenching, and maximal photochemical yield	Tognetti et al. (2010)
Auxin	*Arabidopsis*	Nitrogen deficiency	Seedling stage	Alter root architecture	Zhang et al. (2007)
Auxin	*Helianthus annuus L.*	Copper toxicity	10-day old seedlings	Improve root length root hair formation and water use efficiency	Ouzounidou and Ilias (2005)
Auxin	*Arabidopsis*, *Lupinus albus*	Phosphorus deficiency	Seeding stage	Improve a plant's sensitivity to auxin and promote the production of lateral roots in plants through auxin	Nacry et al. (2005); Neumann et al. (2000)
Auxin	*Phaseolus vulgaris*, *Cucumis sativus*, *Trifolium pratense*	Iron deficiency	Seedlings	Promote iron uptake	Li et al. (2000); Li and Li (2004); Zheng et al. (2003)
GA	*Triticum aestivum*	Drought	7-day old seedlings	Upregulate antioxidant defense mechanisms	Al Mahmud et al. (2019)
GA	*T. aestivum*	Salinity	Seeds	Increase photosynthetic indexes and water use efficiency	Iqbal and Ashraf (2013)
GA	*Z. mays*	Salinity	10–45-day old seedlings	Improve accumulation of chlorophyll and proline content	Tuna et al. (2008)
GA	*H. annuus L.*	Copper toxicity	10-day old seedlings	Maintain chlorophylls and carotenoid contents	Ouzounidou and Ilias (2005)
GA	*Glycine max*	Cadmium toxicity	Young seedlings	Increase leaf area, stem length, and growth rate	Ghorbanli et al. (2000)
JA	*Pennisetum glaucum*	Drought	Seedlings	Boost the levels of chlorophyll, relative amount of water, and antioxidant enzyme activity	Awan et al. (2021)
JA	*Triticum sativum*	Drought	Seedlings	Upregulate reallocation of photoassimilates to vegetative sinks	Abeed et al. (2021)
JA	*O. sativa*	Heat	Seedlings	Boost the leaves' potential for adjusting their osmotic and antioxidant mechanism	Yang et al. (2020)
JA	*Fragaria × ananassa Duch*	Drought	Explants	Boost the quantity of roots and shoots as well as the proportion of water in them	Yosefi et al. (2020)
GA	*Abelmoschus esculentus* L.	Salinity	Seedlings	Increase shoot and root length	Wang et al. (2019)
JA	*A. thaliana*	Far-red	Seedlings	Improve the association between FIN219 and GUS-CCT1 under blue light	Chen et al. (2018)
JA	*Z. mays*	Heavy metal (nickel)	Seedlings	Improve the activity of antioxidant enzymes	Azeem (2018)

(*Continued*)

TABLE 14.1 (*Continued*)
Effects of Exogenous Phytohormones on Plant Growth and Development from 2000 to 2022

Hormone	Species	Stress	Plant Material	Effects	References
JA	*Glycine max*	Drought	Seedlings	Boost the partitioning of cell walls, both saturated and unsaturated fatty acids	Mohamed and Latif (2017)
JA	*Vicia faba*	Heavy metal (cadmium)	Seedlings	Enhance osmolyte and antioxidant activities that reduce oxidative stress	Ahmad et al. (2017)
JA	*N. tabacum*	Imazapic stress	Seedlings	Reduce MDA content, increase phytohormone levels, and antioxidant activity	Kaya and Doganlar (2016)
JA	*Z. mays*	Salt	Seedlings	Improve Na^+ exclusion	Shahzad et al. (2015)
JA	*A. thaliana*	Light	3-week-old wild type and mutant	Modulate violaxanthin de-epoxidase (VDE) expression and increase ROS	Chen and Gallie (2015)
JA	*Glycine max*	Heavy metal (nickel)	Seeds	Protect DNA synthesis of total proteins	Sirhindi et al. (2015)
JA	*T. aestivum*	Salt	Seedlings	Increase the transcript levels and activities of SOD and POD	Qiu et al. (2014)
JA	*T. aestivum*	Ultraviolet-B radiation	Seedlings	Increase photosynthetic electron transport rate and photosystem II quantum yield	Liu et al. (2012)
JA	*Prunus armeniaca*	Drought	90-day-old seedlings (18–20 leaf seedlings)	Increase MDA levels and promote leaf senescence	Ge et al. (2010)
JA	*Eriobotrya japonica*	Chilling	Fruits	Enhance antioxidant enzyme activity	Cao et al. (2009)
JA	*A. thaliana*	Ozone stress	Seedlings	Inhibit the spread of programmed cell death	Rao et al. (2000)
SA	*T. aestivum L.*	Drought	Plants at the vegetative stage	Decrease ROS accumulation and promote enzymatic antioxidant	Shemi et al. (2021)
SA	*O. sativa*	Salt	Seeds	Decrease the activities of antioxidant enzymes such as SOD, CAT, and POD	Jini and Joseph (2017)
SA	*A. thaliana*	Salt	5-week plants	Increase SOD, GST, and GPOX activity	Horváth et al. (2015)
SA	*Linum usitatissimum L.*	Heavy metal (Cd^{2+})	Seeds	Increase antioxidant activities and related enzymes	Belkadhi et al. (2014)
SA	*Cannabis sativa*	Heavy metal (Cd^{2+})	Seeds	Improve photosynthetic capacity and increase SOD and POD activities	Shi et al. (2009)
SA	*Z. mays*	Salt	Seedling	Ameliorate the deterioration of the membrane and reduce the MDA amount	Gunes et al. (2007)
SA	*T. aestivum L.*	Salt	7-day-old seedlings	Boost the rate of absorption and water efficiency	Arfan et al. (2007)

(Continued)

TABLE 14.1 (*Continued*)
Effects of Exogenous Phytohormones on Plant Growth and Development from 2000 to 2022

Hormone	Species	Stress	Plant Material	Effects	References
BR	*A. thaliana*	Drought	1-month old seedlings	Enhance stomatal closure of leaves	Díaz et al. (2021)
BR	*Brassica juncea L.*	Heavy metal (Pb)	Seeds	Increase the antioxidant activity of CAT and POD enzymes	Soares et al. (2020)
BR	*O. sativa*	Salinity (NaCl)	Seeds	Increase chlorophyll and protein contents and antioxidant activities of CAT and SOD	Sharma et al. (2013)
BR	*A. thaliana*	Drought	21-day-old seedlings	Increase the accumulation and expression of drought-responsive genes	Kagale et al. (2007)
BR	*A. thaliana*	Cold	21-day-old seedlings	Enhance the expression of cold-responsive genes	Kagale et al. (2007)
BR	*A. thaliana*	Heat	21-day-old seedlings	Promote thermotolerance of seedlings	Kagale et al. (2007)
BR	*Sorghum vulgare*	Osmotic	5–6-day old seedlings	Enhance the contents of soluble proteins, free proline, and increased catalase activity	Vidya Vardhini and Seeta Ram Rao (2003)
CK	*Solanum lycopersicum*	Salt	8-week-old seedlings	Improve tolerance to salt stress	Žižková et al. (2015)

CAT, catalase; *GPOX*, glutathione peroxidase; *GST*, glutathione *S*-transferase; *MDA*, malondialdehyde; *POD*, peroxidase; *ROS*, reactive oxygen species; *SOD*, superoxide dismutase; *VDE*, violaxanthin de-epoxidase.

14.6 CONCLUSIONS AND FUTURE PERSPECTIVES

The exogenous hormones are pivotal in emerging several abiotic stresses on plant functions. Several findings demonstrate that exogenous hormones are crucial for a plant's survival during stress. Hence, an understanding of the roles of exogenous hormones concerning the growth and development of plant organs under abiotic stress conditions is needed. Consequently, learning more about how exogenous hormones affect plant growth and development is crucial to making more robust varieties, better understanding their interactions with soil resources, and adapting to environmental stresses efficiently to utilize water and nutrients and adapt to environmental stresses efficiently. Despite this, additional research on the crosstalk between exogenous hormones is necessary to understand the regulatory mechanism since these hormones interact in a complicated way in different environments as this is fundamental to enhance crop tolerance to abiotic stresses.

ACKNOWLEDGMENTS

This work was supported by grants from the National Key R & D Program of China (No. 2021YFD1201603), the National Natural Science Foundation of China (Grant Nos. 32171965, 31871646, 31571691), the Core Technology Development for Breeding Program of Jiangsu Province (JBGS-2021-014), and the Program of Collaborative Innovation Center for Modern Crop Production co-sponsored by Province and Ministry (CIC-MCP).

REFERENCES

Abeed, A. H., Eissa, M. A., & Abdel-Wahab, D. A. (2021). Effect of exogenously applied jasmonic acid and kinetin on drought tolerance of wheat cultivars based on morpho-physiological evaluation. *Journal of Soil Science and Plant Nutrition, 21*, 131–144.

Acharya, B. R., & Assmann, S. M. (2009). Hormone interactions in stomatal function. *Plant Molecular Biology, 69*, 451–462.

Adnan, M., Fahad, S., Zamin, M., Shah, S., Mian, I. A., Danish, S., Zafar-ul-Hye, M., Battaglia, M. L., Naz, R. M. M., & Saeed, B. (2020). Coupling phosphate-solubilizing bacteria with phosphorus supplements improve maize phosphorus acquisition and growth under lime induced salinity stress. *Plants, 9*(7), 900.

Agudelo-Romero, P., Bortolloti, C., Pais, M. S., Tiburcio, A. F., & Fortes, A. M. (2013). Study of polyamines during grape ripening indicate an important role of polyamine catabolism. *Plant Physiology and Biochemistry, 67*, 105–119.

Ahmad, P., Alyemeni, M. N., Wijaya, L., Alam, P., Ahanger, M. A., & Alamri, S. A. (2017). Jasmonic acid alleviates negative impacts of cadmium stress by modifying osmolytes and antioxidants in faba bean (Vicia faba L.). *Archives of Agronomy and Soil Science, 63*(13), 1889–1899.

Alagna, F., Cirilli, M., Galla, G., Carbone, F., Daddiego, L., Facella, P., Lopez, L., Colao, C., Maritti, R., Cultrera, N., Rossi, M, Barcaccia, G., Baldoni, L., Muleo, R., Perrotta, G., & Perrotta, G. (2022). Correction: Transcript analysis and regulative events during flower development in olive (*Olea europaea L.*). *Plos One, 17*(1), e0263101.

Alcázar, R., Bueno, M., & Tiburcio, A. F. (2020). Polyamines: Small amines with large effects on plant abiotic stress tolerance. *Cells, 9*(11), 2373.

Ali, A., Pardo, J. M., & Yun, D.-J. (2020). Desensitization of ABA-signaling: the swing from activation to degradation. *Frontiers in Plant Science, 11*, 379.

Al Mahmud, J., Biswas, P. K., Nahar, K., Fujita, M., & Hasanuzzaman, M. (2019). Exogenous application of gibberellic acid mitigates drought-induced damage in spring wheat. *Acta Agrobotanica, 72*(2), 1776.

Alizadeh-Vaskasi, F., Pirdashti, H., Araei, A. C., & Saadatmand, S. (2018). Waterlogging effects on some antioxidant enzymes activities and yield of three wheat promising lines. *Acta Agriculturae Slovenica, 111*(3), 621–631.

Altaf, M. A., Shahid, R., Kumar, R., Altaf, M. M., Kumar, A., Khan, L. U., Saqib, M., Nawaz, M. A., Saddiq, B., & Bahadur, S. (2023). Phytohormones mediated modulation of abiotic stress tolerance and potential crosstalk in horticultural crops. *Journal of Plant Growth Regulation, 42*(8), 4724–4750.

Alvarez, S., Marsh, E. L., Schroeder, S. G., & Schachtman, D. P. (2008). Metabolomic and proteomic changes in the xylem sap of maize under drought. *Plant, Cell & Environment, 31*(3), 325–340.

Amrine, K. C., Blanco-Ulate, B., & Cantu, D. (2015). Discovery of core biotic stress responsive genes in Arabidopsis by weighted gene co-expression network analysis. *PLoS One, 10*(3), e0118731.

Appu, M., & Muthukrishnan, S. (2014). Foliar application of salicylic acid stimulates flowering and induce defense related proteins in finger millet plants. *Universal Journal of Plant Science, 2*(1), 14–18.

Arfan, M., Athar, H. R., & Ashraf, M. (2007). Does exogenous application of salicylic acid through the rooting medium modulate growth and photosynthetic capacity in two differently adapted spring wheat cultivars under salt stress? *Journal of Plant Physiology, 164*(6), 685–694.

Argueso, C. T., Ferreira, F. J., Epple, P., To, J. P., Hutchison, C. E., Schaller, G. E., Dangl, J. L., & Kieber, J. J. (2012). Two-component elements mediate interactions between cytokinin and salicylic acid in plant immunity. *PLoS Genetics, 8*(1), e1002448.

Arkhipova, T., Martynenko, E., Sharipova, G., Kuzmina, L., Ivanov, I., Garipova, M., & Kudoyarova, G. (2020). Effects of plant growth promoting rhizobacteria on the content of abscisic acid and salt resistance of wheat plants. *Plants, 9*(11), 1429.

Arya, H., Singh, M. B., & Bhalla, P. L. (2021). Towards developing drought-smart soybeans. *Frontiers in Plant Science, 12*, 750664.

Asami, T., & Nakagawa, Y. (2018). Preface to the Special Issue: Brief review of plant hormones and their utilization in agriculture. *Journal of Pesticide Science, 43*(3), 154–158.

Ashraf, M., Mukhtar, N., Rehman, S., & Rha, E. (2004). Salt-induced changes in photosynthetic activity and growth in a potential medicinal plant Bishop's weed (Ammi majus L.). *Photosynthetica, 42*, 543–550.

Asami, T., & Nakagawa, Y. (2018). Preface to the Special Issue: Brief review of plant hormones and their utilization in agriculture. *Journal of Pesticide Science, 43*(3), 154–158.

Avestan, S., Ghasemnezhad, M., Esfahani, M., & Barker, A. V. (2021). Effects of nanosilicon dioxide on leaf anatomy, chlorophyll fluorescence, and mineral element composition of strawberry under salinity stress. *Journal of Plant Nutrition, 44*(20), 3005–3019.

Avni, A., Golan, Y., Shirron, N., Shamai, Y., Golumbic, Y., Danin-Poleg, Y., & Gepstein, S. (2020). From survival to productivity mode: Cytokinins allow avoiding the avoidance strategy under stress conditions. *Frontiers in Plant Science, 11*, 879.

Awan, F. K., Khurshid, M. Y., & Mehmood, A. (2017). Plant growth regulators and their role in abiotic stress management. *International Journal of Research in BioSciences, 1*, 9–21.

Awan, S. A., Khan, I., Rizwan, M., Zhang, X., Brestic, M., Khan, A., El-Sheikh, M. A., Alyemeni, M. N., Ali, S., & Huang, L. (2021). Exogenous abscisic acid and jasmonic acid restrain polyethylene glycol-induced drought by improving the growth and antioxidative enzyme activities in pearl millet. *Physiologia Plantarum, 172*(2), 809–819.

Azeem, U. (2018). Ameliorating nickel stress by jasmonic acid treatment in Zea mays L. *Russian Agricultural Sciences, 44*, 209–215.

Bagautdinova, Z. Z., Omelyanchuk, N., Tyapkin, A. V., Kovrizhnykh, V. V., Lavrekha, V. V., & Zemlyanskaya, E. V. (2022). Salicylic acid in root growth and development. *International Journal of Molecular Sciences, 23*(4), 2228.

Balcerowicz, M., Ranjan, A., Rupprecht, L., Fiene, G., & Hoecker, U. (2014). Auxin represses stomatal development in dark-grown seedlings via Aux/IAA proteins. *Development, 141*(16), 3165–3176.

Belkadhi, A., De Haro, A., Soengas, P., Obregon, S., Cartea, M. E., Chaibi, W., & Djebali, W. (2014). Salicylic acid increases tolerance to oxidative stress induced by hydrogen peroxide accumulation in leaves of cadmium-exposed flax (Linum usitatissimum L.). *Journal of Plant Interactions, 9*(1), 647–654.

Bharadwaj, P. S., Sanchez, L., Li, D., Enyi, D., Van de Poel, B., & Chang, C. (2022). The plant hormone ethylene promotes abiotic stress tolerance in the liverwort Marchantia polymorpha. *Frontiers in Plant Science, 13*, 998267.

Bhat, J. A., Basit, F., Alyemeni, M. N., Mansoor, S., Kaya, C., & Ahmad, P. (2023). Gibberellic acid mitigates nickel stress in soybean by cell wall fixation and regulating oxidative stress metabolism and glyoxalase system. *Plant Physiology and Biochemistry, 198*, 107678.

Bielach, A., Hrtyan, M., & Tognetti, V. B. (2017). Plants under stress: involvement of auxin and cytokinin. *International Journal of Molecular Sciences, 18*(7), 1427.

Boroumand, J. S., Lari, Y. H., & Ranjbar, M. (2011). Effect of salicylic acid on some plant growth parameters under lead stress in Brassica napusvar. OKAPI. *Iranian Journal of Plant Physiology, 1*, 177.

Bruno, L., Pacenza, M., Forgione, I., Lamerton, L. R., Greco, M., Chiappetta, A., & Bitonti, M. B. (2017). In Arabidopsis thaliana cadmium impact on the growth of primary root by altering SCR expression and auxin-cytokinin cross-talk. *Frontiers in Plant Science, 8*, 1323.

Bücker-Neto, L., Paiva, A. L. S., Machado, R. D., Arenhart, R. A., & Margis-Pinheiro, M. (2017). Interactions between plant hormones and heavy metals responses. *Genetics and Molecular Biology, 40*, 373–386.

Cai, X.-T., Xu, P., Zhao, P.-X., Liu, R., Yu, L.-H., & Xiang, C.-B. (2014). Arabidopsis ERF109 mediates cross-talk between jasmonic acid and auxin biosynthesis during lateral root formation. *Nature Communications, 5*(1), 5833.

Čanak, P., Jeromela, A. M., Vujošević, B., Kiprovski, B., Mitrović, B., Alberghini, B., Facciolla, E., Monti, A., & Zanetti, F. (2020). Is drought stress tolerance affected by biotypes and seed size in the emerging oilseed crop Camelina? *Agronomy, 10*(12), 1856.

Canavar, Ö. (2015). Effects of water deficit on gas exchange of different age leaves and both stem and leaves solute content of sunflowers. *Agrociencia, 49*(6), 651–667.

Cao, L., Lu, X., Wang, G., Zhang, Q., Zhang, X., Fan, Z., Cao, Y., Wei, L., Wang, T., & Wang, Z. (2021). Maize ZmbZIP33 is involved in drought resistance and recovery ability through an abscisic acid-dependent signaling pathway. *Frontiers in Plant Science, 12*, 629903.

Cao, S., Zheng, Y., Wang, K., Jin, P., & Rui, H. (2009). Methyl jasmonate reduces chilling injury and enhances antioxidant enzyme activity in postharvest loquat fruit. *Food Chemistry, 115*(4), 1458–1463.

Cardinale, B. J., Duffy, J. E., Gonzalez, A., Hooper, D. U., Perrings, C., Venail, P., Narwani, A., Mace, G. M., Tilman, D., & Wardle, D. A. (2012). Biodiversity loss and its impact on humanity. *Nature, 486*(7401), 59–67.

Caverzan, A., Casassola, A., & Brammer, S. P. (2016). Reactive oxygen species and antioxidant enzymes involved in plant tolerance to stress. *Abiotic and Biotic Stress in Plants-Recent Advances and Future Perspectives, 17*, 463–480.

Chattopadhyay, K., Chakraborty, K., Samal, P., & Sarkar, R. K. (2021). Identification of QTLs for stagnant flooding tolerance in rice employing genotyping by sequencing of a RIL population derived from Swarna× Rashpanjor. *Physiology and Molecular Biology of Plants, 27*(12), 2893–2909

Chaudhry, S., & Sidhu, G. P. S. (2022). Climate change regulated abiotic stress mechanisms in plants: A comprehensive review. *Plant Cell Reports, 41*(1), 1–31.

Chen, H., Bullock Jr, D. A., Alonso, J. M., & Stepanova, A. N. (2021). To fight or to grow: The balancing role of ethylene in plant abiotic stress responses. *Plants, 11*(1), 33.

Chen, H., Xue, L., Chintamanani, S., Germain, H., Lin, H., Cui, H., Cai, R., Zuo, J., Tang, X., & Li, X. (2009). Ethylene insensitive3 and ethylene insensitive3-like1 repress salicylic acid induction deficient2 expression to negatively regulate plant innate immunity in Arabidopsis. *The Plant Cell, 21*(8), 2527–2540.

Chen, H.-J., Fu, T.-Y., Yang, S.-L., & Hsieh, H.-L. (2018). FIN219/JAR1 and cryptochrome1 antagonize each other to modulate photomorphogenesis under blue light in Arabidopsis. *PLoS Genetics, 14*(3), e1007248.

Chen, Y., Wang, X.-M., Zhou, L., He, Y., Wang, D., Qi, Y.-H., & Jiang, D.-A. (2015). Rubisco activase is also a multiple responder to abiotic stresses in rice. *PLoS One, 10*(10), e0140934.

Chen, Z., & Gallie, D. R. (2015). Ethylene regulates energy-dependent non-photochemical quenching in Arabidopsis through repression of the xanthophyll cycle. *PLoS One, 10*(12), e0144209.

Choudhary, S., Wani, K. I., Naeem, M., Khan, M. M. A., & Aftab, T. (2023). Cellular responses, osmotic adjustments, and role of osmolytes in providing salt stress resilience in higher plants: Polyamines and nitric oxide crosstalk. *Journal of Plant Growth Regulation, 42*(2), 539–553.

Christianson, J. A., Wilson, I. W., Llewellyn, D. J., & Dennis, E. S. (2009). The low-oxygen-induced NAC domain transcription factor ANAC102 affects viability of Arabidopsis seeds following low-oxygen treatment. *Plant Physiology, 149*(4), 1724–1738.

Colebrook, E. H., Thomas, S. G., Phillips, A. L., & Hedden, P. (2014). The role of gibberellin signalling in plant responses to abiotic stress. *Journal of Experimental Biology, 217*(1), 67–75.

Collin, A., Daszkowska-Golec, A., Kurowska, M., & Szarejko, I. (2020). Barley ABI5 (Abscisic Acid INSENSITIVE 5) is involved in abscisic acid-dependent drought response. *Frontiers in Plant Science, 11*, 1138.

Considine, M. J., & Foyer, C. H. (2021). Stress effects on the reactive oxygen species-dependent regulation of plant growth and development. *Journal of Experimental Botany, 72*(16), 5795–5806.

Copeland, L. (2016). Food carbohydrates from plants. In K. Hardy & L. K. Lucy Kubiak-Martens (Eds.), *Wild harvest: Plants in the hominin and pre-agrarian human worlds* (pp. 19–30). Oxbow Books.

Cornforth, J., Milborrow, B., Ryback, G., & Wareing, P. (1965). Chemistry and physiology of 'dormins' in sycamore: Identity of sycamore 'dormin' with abscisin II. *Nature, 205*(4978), 1269–1270.

Cuevas, J. C., López-Cobollo, R., Alcázar, R., Zarza, X., Koncz, C., Altabella, T., Salinas, J., Tiburcio, A. F., & Ferrando, A. (2008). Putrescine is involved in Arabidopsis freezing tolerance and cold acclimation by regulating abscisic acid levels in response to low temperature. *Plant Physiology, 148*(2), 1094–1105.

da Ponte, N. H. T., Santos, R. I. N., Lopes Filho, W. R. L., Cunha, R. L., Magalhães, M. M., & Pinheiro, H. A. (2019). Morphological assessments evidence that higher number of pneumatophores improves tolerance to long-term waterlogging in oil palm (Elaeis guineensis) seedlings. *Flora, 250*, 52–58.

Dar, N. A., Amin, I., Wani, W., Wani, S. A., Shikari, A. B., Wani, S. H., & Masoodi, K. Z. (2017). Abscisic acid: A key regulator of abiotic stress tolerance in plants. *Plant Gene, 11*, 106–111.

Daszkowska-Golec, A., Collin, A., Sitko, K., Janiak, A., Kalaji, H. M., & Szarejko, I. (2019). Genetic and physiological dissection of photosynthesis in barley exposed to drought stress. *International Journal of Molecular Sciences, 20*(24), 6341.

Day, L. (2013). Proteins from land plants–potential resources for human nutrition and food security. *Trends in Food Science & Technology, 32*(1), 25–42.

Devireddy, A. R., Zandalinas, S. I., Gómez-Cadenas, A., Blumwald, E., & Mittler, R. (2018). Coordinating the overall stomatal response of plants: Rapid leaf-to-leaf communication during light stress. *Science Signaling, 11*(518), eaam9514.

Díaz, K., Espinoza, L., Carvajal, R., Silva-Moreno, E., Olea, A. F., & Rubio, J. (2021). Exogenous application of brassinosteroid 24-norcholane 22 (S)-23-dihydroxy type analogs to enhance water deficit stress tolerance in Arabidopsis thaliana. *International Journal of Molecular Sciences, 22*(3), 1158.

Dobra, J., Motyka, V., Dobrev, P., Malbeck, J., Prasil, I. T., Haisel, D., Gaudinova, A., Havlova, M., Gubis, J., & Vankova, R. (2010). Comparison of hormonal responses to heat, drought and combined stress in tobacco plants with elevated proline content. *Journal of Plant Physiology, 167*(16), 1360–1370.

Du, P., Arndt, S. K., & Farrell, C. (2019). Is plant survival on green roofs related to their drought response, water use or climate of origin? *Science of the Total Environment, 667*, 25–32.

Du, H., Wang, N., Cui, F., Li, X., Xiao, J., & Xiong, L. (2010). Characterization of the β-carotene hydroxylase gene DSM2 conferring drought and oxidative stress resistance by increasing xanthophylls and abscisic acid synthesis in rice. *Plant Physiology, 154*(3), 1304–1318.

Du, H., Wu, N., Chang, Y., Li, X., Xiao, J., & Xiong, L. (2013). Carotenoid deficiency impairs ABA and IAA biosynthesis and differentially affects drought and cold tolerance in rice. *Plant Molecular Biology, 83*, 475–488.

Egamberdieva, D. (2009). Alleviation of salt stress by plant growth regulators and IAA producing bacteria in wheat. *Acta Physiologiae Plantarum, 31*(4), 861–864.

Emamverdian, A., Ding, Y., & Mokhberdoran, F. (2020). The role of salicylic acid and gibberellin signaling in plant responses to abiotic stress with an emphasis on heavy metals. *Plant Signaling & Behavior, 15*(7), 1777372.

Engelberth, J., Viswanathan, S., & Engelberth, M. J. (2011). Low concentrations of salicylic acid stimulate insect elicitor responses in Zea mays seedlings. *Journal of Chemical Ecology, 37*, 263–266.

Ekinci, M., Yildirim, E., Dursun, A., & Turan, M. (2012). Mitigation of salt stress in lettuce (Lactuca sativa L. var. Crispa) by seed and foliar 24-epibrassinolide treatments. *HortScience, 47*(5), 631–636.

Fàbregas, N., Lozano-Elena, F., Blasco-Escámez, D., Tohge, T., Martínez-Andújar, C., Albacete, A., Osorio, S., Bustamante, M., Riechmann, J. L., & Nomura, T. (2018). Overexpression of the vascular brassinosteroid receptor BRL3 confers drought resistance without penalizing plant growth. *Nature Communications, 9*(1), 4680.

Fahad, S., Hussain, S., Bano, A., Saud, S., Hassan, S., Shan, D., Khan, F. A., Khan, F., Chen, Y., & Wu, C. (2015). Potential role of phytohormones and plant growth-promoting rhizobacteria in abiotic stresses: Consequences for changing environment. *Environmental Science and Pollution Research, 22*, 4907–4921.

Falkowska, M., Pietryczuk, A., Piotrowska, A., Bajguz, A., Grygoruk, A., & Czerpak, R. (2011). The effect of gibberellic acid (GA3) on growth, metal biosorption and metabolism of the green algae Chlorella vulgaris (Chlorophyceae) Beijerinck exposed to cadmium and lead stress. *Polish Journal of Environmental Studies, 20*(1), 53–59.

Fatma, M., Asgher, M., Iqbal, N., Rasheed, F., Sehar, Z., Sofo, A., & Khan, N. A. (2022). Ethylene signaling under stressful environments: Analyzing collaborative knowledge. *Plants, 11*(17), 2211.

Fatma, M., Iqbal, N., Gautam, H., Sehar, Z., Sofo, A., D'Ippolito, I., & Khan, N. A. (2021). Ethylene and sulfur coordinately modulate the antioxidant system and ABA accumulation in mustard plants under salt stress. *Plants, 10*(1), 180.

Fediuc, E., Lips, S. H., & Erdei, L. (2005). O-Acetylserine (thiol) lyase activity in Phragmites and Typha plants under cadmium and NaCl stress conditions and the involvement of ABA in the stress response. *Journal of Plant Physiology, 162*(8), 865–872.

Ferronato, C., Marinari, S., Francioso, O., Bello, D., Trasar-Cepeda, C., & Antisari, L. V. (2019). Effect of waterlogging on soil biochemical properties and organic matter quality in different salt marsh systems. *Geoderma, 338*, 302–312.

Finkelstein, R. (2013). Abscisic acid synthesis and response. *The Arabidopsis book/American society of plant biologists, 11*.

Foyer, C. H. (2018). Reactive oxygen species, oxidative signaling and the regulation of photosynthesis. *Environmental and Experimental Botany*, *154*, 134–142.

Francis, B., Aravindakumar, C., Brewer, P. B., & Simon, S. (2023). Plant nutrient stress adaptation: A prospect for fertilizer limited agriculture. *Environmental and Experimental Botany*, 105431.

Franco-Zorrilla, J. M., López-Vidriero, I., Carrasco, J. L., Godoy, M., Vera, P., & Solano, R. (2014). DNA-binding specificities of plant transcription factors and their potential to define target genes. *Proceedings of the National Academy of Sciences, 111*(6), 2367–2372.

Frankowski, K., Wilmowicz, E., Kućko, A., Kęsy, J., Świeżawska, B., & Kopcewicz, J. (2014). Ethylene, auxin, and abscisic acid interactions in the control of photoperiodic flower induction in Pharbitis nil. *Biologia Plantarum, 58*, 305–310.

Fukao, T., & Bailey-Serres, J. (2008). Ethylene – A key regulator of submergence responses in rice. *Plant Science*, *175*(1–2), 43–51.

Fukazawa, J., Ohashi, Y., Takahashi, R., Nakai, K., & Takahashi, Y. (2021). DELLA degradation by gibberellin promotes flowering via GAF1-TPR-dependent repression of floral repressors in Arabidopsis. *The Plant Cell, 33*(7), 2258–2272.

Gazzarrini, S., & McCourt, P. (2003). Cross-talk in plant hormone signalling: What Arabidopsis mutants are telling us. *Annals of Botany*, *91*(6), 605–612.

Ge, Y.-X., Zhang, L.-J., Li, F.-H., Chen, Z.-B., Wang, C., Yao, Y.-C., Han, Z.-H., Zhang, J., & Shi, Z.-S. (2010). Relationship between jasmonic acid accumulation and senescence in drought-stress. *African Journal of Agricultural Research*, *5*(15), 1978–1983.

Genva, M., Obounou Akong, F., Andersson, M. X., Deleu, M., Lins, L., & Fauconnier, M.-L. (2019). New insights into the biosynthesis of esterified oxylipins and their involvement in plant defense and developmental mechanisms. *Phytochemistry Reviews, 18*, 343–358.

Ghani, A., Khan, I., Ahmed, I., & Mustafa, I. (2015). Amelioration of lead toxicity in Pisum sativum (L.) by foliar application of salicylic acid. *Journal of Environmental & Analytical Toxicology*, *5*(292), 2161–0525.

Ghorbanli, M., Kaveh, S. H., & Sepehr, M. F. (2000). Effects of cadmium and gibberellin on growth and photosynthesis of Glycine max. *Photosynthetica*, *37*, 627–631.

Gill, S. S., & Tuteja, N. (2010). Polyamines and abiotic stress tolerance in plants. *Plant Signaling & Behavior, 5*(1), 26–33.

Glinski, J. (2018). Soil physical conditions and plant roots. CRC Press.

Goda, H., Sasaki, E., Akiyama, K., Maruyama-Nakashita, A., Nakabayashi, K., Li, W., Ogawa, M., Yamauchi, Y., Preston, J., & Aoki, K. (2008). The AtGenExpress hormone and chemical treatment data set: experimental design, data evaluation, model data analysis and data access. *The Plant Journal*, *55*(3), 526–542.

Gondor, O. K., Pál, M., Darkó, É., Janda, T., & Szalai, G. (2016). Salicylic acid and sodium salicylate alleviate cadmium toxicity to different extents in maize (Zea mays L.). *PLoS One*, *11*(8), e0160157.

Grill, E., Winnacker, E.-L., & Zenk, M. H. (1985). Phytochelatins: The principal heavy-metal complexing peptides of higher plants. *Science*, *230*(4726), 674–676.

Guan, C., Cui, X., Liu, H.-Y., Li, X., Li, M.-Q., & Zhang, Y.-W. (2020). Proline biosynthesis enzyme genes confer salt tolerance to switchgrass (Panicum virgatum L.) in cooperation with polyamines metabolism. *Frontiers in Plant Science*, *11*, 46.

Guan, C., Ji, J., Zhang, X., Li, X., Jin, C., Guan, W., & Wang, G. (2015). Positive feedback regulation of a Lycium chinense-derived VDE gene by drought-induced endogenous ABA, and over-expression of this VDE gene improve drought-induced photo-damage in Arabidopsis. *Journal of Plant Physiology*, *175*, 26–36.

Gunes, A., Inal, A., Alpaslan, M., Eraslan, F., Bagci, E. G., & Cicek, N. (2007). Salicylic acid induced changes on some physiological parameters symptomatic for oxidative stress and mineral nutrition in maize (Zea mays L.) grown under salinity. *Journal of Plant Physiology*, *164*(6), 728–736.

Gupta, K., Wani, S. H., Razzaq, A., Skalicky, M., Samantara, K., Gupta, S., Pandita, D., Goel, S., Grewal, S., & Hejnak, V. (2022). Abscisic acid: Role in fruit development and ripening. *Frontiers in Plant Science*, *13*, 856.

Gupta, S., Goyal, M. R., & Singh, A. (2018). Biochemical, physiological, and molecular approaches for improving salt tolerance in crop plants: A review. In S. K. Gupta, M. R. Goyal, & A. Singh (Eds.), *Engineering practices for management of soil salinity* (pp. 201–250). Apple Academic Press.

Haisel, D., Pospíšilová, J., Synková, H., Schnablová, R., & Baťková, P. (2006). Effects of abscisic acid or benzyladenine on pigment contents, chlorophyll fluorescence, and chloroplast ultrastructure during water stress and after rehydration. *Photosynthetica*, *44*, 606–614.

Hamayun, M., Hussain, A., Khan, S. A., Kim, H.-Y., Khan, A. L., Waqas, M., Irshad, M., Iqbal, A., Rehman, G., & Jan, S. (2017). Gibberellins producing endophytic fungus Porostereum spadiceum AGH786 rescues growth of salt affected soybean. *Frontiers in Microbiology*, *8*, 686.

Hamayun, M., Khan, S. A., Khan, A. L., Shin, J.-H., Ahmad, B., Shin, D.-H., & Lee, I.-J. (2010). Exogenous gibberellic acid reprograms soybean to higher growth and salt stress tolerance. *Journal of Agricultural and Food Chemistry, 58*(12), 7226–7232.

Hansen, H., & Grossmann, K. (2000). Auxin-induced ethylene triggers abscisic acid biosynthesis and growth inhibition. *Plant physiology, 124*(3), 1437–1448.

Hao, J., Yin, Y., & Fei, S.-z. (2013). Brassinosteroid signaling network: implications on yield and stress tolerance. *Plant Cell Reports, 32*, 1017–1030.

Hasanuzzaman, M. (2020). Plant ecophysiology and adaptation under climate change: Mechanisms and perspectives II: Mechanisms of adaptation and stress amelioration. Springer Nature.

Hasanuzzaman, M., Al Mahmud, J., Nahar, K., Anee, T. I., Inafuku, M., Oku, H., & Fujita, M. (2017). Responses, adaptation, and ROS metabolism in plants exposed to waterlogging stress. In I. M. R. Khan & N. A. Khan (Eds.), *Reactive oxygen species and antioxidant systems in plants: Role and regulation under abiotic stress* (pp. 257–281). Springer.

Hasanuzzaman, M., Nahar, K., Alam, M. M., Roychowdhury, R., & Fujita, M. (2013). Physiological, biochemical, and molecular mechanisms of heat stress tolerance in plants. *International Journal of Molecular Sciences*, *14*(5), 9643–9684.

Hayat, S., Hayat, Q., Alyemeni, M. N., Wani, A. S., Pichtel, J., & Ahmad, A. (2012). Role of proline under changing environments: a review. *Plant Signaling & Behavior, 7*(11), 1456–1466.

He, M., He, C.-Q., & Ding, N.-Z. (2018). Abiotic stresses: General defenses of land plants and chances for engineering multistress tolerance. *Frontiers in Plant Science*, *9*, 1771.

He, M., Qin, C.-X., Wang, X., & Ding, N.-Z. (2020). Plant unsaturated fatty acids: Biosynthesis and regulation. *Frontiers in Plant Science*, *11*, 390.

He, X., Jiang, J., Wang, C. Q., & Dehesh, K. (2017). ORA59 and EIN3 interaction couples jasmonate-ethylene synergistic action to antagonistic salicylic acid regulation of PDF expression. *Journal of Integrative Plant Biology, 59*(4), 275–287.

Hendrawan, V. S. A., Kim, W., & Komori, D. (2023). Crop response pattern to several drought timescales and its possible determinants: A global-scale analysis during the last decades. *Anthropocene*, *43*, 100389.

Hisamatsu, T., & King, R. W. (2008). The nature of floral signals in Arabidopsis. II. Roles for FLOWERING LOCUS T (FT) and gibberellin. *Journal of Experimental Botany, 59*(14), 3821–3829.

Horváth, E., Brunner, S., Bela, K., Papdi, C., Szabados, L., Tari, I., & Csiszár, J. (2015). Exogenous salicylic acid-triggered changes in the glutathione transferases and peroxidases are key factors in the successful salt stress acclimation of Arabidopsis thaliana. *Functional Plant Biology*, *42*(12), 1129–1140.

Hossain, A., Raza, A., Maitra, S., Asaduzzaman, M., Islam, M. R., Hossain, M. J., Sabagh, A. E., Garai, S., Mondal, M., & Latef, A. A. H. A. (2021). Strigolactones: A novel carotenoid-derived phytohormone–biosynthesis, transporters, signalling, and mechanisms in abiotic stress. In T. Aftab & K. R. Hakeem (Eds.), *Plant growth regulators: Signalling under stress conditions* (pp. 275–303). Springer.

Hou, X., Ding, L., & Yu, H. (2013). Crosstalk between GA and JA signaling mediates plant growth and defense. *Plant Cell Reports, 32*, 1067–1074.

Hu, S., Ding, Y., & Zhu, C. (2020). Sensitivity and responses of chloroplasts to heat stress in plants. *Frontiers in Plant Science*, *11*, 375.

Hu, Y., Jiang, Y., Han, X., Wang, H., Pan, J., & Yu, D. (2017). Jasmonate regulates leaf senescence and tolerance to cold stress: crosstalk with other phytohormones. *Journal of Experimental Botany, 68*(6), 1361–1369.

Huang, J., Zhao, X., & Chory, J. (2019). The Arabidopsis transcriptome responds specifically and dynamically to high light stress. *Cell Reports*, *29*(12), 4186–4199. e4183.

Huang, L.-Z., Zhou, M., Ding, Y.-F., & Zhu, C. (2022). Gene networks involved in plant heat stress response and tolerance. *International Journal of Molecular Sciences*, *23*(19), 11970.

Huang, X., Hou, L., Meng, J., You, H., Li, Z., Gong, Z., Yang, S., & Shi, Y. (2018). The antagonistic action of abscisic acid and cytokinin signaling mediates drought stress response in Arabidopsis. *Molecular Plant, 11*(7), 970–982.

Hwang, I., Chen, H.-C., & Sheen, J. (2002). Two-component signal transduction pathways in Arabidopsis. *Plant Physiology, 129*(2), 500–515.

Ibañez, C., Delker, C., Martinez, C., Bürstenbinder, K., Janitza, P., Lippmann, R., Ludwig, W., Sun, H., James, G. V., & Klecker, M. (2018). Brassinosteroids dominate hormonal regulation of plant thermomorphogenesis via BZR1. *Current Biology, 28*(2), 303–310. e303.

Iglesias, M. J., Terrile, M. C., Windels, D., Lombardo, M. C., Bartoli, C. G., Vazquez, F., Estelle, M., & Casalongué, C. A. (2014). MiR393 regulation of auxin signaling and redox-related components during acclimation to salinity in Arabidopsis. *PLoS One*, *9*(9), e107678.

Iqbal, M., & Ashraf, M. (2013). Gibberellic acid mediated induction of salt tolerance in wheat plants: Growth, ionic partitioning, photosynthesis, yield and hormonal homeostasis. *Environmental and Experimental Botany*, *86*, 76–85.

Iqbal, N., Umar, S., Khan, N. A., & Khan, M. I. R. (2014). A new perspective of phytohormones in salinity tolerance: regulation of proline metabolism. *Environmental and Experimental Botany, 100*, 34–42.

Jaleel, C. A., Manivannan, P., Wahid, A., Farooq, M., Al-Juburi, H. J., Somasundaram, R., & Panneerselvam, R. (2009). Drought stress in plants: a review on morphological characteristics and pigments composition. *International Journal of Agriculture and Biology, 11*(1), 100–105.

Jamil, M., Charnikhova, T., Verstappen, F., Ali, Z., Wainwright, H., & Bouwmeester, H. (2014). Effect of phosphate-based seed priming on strigolactone production and S triga hermonthica infection in cereals. *Weed Research, 54*(3), 307–313.

Janda, M., & Ruelland, E. (2015). Magical mystery tour: Salicylic acid signalling. *Environmental and Experimental Botany*, *114*, 117–128.

Jang, G., Chang, S. H., Um, T. Y., Lee, S., Kim, J.-K., & Choi, Y. D. (2017). Antagonistic interaction between jasmonic acid and cytokinin in xylem development. *Scientific Reports, 7*(1), 10212.

Jhanji, S., Goyal, E., Chumber, M., & Kaur, G. (2024). Exploring fine tuning between phytohormones and ROS signaling cascade in regulation of seed dormancy, germination and seedling development. *Plant Physiology and Biochemistry*, 108352.

Jiang, K., & Asami, T. (2018). Chemical regulators of plant hormones and their applications in basic research and agriculture. *Bioscience, Biotechnology, and Biochemistry, 82*(8), 1265–1300.

Jiang, Y., Liang, G., Yang, S., & Yu, D. (2014). Arabidopsis WRKY57 functions as a node of convergence for jasmonic acid–and auxin-mediated signaling in jasmonic acid–induced leaf senescence. *The Plant Cell*, *26*(1), 230–245.

Jini, D., & Joseph, B. (2017). Physiological mechanism of salicylic acid for alleviation of salt stress in rice. *Rice Science*, *24*(2), 97–108.

Joshi, R., Sahoo, K. K., Tripathi, A. K., Kumar, R., Gupta, B. K., Pareek, A., & Singla-Pareek, S. L. (2018). Knockdown of an inflorescence meristem-specific cytokinin oxidase–OsCKX2 in rice reduces yield penalty under salinity stress condition. *Plant, Cell & Environment, 41*(5), 936–946.

Junttila, O., Heide, O. M., Lindgård, B., & Ernstsen, A. (1997). Gibberellins and the photoperiodic control of leaf growth in Poa pratensis. *Physiologia Plantarum, 101*(3), 599–605.

Kagale, S., Divi, U. K., Krochko, J. E., Keller, W. A., & Krishna, P. (2007). Brassinosteroid confers tolerance in Arabidopsis thaliana and Brassica napus to a range of abiotic stresses. *Planta*, *225*, 353–364.

Karpets, Y. V., Kolupaev, Y. E., Lugovaya, A., & Oboznyi, A. (2014). Effect of jasmonic acid on the pro-/antioxidant system of wheat coleoptiles as related to hyperthermia tolerance. *Russian Journal of Plant Physiology, 61*, 339–346.

Kaur, G., & Asthir, B. (2017). Molecular responses to drought stress in plants. *Biologia Plantarum, 61*, 201–209.

Kaushal, N., Bhandari, K., Siddique, K. H., & Nayyar, H. (2016). Food crops face rising temperatures: An overview of responses, adaptive mechanisms, and approaches to improve heat tolerance. *Cogent Food & Agriculture*, *2*(1), 1134380.

Kaya, A., & Doganlar, Z. B. (2016). Exogenous jasmonic acid induces stress tolerance in tobacco (Nicotiana tabacum) exposed to imazapic. *Ecotoxicology and Environmental Safety*, *124*, 470–479.

Kazan, K. (2013). Auxin and the integration of environmental signals into plant root development. *Annals of Botany, 112*(9), 1655–1665.

Kazemi, M. (2013). Foliar application of salicylic acid and calcium on yield, yield component and chemical properties of strawberry. *Bulletin of Environment, Pharmacology and Life Sciences, 2*(11), 19–23.

Kermode, A. R. (2005). Role of abscisic acid in seed dormancy. *Journal of Plant Growth Regulation, 24*, 319–344.

Khalil, R., Haroun, S., Bassyoini, F., Nagah, A., & Yusuf, M. (2021). Salicylic acid in combination with kinetin or calcium ameliorates heavy metal stress in Phaseolus vulgaris plant. *Journal of Agriculture and Food Research*, *5*, 100182.

Khan, A., Shen, F., Yang, L., Xing, W., & Clothier, B. (2022). Limited acclimation in leaf morphology and anatomy to experimental drought in temperate forest species. *Biology, 11*(8), 1186.

Khan, M., & Khan, N. (2013). Salicylic acid and jasmonates: Approaches in abiotic stress tolerance. *Journal of Plant Biochemistry and Biotechnology*, *1*(4), e113.

Khan, N. A., Nazar, R., Iqbal, N., & Anjum, N. A. (2012). *Phytohormones and abiotic stress tolerance in plants*. Springer Science & Business Media.

Khandaker, L., Akond, A. M., & Oba, S. (2011). Foliar application of salicylic acid improved the growth, yield and leaf's bioactive compounds in red Amaranth (L.). *Journal of Fruit and Ornamental Plant Research, 74*(1), 77–86.

Kim, T.-H. (2014). Mechanism of ABA signal transduction: agricultural highlights for improving drought tolerance. *Journal of Plant Biology, 57*, 1–8.

Kissoudis, C., van de Wiel, C., Visser, R. G., & van der Linden, G. (2014). Enhancing crop resilience to combined abiotic and biotic stress through the dissection of physiological and molecular crosstalk. *Frontiers in Plant Science, 5*, 207.

Knight, H., & Knight, M. R. (2001). Abiotic stress signalling pathways: specificity and cross-talk. *Trends in Plant Science, 6*(6), 262–267.

Kohli, S. K., Handa, N., Kaur, R., Kumar, V., Khanna, K., Bakshi, P., Singh, R., Arora, S., Kaur, R., & Bhardwaj, R. (2017). Role of salicylic acid in heavy metal stress tolerance: insight into underlying mechanism. In *Salicylic acid: A multifaceted hormone* (pp. 123–144).

Kovács, V., Gondor, O. K., Szalai, G., Darkó, É., Majláth, I., Janda, T., & Pál, M. (2014). Synthesis and role of salicylic acid in wheat varieties with different levels of cadmium tolerance. *Journal of Hazardous Materials, 280*, 12–19.

Kuiper, D. (1993). Sink strength: established and regulated by plant growth regulators. *Plant, Cell & Environment, 16*(9), 1025–1026.

Kumar, R., Banyal, R., Singh, A., Yadav, R. K., & Sharma, P. C. (2021). Salinity stress effects on the growth, morphological, physiological, and biochemical properties of melia (Melia dubia Cav.) plant. *Dendrobiology, 86*, 56–68.

Kurepa, J., & Smalle, J. A. (2022). Auxin/cytokinin antagonistic control of the shoot/root growth ratio and its relevance for adaptation to drought and nutrient deficiency stresses. *International Journal of Molecular Sciences, 23*(4), 1933.

Krishnamurthy, A., & Rathinasabapathi, B. (2013). Auxin and its transport play a role in plant tolerance to arsenite-induced oxidative stress in Arabidopsis thaliana. *Plant, Cell & Environment, 36*(10), 1838–1849.

Lal, M., Kumari, A., Pooja, & Sheokand, S. (2019). Reactive oxygen species, reactive nitrogen species and oxidative metabolism under waterlogging stress. In M. Hasanuzzaman, V. Fotopoulos, K. Nahar, & M. Fujita (Eds.), *Reactive oxygen, nitrogen and sulfur species in plants: Production, metabolism, signaling and defense mechanisms* (pp. 777–812). John Wiley & Sons.

Larkindale, J., Hall, J. D., Knight, M. R., & Vierling, E. (2005). Heat stress phenotypes of Arabidopsis mutants implicate multiple signaling pathways in the acquisition of thermotolerance. *Plant Physiology, 138*(2), 882–897.

Lei, L., Pan, H., Hu, H.-Y., Fan, X.-W., Wu, Z.-B., & Li, Y.-Z. (2023). Characterization of ZmPMP3g function in drought tolerance of maize. *Scientific Reports, 13*(1), 7375.

Lenis, Y. Y., Elmetwally, M. A., Maldonado-Estrada, J. G., & Bazer, F. W. (2017). Physiological importance of polyamines. *Zygote, 25*(3), 244–255.

Li, A., Sun, X., & Liu, L. (2022). Action of salicylic acid on plant growth. *Frontiers in Plant Science, 13*, 878076.

Li, C. J., Zhu, X. P., & Zhang, F. S. (2000). Role of shoot in regulation of iron deficiency responses in cucumber and bean plants. *Journal of Plant Nutrition, 23*(11–12), 1809–1818.

Li, J. (2016). Fuel-optimal low-thrust formation reconfiguration via Radau pseudospectral method. *Advances in Space Research, 58*(1), 1–16.

Li, J., Li, C., & Smith, S. M. (2017). *Hormone metabolism and signaling in plants*. Academic Press.

Li, J., Sima, W., Ouyang, B., Wang, T., Ziaf, K., Luo, Z., Liu, L., Li, H., Chen, M., & Huang, Y. (2012). Tomato SlDREB gene restricts leaf expansion and internode elongation by downregulating key genes for gibberellin biosynthesis. *Journal of Experimental Botany, 63*(18), 6407–6420.

Li, P., Chen, L., Zhou, Y., Xia, X., Shi, K., Chen, Z., & Yu, J. (2013). Brassinosteroids-induced systemic stress tolerance was associated with increased transcripts of several defence-related genes in the phloem in Cucumis sativus. *PLoS One, 8*(6), e66582.

Li, S., Liu, J., Liu, H., Qiu, R., Gao, Y., & Duan, A. (2021). Role of hydraulic signal and ABA in decrease of leaf stomatal and mesophyll conductance in soil drought-stressed tomato. *Frontiers in Plant Science, 12*, 653186.

Li, X., & Li, C. (2004). Is ethylene involved in regulation of root ferric reductase activity of dicotyledonous species under iron deficiency? *Plant and Soil, 261*(1–2), 147–153.

Li, Z., & He, Y. (2020). Roles of brassinosteroids in plant reproduction. *International Journal of Molecular Sciences, 21*(3), 872.

Li, Z., Luo, X., Ou, Y., Jiao, H., Peng, L., Fu, X., Macho, A. P., Liu, R., & He, Y. (2021). JASMONATE-ZIM DOMAIN proteins engage Polycomb chromatin modifiers to modulate Jasmonate signaling in Arabidopsis. *Molecular Plant, 14*(5), 732–747.

Lima-Melo, Y., Alencar, V. T., Lobo, A. K., Sousa, R. H., Tikkanen, M., Aro, E.-M., Silveira, J. A., & Gollan, P. J. (2019). Photoinhibition of photosystem I provides oxidative protection during imbalanced photosynthetic electron transport in Arabidopsis thaliana. *Frontiers in Plant Science, 10*, 916.

Liu, J., Zhang, R., Xu, X., Fowler, J. C., Miller, T. E., & Dong, T. (2020). Effect of summer warming on growth, photosynthesis and water status in female and male Populus cathayana: Implications for sex-specific drought and heat tolerances. *Tree Physiology*, *40*(9), 1178–1191.

Liu, Q., Yan, S., Huang, W., Yang, J., Dong, J., Zhang, S., Zhao, J., Yang, T., Mao, X., & Zhu, X. (2018). NAC transcription factor ONAC066 positively regulates disease resistance by suppressing the ABA signaling pathway in rice. *Plant Molecular Biology*, *98*, 289–302.

Liu, X., Chi, H., Yue, M., Zhang, X., Li, W., & Jia, E. (2012). The regulation of exogenous jasmonic acid on UV-B stress tolerance in wheat. *Journal of Plant Growth Regulation*, *31*, 436–447.

Llugany, M., Martin, S., Barceló, J., & Poschenrieder, C. (2013). Endogenous jasmonic and salicylic acids levels in the Cd-hyperaccumulator Noccaea (Thlaspi) praecox exposed to fungal infection and/or mechanical stress. *Plant Cell Reports*, *32*, 1243–1249.

Lone, A. A., Khan, M. H., Dar, Z. A., & Wani, S. H. (2018). Breeding strategies for improving growth and yield under waterlogging conditions in maize: a review. *Maydica, 61*(1), 11.

Lu, Q., Zhang, T., Zhang, W., Su, C., Yang, Y., Hu, D., & Xu, Q. (2018). Alleviation of cadmium toxicity in Lemna minor by exogenous salicylic acid. *Ecotoxicology and Environmental Safety*, *147*, 500–508.

Lu, Y., Tarkowská, D., Turečková, V., Luo, T., Xin, Y., Li, J., Wang, Q., Jiao, N., Strnad, M., & Xu, J. (2014). Antagonistic roles of abscisic acid and cytokinin during response to nitrogen depletion in oleaginous microalga N annochloropsis oceanica expand the evolutionary breadth of phytohormone function. *The Plant Journal, 80*(1), 52–68.

Luo, Z.-B., & Luo, J. (2017). Uncovering the physiological mechanisms that allow nitrogen availability to affect drought acclimation in Catalpa bungei. *Tree Physiology, 37*(11), 1453–1456.

Malhi, G. S., Kaur, M., & Kaushik, P. (2021). Impact of climate change on agriculture and its mitigation strategies: A review. *Sustainability*, *13*(3), 1318.

Mandal, S., Ghorai, M., Anand, U., Samanta, D., Kant, N., Mishra, T., Rahman, M. H., Jha, N. K., Jha, S. K., & Lal, M. K. (2022). Cytokinin and abiotic stress tolerance-What has been accomplished and the way forward? *Frontiers in Genetics, 13*, 943025.

Manik, S., Pengilley, G., Dean, G., Field, B., Shabala, S., & Zhou, M. (2019). Soil and crop management practices to minimize the impact of waterlogging on crop productivity. *Frontiers in Plant Science*, 140.

Mao, Q., Lu, X., Mo, H., Gundersen, P., & Mo, J. (2018). Effects of simulated N deposition on foliar nutrient status, N metabolism and photosynthetic capacity of three dominant understory plant species in a mature tropical forest. *Science of the Total Environment, 610*, 555–562.

Masood, A., Iqbal, N., & Khan, N. A. (2012). Role of ethylene in alleviation of cadmium-induced photosynthetic capacity inhibition by sulphur in mustard. *Plant, Cell & Environment*, *35*(3), 524–533.

Mason, M. G., Jha, D., Salt, D. E., Tester, M., Hill, K., Kieber, J. J., & Eric Schaller, G. (2010). Type-B response regulators ARR1 and ARR12 regulate expression of AtHKT1; 1 and accumulation of sodium in Arabidopsis shoots. *The Plant Journal, 64*(5), 753–763.

Matiu, M., Ankerst, D. P., & Menzel, A. (2017). Interactions between temperature and drought in global and regional crop yield variability during 1961–2014. *PLoS One*, *12*(5), e0178339.

Matsoukas, I. G. (2014). Interplay between sugar and hormone signaling pathways modulate floral signal transduction. *Frontiers in Genetics, 5*, 218.

Miura, K., & Tada, Y. (2014). Regulation of water, salinity, and cold stress responses by salicylic acid. *Frontiers in Plant Science, 5*, 4.

Mohamed, H. I., & Latif, H. H. (2017). Improvement of drought tolerance of soybean plants by using methyl jasmonate. *Physiology and Molecular Biology of Plants*, *23*, 545–556.

Mostofa, M. G., Rahman, M. M., Ansary, M. M. U., Fujita, M., & Tran, L.-S. P. (2019). Interactive effects of salicylic acid and nitric oxide in enhancing rice tolerance to cadmium stress. *International Journal of Molecular Sciences*, *20*(22), 5798.

Muhlemann, J. K., Younts, T. L., & Muday, G. K. (2018). Flavonols control pollen tube growth and integrity by regulating ROS homeostasis during high-temperature stress. *Proceedings of the National Academy of Sciences*, *115*(47), E11188–E11197.

Murti, G., & Upreti, K. (2007). Plant growth regulators in water stress tolerance. *Journal of Horticultural Sciences*, *2*(2), 73–93.

Nacry, P., Canivenc, G., Muller, B., Azmi, A., Van Onckelen, H., Rossignol, M., & Doumas, P. (2005). A role for auxin redistribution in the responses of the root system architecture to phosphate starvation in Arabidopsis. *Plant Physiology*, *138*(4), 2061–2074.

Nagar, S., Singh, V., Arora, A., Dhakar, R., Singh, N., Singh, G., Meena, S., Kumar, S., & Shiv Ramakrishnan, R. (2021). Understanding the role of gibberellic acid and paclobutrazol in terminal heat stress tolerance in wheat. *Frontiers in Plant Science, 12*, 692252.

Napieraj, N., Janicka, M., & Reda, M. (2023). Interactions of polyamines and phytohormones in plant response to abiotic stress. *Plants*, *12*(5), 1159.

Naumann, G., Alfieri, L., Wyser, K., Mentaschi, L., Betts, R. A., Carrao, H., Spinoni, J., Vogt, J., & Feyen, L. (2018). Global changes in drought conditions under different levels of warming. *Geophysical Research Letters*, *45*(7), 3285–3296.

Netshimbupfe, M. H., Berner, J., & Gouws, C. (2022). The interactive effects of drought and heat stress on photosynthetic efficiency and biochemical defense mechanisms of Amaranthus species. *Plant-Environment Interactions*, *3*(5), 212–225.

Neumann, G., Massonneau, A., Langlade, N., Dinkelaker, B., Hengeler, C., Römheld, V., & Martinoia, E. (2000). Physiological aspects of cluster root function and development in phosphorus-deficient white lupin (Lupinus albus L.). *Annals of Botany*, *85*(6), 909–919.

Neupane, D., Mayer, J. A., Niechayev, N. A., Bishop, C. D., & Cushman, J. C. (2021). Five-year field trial of the biomass productivity and water input response of cactus pear (Opuntia spp.) as a bioenergy feedstock for arid lands. *GCB Bioenergy*, *13*(4), 719–741.

Nguyen, K. H., Ha, C. V., Nishiyama, R., Watanabe, Y., Leyva-González, M. A., Fujita, Y., Tran, U. T., Li, W., Tanaka, M., & Seki, M. (2016). Arabidopsis type B cytokinin response regulators ARR1, ARR10, and ARR12 negatively regulate plant responses to drought. *Proceedings of the National Academy of Sciences, 113*(11), 3090–3095.

Nguyen, L. T., Osanai, Y., Anderson, I. C., Bange, M. P., Tissue, D. T., & Singh, B. K. (2018). Flooding and prolonged drought have differential legacy impacts on soil nitrogen cycling, microbial communities and plant productivity. *Plant and Soil, 431*, 371–387.

Nishiyama, R., Watanabe, Y., Leyva-Gonzalez, M. A., Van Ha, C., Fujita, Y., Tanaka, M., Seki, M., Yamaguchi-Shinozaki, K., Shinozaki, K., & Herrera-Estrella, L. (2013). Arabidopsis AHP2, AHP3, and AHP5 histidine phosphotransfer proteins function as redundant negative regulators of drought stress response. *Proceedings of the National Academy of Sciences, 110*(12), 4840–4845.

Niu, J.-h., Ahmad Anjum, S., Wang, R., Li, J.-h., Liu, M.-r., Song, J.-x., Zohaib, A., Lv, J., Wang, S.-g., & Zong, X.-f. (2016). Exogenous application of brassinolide can alter morphological and physiological traits of Leymus chinensis (Trin.) Tzvelev under room and high temperatures. *Chilean Journal of Agricultural Research, 76*(1), 27–33.

Noriega, G., Cruz, D. S., Batlle, A., Tomaro, M., & Balestrasse, K. (2012). Heme oxygenase is involved in the protection exerted by jasmonic acid against cadmium stress in soybean roots. *Journal of Plant Growth Regulation, 31*, 79–89.

Oguz, M. C., Aycan, M., Oguz, E., Poyraz, I., & Yildiz, M. (2022). Drought stress tolerance in plants: Interplay of molecular, biochemical and physiological responses in important development stages. *Physiologia, 2*(4), 180–197.

Oh, E., Zhu, J.-Y., & Wang, Z.-Y. (2012). Interaction between BZR1 and PIF4 integrates brassinosteroid and environmental responses. *Nature Cell Biology, 14*(8), 802–809.

Ohnishi, T. (2018). Recent advances in brassinosteroid biosynthetic pathway: insight into novel brassinosteroid shortcut pathway. *Journal of Pesticide Science, 43*(3), 159–167.

Örvar, B. L., Sangwan, V., Omann, F., & Dhindsa, R. S. (2000). Early steps in cold sensing by plant cells: The role of actin cytoskeleton and membrane fluidity. *The Plant Journal*, *23*(6), 785–794.

Osakabe, Y., Osakabe, K., Shinozaki, K., & Tran, L.-S. P. (2014). Response of plants to water stress. *Frontiers in Plant Science*, *5*, 86.

Oshino, T., Miura, S., Kikuchi, S., Hamada, K., Yano, K., Watanabe, M., & Higashitani, A. (2011). Auxin depletion in barley plants under high-temperature conditions represses DNA proliferation in organelles and nuclei via transcriptional alterations. *Plant, Cell & Environment*, *34*(2), 284–290.

Ouzounidou, G., & Ilias, I. (2005). Hormone-induced protection of sunflower photosynthetic apparatus against copper toxicity. *Biologia Plantarum*, *49*, 223–228.

Pál, M., Szalai, G., & Janda, T. (2015). Speculation: Polyamines are important in abiotic stress signaling. *Plant Science*, *237*, 16–23.

Pandey, C., & Gupta, M. (2015). Selenium and auxin mitigates arsenic stress in rice (Oryza sativa L.) by combining the role of stress indicators, modulators and genotoxicity assay. *Journal of Hazardous Materials, 287*, 384–391.

Pandey, R., Vengavasi, K., & Hawkesford, M. J. (2021). Plant adaptation to nutrient stress. *Plant Physiology Reports, 26*(4), 583–586.

Parida, A. K., Veerabathini, S. K., Kumari, A., & Agarwal, P. K. (2016). Physiological, anatomical and metabolic implications of salt tolerance in the halophyte Salvadora persica under hydroponic culture condition. *Frontiers in Plant Science, 7*, 351.

Park, H. J., Kim, W.-Y., & Yun, D.-J. (2013). A role for GIGANTEA: keeping the balance between flowering and salinity stress tolerance. *Plant Signaling & Behavior, 8*(7), e24820.

Parwez, R., Aftab, T., Gill, S. S., & Naeem, M. (2022). Abscisic acid signaling and crosstalk with phytohormones in regulation of environmental stress responses. *Environmental and Experimental Botany, 199*, 104885.

Pasternak, T., Groot, E. P., Kazantsev, F. V., Teale, W., Omelyanchuk, N., Kovrizhnykh, V., Palme, K., & Mironova, V. V. (2019). Salicylic acid affects root meristem patterning via auxin distribution in a concentration-dependent manner. *Plant Physiology, 180*(3), 1725–1739.

Pearce, R. S. (2001). Plant freezing and damage. *Annals of Botany, 87*(4), 417–424.

Pearce, S., Kippes, N., Chen, A., Debernardi, J. M., & Dubcovsky, J. (2016). RNA-seq studies using wheat PHYTOCHROME B and PHYTOCHROME C mutants reveal shared and specific functions in the regulation of flowering and shade-avoidance pathways. *BMC Plant Biology, 16*(1), 1–19.

Peleg, Z., & Blumwald, E. (2011). Hormone balance and abiotic stress tolerance in crop plants. *Current Opinion in Plant Biology, 14*(3), 290–295.

Per, T. S., Khan, M. I. R., Anjum, N. A., Masood, A., Hussain, S. J., & Khan, N. A. (2018). Jasmonates in plants under abiotic stresses: Crosstalk with other phytohormones matters. *Environmental and Experimental Botany, 145*, 104–120.

Petersen, M., Brodersen, P., Naested, H., Andreasson, E., Lindhart, U., Johansen, B., Nielsen, H. B., Lacy, M., Austin, M. J., & Parker, J. E. (2000). Arabidopsis MAP kinase 4 negatively regulates systemic acquired resistance. *Cell, 103*(7), 1111–1120.

Pieterse, C. M., Leon-Reyes, A., Van der Ent, S., & Van Wees, S. C. (2009). Networking by small-molecule hormones in plant immunity. *Nature Chemical Biology, 5*(5), 308–316.

Pieterse, C. M., Van der Does, D., Zamioudis, C., Leon-Reyes, A., & Van Wees, S. C. (2012). Hormonal modulation of plant immunity. *Annual Review of Cell and Developmental Biology, 28*, 489–521.

Pinon, V., Prasad, K., Grigg, S. P., Sanchez-Perez, G. F., & Scheres, B. (2013). Local auxin biosynthesis regulation by PLETHORA transcription factors controls phyllotaxis in Arabidopsis. *Proceedings of the National Academy of Sciences, 110*(3), 1107–1112.

Praveen, A. (2022). Nitric oxide mediated alleviation of abiotic challenges in plants. *Nitric Oxide, 128*, 37.

Qadir, M., Sposito, G., Smith, C., & Oster, J. D. (2021). Reassessing irrigation water quality guidelines for sodicity hazard. *Agricultural Water Management, 255*, 107054.

Qiu, Z., Guo, J., Zhu, A., Zhang, L., & Zhang, M. (2014). Exogenous jasmonic acid can enhance tolerance of wheat seedlings to salt stress. *Ecotoxicology and Environmental Safety, 104*, 202–208.

Rady, M. M. (2011). Effect of 24-epibrassinolide on growth, yield, antioxidant system and cadmium content of bean (Phaseolus vulgaris L.) plants under salinity and cadmium stress. *Scientia Horticulturae, 129*(2), 232–237.

Rahneshan, Z., Nasibi, F., & Moghadam, A. A. (2018). Effects of salinity stress on some growth, physiological, biochemical parameters and nutrients in two pistachio (Pistacia vera L.) rootstocks. *Journal of Plant Interactions, 13*(1), 73–82.

Rajewska, I., Talarek, M., & Bajguz, A. (2016). Brassinosteroids and response of plants to heavy metals action. *Frontiers in Plant Science, 7*, 629.

Ramegowda, V., & Senthil-Kumar, M. (2015). The interactive effects of simultaneous biotic and abiotic stresses on plants: mechanistic understanding from drought and pathogen combination. *Journal of Plant Physiology, 176*, 47–54.

Rao, M. V., Lee, H.-I., Creelman, R. A., Mullet, J. E., & Davis, K. R. (2000). Jasmonic acid signaling modulates ozone-induced hypersensitive cell death. *The Plant Cell, 12*(9), 1633–1646.

Rao, K. M., Raghavendra, A. S., & Reddy, K. J. (2006). *Physiology and molecular biology of stress tolerance in plants*. Springer Science & Business Media.

Raza, A., Charagh, S., Najafi-Kakavand, S., Abbas, S., Shoaib, Y., Anwar, S., Sharifi, S., Lu, G., & Siddique, K. H. (2023). Role of phytohormones in regulating cold stress tolerance: physiological and molecular approaches for developing cold-smart crop plants. *Plant Stress*, 100152.

Reinhardt, D., Mandel, T., & Kuhlemeier, C. (2000). Auxin regulates the initiation and radial position of plant lateral organs. *The Plant Cell, 12*(4), 507–518.

Riyazuddin, R., Verma, R., Singh, K., Nisha, N., Keisham, M., Bhati, K. K., Kim, S. T., & Gupta, R. (2020). Ethylene: A master regulator of salinity stress tolerance in plants. *Biomolecules, 10*(6), 959.

Roe, R. M., Burton, J. D., & Kuhr, R. J. (1997). *Herbicide activity: Toxicology, biochemistry and molecular biology*. IOS Press.

Romera, F. J., Alcantara, E., & De La Guardia, M. D. (1999). Ethylene production by Fe-deficient roots and its involvement in the regulation of Fe-deficiency stress responses by strategy I plants. *Annals of Botany, 83*(1), 51–55.

Rostás, M., Winter, T. R., Borkowski, L., & Zeier, J. (2013). Copper and herbivory lead to priming and synergism in phytohormones and plant volatiles in the absence of salicylate-jasmonate antagonism. *Plant Signaling & Behavior, 8*(6), e24264.

Ruan, S., Xue, Q., & Tylkowska, K. (2002). The influence of priming on germination of rice (Oryza sativa L.) seeds and seedling emergence and performance in flooded soil. *Seed Science and Technology, 30*(1), 61–67.

Ruelland, E., Vaultier, M.-N., Zachowski, A., & Hurry, V. (2009). Cold signalling and cold acclimation in plants. *Advances in Botanical Research, 49*, 35–150.

Sakakibara, H., Takei, K., & Hirose, N. (2006). Interactions between nitrogen and cytokinin in the regulation of metabolism and development. *Trends in Plant Science, 11*(9), 440–448.

Salazar-Iribe, A., & De-la-Pena, C. (2020). Auxins, the hidden player in chloroplast development. *Plant Cell Reports, 39*(12), 1595–1608.

Schopfer, P., & Plachy, C. (1984). Control of seed germination by abscisic acid: II. Effect on embryo water uptake in Brassica napus L. *Plant Physiology, 76*(1), 155–160.

Schulze-Lefert, P., & Robatzek, S. (2006). Plant pathogens trick guard cells into opening the gates. *Cell, 126*(5), 831–834.

Shah, S. H., Islam, S., Parrey, Z. A., & Mohammad, F. (2021). Role of exogenously applied plant growth regulators in growth and development of edible oilseed crops under variable environmental conditions: a review. *Journal of Soil Science and Plant Nutrition, 21*(4), 3284–3308.

Shahzad, A., Pitann, B., Ali, H., Qayyum, M., Fatima, A., & Bakhat, H. (2015). Maize genotypes differing in salt resistance vary in jasmonic acid accumulation during the first phase of salt stress. *Journal of Agronomy and Crop Science, 201*(6), 443–451.

Shao, H.-B., Chu, L.-Y., Jaleel, C. A., & Zhao, C.-X. (2008). Water-deficit stress-induced anatomical changes in higher plants. *Comptes Rendus Biologies, 331*(3), 215–225.

Sharma, A., Kumar, V., Shahzad, B., Ramakrishnan, M., Singh Sidhu, G. P., Bali, A. S., Handa, N., Kapoor, D., Yadav, P., & Khanna, K. (2020). Photosynthetic response of plants under different abiotic stresses: A review. *Journal of Plant Growth Regulation, 39*, 509–531.

Sharma, I., Ching, E., Saini, S., Bhardwaj, R., & Pati, P. K. (2013). Exogenous application of brassinosteroid offers tolerance to salinity by altering stress responses in rice variety Pusa Basmati-1. *Plant Physiology and Biochemistry, 69*, 17–26.

Sharma, P., Jha, A. B., Dubey, R. S., & Pessarakli, M. (2012). Reactive oxygen species, oxidative damage, and antioxidative defense mechanism in plants under stressful conditions. *Journal of Botany*, 2012.

Sharma, S., Kaur, P., & Gaikwad, K. (2022). Role of cytokinins in seed development in pulses and oilseed crops: Current status and future perspective. *Frontiers in Genetics, 13*, 940660.

Shah, S. H., Islam, S., Parrey, Z. A., & Mohammad, F. (2021). Role of exogenously applied plant growth regulators in growth and development of edible oilseed crops under variable environmental conditions: a review. *Journal of Soil Science and Plant Nutrition, 21*(4), 3284–3308.

Sharma, S. K., Kulshreshtha, N., Kumar, A., Yaduvanshi, N., Singh, M., Prasad, K., & Basak, N. (2018). Waterlogging effects on elemental composition of wheat genotypes in sodic soils. *Journal of Plant Nutrition, 41*(10), 1252–1262.

Shemi, R., Wang, R., Gheith, E.-S. M., Hussain, H. A., Cholidah, L., Zhang, K., Zhang, S., & Wang, L. (2021). Role of exogenous-applied salicylic acid, zinc and glycine betaine to improve drought-tolerance in wheat during reproductive growth stages. *BMC Plant Biology, 21*, 1–15.

Sheng, X.-F., & Xia, J.-J. (2006). Improvement of rape (Brassica napus) plant growth and cadmium uptake by cadmium-resistant bacteria. *Chemosphere, 64*(6), 1036–1042.

Shi, G., Cai, Q., Liu, Q., & Wu, L. (2009). Salicylic acid-mediated alleviation of cadmium toxicity in hemp plants in relation to cadmium uptake, photosynthesis, and antioxidant enzymes. *Acta Physiologiae Plantarum, 31*, 969–977.

Shumilina, J., Kusnetsova, A., Tsarev, A., Janse van Rensburg, H. C., Medvedev, S., Demidchik, V., Van den Ende, W., & Frolov, A. (2019). Glycation of plant proteins: Regulatory roles and interplay with sugar signalling? *International Journal of Molecular Sciences*, *20*(9), 2366.

Siddiqui, H., Hayat, S., & Bajguz, A. (2018). Regulation of photosynthesis by brassinosteroids in plants. *Acta Physiologiae Plantarum, 40*, 1–15.

Siemens, J. A., & Zwiazek, J. J. (2004). Changes in root water flow properties of solution culture-grown trembling aspen (Populus tremuloides) seedlings under different intensities of water-deficit stress. *Physiologia Plantarum, 121*(1), 44–49.

Singh, A., & Roychoudhury, A. (2023). Abscisic acid in plants under abiotic stress: Crosstalk with major phytohormones. *Plant Cell Reports*, *42*, 1–14.

Singh, H., Bhat, J. A., Singh, V. P., Corpas, F. J., & Yadav, S. R. (2021). Auxin metabolic network regulates the plant response to metalloids stress. *Journal of Hazardous Materials*, *405*, 124250.

Sirhindi, G., Mir, M. A., Sharma, P., Gill, S. S., Kaur, H., & Mushtaq, R. (2015). Modulatory role of jasmonic acid on photosynthetic pigments, antioxidants and stress markers of Glycine max L. under nickel stress. *Physiology and Molecular Biology of Plants*, *21*, 559–565.

Soares, T. F. S. N., dos Santos Dias, D. C. F., Oliveira, A. M. S., Ribeiro, D. M., & dos Santos Dias, L. A. (2020). Exogenous brassinosteroids increase lead stress tolerance in seed germination and seedling growth of Brassica juncea L. *Ecotoxicology and Environmental Safety*, *193*, 110296.

Song, Y., Dou, L., & Zhang, H. (2014). Molecular and genetic mechanisms of control of floral induction in higher plants. *Plant Physiology Journal, 50*(10), 1459–1468.

Song, W., Liu, S., Meng, L., Xue, R., Wang, C., Liu, G., Dong, C., Wang, S., Dong, J., & Zhang, Y. (2015). Potassium deficiency inhibits lateral root development in tobacco seedlings by changing auxin distribution. *Plant and Soil*, *396*, 163–173.

Sreenivasulu, N., Harshavardhan, V. T., Govind, G., Seiler, C., & Kohli, A. (2012). Contrapuntal role of ABA: does it mediate stress tolerance or plant growth retardation under long-term drought stress? *Gene, 506*(2), 265–273.

Stegner, M., Buchner, O., Schäfernolte, T., Holzinger, A., & Neuner, G. (2022). Responses to ice formation and reasons of frost injury in potato leaves. *Crops, 2*(4), 378–389.

Stintzi, A., & Browse, J. (2000). The Arabidopsis male-sterile mutant, opr3, lacks the 12-oxophytodienoic acid reductase required for jasmonate synthesis. *Proceedings of the National Academy of Sciences, 97*(19), 10625–10630.

Sun, J., Xu, Y., Ye, S., Jiang, H., Chen, Q., Liu, F., Zhou, W., Chen, R., Li, X., & Tietz, O. (2009). Arabidopsis ASA1 is important for jasmonate-mediated regulation of auxin biosynthesis and transport during lateral root formation. *The Plant Cell*, *21*(5), 1495–1511.

Suzuki, N., Bassil, E., Hamilton, J. S., Inupakutika, M. A., Zandalinas, S. I., Tripathy, D., Luo, Y., Dion, E., Fukui, G., & Kumazaki, A. (2016). ABA is required for plant acclimation to a combination of salt and heat stress. *PLoS One*, *11*(1), e0147625.

Suzuki, N., Devireddy, A. R., Inupakutika, M. A., Baxter, A., Miller, G., Song, L., Shulaev, E., Azad, R. K., Shulaev, V., & Mittler, R. (2015). Ultra-fast alterations in mRNA levels uncover multiple players in light stress acclimation in plants. *The Plant Journal*, *84*(4), 760–772.

Swain, R., Sahoo, S., Behera, M., & Rout, G. R. (2023). Instigating prevalent abiotic stress resilience in crop by exogenous application of phytohormones and nutrient. *Frontiers in Plant Science, 14*, 1104874.

Szalai, G., Pál, M., Árendás, T., & Janda, T. (2016). Priming seed with salicylic acid increases grain yield and modifies polyamine levels in maize. *Cereal Research Communications*, *44*(4), 537–548.

Taghvaei, M., Nasrolahizadehi, A., & Mastinu, A. (2022). Effect of Light, Temperature, Salinity, and Halopriming on Seed Germination and Seedling Growth of Hibiscus sabdariffa under Salinity Stress. *Agronomy, 12*(10), 2491.

Tanaka, Y., Sano, T., Tamaoki, M., Nakajima, N., Kondo, N., & Hasezawa, S. (2006). Cytokinin and auxin inhibit abscisic acid-induced stomatal closure by enhancing ethylene production in Arabidopsis. *Journal of Experimental Botany*, *57*(10), 2259–2266.

Tang, B., Xie, L., Yang, H., Li, X., Chen, Y., Zou, X., Liu, F., & Dai, X. (2022). Analysis of the expression and function of key genes in pepper under low-temperature stress. *Frontiers in Plant Science*, *13*, 852511.

Thomas, A., Beena, R., Laksmi, G., Soni, K., Alex, S., & Viji, M. (2022). Changes in sucrose metabolic enzymes to water stress in contrasting rice genotypes. *Plant Stress, 5*, 100088.

Tognetti, V. B., Van Aken, O., Morreel, K., Vandenbroucke, K., Van De Cotte, B., De Clercq, I., Chiwocha, S., Fenske, R., Prinsen, E., & Boerjan, W. (2010). Perturbation of indole-3-butyric acid homeostasis by the UDP-glucosyltransferase UGT74E2 modulates Arabidopsis architecture and water stress tolerance. *The Plant Cell*, *22*(8), 2660–2679.

Trevisan, S., Forestan, C., Brojanigo, S., Quaggiotti, S., & Varotto, S. (2020). Brassinosteroid application affects the growth and gravitropic response of maize by regulating gene expression in the roots, shoots and leaves. *Plant Growth Regulation, 92*, 117–130.

Tromas, A., & Perrot-Rechenmann, C. (2010). Recent progress in auxin biology. *Comptes Rendus Biologies, 333*(4), 297–306.

Tuna, A. L., Kaya, C., Dikilitas, M., & Higgs, D. (2008). The combined effects of gibberellic acid and salinity on some antioxidant enzyme activities, plant growth parameters and nutritional status in maize plants. *Environmental and Experimental Botany, 62*(1), 1–9.

Ullah, A., Tian, Z., Xu, L., Abid, M., Lei, K., Khanzada, A., Zeeshan, M., Sun, C., Yu, J., & Dai, T. (2022). Improving the effects of drought priming against post-anthesis drought stress in wheat (Triticum aestivum L.) using nitrogen. *Frontiers in Plant Science, 13*, 965996.

van der Fits, L., & Memelink, J. (2000). ORCA3, a jasmonate-responsive transcriptional regulator of plant primary and secondary metabolism. *Science, 289*(5477), 295–297.

Van Zelm, E., Zhang, Y., & Testerink, C. (2020). Salt tolerance mechanisms of plants. *Annual Review of Plant Biology, 71*, 403–433.

Vaughan, M. M., Block, A., Christensen, S. A., Allen, L. H., & Schmelz, E. A. (2018). The effects of climate change associated abiotic stresses on maize phytochemical defenses. *Phytochemistry Reviews, 17*, 37–49.

Verma, V., Ravindran, P., & Kumar, P. P. (2016). Plant hormone-mediated regulation of stress responses. *BMC Plant Biology, 16*, 1–10.

Vidya Vardhini, B., & Seeta Ram Rao, S. (2003). Amelioration of osmotic stress by brassinosteroids on seed germination and seedling growth of three varieties of sorghum. *Plant Growth Regulation, 41*, 25–31.

Villano, C., Esposito, S., D'Amelia, V., Garramone, R., Alioto, D., Zoina, A., Aversano, R., & Carputo, D. (2020). WRKY genes family study reveals tissue-specific and stress-responsive TFs in wild potato species. *Scientific Reports, 10*(1), 7196.

Visentin, I., Vitali, M., Ferrero, M., Zhang, Y., Ruyter-Spira, C., Novák, O., Strnad, M., Lovisolo, C., Schubert, A., & Cardinale, F. (2016). Low levels of strigolactones in roots as a component of the systemic signal of drought stress in tomato. *New Phytologist, 212*(4), 954–963.

Waadt, R., Seller, C. A., Hsu, P.-K., Takahashi, Y., Munemasa, S., & Schroeder, J. I. (2022). Plant hormone regulation of abiotic stress responses. *Nature Reviews Molecular Cell Biology, 23*(10), 680–694.

Wang, J., Song, L., Gong, X., Xu, J., & Li, M. (2020). Functions of jasmonic acid in plant regulation and response to abiotic stress. *International Journal of Molecular Sciences, 21*(4), 1446.

Wang, X., & Komatsu, S. (2022). The role of phytohormones in plant response to flooding. *International Journal of Molecular Sciences, 23*(12), 6383.

Wang, Y., Mostafa, S., Zeng, W., & Jin, B. (2021). Function and mechanism of jasmonic acid in plant responses to abiotic and biotic stresses. *International Journal of Molecular Sciences, 22*(16), 8568.

Wang, Y.-H., Zhang, G., Chen, Y., Gao, J., Sun, Y.-R., Sun, M.-F., & Chen, J.-P. (2019). Exogenous application of gibberellic acid and ascorbic acid improved tolerance of okra seedlings to NaCl stress. *Acta Physiologiae Plantarum, 41*, 1–10.

Wani, S. H., & Kumar, V. (2015). Plant stress tolerance: engineering ABA: a potent phytohormone. *Transcriptomics, 3*(2), 1000113.

Wani, S. H., Kumar, V., Shriram, V., & Sah, S. K. (2016). Phytohormones and their metabolic engineering for abiotic stress tolerance in crop plants. *The Crop Journal, 4*(3), 162–176.

Wasternack, C. (2014). Action of jasmonates in plant stress responses and development—applied aspects. *Biotechnology Advances, 32*(1), 31–39.

Wei, L., Wang, L., Yang, Y., Wang, P., Guo, T., & Kang, G. (2015). Abscisic acid enhances tolerance of wheat seedlings to drought and regulates transcript levels of genes encoding ascorbate-glutathione biosynthesis. *Frontiers in Plant Science, 6*, 458.

Woodward, A. W., & Bartel, B. (2005). Auxin: regulation, action, and interaction. *Annals of Botany, 95*(5), 707–735.

Wu, C., Cui, K., Wang, W., Li, Q., Fahad, S., Hu, Q., Huang, J., Nie, L., Mohapatra, P. K., & Peng, S. (2017). Heat-induced cytokinin transportation and degradation are associated with reduced panicle cytokinin expression and fewer spikelets per panicle in rice. *Frontiers in Plant Science, 8*, 371.

Xin, Z., Zhao, Y., & Zheng, Z.-L. (2005). Transcriptome analysis reveals specific modulation of abscisic acid signaling by ROP10 small GTPase in Arabidopsis. *Plant Physiology, 139*(3), 1350–1365.

Xuewen, X., Huihui, W., Xiaohua, Q., Qiang, X., & Xuehao, C. (2014). Waterlogging-induced increase in fermentation and related gene expression in the root of cucumber (Cucumis sativus L.). *Scientia Horticulturae, 179*, 388–395.

Yadav, S., Modi, P., Dave, A., Vijapura, A., Patel, D., & Patel, M. (2020). Effect of abiotic stress on crops. In M. Hasanuzzaman, M. Filho, M. Fujita, & T. Nogueira (Eds.), *Sustainable crop production, 3*. IntechOpen.

Yamaguchi, K., Takahashi, Y., Berberich, T., Imai, A., Takahashi, T., Michael, A. J., & Kusano, T. (2007). A protective role for the polyamine spermine against drought stress in Arabidopsis. *Biochemical and Biophysical Research Communications, 352*(2), 486–490.

Yan, S., Zhang, T., Zhang, B., Zhang, T., Cheng, Y., Wang, C., Luo, M., Feng, H., & Siddique, K. H. (2023). The higher relative concentration of K+ to Na+ in saline water improves soil hydraulic conductivity, salt-leaching efficiency and structural stability. *SOIL, 9*(1), 339–349.

Yang, C., Huang, Y., Lv, P., Antwi-Boasiako, A., Begum, N., Zhao, T., & Zhao, J. (2022). NAC transcription factor GmNAC12 improved drought stress tolerance in soybean. *International Journal of Molecular Sciences, 23*(19), 12029.

Yang, J., Duan, G., Li, C., Liu, L., Han, G., Zhang, Y., & Wang, C. (2019). The crosstalks between jasmonic acid and other plant hormone signaling highlight the involvement of jasmonic acid as a core component in plant response to biotic and abiotic stresses. *Frontiers in Plant Science, 10*, 1349.

Yang, J., Fei, K., Chen, J., Wang, Z., Zhang, W., & Zhang, J. (2020). Jasmonates alleviate spikelet-opening impairment caused by high temperature stress during anthesis of photo-thermo-sensitive genic male sterile rice lines. *Food and Energy Security*, *9*(4), e233.

Yoneyama, K., Xie, X., Kusumoto, D., Sekimoto, H., Sugimoto, Y., Takeuchi, Y., & Yoneyama, K. (2007). Nitrogen deficiency as well as phosphorus deficiency in sorghum promotes the production and exudation of 5-deoxystrigol, the host recognition signal for arbuscular mycorrhizal fungi and root parasites. *Planta, 227*, 125–132.

Yoon, Y., Seo, D. H., Shin, H., Kim, H. J., Kim, C. M., & Jang, G. (2020). The role of stress-responsive transcription factors in modulating abiotic stress tolerance in plants. *Agronomy, 10*(6), 788.

Yosefi, A., Mozafari, A. A., & Javadi, T. (2020). Jasmonic acid improved in vitro strawberry (Fragaria × ananassa Duch.) resistance to PEG-induced water stress. *Plant Cell, Tissue and Organ Culture (PCTOC), 142*, 549–558.

Yoshida, S., Tamaoki, M., Ioki, M., Ogawa, D., Sato, Y., Aono, M., Kubo, A., Saji, S., Saji, H., & Satoh, S. (2009). Ethylene and salicylic acid control glutathione biosynthesis in ozone-exposed Arabidopsis thaliana. *Physiologia Plantarum, 136*(3), 284–298.

Yu, F., Wu, Y., & Xie, Q. (2016). Ubiquitin–proteasome system in ABA signaling: from perception to action. *Molecular Plant, 9*(1), 21–33.

Zaman, M., Shahid, S. A., Heng, L., Shahid, S. A., Zaman, M., & Heng, L. (2018). Soil salinity: Historical perspectives and a world overview of the problem. In *Guideline for salinity assessment, mitigation and adaptation using nuclear and related techniques* (pp. 43–53).

Zhai, Y., Wang, Y., Li, Y., Lei, T., Yan, F., Su, L., Li, X., Zhao, Y., Sun, X., & Li, J. (2013). Isolation and molecular characterization of GmERF7, a soybean ethylene-response factor that increases salt stress tolerance in tobacco. *Gene, 513*(1), 174–183.

Zahra, N., Hafeez, M. B., Ghaffar, A., Kausar, A., Al Zeidi, M., Siddique, K. H., & Farooq, M. (2023). Plant photosynthesis under heat stress: Effects and management. *Environmental and Experimental Botany, 206*, 105178.

Zhang, G., Zhao, F., Chen, L., Pan, Y., Sun, L., Bao, N., Zhang, T., Cui, C.-X., Qiu, Z., & Zhang, Y. (2019a). Jasmonate-mediated wound signalling promotes plant regeneration. *Nature Plants*, *5*(5), 491–497.

Zhang, H., Rong, H., & Pilbeam, D. (2007). Signalling mechanisms underlying the morphological responses of the root system to nitrogen in Arabidopsis thaliana. *Journal of Experimental Botany*, *58*(9), 2329–2338.

Zhang, H., Wang, W., Huang, J., Wang, Y., Hu, L., Yuan, Y., Lyu, M., & Wu, B. (2022). Role of gibberellin and its three GID1 receptors in Jasminum sambac stem elongation and flowering. *Planta, 255*, 1–16.

Zhang, J.-Y., He, S.-B., Li, L., & Yang, H.-Q. (2014). Auxin inhibits stomatal development through MONOPTEROS repression of a mobile peptide gene STOMAGEN in mesophyll. *Proceedings of the National Academy of Sciences, 111*(29), E3015–E3023.

Zhang, X., Long, Y., Chen, X., Zhang, B., Xin, Y., Li, L., Cao, S., Liu, F., Wang, Z., & Huang, H. (2021). A NAC transcription factor OsNAC3 positively regulates ABA response and salt tolerance in rice. *BMC Plant Biology*, *21*, 1–13.

Zhang, X., Wang, X., Zhuang, L., Gao, Y., & Huang, B. (2019b). Abscisic acid mediation of drought priming-enhanced heat tolerance in tall fescue (Festuca arundinacea) and Arabidopsis. *Physiologia Plantarum, 167*(4), 488–501.

Zhang, Y., Li, Y., Hassan, M. J., Li, Z., & Peng, Y. (2020). Indole-3-acetic acid improves drought tolerance of white clover via activating auxin, abscisic acid and jasmonic acid related genes and inhibiting senescence genes. *BMC Plant Biology*, *20*(1), 1–12.

Zhang, X., Takano, T., & Liu, S. (2006). Identification of a mitochondrial ATP synthase small subunit gene (RMtATP6) expressed in response to salts and osmotic stresses in rice (Oryza sativa L.). *Journal of Experimental Botany, 57*(1), 193–200.

Zhao, G., Xu, H., Zhang, P., Su, X., & Zhao, H. (2017). Effects of 2, 4-epibrassinolide on photosynthesis and Rubisco activase gene expression in Triticum aestivum L. seedlings under a combination of drought and heat stress. *Plant Growth Regulation, 81*, 377–384.

Zhao, M. L., Wang, J. N., Shan, W., Fan, J. G., Kuang, J. F., Wu, K. Q., Li, X. P., Chen, W. X., He, F. Y., & Chen, J. Y. (2013). Induction of jasmonate signalling regulators MaMYC2s and their physical interactions with MaICE1 in methyl jasmonate-induced chilling tolerance in banana fruit. *Plant, Cell & Environment, 36*(1), 30–51.

Zheng, C., Jiang, D., Liu, F., Dai, T., Jing, Q., & Cao, W. (2009). Effects of salt and waterlogging stresses and their combination on leaf photosynthesis, chloroplast ATP synthesis, and antioxidant capacity in wheat. *Plant Science, 176*(4), 575–582.

Zheng, Y., Wang, X., Cui, X., Wang, K., Wang, Y., & He, Y. (2023). Phytohormones regulate the abiotic stress: An overview of physiological, biochemical, and molecular responses in horticultural crops. *Frontiers in Plant Science, 13*, 1095363.

Zheng, S. J., Tang, C., Arakawa, Y., & Masaoka, Y. (2003). The responses of red clover (Trifolium pratense L.) to iron deficiency: A root Fe (III) chelate reductase. *Plant Science, 164*(5), 679–687.

Zhou, R., Hyldgaard, B., Yu, X., Rosenqvist, E., Ugarte, R. M., Yu, S., Wu, Z., Ottosen, C.-O., & Zhao, T. (2018). Phenotyping of faba beans (Vicia faba L.) under cold and heat stresses using chlorophyll fluorescence. *Euphytica, 214*, 1–13.

Zhou, Z. S., Guo, K., Elbaz, A. A., & Yang, Z. M. (2009). Salicylic acid alleviates mercury toxicity by preventing oxidative stress in roots of Medicago sativa. *Environmental and Experimental Botany, 65*(1), 27–34.

Zhu, T., Deng, X., Zhou, X., Zhu, L., Zou, L., Li, P., Zhang, D., & Lin, H. (2016). Ethylene and hydrogen peroxide are involved in brassinosteroid-induced salt tolerance in tomato. *Scientific Reports, 6*(1), 35392.

Zhu, X. F., Wang, Z. W., Dong, F., Lei, G. J., Shi, Y. Z., Li, G. X., & Zheng, S. J. (2013). Exogenous auxin alleviates cadmium toxicity in Arabidopsis thaliana by stimulating synthesis of hemicellulose 1 and increasing the cadmium fixation capacity of root cell walls. *Journal of Hazardous Materials, 263*, 398–403.

Žižková, E., Dobrev, P. I., Muhovski, Y., Hošek, P., Hoyerová, K., Haisel, D., Procházková, D., Lutts, S., Motyka, V., & Hichri, I. (2015). Tomato (Solanum lycopersicum L.) SlIPT3 and Solanum lycopersicum SlIPT4 isopentenyltransferases mediate salt stress response in tomato. *BMC Plant Biology, 15*(1), 85.

Zörb, C., Geilfus, C. M., & Dietz, K. J. (2019). Salinity and crop yield. *Plant Biology, 21*, 31–38.

15 Plant Tolerance Mechanisms to Heavy Metal Stress and Breeding Strategies

Battana Swapna, Srinivasan Kameswaran, and Mandala Ramakrishna

15.1 INTRODUCTION

One of the main environmental contaminants that can find their way into soil, water, and the atmosphere is heavy metal (HM). This can happen naturally, as in the case of weathering rocks and volcanic activity or by anthropogenic activity like metallurgical, chemical processing, transportation, and the use of mineral fertilizers. Lead, arsenic, aluminum, mercury, copper, zinc, and manganese are examples of HMs. Through coordinated molecular pathways, plants adapt morphologically and physiologically to offset the stresses imposed by HMs. Novel methods such as transcriptomics, genomes, proteomics, and metabolomics investigations have made it possible to comprehend these intricate networks.

Under severe metal stress, plants use intricate generegulatory mechanisms. MicroRNAs are short RNAs with 21 nucleotides that do not code and are significant post-transcriptional modulators of gene expression. A lot of plant 'heavy-metal-responsive microRNAs' have been discovered recently, thanks to high-throughput sequencing. Metal chelation, HM uptake and transport, antioxidative metabolism, hormone signaling, and biogenesis of microRNA are just a few of the biological processes that are governed by the intricate regulatory network that includes microRNAs regulated by metals.

The molecular mechanisms underlying plant defense and adaptation responses to metal stress are clarified in this chapter. It emphases on the introduction of HMs into plants, HM movement and buildup, their impact on key physiological functions, and the gene expressions involved in plant resistance to HM stress. Extracellular and intracellular HM sequestration is the primary mechanism of HM tolerance. The extracellular sequestration of HM ions is accomplished by organic anions such as malate and oxalate. Cellular receptors and transmembrane transporters play a role in intracellular HM sequestration in addition to a binding technique by means of various polymers like pectin, cellulose, lignin, and hemicellulose.

Root exudates, transport via the membrane, sequestration in the xylem, phloem, and vacuole, and chelation of different ligands are the principal mechanisms of tolerance.

Plants are capable of detoxifying and tolerating HMs through a variety of unique and efficient mechanisms, such as regulating metal influx, speeding up metal efflux, chelating and sequestering metals, remobilizing metal, and scavenging metal-induced reactive oxygen species (ROS) like OH, $1O_2$, and H_2O_2, which are activated forms of oxygen. Certain metals like Pb, Ni, Zn, Al, and Mn may only produce ROS indirectly through various mechanisms; they cannot directly produce ROS, in contrast to iron and copper. Antioxidative enzymes like superoxide dismutase, ascorbate peroxidase, monodehydroascorbate reductase, dehydroascorbate reductase, glutathione reductase, glutathione peroxidase, guaicol peroxidase, glutathione-*S*-transferase, and catalase in addition to non-enzymatic metabolites like proline, ascorbic acid, alpha tocopherol, carotenoids, glutathione, and flavonoids scavenge metal-induced ROS accumulation. Most organisms use ROS

 DOI: 10.1201/9781003457237-15

as essential signal transduction molecules that are necessary for life at basal levels. Intense production of ROS is thought as advantageous to trigger the molecular plant tolerance mechanisms to HM toxicity.

Plants can adapt to harsh environmental conditions, such as biotic and abiotic stressors, by transferring beneficial traits through breeding. Plant growth and development are severely limited by HMT. A successful approach to addressing the issue of HMT in agriculture may involve creating crops that are resilient to HMs. By using anther culture to create a cross between Prabhavati and Basmati, the improved variety for grain quality, Parag 401 was created to help rice resist Fe chlorosis (Patil et al., 1997). On the basis of root and shoot ratio, Pb-tolerant watermelon varieties were found (Khan et al., 2020). Tolerant brown rice in Bangladesh (BRRI-29, BR-26, and BR-14) was developed and commercialized (Halim et al., 2016). Due to its great produce, highly proteinaceous grain, and very low concentration of Cd accumulation in grains, Strongfield, a wheat variety, is highly famous in Canada (Clarke et al., 2005).

15.2 PLANT TOLERANCE MECHANISMS FOR HM STRESS

Once HMs build up in the plant body, a variety of tolerance and defense mechanisms begin in physiological and molecular stages. According to Shahid et al. (2017), a network of interconnected physiological and molecular processes controls plants' tolerance to specific HMs. For various forms of HM uptake control, the plant body uses two fundamental functional strategies: metal exclusion as an avoidance mechanism and metal accumulation as a tolerance mechanism (Sameena and Puthur, 2021). Plant species have been divided into three types based on the accumulation of HMs: excludator metal accumulator, hyperaccumulator, and accumulator (Greger, 2004). Strong resistance to HM buildup is exhibited by metal excluder-type plant species, which also lessen the transport of HMs from the surrounding medium. Concentrating accumulated HM within their bodies, accumulator plants focus on resistance. Rascio and Navari-Izzo (2011), Sameena and Puthur (2021), and other observations showed that a particular hyperaccumulator species has the capacity to absorb significantly higher amounts – between 10 and 1,000 times higher than non-hyperaccumulating plant species. Following are some of the tolerance mechanisms for HM stress.

15.2.1 Root Exudates

The plant components' root zone served as the first line of defense against the buildup of HMs. HM detoxification in the root sections may take place internally or externally. Plants can complex HMs as part of detoxification inside the plant root, but they can also complex the metals outside the root by forming a metal–oxalate complex and continuing absorption (Sameena and Puthur, 2021). Root exudate production has the potential to modify pH equilibrium within the plant rhizosphere and limit the accumulation of HMs within the apoplast zone (Ghori et al., 2019). Additionally, the release of root exudates varies greatly throughout plant species. In response to Al toxicity, for instance, a small number of resistant wheat cultivars were able to secrete malic acid more effectively than sensitive counterparts. As part of metal detoxification processes, root exudates may repeatedly bind HMs to decrease HMs movement in the plant (Ghori et al., 2019).

15.2.2 Physical Barriers

Plants shield themselves from metals mostly with physical barriers. When plants encounter HM stress, certain anatomical characteristics including the thick cuticle, structures like trichomes, cell walls, and mycorrhizae association might operate as barriers (Jia et al., 2021). Trichomes can either store HMs during the detoxification process or release a number of metabolites that mitigate the negative effects of metals (Hauser, 2014; Koul et al., 2021).

15.2.3 Cellular Mechanisms

15.2.3.1 Vacuolar Compartmentation

The tolerance of non-essential metals in the plant body is significantly influenced by the cell vacuole found inside plant cells. Vacuolar ATPase (V-ATPase) and vacuolar pyrophosphatase (V-PPase) are two examples of the transporters that are important for the HM removal process inside the cell vacuole, where proton pumps and metal/H^+ antiporters are key players (Wei et al., 2021).

15.2.3.2 Cellular Macromolecules

According to Ghori et al. (2019) and Jalmi et al. (2018), one of the most frequent processes brought on by HM stress in plants is the creation of ROS. Mittler (2002) and Srivastava et al. (2011) describe how the plant uses a variety of chemicals and enzymes as well as antioxidant properties to combat free radicals in plants. Several plants launched multiple measures, including the development of an organic molecular complex to lower the availability of HMs, as part of its detoxifying mechanisms, and to minimize HM stress. Many chemical compounds, including glutathione (GSH), low-molecular-weight peptides, metallothionein (MT), malate, citrate, and phytochelatins (PCs), may help to construct complexes with metals cumulated in plant bodies (Filiz et al., 2019). PCs remained inactive till the plants were subjected to particular metal stress (Ejaz et al., 2023). According to Rascio and Navari-Izzo (2011), hyperaccumulator plant species exhibit remarkable strategies for both HM accumulation and detoxification. Specifically, a large number of accumulated HMs were stored in aerial parts of the plant. Meanwhile, the removal of HMs primarily takes place in the cuticles, trichomes, and epidermis of the plant, which has less photosynthetic activity. Succulent plant species (halophyte) were abundant in arid environments, although some species have been capable of withstanding HMs in their succulent sections (Li et al., 2019).

Conversely, after HMs get through biophysical barriers and reach tissues and cells, plants launch a series of defensive processes to counteract and lessen the harmful effects of HMs. The main defense against or tolerance of HM stress was the macromolecule synthesis in cells. It includes the initiation of a wide range of chelators and 'low-molecular-weight protein metallochaperones' including putrescine, nicotinamide, spermine, glutathione, organic acids, MTs, and PCs; or exudations in cells, including heat shock proteins, proline, histidine, flavonoids, and phenolic compounds; hormones, including ethylene, jasmonic acid, and salicylic acid.

15.2.3.3 Antioxidant Defense

Redox homeostasis in plants is distressed when the aforementioned tactics fail to prevent metal poisoning, which increases the induction of free radicals. Plants progressed an antioxidative system to counteract the harmful effects of ROS. This defense mechanism is made up of both non-enzymatic antioxidants and enzymatic antioxidants that act as radical quenchers. Some compounds engaged in restoration can comprise many functions, such as antioxidant, chelating, or antiradical properties, as previously mentioned. Any of these biomolecules and mechanisms that are exploited and upregulated may be controlled by the metal to which it is exposed, the plant developmental phase, the metal tolerance level (Raza et al., 2021), and the species of the plant.

15.2.3.4 Phytochelatins

Plants use a mechanism to get rid of HMs that involves creating short-chain thiol-rich peptides. These peptides bind to HMs at toxic levels and help to scavenge them. The peptides are made from GSH, which is sulfur-rich, by an enzyme called PC synthase (PCS). PCs are used by plants as a defense mechanism against a variety of stresses, including heat, salt, UV-B rays, herbicides, and metal-related stresses (Zagorchev et al., 2013). PCs produced in the cytosol were transferred to vacuoles like metal PC aggregates (Song et al., 2014).

Sulfhydryl and carboxyl groups on PCs allow them to connect to certain HM ions, including Cd, Ag, Cu, Zn, Ni, and Pb (Rauser, 1995; Cobbett, 2000). This leads to the abrupt increase in cellular

PCs due to different amounts of certain HM types. Among them, cadmium ions are proved as efficient activators of PC production, inducing four to sixfold more PCs than copper and zinc ions (Thangavel et al., 2007). In both roots and aerial organs, PCs can collect and be created. However, most research indicates that they often start as roots, where they accumulate through biosynthesis. Research has demonstrated that when sunflowers were exposed to Cd poisoning, the amounts of PCs in their leaves were lower than in their roots (ÇIKILI et al., 2020). Fidalgo et al. (2013) demonstrated in *Solanum nigrum* L. that when subjected to 200 µmol/L copper concentration, PC generation in roots was increased. This prevented the excessive Cu in the roots from migrating toward the stem and immobilized it. According to the findings of Batista et al. (2014), Arsenic transportation from below-ground parts to the above-ground parts of plants was decreased by stimulating distinct arsenic–PC complex in the rice roots exposed to severe arsenic stress. When it comes to keeping hazardous metals out of the edible portions of crops, these tactics may work.

However, other studies demonstrate that PCs-related activity and concentration in leaves increase when time is included along with prolonged HM stress. Heiss et al. (2003) showed that *Brassica juncea* accumulated three times more PCs in its leaves than in its roots when exposed to Cd for an extended period of time. Longer-term Cd treatment of maize plants resulted in reduced PC activity in the roots and increased levels of PCS in the leaves, according to Szalai et al. (2013). They proposed that this phenomenon might be explained by the substrate reduction or feedback control mechanism. Along with the previously listed variables, it appears that the kind of plants and how tolerant they are to HM excess affects PC formation, accumulation, transit, and preferred movement path in plants. Along with HM kinds, PCs vary between plants. The most prevalent phytochelatin was PC4 in *Nicotiana tabacum* L., according to Brunetti et al. (2011), while PC3 was more concentrated in *Arabidopsis*. Extended-chain PCs have been found to bind Pb more strongly than shorter-chain PCs in legumes (Piechalak et al., 2002).

Under high metal stress, phytochelatins and antioxidant enzymes can work together to create a synergistic defensive regime in plants that can bolster those plants'defenses against metal intoxication. The combined action of *Brassica chinensis* L.'s antioxidative system and enhanced PC enzymatic production resulted in efficient Cd detoxification, as Chen et al. (2008) presented. The PCS genes that cause PCs to be produced have been identified and cloned with much effort. The earliest plant genes for PCS isolated were PC synthase (TaPCS1) from *Triticum aestivum* L. and PC synthase (AtPCS1) from *Arabidopsis thaliana* (Cobbett, 2000). As a result of the continued research in this field, different PCS genes turned out to be recognized in other plants, including *Oryza sativa* L. and *Brassica juncea* (Liu et al., 2021; Shen et al., 2010).

A very bright future for the potential to increase plant tolerance to HMs is offered by the expression of PCS genes in hyperaccumulator plants. The artificial synthesis of the PC gene improved the resilience of transgenic tobacco plants to different cadmium concentrations (Postrigan et al., 2012). Because of synthetic PC expression (ECs), Shukla et al. (2013), demonstrated that transgenic *Arabidopsis* plants significantly accumulate stronger HM than wild *Arabidopsis*. Guo et al. (2012) showed that overexpressing AsPCS1 (arsenic–PC synthase 1) and YCF1 (yeast cadmium factor 1), which were derived from baking yeast and garlic, augmented the plant's forbearance to Cd and Asas well as its capacity to hyperaccumulate the metals in *A. thaliana*.

15.2.3.5 Metallothioneins

Another group of small, cysteine-rich cytoplasmic metal-binding proteins known as MTs were isolated from prokaryotes, fungi, and also plants. MTs were produced by virtue of mRNA translation, whereas PCs were the outcome of enzymatic peptide synthesis (Verkleij et al., 2003). These ligands help plants counteract the harmful effects of HMs by removing excessive metal ions, maintaining ion homeostasis, and adjusting HM transportation (Guo et al., 2013; Kohler et al., 2004). MTs were recognized as vigorous substances in several processes, such as ROS scavenging (Khalid et al., 2020), redox level maintenance (Macovei et al., 2010), plasma membrane repair (Mishra and Dubey, 2006), and DNA repair (Grennan, 2011). These are additional functions to their part in HM sequestration. HMs were not the only endogenous or exogenous factor that could stimulate the creation and expression of MTs.

Depending on how the Cys residues are arranged, plants can have up to four different subgroups of MT types (Huang and Wang, 2009). MTs show distinctiveness toward plant part and growth stage patterns, such as the following: type 1 MTs expression is shown in root system, type 2 in stems, type 3 in foliage as well as at the time of fruit ripening, and type 4 during seed development. All plant MTs appear to be capable of binding HMs with function in metal chelation and stores. Nevertheless, accumulating data indicates that plant MTs might behave differently depending on the type of metal they are exposed to. Additionally, the role of plant MTs, including tissue localization and metal chelation properties, may differ within or between plant species.

According to Grennan (2011), there is a high probability of participation in chelating copper ions in *Arabidopsis* by types 1, 2, and 3 MT isoforms, whereas type 4 MT isoforms bind zinc. Küpper and Andresen (2016) demonstrated that MT1 might be more crucial for removing copper ions in leaf venation in certain *Arabidopsis* mutants. According to Yang et al. (2009), OsMT1a maintains zinc homeostasis in rice roots. A study on MTs in soybeans demonstrated that types 1, 2, and 3 appeared to be engaged in the sequestration of harmful levels of cadmium, whereas MT4 serves as a zinc binder (Pagani et al., 2012).

A study by Zhou et al. (2019) showed TaMT3, anMT in *Tamarixandrossowii*, when introduced into tobacco, enhanced resistance to cadmium toxicity by significantly increasing SOD functionality, which improved the transgenic plant's ability to clean up ROS; it also resulted in decreased POD activity. Further research is necessary as it appears that the produced MT of transgenic plants affects different constituents of the antioxidative machinery in different ways. Transgenic plants with ectopically produced MTs exhibit increased resistance to metal poisoning. When transgenic tobacco was ectopically expressed with MT1 OSMT1e-p, derived from *O. sativa* L. cv. Pokkali, conferred tolerance to zinc and copper stress (Kumar et al., 2012). They found that tobacco plants carrying the gene showed the tendency to hold on to higher levels of Zn^{+2} and Cu^{2+} in lower leaves or else roots, which greatly decreased the movement and content of HMs ions to and within upper foliage and harvestable organs.

15.2.3.6 Proline (Pro)

As a suitable and metabolic osmolyte, radical quencher, macromolecule stabilizer,and antioxidizing substance, pro is an amino acid (AA) with fivecarbons (Zarei et al., 2012). The normal nonenzymic stress response brought through a variety of biotic and abiotic stresses, including salt, drought, UV radiation, HMs, and oxidative stress, is the synthesis of heightened amounts of Pro by higher plants (Szabados and Savoure, 2010). When it comes to helping plants withstand stress, Pro really performs a variety of functions, like signaling, adaptation, and recuperation (Fidalgo et al., 2013).

According to the review of the literature, plants' activation of proline in response to HMs is largely concentration-, organ-, and metal-specific. Numerous studies involving various plants have shown similar specificity in proline accumulation, where shoots have less proline than roots. These experiments include *Solanum nigrum* L. under Cu stress (Fidalgo et al., 2013), Cd stress in wheat (Lesko and Simon-Sarkadi, 2002), and *Cymbopogon flexuosus* Stapf under Hg and Cd stress (Handique and Handique, 2009). Instead of letting metals penetrate to above-ground areas, fast generation of proline content in roots along with the formation of proline–metal complexes can provide a more efficient means of neutralizing metal toxicity.

Metal-dependent proline concentration was found to be increased in *T. aestivum* under Cd-induced stress. The increase in proline concentration helped to reverse the oxidative damage caused by Cd. Proline mended tissue damage, stimulated enzymes for DNA repair, stabilized protein structure, and enhanced the production of PCs (Meng et al., 2019). The application of jasmonic acid and gibberellic acids in *Cicer arietinum* under Cdstress showed proline regulation of cadmium stress (Ahmad et al., 2021). Howladar et al. (2018) stated that proline content in silicon-stressed *T. aestivum* helped to reduce osmotic stress caused by cadmium stress. Also, they saw that when the plant got Pro from exogenous sources, its production of Pro increased significantly, which helped it handle the metal stress better. It provides a promise in further research about priming seeds with Pro to increase plant resistance to HM toxicity.

15.3 ARBUSCULAR MYCORRHIZAE

In a mutualistic symbiosis formed by symbiotic mycorrhizal fungi like arbuscular mycorrhizae (AM) with plant roots, these fungi benefit from food given by the plants, which in turn improves plant nutritional eminence and increases resistance to certain stressors (Upadhyaya et al., 2010). Even though mycorrhizae in relation to HMs is primarily discussed in the context of bioremediation techniques, it is essential to analyze this association from a defensive perspective due to its numerous and vital services to plants. This helps to strengthen plants' overall defensive systems, particularly against external stressors like HMs.

Mycorrhizal fungi use three main strategies to mitigate the effects of HM stress in plants: they operate as a barrier by sequestering metals in their vacuoles or other organelles, attaching metals to cell walls or mycelium, or serving as an obstruction by deposition of metals in cells of the cortex. Heatshock protein and glutathione are released, metals are precipitated or chelated in the soil by synthesizing glycoprotein or forming complexes in the mycelium (Aloui et al., 2011), and metal strength is decreased by increased growth in plants (Mohammadi et al., 2011). The immobilization of HMs in soil was demonstrated to drastically inhibit the movement of Zn, Cd, and Ni from soil to various areas of the *Lolium perenne* L., associated with mycorrhizae (Takacs et al., 2001). Metal immobilization in the mycorrhizosphere region is said to be mostly attributed to changes in soil pH brought on by AM fungus activities (Bano and Ashfaq, 2013).

Abad and Khara's study (2007) revealed that,in comparison to non-AM wheat, wheat plants colonized by various AM fungal species and exposed to hazardous Cd levels exhibited noticeably higher levels of protective antioxidant activity in their roots and shoots, namely in APX and GPX. Additionally, they noted that *Glomus etunicatum* and *Glomus veruciforme* were potent stimulators of the defense proteins among the fungal species. Mycorrhizal *Cajanus cajan* L. plants grown in cadmium and lead-polluted soils exhibit decreased lipid peroxidation and higher activity of the antioxidative system (Garg and Aggarwal, 2012). Due to the AM fungus's ability to sequester zinc in its hyphae, Farshian et al. (2007) showed that *Lactuca sativa* L. plants associated with mycorrhizae subjected to $ZnSO_4$ stress have high protein content consequently, elevated levels of antioxidants.

There is a paucity of clear knowledge regarding the mechanisms of mycorrhizae affecting the synthesis and enhancement of metabolites, including PCs, MTs, and proline in plants under HM stress. Using maize plants under copper and cadmium stress treated with two species of mycorrhizae (*Acaulospora laevis* and *Glomus mosseae*), Abdelmoneim et al. (2014) observed a reduction of proline accumulation in AM-treated maize compared to nontreated maize, and other factors were thought to be responsible for the fungal and plant association's success in reducing the harmful impact of HMs. However, as Ruscitti et al. (2011) show, the combination of Cr-stressed pepper plants and mycorrhizae treatment with *Glomus mosseae* and *Glomus intraradices* led to a decrease in root Pro concentration but an increase in leaf Pro content. To find out whether AM fungus and the defensive compounds that metal-stressed plants abundantly produce work in opposition to or in concert with each other, more research is required. The hosts of AM fungi have been shown to include major crops including wheat, maize, and rice (Bhalerao, 2013). As a result, it is important to identify and propagate the right fungus species to increase plant resistance to HM toxicity.

15.4 OVEREXPRESSION OF GENES PRODUCING ANTIOXIDANT MACHINERY COMPONENTS

Using a variety of transgenes and host plants, transformation experiments were carried out to enhance general tolerance to abiotic stimuli by overexpressing genes encoding antioxidant enzymes. On the contrary, the findings concerning the reaction to HMs were a little unclear. It was discovered that transgenic plants overexpressing the genes for SOD, APX, and CAT produced fewer ROS following exposure to HMs than wild-type plants did and that antioxidant enzyme activity was significantly increased during stress conditions (Lee et al., 2007; Guan et al., 2009; Gao et al., 2016).

Catalase may not have been entirely necessary to shield tobacco plants against Cd stress, according to Iannone et al. (2015). However, when exposed to Cd stress, transgenic plants with reduced constitutive CAT activity (line CAT1AS) were able to activate alternative defense mechanisms, primarily through increased ascorbate peroxidase and guaiacol peroxidase activity. Transgenic lucerne plants that co-express the genes for human cytochrome P450 2E1 and GST proved very productive. Plants exhibited stronger resistance against the harmful effects of organic xenobiotic trichloroethylene and HMs like cadmium and mercury and also accumulated higher levels of these pollutants compared to the control (Zhang and Liu, 2011; Zhang et al., 2013). According to Alvarez et al. (2009), antioxidant enzymes can also be recognized as distinct isoforms of the same proteins, some of which may be elevated and others downregulated in reaction to HMs. According to Ovečka and Takáč (2014), stress exposure also affects the regulation of genes involved in antioxidant responses, which means that over time, different antioxidant molecules become activated.

According to recent research, overexpressing genes involved in transcription and DNA repair may help plants become more resistant to HM doses. This time, the mechanism of tolerance is associated with the genes that encode antioxidative enzymes, like the DNA repair system enzymes, MTs, metal transporters, catalase, and superoxide dismutase (Faè et al., 2014; Charfeddine et al., 2017). Transgenic *Medicago truncatula* overexpressed the DNA repair gene MtTdp2α, which encodes tyrosyl-DNA phosphodiesterase 2, was able to achieve increased tolerance to Cu (Faè et al., 2014), as well as to copper and cadmium in potatoes with overexpression of genes encoding TFs (Ban et al., 2011; Charfeddine et al., 2017). The aforementioned instances demonstrated the intricate nature of plant stress tolerance and the necessity of processes involving the interaction between HMs and plants for effective plant engineering. These mechanisms will need to be further investigated in subsequent research.

15.5 TRADITIONAL BREEDING TO GAIN TOLERANCE TO HMs

Enhancing stress tolerance is more effectively and powerfully accomplished through conventional breeding approaches. Conventional breeding methods, however, have a lower success rate. The intricate process of plant tolerance is reliant on both genetic and physiological reactions. These procedures include phenotypic assessment in addition to finding quantitative trait loci (QTLs) that are tightly linked to molecular markers. Traditional plant breeding comprises selecting, hybridizing, polyploidizing, and introducing events to improve agricultural plants, which helps screen genotypes for resistance to HMs. Furthermore, effective current breeding techniques supported by biotechnology are applied to generate metal-tolerant cultivars that may be sustainably used in HM-contaminated regions (Figure 15.1).

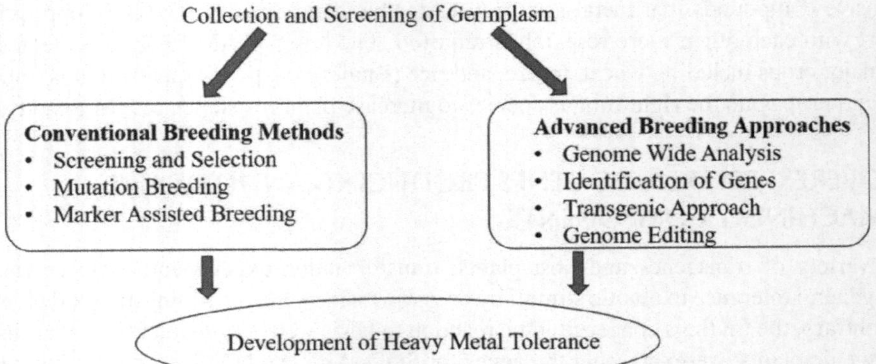

FIGURE 15.1 Different breeding tools to develop tolerance in plants.

15.5.1 Screening for HM Tolerance

For the goal of creating new crop varieties, the presence of natural variation for the desired characteristic is essential. Previous studies using physiological parameters like relative shoot biomass, root biomass, and total chlorophyll content have found a large genetic variety for several HMs such as Cu, Zn, Cd, Ni, Pb, Sn, and As. To check for Zn tolerance in wheat and barley cultivars, relative shoot-to-root ratios were employed (Lombnaesand Singh, 2003). Hydroponics has proven to be a highly efficacious method for assessing tolerance to HMs. In the hydroponics setup, 16 different rice cultivars were tested for Cu tolerance. It has been suggested that four varieties – Xiushui 123, Xiushui 134, Jiahe 991, and Xianghu 301 – be grown locally after they showed improved morphological and enzyme activity of roots when exposed to copper stress (Lin et al., 2017). In a different investigation, 'the rice variety BRRI-29 from the Bangladesh Rice Research Institute (BRRI)' exhibited the highest biomass, the longest root length, and the best germination percentage (82%) among all the entries examined (Halim et al., 2016). The greater levels of proline, and chlorophyll in the leaves, and the area of the flag leaf as markers of lead tolerance.

To check for Cd and Zn buildup, shoot tip cultures of *Linum usitatissimum* are utilized (Smykalova et al., 2010). On the basis of germination rate, root, and epicotyl length, the wheat variety Pirsabak 2004 that is tolerant to lead was identified (Shafiq et al., 2019). Fifty-four additive QTLs were found using two-locus analysis in wheat aluminum (Al) tolerance screening utilizing recombinant inbred lines (RILs) from SeriM82×Babax (Farokhzadeh et al., 2019). Based on observed kernel traits, various wheat genotypes were analyzedagainst Sn and Pb toxicity. High Pb tolerance was demonstrated by the genotype Benisweif-7. Sn had a major effect on kernel diameter. Breeding programs could benefit from using genotype Gimmiza-11 as a parent because of its high tolerance to sulfur (Mourad et al., 2021). Metal resistance in *Agrostis tenuis* from quarry and meadows was assessed in north and central Wales (Gregory and Bradshaw, 1965). Growing populations on mining sites are more resistant to HMs than meadow populations. Even though the 'clone L(II)' was far from a Cu mine, it demonstrated a high tolerance to Cu (Gregory and Bradshaw, 1965). Clone Vil of Poplar alba has demonstrated tolerance to As, Cu, and Zn through phytoremediation (Lonardo et al., 2011).

For breeding and developing new crop varieties, the study of natural variation in related qualities is crucial. Based on physiological parameters such as relative shoot weight, root weight, and chlorophyll, Tang et al. (2021) found that a large genetic variety was present for several HMs (HMs) such as Zn, Cu, Cd, Pb, Ni, As, and Sn. To produce rice genotypes that are tolerant to HM, screening strategies such as hydroponic systems are used to assess the tolerance level (Mishra et al., 2018).

15.5.2 Genetics of HM Tolerance

When the Japonica rice cultivar Koshihikari was exposed to carbon ion-beam radiation, mutants with low Cd content in seeds were found and Cd-free grains were the outcome. These mutants have reduced root absorption of Cd due to a deficiency in the OsNRAMP5transporter gene, as demonstrated by functional investigation (Ishikawa et al., 2012). Zinc and copper are transported to the vacuole via the HMA3 gene, which is located in the vacuolar membrane. Little cadmium uptake in durum wheat is regulated by a dominant allele on chromosome 5B. The production of F_1 from the crossbreeding of hyperaccumulator *Arabidopsis halleri* and non-accumulator *Arabidopsis petraea* shows certain Cd, Cu, and Zn tolerance is regulated by a single gene (Kulaevaand Tsyganov, 2011). In both *Silene vulgaris* (Kulaevaand Tsyganov, 2011) and *Mimulus guttatus* (Macnair, 1983), a particular gene controls Cu tolerance. AsTol is an arsenic tolerance gene found in rice RILs originating from Bala×Azucena (Dasgupta et al., 2004). To investigate the genetics of Cd tolerance, two types of barley were used: tolerant cultivar Weisuobuzhi and sensitive cultivar Dong 17 (Cao et al., 2014). After being exposed to mercury, six genes in *Pisum sativum* roots were elevated. Among them, PsHMIP6B is the novel gene. According to Savenstrand and Strid (2004), several genes were engaged in stimulating antioxidative responses in challenging environments. Two cultivars of hemp, Ym (tolerant) and Nx (sensitive), have been found to have different levels of Cd tolerance. Ym has a sequestration mechanism aided by an antioxidative system, while Nx lacks these genes (Huang et al., 2019).

15.5.3 Mutation Breeding

Using mutants with modified interaction with HMs is one of the key ways to explore plant responses to HMs. To create genetic variety for desired features, such as the creation of novel kinds with greater resistance to HMs, mutation breeding is a crucial technique (Kumagai et al., 2014). According to Shen et al. (2017), several mutagenesis investigations yielded functionally defective mutants that considerably decreased the uptake of related HMs in various rice plant sections. Mutagenesis techniques that are frequently used include irradiation, space, and EMS mutagenesis (Qi et al., 2009). To examine the impact of these genes on HM uptake or displacement in plants, their expression was muted or otherwise altered by knocking out their alleles. This has the potential to create genotypes that are resistant to HM toxicity due to their lower HM accumulation. In rice plants, HM absorption, transport, or distribution is mediated by a number of transporter genes that have been discovered (Chen et al., 2019).

15.5.4 Molecular Markers

Many crops have discovered genomic areas linked to their ability to withstand HMs. Rice QTLs for Fe and Zn (Liu et al., 2017) and Cd toxicity tolerance (Shilin et al., 2021) have been identified. A single-nucleotide polymorphism marker on chromosome 9 (TBGI380435) has been identified to be linked to HM transport detoxification (HTDT) in rice, which can help in developing ferrous toxicity tolerance. This was discovered by Nugraha et al. (2016). Gupta et al. (2013) found RAPD and SCAR markers that can be used for HM screening in *Ceratophyllum* and *Hydrilla*. Additionally, another study conducted by Salarizadeh and Kavousi in 2015 identified RAPD markers that can be helpful in screening for Cd tolerance in *Cichorium pumilum* and cumin (Salarizadeh and Kavousi, 2015).

15.6 ADVANCED BREEDING METHODS FOR HM TOLERANCE

The OMICS technologies are helpful in characterizing the genes through genomics, mRNA by transcriptomics, metabolites via metabolomics, proteins with proteomics, and mineral nutrients by ionomics of the biological system in a comprehensive way (Razzaq et al., 2021). With the intention of improving the HMT problem in agriculture, we have compiled the main conclusions from the omics approaches below.

15.6.1 Metabolomics

15.6.1.1 Amino Acids

AAs are cellular communicators that play a vital role in regulating plant progression. They also help reduce HM toxicity by metal chelation in various plant parts. Histidine and proline, in particular, found binding metals within cells, as well as xylem (Wang et al., 2021a, b, c). Studies have shown that L-histidine plays a significant role in the impairment of nickel mobilization in the root cell sap of Ni-stressed *N. caerulescens*, highlighting its ability to scavenge HMs (Kozhevnikova et al., 2020). Studies in the laboratory show that one of the precursors of AA, aminocarboxylate (NA), effectively chelate metal ions. NAs play a vital role in the translocation of nutrients, maintaining plant nutritional status during HM toxicity (Rahale et al., 2021). Additionally, NA immobilizes certain metals such as Cu, Fe, Cd, Zn, and Mn through adsorption (Prasad et al., 2022).

15.6.1.2 Organic Acids

Organic acids such as oxalate, malate, and citrate, are known for their ability to transport metals in the xylem sap. They have also been found to play a role in sequestering vacuolar metals. Studies by Kosakivska et al. (2021) and Osmolovskaya et al. (2018) have shown that these organic acids can help in metal sequestration. Citrate, derived from citrate synthase, has a high inclination toward binding with cadmium, iron, and nickel in cell organelles, especially iron (Qureashi et al., 2021). On the contrary, malate binds with Zn, and it has the nature of maximum binding with zinc. Its primary function was to solubilize ions in the cytoplasm (Podar and Maathuis, 2022).

15.6.1.3 GSH and α-Tocopherol

A tripeptide thiol, glutathione (c-Glu–Cys–Gly), attracts water and has low molecular weight. It was very important in detoxifying HMs through its defense mechanism (Li et al., 2021; Nguyen et al., 2021; Souri et al., 2021). Glutathione regulates HM stress in plants by preventing the chelation of proteins with metals (Liu et al., 2017). Souri et al. (2020) reported a synergistic correlation between the expression of glutathione reductase and arsenic stress in *Isatiscappadocica*Desv shoots under metal stress. This emphasizes glutathione reductase's defensive activity on HM toxicity. Several researchers have noticed that α-tocopherol expression changes with various HM concentrations. Mishra et al. (2019)noticed an increase in the concentration of α-tocopherol with cadmium stress and also revealed that the upward trend in expressing genes is straightly linked to metabolite generation during HM stress.

15.6.1.4 Phenols

The production of phenol, a type of organic compound, is directly linked to its capability to bind metals when exposed to HMs. Many studies have shown that phenolic compounds present in different plants like *Vitis vinifera* (Zhao et al., 2021) and *Solanum lycopersicum* (Pappi et al., 2021), become activated in the presence of HMs. Tannin, a specific phenolic compound found in large quantities in *Camellia sinensis*, was more tolerant to manganese, and furthermore it helps protect tea plants from manganese stress by forming complexes and by being deposited in various plant organs (Farzadfar et al., 2017).

15.6.2 Ionomics

Ionome has depicted the importance of both micro- and macronutrients, in providing tolerance to plant HM stress. While excessive nutrients can be toxic, optimal usage can be defensive against HM stress (He et al., 2021; Imran et al., 2021). Kerdsomboon et al. (2022) found that 10 mM of phosphorus considerably improves low cadmium stress by enhancing the antioxidant mechanism in *Saccharomyces cerevisiae.* This is achieved by regulating the overproduction of ROS. Similarly, Huang et al. (2022) affirmed that adding $CaSO_4$ to pak choi can lessen cadmium stress by enhancing the uptake of sulfate, synthesis, and accumulation of PCs along with activating some sequestering agents in vacuoles.

15.6.3 Transcriptomics

Transcriptomics is a collection of methods that are used to study the transcription in living beings. Transcriptomics has proven to be most effective in increasing plant HM tolerance through the assessment of signaling cascade and gene ontology. A molecular study on transcriptome profiling of chickpea-subjected As, Cr, and Cd stress was conducted. The results showed that cellular metabolism was stimulated during HM stress and some genes played a significant role in the regulation of biochemical pathways through differential expression (Yadav et al., 2019).

Transcription factors (TFs) play a critical role in regulating gene expression in various plants, particularly under stress. Earlier investigations showed that an individual TF could control several key gene transcription through interaction with the *Cis*-acting region of the desired gene primers (Statello et al., 2021). When exposed to Cd stress, *A. thaliana* showed upregulation of two TFs, namely ERF1 and ERF5, belonging to the AP2/ERF superfamily, for Cd detoxification (Ramegowda et al., 2020). Identical upregulation of TFs was noticed in *A. halleri* under cadmium stress (Corso et al., 2018).

15.6.4 Proteomics

Another study in the recent past by Mehmood et al. (2021) has shed light on identifying protein targets concerned with the removal of HMs. The study found that proteomic modifications in xylem sap, accountable for the translocation of metal ions in addition to water and minerals from root to shoot, considerably eliminate cadmium ions. In the other study, *Brassica* seedlings exposed to 10 μM cadmium showed differential expression of proteins in the xylem sap that involves severe harm of biochemical reactions through interaction with various metabolic activities. The study also identified BnPDFL, a protective factor from the xylem, which acts as a moderator for cadmium chelation triggering cadmium tolerance (Luo and Zhang, 2019). Furthermore, a root proteomic study in maize using the iTRAQ technique by Wen et al. (2019) confirmed the usual modifications in the proteome. After exposing *Eucalyptus camaldulensis* seedlings to high concentrations of copper, proteomic analysis revealed that 11 proteins were upregulated, while 15 proteins were downregulated. The recognized proteins involved photosynthesis, antioxidative metabolism, translation, and transcription (Alotaibi et al., 2019).

15.6.5 Genomics

15.6.5.1 QTL Mapping

Several approaches were designed to fine-tune the mapping sequence of traits, such as quantitative trait locus mapping, QTL-sequencing, and RNAsequencing. These methods can quickly differentiate identical genes within primary QTLs. For example, in *Hordeum vulgare*, two QTLs for cadmium tolerance were identified. These QTLs were found to be correlated with Cd resistance, with 47.24% and 38.59% of the phenotypical variance, respectively. This study also identified 16 Cd resistance genes. Another study by HongYu et al. (2020) showed three QTLs for arsenic toxicity, one for copper and mercury toxicity, and two for Fe and Zn contents in *O. sativa* under HM stress.

15.6.5.2 Genome-Wide Association

Utilizing various molecular investigations, such as genome-wide association studies (GWAS), has proven to be an efficient method for leveraging elite alleles that regulate significant agronomic variables in rice germplasm. Additionally, GWAS has been utilized in a number of molecular screening experiments by researchers to find a significant amount of QTLs for HM uptake in rice (Zhang et al., 2023). Numerous molecular studies have looked into how crop species react to elevated levels of HM stress (Wu et al., 2021). Superior natural variations linked to HM resistance were found by an investigation of tolerance germplasm resources based on GWAS and transcriptome analysis (Chen et al., 2020). All the discovered QTLs offer a valuable resource for enhancing resistance to HM toxicity in current primary cultivars and offer insightful data for utilizing these QTLs/genes in subsequent breeding initiatives.

15.6.5.3 Gene Silencing

Using various gene silencing techniques, including RNA interference (RNAi), the clustered regularly interspaced short palindromic repeats (CRISPR)/Cas9 system (Zhao et al., 2018), and transgenics, new lines of rice with little HM deposition in paddy have been developed. By utilizing RNA interference (RNAi) to block the expression of the PCS gene OsPCS1, Li et al. (2007) were able to effectively limit the cumulation of cadmium ions in rice grains. Using CRISPR/Cas9 editing technology, two mutant lines pmei12-J3 and pmei12-J8 were created through modification of the *OsPMEI12* gene. The pmei12 line's resistance to cadmium stress was shown to be greatly enhanced.

Through genetic transformation, it is possible to produce plants that are resistant to HMs by modifying the metabolic pathways that underlie stress defense mechanisms. Tolerance is typically demonstrated at the cellular level by the activation of processes related to radical scavenging and DNA repair. Plant resistance to harmful substances can be further increased by genetically modifying these two characteristics. The most often used approach in research on HM stress is to increase the antioxidative metabolism of plants.

15.7 DIFFICULTIES IN HM BREEDING

Tolerance to HM toxicity is genetically more complex than tolerance to other abiotic stimuli, involving many polygenic components of translocation and signal transduction pathways. As such, it is comparatively harder to regulate and replicate. However, because of complex genetics and a lack of knowledge about their epigenetic behavior, the promise of genetic modification of HM-responsive genes was not been completely realized despite unexpected results (Singh et al., 2016). Fruitful breeding for the production of HM-tolerant plant is negatively impacted by a number of other factors, including environmental impacts and the important interrelationships among different HMs (Shephard et al., 2021).

15.8 CONCLUSION

One prospective solution to the HMT issue in agriculture is the development of crop varieties that are resistant to HMs because HMs can build up in edibles, potentially entering the bodies of humans and animals. For a variety of commercially significant food crops, a number of metal-resistant crops were created and released into the market. It might be possible to expedite the cultivar development process by utilizing omics like genomics, transcriptomics, proteomics, metabolomics, and ionomics and gene-editing techniques like CRISPR and others. Recently, crop species tolerant to HMs have progressed using the CRISPR technology. The process of identification and breeding cultivars for metal tolerance may be sped up by creating a high throughput, reasonably priced HM screening equipment. The development of effective real-time sensors for identifying HMs in soil, water,and plants must be the main focus of future research. Developing crop management strategies, cultivar development, gene target identification, and effective HM detection calls for an interdisciplinary team comprising agronomists, breeders, microbiologists, molecular biologists, and electrical engineers.

On the contrary, cultivating plants that are naturally tolerant to HM is more economical, environmentally benign, and sustainable. Therefore, to prevent a sharp fall in agricultural output owing to soil HM intensification, it is imperative to find novel genes and produce HM stress-tolerant crop species. It has been demonstrated thus far that tolerance to some HM is only limited and inadequate to diminish the potential effects of HMs on human health and food safety. To lessen the absorption and buildup of HM in the grains, efforts have also been made to improve crops' resistance to HM stress. The prime regulators that initiate the complete signaling network have yet to be fully identified, though.

The efficacy of HM detoxification has been enhanced recently by the application of more sophisticated genome engineering techniques including base editing, CRISPR/Cas9, CRISPR/Cas12, and others that alter different biochemical pathways. In the future, a coordinated use of molecular biology and conventional breeding methods may prove beneficial. Improved HM tolerance would reduce human food exposure while allowing metal-contaminated fields to be used continuously for crop production.

15.9 FUTURE PROSPECTS

(1) To improve food security, efforts should be made to build a standardized technological system for identifying cultivars that are secure from contamination or stable when it comes to accumulating metals. (2) To completely comprehend the influence of the environment on PSCs as well as the mechanism underlying the translocation of HMs in plants, more research is necessary. (3)

Safe farming in the field might be made a lot simpler with a framework based on biotechnology. (4) Research on the simultaneous use of contemporary and traditional breeding techniques is necessary because breeding programs will function more efficiently if molecular breeding and conventional breeding techniques are successfully integrated. Gene pyramiding, for instance, can be used to integrate several genes/QTLs into a single genotype using DNA marker assays. This is extremely difficult to accomplish through conventional breeding. The methods by which various amendments lessen metal toxicity in food crops need to be thoroughly studied. To stabilize soil HM levels, it is also necessary to locate and evaluate inexpensive, effective amendments.

REFERENCES

Abad, A. K. J., & Khara, J. (2007). Effect of cadmium toxicity on the level of lipid peroxidation and antioxidative enzymes activity in wheat plants colonized by arbuscular mycorrhizal fungi. Pakistan Jour*nal of Biological Sciences, 10*(14), 2413–2417.

Abdelmoneim, T. S., Moussa, T. A. A., Almaghrabi, O. A., & Abdelbagi, I. (2014). Investigation the effect of arbuscular mycorrhizal fungi on the tolerance of maize plant to heavy metals stress. *Life Science Journal, 11*(4), 255–263.

Ahmad, P., Raja, V., Ashraf, M., Wijaya, L., Bajguz, A., &Alyemeni, M. N. (2021). Jasmonic acid (JA) and gibberellic acid (GA3) mitigated Cd-toxicity in chickpea plants through restricted cd uptake and oxidative stress management. *Scientific Reports, 11*(1), 19768.

Alotaibi, M. O., Mohammed, A. E., Almutairi, T. A., & Elobeid, M. M. (2019). Morphophysiological and proteomic analyses of Eucalyptus camaldulensis as a bioremediator in copper-polluted soil in Saudi Arabia. *Plants, 8*(2), 43.

Aloui, A., Recorbet, G., Robert, F., Schoefs, B., Bertrand, M., Henry, C., Gianinazzi-Pearson, V., Dumas-Gaudot, E., & Aschi-Smiti, S. (2011). Arbuscular mycorrhizal symbiosis elicits shoot proteome changes that are modified during cadmium stress alleviate on in Medicago truncatula. *BMC Plant Biology, 11*, 75.

Alvarez, S., Berla, M. B., Sheffield, J., Cahoon, R. E., Jez, J. M., & Hicks, L. M. (2009). Comprehensive analysis of the *Brassica juncea* root proteome in response to cadmium exposure by complementary proteomic approaches. *Proteomics, 9*, 2419–2431.

Ban, Q., Liu, G., & Wang, Y. (2011). A DREB gene from Limonium bicolor mediates molecular and physiological responses to copper stress in transgenic tobacco. *Journal of Plant Physiology, 168*, 449–458.

Bano, S. A., & Ashfaq, D. (2013). Role of mycorrhiza to reduce heavy metal stress. *Natural Science, 5*(12), 16–20.

Batista, B. L., Nigar, M., Mestrot, A., Rocha, B. A., Júnior, F. B., Price, A. H., Raab, A., & Feldmann, J. (2014). Identification and quantification of phytochelatins in roots of rice to long-term exposure: Evidence of individual role on arsenic accumulation and translocation. The Jour*nal of Experimental Botany, 65*(6), 1467–1479.

Bhalerao, S. A. (2013). Arbuscular mycorrhizal fungi: A potential biotechnology tool for phytoremediation of heavy metal contaminated soils. *International Journal of Science and Nature, 4*(1), 1–15.

Brunetti, P., Zanella, L., Proia, A., Paolis, A. D., Falasca, G., Altamura, M. M., Toppi, L. S. D., Costantino, P., & Cardarelli, M. (2011). Cadmium tolerance and phytochelatin content of Arabidopsis seedlings over-expressing the phytochelatin synthase gene AtPCS1. *Journal of Experimental Botany, 62*(15), 5509–5519.

Cao, F., Chen, F., Sun, H., Zhang, G., Chen, Z. H., & Wu, F. (2014). Genome-wide transcriptome and functional analysis of two contrasting genotypes reveals key genes for cadmium tolerance in barley. *BMC Genomics, 15*, 1–14.

Charfeddine, M., Charfeddine, S., Bouaziz, D., Messaoud, R. B., & Bouzid, R. G. (2017). The effect of cadmium on transgenic potato (*Solanumtuberosum*) plants overexpressing the StDREB transcription factors. *Plant Cell Tissue and Organ Culture, 128*(3), 521–525.

Chen, J., Zou, W., Meng, L., Fan, X., Xu, G., & Ye, G. (2019). Advances in the uptake and transport mechanisms and QTLs mapping of cadmium in rice. *International Journal of Molecular Sciences, 20*, 3417.

Chen, L., Guo, Y., Yang, L., & Wang, Q. (2008). Synergistic defensive mechanism of phytochelatins and antioxidative enzymes in Brassica chinensis L. against Cd stress. *Chinese Science Bulletin, 53*(10), 1503–1511.

Chen, X.-X., Liu, Y.-M., Zhao, Q.-Y., Cao, W.-Q., Chen, X.-P., & Zou, C.-Q. (2020). Health risk assessment associated with heavy metal accumulation in wheat after long-term phosphorus fertilizer application. *Environmental Pollution, 262*, 114348.

Çikili, Y., Samet, H., & Çiçek, N. (2020). Response of sunflower (Helianthus annuus L.) at early growth stage to cadmium exposure. *Türk TarımveDoğaBilimleriDergisi*, *7*(4), 1098–1107.

Clarke, J. M., McCaig, T. N., DePauw, R. M., Knox, R. E., Clarke, F. R., Fernandez, M. R., & Ames, N. P. (2005). Strongfield durum wheat. *Canadian Journal of Plant Science*, *85*, 651–654.

Cobbett, C. S. (2000). Phytochelatins and their roles in heavy metal detoxification. *Plant Physiology*, *123*(3), 825–832.

Corso, M., Schvartzman, M. S., Guzzo, F., Souard, F., Malkowski, E., Hanikenne, M., & Verbruggen, N. (2018). Contrasting cadmium resistance strategies in two metallicolous populations of *Arabidopsis halleri*. *The New Phytologist, 218*(1), 283–297.

Dasgupta, T., Hossain, S. A., Meharg, A. A., & Price, A. H. (2004). An arsenate tolerance gene on chromosome 6 of rice. *The New Phytologist, 163*, 45–49.

di Lonardo, S., Capuana, M., Arnetoli, M., Gabbrielli, R., & Gonnelli, C. (2011). Exploring the metal phytoremediation potential of three Populus alba L. clones using an in vitro screening. *Environmental Science and Pollution Research International, 18*, 82–90.

Ejaz, U., Khan, S. M., Khalid, N., Ahmad, Z., Jehangir, S., Fatima Rizvi, Z., Lho, L. H., Han, H., & Raposo, A. (2023). Detoxifying the heavy metals: A multipronged study of tolerance strategies against heavy metals toxicity in plants. *Frontiers in Plant Science*, *14*, 1154571.

Faè, M., Balestrazzi, A., Confalonieri, M., Donà, M., Macovei, A., Valassi, A., & Carbonera, D. (2014). Copper-mediated genotoxic stress is attenuated by the overexpression of the DNA repair gene MtTdp2α (tyrosyl-DNA phosphodiesterase 2) in *Medicago truncatula* plants. *Plant Cell Reports*, *33*(7), 1071–1080.

Farokhzadeh, S., Fakheri, B. A., Nezhad, N. M., Tahmasebi, S., & Mirsoleimani, A. (2019). Mapping QTLs of flag leaf morphological and physiological traits related to aluminum tolerance in wheat (*Triticum aestivum* L.). *Physiology and Molecular Biology of Plants*, *25*, 975.

Farshian, S., Khara, J., & Malekzadeh, P. (2007). Influence of arbuscular mycorrhizal fungus (*Glomus etunicatum*) with lettuce plants under zinc toxicity in nutrient solution. *Pakistan Journal of Biological Sciences*, *10*(14), 2363–2367.

Farzadfar, S., Zarinkamar, F., & Hojati, M. (2017). Magnesium and manganese affect photosynthesis, essential oil composition and phenolic compounds of Tanacetum parthenium. *Plant Physiology and Biochemistry*, *112*, 207–217.

Fidalgo, F., Azenha, M., Silva, A. F., de Sousa, A., Santiago, A., Ferraz, P., & Teixeira, J. (2013). Copper-induced stress in *Solanum nigrum* L. and antioxidant defense system response. *Food and Energy Security*, *2*(1), 70–80.

Filiz, E., Saracoglu, I. A., Ozyigit, I. I., & Yalcin, B. (2019). Comparative analyses of phytochelatin synthase (PCS) genes in higher plants. *Biotechnology & Biotechnological Equipment*, *33*(1), 178–194.

Gao, X., Ai, W. L., Gong, H., Cui, L. J., Chen, B. X., Luo, H. Y., & Qiu, B. S. (2016). Transgenic NfFeSOD *Sedumal fredii* plants exhibited profound growth impairments and better relative tolerance to long-term abiotic stresses. *Plant Biotechnology Reports*, *10*(2), 117–128.

Garg, N., & Aggarwal, N. (2012). Effect of mycorrhizal inoculations on heavy metal uptake and stress alleviation of Cajanus cajan (L.) Millsp. genotypes grown in cadmium and lead contaminated soils. *Plant Growth Regulation*, *66*(1), 9–26.

Ghori, N. H., Ghori, T., Hayat, M. Q., Imadi, S. R., Gul, A., Altay, V., & Ozturk, M. (2019). Heavy metal stress and responses in plants. *International Journal of Environmental Science and Technology*, *16*, 1807–1828.

Greger, M. (2004). Metal availability, uptake, transport and accumulation in plants. In M. N. V. Prasad (Ed.), *Heavy metal stress in plants* (pp. 1–27). Springer.

Gregory, R. P. G., & Bradshaw, A. D. (1965). Heavy metal tolerance in populations of *agrostis tenuis* sibth. and other grasses. *The New Phytologist, 64*, 131–143.

Grennan, A. K. (2011). Metallothioneins, a diverse protein family. *Plant Physiology*, *155*(4), 1750–1751.

Guan, Z., Chai, T., Zhang, Y., Xu, J., & Wei, W. (2009). Enhancement of Cd tolerance in transgenic tobacco plants overexpressing a Cd-induced catalase cDNA. *Chemosphere*, *76*(5), 623–630.

Guo, J., Xu, W., & Ma, M. (2012). The assembly of metals chelation by thiols and vacuolar compartmentalization conferred increased tolerance to and accumulation of cadmium and arsenic in transgenic *Arabidopsis thaliana*. *Journal of Hazardous Materials*, *199–200*, 309–313.

Guo, J. L., Xu, L. P., Su, Y. C., Wang, H., Gao, S., Xu, J., & Que, Y. (2013). ScMT2-1-3, a metallothionein gene of sugarcane, plays an important role in the regulation of heavy metal tolerance/accumulation. *BioMed Research International*, *904769*, 12.

Gupta, D., Vandenhove, H., & Inouhe, M. (2013). Role of phytochelatins in heavy metal stress and detoxification mechanisms in plants. In M. N. V. Prasad (Ed.), *Heavy metal stress in plants* (pp. 73–94). Springer.

Halim, M. A., Ghosh, M., Nigar, M., Hossain, F., & Akhter, N. (2016). Screening of arsenic tolerance in rice at germination and early seedling stage as influenced by sodium arsenate. *Jahangirnagar University Journal of Biological Sciences*, *3*, 17–26.

Handique, G. K., & Handique, A. K. (2009). Proline accumulation in lemongrass (*Cymbopogon flexuosus* Stapf.) due to heavy metal stress. *Journal of Environmental Biology*, *30*(2), 299–302.

Hauser, M. T. (2014). Molecular basis of natural variation and environmental control of trichome patterning. *Frontiers in Plant Science*, *5*(320), 1–7.

He, F., Zhao, Q., Huang, J.-L., Niu, M.-X., Feng, H.-C., Shi, Y.-J., Zhao, K.-J., Cui, X.-L., Wu, X.-L., Mi, J.-X., Zhong, Y., Liu, Q.-L., Chen, L.-H., Wan, X.-Q., & Zhang, F. (2021). External application of nitrogen alleviates toxicity of cadmium on poplars via starch and sucrose metabolism. *Tree Physiology*, *41*(11), 2126–2141.

Heiss, S., Wachter, A., Bogs, J., Cobbett, C., & Rausch, T. (2003). Phytochelatin synthase (PCS) protein is induced in Brassica juncea leaves after prolonged Cd exposure. *Journal of Experimental Botany*, *54*(389), 1833–1839.

HongYu, Z., Sheng, W., Zhang, Y., Zhu, H, Zhu, X., Zhang, Y., Wu, X., Rao, Y., Wang, Y., & Lin, H. (2020). QTL excavation and analysis of candidate genes in contents of As, Cu, Fe, Hg and Zn in rice grain. *Scientia Sinica Vitae*, *50*(6), 623–632.

Howladar, S. M., Al-Robai, S. A., Al-Zahrani, F. S., Howladar, M. M., &Aldhebiani, A. Y. (2018). Silicon and its application method effects on modulation of cadmium stress responses in Triticum aestivum (L.) through improving the antioxidative defense system and polyamine gene expression. *Ecotoxicology and environmental safety*, *159*, 143–152.

Huang, G. Y., & Wang, Y. S. (2009). Expression analysis of type 2 metallothionein gene in mangrove species (*Bruguiera gymnorrhiza*) under heavy metal stress. *Chemosphere*, *77*(7), 1026–1029.

Huang, Y., Chen, J., Sun, Y., Wang, H., Zhan, J., Huang, Y., Zou, J., Wang, L., Su, N., & Cui, J. (2022). Mechanisms of calcium sulfate in alleviating cadmium toxicity and accumulation in pak choi seedlings. *The Science of the Total Environment*, *805*, 150115.

Huang, Y., Li, D., Zhao, L., Chen, A., Li, J., Tang, H., Pan, G., Chang, L., Deng, Y., & Huang, S. (2019). Comparative transcriptome combined with physiological analyses revealed key factors for differential cadmium tolerance in two contrasting hemp (*Cannabis sativa* L.) cultivars. *Industrial Crops and Products*, *140*, 111638.

Iannone, M. F., Groppa, M. D., & Benavides, M. P. (2015). Cadmium induces different biochemical responses in wild type and catalase-deficient tobacco plants. *Environmental and Experimental Botany*, *109*, 201–211.

Imran, M., Hussain, S., Rana, M. S., Saleem, M. H., Rasul, F., Ali, K. H., Potcho, M. P., Pan, S., Duan, M., & Tang, X. (2021). Molybdenum improves 2-acetyl-1-pyrroline, grain quality traits and yield attributes in fragrant rice through efficient nitrogen assimilation under cadmium toxicity. *Ecotoxicology and Environmental Safety*, *211*, 111911.

Ishikawa, S., Ishimaru, Y., Igura, M., Kuramata, M., Abe, T., Senoura, T., Hase, Y., Arao, T., Nishizawa, N. K., & Nakanishi, H. (2012). Ion-beam irradiation, gene identification, and marker-assisted breeding in the development of low-cadmium rice. *Proceedings of the National Academy of Sciences of the United States of America*, *109*, 19166.

Jalmi, S. K., Bhagat, P. K., Verma, D., Noryang, S., Tayyeba, S., Singh, K., & Sinha, A. K. (2018). Traversing the links between heavy metal stress and plant signaling. *Frontiers in Plant Science*, *9*, 12.

Jia, H., Wang, X., Wei, T., Wang, M., Liu, X., Hua, L., Ren, X., Guo, J., & Li, J. (2021). Exogenous salicylic acid regulates cell wall polysaccharides synthesis and pectin methylation to reduce Cd accumulation of tomato. *Ecotoxicology and Environmental Safety*, *207*, 111550.

Kerdsomboon, K., Techo, T., Limcharoensuk, T., Tatip, S., & Auesukaree, C. (2022). Low phosphate mitigates cadmium-induced oxidative stress in Saccharomyces cerevisiae by enhancing endogenous antioxidant defence system. *Environmental Microbiology*, *24*(2), 707–720.

Khalid, M., Nan, H. U. I., Kayani, S. I., & Kexuan, T. A. N. G. (2020). Diversity and versatile functions of metallothioneins produced by plants: A review. *Pedosphere*, *30*(5), 577–588.

Khan, J., Yang, Y., Fu, Q., Shao, W., Wang, J., Shen, L., Huai, Y., Malangisha, G. K., Ali, A., Mahmoud, A., Lin, Y., Ren, Y., Yang, J., Hu, Z., & Zhang, M. (2020). Screening of watermelon varieties for lead tolerance at the seedling stage. *Hortscience*, *55*, 858–869.

Kohler, A., Blaudez, D., Chalot, M., & Martin, F. (2004). Cloning and expression of multiple metallothioneins from hybrid poplar. New Phytologist, *164*(1), 83–93.

Kosakivska, I. V., Babenko, L. M., Romanenko, K. O., Korotka, I. Y., & Potters, G. (2021). Molecular mechanisms of plant adaptive responses to heavy metals stress. *Cell Biology International*, *45*(2), 258–272.

Koul, M., Thomas, L. & Karmakar, K. (2021), Functional aspects of solanaceae trichomes in heavy metal detoxification. *Nordic Journal of Botany, 39*(5).

Kozhevnikova, A., Seregin, I., & Schat, H. (2020). Accumulation of nickel by excluder Thlaspiarvense and hyperaccumulator *Noccaea caerulescens* upon short-term and long-term exposure. *Russian Journal of Plant Physiology, 67*(2), 303–311.

Kulaeva, O. A., & Tsyganov, V. E. (2011). Molecular-genetic basis of cadmium tolerance and accumulation in higher plants. *Russian Journal of Genetics: Applied Research, 1*, 349–360.

Kumagai, S., Suzuki, T., Tezuka, K., Satoh-Nagasawa, N., Takahashi, H., Sakurai, K., Watanabe, A., Fujimura, T., & Akagi, H. (2014). Functional analysis of the C-terminal region of the vacuolar cadmium-transporting rice OsHMA3. *FEBS Letters, 588*, 789–794.

Kumar, G., Kushwaha, H. R., Panjabi-Sabharwal, V., Kumari, S., Joshi, R., Karan, R., Mittal, S., Pareek, S. L. S., & Pareek, A. (2012). Clustered metallothionein genes are co-regulated in rice and ectopic expression of OsMT1e-P confers multiple abiotic stress tolerance in tobacco via ROS scavenging. *BMC Plant Biology, 12*, 107.

Küpper, H., & Andresen, E. (2016). Mechanisms of metal toxicity in plants. *Metallomics, 8*(3), 269–285.

Lee, S. H., Ahsan, N., Lee, K. W., Kim, D. H., Lee, D. G., Kwak, S. S., & Lee, B. H. (2007). Simultaneous overexpression of both Cu Zn superoxide dismutase and ascorbate peroxidase in transgenic tall fescue plants confers increased tolerance to a wide range of abiotic stresses. *Journal of Plant Physiology, 164*(12), 1626–1638.

Lesko, K., & Simon-Sarkadi, L. (2002). Effect of cadmium stress on amino acid and polyamine content of wheat seedlings. *Periodica Polytechnica: Chemical Engineering, 46* (1–2), 65–71.

Li, J. C., Guo, J. B., Xu, W. Z., & Ma, M. (2007). RNA interference mediated silencing of phytochelatin synthase gene reduce cadmium accumulation in rice seeds. *Journal of Integrative Plant Biology, 49*(7), 1032–1037.

Li, B., Wang, J., Yao, L., Meng, Y., Ma, X., Si, E., Ren, P., Yang, K., Shang, X., & Wang, H. (2019). Halophyte *Halogeton glomeratus*, a promising candidate for phytoremediation of heavy metal-contaminated saline soils. *Plant and Soil, 442*, 323–331.

Li, G.-Z., Chen, S.-J., Li, N.-Y., Wang, Y.-Y., & Kang, G.-Z. (2021). Exogenous glutathione alleviates cadmium toxicity in wheat by influencing the absorption and translocation of cadmium. *Bulletion of Environmental Contamination and Toxicology, 107*(2), 320–326.

Lin, S., Yang, T., Zhu, S., Wang, J., & Ni, W. (2017). A method for screening copper-tolerant rice (*Oryza sativa* L.) cultivars based on hydroponic experiments and cluster analysis. *International Journal of Phytoremediation, 19*, 1093–1099.

Liu, J., Zhang, J., Kim, S. H., Lee, H. S., Marinoia, E., & Song, W. Y. (2021). Characterization of Brassica rapa metallothionein and phytochelatin synthase genes potentially involved in heavy metal detoxification. *PLoS One, 16*(6), e0252899.

Liu, X., Zhang, H., Jiao, H., Li, L., Qiao, X., Fabrice, M. R., Wu, J., & Zhang, S. (2017). Expansion and evolutionary patterns of cysteine-rich peptides in plants. *BMC Genomics, 18*(1), 1–14.

Lombnaes, P., & Singh, B. R. (2003). Varietal tolerance to zinc deficiency in wheat and barley grown in chelator-buffered nutrient solution and its effect on uptake of Cu, Fe, and Mn. *Journal of Plant Nutrition and Soil Science, 166*, 76–83.

Luo, J.-S., & Zhang, Z. (2019). Proteomic changes in the xylem sap of *Brassica napus* under cadmium stress and functional validation. *BMC Plant Biology, 19*(1), 1–14.

Macnair, M. R. (1983). The genetic control of copper tolerance in the yellow monkey flower, Mimulus guttatus. *Heredity, 50*, 283–293.

Mehmood, S. S., Lu, G., Luo, D., Hussain, M. A., Raza, A., Zafar, Z., Zhang, X., Cheng, Y., Zou, X., & Lv, Y. (2021). Integrated analysis of transcriptomics and proteomics provides insights into the molecular regulation of cold response in *Brassica napus*. *Environmental and Experimnental Botany, 187*, 104480.

Meng, Y., Zhang, L., Wang, L., Zhou, C., Shangguan, Y., & Yang, Y. (2019). Antioxidative enzymes activity and thiol metabolism in three leafy vegetables under Cd stress. *Ecotoxicology and Environmental Safety, 173*, 214–224.

Mishra, B., Chand, S., & Sangwan, N. S. (2019). ROS management is mediated by ascorbateglutathione-α-t ocopherol triad in co-ordination with secondary metabolic pathway under cadmium stress in *Withania somnifera*. *Plant Physiology and Biochemistry, 139*, 620–629.

Mishra, N., Srivastava, A. P., Esmaeili, N., Hu, W., & Shen, G. (2018). Overexpression of the rice gene OsSIZ1 in Arabidopsis improves drought-, heat-, and salt-tolerance simultaneously. *PLoS One, 13*, e0201716.

Mishra, S., & Dubey, R. S. (2006). Heavy metal uptake and detoxification mechanisms in plants. *International Journal of Agricultural Research, 1*(2), 122–141.

Mittler, R. (2002). Oxidative stress, antioxidants and stress tolerance. *Trends in Plant Science*, *7*, 405–410.

Mohammadi, K., Khalesro, S., Sohrabi, Y., & Heidari, G. (2011). A review: Beneficial effects of the mycorrhizal fungi for plant growth. *Journal of Applied Environmental and Biological Sciences*, *1*(9), 310–319.

Mourad, A. M. I., Amin, A. E. E. A. Z., & Dawood, M. F. A. (2021). Genetic variation in kernel traits under lead and tin stresses in spring wheat diverse collection. *Environmenal and Experimental Botany*, *192*, 104646.

Nguyen, T. Q., Sesin, V., Kisiala, A., & Emery, R. N. (2021). Phytohormonal roles in plant responses to heavy metal stress: Implications for using macrophytes in phytoremediation of aquatic ecosystems. *Environmental Toxicology and Chemistry*, *40*(1), 7–22.

Nugraha, Y., Utami, D. W., Rosdianti, I., Ardie, S. W., Ghulammahdi, M., Suwarno, S., & Aswidinnoor, H. (2016). Markers-traits association for iron toxicity tolerance in selected Indonesian rice varieties. *Biodiversitas*, *17*, 753–763.

Osmolovskaya, N., Dung, V. V., & Kuchaeva, L. (2018). The role of organic acids in heavy metal tolerance in plants. *Biology Communications*, *63*(1), 9–16.

Ovečka, M., & Takáč, T. (2014). Managing heavy metal toxicity stress in plants: Biological and biotechnological tools. *Biotechnology Advances*, *32*, 73–86.

Pagani, M. A., Tomas, M., Carrillo, J., Bofill, R., Capdevila, M., Atrian, S., & Andreo, C. S. (2012). The response of the different soybean metallothionein isoforms to cadmium intoxication. *Journal of Inorganic Biochemistry*, *117*, 306–315.

Pappi, P., Nikoloudakis, N., Fanourakis, D., Zambounis, A., Delis, C., & Tsaniklidis, G. (2021). Differential triggering of the phenylpropanoid biosynthetic pathway key genes transcription upon cold stress and viral infection in tomato leaves. *Horticulturae*, *7*(11), 448.

Patil, V. D., Nerkar, Y. S., Misal, M. B., & Harkal, S. R. (1997). Parag 401, a semidwarf rice variety developed through anther culture. *International Rice Research Notes*, *22*, 1.

Piechalak, A., Tomaszewska, B., Baralkiewicz, D., & Malecka, A.(2002). Accumulation and detoxification of lead ions in legumes. *Phytochemistry*, *60*(2), 153–162.

Podar, D., & Maathuis, F. J. (2022). The role of roots and rhizosphere in providing tolerance to toxic metals and metalloids. *Plant, Cell & Environment*, *45*(3), 719–736.

Postrigan, B. N., Knyazev, A. B., Kuluev, B. R., Yakhin, O. I., & Chemeris, A. V. (2012). Expression of the synthetic phytochelatin gene in tobacco. *Russian Journal of Plant Physiology*, *59*(2), 275–280.

Prasad, R., Shivay, Y. S., & Mandi, S. (2022). Phytosiderophores and absorption of iron and other cations by plants. In S. K. Upadhyay (Ed.), *Cation transporters in plants* (pp. 385–399). Academic Press.

Qi, D.-L., Lee, J.-R., Yang, C.-G., Lee, M.-C., Cao, G.-L., Zhang, J.-G., Zhou, Q.-Y., Suh, S.-C., Zhang, S.-Y., & Han, L.-Z. (2009). Detection of QTL for alkali tolerance at the germination stage in japonica rice. *Chinese Journal of Rice Science*, *23*, 589.

Qureashi, A., Pandith, A. H., Bashir, A., Manzoor, T., Malik, L. A., & Sheikh, F. A. (2021). Citrate coated magnetite: A complete magneto dielectric, electrochemical and DFT study for detection and removal of heavy metal ions. *Surfaces and Interfaces*, *23*, 101004.

Rahale, C. S., Lakshmanan, A., Sumithra, M., & Kumar, E. R. (2021). Humic acid involved chelation of ZnO nanoparticles for enhancing mineral nutrition in plants. *Solid State Communications*, *333*, 114355.

Ramegowda, V., Da Costa, M. V. J., Harihar, S., Karaba, N. N., & Sreeman, S. M. (2020). Abiotic and biotic stress interactions in plants: A cross-tolerance perspective. In M. A. Hossain, F. Liu, D. J. Burritt, M. Fujita, & B. Huang (Eds.), *Priming-mediated stress and cross-stress tolerance in crop plants* (pp. 267–302). Academic Press

Rascio, N., & Navari-Izzo, F. (2011). Heavy metal hyperaccumulating plants: How and why do they do it? And what makes them so interesting? *Plant Science*, *180*(2), 169–181.

Rauser, W. E. (1995). Phytochelatins and related peptides. Structure, biosynthesis, and function. *Plant Physiology*, *109*(4), 1141–1149.

Raza, A., Hussain, S., Javed, R., Hafeez, M. B., & Hasanuzzaman, M. (2021). Antioxidant defense systems and remediation of metal toxicity in plants. In M. Hasanuzzaman (Ed.), *Approaches to the remediation of inorganic pollutants* (pp. 91–124). Springer.

Razzaq, M. K., Aleem, M., Mansoor, S., Khan, M. A., Rauf, S., Iqbal, S., & Siddique, K. H. (2021). Omics and CRISPR-Cas9 approaches for molecular insight, functional gene analysis, and stress tolerance development in crops. *International Journal of Molecular Sciences*, *22*(3), 1292.

Ruscitti, M., Arango, M., Ronco, M., & Beltrano, J. (2011). Inoculation with mycorrhizal fungi modifies praline metabolism and increases chromium tolerance in pepper plants (*Capsicum annuum* L.). *Brazilian Journal of Plant Physiology*, *23*(1), 15–25.

Salarizadeh, S., & Kavousi, H. R. (2015). Application of random amplified polymorphic DNA (RAPD) to detect the genotoxic effect of cadmium on tow Iranian ecotypes of cumin (*Cuminum cyminum*). *Journal of Cell and Molecular Research*, *7*, 38–46.

Sameena, P. P., & Puthur, J. T. (2021). Heavy metal phytoremediation by bioenergy plants and associated tolerance mechanisms. *Soil and Sediment Contamination: An International Journal*, *30*(3), 253–274.

Savenstrand, H., & Strid, Å. (2004). Six genes strongly regulated by mercury in *Pisum sativum* roots. *Plant Physiology and Biochemistry*, *42*, 135–142.

Shafiq, S., Zeb, Q., Ali, A., Sajjad, Y., Nazir, R., Widemann, E., & Liu, L. (2019). Lead, cadmium and zinc phytotoxicity alter DNA methylation levels to confer heavy metal tolerance in wheat. *International Journal of Molecular Science*, *20*(19), 4676.

Shahid, M., Dumat, C., Khalid, S., Schreck, E., Xiong, T., & Niazi, N. K. (2017). Foliar heavy metal uptake, toxicity and detoxification in plants: A comparison of foliar and root metal uptake. *Journal of Hazardous Materials*, *325*, 36–58.

Shen, G. M., Zhu, C., & Du, Q. Z. (2010). Genome-wide identification of PHYTOCHELATIN and PHYTOCH SYNTH domaincontaining phytochelatin family from rice. *Electronic Journal of Biology*, *6*, 373–379.

Shen, J., Lv, B., Luo, L., He, J., Mao, C., Xi, D., & Ming, F. (2017). The NAC-type transcription factor OsNAC2 regulates ABA-dependent genes and abiotic stress tolerance in rice. *Scientific Reports*, *7*, 40641.

Shephard, A. M., Zambre, A. M., & Snell-Rood, E. C. (2021). Evaluating costs of heavy metal tolerance in a widely distributed, invasive butterfly. *Evolutionary Applications*, *14*, 1390–1402.

Shilin, D., Chaolei, L., Lianguang, S., Shenglong, Y., Anpeng, Z., Hongzhen, J., Banpu, R., Guonan, F., Biao, T., Guoyou, Y., Longbiao, G., Qian, Q., & Zhenyu, G. (2021). Identification of QTLs for cadmium tolerance during seedling stage and validation of qCDSL1 in rice. Rice Science, *28*, 81–88.

Shukla, D., Tiwari, M., Tripathi, R. D., Nath, P., & Trivedi, P. K. (2013). Synthetic phytochelatins complement a phytochelatindeficient *Arabidopsis* mutant and enhance the accumulation of heavy metal(loid)s. *Biochemical and Biophysical Research Communications*, *434*(3), 664–669.

Singh, S., Parihar, P., Singh, R., Singh, V. P., & Prasad, S. M. (2016). Heavy metal tolerance in plants: Role of transcriptomics, proteomics, metabolomics, and ionomics. *Frontiers in Plant Science*, *6*, 1143.

Smykalova, I., Vrbova, M.,Tejklova, E., Vetrovcova, M., & Griga, M. (2010). Large scale screening of heavy metal tolerance in flax/linseed (*Linum usitatissimum* L.) tested in vitro. *Industrial Crops & Products*, *32*, 527–533.

Song, W.-Y., Mendoza-Cozatl, D. G., Lee, Y., Schroeder, J. I., Ahn, S.-N., Lee, H.-S., Wicker, T., & Martinoia, E. (2014). Phytochelatin-metal(loid) transport into vacuoles shows different substrate preferences in barley and Arabidopsis. *Plant, Cell and Environment*, *37*(5), 1192–1201.

Souri, Z., Karimi, N., & Ahmad, P. (2021). The effect of NADPH oxidase inhibitor diphenyleneiodonium (DPI) and glutathione (GSH) on *Isatis cappadocica*, under Arsenic (As) toxicity. *International Journal of Phytoremediation*, *23*(9), 945–957.

Srivastava, S., Shrivastava, M., Suprasanna, P., & D'souza, S. F. (2011). Phytofiltration of arsenic from simulated contaminated water using *Hydrilla verticillata* in field conditions. *Ecological Engineering*, *37*(11), 1937–1941.

Statello, L., Guo, C.-J., Chen, L.-L., & Huarte, M. (2021). Gene regulation by long non-coding RNAs and its biological functions. *Nature Reviews Molecular Cell Biology*, *22*(2), 96–118.

Szabados, L., & Savoure, A. (2010). Proline: A multifunctional amino acid. *Trends in Plant Science*, *15*(2), 89–97.

Szalai, G., Krantev, A., Yordanova R., Popova, L. P., & Janda, T. (2013). Influence of salicylic acid on phytochelatin synthesis in Zea mays during Cd stress. *Turkish Journal of Botany*, *37*(4), 708–714.

Takacs, T., Biró, B., & Vörös, I. (2001). Arbuscular mycorrhizal effect on heavy metal uptake of ryegrass (*Lolium perenne* L.) in pot culture with polluted soil. In W. W. J. Horst, M. K. Scheck, A. Burkert, & N. Classsen (Eds.), *Plant nutrition: Food security and sustainability of agro-ecosystems through basic and applied research, developments in plant and soil sciences* (pp. 480–481). Kluwer Academic Publishers.

Tang, L., Dong, J., Tan, L., Ji, Z., Li, Y., Sun, Y., Chen, C., Lv, Q., Mao, B., & Hu, Y. (2021). Overexpression of OsLCT2, a low-affinity cation transporter gene, reduces cadmium accumulation in shoots and grains of rice. *Rice*, *14*, 1–15.

Thangavel, P., Long, S., & Minocha, R. (2007). Changes in phytochelatins and their biosynthetic intermediates in red sp (Picea rubens Sarg.) cell suspension cultures under cadmium and zinc stress. *Plant Cell, Tissue and Organ Culture*, *88*(2), 201–216.

Upadhyaya, H., Panda, S. K.,Bhattacharjee, M. K., & Dutta, S. (2010). Role of arbuscular mycorrhiza in heavy metal tolerance in plants: Prospects for phytoremediation. *Journal of Phytology*, *2*(7), 16–27.

Verkleij, J. A. C., Sneller, F. E. C., & Schat, H. (2003). Metallothioneins and phytochelatins: Ecophysiological aspects. In Y. P. Abrol & A. Ahmad (Eds.), *Sulphur in plants* (pp. 163–176). Springer.

Wang, H.-R., Che, Y.-H., Wang, Z.-H., Zhang, B.-N., Huang, D., Feng, F., & Ao, H. (2021c). The multiple effects of hydrogen sulfide on cadmium toxicity in tobacco may be interacted with CaM signal transduction. *Journal of Hazardous Materials*, *403*, 123651.

Wang, S., He, T., Xu, F., Li, X., Yuan, L., Wang, Q., & Liu, H. (2021b). Analysis of physiological and metabolite response of *Celosia argentea* to copper stress. *Plant Biology*, *23*(2), 391–399.

Wang, S., Huang, D.-Y., Zhu, Q.-H., Li, B.-Z., Xu, C., Zhu, H.-H., & Zhang, Q. (2021c). Agronomic traits and ionomics influence on Cd accumulation in various sorghum (*Sorghum bicolor* (L.) Moench) genotypes. *Ecotoxicology and Environmental Safety*, *214*, 112019.

Wei, W., Peng, H., Xie, Y., Wang, X., Huang, R., Chen, H., & Ji, X. (2021). The role of silicon in cadmium alleviation by rice root cell wall retention and vacuole compartmentalization under different durations of Cd exposure. *Ecotoxicology and Environmental Safety*, *226*, 112810.

Wen, R. E. N., Ya, L. I. U., Zhou, M. Y., Zi, S. H. I., Wang, T. Y., Zhao, J. R., & Yu, L. I. (2019). Dynamic changes of root proteome reveal diverse responsive proteins in maize subjected to cadmium stress. *Journal of Integrative Agriculture*, *18*(10), 2193–2204.

Wu, D., He, G., Tian, W., Saleem, M., Li, D., Huang, Y. & He, T. (2021). OPT gene family analysis of potato (Solanum tuberosum) responding to heavy metal stress: Comparative omics and co-expression networks revealed the underlying core templates and specific response patterns. *International Journal of Biological Macromolecules*, *188*, 892–903.

Yadav, B. S., Singh, S., Srivastava, S., Singh, N. K., & Mani, A. (2019). Whole transcriptome expression profiling and biological network analysis of chickpea during heavy metal stress. *Journal of Plant Biochemistry and Biotechnology*, *28*(3), 345–352.

Yang, Z., Wu, Y., Li, Y., Ling, H.-Q., & Chu, C. (2009). OsMT1a, a type 1 metallothionein, plays the pivotal role in zinc homeostasis and drought tolerance in rice. *Plant Molecular Biology*, *70*(1–2), 219–229.

Zagorchev, L., Seal, C. E., Kranner, I., & Odjakova, M. (2013). A central role for thiols in plant tolerance to abiotic stress. *International Journal of Molecular Sciences*, *14*(4), 7405–7432.

Zarei, S., Ehsanpour, A. A., & Abbaspour, J. (2012). The role of over expression of P5CS gene on proline, catalase, ascorbate peroxidase activity and lipid peroxidation of transgenic tobacco (*Nicotiana tabacum* L.) plant under in vitro drought stress. *Journal of Cell and Molecular Research*, *4*(1), 43–49.

Zhang, Y., & Liu, J. (2011). Transgenic alfalfa plants co-expressing glutathione S-transferase (GST) and human CYP2E1 show enhanced resistance to mixed contaminates of heavy metals and organic pollutants. *Journal of Hazard Materials*, *189*(1), 357–362.

Zhang, Y., Liu, J., Zhou, Y., Gong, T., Wang, J., & Ge, Y. (2013). Enhanced phytoremediation of mixed heavy metal (mercury)–organic pollutants (trichloroethylene) with transgenic alfalfa co-expressing glutathione S-transferase and human P450 2E1. *Journal of Hazard Materials*, *260*, 1100–1107.

Zhang, Y., Zhang, M., Ye, J., Xu, Q., Feng, Y., Xu, S., ... & Yang, Y. (2023). Integrating genome-wide association study into genomic selection for the prediction of agronomic traits in rice (Oryza sativa L.). *Molecular Breeding*, *43*(11), 81.

Zhao, T., Li, R., Yao, W., Wang, Y., Zhang, C., & Li, Y. (2021). Genome-wide identification and characterisation of phenylalanine ammonia-lyase gene family in grapevine. *The Journal of Horticultural Science and Biotechnology*, *96*(4), 456–468.

Zhao, Z., Zhang, H., Fu, Z., Chen, H., Lin, Y., Yan, P., Li, W., Xie, H., Guo, Z., & Zhang, X. (2018). Genetic-based dissection of arsenic accumulation in maize using a genome wide association analysis method. *Plant Biotechnology Journal*, *16*(5), 1085–1093.

Zhou, M., Zheng, S., Liu, R., Lu, J., Lu, L., Zhang, C., Liu, Z., Luo, C., Zhang, L., & Wu, Y. (2019). Comparative analysis of root transcriptome profiles between low- and high cadmium-accumulating genotypes of wheat in response to cadmium stress. *Functional & Integrative Genomics*, *19*, 281–294.

16 Reducing the Toxicity of Heavy Metals and Metalloids on Grain Crops by Breeding

Current Achievements on Conventional and Molecular Breeding Approaches

Fatima El Amerany, Samia Elouali, Youssef Ait Hamdan, Iman Janah, and Mohammed Rhazi

16.1 INTRODUCTION

The agricultural land resources play a crucial role as strategic reserves for the sustainable development of human society, ensuring security, and maintaining stability (Chen et al., 2023). To this extent, increasing the production of grain crops, such as rice, wheat, and maize, is of strategic importance in ensuring a sustainable global food supply. However, this production can be affected due to several environmental problems including soil degradation and agricultural pollution (Smith and Siciliano, 2015).

The world's population is currently growing significantly, leading to a substantial increase in food demand. According to the United Nations World Population Prospects Report (2019), the world's population is expected to reach 9.7 billion by 2050, presenting a major challenge for food security (Chen et al., 2023). However, the contamination of water and soil resources by metals and metalloids, specifically arsenic and selenium, as a result of natural and human-made activities will pose a threat to food safety (Ahmad et al., 2020).

Grain crops such as wheat, maize, and rice, grown over large areas, are more exposed to metal and metalloid toxicity, which affects not only their growth and physiology but also their metabolic processes (Ahmad et al., 2020). To face this challenge, scientists use a molecular approach, which has become an essential tool for improving tolerant cultivars against metals and metalloids (Raza et al., 2022; Haider et al., 2023).

The current chapter highlights grain production worldwide and its main producer. It specifies the main metals and metalloids that affect grain crop production and illustrates plant mechanisms to deal with their toxicity. It also looks at the different strategies applied in the agricultural fields to mitigate metals and metalloids' effect on grain crops with an eye on breeding technology.

16.2 GRAIN CROP PRODUCTION WORLDWIDE

Grain crops have been considered the principal component of the human diet for thousands of years and have played a major role in shaping human civilization. Around the world, grain crops that are to a greater extent (i.e., rice, wheat, and maize) and a lesser extent (i.e., sorghum and millets) are important staples critical to the daily survival of billions of people. More than 50% of the world's

DOI: 10.1201/9781003457237-16

daily caloric intake is derived directly from grain crop consumption (Awika, 2011). In 1992, the US Department of Agriculture created the food pyramid, whose main purpose was to represent the share held by each food group in all foods eaten in a single day. According to this model, the consumption of grain crops was positioned at the bottom with bread and pasta (Chiran et al., 2007).

To date, the top four countries that produce grain crops are China, India, the United States, and Brazil (Akibode and Maredia, 2012). China was the world's leading agricultural producer because it produced a quarter of the global grain output with an annual output that reached $1.56 trillion in 2020 (Erokhin et al., 2021). The second, third, and fourth producers of grain crops were India, the United States, and Brazil with an output of up to $403.5 billion, $306.4 billion, and $135.8 billion, respectively, in 2020.

Various categories of grain crops are extremely popular. Among them are maize, wheat, and rice, which occupy a significant place in the eating habits of high-consumption regions, such as Latin America, sub-Saharan Africa, China, India, and Italy. Moreover, the average annual consumption of maize in the world is estimated to be 19.6 kg/capita. For wheat and rice, the average annual consumption per capita is 65.6 and 81.4 kg, respectively (Milani et al., 2022). In addition, other grain crops such as sorghum, millets, teff, oats, rye, fonio, quinoa, amaranth, and buckwheat are crops of regional and local importance. Sorghum is predominant in Africa and India, with an average annual consumption of 3.3 kg/capita (Noort et al., 2022). Moreover, millets, such as pearl millet, foxtail, proso, and finger millet, are widely consumed in Africa, Eurasia, China, India, and Pakistan, with an average annual of 2.8 kg/capita (Hussain et al., 2019). Teff, also classified as a millet, is an essential grain crop in Africa, particularly in the Horn of Africa, and its consumption can be more than 29.0 kg/capita (Taylor, 2019). Oats are mainly found in Northern Europe and Western countries, with an average consumption of 3.2 kg for each person (Smulders et al., 2018). The average annual consumption of the remaining grain (rye, fonio, quinoa, amaranth, and buckwheat) in Central and Eastern Europe, South America, and Africa was less than 1.9 (Milani et al., 2022). The difference in grain crops consumption reflects the diversity of diets and cultures worldwide.

The evolution of grain crop production over 10 years, mainly between 2012/2013 and 2022/2023, is presented in Table 16.1. A significant increase of about 37.39% (equal to 313 million metric tons) in maize production was observed over this decade. In addition, wheat and rice production recorded an increase of 9.24% (equal to 66 million metric tons) and 6.57% (equal to 31 million metric tons), respectively (Table 16.1). The yield of barley, sorghum, and oats is also induced, but it is still lower in comparison to that of maize (Table 16.1). However, rye production was reduced by 11.97% over this decade (Table 16.1), and this decrease might be due to an increase in climate change and pollution by metals and metalloids.

TABLE 16.1
The Evolution of Grain Crop Production Over a Decade (between 2012/2013 and 2022/2023)

	Production Worldwide/Year			
	Million Metric Tons			
Grain Crops	2012–2013	2022–2023	10-Year Gap	% of Production
Maize	838.00	1151.36	313.36	37.39
Wheat	717.00	783.30	66.30	9.24
Rice	471.90	502.92	31.02	6.57
Barley	148.33	150.48	2.15	1.44
Sorghum	57.60	58.54	0.94	1.63
Oats	20.73	25.05	4.32	20.83
Rye	13.70	12.06	−1.64	−11.97

16.3 EFFECTS OF HEAVY METALS AND METALLOIDS ON GRAIN CROPS

16.3.1 Overview of Heavy Metals and Metalloids

Heavy metals (i.e., cadmium (Cd), copper (Cu), zinc (Zn), mercury (Hg)) and metalloids (i.e., arsenic (As), germanium (Ge), tellurium (Te)) are metallic and semi-metallic elements, respectively, that pollute the environment and influence the grain crop health and its productivity. When these harmful elements build up in the soil, they can be taken up by plants, causing negative effects. These elements disrupt physiological processes in plants, resulting in stunted growth, decreased photosynthesis efficiency, and reduced absorption of nutrients and crop productivity (Rashid et al., 2023; Philomene and Komariah, 2021; Vasilachi et al., 2023). Contrary to that, there are heavy metals and alkaline earth metals that have a positive impact on grain crops with low concentration (Vasilachi et al., 2023), and some heavy metals, such as Cu, Zn, iron (Fe), manganese (Mn), molybdenum (Mo), nickel (Ni), and cobalt (Co), are considered essential micronutrients for plants (Reeves and Baker, 2000).

As shown in Figure 16.1, various sources of heavy metals and metalloids contribute to their presence in the soil. Natural sources such as the weathering of rocks and volcanic activity could increase the contamination of agricultural land by metals (Bradl, 2005). In addition, different industrial activities, specifically mining and refinement processes, are significant sources of heavy metals and metalloids (Srivastava et al., 2017). Also, the application of sewage sludge and wastewater irrigation, and the use of inorganic fertilizers and pesticides can introduce toxic metals into the soil, impacting plant health, soil's biodiversity, and crop productivity (Srivastava et al, 2017; Alengebawy et al., 2021). Additionally, soil quality might be affected by dumping batteries and medical waste (Adama et al., 2016; Ephraim et al., 2013).

16.3.2 Positive and Negative Effects of Heavy Metals and Metalloids on Grain Crops

Grain crops such as maize, rice, wheat, barley, and sorghum could be both positively and negatively (Figure 16.2) impacted by the presence of heavy metals and metalloids in the soil. For instance, rice and wheat were biofortified with Zn to improve their productivity and to enhance the quality and germination of their grain (Rashid et al., 2019; Hafeez et al., 2021; Zarea and Karimi, 2023). Also, several studies mentioned that Zn is considered an important micronutrient for improving the

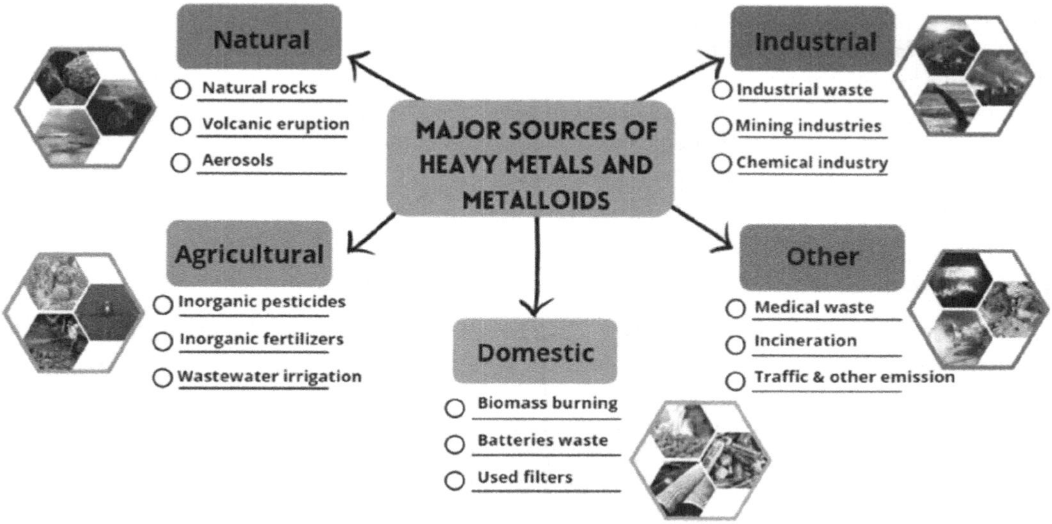

FIGURE 16.1 Major sources of heavy metals and metalloids.

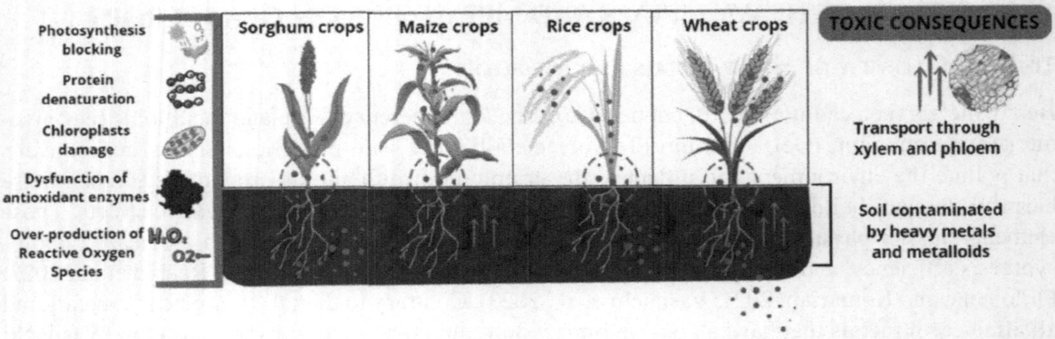

FIGURE 16.2 Toxic effects of the translocation of heavy metals and metalloids in grain crops.

growth attributes in wheat and other grain crops (Basha and Selvaraju, 2015; Khan, 2023; Tao et al., 2018; Zhang and Wang, 2021; Praharaj et al., 2021; Hafeez et al., 2021). Research done by Rashid et al. (2019) showed that when seeds contain Zn (40–50 mg/kg), their germination rate is induced. However, the use of Zn with a high concentration (100–300 mg/kg) can decrease the number of spikes and grains as well as their weight (Dawar et al., 2022).

Magnesium (Mg) is an alkaline earth metal (Phillips et al., 2023), which is also important for producing chlorophylls and activating the specific enzyme needed during carbohydrate metabolism and other metabolic processes (Cakmak, 2013). Moreover, it can be involved in the mitigation of plants to the negative effects of other heavy metals, such as aluminum (Al) and Mn (Kaiser, 2023).

Regarding the negative effects of heavy metals, in a pot experiment done by Xu et al. (2008), three different levels of Cu were found to be toxic to rice seedlings. The results show that the yield of rice was decreased by 10%, 50%, and 90% with soil containing 100, 300–500, and 1,000 mg/kg Cu, respectively. In addition, other heavy metals such as Cd could reduce seed germination, inhibit photosynthesis, degrade the ultrastructure of chloroplasts (Munir et al., 2023), and negatively alter the root elongation of barley seedlings (Jawad Hassan et al., 2020). It could also cause oxidative stress in plants by decreasing the activity of antioxidant enzymes such as superoxide dismutase (SOD), catalase (CAT), and glutathione reductase (GR) (Kintlová et al., 2021). Most research showed that Hg is a toxic heavy metal to grain crops; even at low concentrations (Hu et al., 2016; Feng Shao et al., 2022), it can affect the germination of cereal seeds and induce the hypertrophy of the roots and coleoptile (Patra et al., 2000). In rice, wheat, and sorghum, Jawad Hassan et al. (2020) reported that Hg could directly induce stress-responsive mechanisms and oxidative damage, which affect membrane permeability.

Concerning metalloid effects, Chandrakar et al. (2016) and Begum et al. (2016) showed that As can induce several modifications in the morphology, physiology, and biochemistry of cereals due to the generation of reactive oxygen species (ROS), such as hydrogen peroxide (H_2O_2), hydroxyl radical (OH·), and superoxide radical (O^{2-}) (Rafiq et al., 2017; Shahid et al., 2017).

16.3.3 Grain Crops'Mechanisms against Heavy Metals Stress

To lower stress induced by heavy metals and metalloids, grain crops have developed several mechanisms of detoxification. For example, different antioxidants (i.e., ascorbate peroxidase (APX), CAT, GR, peroxidase (POD), SOD, phenolics) can be produced and used by plants to detoxify metal-induced ROS. The work done by Khan et al. (2018) showed that As stress in rice is reduced due to an increase in antioxidants (CAT, SOD, glutathione peroxidase, glutathione S-transferase) levels in shoots and roots. Moreover, the sequestration of metals and metalloids in the vacuole is considered the most applied strategy by plants, especially by grain crops, to reduce metal and metalloid stresses. This mechanism requires the involvement of two proteins (phytochelatins and metallothionein) that are systematically synthesized when there is an excess of heavy metals and metalloids (Zagorchev et al., 2013). Phytochelatins serve to chelate metal ions (Ahmad et al., 2020), while metallothionein proteins continue to sequester them in the vacuole (Hassan et al., 2017). In addition, a study in rice showed that stress induced by metals and metalloids was decreased due to the accumulation of cysteine and proline (Khan et al, 2018). In wheat, it has been demonstrated that chelating metal and semi-metal ions can also be possible by organic ligands such as citric acid (Tahir et al, 2015).

16.4 STRATEGIES APPLIED IN THE AGRICULTURAL FIELDS TO MITIGATE HEAVY METALS AND METALLOIDS' EFFECT ON GRAIN CROPS

The presence of heavy metals and metalloids in agricultural ecosystems has emerged as a significant concern due to their potential impact on the growth, development, and safety of grain crops. They can exert diverse effects on plant systems depending on their type, their concentration in the soil, and the specific crops involved. Thus, understanding the consequences of heavy metal and metalloid exposure is crucial for sustainable agriculture and food safety.

In agricultural settings, exploring strategies to mitigate the effect of toxic metals on grain crops is pivotal for sustaining both productivity and environmental health. The top four strategies applied were using organic and inorganic fertilizers, managing water use, applying microorganisms, and employing breeding (Figure 16.3). Through the application of soil amendments (Table 16.2), such as organic and inorganic fertilizers, the bioavailability of heavy metals and metalloids can be reduced, thus promoting a healthier soil structure. Additionally, applying water management strategies will contribute to minimizing the impact of metals on grain crops. Moreover, the use of metals-resistant microorganisms (bacteria, fungi, and algae) has also emerged as a pivotal strategy to counteract the detrimental effects of toxic metals. Plant breeding is a process that imparts beneficial traits to grain crops, enabling them to thrive in challenging environmental conditions.

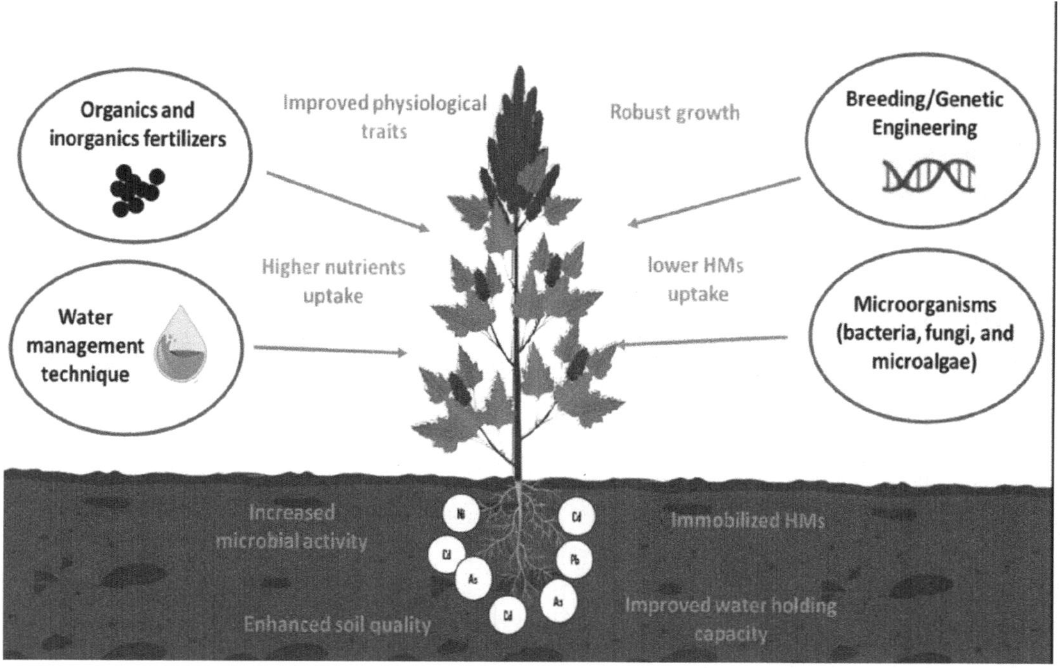

FIGURE 16.3 Strategies for mitigating heavy metals and metalloid stresses.

TABLE 16.2
Effect of Organic and Inorganic Fertilizers on Grain Crops Exposed to Heavy Metal and Metalloid Stresses

Organic and Inorganic Fertilizers	Application Rate	Crops	Heavy Metals/ Metalloids	Results	References
Biochar	20 tons/haor 1.5%	Rice	Cd, Zn, Pb, As, and Al	A reduction in ions uptake and oxidative stress;an increase in growth	Zheng et al. (2015), Irshad et al. (2020), Shetty and Prakash (2020)
	1% or 3% or 3 g/L	Wheat	Cr, Cd, Cu, Pb, Ni, Zn, Mg, Co, and Fe	A decrease in heavy metals content and an increase in relative growth rate, relative water content, chlorophyll fluorescence, germination rate, and proline content	Pandey et al. (2022), Mazhar et al. (2020), Yildiztugay et al. (2019)
	2%, 4%, 5%, and 10%	Maize	Pb, Ni, Cd, and Cr	An increase in shoot and dry root weight, total chlorophyll contents, gas exchange, and soil characteristics. A decrease in Pb, Cr, and Cd toxicity	Irfan et al. (2021), Sayyadian et al. (2019), Alaboudi et al. (2019)
	5, 10, 20, and 30 g/ kgor 20 t/ha	Barley	As, Pb, Cr, and Cd	An increase in growth and oxidative stress;a reduction in heavy metals levels in grain	Moreno-Jiménez et al. (2016), Dai et al. (2022)
Compost	2.5%	Maize	As	A reduction in the levels of As in shoots and an improvement in dry biomass and physiological attributes	Mehmood et al. (2017)
Silicon	1 mM	Wheat	Cd	An improvement in growth and a reduction in Cd uptake	Liu et al. (2023)
	1 mM	Rice	Cd	A decrease in the expression of *OSNRAMP5* (a major transporter for Cd uptake in rice roots) and *OSHMA2* (gene controlling the translocation of Cd from root to shoot)	Feng Shao et al. (2017)
Nano-Se, selenite, and selenate	1 μmol/L	Rice	Cd	A decrease in the accumulation and translocation of Cd	Xu et al. (2020)
Sodium nitroprusside	100 μm	Rice	Pb	A reduction in electrolyte leakage and H_2O_2	Rahim et al. (2022)
Zinc oxide nanoparticles	100 mg/L	Maize	Cd	An increase in growth, chlorophyll concentrations, and gas exchange. A reduction in electrolyte leakage and H_2O_2	Rizwan et al. (2019)
	100 mg/L	Wheat	Cd	An increase in growth and chlorophyll contents	Adrees et al. (2021)
TiO_2 nanoparticles	100 mg/L	Maize	Cd	A reduction in the uptake and translocation of Cd in plants and an improvement in plant physiological status	Lian et al. (2020)

Al, aluminum; As, arsenic; Cd, cadmium; Co, cobalt; Cr, chromium; Cu, copper; Fe, iron; Mg, magnesium; Ni, nickel; Pb, lead; Zn, zinc.

16.4.1 Organic and Inorganic Fertilizers

It has been demonstrated that an increase in a plant's mineral ions exchange capacity and the application of organic fertilizers, such as compost, manure, and biochar to provide humic substances and beneficial microbes, can reduce the effects of heavy metals and semi-metals on plants (Udovic and Mcbride, 2012). A work done by Pandey et al. (2022) revealed significant improvement in wheat growth and germination rate, following the application of biochar. The incorporation of this fertilizer into the soil leads to a substantial decrease (49.29–78.11%) in metal uptake rate. Similar results have been found in maize (Irfan et al., 2021), rice (Zheng et al., 2015), and sorghum (De Almeida et al., 2020). Moreover, studies conducted on barley have shown that the use of compost (50 tons/ha) may lead to a substantial reduction in the bioaccumulation of Cu (by 29%), chromium (Cr by 49%), and Fe (by 77.5%) (Masu et al., 2012). In addition, the application of sulfur fertilization (500 mg/kg) in Cu-polluted paddy soils can positively influence the biomass of rice (Sun et al., 2016). Another work demonstrated that foliar and soil application of silicon (Si) nanoparticles can enhance the growth of wheat plants and their photosynthesis and reduce the accumulation of Cd in their shoots, roots, and grains by 11–53%, 10–59%, and 22–83%, respectively (Ali et al., 2019). Furthermore, we have provided a summary outlining the impact of various fertilizers on grain crops subjected to heavy metals stress (Table 16.2).

16.4.2 Water Management Technique

Alternate wetting and drying (AWD) is a water management technique that is mostly used for cultivating rice, maintaining its yield, and reducing greenhouse gas emissions (Martínez-Eixarch et al., 2021). This water-saving practice might also be used to mitigate the detrimental effects of metals and metalloids on rice. A study showed that the contents of As in rice grains can be reduced by 23% after using the AWD irrigation practice, in comparison to incessant flooding (Islam et al., 2019). In addition, it keeps the concentration of Zn, selenium (Se), Cu, and Cd below the desired thresholds (Martínez-Eixarch et al., 2021).

16.4.3 Microorganisms

The majority of microorganisms are affected by heavy metals and metalloids, except those isolated from metal-polluted soils (Ayangbenro and Babalola, 2017). In grain crops, the use of plant growth-promoting rhizobacteria (PGPR) has been shown to confer resistance to metal and metalloid ions through several processes, such as immobilization and transformation of these ions (Hassan et al., 2017). For instance, PGPRs like *Ochrobactrum* sp. and *Bacillus* sp. can boost rice growth and mitigate the toxic effects of the metal by producing siderophores, indole acetic acid, and 1-aminocyclopropane-1-carboxylate deaminase; fixing the atmospheric nitrogen; and solubilizing minerals (i.e., phosphate, Zn, and potassium); as well as changing bioavailability of heavy metal (Pandey, 2012). In addition, *Azospirillum brasilense* can mitigate the adverse effects of Cd stress on wheat due to an increase in the transcript level of *Triticum aestivum* salt overly sensitive 1 (*TaSOS1*) gene, as well as an improvement in the level of pigments, proline, and antioxidant enzymes, such as APX and CAT (Ghassemi and Mostajeran, 2018).

Moreover, several reports have indicated that arbuscular mycorrhizal fungi play a pivotal role in reducing the phytoavailability of heavy metals through their immobilization via hyphae (Wu et al., 2015) and glomalin (Wu et al., 2014). They can also alter the expression of As and Cd transporter genes (Li et al., 2020). The cell wall of *Diversispora spurcum* and *Funneliformis mosseae* can bind Zn, lead (Pb), and Cd ions, which reduces their translocation to maize plants (Abdelmoneim et al., 2014; Zhan et al., 2018).

16.4.4 Current Achievements in Breeding

Over the past few decades, an increase in the accumulation of heavy metals and metalloids in the soil has caused increased damage to grain crops. Therefore, agricultural experts have been focused on achieving toxicity resistance to heavy metals and metalloids by removing them and reducing their absorption and transport in plant tissues. Previous studies have proposed cultivating rice in non-flooded conditions for reducing the uptake and accumulation of As in grain (Xu et al., 2008); however, non-flooded rice production might be susceptible to insects and may not provide the same yield as anaerobic rice. Other studies demonstrated that producing grain crops without affecting their yield and with less heavy metals and metalloid contents can be possible by selecting tolerant varieties and using interbreeding (or crossing) and molecular breeding, which is based on the discovery of candidate genes and identifying reliable loci.

16.4.4.1 Tolerant Varieties

Most of the time, the rivers that originate in the mountains are laden with rock sediments, heavy metals, and metalloids, making grain crop cultivation vulnerable to pollution. Thus, to reduce their negative impacts, it would be necessary to identify grain varieties that can survive and adapt to polluting land. The selection of different tolerant cultivars was done based on the phenotypical and biochemical characteristics such as kernel trait, grain yield, flowering time, and the reduced levels of heavy metals and metalloids in roots, shoots, and seeds. In addition, employing molecular tools to identify genes that are related to plant resistance to heavy metal and metalloid ions contamination is the main step that allows breeders to select tolerant cultivars with excellent grain quality and to develop safer varieties with low toxic ion accumulation.

The selection of candidate genes is carried out according to the following steps (Figure 16.4): the first step requires the analysis of the chromosome profile of a plant species and the determination of the number of chromosomes, their size, and their morphology. The next step is the identification of

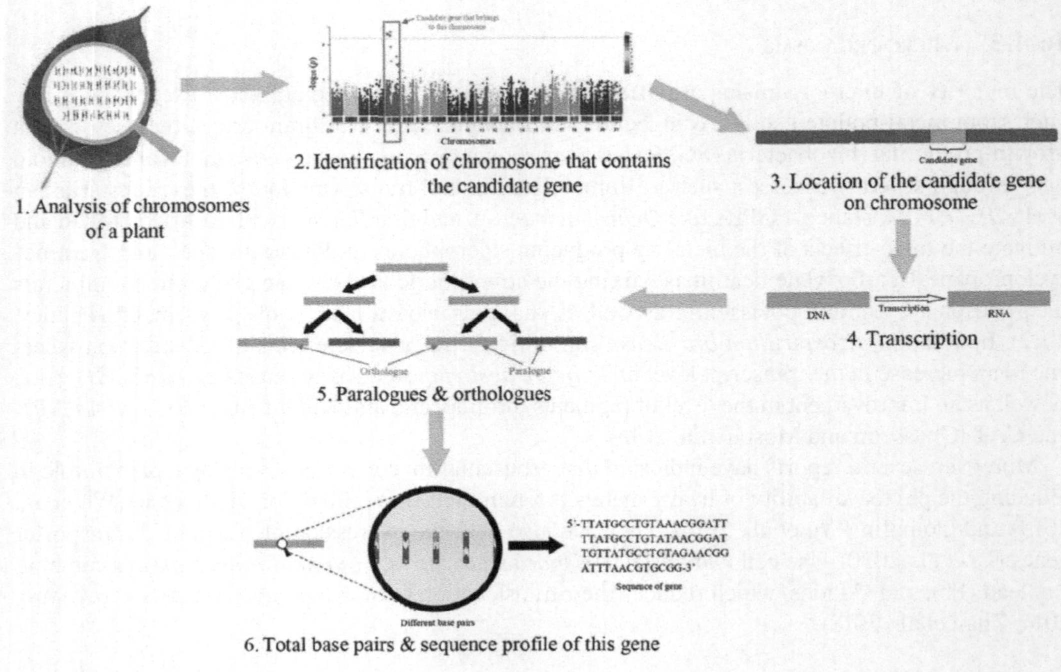

FIGURE 16.4 Steps for the identification of the candidate genes.

the chromosome that contains the candidate gene and where it is localized. Then, make RNA copies and determine the total base pairs and amino acids and the sequence profile of the candidate gene.

Different studies have shown significant differences in the accumulation of As, tin (Sn), Pb, and Cd in grain crops. Grain cultivars with low metalloid and heavy metals accumulating capability are summarized in Table 16.3. Every tolerant grain cultivar has a unique gene responsible for its resistance. In rice tolerant cultivar (*Oryza sativa* var. Yueguang), *OsCdl*gene can disrupt the absorption and the uptake of Cd in roots (Li et al., 2023), while in another rice cultivar, such as IRAT129, *qCddis8* gene can limit the distribution of Cd from roots to shoots and between leaves (Chen et al., 2021). Another study demonstrated that the introduction of a specific region of DNA that influences As accumulation (the As QTL) on chromosome 5 could lead to a decrease in As levels in grains of high-yielding cultivars (Norton et al., 2019). In addition, the negative impact of As on wheat growth can be reduced due to the expression of the *Os10g0529800* gene that controls the glutathione metabolic process (Saeed et al., 2022). Also, in wheat genotypes, a single-nucleotide polymorphism (Ku_c269_2643) located with the *TraesCS2A02G080700* gene was considered to be important for improving kernel trait under Pb and Sn application (Mourad et al., 2023).

Thus, the tolerant varieties might be used as candidate parents for crossing and developing new genotypes in the future.

TABLE 16.3
Tolerant Varieties of Grain Crops against Heavy Metals and Metalloids

Tolerant Varieties	Heavy Metals/ Metalloids	Candidate Genes Responsible for Plant Tolerance	References
Wheat (*Triticum aestivum* L.)			
• Kohistan-97 • As-2002 • Barani-70 • Pari-73 • D-97	As	*Os10g0529800*	Saeed et al. (2022)
• Gimmiza_11	Sn	Not identified	Mourad et al. (2021)
• BeniSweif_7	Pb	Not identified	
Rice (*Oryza sativa* L.)			
• Soribhog • Jhona 349	As	Not identified	Norton et al. (2019)
• Kahawanu	As	Not identified	Manawasinghe and Chandrajith (2023)
• Tianyou 998	Cd	Not identified	Wang et al. (2011)
• Daganzaogu • Hongsanlicun • Changgu • Zaomagu • Hongtaonuo	Cd	*qCd1–3/OsABCB24*	Pan et al. (2020)
• IRAT129	Cd	*qCddis8*	Chen et al. (2021)
• Yueguang • Yunlu 103 • Minghui 63	Cd	*OsPCR1*, *CAL1*, and *Oscd1*	Li et al. (2023)
Maize (*Zea mays* L.)			
• Cateto	Al	Not identified	Doncheva et al. (2005)
• 203B-14	Al	*ZmNrat1*	Matonyei et al. (2020)
• B73	Cd	*GRMZM2G175576*	Zhao et al. (2018)

Al, aluminum;As, arsenic; Cd, cadmium; Pb, lead; Sn, tin.

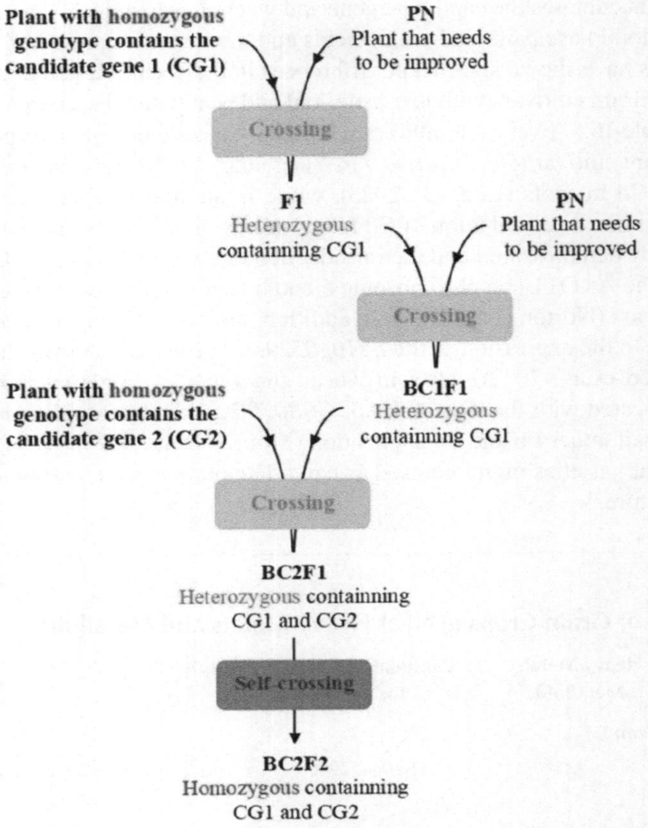

FIGURE 16.5 Steps for developing interbreeding lines carrying two genes.

16.4.4.2 Crossing Plants

Two or more plants from the same family can be crossed either to introduce new traits or to transfer a gene from one of them to another. Steps for crossing plants to develop lines carrying two genes were summarized in Figure 16.5. For instance, crossing a sensitive cultivar of rice (*Oryza sativa* L. var. Chuang-5S) with a tolerant cultivar of low Cd accumulation-related gene (*OsHMA3* or *qlGCd3*) then with rice blast resistance genes (*Pi49* or *Pi48*, respectively) can maintain plant growth and decrease Cd concentration (by 50–53%) and disease levels (by 60%) (Chen et al., 2020).

16.4.4.3 Molecular Breeding

Molecular breeding is a genetic engineering approach that produces transgenic lines with improved resistance to heavy metals and metalloids due to mutation, deletion (knockout), or overexpression of candidate genes. The suitable grain crop varieties for As-, Cd-, and aluminum (Al)-contaminated ecosystems that are genetically developed are presented in Table 16.4. Du et al. (2020) reported that the incorporation of the *ZmAT6* gene into the genome of immature maize callus by *Agrobacterium tumefaciens* reduces the accumulation of Al and the levels of ROS. In addition, Ueno et al. demonstrated that an overexpression of the *OsHMA3* gene that encodes a metal transporter P_{1B}-type ATPase regulates vacuolar sequestration capacity, specifically, for Cd and reduces its translocation from roots to shoots of rice seedlings (Ueno et al., 2010). Overexpression of another gene such as *GRMZM2G175576* encodes Cd/Zn-transporting ATPase, in shoots, stems, and roots, and reduces

TABLE 16.4
Resistant Grain Crops Developed by Molecular Breeding

Heavy Metals/ Metalloids	Grain Species	QTLs/Genes Mapped	Chromosomes	Effects	References
As	Rice (*Oryza sativa* L.)	*qAsR8.1* and *qAsR8.2* *qAsS2*, *qAsS5.1*, *qAsS5.2*, *qAsS6*, *qAsS9.1*, and *qAsS9.2*	Chromosome 8 in root Chromosomes 2, 5, 6, and 9 in shoots	Increase in the transcript levels of heat shock protein, heavy metal associated domain-containing protein, and inorganic H^+ pyrophosphatase gene. Decrease in the expression of some genes, especially IAA gene family member and auxin efflux carrier component.	Murugaiyan et al. (2019)
Cd	Rice (*Oryza sativa* L.)	*OsNRAMP1*	Chromosome 7	Decrease in grain Cd.	Norton et al. (2021)
		OsHMA3	Not identified	Limitation in the translocation of Cd from roots to shoots.	Ueno et al. (2010)
		OsHMA3-OsNramp5-OsNramp1	Chromosome 7	Decrease in grain Cd.	Wang et al. (2021)
	Maize (*Zea mays* L.)	*GRMZM2G175576*	Chromosome 2	Decrease in Cd accumulation in leaves, stems, and roots.	Zhao et al. (2018)
Al	Maize (*Zea mays* L.)	*ZmAT6*	Not identified	Increase in root length, shoot growth, and plant tolerance to Al by reducing malondialdehyde contents and peroxide hydrogen levels.	Du et al. (2020)

Al, aluminum; As, arsenic; Cd, cadmium.

Cd assimilation and accumulation in these tissues (Zhao et al., 2018). Also, Norton et al. (2021) found that there is a link between the *OsNRAMP1* gene and Cd accumulation in rice grains. When the *OsNRAMP1* gene is deleted, the rice cultivars that have been grown in two irrigation systems (continuous flooding and AWD) have reduced levels of Cd in grain.

Despite the positive effects of breeding on plants, it takes a long time, up to several years, to test new variants that have resistance to a specific dose of heavy metals.

16.5 CONCLUSION AND PERSPECTIVES

Various kinds of heavy metals such as Cd, Pb, Hg, and Al can pose a serious threat to grain crops when present in low and elevated concentrations. The uptake of these metals can trigger oxidative stress, ultimately resulting in a reduction in biomass, photosynthesis, and grain yield. For handling this type of stress many approaches have been proposed and the most effective ones are applying organic and inorganic amendments, beneficial bacteria and fungi, water-saving practices (AWD), and breeding techniques. Even though breeding is a complex process, it is still the fastest means to ensure getting the desired result, such as inducing the resistance and grain yield. However, from the current data available in the literature, rice and maize, are the two-grain crop types that were genetically modified to treat As, Cd, and Al. In addition, no work was done on grafted plants.

REFERENCES

Abdelmoneim, T. S., Moussa, T. A., Almaghrabi, O. A., & Abdelbagi, I. (2014). Investigation the effect of arbuscular mycorrhizal fungi on the tolerance of maize plant to heavy metals stress. *Life Science Journal, 11*(4), 255–263.

Adama, M., Esena, R., Fosu-Mensah, B., & Yirenya-Tawiah, D. (2016). Heavy metal contamination of soils around a hospital waste incinerator bottom ash dumps site. *Journl of Environmental and Public Health, 2016*, 8926453. https://doi.org/10.1155/2016/8926453

Adrees, M., Khan, Z. S., Hafeez, M., Rizwan, M., Hussain, K., Asrar, M., & Ali, S. (2021). Foliar exposure of zinc oxide nanoparticles improved the growth of wheat (Triticum aestivum L.) and decreased cadmium concentration in grains under simultaneous Cd and water deficient stress. *Ecotoxicology and Environmental Safety, 208*, 111627.

Ahmad, I., Tahir, M., Daraz, U., Ditta, A., Hussain, M. B., & Khan, Z. U. H. (2020). Responses and tolerance of cereal crops to metal and metalloid toxicity. In M. Hasanuzzaman (Ed.), *Agronomic crops*. Springer. https://doi.org/10.1007/978-981-15-0025-1_14

Akibode, C. S., & Maredia, M. K. (2012). *Global and regional trends in production, trade and consumption of food legume crops* (No. 1099-2016-89132).

Alaboudi, K. A., Ahmed, B., & Brodie, G. (2019). Effect of biochar on Pb, Cd and Cr availability and maize growth in artificial contaminated soil. *Annals of Agricultural Sciences, 64*(1), 95–102.

Alengebawy, A., Abdelkhalek, S. T., Qureshi, S. R., & Wang, M. Q. (2021). Heavy metals and pesticides toxicity in agricultural soil and plants: Ecological risks and human health implications. *Toxics, 9*(3), 42. https://doi.org/10.3390/toxics9030042

Ali, S., Rizwan, M., Hussain, A., urRehman, M. Z., Ali, B., Yousaf, B., & Ahmad, P. (2019). Silicon nanoparticles enhanced the growth and reduced the cadmium accumulation in grains of wheat (Triticum aestivum L.). *Plant Physiology and Biochemistry, 140*, 1–8.

Awika, J. M. (2011). Major cereal grains production and use around the world. In J. M. Awika, V. Piironen, & S. Bean (Eds.), *Advances in cereal science: Implications to food processing and health promotion* (pp. 1–13). American Chemical Society.

Ayangbenro, A. S., & Babalola, O. O. (2017). A new strategy for heavy metal polluted environments: A review of microbial biosorbents. *International Journal of Environmental Research and Public Health, 14*(1), 94.

Basha, S. A., & Selvaraju, M. (2015). Toxic effect of zinc on growth and nutrient accumulation of cow pea. *International Letters of Natural Sciences, 43*, 48–53. https://doi.org/10.56431/p-6tf03z

Begum, M. C., Islam, M. S., Islam, M., Amin, R., Parvez, M. S., & Kabir, A. H. (2016). Biochemical and molecular responses underlying differential arsenic tolerance in rice (Oryza sativa L.). *Plant Physiology and Biochemistry, 104*, 266–277. https://doi.org/10.1016/j.plaphy.2016.03.034

Bradl, H. B. (2005). Sources and origins of heavy metals. *Interface Science and Technology, 6*, 1–27. https://doi.org/10.1016/S1573-4285(05)80020-1

Cakmak, I. (2013). Magnesium in crop production, food quality and human health. *Plant Soil, 368*, 1–4. https://doi.org/10.1007/s11104-013-1781-2

Chandrakar, V., Naithani, S. C., & Keshavkant, S. (2016). Arsenic-induced metabolic disturbances and their mitigation mechanisms in crop plants: A review. *Biologia, 71*, 367–377. https://doi.org/10.1515/biolog-2016-0052

Chen, Q., Tang, W., Zeng, G., Sheng, H., Shi, W., & Xiao, Y. (2020). Reduction of cadmium accumulation in the grains of male sterile rice Chuang-5S carrying Pi48 or Pi49 through marker-assisted selection. *3 Biotech, 10*, 1–10.

Chen, W., Wang, G., Cai, W., Che, X., Zhou, W., Zhang, C., & Zeng, J. (2023). Inadéquation spatio-temporelle de la production céréalière mondiale et des terres agricoles et ses facteurs d'influence. *Ressources, Conservation et Recyclage, 194*, 107008.

Chen, Y. B., Chen, Y. C., Zhu, Y. X., Li, S., Deng, H. B., Wang, J. R., & Sun, L. (2021). Genetic control diversity drives differences between cadmium distribution and tolerance in rice. *Frontiers in Plant Science, 12*, 638095.

Chiran, A., Drobotă, B., & Gîndu, E. (2007). Prospects on the world cereal market. *Cercet A ri Agronomice în Moldova, 1*

Dai, L., Chen, Y., Liu, L., Sun, P., Liu, J., Wang, B., & Yang, S. (2022). Effect of biochar on the uptake, translocation and phytotoxicity of chromium in a soil-barley pot system. *Science of the Total Environment, 826*, 153905.

Dawar, K., Ali, W., Bibi, H., Mian, I. A., Ahmad, M. A., Hussain, M. B., Ali, M., Ali, S., Fahad, S., & Rehman, S. U. (2022). Effect of different levels of zinc and compost on yield and yield components of wheat. *Agronomy*, *12*(7), 1562. https://doi.org/10.3390/agronomy12071562

De Almeida, J. B., Santos, L. S., Pandey, S. D., Nunes, C. F., Colen, F., Sampaio, R. A., & Fernandes, L. A. (2020). Alleviation of aluminum and cadmium toxicity by biochar and its potential toxicity to sorghum. *Semina: CiênciasAgrárias*, *41*(1), 95–108.

Doncheva, S., Amenos, M., Poschenrieder, C., & Barcelo, J. (2005). Root cell patterning: A primary target for aluminium toxicity in maize. *Journal of Experimental Botany*, *56*(414), 1213–1220.

Du, H., Huang, Y., Qu, M., Li, Y., Hu, X., Yang, W., & Zhang, S. (2020). A maize ZmAT6 gene confers aluminum tolerance via reactive oxygen species scavenging. *Frontiers in Plant Science*, *11*, 1016.

Ephraim, P. I., Ita, A., & Eusebius, I. O. (2013). Investigation of soils affected by burnt hospital wastes in Nigeria using PIXE. *Springerplus*, *2*(1), 208. https://doi.org/10.1186/2193-1801-2-208

Erokhin, V., Tianming, G., & Ivolga, A. (2021). International agricultural trade: Exporters and importers. In V. Erokhin, G. Tianming, & J. V. Andrei (Eds.), *Shifting patterns of agricultural trade: The protectionism outbreak and food security* (pp. 21–51). Springer.

Feng Shao, J., Che, J., Yamaji, N., Fang Shen, R., & Feng Ma, J. (2017). Silicon reduces cadmium accumulation by suppressing expression of transporter genes involved in cadmium uptake and translocation in rice. *Journal of Experimental Botany*, *68*(20), 5641–5651.

Ghassemi, H. R., & Mostajeran, A. (2018). TaSOS1 and TaTM20 genes expression and nutrient uptake in wheat seedlings may be altered via excess cadmium exposure and inoculation with Azospirillumbrasilense Sp7 under saline condition. *Applied Ecology & Environmental Research*, *16*(2), 1797–1817.

Hafeez, M. B., Ramzan, Y., Khan, S., Ibrar, D., Bashir, S., Zahra, N., Rashid, N., Nadeem, M., Rahman, S. U., Shair, H., Ahmad, J., Hussain, M., Irshad, S., Al-Hashimi, A., Alfagham, A., & Diao, Z.-H. (2021). Application of zinc and iron-based fertilizers improves the growth attributes, productivity, and grain quality of two wheat (Triticum aestivum) cultivars. *Frontiers Nutrition*, *8*, 779595. https://doi.org/10.3389/fnut.2021.779595

Haider, Z., Ahmad, I., Zia, S., & Gan, Y. (2023). Recent developments in rice molecular breeding for tolerance to heavy metal toxicity. *Agriculture*, *13*(5), 944.

Hanusa, T. P., & Phillips, C. S. G. (2023). *Alkaline-Earth Metal*. Encyclopedia Britannica. https://www.britannica.com/science/alkaline-earth-metal

Hassan, T. U., Bano, A., & Naz, I. (2017). Alleviation of heavy metals toxicity by the application of plant growth promoting rhizobacteria and effects on wheat grown in saline sodic field. *International Journal of Phytoremediation*, *19*, 522–529. https://doi.org/10.1080/15226514.2016.1267696

Hu, H. Y., Li, Z. J., Yao, F. E. N. G., Liu, Y. W., Xue, J. M., Davis, M., & Liang, Y. C. (2016). Prediction model for mercury transfer from soil to corn grain and its cross-species extrapolation. *Journal of Integrative Agriculture*, *15*(10), 2393–2402.

Hussain, S., Ahmad, S., Ahmad, N., Muhammad, S., Muhammad, I., Sami, U., & Abdul, S. (2019). *Agronomic crops: Volume 1: Production technologies* (p. 273). Springer.

Irfan, M., Mudassir, M., Khan, M. J., Dawar, K. M., Muhammad, D., Mian, I. A., & Dewil, R. (2021). Heavy metals immobilization and improvement in maize (Zea mays L.) growth amended with biochar and compost. *Scientific Reports*, *11*(1), 18416.

Irshad, M. K., Noman, A., Alhaithloul, H. A., Adeel, M., Rui, Y., Shah, T., & Shang, J. (2020). Goethite-modified biochar ameliorates the growth of rice (Oryza sativa L.) plants by suppressing Cd and As-induced oxidative stress in Cd and As co-contaminated paddy soil. *Science of the Total Environment*, *717*, 137086.

Islam, S., Rahman, M. M., & Naidu, R. (2019). Impact of water and fertilizer management on arsenic bioaccumulation and speciation in rice plants grown under greenhouse conditions. *Chemosphere*, *214*, 606–613.

Jawad Hassan, M., Ali Raza, M., Ur Rehman, S., Ansar, M., Gitari, H., Khan, I., Wajid, M., Ahmed, M., Abbas Shah, G., Peng, Y., & Li, Z. (2020). Effect of cadmium toxicity on growth, oxidative damage, antioxidant defense system and cadmium accumulation in two sorghum cultivars. *Plants (Basel)*, *9*(11), 1575. https://doi.org/10.3390/plants9111575

Khan, F., Hussain, S., Tanveer, M., Khan, S., Hussain, H. A., Iqbal, B., & Geng, M. (2018). Coordinated effects of lead toxicity and nutrient deprivation on growth, oxidative status, and elemental composition of primed and non-primed rice seedlings. *Environmental Science and Pollution Research International*, *25*, 21185. https://doi.org/10.1007/s11356-018-2262-1

Khan, A., & Shahwaar, S. (2023). Assessment of soil, water and vegetable crops for zinc deficiency in gadap basin for agricultural solution. *International Journal of Ground Sediment & Water*, *18*. 1215–1249.

Kintlová, M., Vrána, J., Hobza, R., Blavet, N., & Hudzieczek, V. (2021). Transcriptome response to cadmium exposure in barley (Hordeum vulgare L.). *Frontiers in Plant Science*, *12*, 629089. https://doi.org/10.3389/fpls.2021.629089

Li, H., Chen, X. W., & Wu, L. (2020). Effects of arbuscular mycorrhizal fungi on redox homeostasis of rice under Cd stress. *Plant and Soil*, *445*, 121–138.

Li, W., Xu, F., Cai, T., Zhao, W., Lin, J., Huang, J., & Zhu, C. (2023). The relationship between cadmium-related gene sequence variations in rice and cadmium accumulation. *Agronomy*, *13*(3), 800.

Lian, J., Zhao, L., Wu, J., Xiong, H., Bao, Y., Zeb, A., & Liu, W. (2020). Foliar spray of TiO_2 nanoparticles prevails over root application in reducing Cd accumulation and mitigating Cd-induced phytotoxicity in maize (Zea mays L.). *Chemosphere*, *239*, 124794.

Liu, H., Jiao, Q., Fan, L., Jiang, Y., Alyemeni, M. N., Ahmad, P., & Li, G. (2023). Integrated physio-biochemical and transcriptomic analysis revealed mechanism underlying of Si-mediated alleviation to cadmium toxicity in wheat. *Journal of Hazardous Materials*, *452*, 131366.

Manawasinghe, S., & Chandrajith, R. (2023). Growth responses of genetically diverse rice (Oryza sativa L.) cultivars to Arsenic stress. *Ceylon Journal of Science*, *52*(3), 381–389.

Martínez-Eixarch, M., Alcaraz, C., Guàrdia, M., Català-Forner, M., Bertomeu, A., Monaco, S., & Price, A. H. (2021). Multiple environmental benefits of alternate wetting and drying irrigation system with limited yield impact on European rice cultivation: The Ebre Delta case. *Agricultural Water Management*, *258*, 107164.

Masu, S., Dragomir, N., Morariu, F., Jurj, L., Luminita, N., & Popescu, D. (2012). The bioaccumulation of heavy metals in barley (Hordeum vulgare L) cultivated on a fly ash dump mixed with compost and natural zeolite materials. *Scientific Papers: Animal Science and Biotechnologies*, *45*(2), 237.

Matonyei, T. K., Barros, B. A., Guimaraes, R. G., Ouma, E. O., Cheprot, R. K., Apolinário, L. C., & Guimaraes, C. T. (2020). Aluminum tolerance mechanisms in Kenyan maize germplasm are independent from the citrate transporter ZmMATE1. *Scientific Reports*, *10*(1), 7320.

Mazhar, R., Ilyas, N., Arshad, M., & Khalid, A. (2020). Amelioration potential of biochar for chromium stress in wheat. *Pakistan Journal of Botany*, *52*(4), 1159–1168.

Mehmood, T., Bibi, I., Shahid, M., Niazi, N. K., Murtaza, B., Wang, H., Ok, Y. S., Sarkar, B., Javed, M. T., & Murtaza, G. (2017). Effect of compost addition on arsenic uptake, morphological and physiological attributes of maize plants grown in contrasting soils. *Journal of Geochemical Exploration*, *178*, 83–91.

Milani, P., Torres-Aguilar, P., Hamaker, B., Manary, M., Abushamma, S., Laar, A., Steiner, R., Ehsani, M., de la Parra, J., Skaven-Ruben, D., de Kock, H., Hawkes, C., Covic, N., Mitchell, C., & Taylor, J. (2022). The whole grain manifesto: From green revolution to grain evolution. *Global Food Security*, *34*, 100649.

Moreno-Jiménez, E., Fernández, J. M., Puschenreiter, M., Williams, P. N., & Plaza, C. (2016). Availability and transfer to grain of As, Cd, Cu, Ni, Pb and Zn in a barley agri-system: Impact of biochar, organic and mineral fertilizers. *Agriculture*, Ecosystems & Environment, *219*, 171–178.

Mourad, A. M., Amin, A. E. E. A. Z., & Dawood, M. F. (2021). Genetic variation in kernel traits under lead and tin stresses in spring wheat diverse collection. *Environmental and Experimental Botany*, *192*, 104646.

Mourad, A. M., Eltaher, S., Börner, A., & Sallam, A. (2023). Unlocking the genetic control of spring wheat kernel traits under normal and heavy metals stress conditions. *Plant and Soil*, *484*(1–2), 257–278.

Munir, R., Jan, M., Muhammad, S., Afzal, M., Jan, N., Yasin, M. U., Munir, I., Iqbal, A., Yang, S., Zhou, W., & Gan, Y. (2023). Detrimental effects of Cd and temperature on rice and functions of microbial community in paddy soils. *Environmental Pollution*, *324*, 121371.

Murugaiyan, V., Ali, J., Mahender, A., Aslam, U. M., Jewel, Z. A., Pang, Y., & Li, Z. (2019). Mapping of genomic regions associated with arsenic toxicity stress in a backcross breeding populations of rice (Oryza sativa L.). *Rice*, *12*, 1–14.

Noort, M. W., Renzetti, S., Linderhof, V., du Rand, G. E., Marx-Pienaar, N. J., de Kock, H. L., & Taylor, J. R. (2022). Towardssustainable shifts to healthydiets and foodsecurity in sub Saharan Africa with climate-resilient crops in bread-type products: A food system analysis. *Foods*, *11*(2), 135.

Norton, G. J., Travis, A., Ruang-Areerate, P., Nicol, G. W., Adeosun, A. A., Hossain, M., & Price, A. H. (2021). Genetic loci regulating cadmium content in rice grains. *Euphytica*, *217*, 1–16.

Norton, G. J., Travis, A. J., Talukdar, P., Hossain, M., Islam, M. R., Douglas, A., & Price, A. H. (2019). Genetic loci regulating arsenic content in rice grains when grown flooded or under alternative wetting and drying irrigation. *Rice*, *12*(1), 1–15.

Pan, X., Li, Y., Liu, W., Liu, S., Min, J., Xiong, H., & Li, X. (2020). QTL mapping and candidate gene analysis of cadmium accumulation in polished rice by genome-wide association study. *Scientific Reports*, *10*(1), 11791.

Pandey, B., Suthar, S., & Chand, N. (2022). Effect of biochar amendment on metal mobility, phytotoxicity, soil enzymes, and metal-uptakes by wheat (Triticum aestivum) in contaminated soils. *Chemosphere*, *307*, 135889.

Pandey, V. C. (2012). Phytoremediation of heavy metals from fly ash pond by Azolla caroliniana. *Ecotoxicology and Environmental Safety*, *82*, 8–12.

Patra, M., & Archana, S. (2000). Mercury toxicity in plants. *The Botanical Review*, *66*, 379–422. https://doi.org/10.1007/BF02868923

Philomene, N., & Komariah, S. (2021). Harmful impacts of heavy metal contamination in the soil and crops grown around dumpsites. *Reviews in Agricultural Science*, *9*, 271–282, https://doi.org/10.7831/ras.9.0_271

Praharaj, S., Skalicky, M., Maitra, S., Bhadra, P., Shankar, T., Brestic, M., Hejnak, V., Vachova, P., & Hossain, A. (2021). Zinc biofortification in food crops could alleviate the zinc malnutrition in human health. *Molecules*, *26*(12), 3509. https://doi.org/10.3390/molecules26123509

Rafiq, M., Shahid, M., Abbas, G., Shamshad, S., Khalid, S., Niazi, N. K., & Dumat, C. (2017). Comparative effect of calcium and EDTA on arsenic uptake and physiological attributes of Pisum sativum. *International Journal of Phytoremediation*, *19*, 662–669. https://doi.org/10.1080/15226514.2016.1278426

Rahim, W., Khan, M., Al Azzawi, T. N. I., Pande, A., Methela, N. J., Ali, S., & Yun, B. W. (2022). Exogenously applied sodium nitroprusside mitigates lead toxicity in rice by regulating antioxidants and metal stress-related transcripts. *International Journal of Molecular Sciences*, *23*(17), 9729.

Rashid, A., Ram, H., Zou, C. Q., Rerkasem, B., Duarte, A. P., Simunji, S., Yazici, A., Guo, S., Rizwan, M., Bal, R. S., Wang, Z., Malik, S. S., Phattarakul, N., de Freitas, R. S., Lungu, O., Barros, V., & Cakmak, I. (2019). Effect of zinc-biofortified seeds on grain yield of wheat, rice, and common bean grown in six countries. Journal of Plant Nutrition and Soil Science, *182*, 791–804. https://doi.org/10.1002/jpln.201800577

Rashid, A., Schutte, B. J., Ulery, A., Deyholos, M. K., Sanogo, S., Lehnhoff, E. A., & Beck, L. (2023). Heavy metal contamination in agricultural soil: Environmental pollutants affecting crop health. *Agronomy*, *13*, 1521. https://doi.org/10.3390/agronomy13061521

Raza, A., Tabassum, J., Zahid, Z., Charagh, S., Bashir, S., Barmukh, R., & Varshney, R. K. (2022). Advances in "omics" approaches for improving toxic metals/metalloids tolerance in plants. *Frontiers in Plant Science*, *12*, 794373.

Reeves, R. D., & Baker, A. J. M. (2000). Metal-accumulating plants. In I. Raskin & B. D. Ensley (Eds.), *Phytoremediation of toxic metals: Using plants to clean up the environment* (pp. 193–229). Wiley.

Rizwan, M., Ali, S., Ur Rehman, M. Z., Adrees, M., Arshad, M., Qayyum, M. F., & Imran, M. (2019). Alleviation of cadmium accumulation in maize (Zea mays L.) by foliar spray of zinc oxide nanoparticles and biochar to contaminated soil. *Environmental Pollution*, *248*, 358–367.

Saeed, M., Quraishi, U. M., & Malik, R. N. (2022). Identification of arsenic-tolerant varieties and candidate genes of tolerance in spring wheat (Triticum aestivum L.). *Chemosphere*, *308*, 136380.

Sayyadian, K., Moezzi, A., Gholami, A., Panahpour, E., & Mohsenifar, K. (2019). Effect of biochar on cadmium, nickel and lead uptake and translocation in maize irrigated with heavy metal contaminated water. *Applied Ecology & Environmental Research*, *17*(1), 969–982.

Shahid, M., Dumat, C., Khalid, S., Schreck, E., Xiong, T., & Niazi, N. K. (2017). Foliar heavy metal uptake, toxicity and detoxification in plants: A comparison of foliar and root metal uptake. *Journal of Hazardous Materials*, *325*, 36–58. https://doi.org/10.1016/j.jhazmat.2016.11.063

Shetty, R., & Prakash, N. B. (2020). Effect of different biochars on acid soil and growth parameters of rice plants under aluminium toxicity. *Scientific Reports*, *10*(1), 12249.

Smith, L. E., & Siciliano, G. (2015). A comprehensive review of constraints to improved management of fertilizers in China and mitigation of diffuse water pollution from agriculture. *Agriculture, Ecosystems & Environment*, 209, 15–25.

Smulders, M. J., van de Wiel, C. C., van den Broeck, H. C., van der Meer, I. M., Israel-Hoevelaken, T. P. M., Timmer, R. D., &Gilissen, L. J. (2018). Oats in healthy gluten-free and regulardiets: A perspective. *Food Research International*, *110*, 3–10.

Srivastava, V., Sarkar, A., Singh, S., Singh, P., de Araujo Ademir, S. F., & Singh Rajeev, P. (2017). Agroecological responses of heavy metal pollution with special emphasis on soil health and plant performances. *Frontiers in Environmental Science*, *5*. https://doi.org/10.3389/fenvs.2017.00064

Sun, L., Zheng, C., Yang, J., Peng, C., Xu, C., Wang, Y., & Shi, J. (2016). Impact of sulfur (S) fertilization in paddy soils on copper (Cu) accumulation in rice (Oryza sativa L.) plants under flooding conditions. *Biology and Fertility of Soils*, *52*, 31–39.

Tahir, M., Mirza, M. S., Hameed, S., Dimitrov, M. R., & Smidt, H. (2015). Cultivation-based and molecular assessment of bacterial diversity in the rhizosheath of wheat under different crop rotations. *PLoS One*, *10*, 0130030.

Tao, Z., Wang, D., Chang, X., Wang, Y., Yang, Y., & Zhao, G. (2018). Effects of zinc fertilizer and short-term high temperature stress on wheat grain production and wheat flour proteins. *Journal of Integrative Agriculture*, *17*(9), 1979–1990. https://doi.org/10.1016/S2095-3119(18)61911-2

Taylor, J. R. (2019). Sorghum and millets: Taxonomy, history, distribution, and production. In J. R. N. Taylor & K. G. Duodu (Eds.), *Sorghum and millets* (pp. 1–21). AACC International Press.

Udovic, M., & McBride, M. B. (2012). Influence of compost addition on lead and arsenic bioavailability in reclaimed orchard soil assessed using Porcellio scaber bioaccumulation test. *Journal of Hazardous Materials*, *205*, 144–149.

Ueno, D., Yamaji, N., Kono, I., Huang, C. F., Ando, T., Yano, M., & Ma, J. F. (2010). Gene limiting cadmium accumulation in rice. *Proceedings of the National Academy of Sciences*, *107*(38), 16500–16505.

Vasilachi, I. C., Stoleru, V., & Gavrilescu, M. (2023). Analysis of heavy metal impacts on cereal crop growth and development in contaminated soils. *Agriculture*, *13*, 1983. https://doi.org/10.3390/agriculture13101983

Wang, K., Yan, T. Z., Xu, S. L., Yan, X., Zhou, Q. F., Zhao, X. H., & Yang, Y. Z. (2021). Validating a segment on chromosome 7 of japonica for establishing low-cadmium accumulating indica rice variety. *Scientific Reports*, *11*(1), 6053.

Wang, M. Y., Chen, A. K., Wong, M. H., Qiu, R. L., Cheng, H., & Ye, Z. H. (2011). Cadmium accumulation in and tolerance of rice (Oryza sativa L.) varieties with different rates of radial oxygen loss. *Environmental Pollution*, *159*(6), 1730–1736.

Wu, S., Zhang, X., Sun, Y., Wu, Z., Li, T., Hu, Y., & Chen, B. (2015). Transformation and immobilization of chromium by arbuscular mycorrhizal fungi as revealed by SEM–EDS, TEM–EDS, and XAFS. *Environmental Science & Technology*, *49*(24), 14036–14047.

Wu, Z., McGrouther, K., Huang, J., Wu, P., Wu, W., & Wang, H. (2014). Decomposition and the contribution of glomalin-related soil protein (GRSP) in heavy metal sequestration: Field experiment. *Soil Biology and Biochemistry*, *68*, 283–290.

Xu, H., Yan, J., Qin, Y., Xu, J., Shohag, M. J. I., Wei, Y., & Gu, M. (2020). Effect of different forms of selenium on the physiological response and the cadmium uptake by rice under cadmium stress. *International Journal of Environmental Research and Public Health*, *17*(19), 6991.

Xu, X. Y., McGrath, S. P., Meharg, A. A., & Zhao, F. J. (2008). Growing rice aerobically markedly decreases arsenic accumulation. *Environmental Science & Technology*, *42*(15), 5574–5579.

Yildiztugay, A., Ozfidan-Konakci, C., Yildiztugay, E., & Kucukoduk, M. (2019). Biochar triggers systemic tolerance against cobalt stress in wheat leaves through regulation of water status and antioxidant metabolism. *Journal of Soil Science and Plant Nutrition*, *19*, 935–947.

Zagorchev, L., Seal, C. E., Kranner, I., & Odjakova, M. (2013). A central role for thiols in plant tolerance to abiotic stress. *International Journal of Molecular Sciences*, *14*, 7405–7432.

Zarea, M. J., & Karimi, N. (2023) Grain yield and quality of wheat are improved through post-flowering foliar application of zinc and 6-benzylaminopurine under water deficit condition. *Frontiers in Plant Science*, *13*, 1068649. https://doi.org/10.3389/fpls.2022.1068649

Zhan, F., Li, B., Jiang, M., Yue, X., He, Y., Xia, Y., & Wang, Y. (2018). Arbuscular mycorrhizal fungi enhance antioxidant defense in the leaves and the retention of heavy metals in the roots of maize. *Environmental Science and Pollution Research*, *25*, 24338–24347.

Zhang, W., & Wang, X. (2021). Quantitative evaluation of the grain zinc in cereal crops caused by phosphorus fertilization. A meta-analysis. *Agronomy for Sustainable Development*, *41*, 6. https://doi.org/10.1007/s13593-020-00661-0

Zhao, X., Luo, L., Cao, Y., Liu, Y., Li, Y., Wu, W., & Lin, H. (2018). Genome-wide association analysis and QTL mapping reveal the genetic control of cadmium accumulation in maize leaf. *BMC Genomics*, *19*(1), 1–13.

Zheng, R., Chen, Z., Cai, C., Tie, B., Liu, X., Reid, B. J., & Baltrėnaitė, E. (2015). Mitigating heavy metal accumulation into rice (Oryza sativa L.) using biochar amendment – A field experiment in Hunan, China. *Environmental Science and Pollution Research*, *22*, 11097–11108.

17 Molecular Aspects of Abiotic Stress Tolerance in Horticultural Crops

Challenges and Prospects

Lokesh Thakur, Vinay Kumar Dhiman, Devendra Singh, Vivek Kumar Dhiman, Himanshu Pandey, and Avinash Sharma

17.1 INTRODUCTION

Biological stress is described as situations that create undesirable or harmful effects in organisms or populations, resulting in organisms being unable to completely express their genetic potential for growth, development, and reproduction. From an agricultural standpoint, it may be described as any event or process that results in either decreased crop production and nutritional content or the loss of crop biomass. Plant stresses may be classified into two categories depending on the elements that cause them, namely biotic and abiotic stresses. Biotic factors include live creatures for instance pests and diseases, whereas abiotic factors include environmental elements like radiation, temperature, and water (Kashyap et al., 2023). Plants are regularly subjected to abiotic stresses such as high temperature (hot), low temperature (cold), flood, drought, soil and water salinity, and oxidative stress, which all adversely impact plant production (Prasad et al., 2023). In important crops, adverse climatic and soil conditions might result in a 70% yield loss (Boyer, 1982).

17.2 ABIOTIC STRESSES ON HORTICULTURAL CROPS

Horticulture crops are regularly subjected to a broad range of abiotic stresses, which are among the primary causes of production and yield loss (Francini and Sebastiani, 2019). These stresses include salt, water availability extremes (flood and drought), temperature extremes (low and high), and others like wind and hail, nutrition shortage, metal toxicity, air pollution, UV exposure, photoinhibition, and gaseous deprivation (Yadav et al., 2023).

17.2.1 Salinity Stress

Salts are a necessary component of soil, in general, and the majority of them (such as potassium and nitrates) are needed plant nutrients. Inorganic fertilizers, mineral weathering, soil changes (e.g., manures and composts), and irrigation fluids all contribute to soil salts (Kotuby-Amacher et al., 2000). Irrigation of crops, in particular, greatly increases and intensifies the way of soil salinization. As a result, from the standpoint of salinity, irrigation has the effect of introducing a considerable quantity of additional salts to the soil that were not previously there (Munns et al., 2004). Soil salinity is a major limiting factor in crop productivity. Arid and semi-arid areas cover over 40% of the earth's land area, and soil salinity is a developing concern in these places (Fischer and Turner, 1978).

DOI: 10.1201/9781003457237-17

17.2.2 Drought Stress

Water stress is an extremely important environmental stress for plant production globally, whether it is in short supply (drought) or abundant (flood). Drought can be described as a lengthy period of low rainfall with major economic, social, and ecological consequences (Folger and Cody, 2014). Extreme droughts occur on occasion in various food-experiencing nations, having a significant impact on food production and world supplies. Due to severe droughts in many regions, worldwide grain output has decreased by 5% or more for several years (Ashraf and Harris, 2005). Water shortages in plants are caused by factors such as decreased rainfall, the presence of salt, harsh temperatures, and excessive light intensity.

17.2.3 Flood Stress

Flood stress is a challenging global environmental issue that happens all over the planet and produces changes within the tissues of plants (Perata et al., 2011). Flooding is an anoxic condition in which the soil is completely saturated with water (Jackson and Colmer, 2005). Worldwide, flooding has impacted almost 33% of the irrigated land. Flooding is majorly caused by severe rainfall, heavy soil texture, poor drainage, excessive irrigation, river overflow, etc. (Rao et al., 2016). Flood stress caused evident damage to the shoots, causing withering and abscission, especially in the lower plant portions.

17.2.4 Temperature Extremes

Temperature stress is a phenomenon in which prolonged exposure to temperature in excess of or below the set threshold limits has a negative impact on cellular and tissue metabolism, resulting in retardation in growth or injury and, ultimately, restricting yields based on genetic potential (Greaves, 1996). Horticultural crops are divided into three categories based on temperature requirements: tropical crops (e.g., guava, pineapple, banana, coconut, and papaya), subtropical crops (e.g., litchi, citrus, pomegranate, mango, and grape), and temperate crops (e.g., walnut, almond, cherry, apple, pear, and strawberry). Stress caused by temperature extremes in plants is usually divided into two distinct categories:

a. **High-Temperature Stress**

High-temperature damage begins with an increased respiration process caused by photosynthesis and progresses to secondary impacts such as stored food depletion and related stress factors such as famine (Levitt, 1980). High temperatures can affect plants on a daily or seasonal basis. Plants may be harmed by both high temperatures and prolonged periods of moderate temperature, however, how they deal with these stresses varies. Heat stress causes aggregation and denaturation of proteins while also increasing the flexibility of membrane lipids, affecting membrane structure (Uprety and Reddy, 2016).

b. **Low-Temperature Stress**

Plant injury caused by low temperatures is classified into two categories: chilling or freezing stress. Raison and Lyons (1986) define chilling stress in terms of two critical components: temperature and time. Chilling stress is defined as a reduction in the outside temperature that does not cause ice formation inside the cells of plants and typically occurs between 0°C and 15°C (Mckersie et al., 1994). Freezing stress, on the contrary, takes place when the temperature falls below 0°C or the freezing level of water, causing ice nucleation and consequent ice accumulation inside the tissues of plants. Frequently, the nucleation process begins in locations with the least negative osmotic potential. These sections have precise form and size constraints, which are frequently related to cell wall components.

As soon as the process of ice nucleation begins, additional nucleation can occur on the surface of existing ice crystals. The shape and chemistry of the plant's cell wall play an important role in defining the location of ice nucleation sites within the tissue. In the absence of accessible nucleation sites, water can cool below its freezing point without crystallizing into ice.

17.2.5 Nutritional Deficiency Stresses

A common reason for nutrient deficiency in crops consists of decreased dependence on organic manure, reluctance to apply high-density planting, insufficient use of rootstocks to develop disease and salt-resistant varieties, disproportionate usage of nitrogenous, phosphatic, and potassium fertilizers, as well as insufficient outreach programs for distributing horticultural innovations to marginal lands. To maximize crop productivity and generate high-quality agricultural products, it is critical to diagnose shortages of nutrients early on, before the appearance of symptoms (Dhiman et al., 2022b; Sharma et al., 2023b). To deal with prominent and hidden micronutrient deficits, spraying on the foliage or direct soil administration of micronutrients is critical. Plants can suffer nutritional stress not solely from nutrient deficits but also from nutrient surpluses, emphasizing the significance of keeping optimum nutrient levels for ensuring productive and healthy crops.

17.2.6 Metal Toxicity

Several human activities, such as mining, modern farming practices, and industrialization, have long-term negative effects on our natural environment (Singh and Verma, 2023). All of these causes contribute to a rise in heavy metals such as cobalt (Co), cadmium (Cd), nickel (Ni), mercury (Hg), lead (Pb), copper (Cu), chromium (Cr), zinc (Zn), and arsenic (As) concentrations in soil, water, and air. Heavy metal poisoning in soil causes a number of environmental issues as well as a harmful effect on plants (Tiwari and Lata, 2018). Beyond a certain concentration, these can reduce agricultural production, induce morphological changes, increase reactive oxygen species (ROS), and cause metabolic problems (Sachdev et al., 2021).

17.3 MOLECULAR MECHANISMS OF ABIOTIC STRESS RESPONSES

17.3.1 Perception of Stress

Plants have evolved elaborate systems to recognize and detect diverse abiotic stresses in their environment, kicking off a complicated chain of biochemical reactions that eventually result in adaptive responses. This perceptual capacity comprises many fundamental components: To begin, plant cells have specialized receptors on their membranes that can detect certain stress signals. For example, the formation of ROS such as H_2O_2 and superoxide radicals is widespread in response to numerous abiotic conditions like harsh temperatures, drought, or excessive salt (Sachdev et al., 2021). Surprisingly, plants have receptors that detect changes in ROS levels and convert them into signaling molecules that activate specialized stress responses. Furthermore, osmoreceptors are activated under instances of osmotic stress like dehydration or salinity stress (Ma et al., 2022). Osmoreceptors are capable of responding to changes in cell turgor pressure or ion concentrations, allowing plants to detect changes in their osmotic environment. Second, protein kinases are important in the plant's capacity to detect stress. Specific protein kinases are activated when a stressor is detected, triggering a cascade of downstream signaling pathways. In reaction to stress, mitogen-activated protein kinases are commonly activated (Soares-Silva et al., 2016). These kinases perform an important part in phosphorylating target proteins, such as transcription factors. This phosphorylation process acts as an activation signal for stress-responsive genes, directing a cellular response to the stressor encountered. Another group of important participants in stress perception is the calcium-dependent protein kinases (CDPKs). They act as calcium sensors, activating stress-signaling pathways in response to increasing cytosolic calcium levels. These pathways are crucial in connecting the experience of stress to the subsequent cellular responses. Finally, plant hormones like abscisic acid (ABA) play an important part in stress perception. ABA, for instance, is important in the plant's reaction to many stresses, including drought stress. Plants synthesize and accumulate ABA in response to decreasing soil moisture levels. Once produced, this hormone is responsible for initiating downstream reactions such as the closing of stomata, which are tiny holes located on the surface of leaves. This stomatal closure is an important adaptive step done by the plant during drought circumstances to control water loss, conserve valuable water resources, and prevent desiccation. As a result, these sophisticated stress-sensing mechanisms in plants provide the groundwork for their eventual reactions to a dynamic and frequently demanding environment (Muhammad Aslam et al., 2022) (Figure 17.1).

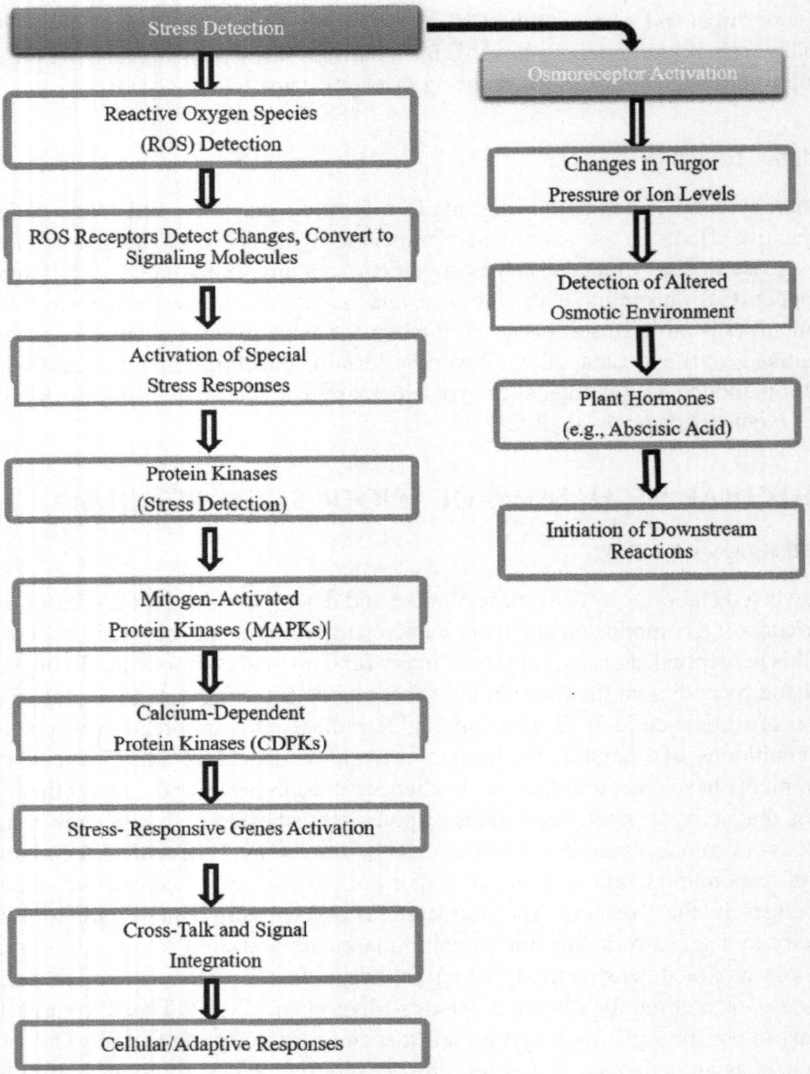

FIGURE 17.1 Flowchart showing the perception of stress.

17.3.2 Stress-Responsive Genes

Plants go through a complex biochemical journey in reaction to abiotic stress, which is primarily controlled by the stimulation of stress-responsive genes. These genes act as frontline defenders, launching a slew of protective and adaptive processes that are carefully designed to address the unique problems provided by the encountered stressor. In this regulatory network, transcription factors emerge as critical participants. These proteins function as master regulators, attaching to particular DNA sequences inside target gene promoter regions and either activating or inhibiting transcription (Chauhan et al., 2023a). Drought-responsive transcription factors, or DREBs, become increasingly prominent as a result of drought and cold stress. They control the expression of genes related to osmoregulation, letting the plant adapt to changes in water availability. These transcription factors regulate ABA-responsive gene expression, enhancing the ability of plants to respond to stress. Stress-responsive genes encode effector proteins, which act as direct protectors against the stressor's negative consequences. Dehydrins, for example, build in response to drought stress, acting as molecular shields to protect cellular structures from desiccation and water loss harm.

Antioxidant enzymes, such as peroxidases and superoxide dismutase (SOD), are essential in combating oxidative stress caused by the stressor. They neutralize potentially hazardous ROS, protecting cellular redox equilibrium (Sharma et al., 2022). Furthermore, these genes have altered metabolic pathways, allowing plants to change their biochemistry to better cope with stress. Osmoprotectant-producing genes are activated in reaction to osmotic stress. They regulate the synthesis of osmoprotectants like soluble carbohydrates and proline, which aid in maintaining cellular osmotic equilibrium and avoiding excessive water loss.

Heat shock protein (HSP) genes come to the fore in conditions of heat stress. These proteins help in protein folding, stability, and refolding, ensuring that vital cellular functions continue even as temperatures rise. Transporter proteins, which are frequently produced by stress-responsive genes, allow ions or metabolites to flow within plant cells. They are essential for maintaining ion homeostasis and aiding with osmotic adjustment, both of which are important components of stress tolerance (Singh et al., 2022b). Furthermore, epigenetic regulation, which involves changes such as DNA methylation and histone alterations, adds another degree of complexity to the control of stress-responsive genes. These epigenetic changes can impact chromatin accessibility, influencing the expression of genes important for stress response.

17.4 MOLECULAR MECHANISMS OF STRESS RESPONSE

17.4.1 Water Uptake and Transport: Regulation of Water Stress

Plants use a complex collection of molecular processes to secure their survival and retain key functioning in the face of drought stress. The mechanism of water intake and transport, which begins with root uptake, is important to this reaction. Water absorption begins in the roots, where the coordinated activity of root hairs and specialized transport proteins allows water from the soil to enter the root cells. This mechanism is primarily driven by osmotic potential, with water migrating from lower solute concentration locations in the soil to greater solute concentration regions within the root cells. When water is absorbed by the roots, it begins an amazing trip upward through the plant, aided by xylem tissue. The cohesion–tension hypothesis explains how transpiration, or the loss of water from leaves, causes tension inside the xylem. This tension, together with the cohesive qualities of water molecules and their attachment to the inner walls of xylem vessels, drives water upward, allowing a continuous flow to occur throughout the plant. The modulation of aquaporins, integral membrane proteins present in a variety of plant tissues including root cells and leaf stomata, is a critical component of drought response. These proteins function as specialized water channels, allowing water molecules to travel quickly across cell membranes while effectively inhibiting solute passage. Plants have the ability to adjust the activity of aquaporins under drought circumstances. This regulation provides two functions: managing water absorption and reducing water loss through transpiration. Plants optimize their water balance by fine-tuning aquaporin activity, increasing their chances of survival in the face of restricted water supplies (Singh et al., 2022d). Furthermore, plants use osmotic adjustment mechanisms to maintain cellular turgor pressure during drought stress, which is an important step in preventing wilting and cellular damage. This adaptation involves the buildup of osmoprotectants within plant cells, such as proline and soluble carbohydrates. These organic compounds enhance the solute content in the cell, lowering the osmotic gradient between the cell and the surrounding parched soil and therefore minimizing water loss. Furthermore, several plant species can change the characteristics of their cell walls, increasing cell wall flexibility. This adaptation allows cells to withstand turgor pressure variations without collapsing. Stomatal regulatory systems also contribute to drought resistance. Stomata, or microscopic holes on leaf surfaces, regulate plant gas exchange and water loss. Plants trigger distinct mechanisms to preserve water during drought stress. ABA, a phytohormone crucial in drought response, is crucial. Plants synthesize and collect ABA when soil moisture drops, which is transferred from roots to leaves via the xylem. ABA causes a number of reactions in the leaf, including stomatal closure. Guard cells that encircle stomatal holes lose turgor pressure as a result of ABA signaling, prompting them to contract and seal the stomatal apertures. ABA also increases the production of osmoprotectants like proline and soluble sugars, which aids in osmotic adjustment and cellular protection. Other elements, such as humidity, light, and internal CO_2 concentrations, interact with ABA signaling to fine-tune stomatal responses, improving the plant's capacity to deal with drought.

17.4.2 Ion Transporters: Regulation of Salt Stress

Salt stress is a major concern for plants because it upsets the delicate ion balance within their cells, mostly as a result of high amounts of salt ions like sodium (Na^+) and chloride (Cl^-) in salty soils. Plants use various sophisticated techniques to combat the adverse effects of salt stress and maintain ion homeostasis. Selective ion absorption is an important process that allows plants to absorb vital ions like potassium (K^+) while avoiding detrimental ions like sodium. This selective absorption is accomplished by specialized ion transporters found in root cell membranes. These transporters guarantee that important nutrients are received while also preventing harmful ions from accumulating inside the plant. Some plant species have evolved specialized ion exclusion mechanisms in addition to selective absorption. These methods involve sodium transporters that actively pump sodium out of root cells, preventing it from being transported to above-ground tissues. This activity lowers the likelihood of salt toxicity in plant cells. Ion transporters such as HKT (high-affinity K^+ transporter) and SOS1 (salt overly sensitive 1) play crucial roles in controlling ion homeostasis under salt stress. HKT transporters perform a crucial part in maintaining the equilibrium of sodium and potassium inside plant cells. They promote potassium intake while excluding sodium, reducing excessive salt buildup. SOS1 transporters, on the contrary, are in charge of aggressively moving sodium ions out of root cells and into the soil, dramatically lowering sodium concentrations within plant cells. By reducing the water potential in the surrounding soil, salt stress also causes osmotic stress in plant cells. This osmotic action has numerous repercussions, including water loss from plant cells, which causes cellular dryness and lower turgor pressure. As a result of this osmotic stress, cell expansion and growth are inhibited, resulting in decreased overall plant development. Plants collect suitable solutes like proline and soluble carbohydrates like sucrose within their cells to counteract the effects of osmotic stress. These compatible solutes contribute to osmotic equilibrium by boosting solute concentration inside the cell, preventing excessive water loss from the cell. Furthermore, as previously stated, ion transporters play an essential part in the osmotic stress response by managing the intake and transportation of ions such as potassium and sodium. These ions help in osmotic adjustment, allowing the plant to adapt and survive under salt stress circumstances.

17.4.3 Plant Antifreeze Proteins: Regulation of Cold Stress

Plants go through a process referred to as cold acclimation in response to cold stress, which equips them with enhanced cold tolerance to live and thrive in low-temperature environments. This adaptation incorporates numerous chemical pathways that work together to improve a plant's capacity to endure cold settings. The shift in gene expression patterns is a key feature of cold adaptation. Plants respond to cold stress by upregulating certain cold-responsive genes. Among these genes are those that code for proteins that play critical roles in cold tolerance. Antifreeze proteins, which restrict the production of ice crystals inside plant tissues, and dehydrins, which operate as protective agents insulating cells from harm caused by freezing temperatures, are two notable examples. The alteration of cell membrane characteristics is another key aspect of cold adaptation. Plants adjust the composition and fluidity of their cell membranes to operate well under cooler circumstances. This adaptation involves changes in lipid content, which allows membranes to maintain structural integrity and performance at lower temperatures. The CBF/DREB transcription factors are essential for orchestrating the response against cold stress. These transcription factors play a critical part in the plant's defense against cold injury. CBF/DREB transcription factors express the cold-responsive genes by binding to their promoters. As a result, the creation of protective proteins increases, and many cellular processes are modified to better tolerate the rigors of cold settings. In summary, CBF/DREB transcription factors play a critical part in allowing plants to generate a robust and effective response to cold stress, guaranteeing their survival and adaptability in cold climates.

17.4.4 HSPs: Regulation of Heat Stress

Plants have a sophisticated range of mechanisms to safeguard their cellular structures and maintain critical physiological processes when exposed to high temperatures. These reactions combine to form the plant's heat stress response, which is critical for survival and adaptability to high temperatures. The creation of HSPs, also known as chaperones, is an important component of the heat stress response. When subjected to heat stress, these specialized proteins serve a critical role in ensuring protein stability. HSPs act as molecular carers, helping in the proper folding of proteins and avoiding denaturation, which is a frequent result of high temperatures. HSPs guarantee that important molecules may continue to function efficiently by aiding the refolding of damaged proteins. In essence, HSPs and chaperones protect protein integrity, eventually boosting plant cell viability under heat stress conditions. Another crucial element of the heat stress response is the increase in antioxidant enzyme synthesis. High temperatures can enhance oxidative stress inside plant cells by causing the formation of damaging ROS. Plants increase the production of antioxidant enzymes such as glutathione peroxidase and SOD to offset the adverse effects of oxidative damage. These enzymes serve to maintain cellular redox equilibrium and safeguard plant cells from oxidative injury by neutralizing ROS (Dhiman et al., 2021; Chauhan et al., 2022).

17.4.5 Phytochelation: Regulation of Heavy Metal Stress

Plants have evolved complex ways to deal with heavy metal stress, which is a severe environmental concern in many areas. These processes include metal absorption, detoxification, and antioxidant defense, all of which strive to preserve cellular homeostasis and protect against the harmful effects of heavy metals. Plants absorb harmful substances from the soil via their root systems, which is the first step in heavy metal absorption. This mechanism is dependent on specialized transporters found on root cell membranes that aid in the absorption of heavy metals like zinc (Zn), lead (Pb), and cadmium (Cd). Heavy metals, once within the plant, can endanger cellular integrity and function. Plants, on the contrary, have developed chelation processes as a defense measure. Chelators, such as phytochelatins, bind firmly to heavy metals and form complexes. The harmful metals are successfully sequestered by these chelates, preventing them from interfering with important cellular functions (Singh et al.,

2022c). Metal transporters and chelation processes play an essential part in the plant's heavy metal stress response. These transporters aid in the selective intake of necessary ions while actively rejecting or sequestering toxic heavy metals (Dhiman et al., 2022a). Some vacuolar membrane transporters aid in the sequestration of heavy metals inside vacuoles, restricting their entrance into sensitive cellular compartments. Exposure to heavy metals can also cause oxidative stress within plant cells. Heavy metals can cause the formation of ROS, such as superoxide radicals (O^{2-}) and hydrogen peroxide (H_2O_2), which can induce cellular damage. Plants use an arsenal of antioxidants to resist this harm. SOD enzymes serve a critical role in minimizing the detrimental effects of superoxide buildup by converting superoxide radicals to hydrogen peroxide. Furthermore, glutathione peroxidase functions as an important antioxidant enzyme by detoxifying hydrogen peroxide and converting it to harmless water molecules, reducing cellular damage induced by excessive ROS.

17.4.6 Physiological and Biochemical Regulation

Plants have a broad range of physiological and biochemical reactions to deal with abiotic stresses in their environment. These responses are complex and interrelated systems that allow plants to survive and grow in harsh environments. Plants, for example, use the hormone ABA to control the opening and shutting of stomata, or small holes on their leaves, in response to drought stress. This reduces water loss via transpiration but may restrict carbon dioxide intake for photosynthesis. Furthermore, in response to water constraints, plants frequently store suitable solutes such as proline and soluble carbohydrates to maintain cell turgor and safeguard cellular structures. Abiotic stresses also stimulate the development of antioxidants such as catalase and SOD, which detoxify toxic ROS created during stress.

HSPs are proteins that help denatured proteins refold and avoid protein aggregation when exposed to high temperatures. Similarly, cold stress causes the development of antifreeze proteins as well as changes in membrane lipid composition to protect against ice crystal formation. Plants respond to nutrient scarcity by changing their root architecture, secreting organic acids, and creating symbiotic relationships with mycorrhizal fungi. Plants regulate soil pH in response to alkaline or acidic circumstances by adjusting root exudation and proton pump activity, which improves nutrient absorption. Aside from its involvement in stomatal control, the hormone in many abiotic stress reactions, ABA, functions as a key signaling molecule, activating stress-responsive genes. Excessive light can impair the photosynthetic machinery, causing plants to develop photoprotection mechanisms to remove excess light energy and protect photosystems from damage. Plants may also vary their cell wall composition, fortifying it through enhanced lignification or changing the structure of pectin to retain cell integrity under stress situations (Negi et al., 2021). These complicated biochemical and physiological responses are highly adapted to each plant species at different developmental stages. Understanding these reactions is critical for designing methods to improve agricultural resilience and production in the face of adversity, eventually contributing to global food security (Sharma et al., 2023a).

17.4.7 Phyto-hormones in Stress Regulation

Plant hormones play a vital part in coordinating abiotic stress responses, and the three main hormones implicated in this sophisticated regulation network are ABA, cytokinins, and ethylene (Singh et al., 2022b).

ABA is a key participant in stress reactions, particularly in water-stressed circumstances such as drought and excessive salinity. When plants detect a decrease in soil moisture, their ABA levels rise, activating a series of adaptive reactions. One of the most noticeable impacts is the closing of stomata, which are small openings on leaves that minimize water loss through transpiration. Additionally, ABA encourages the buildup of osmoprotectants like proline and soluble carbohydrates, which assists in cellular osmotic adjustment. Furthermore, ABA increases the stress-responsive gene's expression, resulting in the enhanced ability of plants to tolerate harsh circumstances.

Cytokinins, on the contrary, have the opposite effect. These hormones are linked to the promotion of cell division and growth. Cytokinin levels frequently fall in stressful environments, such as drought or severe heat. This decrease in cytokinin levels assists plants in prioritizing resource allocation for stress tolerance over development, resulting in decreased shoot and root growth. Plants can improve their odds of surviving under adverse environmental circumstances by reallocating resources.

Ethylene, sometimes known as the 'stress hormone', has multiple roles in abiotic stress reactions. Stresses such as floods, heat, and mechanical damage generally boost ethylene production. Ethylene may cause a variety of reactions, including fruit ripening and senescence, but it can also trigger adaptive changes in the setting of abiotic stress. Under waterlogged conditions, for example, ethylene promotes the growth of adventitious roots, allowing the plant to acquire oxygen more efficiently. Furthermore, ethylene, like ABA, can modulate stomatal closure, assisting in the reduction of water loss during stress.

17.4.7.1 Hormonal Crosstalk during Abiotic Stress Responses

Hormone crosstalk is an important regulatory mechanism in horticulture crop response to abiotic stresses such as severe temperatures, drought, salt, and nutritional deficits. Among the primary hormones involved, ABA is critical for stress response. ABA frequently works in tandem with ethylene (ET), especially under drought and salt stress, where they work synergistically to improve stress tolerance. Gibberellins (GA), on the contrary, can offset ABA effects by stimulating development, emphasizing their antagonistic connection. Auxins, which are predominantly growth regulators, also participate in stress responses, sometimes counteracting the impacts of ABA during drought. Cross talk (CTs), which have a role in cell division, and brassinosteroids, which have a part in growth and stress response, interact with ABA and other hormones to fine-tune stress adaptation. Furthermore, ethylene can regulate auxin production and transport, influencing root development, which is critical for stress tolerance. The interaction of these hormones affects crops' complicated responses to abiotic stress, providing insights for crop development techniques that optimize growth and stress tolerance in tough environmental circumstances (Figure 17.2).

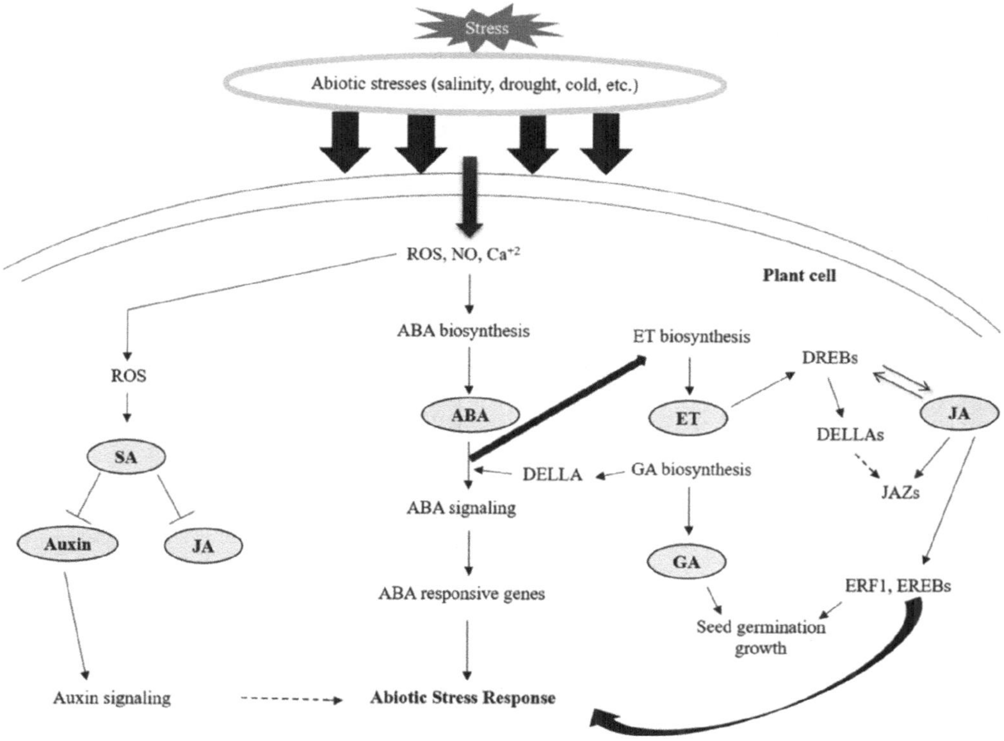

FIGURE 17.2 Schematic representation of stress response.

17.4.7.1.1 Breeding for Stress Tolerance

Crop breeding for abiotic stress resistance is an important strategy for addressing the mounting problems provided by climate change and poor environmental conditions. This method entails the creation of crop types that can endure and grow in the face of environmental challenges like drought, salt, severe temperatures, and nutrient deficiencies (Sunya et al., 2022). Plant breeders use a variety of tactics and technology to accomplish this. Traditional breeding is a fundamental strategy that involves selecting and crossbreeding plants with optimal stress tolerance features across numerous generations. This strategy is based on crop species' genetic diversity and tries to generate superior varieties through natural genetic variation. Traditional breeding, on the contrary, is typically time-consuming and may not give quick answers to rising stress-related disorders. Breeding for abiotic stress tolerance has been transformed by modern biotechnological technologies. Specific genes responsible for stress responses can be introduced or modified using genetic engineering techniques. For example, introducing genes encoding osmoprotectants like proline or overexpressing stress-responsive genes might improve a plant's ability to withstand drought or salt. Furthermore, genome-editing technologies such as CRISPR/Cas9 allow for precision alteration of a plant's DNA to improve its tolerance against stress. Marker-assisted selection (MAS) is another strategy that discovers particular genetic markers related to stress tolerance features. This allows breeders to more quickly screen and select plants with desirable features, decreasing the time necessary to generate stress-tolerant cultivars. Interdisciplinary research combining plant physiology, genomics, and computational biology helps breeding programs better understand the complex genetic and molecular pathways driving stress responses. This knowledge is used to build focused breeding techniques. Modern breeding programs attempt to speed the generation of stress-tolerant crop types that can adapt to changing climatic circumstances by incorporating varied germplasm, researching wild crop relatives, and utilizing the power of genomics (Table 17.1).

17.4.8 Biotechnological Interventions for Stress Tolerance

Biotechnological treatments have transformed attempts to enhance abiotic stress tolerance in horticulture crops by tackling issues like drought, severe temperatures, salt, and nutrient deficits. These novel techniques have greatly improved crop resilience and productivity. Scientists may now add or change certain genes related to stress tolerance using genetic engineering techniques. For example,

TABLE 17.1
Different Breeding Techniques Used to Induce Abiotic Stress Resistance in Horticultural Crops

S. No.	Horticultural Crop	Abiotic Stress	Breeding Techniques	References
1.	Safflower (cultivar: Morlin)	Drought stress	Derived from a single plant selection from the F11 population	Singh et al. (2022a)
2.	Chickpea (cultivar: FLIP 87-59C)	Drought stress	Cross of ILC3843 with FLIP87	Singh et al. (2022a)
3.	Soya bean	Drought stress	Mutation breeding	Chauhan et al (2023b)
4.	Faba bean	Drought stress	Genome-wide association analysis	Gutiérrez et al (2023)
5.	Cowpea	Salinity stress	QTL analysis	Chankaew et al. (2014)
6.	Common bean	Drought stress	QTL analysis	Mukeshimana et al. (2014)
7.	Pea	Frost	QTL analysis	Klein et al. (2014)
8.	Mungbean	Salinity stress	MAS	Sehrawat et al. (2014)
9.	Potato	Freezing stress	MAS	Vega et al. (2003)
10.	Pea	Cold stress	MAS	Yang et al. (2015)

genes responsible for the generation of osmoprotectants such as glycine betaine have been inserted into crop genomes to improve salinity and drought tolerance. Stress-responsive genes that govern antioxidant defense mechanisms have also been overexpressed, reducing oxidative damage caused by stresses. Transgenic plants, which are formed by inserting foreign genes into crop genomes, have been designed to have better stress tolerance. Bt cotton, for example, is developed to resist specific insect pests, reducing stress caused by infestations. CRISPR/Cas9 genome-editing tools, for example, have permitted precise changes in particular genes connected to stress responses. This method allows for the improvement of a plant's adaptation to harsh circumstances while avoiding the introduction of alien genetic material. RNA interference (RNAi) has been used to inhibit stress-responsive gene expression, reducing the negative effect of abiotic stresses on crops. Biotechnology has also accelerated the breeding process making use of MAS. Molecular markers related to stress tolerance features are found and used for more effective plant selection with desired qualities. Omics technologies, which include genomes, proteomics, and metabolomics, have helped us better comprehend the complicated molecular and metabolic networks that drive stress reactions (Chauhan et al., 2023a). This understanding contributes to the identification of critical genes and proteins that may be targeted for biotechnological treatments. The burgeoning discipline of synthetic biology has ushered in the creation and assembly of novel genomic circuits and pathways inside plants, providing many levels of resistance to a variety of environmental events (Table 17.2).

TABLE 17.2
The Technology Used to Develop Resistance in Different Horticulture Crops

S. No.	Horticultural Crop	Abiotic Stress	Technology for Resistance/Tolerance	References
1.	Tomato	Drought	Overexpression of *miRNA169*	Zhang et al. (2011)
2.	Apple	Cold	Overexpression of *MdERF1B*	Wang et al. (2021)
3.	Rice	Salinity	Overexpression of genes associated with programmed cell death	Hoang et al. (2015)
4.	Potato	Drought	In vitro selection	Sabbah and Tal (1990)
5.	Brown mustard	Salinity	Transgenics	Singh et al. (2022a)
6.	Strawberry	Waterlogging	RNAi	Luo et al. (2022)
7.	Pepper	High light intensity	Application of 5 aminolevulinic acid	Korkmaz et al. (2010)
8.	Pineapple	Bialaphos	Bar gene; gene gun-mediated gene transfer	Sripaoraya et al. (2006)
9.	Citrus	Drought	Transgenics	De Carvalho et al. (2013)
10.	Chrysanthemum	Drought	Transgenics	Fan et al. (2016)
11.	Mulberry	Heat	Drought and salinity stress	Das et al. (2011)
12.	China rose (*Rosa chinensis*)	Salinity	Overexpression of AtDREB2A-CA gene	Josine et al. (2015)
13.	Kiwi	Salinity	Transgenics	Tian et al. (2011)
14.	Sweet potato	Salinity, chilling/cold stress, and oxidative stress	Transgenics	Fan et al. (2012)
15.	Bottle gourd	Salinity	AVP1 gene transfer	Han et al. (2015)
16.	Cherry tomato	Salinity	Transgenics	Lim et al. (2016)
17.	Lotus	Salinity	Transgenics	Liu et al. (2014)
18.	Petunia	Cold	CBF3 gene transferred from *Arabidopsis*	Warner (2010)
19.	Rhododendron	Iron stress	Fro2 gene transferred from *Arabidopsis*	Dunemann et al. (2001)
20.	Orchid (*Phalaenopsis amabilis*)	Cold	Transgenics	Qin et al. (2011)

17.5 EMERGING STRATEGIES, CHALLENGES, AND FUTURE PROSPECTS

Emerging solutions for improving crop stress tolerance pave the path for more resilient and sustainable agriculture. A cutting-edge technology, synthetic biology, permits the building of plants with custom-designed genetic circuits that can sense and respond to specific stress stimuli. This kind of accuracy has the ability to produce crops that flourish in a variety of difficult situations. In the meanwhile, epigenetic modifications are being studied as a means of producing heritable alterations in gene expression without affecting the basic sequence of DNA, improving stress responses. Another interesting path is to use the plant microbiome, which is made up of helpful bacteria. Researchers are deciphering the complicated connections between plants and their microbial partners to find species that improve stress tolerance. This knowledge may be used to create microbial inoculants for crops (Pandey et al., 2022). Multi-omics methods are helping us comprehend plant stress adaptation at the molecular level, offering significant insights for targeted therapies. As we go forward with these novel solutions, it is critical to perform thorough research to assure their safety, efficacy, and ethical application, while also providing hope for a more sustainable and food-secure future.

Abiotic stress-resistant crop development is a complicated and multi-dimensional endeavor fraught with substantial obstacles. These problems provide impediments to attaining the objective of guaranteeing sustainable agriculture and food security in the context of growing environmental stresses. The genetic complexity of abiotic stress tolerance is one of the most significant difficulties. This feature includes several genes and regulatory pathways, making it difficult to discover and control all of the essential genetic components. Furthermore, various horticulture crops demonstrate variable degrees of susceptibility to abiotic stresses, and each may necessitate specific remedies, complicating breeding efforts. Environmental variability complicates matters further since stresses vary in severity and frequency across areas and seasons. Another challenging obstacle is the lengthy breeding cycles associated with traditional breeding procedures. It can take several years to decades to establish a single crop variety, making it difficult to adapt swiftly to growing environmental concerns. Climate change, with its unexpected fluctuations in weather patterns and new pressures, adds another degree of complication. Furthermore, worries about the environmental effect of stress tolerance characteristics, regulatory barriers for genetically modified crops, and limited resources in many agricultural regions offer significant problems. Ethical concerns about genetic manipulation and the expansion of profitable crop types also play a part in the creation and adoption of stress-resistant crops. Despite these obstacles, the need to create abiotic stress-resistant crops remains critical, given the growing dangers presented by climate change and the need to secure global food security. Addressing these difficulties would need long-term efforts, multidisciplinary collaboration, breeding method innovation, and a commitment to ethical and sustainable agriculture practices.

17.6 CONCLUSION

The investigation of molecular characteristics of tolerance against abiotic stress in horticulture crops has shown both the difficulties and the enticing potential that lay ahead. Understanding the complicated signaling networks, hormone crosstalk, and genetic regulatory systems involved in stress responses has offered vital insights into how these crops may adapt to adversity. Many obstacles remain, however, including the need for in-depth functional genomics research, the creation of crop-specific stress-tolerant cultivars, and the practical use of biotechnological and synthetic biology methods in agricultural contexts. As global agriculture faces enormous problems as a result of climate change, the development of new solutions is critical. While the path ahead may be complicated, the potential advantages for improving crop resilience, guaranteeing food security, and minimizing the impact of abiotic stress on horticulture crops are compelling and necessary. We may look forward to a future where horticulture crops are better adapted to survive in a changing environment if we address these difficulties with dedication and multidisciplinary teamwork.

REFERENCES

Ashraf, M., Harris, P. J. C. (Eds.) (2005). *Abiotic Stresses: Plant Resistance through Breeding and Molecular Approaches*. New York: CRC Press.

Boyer, J. S. (1982). Plant productivity and environment. *Science*, *218*(4571), 443–448.

Chankaew, S., Isemura, T., Naito, K., Ogiso-Tanaka, E., Tomooka, N., Somta, P., ... & Srinives, P. (2014). QTL mapping for salt tolerance and domestication-related traits in Vigna marina subsp. oblonga, a halophytic species. *Theoretical and Applied Genetics*, *127*, 691–702.

Chauhan, D., Pandey, H., Prasad, G., & Singh, D. (2023a). Role of next-generation sequencing technology in crop improvement. In: K. Chandra, S. Chand, R. P. Saini, & R. Sharma (Eds.), *Smart Breeding :Molecular Interventions and Advancements for Crop Improvement* (pp. 371–390). Milton: Apple Academic Press.

Chauhan, D., Singh, D., Pandey, H., Khan, Z., Srivastava, R., Dhiman, V. K., & Dhiman, V. K. (2023b). Impact of transcription factors in plant abiotic stress: a recent advancement for crop improvement. In: D. H. González (Ed.), *Plant Transcription Factors* (pp. 271–286). London: Academic Press.

Chauhan, V., Kaushal, D., Dhiman, V. K., Kanwar, S. S., Singh, D., Dhiman, V. K., & Pandey, H. (2022). An insight in developing carrier-free immobilized enzymes. *Frontiers in Bioengineering and Biotechnology*, *10*, 794411.

Das, M., Chauhan, H., Chhibbar, A., Rizwanul Haq, Q. M., & Khurana, P. (2011). High-efficiency transformation and selective tolerance against biotic and abiotic stress in mulberry, Morus indica cv. K2, by constitutive and inducible expression of tobacco osmotin. *Transgenic Research*, *20*, 231–246.

De Carvalho, K., de Campos, M. K. F., Domingues, D. S., Pereira, L. F. P., & Vieira, L. G. E. (2013). The accumulation of endogenous proline induces changes in gene expression of several antioxidant enzymes in leaves of transgenic Swingle citrumelo. *Molecular Biology Reports: An International Journal on Molecular and Cellular Biology*, *40*, 3269 doi:10.1007/s11033-012-2402–5.

Dhiman, V. K., Chauhan, V., Kanwar, S. S., Singh, D., & Pandey, H. (2021). Purification and characterization of actinidin from Actinidia deliciosa and its utilization in inactivation of α-amylase. *Bulletin of the National Research Centre*, *45*, 1–9.

Dhiman, V. K., Rana, N., Dhiman, V. K., Pandey, H., Verma, P., & Singh, D. (2022a). Effect of rhizobial isolates and nitrogen fertilizers on nursery performance, nodulation behavior and nitrogenase activity of Dalbergia sissoo Roxb. seedlings. *Plant Stress*, *4*, 100080.

Dhiman, V. K., Singh, D., Pandey, H., Chauhan, D., Dhiman, V. K., & Pandey, D. (2022b). Mycoremediation of agricultural waste for the cultivation of edible mushroom. In: T. Aftab (Ed.), *Sustainable Management of Environmental Contaminants: Eco-friendly Remediation Approaches* (pp. 471–482). Cham: Springer International Publishing.

Dunemann, F., Illgner, R., & Stange, I. (2001, July). Transformation of Rhododendron with genes for abiotic stress tolerance. In: Eucarpia, E. van Bockstaele, J. van Huylenbroeck, & P. Debergh (Eds.), XX International Eucarpia Symposium, Section Ornamentals, Strategies for New Ornamentals-Part II : Melle, Belgium, 3-6 July, 2001 572 (pp. 113–120). Leuven: ISHS.

Fan, Q., Song, A., Jiang Zhang, T., Sun, H., & Wang, Y. (2016). CmWRKY1 enhances the dehydration tolerance of chrysanthemum through the regulation of ABA-associated genes. *PLoS One*, *11*, e0150572

Fan, W., Zhang, M., Zhang, H., & Zhang, P. (2012). Improved tolerance to various abiotic stresses in transgenic sweet potato (Ipomoea batatas) expressing spinach betaine aldehyde dehydrogenase. *PLoS One*, *7*, e37344

Fischer, R. A., & Turner, N. C. (1978). Plant productivity in the arid and semiarid zones. *Annual Review of Plant Physiology*, *29*(1), 277–317.

Folger, P. F., & Cody, B. A. (2014). Drought in the United States: causes and current understanding. CRS Report.

Francini, A., & Sebastiani, L. (2019). Abiotic stress effects on performance of horticultural crops. *Horticulturae*, *5*(4), 67.

Greaves, J. A. (1996). Improving suboptimal temperature tolerance in maize-the search for variation. *Journal of Experimental Botany*, *47*(3), 307–323.

Gutiérrez, N., Pégard, M., Balko, C., & Torres, A. M. (2023). Genome-wide association analysis for drought tolerance and associated traits in faba bean (Vicia faba L.). *Frontiers in Plant Science*, *14*, 1091875.

Han, J. S., Park, K. I., Jeon, S. M., Park, S., Naing, A. H., & Kim, C. K. (2015). Assessments of salt tolerance in a bottle gourd line expressing the Arabidopsis H^+-pyrophosphatase AVP 1 gene and in a watermelon plant grafted onto a transgenic bottle gourd rootstock. *Plant Breeding*, *134*(2), 233–238.

Hoang, T. M., Moghaddam, L., Williams, B., Khanna, H., Dale, J., & Mundree, S. G. (2015). Development of salinity tolerance in rice by constitutive-overexpression of genes involved in the regulation of programmed cell death. *Frontiers in Plant Science*, *6*, 175.

Jackson, M. B., & Colmer, T. (2005). Response and adaptation by plants to flooding stress. *Annals of Botany, 96*(4), 501–505.

Josine, T. L., Ji, J., Wang, G., Zhao, Q., Yang, H., Wang, Y., & Wu, W. (2015). AtDREB2A-CA gene over-expression in Rosa chinensis Jacq. affect leaf ultrastructure response to salt stress. *International Journal of Agriculture and Crop Sciences (IJACS), 8*(3), 463–476.

Kashyap, P., Shirkot, P., Das, R., Pandey, H., & Singh, D. (2023). Biosynthesis and characterization of copper nanoparticles from Stenotrophomonas maltophilia and its effect on plant pathogens and pesticide degradation. *Journal of Agriculture and Food Research, 13*, 100654.

Klein, A., Houtin, H., Rond, C., Marget, P., Jacquin, F., Boucherot, K., … & Burstin, J. (2014). QTL analysis of frost damage in pea suggests different mechanisms involved in frost tolerance. *Theoretical and Applied Genetics, 127*, 1319–1330.

Korkmaz, A., Korkmaz, Y., & Demirkıran, A. R. (2010). Enhancing chilling stress tolerance of pepper seedlings by exogenous application of 5-aminolevulinic acid. *Environmental and Experimental Botany, 67*(3), 495–501.

Kotuby-Amacher, J., Koenig, R., & Kitchen, B. (2000). *Salinity and Plant Tolerance*. Electronic Publication AG-SO-03. Logan: Utah State University Extension.

Levitt, J. (1980). *Responses of Plants to Environmental Stress, Volume 1: Chilling, Freezing, and High Temperature Stresses*. Orlando: Academic Press.

Lim, M. Y., Jeong, B. R., Jung, M., & Harn, C. H. (2016). Transgenic tomato plants expressing strawberry D-galacturonic acid reductase gene display enhanced tolerance to abiotic stresses. *Plant Biotechnology Reports, 10*, 105–116.

Liu, R., Shi, H., Wang, Y., Chen, S., Deng, J., Liu, Y., … & Chan, Z. (2014). Comparative physiological analysis of lotus (Nelumbo nucifera) cultivars in response to salt stress and cloning of NnCIPK genes. *Scientia Horticulturae, 173*, 29–36.

Luo, J., Yu, W., Xiao, Y., Zhang, Y., & Peng, F. (2022). Strawberry FaSnRK1α regulates anaerobic respiratory metabolism under waterlogging. *International Journal of Molecular Sciences, 23*(9), 4914.

Ma, L., Liu, X., Lv, W., & Yang, Y. (2022). Molecular mechanisms of plant responses to salt stress. *Frontiers in Plant Science, 13*, 934877.

Mckersie, B. D., Leshem, Y. A. Y., Mckersie, B. D., & Leshem, Y. A. Y. (1994). Freezing stress. In: B. D. McKersie, & Y. Y. Leshem (Eds.), *Stress and Stress Coping in Cultivated Plants* (pp. 104–131). Dordrecht: Kluwer Academic Publishers.

Muhammad Aslam, M., Waseem, M., Jakada, B. H., Okal, E. J., Lei, Z., Saqib, H. S. A., … & Zhang, Q. (2022). Mechanisms of abscisic acid-mediated drought stress responses in plants. *International Journal of Molecular Sciences, 23*(3), 1084.

Mukeshimana, G., Butare, L., Cregan, P. B., Blair, M. W., & Kelly, J. D. (2014). Quantitative trait loci associated with drought tolerance in common bean. *Crop Science, 54*(3), 923–938.

Munns, R., Goyal, S., & Passioura, J. (2004). *Salinity Stress and Its Mitigation*. Davis: University of California.

Negi, S., Singh, D., Pandey, H., & Singhal, R. K. (2021). Biofortification for high Fe and Zn in various Poaceae crops by using different molecular breeding and biotechnological approaches. *Plant Physiology Reports, 26*, 1–11.

Pandey, H., Singh, D., Dhiman, V. K., Dhiman, V. K., & Pandey, D. (2022). Extremophiles and their application in bioremediation. In: A. B. Gunjal, R. Thombre, & J. A. Parray (Eds.), *Physiology, Genomics, and Biotechnological Applications of Extremophiles* (pp. 188–206). Hershey: IGI Global.

Perata, P., Armstrong, W., & Voesenek, L.A. (2011). Plants and flooding stress. *New Phytologist, 190*(2), 269–273.

Prasad, G., Chauhan, D., Pandey, H., Singh, D., Dhiman, V., & Dhiman, V. (2023). Recent advances in CRISPR-Cas for climate-resilient horticulture in fruits crops. In: A. Singh, & S. Pandey (Eds.), *Climate-Resilient Agriculture: A Molecular Perspective*. Milton: Apple Academic Press, Incorporated.

Qin, X., Liu, Y., Mao, S., Li, T., Wu, H., Chu, C., & Wang, Y. (2011). Genetic transformation of lipid transfer protein encoding gene in Phalaenopsis amabilis to enhance cold resistance. *Euphytica, 177*, 33–43.

Raison, J. K., & Lyons, J. M. (1986). Chilling injury: a plea for uniform terminology. *Plant, Cell & Environment, 9*(9), 685–686.

Rao, N. S., Laxman, R. H., & Shivashankara, K. S. (2016). Physiological and morphological responses of horticultural crops to abiotic stresses. In: N. K. Srinivasa Rao, K. S. Shivashankara, & R. H. Laxman (Eds.), *Abiotic Stress Physiology of Horticultural Crops* (pp. 3–17). India: Springer.

Sabbah, S., & Tal, M. (1990). Development of callus and suspension cultures of potato resistant to NaCl and mannitol and their response to stress. *Plant Cell, Tissue and Organ Culture, 21*, 119–128.

Sachdev, S., Ansari, S. A., Ansari, M. I., Fujita, M., & Hasanuzzaman, M. (2021). Abiotic stress and reactive oxygen species: generation, signaling, and defense mechanisms. *Antioxidants, 10*(2), 277.

Sehrawat, N., Bhat, K. V., Kaga, A., Tomooka, N., Yaday, M., & Jaiwal, P. K. (2014). Development of new gene-specific markers associated with salt tolerance for mungbean (Vigna radiata L. Wilczek). *Spanish Journal of Agricultural Research*, (3), 732–741.

Sharma, A., Pandey, H., Manpoong, C., Vashishth, A., Singh, D., & Bang, N. T. H. (2023b). Biometric analysis, biosynthetic pathway and multipurpose uses of buckwheat local varieties in Namsai district of eastern Himalayas in India. *Trends in Food Science & Technology*, 136, 251–267.

Sharma, A., Pandey, H., Nampoothiri Devadas, V. A. S., Singh, D., Kartha, B. D., & Wangsu, M. (2023a). Regulations and factors affecting of light emitting diodes artificial light in crop production under soilless culture. *ACS Agricultural Science & Technology*, *3*(8), 616–630.

Sharma, S. K., Singh, D., Pandey, H., Jatav, R. B., Singh, V., & Pandey, D. (2022). An overview of roles of enzymatic and nonenzymatic antioxidants in plant. In: T. Aftab, & K. R. Hakeem (Eds.), *Antioxidant Defense in Plants: Molecular Basis of Regulation* (pp. 1–13). Singapore: Springer.

Singh, D., Dhiman, V. K., Pandey, H., Dhiman, V. K., & Pandey, D. (2022a). Crosstalk between salicylic acid and auxins, cytokinins and gibberellins under biotic stress. In: T. Aftab (Ed.), *Auxins, Cytokinins and Gibberellins Signaling in Plants* (pp. 249–262). Cham: Springer International Publishing.

Singh, D., Pandey, H., Dhiman, V. K., Bang, N. T. H., Dhiman, V. K., Singh, V., & Pandey, D. (2022b). Secondary metabolite engineering for plant immunity against various pathogens. In: T. Aftab, & K. R. Hakeem (Eds.), *Metabolic Engineering in Plants* (pp. 123–143). Singapore: Springer Nature Singapore.

Singh, D., Pandey, H., Dhiman, V. K., Thakur, K., Sharma, I., Dhiman, V. K., … & Pandey, D. (2022c). Assessment of environmental impacts of metal/metalloid pollution on plants. In T. Aftab, & K. Hakeem (Eds.), Metals Metalloids Soil Plant Water Systems (pp. 217–232). London: Academic Press.

Singh, D., Pandey, H., Thakur, K., Sharma, I., & Naraian, R. (2022d). Aquaporins and their functions in water transportation in different plant species. *Environmental Sustainability*, *5*(4), 443–456.

Singh, D., & Verma, S. K. (2023). Impacts of particulate matter pollution on plants. In: T. Aftab (Ed.), *New Frontiers in Plant-Environment Interactions: Innovative Technologies and Developments* (pp. 483–500). Cham: Springer Nature Switzerland.

Soares-Silva, M., Diniz, F. F., Gomes, G. N., & Bahia, D. (2016). The mitogen-activated protein kinase (MAPK) pathway: role in immune evasion by trypanosomatids. *Frontiers in Microbiology*, *7*, 183.

Sripaoraya, S., Keawsompong, S., Insupa, P., Power, J. B., Davey, M. R., & Srinives, P. (2006). Genetically manipulated pineapple: transgene stability, gene expression and herbicide tolerance under field conditions. *Plant Breeding*, *125*(4), 411–413.

Sunya, H., Sharma, A., Manpoong, C., Devadas, V. S., Pandey, H., & Singh, D. (2022). Evaluation of improved varieties of rice (Oryza sativa L.) under environmental condition of Namsai, Arunachal Pradesh, India. *Ecology, Environment and Conservation*, *28*, 907.

Tian, N., Wang, J., & Xu, Z. Q. (2011). Overexpression of Na^+/H^+ antiporter gene AtNHX1 from Arabidopsis thaliana improves the salt tolerance of kiwifruit (Actinidia deliciosa). *South African Journal of Botany*, *77*(1), 160–169.

Tiwari, S., & Lata, C. (2018). Heavy metal stress, signaling, and tolerance due to plant-associated microbes: an overview. *Frontiers in Plant Science*, *9*, 452.

Uprety, D. C., Reddy, V. R., Uprety, D. C., & Reddy, V. R. (2016). Mitigation technologies to control high-temperature stress in crop plants. In: D. C. Uprety, & V. R. Reddy (Eds.), *Crop Responses to Global Warming* (pp. 117–125). Singapore: Springer.

Vega, S. E., Del Rio, A. H., Jung, G., Bamberg, J. B., & Palta, J. P. (2003). Marker-assisted genetic analysis of non-acclimated freezing tolerance and cold acclimation capacity in a backcross Solarium population. *American Journal of Potato Research*, *80*, 359–369.

Wang, Y., Jiang, H., Mao, Z., Liu, W., Jiang, S., Xu, H., … & Chen, X. (2021). Ethylene increases the cold tolerance of apple via the MdERF1B–MdCIbHLH1 regulatory module. *The Plant Journal*, *106*(2), 379–393.

Warner, R. (2010). Genetic approaches to improve cold tolerance of petunia. The American Floral Endowment, 306.

Yadav, G., Prajapati, P., Singh, D., Hora, S., Singh, S., Vishwakarma, K., & Ahmad, I. Z. (2023). Emerging trends in plant metabolomics and hormonomics to study abiotic stress tolerance associated with rhizospheric probiotics. In M. I. R. Khan, A. Singh, & P. Poór (Eds.), *Plant Hormones in Crop Improvement* (pp. 283–306). USA: Academic Press.

Yang, T., Fang, L., Zhang, X., Hu, J., Bao, S., Hao, J., … & Zong, X. (2015). High-throughput development of SSR markers from pea (Pisum sativum L.) based on next generation sequencing of a purified Chinese commercial variety. *PLoS One*, *10*(10), e0139775.

Zhang, X., Zou, Z., Gong, P., Zhang, J., Ziaf, K., Li, H., … & Ye, Z. (2011). Over-expression of microRNA169 confers enhanced drought tolerance to tomato. *Biotechnology Letters*, *33*, 403–409.

18 Brassicaceae

Stress Responses and Breeding Strategies in Unfriendly Environments

Branka Salopek-Sondi, Mirta Tkalec, Nataša Bauer, Nenad Jasprica, and Smiljana Goreta Ban

18.1 INTRODUCTION

Species from the Brassicaceae family are highly widespread involving wild representatives but also many crop plants important for human consumption and industrial uses (Raza et al., 2020). Cruciferous vegetables have been popular from ancient times until the present because of their relatively easy cultivation, affordable price and availability at local markets, and their nutritional and health potential. The consumption of *Brassica* vegetables has been reported in a number of epidemiological and meta-analyses due to its beneficial effects (antioxidant, antimicrobial, anticancer, etc.) against various diseases. The advantages of cruciferous vegetables for human health are mainly linked to the richness in 'healthy phytochemicals' (glucosinolates, carotenoids, phenols, vitamins) that make these vegetables recognized as a functional food (Biondi et al., 2021; Šamec et al., 2017, 2019).

Global warming and accompanying climate changes cause unfriendly environments for plants in general (high temperatures, increased salinity, long-term drought periods, extreme events of temperature waves or flood, etc.) and may lead to biodiversity disturbance, quality decrease, and productivity loss of many plants species, including Brassicaceae. Drought has already affected about 40% of the world's land area. From 1990 to the present, average surface temperatures have increased by about 1°C and are expected to increase up to 4.8°C by the end of the century (Arias et al., 2021; Leisner, 2020). Increased temperatures and drought occur together and present the major obstruction to agricultural production (Hassan et al., 2021). Semi-arid and arid regions in the Mediterranean basin are the most sensitive and therefore more vulnerable to global warming and the problem is expected to worsen in the future (Corwin, 2021; Guiot & Cramer, 2016). Besides drought and heat stress, increased soil salinity as a consequence of natural but also anthropogenic activities is highly abundant. Salinity stress has already affected about 7% of the world's total land and about 20% of irrigated land and shows an increasing tendency in the future (Munns & Gilliham, 2015). Increased salinity may be particularly harmful to plants because it induces three kinds of interacting stresses: ionic, osmotic, and oxidative stresses (Liang et al., 2018).

Climate change strongly affects the cultivation of crops in general, including brassicas as well, particularly in the Mediterranean region. Since Brassicaceae are reported to be sensitive to abiotic stresses, particularly increased temperatures, continued global warming is expected to impact crop yield and quality all over the world (Lobell & Gourdji, 2012). For example, a 17% yield reduction was reported in brassicas exposed to a temperature increase of 1°C (Kaushal et al., 2016). Elferjani and Soolanayakanahally (2018) reported a reduction in canola seed yield by 85.3% and 31% under the heat and drought treatments, respectively. Considering the salinity tolerance of *Brassica* crops, the amphidiploid species (*Brassica juncea*, *Brassica napus*,

 DOI: 10.1201/9781003457237-18

and *Brassica carinata*) are reported to be more tolerant than the diploids (*Brassica oleracea*, *Brassica nigra*, and *Brassica rapa*) (Purty et al., 2008).

Abiotic stress responses in plants, including the Brassicaceae family, involve a range of mechanisms at both macro- and micro-levels. These mechanisms are often diverse, including changes at the molecular (gene expression, protein synthesis, antioxidant accumulation, etc.), physiological (photosynthesis performance), and morphological (growth retardation) levels. Mechanisms of tolerance are a complex network of various players including transcription factors, plant hormones, osmoprotectants, and non-enzymatic and enzymatic antioxidants (Singh et al., 2019). Abiotic stress response and mechanisms of tolerance cannot be generalized; they are species-, even varieties- or accession-specific among Brassicaceae (Bauer et al., 2022; Orsini et al., 2010; Pavlović et al., 2018, 2019). Thus, knowledge of abiotic tolerance mechanisms revealed in the widely used plant model *Arabidopsis thaliana* is highly appreciated but cannot strictly be applied to other plant species from the Brassicaceae family. Recent interests are focused on wild relatives and related species to *Brassica* crops such as *Anastatica hierochuntica*, *Eutrema salsugineum*, *Schrenkiella parvula*, and *Cakile maritima* as potential model plants in evaluating stress responses and investigating mechanisms of stress tolerance to build crop resilience and resistances in the face of global climate change (Quezada-Martinez et al., 2021). Due to genetic diversity, crop wild relatives (CWRs) are a great source of genes with the potential to improve crop tolerance to often unfriendly environmental conditions (Razzaq et al., 2021).

The selection of more tolerant species to abiotic stressors and the application of breeding strategies in the creation of more resilient crops are of great interest in agriculture. Genetically based strategies including classical crossing and mutagenesis and modern transgene manipulations and genome editing are more or less time-consuming methods. On the contrary, exogenous plant and seed treatments with various agents, bio-based or synthesized, are also promising practices to help plants cope with the negative effects of environmental stress (Patel et al., 2020).

18.2 ABOUT BRASSICACEAE

18.2.1 Taxonomy, Morphology, and Botanical Overview

The mustard family (Brassicaceae) consists of ca. 4,140 species belonging to 346 accepted genera (POWO, 2023). According to the most complete study to date on the generic phylogeny of the Brassicaceae (Hendriks et al., 2023), based on nuclear (1,081) and plastome (60) genes, 319 and 265 genera were found, respectively. The species are distributed all over the globe except Antarctica, mostly in temperate areas, with the highest diversity in the Irano-Turanian and Mediterranean floristic regions as well as in the Himalayas, Cape floristic region, Andes, western United States, Australia, and New Zealand (German et al., 2023; Nikolov, 2019; Nikolov et al., 2019). The radiation of the core Brassicaceae most likely began in the Irano-Turanian region (Franzke et al., 2011). Although most molecular data point to a pre-Miocene origin of the family, evidence from paleobotany and palaeoecology suggests a Miocene radiation; i.e. newly developed habitats and/or niches created by climatic changes in the Miocene acted as an 'extrinsic' factor for the radiation of the family (Franzke et al., 2009). However, the results of recent studies on Brassicaceae converge to a crown age of 37.1–32.4 Ma (Eocene-Oligocene) (Hendriks et al., 2023).

The family includes annual, biennial, or perennial species, sometimes subshrubs or shrubs. Despite their species diversity, most species are easily recognized by their conserved flowering plan with a cruciform appearance that led to their classic botanical name, the Cruciferae. The Brassicaceae belongs to the order Brassicales, which comprises 18 plant families characterized by the accumulation of glucosinolates, specialized metabolites with a protective role against biotic and abiotic stresses (Edger et al., 2015, 2018). Phylogenetic studies over the last two and a half decades support the recognition of two highly disparate subfamilies, the new unigeneric Aethionemoideae (type genus *Aethionema*) with 58 species and the much larger Brassicoideae (type genus *Brassica*), which comprises the other 98.6% of species and the remainder of the family's generic and tribal diversity (German et al., 2023). The Brassicoideae are divided into five supertribes (Arabodae, Brassicodae, Camelinodae, Heliophilodae, and Hesperodae) corresponding to the major evolutionary lineages (Hendriks et al., 2023). Updating the genus boundaries and number of species at the tribal level revealed that the tribes with the highest number of genera and species include Thelypodieae (34 genera; 235 species), Euclidieae (30; 155), Cardamineae (14; 344), Lepidieae (1; 268), and Coluteocarpeae (1–12; 130) (German et al., 2023; Hendriks et al., 2023).

The Brassicaceae exhibits considerable morphological diversity in flowers, leaves, trichomes, and fruit (Nikolov, 2019). In general, the delimitation of genera is often difficult, and some authors emphasize that mature fruits are often needed for reliable identification (Warwick & Hall, 2009). In fact, many identification keys, books, and guides are based on the morphological characteristics of the fruits. According to Al-Shehbaz et al. (2006), important points in the delimitation of genera are emphasized here:

> (i) monotypic or oligotypic genera should not be established without prior molecular studies and a critical evaluation of morphology; (ii) due to widespread convergence in most morphological characters, especially fruit types and embryo position, these characters should be used with extreme caution when establishing genus boundaries; (iii) large differences in fruit morphology can be misleading, and the examples of *Heliophila*, *Tropidocarpum* and *Vella* should be a constant reminder against drawing incorrect taxonomic conclusions by overemphasizing fruit morphology at the expense of other potentially very useful vegetative and floral characters.

18.2.2 Species and Their Uses

Many Brassicaceae plant species have been cultivated since the earliest beginnings of human civilization, even as far back as 8,000 years ago (Blondel & Aronson, 1999). The Brassicaceae includes important crops grown as vegetables (*B. rapa*, *B. oleracea*, *Raphanus sativus*, *Lepidium sativum*, *Nasturtium officinale*), oilseeds as source of industrial and edible oils (*B. napus*, *B. juncea*, *B. rapa*, *B. carinata*), and spices (*B. nigra*, *Sinapis alba*, *Armoracia rusticana*, *Eutrena japonica*). Additionally, there are more than 120 weedy species that have the potential to be used as feed or organic fertilizers. The Brassicaceae also include ornamental species belonging to the genera *Aethionema*, *Alyssum*, *Arabis*, *Aubrieta*, *Aurinia*, *Cheiranthus*, *Erysimum*, *Hesperis*, *Iberis*, *Lobularia*, *Lunaria*, *Malcolmia*, and *Matthiola*.

Brassica oilseeds are the third most important source of vegetable oils in the world (Cartea et al., 2019). In addition, the increasing importance of renewable energy, feedstocks, industrial oils, and novel uses of vegetable oils, as well as the evergrowing bioeconomy, offers significant opportunities for industrial (non-edible) oils (McVetty et al., 2016).

Cabbages (*B. oleracea*) have been an important food source for centuries, as leafy and root vegetables, fodder, and forage (Shyam et al., 2012; Šamec et al., 2017). One of the representative crops derived from *B. oleracea* L. is kale (*B. oleracea* var. *acephala* DC.), a traditional vegetable crop in the coastal region of Croatia, the northeast of the Mediterranean (Figure 18.1).

Brassicaceae species evolved after multiple whole-genome duplication events (Kagale et al., 2014) and are formed by the natural hybridization of their ancestor diploids. U-triangle (Nagaharu, 1935) describes the genetic relationship among different *Brassica* species (Figure 18.2). For example, *B. napus* (n=19 chromosomes, AC-genome) is domesticated allotetraploid originated from *B. rapa* (n=10 chromosomes, A-genome) and *B. oleracea* (n=9 chromosomes, C-genome) about 7,500 years ago (Chalhoub et al., 2014; Lv et al., 2020; Song et al., 2020; Zhang et al., 2018) and have been further selected and bred by humans as part of the process of domestication and crop improvement.

Well-known and widely grown vegetables such as broccoli (var. *italica*), Brussels sprouts (var. *gemmifera*), cabbage (var. *capitata*), cauliflower (var. *botrytis*), kale (var. *acephala*), and kohlrabi (var. *gongylodes*) belong to the same species, *B. oleracea*. Today, little is known about the origin species of domesticated *B. oleracea*. Attempts to identify the ancestor species are focused

FIGURE 18.1 Among the various crops derived from *Brassica oleracea* L., kale (*B. oleracea* var. *acephala* DC.) is a common vegetable crop in the coastal region of Croatia, the north-east of the Mediterranean. Kale is able to survive dry and hot climates but also low temperatures, snow, and frost in the continental region during wintertime. Photo: N. Jasprica, B. Salopek-Sondi.

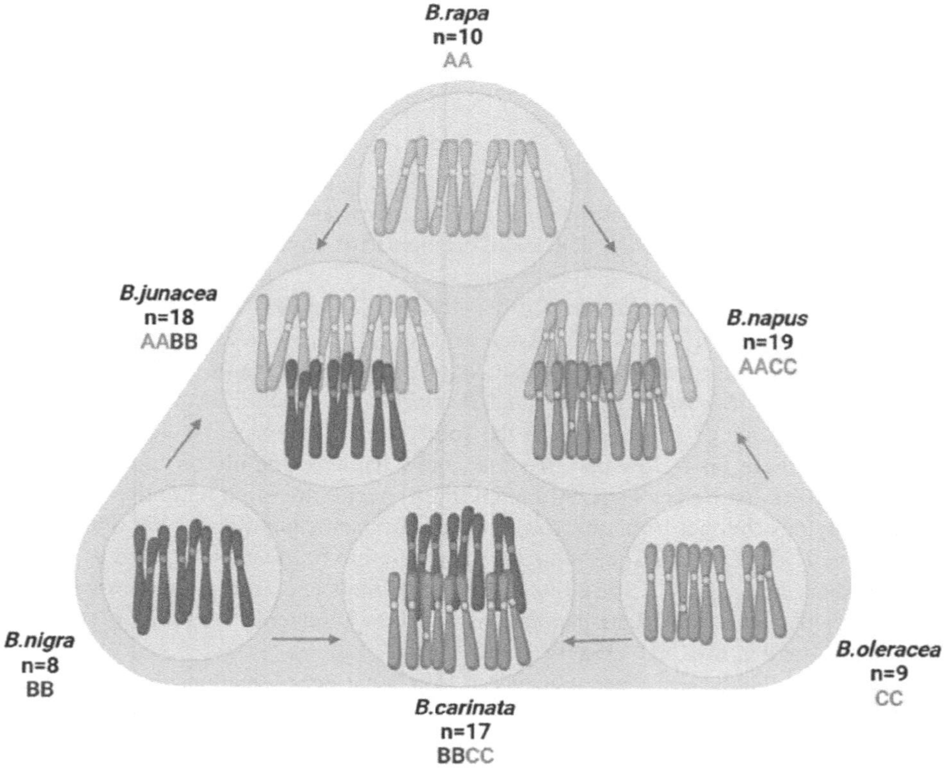

FIGURE 18.2 U-triangle explaining the relationship of the six major *Brassica* species. Allotetraploids *Brassica napus*, *Brassica junacea*, and *Brassica carinata* are formed by the natural hybridization of their ancestor diploids *Brassica nigra*, *Brassica oleracea*, and *Brassica rapa*. Prepared by N. Bauer, according to Nagaharu (1935) and Song et al. (2021) using BioRender (https://app.biorender.com).

on wild crop relative species capable of interbreeding with *B. oleracea*. Wild species such as *Brassica bourgeaui*, *Brassica cretica*, *Brassica hilarionis*, *Brassica incana*, *Brassica insularis*, *Brassica macrocarpa*, *Brassica montana*, and *Brassica rupestris* share the same type of genome with domesticated *B. oleracea* (C-genome) and, therefore, can hybridize with agricultural important varieties to improve cultivated Brassicaceae varieties (cultivars) (e.g. Maggioni et al., 2014). The relationship between these weedy populations, and the domesticated *B. oleracea* is unclear. Some studies suggest that these wild relatives represent the progenitors, while others identify these wild forms as ones escaping cultivation (Mabry et al., 2021). Taxonomy of wild species is still under consideration, and many of them are referred to by alternative names or have multiple subspecies. Tatout et al. (1999), for example, proposed that *Brassica villosa* subsp. *drepanensis* should be considered a separate species based on molecular analyses, suggesting a hybrid origin of this taxon from *B. incana* and *B. villosa*. Although the usage of advanced genetic techniques and chemical analyses has clarified the relationships, there is still general disagreement among authorities as to whether a particular form should be considered a species, a subspecies, or a variety within a species.

Besides their culinary use, Brassicaceae species have been extensively used in traditional medicine from ancient times until the present. They exhibit biological activities, such as antioxidant, antibacterial, anticancer, and antiviral effects, and for the innate immune system, these vegetables act as potent modulators (for details see Lučić et al., 2023; Shankar et al., 2019; Šamec et al., 2017, 2019). In the last couple of decades, epidemiological studies have provided evidence about the beneficial role of the consumption of *Brassica* vegetables, in the prevention of various chronic diseases (see for review Šamec & Salopek-Sondi, 2019). Many epidemiological studies reported a lower risk of different types of cancer linked to a diet including cruciferous vegetables (see for review Mandrich & Caputo, 2020).

18.2.3 Ecology

Brassicaceae plant species occupy a wide range of habitats in cosmopolitan and extreme environments (Trautwig et al., 2023). Some of those representatives are shown in Figure 18.3. Multiple genome duplications among Brassicaceae are likely liable for high diversification and adaptation processes.

Generic and specific endemism of the family is the most abundant in the Irano-Turanian region and the Mediterranean region (subtribe *Brassicinae*) (Hedge, 1976). Sicily, the largest island in the Mediterranean, is one of the centers of diversification of the wild taxa of *Brassica* sect. *Brassica* (Snogerup et al., 1990). *Brassica rupestris* and *B. villosa* are endemic to western Sicily, with a single disjunct population of *B. rupestris* in the south of the Italian peninsula (Calabria region). All populations of these taxa grow on cliffs and rocky habitats (mainly on limestone) from sea level to 1,000–1,200 m above sea level. Although these species are often naturally protected from most human activities (but not from quarries, for example), some populations are often exposed to anthropogenic activities or stochastic events and can therefore become endangered or even extinct within a very short time (IUCN, 2012). In some other habitats (e.g. in the Mediterranean maquis; Jasprica & Stamenković, 2023), their populations are often threatened by limited or highly fragmented habitats, grazing, fires, etc. (Figure 18.4).

It is also reported that Brassicaceae species can form a variety of relationships with fungi, from harmful interactions with pathogens to beneficial associations with endophytic fungi (Mourou et al., 2023). Although the Brassicaceae family has been consistently identified as non-mycorrhizal, there are some exceptions (*Arabidopsis thaliana*, *Alliaria petiolata*, *B. napus*, *B. oleracea*, *C. maritima*, *Capsella bursa-pastoris*, *Cardamine bulbifera*, etc.) to this general observation (e.g. Trautwig et al., 2023). However, the molecular mechanisms of mycorrhizal association in the Brassicaceae, as well as their specific role remain unresolved and require additional investigation.

FIGURE 18.3 Brassicaceae plant species occupy a wide range of ecological niches in cosmopolitan and extreme environments. The Appenine-Blakanic species *Brassica incana* Ten. grows on the low rocky coast (a), on the high vertical coastal cliffs (b), and in the low evergreen maquis (c). Other species (e.g. the southern European species *Peltaria alliacea* Jacq.) can also be found on the scree slopes (d). Other Brassicaceae plant species (e.g. *Arabis alpina* L. subsp. *alpina*) are found in alpine and subalpine grassland (e) or in rock crevices (e.g. *Resetnikia triquetra*) (f). Photo: N. Jasprica.

Last but not least, among the naturalized alien species worldwide, the Brassicaceae is one of the families that are not only rich in naturalized species but also quite widespread (i.e. representatives can be found in many regions). Among the top 30 families ranked by the total number of naturalized species in the world, Brassicaceae ranks seventh with 311 species (Pyšek et al., 2017). This family also contains disproportionately more naturalized alien species in mainland regions than in

FIGURE 18.4 The Brassicaceae species occurring in the Mediterranean region: (a) *Sisymbrium irio* L. colonizes nithrophilous ruderal sites, fallow land, in and near settlements, and along roadsides. Fruit is siliqua. (b) *Alyssoides utriculata* (L.) Medik. occurs in calcareous, rocky places. (c) *Aethionema saxatile* (L.) W.T. Aiton grows mainly on dry grasslands. Fruit is a short and broad silicle (silicula). (d) *Cakile maritima* Scop. occupies sandy and pebble beaches. (e) *Cheiranthus cheiri* L. is a widely cultivated ornamental plant of European origin. (f) *Resetnikia triquetra* (DC.) Španiel et al. are a rare stenoendemic species restricted to a few locations along the Croatian Adriatic coast. Photo: N. Jasprica.

island regions, and, even in these cases, this imbalance is less pronounced (Pyšek et al., 2017). Key findings to date include the fact that climate change will increase the risk of naturalization of alien plants in Europe (Dullinger et al., 2017) and that emerging economies in countries with high biodiversity are the regions most at risk of future plant invasions due to the interaction of global trade and climate change (Seebens et al., 2015).

18.3 HOW DO BRASSICACEAE RESPOND TO ABIOTIC STRESS?

18.3.1 Morphological Adaptation

The most adverse effect of abiotic stressors like salinity, drought or high temperatures on various Brassicaceae species is the reduction in growth and yield. Growth inhibition is a common response to the stress of most crops including brassicas because instead of growing, plants redirect their resources and energy into defense mechanisms that enable them to survive (Eziz et al., 2017; Karasov et al., 2017). There are significant inter- and intraspecific differences within Brassicaceae species in tolerance to various stress factors. Different morphological parameters can be used to screen tolerance, but the results are not always unambiguous and depend on the investigated parameter, developmental stages as well as species. All growth stages can be sensitive to stress, and the effects depend on the duration, intensity, and timing (Wu et al., 2018). In *Brassica* crops exposure to salinity, drought or high temperatures at the seedling stage can inhibit germination, and seedling vigor, resulting in decreased plant densities and reduced yields (Wu et al., 2018; Zhang et al., 2015). Early seedling growth has been proposed as a good marker for stress tolerance screening because it is more rapid and allows a high-throughput screening in a laboratory reducing the time and labor for field testing at a later developmental stage (Bakirov et al., 2022; Bouchyoua et al., 2024; Channaoui et al., 2019). Among growth-inhibition changes in various brassicas during drought and heat stress are reduction of plant height, leaf area and number, shoot/root biomass, length, and fresh or dry weight with above-ground organs being more sensitive than roots (Akram et al., 2016; Ben Ammar et al., 2022; Gunasekera et al., 2006; Husen et al., 2014; Pokharel et al., 2020; Rodríguez et al., 2015). Since most stress factors reduce the availability of water to the plant, the reduction of growth of above-ground organs is considered an adaptation that can help the plants reduce transpiration and conserve water (Munns & Millar, 2023). In general, less inhibited or unchanged shoot growth and unchanged or even stimulated root growth are characteristics of more tolerant species/cultivars (Ashraf & Sharif, 1998; Pokharel et al., 2020). Bakirov et al. (2022) screened 200 germplasms of rapeseed *B. napus* L. and found that salt tolerance correlated with total fresh weight thus considering it as the reliable parameter to evaluate salt tolerance. Root biomass and especially length are usually less affected by salt or drought stress increasing root-to-shoot ratio (Ben Ammar et al., 2022; Bouchyoua et al., 2024; Wu et al., 2018). In tolerant species/cultivars roots can continue to grow and elongate enhancing water uptake and maintaining water balance, for example in *B. napus* (Bouchyoua et al., 2024; Channaoui et al., 2019; Li et al., 2018; Sharif et al., 2018;), *B. oleracea* (Kage et al., 2004), and *B. juncea* var. CS52 (Kumar et al., 2009). In a study comparing drought tolerance in *B. carinata*, *B. napus* and various lines of *Eruca vesicaria* subsp. *sativa*, one of the drought-resistant species of Brassicaceae, water stress significantly inhibited the root and shoot growth and biomass of seedlings in all species but much higher in drought-tolerant *B. napus* cultivar Zhongshuang 9 than in *Eruca* lines, while *B. carinata* cultivar XB1 exhibited high drought sensitivity (Huang et al., 2017). In screening *B. napus* accessions under osmotic treatments with polyethylene glycol the root-to-shoot length ratio was found to be a reliable indicator of water stress tolerance (Zhang et al., 2015). The importance of roots in adaptation to salinity has been confirmed in *A. hierochuntica*, a heat and salt-tolerant member of Brassicaceae, whose roots continue to elongate at 200 mM NaCl while ceased in *Arabidopsis* seedlings (Eshel et al., 2017). Moreover, in *S. parvula*, one of several extremophile relatives of *Arabidopsis*, the increase in primary root length enables deeper growing to lower salt concentrations (Tran et al., 2023).

In addition to growth changes, anatomical features like thicker leaves were found in drought-tolerant *B. napus* genotypes, contributing to lower water loss through reduced transpiration, and higher leaf water transport efficiency (Zhu et al., 2011b). *Schrenkiella parvula* leaves under long-term salt stress showed increased leaf thickness and wax content which was associated with higher relative water content and growth under salt stress (Inan et al., 2004; Tran et al., 2023). Also, *S. parvula* was shown to increase the xylem vessel area allowing a higher bulk flow through the xylem during the long-term salt stress, thus optimizing the effective use of water instead of just reducing water loss (Tran et al., 2023).

18.3.2 Photosynthesis

Photosynthesis is markedly affected by heat or lack of water due to increased salinity or drought, particularly in C3 metabolic pathway crops like Brassicaceae. The effects depend on the intensity and duration of stressors and whether the stressors occur individually or are combined (Elferjani & Soolanayakanahally, 2018). The most investigated stress parameter regarding the photosynthetic process is chlorophyll content which has almost always been reduced in various *Brassica* crops (Akram et al., 2016; Bhuiyan et al., 2019; Hajiboland & Amirazad, 2010; Husen et al., 2014; Jan et al., 2016; Latif et al., 2016; Morsi et al., 2023; Munawar et al., 2019; Ramzan et al., 2023; Yusuf et al., 2010) and in their relatives such as *Arabidopsis* (Eshel et al., 2017) and arugula *E. sativa* (Nikzad et al., 2023). Water stress in the cell can induce oxidative damage to membranes and negatively influence chlorophyll and its biosynthesis (Tefera et al., 2021). Higher chlorophyll content has been associated with more tolerant species and genotypes. For example, *B. napus* exposed to drought showed greater fresh and dry biomass than *B. rapa*, *B. juncea*, and *B. carinata* which correlated with higher water and chlorophyll content (Ashraf & Mehmood, 1990). In *B. juncea* plants chlorophyll and carotenoid content declined with increasing levels of temperature but more tolerant genotypes maintain high water content accompanied with higher chlorophyll and carotenoid contents and preserved membrane integrity (Kumar et al., 2013b). Heat-tolerant brassicas relative *A. hierochuntica* after 4 hours of exposure to heat stress maintained chlorophyll levels around 65% of control levels with no effect on carotenoid content while *Arabidopsis* seedlings were completely bleached (Eshel et al., 2017). Similarly, in salt-tolerant *E. salsugineum* exposure to up to 200 mM NaCl did not cause any change in chlorophyll content contrary to *Arabidopsis* where it significantly dropped (Stępień & Johnson, 2009; Sui & Han, 2014).

Exposure to drought and heat stress, especially if combined also with water deficit, markedly reduced net photosynthetic assimilation rate in sensitive *Brassica* crops, which was accompanied by lower CO_2 uptake due to negatively influenced stomatal conductance and transpiration rate (Elferjani & Soolanayakanahally, 2018; Hajiboland & Amirazad, 2010; Husen et al., 2014; Xiong et al., 2018). Plants under stress tend to conserve water lost through transpiration by closing their stomata, which consequently reduces the carboxylation rate of Rubisco (Elferjani & Soolanayakanahally, 2018). Accumulation of phytohormone abscisic acid (ABA), known to be involved in activating stomatal closure, has been confirmed in drought and salinity-stressed brassicas such as stressed *B. napus* seedlings (Qaderi et al., 2006), as well as *B. oleracea* var. *capitata*, *B. rapa*, and *B.oleracea* var. *acephala* (Pavlović et al., 2018, 2019). ABA accumulation usually negatively correlates with photosynthetic performance. Inhibition of CO_2 fixation often leads to photoinhibition and inhibition of electron transport through photosystem II (PSII), which has also been reported from chlorophyll *a* measurement in *Brassica* crops during drought (Fariduddin et al., 2009; Husen et al., 2014). Similarly, under heat stress *B. napus* plants showed an impairment of the photosynthetic activity (Rodríguez et al., 2015), while in *B. oleracea* a reduction in the capacity of the photosynthetic electron transport, measured as quantum yield of PSII (ΦPSII), was reported but in the combination with high light exposure (Díaz et al., 2007), which is often seen in natural conditions. The decreased photochemical efficiency of PSII, as indicated by a decline in the ratio of variable to maximum fluorescence (F_v/F_m), was accompanied by a significant increase in thylakoid membrane damage in sensitive cultivars of *B. napus* under high night-time temperatures (Pokharel et al., 2020).

Contrasting photosynthetic responses to salt stress have been observed between sensitive and tolerant Brassicaceae species. Sensitive *Arabidopsis* showed rapid stomatal closure and a drop in CO_2 assimilation (Stępień & Johnson, 2009), which contributed to the cessation of growth. This was accompanied by a lowering of F_v/F_m, inhibition of PSII electron transport, and increased non-photochemical quenching (NPQ), a mechanism for dissipating excess energy as heat. Moreover, the electron transport rate (ETR) through PSI increased implying an increase in cyclic electron flow due to stress (Stępień & Johnson, 2009). In salt-tolerant extremophytes related to *Arabidopsis*, *S. parvula*, and *E. salsugineum*, stomatal conductance and CO_2 assimilation were relatively stable with unaffected net photosynthetic rate and F_v/F_m, thus allowing growth to be retained under salinity (Li et al., 2022; Sui & Han, 2014; Tran et al., 2023; Wiciarz et al., 2015). Furthermore, NPQ values in *E. salsugineum* did not change significantly even at 500 mM NaCl while ΦPSII increased (Stępień & Johnson, 2009), suggesting that more energy was used for photochemistry. Increased concentrations of unsaturated fatty acids in membrane lipids found in *E. salsugineum* under salt stress could enhance the resistance of PSII to salt-induced photoinhibition (Sui & Han, 2014). Similarly, in tolerant *A. hierochuntica* salt stress caused a lesser decline in PSII electron transport, which was accompanied by lower NPQ, thus exhibiting salt-resilient photochemistry (Eshel et al., 2017). Moreover, even in control conditions *A. hierochuntica* had higher PSII ETR and lower NPQ than *Arabidopsis* (Eshel et al., 2017) which indicates the importance of the resistance of the photosynthetic apparatus in the adaptive stress tolerance.

18.3.3 Biochemical Stress Markers

In most Brassicaceae stressors like high temperature, salinity or drought induce inhibition of growth which often correlates with reduced relative water content (RWC) in leaves or shoots (Akram et al., 2016, 2018; Bhuiyan et al., 2019; Hasanuzzaman et al., 2018; Husen et al., 2014; Jan et al., 2016; Kumar et al., 2013b; Latif et al., 2016; Nikzad et al., 2023). An increase in free proline was found to be a common metabolic strategy of Brassicaceae species to cope with reduced water potential in plant cells, although an increase in glycine betaine was also observed (Akram et al., 2016, 2018; Bhuiyan et al., 2019; Latif et al., 2016; Munawar et al., 2019). Both compounds can serve for osmotic adjustment but may also act as protectants of membrane proteins during stress. Some studies reported an increase in the level of soluble sugars, which might also contribute to the osmotic adjustment (Akram et al., 2016; Xie et al., 2013). Concomitantly, stress-induced metabolic disturbances increase the production of reactive oxygen species (ROS) resulting in lipid peroxidation and membrane damage. Overgeneration of ROS, mostly measured as superoxide ($O_2^{\bullet-}$) or H_2O_2 content and severe oxidative damage as indicated by malondialdehyde (MDA) content or electrolyte leakage, was observed in various Brassicaceae species under stress including *B. rapa*, *B. oleracea*, and *B. napus* under drought (Bhuiyan et al., 2019; Hasanuzzaman et al., 2018; Latif et al., 2016), *B. juncea* exposed to increased temperature (Rai et al., 2020) and drought (Kumar et al., 2013a), *B. rapa* exposed to salt and drought stress (Zhang et al., 2019), and radish under water stress (Mahesh et al., 2013). To overcome oxidative stress, plants activate antioxidative defense including antioxidative enzymes such as superoxide dismutase (SOD), catalase (CAT) ascorbate peroxidase (APX), guaiacol peroxidase (POD) and glutathione peroxidase (GPX), glutathione reductase (GR) as well as antioxidants like glutathione (GSH) and ascorbate (AsA). In *B. oleracea*, *B. juncea*, and *B. rapa* exposed to either drought or salt stress, AsA increased as well as activities of SOD, CAT, GR, and APX (Hajiboland & Amirazad, 2010; Kumar et al., 2013a; Latif et al., 2016; Munawar et al., 2019; Zhang et al., 2019). Similarly, water-stressed *R. sativus* cultivars showed a significant increase in the activity of SOD, as well as increased activity of CAT and APX reflecting enhanced O_2 removal and H_2O_2 detoxification (Akram et al., 2016; Mahesh et al., 2013). However, depending on species some differences in responses were observed probably reflecting genotype specificity. For example, in *B. napus* exposed to drought glutathione (GSH) content increased together with APX and GPX activity while CAT activity decreased as well as activities of enzymes involved in the AsA–GSH cycle resulting also in decreased AsA level (Hasanuzzaman et al., 2018). Among more specific metabolites it was observed that the amount of total tocopherol increased in *B. juncea* plants under salinity and osmotic stress (Yusuf et al., 2010). While water-soluble AsA and GSH scavenge the oxygen radicals in the aqueous phase, lipophilic tocopherols act as ROS scavengers in the membrane directly preventing its peroxidation (Kumar et al., 2013a). Moreover, α-tocopherol-enriched transgenic *B. juncea* plants were more tolerant to salt and osmotic stress due to higher relative water content and lower membrane damage and ROS content as well as enhanced activities and transcript levels of antioxidant enzymes SOD, CAT, APX, and GR (Kumar et al., 2013a).

Comparison of sensitive and tolerant cultivars/genotypes in various Brassicaceae species revealed that higher accumulation of proline and enhanced antioxidant enzyme activity are responsible for stress tolerance due to increased ROS-scavenging ability and lower membrane damage (Bailey-Serres & Mittler, 2006). For example, cluster analysis of ten varieties of *B. napus* with various drought resistance showed that MDA content as a marker of membrane damage was negatively correlated with drought resistance, while the increase of proline and soluble sugar content and the activity of POD were positively correlated (Xie et al., 2013). Moreover, the enhanced activity of SOD and POD and increased expression of Mn-SOD isoenzyme was found in more drought-tolerant *B. napus* (Abedi & Pakniyat, 2010). Kumar et al. (2009) showed that the amphidiploid species, especially *B. juncea* var. CS52 maintained higher levels of proline and efficiently restricted the membrane damage under extended salinity stress contributing to their better salt tolerance compared to diploids. Similarly, *B. juncea* genotypes CS-52 and NDR 8801 maintained higher RWC and preserved membrane integrity during heat stress (Kumar et al., 2013a). Moreover, screening of *B. juncea* genotypes after exposure of seedlings to 4-day cycle of heat stress (45°C, 4 hours) showed lower levels of ROS and MDA in heat-tolerant plants than in sensitive ones due to improved ROS scavenging as a result of enhanced SOD and CAT activity as well as upregulation of heat stress-related genes (Rai et al., 2020).

The importance of an activated antioxidative system for achieving tolerance has been confirmed in different stress-resistant *Brassica* relatives. A comprehensive study on genotypes of *E. vesicaria sativa* showed that the drought-tolerant genotypes were characterized by high activities of APX, CAT, and POD as well as higher levels of proline (Nikzad et al., 2023). Increased antioxidative response through activation of the AsA-GSH cycle was found to have a significant role in achieving salt tolerance in *C. maritima* (Amor et al., 2006), *S. parvula* (Uzilday et al., 2015) and *E. salsugineum* (Wiciarz et al., 2018). Moreover, the upregulated antioxidative system was associated with higher H_2O_2 content in leaves of *E. salsugineum* without inducing oxidative damage as indicated by low MDA level (Li et al., 2022), thus confirming the important signaling role of H_2O_2 in plant stress responses (Pilarska et al., 2021). Since APX activity was decreased in *E. salsugineum* (Pilarska et al., 2016) and GPX activity was enhanced (Li et al., 2022), the latter probably contributed to precise control of H_2O_2 accumulation. Investigations of extremophyte Brassicaceae relatives *A. hierochuntica, E. salsugineum and S. parvula* which are tolerant to multiple stressors including heat and salt stress revealed that all species had greater proline content than stress-sensitive *Arabidopsis* under normal growth conditions (Eshel et al., 2017; Kant et al., 2006; Taji et al., 2004) which were interpreted as metabolic stress-anticipation (Gong et al., 2005; Lee et al., 2012). Interestingly P5CS1, the key gene involved in proline biosynthesis, was induced by salt in *A. hierochuntica* and *E. salsugineum* as well as in *Arabidopsis* but *ProDH1* gene, which encodes an enzyme that degrades proline, was suppressed only in the extremophytes under salt stress, resulting in proline accumulation (Eshel et al., 2017). Moreover, *A. hierochuntica* and *E. salsugineum* under saline conditions exhibit tight control of Na^+ accumulation compared to *Arabidopsis* but have higher shoot Na^+ content under control conditions (Eshel et al., 2017; Kant et al., 2006) which probably helps them to maintain a constitutively more negative leaf osmotic potential and retain water (Inan et al., 2004). Furthermore, the levels of sugars such as glucose and fructose and their derivatives fructose-6-phosphate and glucose-6-phosphate as well as antioxidants ascorbate and dehydroascorbate were higher in *A. hierochuntica* and *E. salsugineum* than in *Arabidopsis* (Eshel et al., 2017). Much higher basal levels of sugars and proline can either be used as resources to maintain growth in stress conditions when photosynthesis is impaired or serve as osmoprotectants or antioxidants effectively reducing membrane damage (M'rah et al., 2007). However, the allocation of more resources to storage, structural, or defense traits is costly (Munns, 2002) and therefore associated with slow plant growth as was observed for *E. salsugineum* roots grown under control conditions (Lugan et al., 2010; Tran et al., 2023). On the contrary, a different mechanism was observed in faster-growing *S. parvula*, which instead of having very high basal levels of sugars, and proline showed more dynamic accumulation of these metabolites during salt treatments (Tran et al., 2023). Furthermore, *A. hierochuntica* by accumulating antioxidants and to a lesser extent sugars and amino acids evolved a stress tolerance strategy with lower metabolic costs (Eshel et al., 2017).

18.3.4 Specialized Metabolites as Abiotic Stress Protectors

Brassicaceae are well-known as rich in specialized or secondary metabolites such as polyphenols, glucosinolates, and carotenoids which are often species-specific, and their accumulation is affected by environmental factors (Neugart et al., 2018; Šamec et al., 2019). Plants use specialized metabolites to adapt to the changing environment. Generally, under stress conditions, accumulation of specialized metabolism in plants can be induced, because growth is often limited more than photosynthesis performance, and assimilate sink is directed to specialized metabolites production (Martínez-Ballesta et al., 2013). Thus, it is important to understand the roles of particular specialized metabolites in stress response, with a focus on the development of abiotic stress-tolerant crops (Muthusamy & Lee, 2024).

Polyphenols and carotenoids are mostly colored compounds that give an attractive look to plant flowers and fruits and participate in pollination and seed dispersion. They have also protective roles in various abiotic and biotic stresses acting mostly as antioxidants that may neutralize harmful ROS accumulated in plants under stress conditions. Unfriendly environmental conditions such as drought, extreme temperatures, and salinity usually activate the phenylpropanoid biosynthetic pathway resulting in the accumulation of various phenolic compounds (Šamec et al., 2021). Phenolic compounds widely studied for their role in stress response are flavonoids and phenolic acids. There is a positive correlation between the level of tolerance of certain *Brassica* species to abiotic stressors and the ability to accumulate higher levels of polyphenols. A comparative study of yellow- and black-coated rape seeds of *B. rapa* demonstrated that a higher flavonoid content in black-seeded rapes was in agreement with its higher tolerance to salinity stress (Xuan et al., 2015). An increase in flavonoid content in mustard (*B. juncea*) was shown to improve its stress tolerance (Ahmad et al., 2017). Similarly, the flavonol and total flavonoid contents were increased in canola cultivars under mild drought (Rezayian et al., 2018). Comparative analysis of three brassicas (Chinese cabbage, white cabbage, and kale) showed that more tolerant species/varieties accumulate specific phenolic acids and perform better under salinity (Linić et al., 2019; Pavlović et al., 2019) and drought (Pavlović et al., 2018). Rapeseed also accumulated phenolic compounds, particularly flavonoids during exposure to drought (Salami et al., 2023).

Good examples of successful survivors in extreme conditions are halophytes facing a combination of extreme conditions (high temperatures, salinity, drought or flood, anoxia, etc.) in natural habitats such as salt marshes and mangroves, salt flats and alkaline and limestone soils (Lopes et al., 2023). They accumulate extremely high phenolic content and show corresponding antioxidant activity. Examples of halophytes from the Brassicaceae family with reported high levels of polyphenols are *C. maritima* (Meot-Duros et al., 2008), *Lepidium ruderale* (Stanković et al., 2015), and *Raphanus raphanistrum* subsp. *landra* (Jdey et al., 2017).

Glucosinolates (GLs) are metabolites specifically found in Brassicaceae and have protective roles in abiotic and biotic stress (Chowdhury, 2022). There are three groups of GLs depending on the origin amino acids that they are derived from aliphatic (from methionine, alanine, valine, or leucine); aromatic (from phenylalanine or tyrosine); and indolic (from tryptophan). The types and abundances of glucosinolates among different genera are significantly different, suggesting that each genus developed specific biosynthetic pathways which are also affected by environmental conditions (Liu et al., 2020). A comparison of GLs profile in two broccoli varieties under salinity stress showed that the indole GLs were not affected in the more tolerant broccoli variety (cv. Parthenon) while it decreased significantly in the sensitive one (cv. Naxos), The opposite situation was observed for aliphatic GLs suggesting that each broccoli cultivar adopted distinct strategy for coping with salinity stress (Zaghdoud et al., 2012). Besides being species-specific, GL accumulation was found to be tissue-specific as well. Thus, *Brassica* crops (broccoli, cauliflower, kale) grown under drought stress showed a higher amount of GLs in their leaves compared to their roots (Ben Ammar et al., 2023). In general, moderate and/or short-time abiotic stresses (salinity, heat, and drought) caused an increase in GLs content in various Brassicaceae helping to improve stress tolerance along with other factors. More severe or long-term stresses usually decrease GL content, particularly in more sensitive species (Chowdhury, 2022; Cocetta et al., 2018; Martínez-Ballesta et al., 2013). Experiments with *B. rapa* BrPP5.2 transgenic line demonstrated a positive correlation between high GLs content and heat tolerance (Muthusamy et al., 2021). On the contrary, the *Arabidopsis* TU8 mutant deficient in GLs metabolism was found to be remarkably less tolerant to heat stress in comparison to the wild-type (WT) plants (Ludwig-Müller et al., 2000). Screening of different heat stress proteins (Hsps) in TU8 mutants showed that Hsp90 may be responsible for heat tolerance. Furthermore, the results obtained with *A. thaliana* single *myb28* and double *myb28myb29* mutants, defective partly and completely in aliphatic GLs biosynthesis, respectively, indicate that short-chain aliphatic GLs may help to water saving at salinity treatment (Martínez-Ballesta et al., 2015). A higher sensitivity of *myb28myb29* mutant to salinity compared to the WT and *myb28* mutant correlated with the abundance of PIP1 and PIP2 aquaporins. Essoh et al. (2020) recognized *Diplotaxis*, the wild *Brassica* relatives, as a new source of genes associated with GLs accumulation that may be of great potential in building abiotic stress tolerance in Brassica crops.

18.3.5 Natural Compounds as Useful Agents for Bursting the Abiotic Stress Tolerance

Intensive research of plant response to abiotic stresses revealed certain metabolites (amino acids, phenolic compounds, plant hormones, etc.) as the main players involved in plant tolerance mechanisms. Based on that knowledge, the exogenous application of a variety of compounds has been tested to find additional tools to ameliorate plant growth under abiotic stress. The effect of serotonin application on *B. napus* under salt stress showed its ability to alleviate stress by scavenging the ROS, regulating osmotic pressure, and promoting growth (Liu et al., 2021). The study on *B. napus* seedlings exposed to drought stress suggests that exogenous melatonin application improves growth and enhanced antioxidative system reaction and osmotic activity-adjustment and growth-inducing properties, thus making it beneficial for drought acclimatization (Li et al., 2018). Priming of *B. napus* seeds with melatonin and a combination of melatonin priming and salicylic acid (SA) foliar spray were the most effective treatments in counteracting drought stress at the plant reproductive stage, showing the increase in growth and yield compared to untreated plants (Rafique et al., 2023). The ability of melatonin to induce stress resistance and enhance growth under drought stress was confirmed in *B. rapa* plants as well (Hasnain et al., 2023). During drought, foliar application of proline, methionine, and melatonin enhanced cauliflower (*B. oleracea* var. *botrytis* L.) leaf area, leaf chlorophyll content, leaf relative water content, vitamin C, proline, total soluble sugar, reducing sugar, and nonreducing sugar, and yield compared to the corresponding control. However, the phenolic-related enzymes, glucosinolates, ABA, MDA, and total phenols were reduced compared to untreated plants (EL-Bauome et al., 2022). In addition to the efficacy of exogenously applied melatonin in the alleviation of salinity stress, the recovery of salinity-damaged seedlings by improvement of physiologic and biochemical parameters was demonstrated in *B. juncea* plants (Park et al., 2021).

Foliar treatments of *B. rapa* with phenolic acids, particularly ferulic acid, showed attenuated effects of salinity stress, causing a decrease in proline, increased levels of polyphenolic compounds especially sinapic acid, kaempferol, and quercetin, as well as antioxidant activity (Linić et al., 2021). Treatment of Chinese cabbage (*B. rapa* subsp. *parachinensis*) with liquiritoside resulted in a higher level of flavonoids, which helped mitigate the adverse stress effects (Akram et al., 2021).

Application of SA, which is phenolic acid but also a stress hormone, also showed mitigating effect against abiotic stressors depending on applied concentration, plant species, and stress conditions. Thus, SA application was reported to improve salt stress tolerance by activating the antioxidant enzymes in *B. napus* (Hasanuzzaman et al., 2014) as well as pigment levels and photosynthetic efficacy in *B. juncea* (Yusuf et al., 2012). Seed priming of *B. napus* with SA was shown to improve germination and growth during increased salinity and drought (Raees et al., 2022). As already mentioned, the plant hormone ABA has been proven as a protector against abiotic stresses. The application of ABA at appropriate concentrations has been also reported to mitigate the effects of various abiotic stresses (Zhu et al. 2011a; Liao et al., 2023). Metabolome profiling of *B. napus* after ABA treatments identified 77 primary and specialized metabolites (carbohydrates, fatty acids, glucosinolates, and flavonoids). Specialized metabolites such as sinigrin, quercetin, campesterol, and sitosterol were reported to regulate stomatal closure in *A. thaliana*, and *B. napus* (Zhu & Assmann, 2017).

Other biostimulants of organic or inorganic origin have been tested as well. So, foliar application of seaweed extract improved growth, photosynthetic pigments' content, free radical scavenging, and SOD activity of *B. napus* seedlings exposed to drought stress. The level of proline was significantly increased in plants after seaweed extract application as a result of enhanced expression of the *P5CS* gene involved in proline biosynthesis and decreased expression of PRODH responsible for proline degradation (Shahriari et al., 2021). In *B. napus* plants grown hydroponically, salinity stress was ameliorated by the silicon addition in a nutrient solution, and Si-treated plants maintained growth, had lower tissue Na^+ contents, increased ROS-scavenging capacity, and maintained the membrane integrity of root cells (Hashemi et al., 2010). Foliar application of silicon and seaweed extract under irrigated and rainfed conditions and at high temperatures enhanced plant growth and yield, as well as morphophysiological and gas exchange parameters while decreasing membrane injury percentage and MDA content and maintaining cooler canopy temperature of the *B. juncea* (Sujata Goyal et al., 2023).

18.3.6 Stress-Responsive Genes and Their Regulation under Stressful Environment

In response to stress, genes involved in growth and development are downregulated, and stress-associated genes involved in plant defense response are upregulated. These genome-wide reprograming optimize resource allocation, modulate growth, and activate protective mechanisms to ensure survival in stressful environmental conditions.

Perturbation of Ca^{2+} concentration and distribution, mitogen-activated protein kinases phosphorylation and signaling, ROS perturbation, harmful effects and signaling, membrane and cell wall disturbance, and phospholipid signaling, as well as hormonal signaling, are all implicated in gene expression regulation and protein turnover (Ghorbel et al., 2020; Kajal et al., 2023; Shahzad et al., 2022; Yan et al., 2023). Depending on the plant species, developmental stage, type and extent of applied stress, the changes in cell signaling ultimately influence transcriptional, post-transcriptional, translational, and post-translational regulation of many genes and proteins (Guerra et al., 2015; Kourani et al., 2022; Martín et al., 2021; Sahil et al., 2021; Salomé, 2017; Wang et al., 2013; Yang et al., 2022). Recently, our knowledge about key genes involved in stress response extended significantly by quantifying gene expression between control and stressed individuals or between stress-susceptible and resistant varieties or related species (Joshi et al., 2016; Wang et al., 2018). When environmental conditions are perceived as harsh or stressful, cells switch from performing their specific function within an organ/tissue to a more generalized stress response (Berkowitz et al., 2021; Jean-Baptiste et al., 2019; Mustroph et al., 2009). This reprogramming has a critical function in maintaining homeostasis under stressful conditions and involves the degradation or sequestration of currently unnecessary proteins and triggers refold, accumulation and activation of proteins necessary to alleviate the harmful effects of stress. Besides the unspecific degradation of proteins affected and damaged by ROS, targeted protein degradation is implicated in metabolism reprograming (Al-Saharin et al., 2022). Transcriptional reprogramming under stress conditions is mediated through the activation of stress-responsive transcription factors that facilitate massive activation of RNA polymerase II and gene transcription upregulation, while, in parallel, under stress conditions many genes are downregulated after binding of corepressors that disturb transcription or initiate chromatin modification (DNA methylation and histone proteins), impacting the establishment of repressive chromatin structure (Gómez-Zambrano et al., 2019; Lim et al., 2020; Plant et al., 2021; Ueda & Seki, 2019). Under stress, newly synthesized mRNAs appear in minutes and the accumulation of protective proteins and metabolites in a few hours (Guihur et al., 2022). This fast gene expression reprogramming relies on chromatin reorganization and ensures fast expression of master regulators that further impact many stress-related genes (Cortijo et al., 2017). Exposure to stress may perturb the expression of thousands of genes (Liu et al., 2022; Wang et al., 2018). In Brassicaceae species, the temperature stress (cold and heat) causes much broader changes in gene expression then drought, salt, or application of chemical substances (Rasmussen et al., 2013; Wang et al., 2018). Comprehensive abiotic stress response analyses reveal broad gene expression changes of transcription factors from basic helix-loop-helix (bHLH), APETALA2 and ethylene-responsive element binding proteins (AP2/EREBP), basic leucine zipper (bZIP), myeloblastosis (MYB), NAM, ATAF1/2, and CUC2 (NAC) and WRKY families (Liu et al., 2022; Wang et al., 2018), genes for antioxidative enzymes such as SOD (Verma et al., 2019), APX (Ma et al., 2019), and many other genes. The study of cold-resistant and sensitive cultivars of *B. napus* (Qi et al., 2020) and *B. rapa* (Ma et al., 2019) high-throughput gene expression reveals quick adaptation to prepare for ROS detoxification and soluble sugar and protein accumulation. Similar gene expression changes are important for the synthesis of protective organic osmolytes, and a tight balance between ROS production and scavenging in *Brassica* species is highlighted in salt stress tolerance (Shahzad et al., 2022).

Not only induction or repression of gene expression is influenced by abiotic stress. Stress impacts differential alternative splicing and augments proteome diversity by generating diverse protein isoforms or may reduce protein synthesis by imposing premature termination codons, which lead to mRNA degradation by nonsense-mediated RNA decay mechanism (Lee & Adams, 2020; Liu et al., 2022; Martín et al., 2021; Yang et al., 2022). So, alternative splicing has a role not only in specific protein isoform synthesis but also in the prevention of protein synthesis (Liu et al., 2013). Furthermore, stress intensity may trigger diverse mechanisms of alternative splicing and additionally modulate the stress response. This is well studied, especially for heat stress where many heat stress factor genes are alternatively spliced (Liu et al., 2022; Martín et al., 2021). Fine-tuning of gene expression regulation coupled with alternative splicing has thus important role in ecological diversity, species distribution, and performance of field crops. Recent evidence points to the importance of cis-regulating elements repertoire upstream of the transcription initiation site. It has been shown that more resistant species/varieties have higher basal stress-related gene expression levels and upon stress application moderately increase the expression of stress-responsive genes, and respond more specifically to certain stress types (Bauer et al., 2022; Ghorbel et al., 2020). Even more, recent analyses revealed subtle but significant differences in the stress response of different cells and tissues of an organism (Berkowitz et al., 2021; Jean-Baptiste et al., 2019), indicating that not all cells and tissues are equally competent to respond to stress. Currently, our measurements of gene expression mostly rely on the analysis of whole organisms or tissues, and unlike complex environmental conditions that in nature simultaneously affect plants (light intensity, drought, heat, wind, etc.), in controlled laboratory conditions we usually apply one, two, or a maximum of three stress parameters simultaneously. The application of different individual stress types reveals no evidence for a significant common stress response of any given tissue nor common stress-responsive genes across all tissues (Berkowitz et al., 2021). In future investigations, comparative analyses of stress-tolerant and sensitive plant relatives and the response of their specific cell types have the potential to disclose specialized cellular mechanisms of stress response that are until now possibly masked using complex tissues or whole organisms for stress response analysis.

18.4 EXPERIMENTAL MODELS IN REVELING ABIOTIC STRESS TOLERANCE MECHANISMS

18.4.1 *ARABIDOPSIS THALIANA*: CAN WE PROJECT TOLERANCE MECHANISMS TO BRASSICACEAE CROPS?

Due to the rapidity and magnitude of growth, small genome size, and a large collection of mutants, *A. thaliana* become a popular model organism since the 1970s. *Arabidopsis thaliana* belongs to the Brassicaceae family and has a much simpler genome but similar lifestyle, environmental challenges and stress response, modes of reproduction, and body plan. *Arabidopsis thaliana* is the first among plants to have a complete genome sequenced and provides the foundations for a comprehensive comparison of conserved processes and identification of genes valuable for crop improvement (Initiative, 2000).

Arabidopsis thaliana genome is one of the smallest (135 Mb, $n=5$) though *A. thaliana* shares more than 14,150 genes with the other Brassicaceae species (Lv et al., 2020). Deciphering the complex mechanisms of stress response in *A. thaliana* and the recognition of genes involved in stress resistance are thus simpler than in species having more complex genomes or lacking high-quality genome sequence and/or gene expression data.

Plants exposed to environmental cues rarely achieve optimal growth. Response to abiotic stress conditions such as nutrient imbalance, light and temperature extremes, water deficit, or flooding include many changes in cell signaling and lead to the activation and regulation of multiple genes and metabolic pathways. The transcription factors belonging to AP2/EREBP, bZIP, MYB, NAC, and WRKY families (AP2/EREBP, NAC, and WRKY are unique in plants) are master regulators of abiotic stress response and act as molecular switches in the regulation of other genes that respond to abiotic stress (Wang et al., 2018). In *B. napus* genome there are 315 highly co-expressed, denoted as crucial genes in abiotic stress response, while in *A. thaliana* 133 homologous of these crucial genes are identified. In species with more complex genomes, gene duplication may enable subtle sub-specialization for gene expression or in the adaptation of coding sequence to ensure survival in more extreme environmental conditions (Proulx, 2012). So, in polyploid species (like *B. napus*) or in species where gene duplication was extensive during evolution, many traits are under the control of multiple-copy genes with redundant functions (Birchler & Yang, 2022). It is very inefficient to manipulate such traits by particular gene modification, and using *A. thaliana* as a model plant that has much simple genome, well-established methods for gene manipulation and mutant regeneration, has a great potential implication in gene(s) functional characterization and breeding of *Brassica* varieties with improved stress tolerance. In addition, if we take into account CRISPR/Cas technic that is already used for simultaneous manipulation of multiple genes in *Brassica* species (Boniecka, 2023; Huang et al., 2023; Kim et al., 2022) and increasingly available genome and expression sequence data (Moreno-Ger et al., 2021; Wang et al., 2018), using *A. thaliana*, recognizing of valuable traits in wild species has a great potential for breeding Brassicaceae species in future.

18.4.2 CWRS OF BRASSICACEAE

'Crop wild relatives (CWRs) are the weedy relatives of domesticated plant species which are left unutilized in their natural habitat due to the presence of certain undesirable alleles which hamper their yield potential, and, thus, their further domestication' (Verma et al., 2023). Over the past thousands of years, CWRs have survived various biotic and abiotic stresses, many of them extreme, and accumulated many useful traits contributing to tolerance to various environmental stresses. Thus wild species may be useful experimental models for revealing tolerance mechanisms to abiotic stress factors and could serve as useful gene pools in breeding strategies to obtain more stress resilience crops.

Some of the CWRs have been already proposed as 'third-generation model plants' to cover research areas such as plant evolution, plant parasitism, and specific adaptations to environmental conditions (Cesarino et al., 2020). Among them, Brassicaceae species *Eutrema salsugineum* (formerly *Thellungiella salsuginea* or *Thellungiella halophile*) was successfully introduced in abiotic stress (mostly salinity/osmotic) studies revealing stress tolerance mechanisms as described earlier in the text.

The salt cress (*E. salsugineum*) was chosen because of its ability to cope successfully with extreme conditions, such as drought, salinity, and frost, as well as by its morphological similarity to *A. thaliana*. It meets numerous useful criteria for being advantageous as an experimental model; it has a small genome ($2n = 14$) with sequence identity of ~92% to *A. thaliana*; it has a short life cycle, is self-fertile, produces plenty of seeds, and can be efficiently genetically manipulated (Bressan et al., 2001). Furthermore, *E. salsugineum* is tolerant to salinity stress of up to 500 mM NaCl, whereas *A. thaliana* is already showing signs of suffering at 100 mM (Bressan et al., 2001). Metabolomics and transcriptomics data were generated, allowing the indication of candidate genes and metabolites potentially involved in salt/osmotic stress adaptation (Guevara et al., 2012; Inan et al., 2004; Lugan et al., 2010). A review by Li et al. (2022) gives a systematic overview of experimental data published for *E. salsugineum*, shedding light on the mechanism of adaptation of this species to extreme conditions. Recently, a new high-quality genome for *E. salsugineum* has been published, providing valuable information for comparative genomic studies and the functional analysis of plant abiotic stress tolerance in plants (Xiao et al., 2023).

The next Brassicaceae species that is considered a good experimental model for understanding tolerance mechanisms to abiotic stresses is sea rocket (*C. maritima*), an oilseed halophyte naturally growing at sandy sea shores. It also meets useful criteria such as a small diploid genome ($2n = 18$), a short life cycle, and the production of large amounts of seeds (Arbelet-Bonnin et al., 2019). So far, limited data suggest that a general activation of the antioxidant systems, enzymatic and non-enzymatic (mostly polyphenols) antioxidants allow the survival of the halophyte *C. maritima* in extreme conditions (Arbelet-Bonnin et al., 2019; Houmani et al., 2022; Ksouri et al., 2007, 2008).

Wild relatives of modern cultivated cabbages that may give better insight into stress tolerance mechanisms are wild cabbages (*B. incana*, *B. botteri*, etc.) that inhabit marine cliffs and sandy and barren rocks (Figure 18.3a) and are able to survive in extremely unfavorable environmental conditions of increased salinity, high temperature, and drought. They are morphologically similar to non-headed kale (*B. oleracea* var. *acephala*) (Figure 18.1) and exhibit high levels of specialized metabolites (Bootter et al.; 2023; Lučić et al., 2023; Lukić et al., 2023; Miceli et al., 2020). According to their physiological traits, they do not satisfy well-known requirements to be experimental model plants. However, seeds, collected in their natural habitat, germinate easily in laboratory conditions, and plants are successfully cultivated in soil and hydroponic systems in growth chambers and greenhouses to investigate mechanisms of abiotic stress tolerance (Figure 18.5a–c). Seed can be successfully regenerated by using bumblebees for controlled pollination (Figure 18.5d).

Screening of wild species for their potential tolerance against abiotic stress factors is limited so far on a few pioneer works. One example is an analysis of 25 wild species of *B. juncea* for their potential tolerance against heat and drought stress, which resulted in the selection of tolerant germplasms of *B. tournefortii*, *C. bursa-pastoris*, etc. as the potential source of abiotic stress tolerance in the mustard breeding program (Kashyap et al., 2023). As suggested by Essoh et al. (2020), *Diplotaxis*, as a GLs-rich species, may serve as a new source of genes with great agronomic potential, particularly in abiotic stress tolerance. Nevertheless, wild representatives of Brassicaceae are a 'gold mine of good genes' in breeding programs focused on sustainable agriculture and secure food production under climate change.

18.5 BREEDING STRATEGIES

The breading of *Brassica* species has a millennial tradition. New cultivars were created over centuries as a result of classical breeding by crossing individuals with desired phenotypic traits and growing their progenies. Knowledge about the processes that control the inheritance of agronomically important properties and their interaction with the environment has resulted in modern cultivars

FIGURE 18.5 Wild cabbage (*Brassica incana*) is grown in a pot (a) and hydroponically (b) and then replaced in a greenhouse (c), and field where controlled pollination of Brassicaceae can be performed for seed production (d). Photo: B. Salopek-Sondi and S. Goreta Ban.

that have substantially increased yields and improved quality compared to ancestors. Hence, within cultivated species exist a large number of developed cultivars and phenotypes. According to Breseghello and Guedes Coelho (2013), the main eras of plant breeding are

> (i) plant breeding based on the selection of observed variants, disregarding their origin; (ii) generation and selection of expanded variation by controlled mating; (iii) monitoring the inheritance of within-genome variation and selection of specific recombinants; and (iv) the creation and introduction of novel variation.

Classical breeding in *Brassica* includes the crossing and hybridization of pure lines by which breeders create new genetic combinations with the general aim of improving agronomically and economically desirable traits. The process is time-consuming, and it may take years or decades to develop new cultivars or F1 hybrids. Developing the resistant germplasm to abiotic stress is particularly demanding since it requires a comprehensive study of tolerance mechanisms from the seedling stage to harvest and relating them to genotype.

Stress tolerance and yield are difficult to breed using conventional methods because they are polygenic traits highly influenced by the environment and the genotype (Kumar et al., 2015). Therefore, new approaches including conventional breeding and biotechnological approaches have to be adopted to fasten the process and particularly to create germplasm with resistance to main biotic and abiotic stressors.

The molecular markers may be utilized for enhancing selection efficiency via marker-assisted selection (MAS) in *Brassica* breeding for stress tolerance. This approach accelerates crop improvement by integrating genomics and breeding and enables breeders to select plants early in their development for inclusion in breeding programs. The MAS significantly reduces the time needed to identify varieties expressing desired traits and they were mostly utilized for biotic stress resistance.

Inter- and intraspecific variations within *Brassica* species have the potential of molecular breeding to obtain tolerant genotypes to diverse abiotic stresses. Considering various molecular markers, SSR markers are getting into focus, because they are randomly sparsed, are highly variable, and show co-dominant inheritance (Kumar et al., 2015). They were identified to be suitable for separating cold-resistant and cold-susceptible genotypes in *B. rapa* (Amosova et al., 2020) or *B. napus* resistance to salt stress (Cui et al., 2020). Further, the use of quantitative trait loci (QTL) was applied to analyze the trade-off between root development and flowering under drought stress in *B. napus* haplotype. The authors found strong genetic correlations among studied traits and identified 20 QTLs, of which many co-localized to two major QTLs, suggesting that the trade-off is genetically confined (Fletcher et al., 2015). Totally 39 QTLs were identified, of which 18 QTLs affected the percentage of germination, length of root and shoot, and root/shoot ratio in germinating seeds of *B. napus* haploids grown under drought stress (Gad et al., 2021).

The next step in marker-assisted breeding is the application of genome-wide association study (GWAS) as a method developed for the study of associations between a genome-wide set of single-nucleotide polymorphisms (SNPs) and desired phenotypic traits. In the past years, whole-genome sequencing has been largely available in a number of species with phenotype variations promoting the application of GWAS. A GWAS was conducted on 157 of *B. napus* accessions exposed to low temperatures, and a total of 37,454 SNPs were used for the analysis. Ten QTLs were identified to be linked with electrolyte leakage of *B. napus* seedlings of which five were located on the A-genome; all together they explained 43% phenotypic variation (Fiebelkorn et al., 2018). In another study, a total of 37,111 SNPs were used for the analysis of frost tolerance in *B. napus* seedlings, and one QTL located on chromosome C04 was found to explain about 5% of the phenotypic variation and was associated with frost tolerance. Five additional genes were suggested to be associated with frost tolerance and abiotic stress tolerance (Wrucke et al., 2019). In *B. napus* seedlings exposed to drought stress at the seedling stage, 314 marker-trait associations were identified and linked with drought response, and they were distributed to all chromosomes in the A- and C-genomes. Of the identified genes, 85 genes were orthologous to *Arabidopsis* genes known to be associated with drought tolerance (Khanzada et al., 2020). A similar approach was used to conduct association mapping in *B. napus* exposed to salinity stress at the germination stage where authors detected a total of 110 SNPs associated with phenotyped salt tolerance indices and these SNPs were integrated into 44 QTLs. Among them, 56 interesting genes were located in or near the genomic region where the above QTLs were identified (Zhang et al., 2022).

It is evident that up-to-date molecular methods are readily applied in *B. napus* genetics and breeding for abiotic stress resistance; however, significantly less information is available for *B. oleracea* crops regardless of their importance as wide-grown vegetables. Recently, the sensitivity of *B. oleracea* seed germination to hypoxia was studied in 95 doubled haploid progeny lines produced by crossing between Chinese kale and a Calabrese parent. It seems that this trait is under the genetic control of two QTLs, one which influences the rate of germination, and the other one may be oxygen stress specific, which influences the capability of germination (Finch-Savage et al., 2005).

Breeding *Brassica* crops for heat tolerance is particularly important for some *B. oleracea* varieties like broccoli and cauliflower, which are more sensitive to high temperatures, particularly during inflorescence development. The head quality under summer conditions in a broccoli population was related to five QTL and one positive epistatic interaction that explained 62.1% of the variation in heat tolerance, which opens the door to develop markers for MAS (Branham et al., 2017). Uptmoor et al. (2009) developed a two-phase linear function for describing pre-flowering leaf area development, and the plateau function describing the effects of drought on leaf area development in *B. oleracea* L. doubled haploid population. Fourteen QTLs were detected of which nine co-localized to QTL detected as ones linked to osmotic adjustment. However, the predictability of the model was largely reduced when drought stress became more severe. The identification of genes related to abiotic stress provides a genetic resource for the breeding of *Brassica* crops resistant/tolerant to abiotic stress. This has been accelerated in the past 30 years by employing genetic modifications and recently genome editing. The introduction of new traits or the transfer of useful traits from wild relatives to cultivated varieties is now facilitated. Thus, through genetic manipulation, commercial verities of Brassicaceae species are further improved by introducing genes that affect the oil yield and quality, ensure tolerance of plants to various herbicides, ensure better meal quality and usage of phosphate from *Brassica* sp. derived feed, ensure biotic and abiotic stress tolerance, and prevent uncontrolled spread of modified plants due to male sterility (Poveda et al., 2020; Rustagi et al., 2020; Ton et al., 2020). The genetic transformation has been carried out in almost all economically important Brassicaceae species. Many genetically modified Brassicaceae plants are regenerated and show improved performances in laboratory conditions, with the most extensive genetic modifications done on *B. napus* and *B. rapa* including 45 and 4 modification events, respectively, approved for agricultural usage (www.isaaa.org).

As mentioned before, many Brassicaceae species are polyploid species where the introduction of the new trait may work well (such as herbicide tolerance), but breeding programs are significantly complicated in cases when the intention is to destroy property or trait (e.g. eliminate the production of undesired metabolite) or manipulate complex traits such as crop yield or disease tolerance. This requires simultaneous targeting at multiple specific locations in the genome. Dramatic innovations in next-generation sequencing technologies, improvement of our knowledge in functional genomics, manipulation of gene expression, and protein engineering by recently available genome-editing technologies open numerous possibilities for Brassicaceae genome manipulation. Genome editing by CRISPR/Cas technology has already been successfully used to improve stearic acid content in the seeds by manipulating multiple copies of the sensitivity to ABA and drought 2 (*SAD2*) gene in *B. napus* (Huang et al., 2023) or to reduce erucic acid and fiber content in pennycress *Thlaspi arvense* (Phippen et al., 2022), as well as to reduce pod shattering and yield losses in canola and oilseed rape (Walker et al., 2023). An excellent example of breeding by CRISPR/Cas recently came from Kim et al. (2022) where the genome of broccoli protoplasts was modified based on a sequence of wild relative *B. villosa MYB28* gene and glucoraphanin-rich broccoli was regenerated. By using a similar approach based on our knowledge of genetic mechanisms of abiotic stress tolerance in wild extremophile *Brassica* species, we expect extensive usage of genome-editing techniques in the development of abiotic stress-tolerant commercial *Brassica* species in the near future.

REFERENCES

Abedi, T., & Pakniyat, H. (2010). Antioxidant enzyme changes in response to drought stress in ten cultivars of oilseed rape (*Brassica napus* L.). *Czech Journal of Genetics and Plant Breeding*, *46*(1), 27–34.

Ahmad, P., Ahanger, M. A., AlYemeni, M. N., Wijaya, L., Egamberdieva, D., Bhardwaj, R., & Ashraf, M. (2017). Zinc application mitigates the adverse effects of NaCl stress on mustard [*Brassica juncea* (L.) czern & coss] through modulating compatible organic solutes, antioxidant enzymes, and flavonoid content. *Journal of Plant Interactions*, *12*, 429–437. https://doi.org/10.1080/17429145.2017.1385867

Akram, N. A., Iqbal, M., Muhammad, A., Ashraf, M., Al-Qurainy, F., & Shafiq, S. (2018). Aminolevulinic acid and nitric oxide regulate oxidative defense and secondary metabolisms in canola (*Brassica napus* L.) under drought stress. *Protoplasma*, *255*, 163–174.

Akram, N. A., Waseem, M., Ameen, R., & Ashraf, M. (2016). Trehalose pretreatment induces drought tolerance in radish (*Raphanus sativus* L.) plants: Some key physio-biochemical traits. *Acta Physiologiae Plantarum*, *38*(1). https://doi.org/10.1007/s11738-015-2018-1

Akram, W., Khan, W. U., Shah, A. A., Yasin, N. A., & Li, G. (2021). Liquiritoside alleviated Pb induced stress in *Brassica rapa* subsp. parachinensis: Modulations in glucosinolate content and some physiochemical attributes. *Frontiers in Plant Science*, *12*. https://doi.org/10.3389/fpls.2021.722498

Al-Saharin, R., Hellmann, H., & Mooney, S. (2022). Plant E3 ligases and their role in abiotic stress response. *Cells*, *11*(5), 890. https://doi.org/10.3390/cells11050890

Al-Shehbaz, I. A., Beilstein, M. A., & Kellogg, E. A. (2006). Systematics and phylogeny of the *Brassicaceae* (Cruciferae): An overview. *Plant Systematics and* Evolution, *259*, 89–120. https:// doi.org/10.1007/s00606-006-0415-z

Amor, N. B., Jiménez, A., Megdiche, W., Lundqvist, M., Sevilla, F., & Abdelly, C. (2006). Response of antioxidant systems to NaCl stress in the halophyte *Cakile maritima*. *Physiologia Plantarum*, *126*(3), 446–457. https://doi.org/10.1111/j.1399-3054.2006.00620.x

Amosova, A. V., Samatadze, T. E., Mozgova, G. V., Kipen, V. N., Dubovskaya, A. G., Artemyeva, A. M., Yurkevich, O. Y., Zoshchuk, S. A., Lemesh, V. A., & Muravenko, O. V. (2020). Genomic markers associated with cold-hardiness in *Brassica rapa* L. *Molecular Biology*, *54*, 541–552. https://doi.org/10.1134/S0026893320040032

Arbelet-Bonnin, D., Hamed-Louati, I. B., Laurenti, P., Abdelly, C., Ben-Hamed, K., & Bouteau, F. (2019). *Cakile maritima*, a promising model for halophyte studies and a putative cash crop for saline agriculture. *Advances in Agronomy*, *155*, 45–78. https://doi.org/10.1016/bs.agron.2019.01.003

Arias, P. A., Bellouin, N., Coppola, E., Jones, R. G., Krinner, G., Marotzke, J., Naik, V., Palmer, M. D., Plattner, G. K., Rogelj, J., Rojas, M., Sillmann, J., Storelvmo, T., Thorne, P. W., Trewin, B., Achuta Rao, K., Adhikary, B., Allan, R. P., Armour, K., … Zickfeld, K. (2021). Technical summary. In V. Masson-Delmotte, P. Zhai, A. Pirani, S. L. Connors, C. Péan, S. Berger, N. Caud, Y. Chen, L. Goldfarb, M. I. Gomis, M. Huang, K. Leitzell, E. Lonnoy, J. B. R. Matthews, T. K. Maycock, T. Waterfield, O. Yelekçi, R. Yu, & B. Zhou (Eds.), *Climate change 2021: The physical science basis. Contribution of working group I to the sixth assessment report of the intergovernmental panel on climate change* (pp. 33–144). Cambridge University Press.

Ashraf, M., & Mehmood, S. (1990). Response of four *Brassica* species to drought stress. *Environmental and Experimental Botany*, *30*(1), 93–100.

Ashraf, M., & Sharif, R. M. S. (1998). Assessment of inter-cultivar/line variation of drought resistance in a potential oil-seed crop, Ethiopian mustard (*Brassica carinata* A.Br.). *Archives of Agronomy and Soil Science*, *43*(4), 251–265. https://doi.org/10.1080/03650349809366043

Bailey-Serres, J., & Mittler, R. (2006). The roles of reactive oxygen species in plant cells. *Plant Physiology*, *141*(2), 311–311.

Bakirov, A., Zhang, Y., Zhang, Q., Seitahmetovna, S. A., Yu, X., Shii, Y., Xu, Y., Wang, K., Qin, M., Xu, A., & Huang, Z. (2022). Screening of salt tolerance traits and the salt tolerance evaluation method in *Brassica napus* at the seed germination stage. *Italian Journal of Agronomy*, *17*(2). https://doi.org/10.4081/ija.2022.2011

Bauer, N., Tkalec, M., Major, N., Talanga Vasari, A., Tokić, M., Vitko, S., Ban, D., Goreta Ban, S., & Salopek-Sondi, B. (2022). Mechanisms of kale (*Brassica oleracea* var. *acephala*) tolerance to individual and combined stresses of drought and elevated temperature. *International Journal of Molecular Sciences*, *23*(19), 11494. https://doi.org/10.3390/ijms231911494

Ben Ammar, H., Arena, D., Treccarichi, S., Di Bella, M. C., Marghali, S., Ficcadenti, N., Lo Scalzo, R., & Branca, F. (2023). The effect of water stress on the glucosinolate content and profile: A comparative study on roots and leaves of *Brassica oleracea* L. crops. *Agronomy*, *13*, 579. https://doi.org/10.3390/agronomy13020579

Ben Ammar, H., Picchi, V., Arena, D., Treccarichi, S., Bianchi, G., Lo Scalzo, R., Marghali, S., & Branca, F. (2022). Variation of bio-morphometric traits and antioxidant compounds of *Brassica oleracea* L. accessions in relation to drought stress. *Agronomy*, *12*(9), 2016. https://doi.org/10.3390/agronomy12092016

Berkowitz, O., Xu, Y., Liew, L. C., Wang, Y., Zhu, Y., Hurgobin, B., Lewsey, M. G., & Whelan, J. (2021). RNA-seq analysis of laser microdissected *Arabidopsis thaliana* leaf epidermis, mesophyll and vasculature defines tissue-specific transcriptional responses to multiple stress treatments. *The Plant Journal*, *107*(3), 938–955. https://doi.org/10.1111/tpj.15314

Bhuiyan, T. F., Ahamed, K. U., Nahar, K., Al Mahmud, J., Bhuyan, M. B., Anee, T. I., Fujita, M., & Hasanuzzaman, M. (2019). Mitigation of PEG-induced drought stress in rapeseed (*Brassica rapa* L.) by exogenous application of osmolytes. *Biocatalysis and Agricultural Biotechnology*, *20*, 101197.

Biondi, F., Balducci, F., Capocasa, F., Visciglio, M., Mei, E., Vagnoni, M., Mezzetti, B., & Mazzoni, L. (2021). Environmental conditions and agronomical factors influencing the levels of phytochemicals in Brassica vegetables responsible for nutritional and sensorial properties. *Applied Sciences*, *11*, 1927. https://doi.org/10.3390/app11041927

Birchler, J. A., & Yang, H. (2022). The multiple fates of gene duplications: Deletion, hypofunctionalization, subfunctionalization, neofunctionalization, dosage balance constraints, and neutral variation. *The Plant Cell*, *34*(7), 2466–2474. https://doi.org/10.1093/plcell/koac076

Blondel, J., & Aronson, J. (1999). *Biology and wildlife of the Mediterranean region*. Oxford University Press.

Boniecka, J. (2023). CRISPR/Cas-based precision breeding of oilseed rape (*Brassica napus* L.) – Recent improvements. In A. Ricroch, D. Eriksson, D. Miladinović, J. Sweet, K. Van Laere, & E. Woźniak-Gientka (Eds.), *A roadmap for plant genome editing* (pp. 291–307). Cham: Springer. https://doi.org/10.1007/978-3-031-46150-7_18

Bootter, M. B., Li, J., Zhou, W., Edwards, D., & Batley, J. (2023). Diversity of phytosterols in leaves of wild brassicaceae species as compared to Brassica napus cultivars: Potential traits for insect resistance and abiotic stress tolerance. *Plants*, *12*(9), 1866. https://doi.org/10.3390/plants12091866

Bouchyoua, A., Kouighat, M., Hafid, A., Ouardi, L., Khabbach, A., Hammani, K., & Nabloussi, A. (2024). Evaluation of rapeseed (*Brassica napus* L.) genotypes for tolerance to PEG (polyethylene glycol) induced drought at germination and early seedling growth. *Journal of Agriculture and Food Research*, *15*, 100928.

Branham, S. E., Stansell, Z. J., Couillard, D. M. & Farnham, M.W. (2017). Quantitative trait loci mapping of heat tolerance in broccoli (*Brassica oleracea* var. *italica*) using genotyping-by-sequencing. *Theorethical Applied Genettics*, *130*, 529–538. https://doi.org/10.1007/s00122-016-2832-x

Breseghello, F., & Guedes Coelho, A. S. (2013). Traditional and modern plant breeding methods with examples in rice (*Oryza sativa* L.). *Journal of Agricultural and Food Chemistry*, *61*(35), 8277–8286. https://doi.org/10.1021/jf305531j

Bressan, R. A., Zhang, C., Zhang, H., Hasegawa, P. M., Bohnert, H. J., & Zhu, J. K. (2001). Learning from the Arabidopsis experience. The next gene search paradigm. *Plant Physiology*, *127*, 1354–1360.

Cartea, E., De Haro-Bailón, A., Padilla, G., Obregón-Cano, S., Del Rio-Celestino, M., & Ordás, A. (2019). Seed oil quality of *Brassica napus* and *Brassica rapa* germplasm from Northwestern Spain. *Foods*, *8*(8), 292. https://doi.org/10.3390/foods8080292

Cesarino, I., Ioio, R. D., Kirschner, G. K., Ogden, M. S., Picard, K. L., Rast-Somssich, M. I., & Somssich, M. (2020). Plant science's next top models. *Annals of Botany*, *126*, 1–23. https://doi.org/10.1093/aob/mcaa063

Chalhoub, B., Denœud, F., Liu, S., Parkin, I. A. P., Tang, H., Wang, X., Chiquet, J., Belcram, H., Tong, C., Samans, B., Corréa, M., Da Silva, C., Just, J., Falentin, C., Koh, C., Clainche, I. L., Bernard, M., Bento, P., Noël, B., … Wincker, P. (2014). Early allopolyploid evolution in the post-Neolithic *Brassica napus* oilseed genome. *Science*, *345*(6199), 950–953. https://doi.org/10.1126/science.1253435

Channaoui, S., Idrissi, I. S. E., Mazouz, H., & Nabloussi, A. (2019). Reaction of some rapeseed (*Brassica napus* L.) genotypes to different drought stress levels during germination and seedling growth stages. *Oilseeds and Fats, Crops and Lipids*, *26*, 23. https://doi.org/10.1051/ocl/2019020

Chowdhury, P. (2022). Glucosinolates and its role in mitigating abiotic and biotic stress in Brassicaceae. In M. Hasanuzzaman, & K. Nahar (Eds.), *Plant stress physiology – Perspectives in agriculture*. IntechOpen. https://doi.org/10.5772/intechopen.94821

Cocetta, G., Mishra, S., Raffaelli, A., & Ferrante, A. (2018). Effect of heat root stress and high salinity on glucosinolates metabolism in wild rocket. *Journal of Plant Physiology*, *231*, 261–270.

Cortijo, S., Charoensawan, V., Brestovitsky, A., Buning, R., Ravarani, C. N. J., Rhodes, D., Van Noort, J., Jaeger, K. E., & Wigge, P. A. (2017). Transcriptional regulation of the ambient temperature response by H2A.Z nucleosomes and HSF1 transcription factors in Arabidopsis. *Molecular Plant*, *10*(10), 1258–1273. https://doi.org/10.1016/j.molp.2017.08.014

Corwin, D. L. (2021). Climate change impacts on soil salinity in agricultural areas. *European Journal of Soil Science*, *72*(2), 842–862.

Cui, J. Q., Hua, Y. P., Zhou, T., Liu, Y., Huang, J. Y., & Yue C. P. (2020). Global landscapes of the Na^+/H^+ antiporter (NHX) family members uncover their potential roles in regulating the rapeseed resistance to salt stress. *International Journal of Molecular Sciences*, *21*(10), 3429. https://doi.org/10.3390/ijms21103429

Díaz, M., De Haro, V., Muñoz, R., & Quiles, M. J. (2007). Chlororespiration is involved in the adaptation of Brassica plants to heat and high light intensity. *Plant, Cell & Environment*, *30*(12), 1578–1585. https://doi.org/10.1111/j.1365-3040.2007.01735.x

Dullinger, I., Wessely, J., Bossdorf, O., Dawson, W., Essl, F., Gattringer, A., Klonner, G., Kreft, H., Kuttner, M., Moser, D., Pergl, J., Pyšek, P., Thuiller, W., van Kleunen, M., Weigelt, P., Winter, M., & Dullinger, S. (2017). Climate change will increase the naturalization risk from garden plants in Europe. *Global Ecology and Biogeography*, *26*(1), 43–53. https://doi.org/10.1111/geb.12512

Edger, P. P., Hall, J. C., Harkess, A., Tang, M., Coombs, J., Mohammadin, S., Schranz, M. E., Xiong, Z., Leebens-Mack, J., Meyers, B. C., Sytsma, K. J., Koch, M. A., Al-Shehbaz, I. A., & Pires, J. C. (2018). *Brassicales* phylogeny inferred from 72 plastid genes: A reanalysis of the phylogenetic localization of two paleopolyploid events and origin of novel chemical defenses. *American Journal of Botany*, *105*(3), 463–469. https://doi.org/10.1002/ajb2.1040

Edger, P. P., Heidel-Fischer, H. M., Bekaert, M., Rota, J., Glöckner, G., Platts, A. E., Heckel, D. G., Der, J. P., Wafula, E. K., Tang, M., Hofberger, J. A., Smithson, A., Hall, J. C., Blanchette, M., Bureau, T. E., Wright, S. I., dePamphilis, C. W., Schranz, M. E., Conant, G. C., Barker, M. S., Wahlberg, N., Vogel, H., Pires, J. C., & Wheat, C. W. (2015). The butterfly plant arms-race escalated by gene and genome duplications. *Proceedings of the National Academy of Science*, *112*(27), 8362–8366. https://doi.org/10.1073/pnas.1503926112

EL-Bauome, H. A., Abdeldaym, E. A., El-Hady, M. A. M., Darwish, D. B. E., Alsubeie, M. S., El-Mogy, M. M., Basahi, M. A., Al-Qahtani, S. M., Al-Harbi, N. A., Alzuaibr, F. M., Alasmari, A., Ismail, I. A., Dessoky, E. S., & Doklega, S. M. A. (2022). Exogenous proline, methionine, and melatonin stimulate growth, quality, and drought tolerance in cauliflower plants. *Agriculture*, *2*(9), 1301. https://doi.org/10.3390/agriculture12091301

Elferjani, R., & Soolanayakanahally, R. (2018). Canola responses to drought, heat, and combined stress: Shared and specific effects on carbon assimilation, seed yield, and oil composition. *Frontiers in Plant Science*, *9*. https://doi.org/10.3389/fpls.2018.01224

Eshel, G., Shaked, R., Kazachkova, Y., Khan, A. A., Eppel, A., Cisneros, A., Acuña, T., Gutterman, Y., Tel-Zur, N., Rachmilevitch, S., Fait, A., & Barak, S. (2017). *Anastatica hierochuntica*, an Arabidopsis desert relative, is tolerant to multiple abiotic stresses and exhibits species-specific and common stress tolerance strategies with its halophytic relative, *Eutrema* (*Thellungiella*) *salsugineum*. *Frontiers in Plant Science*, *7*. https://doi.org/10.3389/fpls.2016.01992

Essoh, A. P., Monteiro, F., Pena, A. R., Pais, M. S., Moura, M., & Romeiras, M. M. (2020). Exploring glucosinolates diversity in Brassicaceae: A genomic and chemical assessment for deciphering abiotic stress tolerance. *Plant Physiology and Biochemistry, 150*, 151–161

Eziz, A., Yan, Z., Tian, D., Han, W., Tang, Z., & Fang, J. (2017). Drought effect on plant biomass allocation: A meta-analysis. *Ecology and Evolution*, *7*(24), 11002–11010. https://doi.org/10.1002/ece3.3630

Fariduddin, Q., Khanam, S., Hasan, S. A., Ali, B., Hayat, S., & Ahmad, A. (2009). Effect of 28-homobrassinolide on the drought stress-induced changes in photosynthesis and antioxidant system of *Brassica juncea* L. *Acta Physiologiae Plantarum*, *31*, 889–897.

Fiebelkorn, D., Horvath, D., & Rahman, M. (2018). Genome-wide association study for electrolyte leakage in rapeseed/canola (*Brassica napus* L.). *Molecular Breeding*, *38*, 129. https://doi.org/10.1007/s11032-018-0892-0

Finch-Savage, W. E., Côme, D., Lynn, J. R., & Corbineau, F. (2005). Sensitivity of *Brassica oleracea* seed germination to hypoxia: A QTL analysis. *Plant Science*, *169*(4), 753–759. https://doi.org/10.1016/j.plantsci.2005.05.026

Fletcher, R. S., Mullen, J. L., Heiliger, A., & McKay, J. K. (2015). QTL analysis of root morphology, flowering time, and yield reveals trade-offs in response to drought in *Brassica napus*. *Journal of Experimental Botany*, *66*, 245–256. https://doi.org/10.1093/jxb/eru423

Franzke, A., German, D., Al-Shehbaz, I. A., & Mummenhoff, K. (2009). *Arabidopsis* family ties: Molecular phylogeny and age estimates in the Brassicaceae. *Taxon*, *58*(2), 425–437. https://doi.org/10.1002/tax.582009

Franzke, A., Lysak, M. A., Al-Shehbaz, I. A., Koch, M. A., & Mummenhoff, K. (2011). Cabbage family affairs: The evolutionary history of Brassicaceae. *Trends in Plant Science, 16*(2), 108–116. https://doi.org/10.1016/j.tplants.2010.11.005

Gad, M., Chao, H., Li, H., Zhao, W., Lu, G., & Li, M. (2021). Mapping for seed germination response to drought stress in *Brassica napus*. *Frontiers in Plant Science*, *11*. https://doi.org/10.3389/fpls.2020.629970

German, D. A., Hendriks, K. P., Koch, M. A., Lens, F., Lysak, M. A., Bailey, C. D., Mummenhoff, K., & Al-Shehbaz, I. A. (2023). An updated classification of the Brassicaceae (Cruciferae). *PhytoKeys*, *220*, 127–144. https://doi.org/10.3897/phytokeys.220.97724

Ghorbel, M., Saibi, W., & Brini, F. (2020). Abiotic stress signaling in Brassicaceae plants. *Journal of Soil and Plant Biology*, *2020*(1), 138–150.

Gómez-Zambrano, Á., Merini, W., & Calonje, M. (2019). The repressive role of Arabidopsis H2A.Z in transcriptional regulation depends on AtBMI1 activity. *Nature Communications*, *10*(1). https://doi.org/10.1038/s41467-019-10773-1

Gong, Q., Li, P., Ma, S., Rupassara, S. I., & Bohnert, H. J. (2005). Salinity stress adaptation competence in the extremophile *Thellungiella halophila* in comparison with its relative *Arabidopsis thaliana*. *The Plant Journal*, *44*(5), 826–839. https://doi.org/10.1111/j.1365-313x.2005.02587.x

Guerra, D., Crosatti, C., Khoshro, H. H., Mastrangelo, A. M., Mica, E., & Mazzucotelli, E. (2015). Post-transcriptional and post-translational regulations of drought and heat response in plants: A spider's web web of mechanisms. *Frontiers in Plant Science*, *6*. https://doi.org/10.3389/fpls.2015.00057

Guevara, D. R., Champigny, M. J., Tattersall, A., Dedrick, J., Wong, C. E., Li, Y., Labbe, A., Ping, C. L., Wang, Y., Nuin, P., & Golding, G. B. (2012). Transcriptomic and metabolomic analysis of Yukon Thellungiella plants grown in cabinets and their natural habitat show phenotypic plasticity. *BMC Plant Biology*, *12*, 175.

Guihur, A., Rebeaud, M. E., & Goloubinoff, P. (2022). How do plants feel the heat and survive? *Trends in Biochemical Sciences*, *47*(10), 824–838. https://doi.org/10.1016/j.tibs.2022.05.004

Guiot, J., & Cramer, W. (2016). Climate change: The 2015 Paris Agreement thresholds and Mediterranean basin ecosystems. *Science*, *354*(6311), 465–468. https://doi.org/10.1126/science.aah5015

Gunasekera, C., Martin, L., Siddique, K. H. M., & Walton, G. H. (2006). Genotype by environment interactions of Indian mustard (*Brassica juncea* L.) and canola (B. napus L.) in Mediterranean-type environments. *European Journal of Agronomy*, *25*(1), 1–12. https://doi.org/10.1016/j.eja.2005.08.002

Hajiboland, R., & Amirazad, H. (2010). Drought tolerance in Zn-deficient red cabbage (*Brassica oleracea* L. var. capitata f. rubra) plants. *Horticultural Science*, *37*(3), 88–98. https://doi.org/10.17221/64/2009-hortsci

Hasanuzzaman, M., Alam, M. M., Nahar, K., Ahamed, K. U., & Fujita, M. (2014). Exogenous salicylic acid alleviates salt stress-induced oxidative damage in *Brassica napus* by enhancing the antioxidant defense and glyoxalase systems. *Australian Journal of Crop Science*, *8*(4), 631–639.

Hasanuzzaman, M., Nahar, K., Anee, T. I., Khan, M. I. R., & Fujita, M. (2018). Silicon-mediated regulation of antioxidant defense and glyoxalase systems confers drought stress tolerance in *Brassica napus* L. *South African Journal of Botany*, *115*, 50–57.

Hassan, M. U., Chattha, M. U., Khan, I., Chattha, M. B., Barbanti, L., Aamer, M., Iqbal, M. M., Nawaz, M., Mahmood, A., Ali, A., & Aslam, M. T. (2021). Heat stress in cultivated plants: Nature, impact, mechanisms, and mitigation strategies—A review. *Plant Biosystems*, *155*(2), 211–234.

Hashemi, A., Abdolzadeh, A., & Sadeghipour, H. R. (2010). Beneficial effects of silicon nutrition in alleviating salinity stress in hydroponically grown canola, *Brassica napus* L., plants. *Soil Science & Plant Nutrition*, *56*(2), 244–253.

Hasnain, Z., Zafar, S., Usman, S., Zhang, L. H., & Elansary, H. O. (2023). Elucidating role of melatonin foliar spray in ameliorating adverse effects of drought stress on growth and physio-biochemical attributes of *Brassica rapa* plants. *Scientia Horticulturae*, *321*, 112336. https://doi.org/10.1016/j.scienta.2023.112336

Hedge, I. C. (1976). A systematic and geographical survey of the Old World Cruciferae. In J. G. Vaughan, A. J. MacLeod, & B. M. G. Jones (Eds.), *The biology and chemistry of the cruciferae* (pp. 1–45). Academic Press.

Hendriks, K. P., Kiefer, C., Al-Shehbaz, I. A., Bailey, C. D., Hooft van Huysduynen, A., Nikolov, L. A., Nauheimer, L., Zuntini, A. R., German, D. A., Franzke, A., Koch, M. A., Lysak, M. A., Toro-Núñez, Ó., Özüdoğru, B., Invernon, V. R., Walden, N., Maurin, O., Hay, N. M., Shushkov, P., … Lens, F. (2023). Global *Brassicaceae* phylogeny based on filtering of 1,000-gene dataset. *Current Biology*, *33*(19), 4052–4068.e6. https://doi.org/10.1016/j.cub.2023.08.026

Houmani, H., Debez, A., Freitas-Silva, L. D., Abdelly, C., Palma, J. M., & Corpas, F. J. (2022). Potassium (K+) starvation-induced oxidative stress triggers a general boost of antioxidant and NADPH-generating systems in the halophyte *Cakile maritima*. *Antioxidants*, *11*(2), 401. https://doi.org/10.3390/antiox11020401

Huang, B., Su, J., Zhang, G., Luo, X., Wang, H., Gao, Y., Ma, G., Wang, J., Cai, D., Zhang, X., & Huang, B. (2017). Screening for Eruca genotypes tolerant to polyethylene glycol-simulated drought stress based on the principal component and cluster analyses of seed germination and early seedling growth. *Plant Genetic Resources*, *15*(2), 187–193. https://doi.org/10.1017/s1479262115000519

Huang, H., Ahmar, S., Samad, R. A., Pin, Q., Tong, Y., Zhao, Q., Xie, K., Zhang, C., Fan, C., & Zhou, Y. (2023). A novel type of *Brassica napus* with higher stearic acid in seeds developed through genome editing of BnaSAD2 family. *Theoretical and Applied Genetics*, *136*(9). https://doi.org/10.1007/s00122-023-04414-x

Husen, A., Iqbal, M., & Aref, I. M. (2014). Growth, water status, and leaf characteristics of *Brassica carinata* under drought and rehydration conditions. *Brazilian Journal of Botany*, *37*(3), 217–227. https://doi.org/10.1007/s40415-014-0066-1

Inan, G., Zhang, Q., Li, P., Wang, Z., Cao, Z., Zhang, H., Zhang, C., Quist, T. M., Goodwin, S. M., Zhu, J., Shi, H., Damsz, B., Charbaji, T., Gong, Q., Ma, S., Fredricksen, M., Galbraith, D. W., Jenks, M. A., Rhodes, D., … Zhu, J. K. (2004). Salt cress. A halophyte and cryophyte arabidopsis relative model system and its applicability to molecular genetic analyses of growth and development of extremophiles. *Plant Physiology*, *135*(3), 1718–1737. https://doi.org/10.1104/pp.104.041723

Initiative, A. G. (2000). Analysis of the genome sequence of the flowering plant *Arabidopsis thaliana*. *Nature*, *408*(6814), 796–815. https://doi.org/10.1038/35048692

IUCN. (2012). *IUCN red list categories and criteria:* Version 3.1 (2nd ed.). IUCN.

Jan, S. A., Shinwari, Z. K., & Rabbani, M. A. (2016). Morpho-biochemical evaluation of Brassica rapa subspecies for salt tolerance. *Genetika*, *48*(1), 323–338.

Jasprica, N., & Stamenković, V. (2023). Contribution to the syntaxonomy of plant communities with insular endemic species of genus *Brassica* (southern Croatia). *Hacquetia*, *22*(2), 197–214. https://doi.org/10.2478/hacq-2022-0019

Jdey, A., Falleh, H., Jannet, S. B., Hammi, K. M., Dauvergne, X., Ksouri, R., & Magnė, C. (2017). Phytochemical investigation and antioxidant, antibacterial and anti-tyrosinase performances of six medicinal halophytes. *South African Journal of Botany*, *112*, 508–514. https://doi.org/10.1016/j.sajb.2017.05.016

Jean-Baptiste, K., McFaline-Figueroa, J. L., Alexandre, C., Dorrity, M. W., Saunders, L., Bubb, K. L., Trapnell, C., Fields, S., Queitsch, C., & Cuperus, J. T. (2019). Dynamics of gene expression in single root cells of *Arabidopsis thaliana*. *The Plant Cell*, *31*(5), 993–1011. https://doi.org/10.1105/tpc.18.00785

Joshi, R., Wani, S. H., Singh, B., Bohra, A., Dar, Z., Lone, A. A., Pareek, A., & Singla-Pareek, S. L. (2016). Transcription factors and plants response to drought stress: Current understanding and future directions. *Frontiers in Plant Science*, *7*. https://doi.org/10.3389/fpls.2016.01029

Kagale, S., Robinson, S. J., Nixon, J., Xiao, R., Huebert, T., Condie, J., Kessler, D., Clarke, W. E., Edger, P. P., Links, M. G., Sharpe, A., & Parkin, I. A. P. (2014). Polyploid evolution of the Brassicaceae during the cenozoic era. *The Plant Cell*, *26*(7), 2777–2791. https://doi.org/10.1105/tpc.114.126391

Kage, H., Kochler, M., & Stützel, H. (2004). Root growth and dry matter partitioning of cauliflower under drought stress conditions: Measurement and simulation. *European Journal of Agronomy*, *20*(4), 379–394. https://doi.org/10.1016/s1161-0301(03)00061-3

Kajal, R. O., Lohani, P., Deshmukh, R., & Salvi, P. (2023). Engineering the transcriptional regulatory network to improve abiotic stress tolerance in crop plants: Taming the tough time. *Journal of Plant Growth Regulation*. https://doi.org/10.1007/s00344-023-11057-1

Kant, S., Kant, P., Raveh, E., & Barak, S. (2006). Evidence that differential gene expression between the halophyte, *Thellungiella halophila*, and *Arabidopsis thaliana* is responsible for higher levels of the compatible osmolyte proline and tight control of Na^+ uptake in T. halophila. *Plant, Cell & Environment*, *29*(7), 1220–1234. https://doi.org/10.1111/j.1365-3040.2006.01502.x

Karasov, T. L., Chae, E., Herman, J., & Bergelson, J. (2017). Mechanisms to mitigate the trade-off between growth and defense. *The Plant Cell*, *29*(4), 666–680. https://doi.org/10.1105/tpc.16.00931

Kashyap, A., Kumari, S., Garg, P., Kushwaha, R., Tripathi, S., Sharma, J., Gupta, N. C., Kumar, R. R., Yadav, R., Vishwakarma, H., & Rana J. C. (2023). Indexing resilience to heat and drought stress in the wild relatives of rapeseed-mustard. *Life*, *13*(3), 738. https://doi.org/10.3390/life13030738

Kaushal, N., Bhandari, K., Kadambot, H. M., Siddique, K. H. M., & Nayyar, H. (2016). Food crops face rising temperatures: An overview of responses, adaptive mechanisms, and approaches to improve heat tolerance. *Cogent Food & Agriculture*, *2*(1), 1134380. https://doi.org/10.1080/23311932.2015.1134380

Khanzada, H., Wassan, G. M., He, H., Mason, A. S., Keerio, A. A., Khanzada, S., Faheem, M., Solangi, A. M., Zhou, Q., Fu, D., Huang, Y., & Rasheed, A. (2020). Differentially evolved drought stress indices determine the genetic variation of *Brassica napus* at seedling traits by genome-wide association mapping. *Journal of Advanced Research*, *24*, 447–461. https://doi.org/10.1016/j.jare.2020.05.019

Kim, Y., Ahn, W. S., Cha, A., Jie, E. Y., Kim, S. W., Hwang, B., & Lee, S. (2022). Development of glucoraphanin-rich broccoli (*Brassica oleracea* var. italica) by CRISPR/Cas9-mediated DNA-free BolMYB28 editing. *Plant Biotechnology Reports*, *16*(1), 123–132. https://doi.org/10.1007/s11816-021-00732-y

Kourani, M., Mohareb, F., Rezwan, F. I., Anastasiadi, M., & Hammond, J. P. (2022). Genetic and physiological responses to heat stress in Brassica napus. *Frontiers in Plant Science*, *13*. https://doi.org/10.3389/fpls.2022.832147

Ksouri, R., Megdiche, W., Debez, A., Falleh, H., Grignon, C., & Abdelly, C. (2007). Salinity effects on polyphenol content and antioxidant activities in leaves of the halophyte *Cakile maritima*. *Plant Physiology and Biochemistry*, *45*, 244–249. https://doi.org/10.1016/j.plaphy.2007.02.001

Ksouri, R., Megdiche, W., Falleh, H., Trabelsi, N., Boulaaba, M., Smaoui, A., & Abdelly, C. (2008). Influence of biological, environmental and technical factors on phenolic content and antioxidant activities of Tunisian halophytes. *Comptes Rendus Biologies*, *331*(11), 865–873.

Kumar, D., Yusuf, M. A., Singh, P., Sardar, M., & Sarin, N. B. (2013a). Modulation of antioxidant machinery in α-tocopherol-enriched transgenic *Brassica juncea* plants tolerant to abiotic stress conditions. *Protoplasma*, *250*, 1079–1089.

Kumar, G., Purty, R. S., Sharma, M. P., Singla-Pareek, S. L., & Pareek, A. (2009). Physiological responses among Brassica species under salinity stress show strong correlation with transcript abundance for SOS pathway-related genes. *Journal of Plant Physiology*, *166*(5), 507–520. https://doi.org/10.1016/j.jplph.2008.08.001

Kumar, M., Choi, J. Y., Kumari, N., Pareek, A., & Kim, S. R. (2015). Molecular breeding in Brassica for salt tolerance: Importance of microsatellite (SSR) markers for molecular breeding in *Brassica*. *Frontiers in Plant Science*, *6*, 688. https://doi.org/10.3389/fpls.2015.00688

Kumar, S., Sairam, R. K., & Prabhu, K. V. (2013b). Physiological traits for high temperature stress tolerance in *Brassica juncea*. *Indian Journal of Plant Physiology*, *18*(1), 89–93. https://doi.org/10.1007/s40502-013-0015-1

Latif, M., Akram, N. A., & Ashraf, M. (2016). Regulation of some biochemical attributes in drought-stressed cauliflower (*Brassica oleracea* L.) by seed pre-treatment with ascorbic acid. *The Journal of Horticultural Science and Biotechnology*, *91*(2), 129–137.

Lee, J. S., & Adams, K. L. (2020). Global insights into duplicated gene expression and alternative splicing in polyploid *Brassica napus* under heat, cold, and drought stress. *The Plant Genome*, *13*(3). https://doi.org/10.1002/tpg2.20057

Lee, Y. P., Babakov, A., de Boer, B., Zuther, E., & Hincha, D. K. (2012). Comparison of freezing tolerance, compatible solutes and polyamines in geographically diverse collections of *Thellungiella* sp. and *Arabidopsis thaliana* accessions. *BMC Plant Biology*, *12*(1). https://doi.org/10.1186/1471-2229-12-131

Leisner, C. P. (2020). Review: Climate change impacts on food security-focus on perennial cropping systems and nutritional value. *Plant Sciences*, *293*, 110412.

Li, C., Duan, C., Zhang, H., Zhao, Y., Meng, Z., Zhao, Y., & Zhang, Q. (2022). Adaptative mechanisms of halophytic eutrema salsugineum encountering saline environment. *Frontiers in Plant Science*, *13*, 909527. https://doi.org/10.3389/fpls.2022.909527

Li, J., Liu, Z., Cheng, Y., Lu, G., Fu, G., Ma, H., Liu, Q., Zhang, X., Zou, X., & Li, C. (2018). Exogenous melatonin alleviates damage from drought stress in *Brassica napus* L. (rapeseed) seedlings. *Acta Physiologiae Plantarum, 40*(3). https://doi.org/10.1007/s11738-017-2601-8

Liang, W., Ma, X., Wan, P., & Liu, L. (2018). Plant salt-tolerance mechanism: A review. *Biochemical and Biophysical Research Communications*, *495*(1), 286–291. https://doi.org/10.1016/j.bbrc.2017.11.043

Liao, Y., Tang, Y., Wang, S., Su, H., Chen, J., Zhang, D., Wu, J., Zhou, D., Yan, M., & Liu, L. (2023). Abscisic acid modulates differential physiological and biochemical responses to cadmium stress in *Brassica napus*. *Environmental Pollutants and Bioavailability*, *35*, 1. https://doi.org/10.1080/26395940.2023.2168216

Lim, C., Park, J., Shen, M., Park, H., Cheong, M. S., Park, K. S., Baek, D., Bae, M. J., Ali, A., Jan, M., Lee, S. Y., Lee, B., Kim, W., Pardo, J. M., & Yun, D. J. (2020). The histone-modifying complex PWR/HOS15/HD2C epigenetically regulates cold tolerance. *Plant Physiology*, *184*(2), 1097–1111. https://doi.org/10.1104/pp.20.00439

Linić, I., Šamec, D., Grúz, J., Vujčić Bok, V., Strnad, M., & Salopek-Sondi, B. (2019). Involvement of phenolic acids in short-term adaptation to salinity stress is species specific among Brassicaceae. *Plants*, *8*, 155. https://doi.org/10.3390/plants8060155

Linić, I., Mlinarić, S., Brkljačić, L., Pavlović, I., Smolko, A., & Salopek-Sondi, B. (2021). Ferulic acid and salicylic acid foliar treatments reduce short-term salt stress in Chinese cabbage by increasing phenolic compounds accumulation and photosynthetic performance. *Plants*, *10*, 2346. https://doi.org/10.3390/plants10112346

Liu, J., Sun, N., Liu, M., Liu, J., Du, B., Wang, X., & Qi, X. (2013). An autoregulatory loop controlling Arabidopsis HSFA2 expression: Role of heat shock-induced alternative splicing. *Plant Physiology*, *162*(1), 512–521. https://doi.org/10.1104/pp.112.205864

Liu, Y., Cai, Y., Li, Y., Zhang, X., Song, N., Zhao, J., & Yang, H. (2022). Dynamic changes in the transcriptome landscape of Arabidopsis thaliana in response to cold stress. *Frontiers in Plant Science*, *13*. https://doi.org/10.3389/fpls.2022.983460

Liu, Y., Ding, X., Lv, Y., Cheng, Y., Li, C., Yan, L., Tian, S., & Zou, X. (2021). Exogenous serotonin improves salt tolerance in rapeseed (*Brassica napus* L.) seedlings. *Agronomy*, *11*, 400. https://doi.org/10.3390/agronomy11020400

Liu, Y., Rossi, M., Liang, X., Zhang, H., Zou, L., & Ong, C. N. (2020). An integrated metabolomics study of glucosinolate metabolism in different Brassicaceae genera. *Metabolites*, *10*, 313. https://doi.org/10.3390/metabo10080313

Lobell, D., & Gourdji, S. (2012). The influence of climate change on global crop productivity. *Plant Physiology*, *160*, 1686–1697.

Lopes, M., Sanches-Silva, A., Castilho, M., Cavaleiro, C., & Ramos, F. (2023). Halophytes as source of bioactive phenolic compounds and their potential applications. *Critical Reviews in Food Science and Nutrition*, *63*(8), 1078–1101. https://doi.org/10.1080/10408398.2021.1959295

Lučić, D., Pavlović, I., Brkljačić, L., Bogdanović, S., Farkaš, V., Cedilak, A., Nanić, L., Rubelj, I., & Salopek-Sondi, B. (2023). Antioxidant and antiproliferative activities of kale (Brassica oleracea L. var. *acephala* DC.) and wild cabbage (*Brassica incana* Ten.) polyphenolic extracts. *Molecules*, *28*, 1840. https://doi.org/10.3390/molecules28041840

Ludwig-Müller, J., Krishna, P., & Forreiter, C. (2000). A glucosinolate mutant of Arabidopsis is thermosensitive and defective in cytosolic Hsp90 expression after heat stress. *Plant Physiology*, *123*(3), 949–958.

Lugan, R., Niogret, M., Leport, L., Guégan, J., Larher, F., Savouré, A., Kopka, J., & Bouchereau, A. (2010). Metabolome and water homeostasis analysis of *Thellungiella salsuginea* suggests that dehydration tolerance is a key response to osmotic stress in this halophyte. *The Plant Journal*, *64*(2), 215–229. https://doi.org/10.1111/j.1365-313x.2010.04323.x

Lukić, I., Išić, N., Ban, D., Salopek Sondi, B., & Goreta Ban, S. (2023). Comprehensive volatilome signature of various Brassicaceae species. *Plants*, *12*, 177. https://doi.org/10.3390/plants12010177

Lv, H., Wang, Y., Han, F., Ji, J., Fang, Z., Zhuang, M., Li, Z., Zhang, Y., & Yang, L. (2020). A high-quality reference genome for cabbage obtained with SMRT reveals novel genomic features and evolutionary characteristics. *Scientific Reports*, *10*(1). https://doi.org/10.1038/s41598-020-69389-x

Ma, L., Coulter, J. A., Liu, L., Zhao, Y., Chang, Y. M., Pu, Y., Zeng, X., Yao-Zhao, X., Wu, J., Fang, Y., Bai, J., & Sun, W. (2019). Transcriptome analysis reveals key cold-stress-responsive genes in winter rapeseed (*Brassica rapa* L.). *International Journal of Molecular Sciences*, *20*(5), 1071. https://doi.org/10.3390/ijms20051071

Mabry, M. E., Hissong, S. D., Gallagher, E. Y., An, H., Labate, J. A., Gore, M. A., & Pires, C. J. (2021). The evolutionary history of wild, domesticated, and feral *Brassica oleracea* (*Brassicaceae*). *Molecular Biology and Evolution*, *38*(10), 4419–4434. https://doi.org/10.1093/molbev/msab183

Maggioni, L., von Bothmer, R., Poulsen, G., Branca, F., & Bagger Jørgensen, R. (2014). Genetic diversity and population structure of leafy kale and *Brassica rupestris* Raf. in south Italy. *Hereditas*, *151*(6), 145–158. https://doi.org/10.1111/hrd2.00058

Mahesh, K., Balaraju, P., Ramakrishna, B., & Ram Rao, S. (2013). Effect of brassinosteroids on germination and seedling growth of radish (*Raphanus sativus* L.) under PEG-6000 induced water stress. *American Journal of Plant Sciences*, *4*(12), 2305–2313. http://dx.doi.org/10.4236/ajps.2013.412285

Mandrich, L., & Caputo, E. (2020). Brassicaceae-derived anticancer agents: Towards a green approach to beat cancer. *Nutrients*, *12*, 868. https://doi.org/10.3390/nu12030868

Martín, G., Márquez, Y., Mantica, F., Duque, P., & Irimia, M. (2021). Alternative splicing landscapes in *Arabidopsis thaliana* across tissues and stress conditions highlight major functional differences with animals. *Genome Biology*, *22*(1). https://doi.org/10.1186/s13059-020-02258-y

Martínez-Ballesta, M. C., Moreno, D. A., & Carvajal, M. (2013). The physiological importance of glucosinolates on plant response to abiotic stress in Brassica. *International Journal of Molecular Sciences*, *14*, 11607–11625; https://doi.org/10.3390/ijms140611607

Martínez-Ballesta, M., Moreno-Fernández, D. A., Castejón, D., Ochando, C., Morandini, P. A., & Carvajal, M. (2015). The impact of the absence of aliphatic glucosinolates on water transport under salt stress in *Arabidopsis thaliana*. *Frontiers in Plant Science*, *6*, 524. https://doi.org/10.3389/fpls.2015.00524

McVetty, P. B. E., Mietkiewska, E., Omonov, T., Curtis, J., Taylor, D. C., & Weselake, R. J. (2016). Brassica spp. oils. In T. A. McKeon, D. G. Hayes, D. F. Hildebrand, Randall, & R. J. Weselake (Eds.), *Industrial oil crops* (pp. 113–156). Academic Press and AOCS Press. https://doi.org/10.1016/B978-1-893997-98-1.00005-1

Meot-Duros, L., Floch, L. G., & Magnė, C. (2008). Radical scavenging, antioxidant and antimicrobial activities of halophytic species. *Journal of Ethnopharmacology*, *116*(2), 258–262. https://doi.org/10.1016/j.jep.2007.11.024

Miceli, N., Cavò, E., Ragusa, M., Cacciola, F., Mondello, L., Dugo, L., Acquaviva, R., Malfa, G. A., Marino, A., D'Arrigo, M., & Taviano, M. F. (2020). *Brassica incana* Ten. (Brassicaceae): Phenolic constituents, antioxidant and cytotoxic properties of the leaf and flowering top extracts. *Molecules*, *25*(6), 1461. https://doi.org/10.3390/molecules25061461

Moreno-Ger, P., Fexová, S., George, N., Manning, J., Miao, Z., Mohammed, S., Muñoz-Pomer, A., Füllgrabe, A., Bi, Y., Bush, N. G., Iqbal, H., Kumbham, U., Solovyev, A., Zhao, L., Prakash, A., García-Seisdedos, D., Kundu, D. J., Wang, S., Walzer, M., & Papatheodorou, I. (2021). Expression Atlas update: Gene and protein expression in multiple species. *Nucleic Acids Research*, *50*(D1), D129–D140. https://doi.org/10.1093/nar/gkab1030

Morsi, N. A. A., Hashem, O. S. M., ElHady, M., Abd-Elkrem, Y. M., El-Temsah, M. E., Galal, E. G., Gad, K. I., Boudiar, R., Silvar, C., El-Hendawy, S., Mansour, E., & Abdelkader, M. (2023). Assessing drought tolerance of newly developed tissue-cultured canola genotypes under varying irrigation regimes. *Agronomy*, *13*(3), 836. https://doi.org/10.3390/agronomy13030836

Mourou, M., Raimondo, M. L., Lops, F., & Carlucci, A. (2023). Brassicaceae fungi and chromista diseases: Molecular detection and host–plant interaction. *Plants*, 12(5), 1033. https://doi.org/10.3390/plants12051033

M'rah, S., Ouerghi, Z., Eymery, F., Rey, P., Hajji, M., Grignon, C., & Lachaâl, M. (2007). Efficiency of biochemical protection against toxic effects of accumulated salt differentiates *Thellungiella halophila* from *Arabidopsis thaliana*. *Journal of Plant Physiology*, *164*(4), 375–384. https://doi.org/10.1016/j.jplph.2006.07.013

Munawar, A., Akram, N. A., Ahmad, A., & Ashraf, M. (2019). Nitric oxide regulates oxidative defense system, key metabolites and growth of broccoli (*Brassica oleracea* L.) plants under water limited conditions. *Scientia Horticulturae*, *254*, 7–13. https://doi.org/10.1016/j.scienta.2019.04.072

Munns, R. (2002). Comparative physiology of salt and water stress. *Plant, Cell & Environment*, *25*(2), 239–250. https://doi.org/10.1046/j.0016-8025.2001.00808.x

Munns, R., & Gilliham, M. (2015). Salinity tolerance of crops – What is the cost? *New Phytologist*, *208*, 668–673. https://doi.org/10.1111/nph.13519

Munns, R., & Millar, A. H. (2023). Seven plant capacities to adapt to abiotic stress. *Journal of Experimental Botany*, *74*(15), 4308–4323. https://doi.org/10.1093/jxb/erad179

Mustroph, A., Zanetti, M., Jang, C. J. H., Holtan, H., Repetti, P. P., Galbraith, D. W., Girke, T., & Bailey-Serres, J. (2009). Profiling translatomes of discrete cell populations resolves altered cellular priorities during hypoxia in Arabidopsis. *Proceedings of the National Academy of Sciences of the United States of America*, *106*(44), 18843–18848. https://doi.org/10.1073/pnas.0906131106

Muthusamy, M., Kim, J., Kim, S., Park, S., & Lee, S. (2021). BrPP5.2 overexpression confers heat shock tolerance in transgenic *Brassica rapa* through inherent chaperone activity, induced glucosinolate biosynthesis, and differential regulation of abiotic stress response genes. *International Journal of Molecular Sciences*, *22*(12), 6437.

Muthusamy, M., & Lee, S. I. (2024). Abiotic stress-induced secondary metabolite production in Brassica: Opportunities and challenges. *Frontiers in Plant Science*, *14*, 1323085. https://doi.org/10.3389/fpls.2023.1323085

Nagaharu, U. (1935). Genome analysis in Brassica with special reference to the experimental formation of *B. Napus* and peculiar mode of fertilization. *Japanese Journal of Botany*, *7*, 389–452.

Neugart, S., Baldermann, S., Hanschen, F. S., Klopsch, R., Wiesner-Reinhold, M., & Schreiner, M. (2018). The intrinsic quality of brassicaceous vegetables: How secondary plant metabolites are affected by genetic, environmental, and agronomic factors. *Scientia Horticulturae*, *233*, 460–478. https://doi.org/10.1016/j.scienta.2017.12.038

Nikolov, L. A. (2019). Brassicaceae flowers: Diversity amid uniformity. *Journal of Experimental Botany*, *70*(10), 2623–2635. https://doi.org/10.1093/jxb/erz079

Nikolov, L. A., Shushkov, P. S., Nevado, B., Gan, X., Al-Shehbaz, I. A., Filatov, D., Bailey, C. D., & Tsiantis, M. (2019). Resolving the backbone of the *Brassicaceae* phylogeny for investigating trait diversity. *New Phytologist*, *222*(3), 1638–1651. https://doi.org/10.1111/nph.15732

Nikzad, S., Maibody, S. A. M. M., Ehtemam, M. H., Golkar, P., & Mohammadi, S. A. (2023). Response of seed yield and biochemical traits of *Eruca sativa* Mill. to drought stress in a collection study. *Scientific Reports*, *13*(1). https://doi.org/10.1038/s41598-023-38028-6

Orsini, F., D'Urzo, M. P., Inan, G., Serra, S., Oh, D. H., Mickelbart, M. V., Consiglio, F., Li, X., Jeong, J. C., Yun, D., Bohnert, H. J., Bressan, R. A., & Maggio, A. (2010). A comparative study of salt tolerance parameters in 11 wild relatives of Arabidopsis thaliana. *Journal of Experimental Botany*, *61*(13), 3787–3798. https://doi.org/10.1093/jxb/erq188

Park, H. S., Kazerooni, E. A., Kang, S. M.., Al-Sadi, A. M., & Lee, I. J. (2021). Melatonin enhances the tolerance and recovery mechanisms in *Brassica juncea* (L.) Czern. under saline conditions. *Frontiers in Plant Science*, *12*, 593717. https://doi.org/10.3389/fpls.2021.593717

Pavlović, I., Mlinarić, S., Tarkowská, D., Oklestková, J., Novák, O., Lepeduš, H., Vujčić Bok, V., Radić Brkanac, S., Strnad, M., & Salopek-Sondi, B. (2019). Early Brassica crops responses to salinity stress: A comparative analysis between Chinese cabbage, white cabbage and kale. *Frontiers in Plant Science*, *10*, 450. https://doi.org/10.3389/fpls.2019.00450

Pavlović, I., Petřík, I., Tarkowska, D., Lepeduš, H., Vujčić, V., Radić Brkanac, S., Novák, O., & Salopek-Sondi, B. (2018). Correlations between phytohormones and drought tolerance in selected Brassica crops: Chinese cabbage, white cabbage and kale. *International Journal of Molecular Sciences*, *19*, 2866. https://doi.org/10.3390/ijms19102866

Patel, M. K., Kumar, M., Li, W., Luo, Y., Burritt, D. J., Alkan, N., & Tran, L.-S. P. (2020). Enhancing salt tolerance of plants: From metabolic reprogramming to exogenous chemical treatments and molecular approaches. *Cells*, *9*, 2492. https://doi.org/10.3390/cells9112492

Phippen, W. B., Rhykerd, R. L., Sedbrook, J. C., Handel, C., & Csonka, S. (2022). From farm to flight: CoverCreSs as a low carbon intensity cash cover crop for sustainable aviation fuel production. A review of progress towards commercialization. *Frontiers in Energy Research*, 10, 793776. https://doi.org/10.3389/fenrg.2022.793776.

Pilarska, M., Bartels, D., & Niewiadomska, E. (2021). Differential regulation of NAPDH oxidases in salt-tolerant *Eutrema salsugineum* and salt-sensitive *Arabidopsis thaliana*. *International Journal of Molecular Sciences*, *22*(19), 10341. https://doi.org/10.3390/ijms221910341

Pilarska, M., Wiciarz, M., Jajić, I., Kozieradzka-Kiszkurno, M., Dobrev, P., Vanková, R., & Niewiadomska, E. (2016). A different pattern of production and scavenging of reactive oxygen species in halophytic Eutrema salsugineum (*Thellungiella salsuginea*) plants in comparison to *Arabidopsis thaliana* and its relation to salt stress signaling. *Frontiers in Plant Science*, *7*, 1179.

Plant, A. R., Larrieu, A., & Causier, B. (2021). Repressor for hire! The vital roles of TOPLESS-mediated transcriptional repression in plants. *New Phytologist*, *231*(3), 963–973. https://doi.org/10.1111/nph.17428

Pokharel, M., Chiluwal, A., Stamm, M., Deng, M., Rhodes, D. H., & Jagadish, S. V. K. (2020). High night-time temperature during flowering and pod filling affects flower opening, yield and seed fatty acid composition in canola. *Journal of Agronomy and Crop Science*, *206*(5), 579–596. https://doi.org/10.1111/jac.12408

Poveda, J., Francisco, M., Cartea, M. E., & Velasco, P. (2020). Development of transgenic Brassica crops against biotic stresses caused by pathogens and Arthropod pests. *Plants*, *9*(12), 1664. https://doi.org/10.3390/plants9121664

POWO. (2023). *Plants of the world online. Facilitated by the Royal Botanic Gardens, Kew*. https://powo.science.kew.org

Proulx, S. R. (2012). Multiple routes to subfunctionalization and gene duplicate specialization. *Genetics*, *190*(2), 737–751. https://doi.org/10.1534/genetics.111.135590

Purty, R. S., Kumar, G., Singla-Pareek, S. L., & Pareek, A. (2008). Towards salinity tolerance in Brassica: An overview. *Physiology and Molecular Biology of Plants*, *14*, 39–49. https://doi.org/10.1007/s12298-008-0004-4

Pyšek, P., Pergl, J., Essl, F., Lenzner, B., Dawson, W., Kreft, H., Weigelt, P., Winter, M., Kartesz, J., Nishino, M., Antonova, L. A., Barcelona, J. F., Cabezas, F. J., Cárdenas, D., Cárdenas-Toro, J., Castaño, N., Chacón, E., Chatelain, C., Dullinger, S., … van Kleunen, M. (2017). Naturalized alien flora of the world: Species diversity, taxonomic and phylogenetic patterns, geographic distribution and global hotspots of plant invasion. *Preslia*, *89*, 203–274. https://doi.org/10.23855/preslia.2017.203

Qaderi, M. M., Kurepin, L. V., & Reid, D. M. (2006). Growth and physiological responses of canola (*Brassica napus*) to three components of global climate change: Temperature, carbon dioxide and drought. *Physiologia Plantarum*, *128*(4), 710–721. https://doi.org/10.1111/j.1399-3054.2006.00804.x

Qi, W., Wang, F., Ma, L., Qi, Z., Liu, S., Chen, C., Wu, J., Wang, P., Yang, C., Wu, Y., & Sun, W. (2020). Physiological and biochemical mechanisms and cytology of cold tolerance in *Brassica napus*. *Frontiers in Plant Science*, *11*. https://doi.org/10.3389/fpls.2020.01241

Quezada-Martinez, D., Nyarko, C. P. A., Schiessl, S. V., & Mason, A. S. (2021). Using wild relatives and related species to build climate resilience in Brassica crops. *Theoretical and Applied Genetics*, *134*, 1711–1728. https://doi.org/10.1007/s00122-021-03793-3

Raees, N., Ullah, S., & Nafees, M. (2022). Interactive effect of tocopherol, salicylic acid and ascorbic acid on agronomic characters of two genotypes of *Brassica napus* L. under induced drought and salinity stresses. *Gesunde Pflanz*, *75*(5), 1905–1923. https://doi.org/10.1007/s10343-022-00808-x

Rafique, N., Ilyas, N., Aqeel, M., Raja, N. I., Shabbir, G., Ajaib, M., Sayyed, R. Z., Alharbi, S. A., & Ansari, M. J. (2023). Interactive effects of melatonin and salicylic acid on Brassica napus under drought condition. *Plant Soil*. https://doi.org/10.1007/s11104-023-05942-7

Rai, A. N., Saini, N., Yadav, R., & Suprasanna, P. (2020). A potential seedling-stage evaluation method for heat tolerance in Indian mustard (*Brassica juncea* L. Czern and Coss). *3 Biotech*, *10*, 1–10.

Ramzan, T., Shahbaz, M., Maqsood, M. F., Zulfiqar, U., Saman, R. U., Nian, L., Irshad, M., Maqsood, S., Haider, A., Shahzad, M. B., Gaafar, A. Z., & Haider, F. U. (2023). Phenylalanine supply alleviates the drought stress in mustard (*Brassica campestris*) by modulating plant growth, photosynthesis and antioxidant defense system. *Plant Physiology and Biochemistry*, *201*, 107828. https://doi.org/10.1016/j.plaphy.2023.107828

Rasmussen, S., Barah, P., Suarez-Rodriguez, M. C., Bressendorff, S., Friis, P., Costantino, P., Bones, A. M., Nielsen, H. B., & Mundy, J. (2013). Transcriptome responses to combinations of stresses in Arabidopsis. *Plant Physiology*, *161*(4), 1783–1794. https://doi.org/10.1104/pp.112.210773

Raza, A., Hafeez, M. B., Zahra, N., Shaukat, K., Umbreen, S., Tabassum, J., Charagh, S., Ahmad Khan, R. S., & Hasanuzzaman, M. (2020). The plant family Brassicaceae: Introduction, biology, and importance. In M. Hasanuzzaman (Ed.), *The plant family Brassicaceae* (pp. 1–43). Springer.

Razzaq, H., Armstrong, S. J., & Saleem, H. (2021). Brassicas: A complete guide to the potential of their wild relatives. In M. T. Azhar & S. H. Wani (Eds.), *Wild germplasm for genetic improvement in crop plants* (pp. 187–199). Elsevier; Academic Press. https://doi.org/10.1016/B978-0-12-822137-2.00010-2

Rezayian, M., Niknam, V., & Ebrahimzadeh, H. (2018). Differential responses of phenolic compounds of *Brassica napus* under drought stress. *Iranian Journal of Plant Physiology*, *8*(3), 2417–2425.

Rodríguez, V. M., Soengas, P., Alonso-Villaverde, V., Sotelo, T., Cartea, M. E., & Velasco, P. (2015). Effect of temperature stress on the early vegetative development of *Brassica oleracea* L. *BMC Plant Biology*, *15*(1). https://doi.org/10.1186/s12870-015-0535-0

Rustagi, A., Negi, N. P., Choudhury, H. D., Mahajan, A., Verma, S., Kumar, D., Rajwanshi, R., & Sarin, N. B. (2020). Transgenic approaches for improvement of brassica species. In S. H. Wani, A. K. Thakur, & Y. J. Khan (Eds.), *Brassica improvement* (pp. 187–213). Springer.

Sahil, S., Keshan, R., Patra, A., Mehta, S., Abdelmotelb, K. F., Lavale, S. A., Chaudhary, M., Aggarwal, S. K., & Chattopadhyay, A. (2021). Expression and regulation of stress-responsive genes in plants under harsh environmental conditions. In A. Husen (Ed.), *Harsh environment and plant resilience* (pp. 25–44). Springer. https://doi.org/10.1007/978-3-030-65912-7_2

Salami, M., Heidari, B., & Tan, H. (2023). Comparative profiling of polyphenols and antioxidants and analysis of antiglycation activities in rapeseed (*Brassica napus* L.) under different moisture regimes. *Food Chemistry*, *399*, 133946. https://doi.org/10.1016/j.foodchem.2022.133946

Salomé, P. A. (2017). Some like it HOT: Protein translation and heat stress in plants. *The Plant Cell*, *29*(9), 2075. https://doi.org/10.1105/tpc.17.00699

Seebens, H., Essl, F., Dawson, W., Fuentes, N., Moser, D., Pergl, J., Pyšek, P., van Kleunen, M., Weber, E., Winter, M., & Blasius, B. (2015). Global trade will accelerate plant invasions in emerging economies under climate change. *Global Change Biology*, *21*(11), 4128–4140. https://doi.org/10.1111/gcb.13021

Shahriari, A. G., Mohkami, A., Niazi, A., Ghodoum Parizipour, M. H., & Habibi-Pirkoohi, M. (2021). Application of brown algae (*Sargassum angustifolium*) extract for improvement of drought tolerance in canola (*Brassica napus* L.). *Iranian Journal of Biotechnology*, *19*(1), 22–29. https://doi.org/10.30498/IJB.2021.2775

Shahzad, B., Rehman, A., Tanveer, M., Wang, L., Park, S. K., & Ali, A. (2022). Salt stress in Brassica: Effects, tolerance mechanisms, and management. *Journal of Plant Growth Regulation*, *41*(2), 781–795. https://doi.org/10.1007/s00344-021-10338-x

Shankar, S., Segaran, G., Sundar, R. D. V., Settu, S., & Sathiavelu, M. (2019). *Brassicaceae* – A classical review on its pharmacological activities. *International Journal of Pharmaceutical Sciences Review and Research*, *55*(1), 107–113.

Sharif, P., Seyedsalehi, M., Paladino, O., Van Damme, P., Sillanpää, M., & Sharifi, A. A. (2018). Effect of drought and salinity stresses on morphological and physiological characteristics of canola. *International Journal of Environmental Science and Technology*, *15*, 1859–1866. https://doi.org/10.1007/s13762-017-1508-7

Shyam, P., Wu, X. M., & Bhat, S. R. (2012). History, evolution, and domestication of Brassica crops. *Plant Breeding Reviews*, *35*, 19–84. https://doi.org/10.1002/9781118100509.CH2

Singh, S., Upadhyay, S., Pandey, A., & Kumar, S. (Eds.). (2019). *Molecular approaches in plant biology and environmental challenges. Energy, environment, and sustainability* (p. 499). Springer. https://doi.org/10.1007/978-981-15-0690-1

Snogerup, S., Gustafsson, M., & Bothmer von, R. (1990). *Brassica* sect. *Brassica* (Brassicaceae). I. Taxonomy and variation. *Willdenowia*, *19*(2), 271–365.

Song, J., Guan, Z., Hu, J., Guo, C., Yang, Z., Wang, S., Li, D., Wang, B., Lu, S., Zhou, R., Xie, W., YuanFang, C., Zhang, Y., Liu, K., Yang, Q., Chen, L., & Guo, L. (2020). Eight high-quality genomes reveal pan-genome architecture and ecotype differentiation of *Brassica napus*. *Nature Plants*, *6*(1), 34–45. https://doi.org/10.1038/s41477-019-0577-7

Song, X., Wei, Y., Xiao, D., Gong, K., Sun, P., Ren, Y., Yuan, J., Wu, T., Yang, Q., Li, X., Nie, F., Li, N., Feng, S., Pei, Q., Yu, T., Zhang, C., Hou, X., Wang, X., & Yang, J. (2021). *Brassica carinata* genome characterization clarifies U's triangle model of evolution and polyploidy in Brassica. *Plant Physiology*, *186*(1), 388–406. https://doi.org/10.1093/plphys/kiab048

Stanković, M. S., Petrović, M., Godjevac, D., & Stevanović, Z. D. (2015). Screening inland halophytes from the central Balkan for their antioxidant activity in relation to total phenolic compounds and flavonoids: Are there any prospective medicinal plants? *Journal of Arid Environments*, *120*, 26–32. https://doi.org/10.1016/j.jaridenv.2015.04.008

Stępień, P., & Johnson, G. N. (2009). Contrasting responses of photosynthesis to salt stress in the glycophyte Arabidopsis and the halophyte Thellungiella: Role of the plastid terminal oxidase as an alternative electron sink. *Plant Physiology*, *149*(2), 1154–1165. https://doi.org/10.1104/pp.108.132407

Sui, N., & Han, G. (2014). Salt-induced photoinhibition of PSII is alleviated in halophyte *Thellungiella halophila* by increases of unsaturated fatty acids in membrane lipids. *Acta Physiologiae Plantarum*, *36*(4), 983–992. https://doi.org/10.1007/s11738-013-1477-5

Sujata Goyal, V., Baliyan, V., Avtar, R., & Mehrotra, S. (2023). Correction to: Alleviating drought stress in *Brassica juncea* (L.) Czern and Coss. by Foliar application of biostimulants–orthosilicic acid and seaweed extract. *Applied Biochemistry and Biotechnology*, *195*, 722. https://doi.org/10.1007/s12010-022-04110-4

Šamec, D., Karalija, E., Šola, I., Vujčić Bok, V., & Salopek-Sondi, B. (2021). The role of polyphenols in abiotic stress response: The influence of molecular structure. *Plants*, *10*, 118. https://doi.org/10.3390/plants10010118

Šamec, D., Pavlović, I., & Salopek-Sondi, B. (2017). White cabbage (*Brassica oleracea* var. *capitata* f. alba): Botanical, phytochemical and pharmacologicall overwiev. *Phytochemistry Reviews*, *16*, 117–135.

Šamec, D., & Salopek-Sondi, B. (2019). Cruciferous (Brassicaceae) vegetables. In S. M. Nabavi & T. Sanches Silva (Eds.), *Nonvitamin and nonmineral nutritional supplements* (pp. 195–202). Elsevier. https://doi.org/10.1016/B978-0-12-812491-8.00027-8

Šamec, D., Urlić, B., & Salopek-Sondi, B. (2019). Kale (*Brassica oleracea* var. *acephala*) as a superfood: Review of the scientific evidence behind the statement. *Critical Reviews in Food Science and Nutrition*, *59*(15), 2411–2422.

Taji, T., Seki, M., Satou, M., Sakurai, T., Kobayashi, M., Ishiyama, K., Narusaka, Y., Narusaka, M., Zhu, J., & Shinozaki, K. (2004). Comparative genomics in salt tolerance between Arabidopsis and Arabidopsis-related halophyte salt cress using Arabidopsis microarray. *Plant Physiology*, *135*(3), 1697–1709. https://doi.org/10.1104/pp.104.039909

Tatout, C., Warwick, S., Lenoir, A., & Deragon, J.-M. (1999). SINE insertions as clade markers for wild Crucifer species. Molecular Biology and Evolution 16(11), 1614–1621. https://doi.org/10.1093/oxfordjournals.molbev.a026074

Tefera, A., Kebede, M., Tadesse, K., & Getahun, T. (2021). Morphological, physiological, and biochemical characterization of drought-tolerant wheat (*Triticum* spp.) varieties. *International Journal of Agronomy*, 2021, 1–12. https://doi.org/10.1155/2021/8811749

Ton, L. B., Neik, T. X., & Batley, J. (2020). The use of genetic and gene technologies in shaping modern rapeseed cultivars (*Brassica napus* L.). *Genes*, *11*(10), 1161. https://doi.org/10.3390/genes11101161

Tran, K., Pantha, P., Wang, G., Kumar, N., Wijesinghege, C., Oh, D. H., Wimalagunasekara, S., Duppen, N., Li, H., Hong, H., Johnson, J. E., Kelt, R., Matherne, M. G., Nguyen, T. T., Garcia, J., Clement, A., Tran, D., Crain, C., Adhikari, P., … Dassanayake, M. (2023). Balancing growth amidst salt stress – Lifestyle perspectives from the extremophyte model Schrenkiella parvula. *The Plant Journal*, *116*(3), 921–941. https://doi.org/10.1111/tpj.16396

Trautwig, A. N., Jackson, M. R., Kivlin, S. N., & Stinson, K. A. (2023). Reviewing ecological implications of mycorrhizal fungal interactions in the Brassicaceae. *Frontiers in Plant Sciences*, *14*, 1269815. https://doi.org/10.3389/fpls.2023.1269815

Ueda, M., & Seki, M. (2019). Histone modifications form epigenetic regulatory networks to regulate abiotic stress response. *Plant Physiology*, *182*(1), 15–26. https://doi.org/10.1104/pp.19.00988

Uptmoor, R., Osei-Kwarteng, M., Gürtler, S., & Stützel, H. (2009) Modeling the effects of drought stress on leaf development in a *Brassica oleracea* doubled haploid population using two-phase linear functions. *Journal of the American Society for Horticultural Science*, *134*(5), 543–552. https://doi.org/10.21273/JASHS.134.5.543

Uzilday, B., Özgür, R., Sekmen, A. H., Yıldıztugay, E., & Türkan, İ. (2015). Changes in the alternative electron sinks and antioxidant defence in chloroplasts of the extreme halophyte *Eutrema parvulum* (*Thellungiella parvula*) under salinity. *Annals of Botany*, *115*(3), 449–463. https://doi.org/10.1093/aob/mcu184

Verma, D., Lakhanpal, N., & Singh, K. (2019). Genome-wide identification and characterization of abiotic-stress responsive SOD (superoxide dismutase) gene family in *Brassica* j*uncea* and *B. rapa*. *BMC Genomics*, *20*(1). https://doi.org/10.1186/s12864-019-5593-5

Verma, S., Dubey, N., Singh, K. H., Parmar, N., Singh, L., Sharma, D., Rana, D., Thakur, K., Vaidya, D., & Thakur, A. K. (2023). Utilization of crop wild relatives for biotic and abiotic stress management in Indian mustard [*Brassica juncea* (L.) Czern. & Coss.]. *Frontiers in Plant Science*, *14*, 1277922. https://doi.org/10.3389/fpls.2023.1277922

Walker, A., Narváez-Vásquez, J., Mozoruk, J., Niu, Z., Luginbühl, P., Sanders, S. H., Schöpke, C., Sauer, N. J., Radtke, J., Gocal, G. F. W., & Beetham, P. R. (2023). Industrial scale gene editing in Brassica napus. *International Journal of Plant Biology*, *14*(4), 1064–1077. https://doi.org/10.3390/ijpb14040077

Wang, M., Yuan, F., Hao, H., Zhang, Y., Zhao, H., Guo, A., Jing-Jiang, H., Zhou, X., & Xie, C. G. (2013). BolOST1, an ortholog of Open Stomata 1 with alternative splicing products in *Brassica oleracea*, positively modulates drought responses in plants. *Biochemical and Biophysical Research Communications*, *442*(3–4), 214–220. https://doi.org/10.1016/j.bbrc.2013.11.032

Wang, P., Yang, C., Chen, H., Luo, L., Leng, Q., Li, S., Han, Z., Li, X., Song, C., Zhang, X., & Wang, D. (2018). Exploring transcription factors reveals crucial members and regulatory networks involved in different abiotic stresses in Brassica napus L. *BMC Plant Biology*, *18*(1). https://doi.org/10.1186/s12870-018-1417-z

Warwick, S. I., & Hall, J. C. (2009). Phylogeny of *Brassica* and wild relatives. In S. K. Gupta (Ed.), *Biology and breeding of crucifers* (pp. 19–36). CRC Press.

Wiciarz, M., Gubernator, B., Kruk, J., & Niewiadomska, E. (2015). Enhanced chloroplastic generation of H_2O_2 in stress-resistant *Thellungiella salsuginea* in comparison to *Arabidopsis thaliana*. *Physiologia Plantarum*, *153*(3), 467–476.

Wiciarz, M., Niewiadomska, E., & Kruk, J. (2018). Effects of salt stress on low molecular antioxidants and redox state of plastoquinone and P700 in *Arabidopsis thaliana* (glycophyte) and *Eutrema salsugineum* (halophyte). *Photosynthetica*, *56*, 811–819. https://doi.org/10.1007/s11099-017-0733-0

Wrucke, D. F., Mamidi, S. & Rahman, M. (2019). Genome-wide association study for frost tolerance in canola (*Brassica napus* L.) under field conditions. *Journal of Plant Biochemistry and Biotechnology*, *28*(2), 211–222. https://doi.org/10.1007/s13562-018-0472-8

Wu, W., Ma, B. L., & Whalen, J. K. (2018). Enhancing rapeseed tolerance to heat and drought stresses in a changing climate: Perspectives for stress adaptation from root system architecture. *Advances in Agronomy*, *151*, 87–157.

Xiao, M., Hao, G., Guo, X., Feng, L., Lin, H., Yang, W., Chen, Y., Zhao, K., Xiang, L., Jiang, X. & Mei, D. (2023). A high-quality chromosome-level *Eutrema salsugineum* genome, an extremophile plant model. *BMC Genomics*, *24*, 174. https://doi.org/10.1186/s12864-023-09256-x

Xie, X. Y., Zhang, X., & Zhang, B. (2013). Evaluation of drought resistance and analysis of variation of relevant parameters at seedling stage of rapeseed (*Brassica napus* L.). *Scientia Agricultura Sinica*, *46*, 476–485.

Xiong, X., Chang, L., Khalid, M., Jing-Jin, Z., & Huang, D. (2018). Alleviation of drought stress by nitrogen application in *Brassica campestris* ssp. *chinensis* L. *Agronomy*, *8*(5), 66. https://doi.org/10.3390/agronomy8050066

Xuan, L., Hussain, N., Wang, Z., Jiang, Y., Chen, M., & Jiang, L. (2015). Comparison of vitality between seedlings germinated from black-coated and yellow-coated seeds of a turnip rape (*Brassica rapa* L.) subjected to NaCl and $CdCl_2$ stresses. *Plant Growth Regulation*, *76*, 61–70. https://doi.org/10.1007/s10725-014-0019-5

Yan, Z., Xu, J., Li, R., Ge, Y., Li, Y., & Li, R. (2023). Plants' response to abiotic stress: Mechanisms and strategies. *International Journal of Molecular Sciences*, *24*(13), 10915. https://doi.org/10.3390/ijms241310915

Yang, L., Li, Y., Zhao, C., Liu, J., Tong, C., Zhang, Y., Cheng, X., Jiang, H., Shen, J., Xie, M., & S, L. (2022). Differential alternative splicing genes and isoform co-expression networks of Brassica napus under multiple abiotic stresses. *Frontiers in Plant Science*, *13*. https://doi.org/10.3389/fpls.2022.1009998

Yusuf, M., Fariduddin, Q., Varshney, P., & Ahmad, A. (2012). Salicylic acid minimizes nickel and/or salinity-induced toxicity in Indian mustard (*Brassica juncea*) through an improved antioxidant system. *Environmental Science and Pollution Research*, *19*, 8–18. https://doi.org/10.1007/s11356-011-0531-3

Yusuf, M. A., Kumar, D., Rajwanshi, R., Strasser, R. J., Tsimilli-Michael, M., & Sarin, N. B. (2010). Overexpression of γ-tocopherol methyl transferase gene in transgenic *Brassica juncea* plants alleviates abiotic stress: Physiological and chlorophyll a fluorescence measurements. *Biochimica et Biophysica Acta (BBA)-Bioenergetics*, *1797*(8), 1428–1438.

Zaghdoud, C., Alcaraz-Lopez, C., Mota-Cadenas, C., Martínez-Ballesta, M. D., Moreno, D. A., Ferchichi, A., & Carvajal, M. (2012). Differential responses of two broccoli (*Brassica oleracea* L. var. Italica) cultivars to salinity and nutritional quality improvement. *The Scientific World Journal*, 2012, 1–12.

Zhang, J., Mason, A. S., Wu, J., Liu, S., Zhang, X., Luo, T., Redden, R. J., Batley, J., Hu, L., & Yan, G. (2015). Identification of putative candidate genes for water stress tolerance in canola (*Brassica napus*). *Frontiers in Plant Science*, *6*. https://doi.org/10.3389/fpls.2015.01058

Zhang, L., Cai, X., Wu, J., Liu, M., Grob, S., Cheng, F., Liang, J., Cai, C., Liu, Z., Liu, B., Wang, F., Li, S., Liu, F., Li, X., Cheng, L., Yang, W., Li, M., Grossniklaus, U., Zheng, H., & Wang, X. (2018). Improved *Brassica rapa* reference genome by single-molecule sequencing and chromosome conformation capture technologies. *Horticulture Research*, *5*(1). https://doi.org/10.1038/s41438-018-0071-9

Zhang, Y., Li, P., Zhang, J., Li, Y., Xu, A. & Huang, Z. (2022). Genome-wide association studies of salt tolerance at the seed germination stage and yield-related traits in *Brassica napus* L. *International Journal of Molecular Science*, *23*, 15892. https://doi.org/10.3390/ijms232415892

Zhang, T. G, Li, Q. L., Diao, Z. H., Li, P., Wang, J., & Zheng, S. (2019). Effects of salt and drought stresses on antioxidant system and RbohC and RbohF genes expression in *Brassica campestris*. *Ying Yong Sheng Tai Xue Bao*, *30*(3), 969–978. https://doi.org/10.13287/j.1001-9332.201903.017.

Zhu, M., & Assmann, S. M. (2017). Metabolic signatures in response to abscisic acid (ABA) treatment in *Brassica napus* guard cells revealed by metabolomics. *Scientific Reports*, *7*(1), 12875. https://doi.org/10.1038/s41598-017-13166-w

Zhu, S.-Q., Chen, M.-W., Ji, B.-H., De-Mao Jiao, D.-M., & Jian-Sheng Liang, J.-S. (2011a). Roles of xanthophylls and exogenous ABA in protection against NaCl-induced photodamage in rice (*Oryza sativa* L) and cabbage (*Brassica campestris*). *Journal of Experimental Botany*, *62*(13), 4617–4625. https://doi.org/10.1093/jxb/err170

Zhu, Z., Zheng, W., & Zhang, X. (2011b). Principal component analysis and comprehensive evaluation on morphological and agronomic traits of drought tolerance in rapeseed (*Brassica napus* L.). *Scientia Agricultura Sinica*, *44*(9), 1775–1787. https://www.chinaagrisci.com/EN/10.3864/j.issn.0578-1752.2011.09.003

19 Mechanisms of Tolerance to Diverse Environmental Stresses and Adaptive Capacity
Current Achievements in Amaranth

Selene Fragoso-Iñiguez, Maryan Hernández-Jiménez, Angélica Barrales-López, Guillermo Antonio Silva-Martínez, Fabiola Estefania Tristan-Flores, Laura Fonseca-Mendoza, Eduardo Espítia-Rangel, and Gerardo Acosta-García

19.1 INTRODUCTION

Amaranth, taxonomically classified under the genus *Amaranthus* belonging to the family Amaranthaceae, comprehends over 60 distinct species (Waselkov et al., 2018). The most cultivated species for human consumption as grains are *Amaranthus cruentus* L., *Amaranthus hypochondriacus* L., and *Amaranthus caudatus* L. (Aderibigbe et al., 2022). These plants are dicotyledonous, are monoecious, and usually develop well in warm climates around the tropics; they are characterized by their upright growth, striking inflorescences of vibrant colors, and lance-shaped leaves, characteristics highly appreciated for a decorative plant (Figure 19.1).

Amaranth is used mainly as a grain and, therefore, is often referred to as a pseudocereal, because, despite its grain-like qualities and uses, it does not belong to the grass family and does

FIGURE 19.1 *Amaranthus cruentus* var. Amaranteca (a) and *Amaranthus hypochondriacus* var. Nutrisol (b) panicles growing in the greenhouse.

DOI: 10.1201/9781003457237-19

not contain gluten (Martinez-Villaluenga, 2020). Originating from Mesoamerica, amaranth has been a dietary staple for centuries. Beyond its role as sustenance, amaranth held religious significance. The Mexica culture used amaranth during religious ceremonies and rituals, often forming the grains of ceremonial bread into images of their gods. Its use was intertwined with beliefs about the life-giving properties of the plant, further solidifying its cultural importance, but its use was discouraged during colonization. In recent years, amaranth has recovered significance due to its reported high nutritional, particularly rich in protein containing several essential amino acids and minerals which are often deficient in other grains and bioactive compounds such as polyphenols, fatty acids, and squalene (Alvarez-Jubete et al., 2010; Mota et al., 2016). The grain also provides a reliable source of dietary fiber, essential for digestive health and blood sugar regulation. Amaranth is gluten-free, making it an excellent alternative for individuals with gluten sensitivity or celiac disease (Alvarez-Jubete et al., 2010). Additionally, amaranth could be consumed as a vegetable, according to Sarker and Oba (2020); *A. hypochondriacus* leaves have abundant phenolic compounds, flavonoids, pigments such as anthocyanins and vitamin B complex that offer enormous prospects for nourishing people with vitamin and antioxidant; also amaranth leaves and grains are rich in micronutrients such as iron, magnesium, and phosphorus. Besides its nutritious characteristics, amaranth has resilience against diverse abiotic stresses, among those occasioned by climate change as hot and arid conditions.

Although abiotic stresses affect all vegetal species over the planet, plants have evolved traits to cope with those stresses, there are a variety of responses to tolerate those extreme conditions, and there are some species or genera that are notorious. One strategy common to 1–10% of the plants on the earth is C4 metabolism, which represents a remarkable evolutionary solution to optimize carbon fixation in environments characterized by high temperatures and water limitations; in fact, as found by Zhou et al. (2018), water limitation was the primary driver for a C4 advantage, with atmospheric CO_2 as high as 600 ppm.

Plants with C4 metabolism exhibit distinct anatomical and physiological adaptations that set them apart from C3 plants, the more common photosynthetic type. One key feature is the development of Kranz anatomy, characterized by concentric layers of cells surrounding leaf veins, mesophyll cells, and bundle sheath cells, each playing a specific role in the carbon fixation facilitated by the enzyme phosphoenolpyruvate carboxylase (PEPCase), which initially fixes CO_2 in mesophyll cells to oxalacetate (OAA), a four-carbon compound. Depending on the Kranz anatomy of the plant, OAA will be conserved or transformed into another four-carbon compound and then transported to bundle sheath cells, where it will be decarboxylated in mitochondria, for *A. hypochondriacus* L. and *A. cruentus* L., by the NAD-dependent malic enzyme in the photosynthetic process that converts malate into 3-carbon pyruvate and releases CO_2. This CO_2 is then refixed by Rubisco in the Calvin–Benson–Bassham cycle within adjacent chloroplasts (Long et al., 1994; Babayev et al., 2014; Tsutsumi et al., 2017). As a result, these plants capture more organic carbon and accumulate more biomass during photosynthesis than C3 plants. This enhanced efficiency allows C4 plants to better adapt to extreme conditions like heat and drought (Wu et al., 2023). In the case of *A. hypochondriacus* L. and *A. cruentus* L., the NAD-dependent photosynthetic enzyme malic NAD produces 3-carbon pyruvate and releases CO2, for refixation by Rubisco in the Calvin-Benson-Bassham cycle in the adjacent chloroplast (Long et al., 1994; Babayev et al., 2014; Tsutsumi et al., 2017), so they obtain more organic carbon and accumulate more biomass during the photosynthetic process than C3 plants, allowing C4 plants to better adapt to extreme environmental conditions such as heat and drought (Wu et al., 2023).

However, some of the CO_2 fixed in chloroplasts is released from mitochondria during photorespiration, therefore, the positioning of mitochondria within photosynthetic cells plays a crucial role in determining the rate of CO_2 loss during photorespiration (Yoshimura et al., 2004), in those C4 species having a mitochondrial malic enzyme (NAD-ME), as *Amaranthus,* mitochondria are located centripetally to chloroplasts what in turn reduces the leakage of CO_2 into mesophyll cells, compared to the other two possible C4 photosynthesis pathways either NADP-ME or PEP-carboxykinase

(PCK) for decarboxylate C4 compounds (Fan et al., 2022). For those species using a mitochondrial malic enzyme over a chloroplast malic enzyme, the advantage could be the close integration with mitochondrial energy production, redox balance, metabolic flexibility, anaplerotic replenishment of the TCA cycle, and involvement in nitrogen metabolism. Under a wide range of environmental conditions, such as elevated temperatures, dryness, and high light intensities, plants possessing the C4 biochemical pump are more efficient in terms of water and nitrogen use than C3 (Furbank and Hatch, 1987; Singh et al., 2022).

19.2 WATER USE EFFICIENCY

Water use efficiency (WUE) in plants is typically calculated, as the relation between the grain yield or the total biomass produced per unit of water consumed. The quantity of water consumed is estimated according to the total amount of water used from the plant and soil surfaces in addition to that retained within plant structures (Wang et al., 2019).

According to a study by Wei et al. (2022), where they found that amaranth growing in a highly concentrated CO_2 environment (800 ppm) developed a minor stomatal density against that on normal conditions, a mechanism that could be also useful during drought. WUE is defined differently depending on the context, at an agronomic level, is described as the ratio of water used in crop production versus biomass or yield, from a plant physiology perspective, WUE is the amount of CO_2 fixed in photosynthesis compared to the water vapor lost to the atmosphere. Since stomata control both CO_2 intake and water loss, they play a crucial role in regulating WUE (Bertolino et al., 2019).

By adjusting the number of stomata (Figure 19.2), plants control the loss of water in dry environments; therefore, stomatal density decreases. This phenomenon has been reported for *A. hybridus* L. and *A. hypochondriacus* L. growing in low-watered soils compared to those in well-wetted soils (Dada and Ohu, 2013).

Additionally, plant cells need to regulate the opening of stomata, a process overseen by guard cells, the two specialized cells that surround the central pore, which is the stoma itself. By modulating the turgor pressure of their guard cells, plants dynamically adjust their stomatal opening, thereby improving the intake of CO_2 for photosynthesis and the loss of water through a process known as stomatal conductance. There is unanimity regarding the signals that initiate the transduction pathways leading to stomatal opening or closure; these include light intensity, atmospheric CO_2 concentration, humidity, and endogenous plant hormonal stimuli (Araújo et al., 2011; Driesen et al., 2020); among them, it is known that, during elevated ambient CO_2, the phytohormone abscisic acid (ABA) mediates the closure of stomata in response to drought

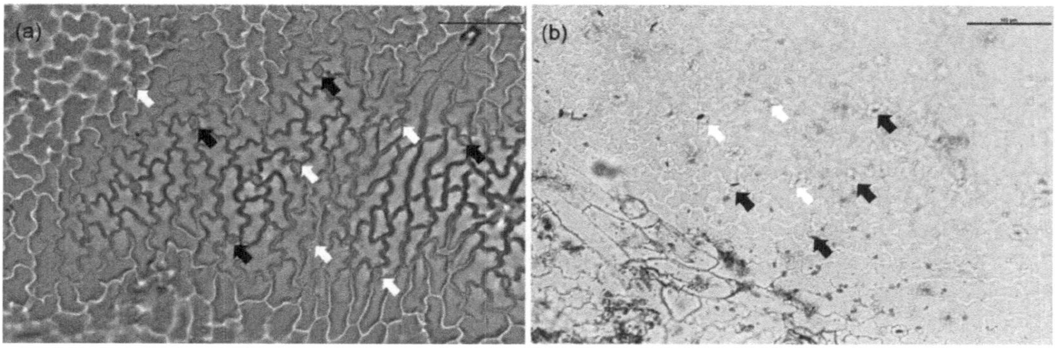

FIGURE 19.2 *Amaranth stomata.* (a) Stomatal density on the leaves of plants grown under normal watering conditions, (b) stomatal density on the leaves of plants grown under drought stress. White arrows point to open stomata, and black arrows point to closed stomata.

(Li et al., 2020), conditions expected during the actual climate change conditions. This mechanism has been studied in plants where SnRK2 kinases are key components. ABA is perceived by receptors known as PYR/PYL/RCAR proteins (pyrabactin resistance 1/PYR1-like/regulatory components of ABA receptors). In the absence of ABA, these receptors interact with protein phosphatases type 2C (PP2Cs) and inhibit the activity of SnRK2 kinases (Lim et al., 2015). When ABA levels increase, the hormone binds to its receptors, causing the receptors to undergo a conformational change. This altered conformation disrupts the interaction between the receptors and PP2Cs, leading to the activation of SnRK2 kinases that phosphorylate downstream targets, such as ion channels or transcriptions factors. This signaling cascade mediated by SnK2 is believed to be functionally conserved among terrestrial plants (Fürst-Jansen et al., 2020) though it has not been reported on *Amaranthus* yet.

ABA is not the only hormone involved in stomata closure; after a transcriptomic analysis of *A. hypochondriacus*, there was found a transcription factor member of APETALA2/ethylene-responsive factor (AP2/ERF) family, AhERF which is reported to be responsive to ethylene and to be induced during water deficit stress (WDS) in amaranth. Moreover, the overexpression of *AhERF* in *Arabidopsis* resulted in increased tolerance to WDS related to increased antioxidant enzymatic activity levels during stress conditions and to reduced water loss rates in detached leaves associated with the significantly increased tendency toward stomatal closure (Massange-Sanchez et al., 2016; Roy et al., 2024). Therefore, the stomatal responses of amaranth during drought highlight the adaptability and resilience of these plants to face water deficits.

19.3 TOLERANCE TO OXIDATIVE STRESS

Drought stress causes the closing of stomata in an attempt to reduce water loss, which results in less absorption of CO_2; therefore, the quantity of harvested light energy and generated reducing power surpasses the rate of its utilization by the Calvin–Benson–Bassham cycle. Thus, the absorbed light energy cannot go to CO_2 fixation, leading to an accumulation of reactive oxygen species (ROS) (Hasanuzzaman et al., 2020). The excess light absorption leads to the inactivation of photosystem II (PSII) and photoinhibition. This process is exacerbated by the excessive ROS that delay the turnover of D1 (Nishiyama et al., 2011). Overproduction of ROS in various organelles causes oxidative damage to macromolecules and activates programmed cell death pathways (Li and Kim, 2022).

Water deficit also leads to salinity in soils, causing osmotic and ionic stress. For these, plants also respond by closing stomata with the concomitant reduction in CO_2 fixation, which leads to ROS production (Hameed et al., 2021) as explained previously. Therefore, plants have developed some strategies that allow them to tolerate the oxidative stress promoted by either salinity or drought. A common physiological mechanism adopted by plants is the synthesis of low-molecular-weight, water-soluble, and non-toxic organic compounds at millimolar concentration termed 'osmolytes'; these include proline, glycine betaine, trehalose, and other sugars (Hossain et al., 2014).

There are reports about the accumulation of solutes as proline and sugars during water-deficient conditions in the grain amaranth (Slabbert and Krüger, 2014; Reyes-Rosales et al., 2023). Proline serves as a compatible solute with antioxidant properties induced in plant cells during osmotic stress; its osmoprotective role helps to maintain cell turgor pressure, preventing cellular dehydration and contributing to the plant's ability to withstand water stress. Proline also works as an effective scavenger of ROS, reducing oxidative damage to cellular structures (Mansour and Ali, 2017)

Proline is known to stabilize proteins by acting as a molecular chaperone; it helps prevent protein denaturation and aggregation, ensuring the structural integrity of key enzymes and proteins essential for plant function (Ghosh et al., 2022). Besides its protective roles, proline participates in floral and seed development. There is a natural increment in proline content during those processes. Its presence in seeds could serve as a readily available source of energy during germination (Trovato et al., 2019). Proline accumulation during drought and salinity stress is a feature character; however, the mechanisms in which this increment benefits plants are yet to be discovered.

Other soluble molecules that may help stabilize redox unbalance are sugars, acting as energy sources, antioxidants, osmoprotectants, signaling molecules, and directly scavenging ROS. The antioxidant character of sugars is based on the redox balance that is achieved through their roles in photosynthesis, mitochondrial respiration, and fatty acid β-oxidation in various organelles (Keunen et al., 2013; Nadarajah, 2020). In several species, e.g. Amaranth, sorghum, maize, and rice (Ogbaga et al., 2016; Voronin et al., 2019; Ichsan et al., 2022), increased levels of soluble sugars have been reported during drought. This increase in simple sugar concentrations is inversely proportional to a decrease in starch concentration (Vargas-Ortiz et al., 2021), a strategy that finally contributes to seed filling during stress conditions such as those derived by heat including water stress.

Finally, the presence of secondary metabolites, such as flavonoids and phenolic compounds, also plays a role in amaranth's oxidative stress tolerance function as antioxidants, but they also have a nutritious value for human consumption (Joshi and Verma, 2020).

19.4 CLIMATE CHANGE RESILIENCE

Climate change refers to long-term shifts in temperature, precipitation, and other atmospheric conditions around the globe. While natural factors contribute to climate variability, human activities have amplified these changes, leading to what is commonly known as anthropogenic climate change. The primary driver of this phenomenon is the increase in greenhouse gas (GHG) concentrations in the atmosphere, primarily carbon dioxide (CO_2) from the combustion of fossil fuels, especially coal, oil, and natural gas; methane (CH_4) emissions that arise from livestock, rice paddies, and the extraction of fossil fuels; and nitrous oxide (N_2O) released from agricultural and industrial activities. The increasing GHG emissions during the past 20 years have accelerated global warming, and climate change is occurring at a rate much faster than anticipated and its impacts can be devastating (IPCC, 2023). Climate change intensifies the frequency and severity of extreme weather events, such as hurricanes, droughts, heat waves, and heavy rainfall. These events can have devastating effects on communities, agriculture, and infrastructure, leading to economic losses and displacement of populations (Atwoli et al., 2021). According to the Intergovernmental Panel on Climate Change (IPCC), highly vulnerable regions, which include approximately 3.3–3.6 billion people, experienced 15 times higher human mortality rates from floods, droughts, and storms compared to regions with very low vulnerability. The time span between the 19th and 21st centuries is considered the period in which the earth had the most significant warming (IPCC, 2023). It is evident that climate change is challenging the continuity of human existence, not only with the increasing number of natural catastrophes but also with the new conditions over which food is produced, condemning humans to famine (Wiebers et al., 2022). Countries must develop new strategies that guarantee food security.

Among food production ways, agriculture is one of the more endangered because of the extreme weather disasters brought on by the impact of climate change on global crop production (Lesk et al., 2016). Traditionally a few crops are utilized as staple food over the world – wheat and maize surpass among others – but there is evidence that climate change could cause a global yield loss of these two main cereals due to the high probability of natural disaster occurrences in breadbasket regions (Kornhuber et al., 2023). This is a conjunctural moment in human history in which, to face new challenges, it is necessary to search for new crops or just reconsider some that were also used by ancient civilizations – this is the case of amaranth.

19.4.1 Amaranth WUE

WUE in plants is typically calculated as the relation between the grain yield and the total biomass produced per unit of water consumed. The quantity of water consumed is estimated according to the total amount of water used from the plant and soil surfaces in addition to that retained within plant structures (Wang et al., 2019). Amaranth has a higher tolerance to drought stress than even other plants with C4 metabolism, such as maize (Figure 19.3).

FIGURE 19.3 Amaranth and maize crops grown in neighboring fields (a) Crops grown under rainfed irrigation conditions. (b) Crops grown in sandy soils.

19.4.2 Salinity Stress Tolerance

Salinity stress on plants is derived from the high accumulation of ions, e.g., calcium (Ca^{2+}), potassium (K^+), and magnesium (Mg^{2+}), although the accumulation of ion sodium (Na^+) also contributes to salinity – it is specifically called soil sodicity. The causes of soil salinity could be natural or anthropogenic, elevated evaporation rates resulting from drought conditions, and irrigation practices in agricultural soils, which contribute to the buildup of ions within the soil. Simultaneously, climate fluctuations, characterized by rising temperatures and a heightened occurrence and severity of droughts in typically moist regions, alongside the exacerbation of aridity and expansion of global drylands, collectively pose soil salinization as a significant global concern (Stavi et al., 2021).

The amaranth responses to soil salinity stress seem to converge with oxidative stress tolerance. In *Amaranthus caudatus* L. it was shown that treatment with 100 mM NaCl improved significantly leaf and root dry biomass with the maintenance of water content; there is also an increase in the activity of antioxidant enzymes – catalase, ascorbate peroxidase, and guaiacol peroxidase. However, the salt tolerance in amaranth is closely associated with maintaining ion balance; it demonstrates its capacity to store Na^+ in the stems as a protective measure for the leaves, while also maintaining a high K^+/Na^+ ratio by reducing Na+ accumulation. This allows the enhancement of potassium mainly in photosynthetic leaves (Tebini et al., 2022). There is also an accumulation of soluble sugars and proline in the roots and shoots of *A. cruentus*, which will help to persist under oxidative stress (Wouyou et al., 2019).

To further comprehend the strategies employed by amaranth to combat salt stress, Estrada et al. (2021), conducted a comparative analysis of the physiological, anatomical, and molecular responses of amaranth (*A. hypochondriacus* L.) and quinoa under saline conditions (100 mM NaCl). High tolerance to saline conditions was attributed to its low basal stomatal conductance, stemming from a reduced stomatal density and limited aperture of stomata on the adaxial leaf surface. This feature enables amaranth to minimize water loss from leaves more efficiently than quinoa under salt stress. Regarding Na^+ regulation, amaranth exhibited a pattern of Na^+ distribution like that of glycophytes, with the highest accumulation in the roots, followed by the stem and the lowest levels detected in the leaves. The analysis of gene expression related to Na^+ homeostasis, plasma membrane Na^+/H^+ transporter SOS1, HKT-type protein, and tonoplast Na^+/H^+ antiporter NHX1 (SOS1, HKT1.1, and NHX1) revealed distinct patterns between amaranth and quinoa, notably with significantly higher basal expression levels in amaranth roots. According to these researchers, amaranth demonstrated tolerance to salt stress throughout its growth cycle, as evidenced by unaffected grain production under the applied saline conditions. These findings were confirmed through an open field trial carried out in Italy over 3 years to evaluate the response of *A. hypochondriacus* L. to various irrigation strategies and salinity, finding that the grain amaranth yield was not negatively affected either by different irrigation volumes or by irrigation time though salinity (same ions and quantities as in seawater) significantly reduced seed yield, but the quality (measured as starch, protein, ashes, and fat content) of amaranth seeds was preserved (Pulvento et al., 2022).

Since roots are the primary organ that faces soil salinity, Huerta-Ocampo et al. (2014) conducted an assay to identify the impact of differentially accumulated proteins in *A. cruentus* L. on hydroponic growth in response to 150 mM NaCl (considered mild-moderate saline stress). The total root protein extract was further separated on 2D gel electrophoresis and liquid chromatography–mass spectrometry/mass spectrometry (LC–MS/MS) technique. A total of 101 protein spots ran differentially in response to salinity stress, of which 77 were successfully identified and correlated with their biological function according to gene ontology (http://www.geneontology.org/). The main function identified of the proteins differentially expressed was carbohydrate metabolism (24%), which suggests that plants adjust their carbohydrate content to maintain growth and development.

To understand the mechanisms underlying amaranth tolerance to saline stress, reverse genetics was performed by Atou et al. (2022) following a recent study, where they developed some amaranth mutant lines (L2, L18 and L23) from the sensitive cultivar locale derived from γ-rays irradiation strategy, which exhibited improved salt resistance compared to the initial cultivar (Atou et al., 2022). Three mutant lines and the control cultivar locale were placed in pots and subjected to three different concentrations of NaCl (0, 100, and 200 mM). After 2 weeks of treatment, assessments were made regarding plant growth, ion levels, and organic solute contents. Notably, the control cultivar locale exhibited earlier and more pronounced growth reduction under salt stress compared to line L2, followed by the salt-resistant lines L18 and L23. The accumulation rates of Na^+, proline, and soluble sugars, as well as the absorption rates of K^+ and Ca^{2+}, and the reduction rates of K^+/Na^+ and Ca^{2+}/Na^+ under salt stress, varied significantly among the populations. The salt resistance observed in line L18 was attributed to Na^+ accumulation in leaves, coupled with the accumulation of proline and soluble sugars, and the maintenance of high absorption levels of Ca^{2+}, along with high K^+/Na^+ and Ca^{2+}/Na^+ ratios. Conversely, the resistance of line L23 was attributed to the exclusion of Na^+ from leaves, while maintaining high absorption levels of K^+ and Ca^{2+} and high K^+/Na^+ and Ca^{2+}/Na^+ ratios. Line L2 exhibited resistance to salt stress through proline accumulation and the maintenance of a high Ca^{2+}/Na^+ ratio. Thus, the different mutant lines developed distinct strategies for salt tolerance (Richard et al., 2023).

19.5 DEFOLIATION TOLERANCE

Defoliation is attributed to several biotic and abiotic factors, such as drought or pathogens, resulting in the loss of leaf area; its severity is related to the developmental stage of the plant; and it could be so grim that the plant cannot support growth or reproduction demands and minimizes the yield of crops (Nezami et al., 2008; Thomason and Battaglia, 2020).

With respect to plant damage by mechanical defoliation, in *A. hypochondriacus* and *A. Cruentus*, it was observed that cutting shoots by 10% and 40% increased the number of leaves and the number of secondary branches, although the number of primary branches was lower than that in the control; in addition, there are differences between species in the distribution of the biomass produced, while in *A. Cruentus* more biomass was distributed in the seed; in *A. hypochondriacus* it was in the leaves. In general, in *A. cruentus* a better overcompensation for damage was observed, but both species are tolerant to defoliation (Martinez-Moreno et al., 1999). In another study on *A. cruentus*, the effect of mechanical damage and defoliation was evaluated, which showed tolerance and resistance mechanisms, modifying growth, carbohydrate accumulation, expression, and activity of trypsin inhibitors (Vargas-Ortíz et al., 2015).

The study of amaranth defoliation tolerance emerges as a promising path for sustainable agriculture, mainly because of its dual-purpose cultivar as both vegetable and grain, a characteristic unbelievably valuable for small producers who can easily consume the leaves and sell the grains.

Some studies demonstrate that amaranth could easily be converted into a staple crop, helping to guarantee food security (Gelaye, 2023). Some other studies have explored the defoliation tolerance of grain amaranth, which is crucial for plant survival, especially in environments where defoliation events are frequent. In 2013 Vargas-Ortiz et al. (2013) cultivated *A. hypochondriacus* L. and *A. cruentus* L. under greenhouse conditions and then simulated the defoliation by perforating gradually the leaves; in these conditions, both species showed to be highly tolerant. However, tolerance was compromised when defoliation was done by gradual cutting of the leaf. Finally, Vargas-Ortiz et al. assessed full defoliation under field conditions where *A. cruentus* L. performed better than *A. hypochondriacus* L. although tolerance tended to decrease in both. One key mechanism in defoliation tolerance is the regrowth of leaves following damage. Amaranth activates dormant buds or develops new vegetative meristems to compensate for lost foliage. *A. cruentus* L. accumulated sufficient C in roots and stem to allow branching and shoot regrowth after defoliation. This rapid response is attributed to the plant's ability to efficiently relocate resources and energy from sink to source organs (Vargas-Ortiz et al., 2015).

In another study, Hoidal et al. (2019) evaluated defoliation at various intensities and frequencies (0%, 25%, 50%, 75%, and 100% defoliation, one, two, and three times) under controlled conditions and Danish and Mexican field conditions. Defoliation tolerance was assessed in a total of seven varieties, spanning the three primary grain amaranth species: *A. cruentus* L., *A. hypochondriacus* L., and *A. caudatus* L. All varieties and environments evaluated had no impact on yield or seed quality from a single defoliation event at intensities of up to 50% leaf removal. Similar responses with two and three consecutive defoliations in which they removed 25% of all leaves were observed. Greater frequency and intensity of defoliation resulted in reduced seed yield, while seed quality did not appear to be affected (Hoidal et al., 2019). These results were confirmed later by the same group of authors in an experiment in which they also measured the nutritious state of the crop. Their findings show that the nutritional value of amaranth leaves differed substantially throughout development, with the highest concentrations of iron midway through vegetative development and the highest levels of vitamin A, magnesium, and copper at the end of the vegetative development stage. They also tested palatability and found that it was highest in young leaves and decreased as plants aged (Hoidal et al., 2020).

In 2021, Cisneros-Hernández et al. (2021) explored deeply the responses of *A. cruentus* L. to defoliation tolerance. In their study, experimental findings reveal that plants exhibit the highest defoliation tolerance during panicle emergence, accompanied by increased cytosolic invertase (CI) activity in all organs and with the extensive induction of various class II trehalose-phosphate synthase genes. This explains the observed changes in sugar metabolism and allocation; inversely, the less defoliation tolerance observed in the flowering phase (anthesis) coincided with a strong depletion of starch reserves and sucrose (Suc) accumulation in leaves and stems, which was consistent with very low levels of CI and vacuolar invertase activities that were not further modified by defoliation. They also found that sucrolytic enzymes have contradictory comportment alongside the organ and phenological development stage. This suggests that a localized sugar starvation response caused by C partitioning is crucial for defoliation tolerance in grain amaranth.

In the context of plant–herbivore interactions and ecosystem dynamics where plants are eaten by the herbivore, they lose leaf area. In these ecosystems, amaranth species with high defoliation tolerance may enjoy a competitive advantage over less-tolerant plants. From a ground study where the identification of several genes overexpressed in different (a)biotic stresses including insect herbivory, the *Ah24* gene of unknown function was identified (Délano-Frier et al., 2011) and posteriorly characterized as a novel stress-inducible gene in a 2015 study (Massange-Sanchez et al., 2015). *Ah24* was expressed in young or developing tissues; it was also strongly induced by mechanical damage, insect herbivory, and methyl jasmonate and in meristems and newly emerging leaves of severely defoliated plants. *Ah24* encodes a cytosolic protein consisting of 164 amino acids. In experimental conditions transgenic *Arabidopsis thaliana*, and *Nicotiana tabacum* overexpressing *Ah24* showed enhanced vegetative growth and increased number of leaves. Additionally, transgenic tobacco plants, which grew and reproduced normally, had increased insect herbivory resistance. Modified vegetative growth in transgenic Arabidopsis coincided with significant changes in the expression of genes controlling phytohormone synthesis or signaling, whereas increased resistance to insect herbivory in transgenic tobacco coincided with higher jasmonic acid and proteinase inhibitor activity levels, plus the accumulation of nicotine and several other putative defense-related metabolites. Therefore the investigators proposed that the primary role of the *Ah24* gene in *A. hypochondriacus* L. is mainly to contribute to a rapid recovery post-wounding or defoliation, although its participation in defense against insect herbivory is also plausible (Massange-Sanchez et al., 2015). In an interesting coincidence, the protein Ah24 was identified in the previously mentioned root proteomic in response to salinity stress (Huerta-Ocampo et al., 2014). In amaranth, C4 metabolism allows high photosynthetic rates at high temperatures and light intensity with low photorespiration and efficient water use. Furthermore, amaranth's photosynthetic pathway, C4 metabolism, contributes to its defoliation tolerance. C4 metabolism enhances the plant's photosynthetic efficiency, allowing it to maintain a stable rate of carbon assimilation even when subjected to leaf loss (Iqbal et al., 2012).

Some crops are grown on deficient soils, and hence they face several constraints due to the stressful conditions. Most of the arable land is classified as semi-arid, so drought is the main limiting factor to producing most of the crops. One of the solutions proposed is practice to capture and conserve soil moisture so that crops can develop, grow, and produce under these drought conditions (Hsiao and Acevedo 1975; Lewis and Christiansen 1981; Seropian and Planchon 1984). In addition, amaranth has mechanisms that allow it to recover quickly from drought stress if irrigated again (Figure 19.4).

FIGURE 19.4 *Amarantus hypochondriacus* var Nutrisol recovering after drought stress condition. (a) Plants after 5 days of treatment. (b) Plants after 15 days of treatment. (c) Recovery 5 days after rewatering. NW, normal watering; DS, drought stress.

19.6 PLANT MORPHOGENESIS IN STRESS CONDITIONS

In addition, plant morphogenesis is strongly influenced by other organisms, being either harmful (pathogens, herbivores, parasites) or beneficial (symbionts). These organisms might evoke plant responses that result in cellular reorganization or new or unusual organ formation.

Cultivated plants generally exhibit three mechanisms of drought resistance: avoidance, tolerance, and escape. Drought-avoiding plants maintain a high water potential when exposed to stress. Two types of plants use this mechanism – water savers and water wasters under water stress conditions – but have mechanisms for the rapid uptake of sufficient water to maintain themselves (Espitia-Rangel et al., 2010).

Amaranth has been reported to have several mechanisms that help maintain its water needs for drought resistance such as stomatal closure, cuticular barriers, cuticle wax accumulation, and leaf area reduction. Leaf area reduction occurs when amaranth leaves curl when exposed to light stress, and amaranth drops leaves under severe stress, thus reducing the area of transpiration (Cavagnaro, 1985).

On the contrary, amaranth has resilience, because it has a root system that can penetrate up to 2 m deep and with a high number of lateral roots; this enables amaranth to recover from stress. It has been reported that amaranth is kept turgid by an osmotic adjustment mechanism. There is another way in which amaranth can resist stress, as there are amaranth genotypes that show precocity, which helps them mature before a water shortage occurs (Kigel and Rubin, 1985). Knowledge of the different factors involved in how drought affects crops and their resistance mechanisms is necessary to identify those characteristics to select the best genotypes for these conditions.

Studies on plant population density provide information on water availability, nutrient availability, and solar radiation interception. In two cultivars of amaranth *Frondosa* and *Tulyehualco* (both *A. hypochondriacus*), grown at two planting densities (62,500 and 375,000 plants/ha), higher yields were obtained at higher densities, as well as higher values for aerial biomass and lower values for stem diameter, percentage lodging and greening, and agronomic characteristics favorable for better seed yield (Torres-Saldaña et al., 2006). Between these two cultivars, the plastic response to density can be observed with a six-fold difference with respect to the traditional sowing of 50,000 and 100,000 plants/ha.

19.7 RECOVERY FROM MECHANICAL DAMAGE

It has also been observed that when amaranth plants are faced with abiotic factors such as frost or hail, amaranth renews its leaf area (depending on the intensity, duration, and phenological stage) (Figure 19.5), although it would have to be investigated whether this response is due to the effect of apical dominance or as a response of the plant for the formation and maintenance of reproductive structures to perpetuate the species or whether both phenomena occur. The recovery of the leaf area represents an energy cost and results in a delay in seed production.

FIGURE 19.5 Recovery of amaranth plants affected by hail. (a) Leaves of amaranth plants damaged by hail. (b) Amaranth plants are damaged by wind and hail. (c) Close-up of the box shown in c. The arrow points to the new shoots.

19.8 CONCLUSIONS AND PERSPECTIVES

The mechanisms described here to cope with water deficit are not completely understood on amaranth. A couple of studies have attempted to detect differences in responses between different *Amaranthus* species, particularly among those utilized as grains. Previous reports compared *A. hypochondriacus* L., *A. cruentus* L., and *A. caudatus* L. with their proposed shared ancestor *A. hybridus* L. (Gónzález-Rodríguez, et al., 2019). Their experiment consisted of an initial screening to detect differences in WDS tolerance between the amaranth species, where they found that *A. hypochondriacus* L., was the most WDS-tolerant species, *A. cruentus* cv. Amaranteca was intermediate and *A. caudatus* was the most sensitive, while *A. hybridus* L. plants became severely desiccated. A more recent paper has explored the differences between *Amaranthus* L. species under WDS; it compares *A. hypocondriacus* cv. Nutrisol and *A. hybridus* L., though both species showed the amelioration of photosynthetic parameters. Again the ancestor shows less efficient photosynthesis, measured among other parameters with the accumulation of chlorophyll, than *A. hypochondriacus* cv. Nutrisol. This variety also presented a higher accumulation of proline (Vargas-Ortiz et al., 2021). These findings are in accordance with an exploratory experiment performed in our group where WDS response was compared between two varieties of *A. hypochondriacus* – Nutrisol and Revancha – and two of *A. cruentus* – Amaranteca and Benito – during two different seasons. The results suggested that *A. hypochondriacus* var Nutrisol is the most tolerant strain maintaining its seed production despite WDS during the winter season. Nevertheless, *A. cruentus* L. seems to tolerate better during springtime. Both strains show an increased chlorophyll, a content probably related to an increase in photosynthetic rate. Similarly, data obtained from Netshimbupfe et al. (2022) found that *A. cruentus* L. showed higher heat and drought tolerance, which in this case is attributed to a more stress-resilient photosystem II and to the accumulation of proline, though the four species evaluated were considered to be tolerant to heat and drought.

Understanding amaranth defoliation tolerance holds great promise for agricultural practices, especially in facing climate change and evolving pest pressures. By identifying and harnessing the genetic factors responsible for defoliation tolerance, researchers aim to develop crop varieties with enhanced resilience to pests and environmental stressors.

With respect to plant damage by mechanical defoliation in *A. hypochondriacus* and *A. Cruentus*, it was observed that cutting shoots by 10% and 40% increased the number of leaves and the number of secondary branches, although the number of primary branches was lower than that in the control; in addition, there are differences between species in the distribution of the biomass produced, while in *A. Cruentus* more biomass was distributed in the seed and in *A. hypochondriacus* it was in the leaves. In general, in *A. cruentus* a better overcompensation for damage was observed, but both species are tolerant to defoliation (Martinez et al., 1999). In another study on *A. cruentus*, the effect of mechanical damage and defoliation was evaluated, which showed tolerance and resistance mechanisms, modifying growth, carbohydrate accumulation, expression, and activity of trypsin inhibitors (Vargas-Ortíz et al., 2012). Because amaranth is a crop with a wide range of tolerance to different stresses and has a high nutritional value, it should be considered within strategies for food security in the face of climate change.

REFERENCES

Aderibigbe, O. R., Ezekiel O. O., Owolade S. O., Korese J. K., Sturm B., & Hensel O. (2022). Exploring the potentials of underutilized grain amaranth (*Amaranthus* spp.) along the value chain for food and nutrition security: A review. *Critical Reviews in Food Science and Nutrition*, *62*, 656–669.

Alvarez-Jubete, L., Arendt, E. K., & Gallagher, E. (2010). Nutritive value of pseudocereals and their increasing use as functional gluten-free ingredients. *Trends in Food Science & Technology*, *21*(2), 106–113. https://doi.org/10.1016/j.tifs.2009.10.014

Araújo, W. L., Fernie, A. R., & Nunes-Nesi, A. (2011). Control of stomatal aperture: A renaissance of the old guard. *Plant Signaling & Behavior*, *6*(9), 1305–1311.

Atou, R., Tonouewa, G., Wouyou, A., Kpochemè, E., Missihoun, A. A., Montcho, D., Ahoton L., Agbangla C., & Gandonou, C. B. (2022). Screening of Amaranth (Amaranthus cruentus L.) mutant lines for salinity tolerance. *International Journal of Plant & Soil Science*, *34*(22), 785–797.

Atwoli, L., Baqui, A. H., Benfield, T., Bosurgi, R., Godlee, F., & Hancocks, S. (2021). Call for emergency action to limit global temperature increases, restore biodiversity, and protect health. *The New England Journal of Medicine*, *385*(12), 1134–1137. https://doi.org/10.1056/NEJMe2113200

Babayev, H., Mehvaliyeva, U., Aliyeva, M., Feyziyev, Y., & Guliyev, N. (2014). The study of NAD-malic enzyme in *Amaranthus cruentus* L. under drought. *Plant Physiology and Biochemistry*, *81*, 84–89.

Bertolino, L. T., Caine, R. S., & Gray, J. E. (2019). Impact of stomatal density and morphology on water-use efficiency in a changing world. *Frontiers in Plant Science*, *10*, 225.

Cavagnaro, J. B. (1985). *Effect of water stress on growth and dry matter partitioning in four species of Amaranthus*. University of California.

Cisneros-Hernández, I., Vargas-Ortiz, E., Sánchez-Martínez, E. S., Martínez-Gallardo, N., Soto González, D., & Délano-Frier, J. P. (2021). Highest defoliation tolerance in *Amaranthus cruentus* plants at panicle development is associated with sugar starvation responses. *Frontiers in Plant Science*, *12*, 658977.

Dada, O., & Ohu, O. (2013). Growth and stomatal response of *Amaranthus* spp. to different watering regimes, organic and inorganic soil amendments. *Nigerian Journal of Science*, *47*, 15–26.

Délano-Frier, J. P., Avilés-Arnaut, H., Casarrubias-Castillo, K., Casique-Arroyo, G., Castrillón-Arbeláez, P. A., Herrera-Estrella, L., Massange-Sánchez, J. M., Martínez-Gallardo, N. A., Parra-Cota, F. I., Vargas-Ortiz, E., & Estrada-Hernández, M. G. (2011). Transcriptomic analysis of grain amaranth (*Amaranthus hypochondriacus*) using 454 pyrosequencing: Comparison with *A. tuberculatus*, expression profiling in stems and in response to biotic and abiotic stress. *BMC Genomics*, *12*(1), 1–18.

Driesen, E., Van den Ende, W., De Proft, M., & Saeys, W. (2020). Influence of environmental factors light, CO_2, temperature, and relative humidity on stomatal opening and development: A review. *Agronomy*, *10*(12), 1975.

Espitia-Rangel, E., Mapes-Sánchez, E. C., Núñez-Colín, C. A., & Escobedo-López, D. (2010). Distribución geográfica de las especies cultivadas de Amaranthus y de sus parientes silvestres en México. *Revista Mexicana de Ciencias Agrícolas*, *1*(3), 427–437.

Estrada, Y., Fernández-Ojeda, A., Morales, B., Egea-Fernández, J. M., Flores, F. B., Bolarín, M. C., & Egea, I. (2021). Unraveling the strategies used by the underexploited amaranth species to confront salt stress: Similarities and differences with quinoa species. *Frontiers in Plant Science*, *12*, 604481.

Fan, Y., Asao, S., Furbank, R. T., von Caemmerer, S., Day, D. A., Tcherkez, G., Sage, T. L., Sage, R. F., & Atkin, O. K. (2022). The crucial roles of mitochondria in supporting C4 photosynthesis. *New Phytologist*, *233*(3), 1083–1096.

Furbank, R. T., & Hatch, M. D. (1987). Mechanism of C4 photosynthesis: The size and composition of the inorganic carbon pool in bundle sheath cells. *Plant Physiology*, *85*, 958–964. https://doi.org/10.1104/pp.85.4.958

Fürst-Jansen, J. M. R., de Vries, S., & de Vries, J. (2020). Evo-physio: On stress responses and the earliest land plants. *Journal of Experimental Botany*, *71*(11), 3254–3269. https://doi.org/10.1093/jxb/eraa007

Gelaye, Y. (2023). A review of Amaranth crop as a potential solution to ethiopia's nutritional crisis. *Nutrition and Dietary Supplements*, *15*, 101–110.

Ghosh, U. K., Islam, M. N., Siddiqui, M. N., Cao, X., & Khan, M. A. R. (2022). Proline, a multifaceted signalling molecule in plant responses to abiotic stress: Understanding the physiological mechanisms. *Plant Biology*, *24*(2), 227–239.

Gónzález-Rodríguez, T., Cisneros-Hernández, I., Acosta Bayona, J., Ramírez-Chavez, E., Martínez-Gallardo, N., Mellado-Mojica, E., López-Pérez, M. G., Molina-Torres, J., & Frier, J. (2019). Identification of factors linked to higher water-deficit stress tolerance in *Amaranthus hypochondriacus* compared to other grain amaranths and *A. hybridus*, their shared ancestor. *Plants*, *8*(7), 239.

Hameed, A., Ahmed, M. Z., Hussain, T., Aziz, I., Ahmad, N., Gul, B., & Nielsen, B. L. (2021). Effects of salinity stress on chloroplast structure and function. *Cells*, *10*(8), 2023.

Hasanuzzaman, M., Bhuyan, M. B., Parvin, K., Bhuiyan, T. F., Anee, T. I., Nahar, K., Hossen, S., Zulfiqar, F., Alam, M. & Fujita, M. (2020). Regulation of ROS metabolism in plants under environmental stress: A review of recent experimental evidence. *International Journal of Molecular Sciences*, *21*(22), 8695.

Hoidal, N., Díaz Gallardo, M., Jacobsen, S. E., & Alandia, G. (2019). Amaranth as a dual-use crop for leafy greens and seeds: Stable responses to leaf harvest across genotypes and environments. *Frontiers in Plant Science*, *10*, 817.

Hoidal, N., Jacobsen, S. E., Odone, A., & Alandia, G. (2020). Defoliation timing for optimal leaf nutrition in dual-use amaranth production systems. *Journal of the Science of Food and Agriculture*, *100*(13), 4745–4755.

Hossain, M. A., Hoque, M. A., Burritt, D. J., & Fujita, M. (2014). Proline protects plants against abiotic oxidative stress: Biochemical and molecular mechanisms. In P. Ahmad (Ed.), *Oxidative damage to plants* (pp. 477–522). Academic Press.

Hsiao, T. C., & Acevedo, E. (1975). Plant responses to water deficits, water-use efficiency, and drought resistance. *Developments in Agricultural and Managed Forest Ecology*, *1*, 59–84.

Huerta-Ocampo, J. A., Barrera-Pacheco, A., Mendoza-Hernandez, C. S., Espitia-Rangel, E., Mock, H. P., & Barba de la Rosa, A. P. (2014). Salt stress-induced alterations in the root proteome of *Amaranthus cruentus* L. *Journal of Proteome Research*, *13*(8), 3607–3627.

Ichsan, C. N., Andani, R., Basyah, B., Zakaria, S., & Efendi, E. (2022). The relationship between relative water content of leaves, soluble sugars, accumulation of dry matter, and yield components of rice (*Oryza sativa* L.) under water-stress condition during the generative stage. *International Journal on Advanced Science, Engineering, and Information Technology, 12*(3), 899–907.

IPCC. (2023). Summary for policymakers. In Core Writing Team, H. Lee, and J. Romero (Eds.), *Climate change 2023: Synthesis report. Contribution of working groups I, II and III to the sixth assessment report of the intergovernmental panel on climate change* (pp. 1–34). IPCC. https://doi.org/10.59327/IPCC/AR6-9789291691647.001

Iqbal, N., Masood, A., & Khan, N. A. (2012). Analyzing the significance of defoliation in growth, photosynthetic compensation, and source-sink relations. *Photosynthetica*, *50*, 161–170.

Joshi, N., & Verma, K. C. (2020). A review on nutrition value of Amaranth (*Amaranthus caudatus* L.): The crop of future. *Journal of Pharmacognosy and Phytochemistry*, *9*(4), 1111–1113.

Keunen, E. L. S., Peshev, D., Vangronsveld, J., Van Den Ende, W. I. M., & Cuypers, A. N. N. (2013). Plant sugars are crucial players in the oxidative challenge during abiotic stress: Extending the traditional concept. *Plant, Cell & Environment*, *36*(7), 1242–1255.

Kigel, J., & Rubin, B. (1985). Amaranthus. In A. H. Halevy (Ed.), *Handbook of flowering* (Vol. 1, pp. 434–449). CRC Press.

Kornhuber, K., Lesk, C., Schleussner, C. F. Jägermeyr, J., Pfleiderer, P., & Horton, R. M. (2023). Risks of synchronized low yields are underestimated in climate and crop model projections. *Nature Communications*, *14*, 3528.

Lesk, C., Rowhani, P., & Ramankutty, N. (2016). Influence of extreme weather disasters on global crop production. *Nature*, *529*, 84–87.

Lewis, C. F., Christiansen, M. N., Jensen, S. D., & Shannon, M. C. (1981). *Breeding plants for stress environments*. In *Plant breeding symposium* (pp. 151–177). Iowa State University. (Original work presented at Plant Breeding Symposium 2, 1979, Ames, IA)

Li, M., & Kim, C. (2022). Chloroplast ROS and stress signaling. *Plant Communications*, *3*, 100264. doi: 10.1016/j.xplc.2021.100264.

Li, S., Li, X., Wei, Z., & Liu, F. (2020). ABA-mediated modulation of elevated CO_2 on stomatal response to drought. *Current Opinion in Plant Biology*, *56*, 174–180.

Lim, C. W., Baek, W., Jung, J., Kim, J. H., & Lee, S. C. (2015). Function of ABA in stomatal defense against biotic and drought stresses. *International Journal of Molecular Sciences*, *16*(7), 15251–15270. https://doi.org/10.3390/ijms160715251

Long, J. J., Wang, J. L., & Berry, J. O. (1994). Cloning and analysis of the C4 photosynthetic NAD-dependent malic enzyme of *Amaranth mitochondria*. *Journal of Biological Chemistry*, *269*(4), 2827–2833.

Mansour, M. M. F., & Ali, E. F. (2017). Evaluation of proline functions in saline conditions. *Phytochemistry*, *140*, 52–68.

Martinez-Villaluenga, C. (2020). Pseudocereal grains: Nutritional value, health benefits and current applications for the development of gluten-free foods. *Food, and Chemical Toxicology*, *137* , 111178. doi:10.1016/j.fct.2020.111178.

Massange-Sanchez, J. A., Palmeros-Suarez, P. A., Espitia-Rangel, E., Rodriguez-Arevalo, I., Sanchez-Segura, L., Martinez-Gallardo, N. A., Alatorre-Cobos, F., Tiessen, A., & Delano-Frier, J. P. (2016). Overexpression of grain amaranth (*Amaranthus hypochondriacus*) AhERF or AhDOF transcription factors in *Arabidopsis thaliana* increases water deficit-and salt-stress tolerance, respectively, via contrasting stress-amelioration mechanisms. *PLoS ONE*, *11*(10), e0164280.

Massange-Sanchez, J. A., Palmeros-Suarez, P. A., Martinez-Gallardo, N. A., Castrillon-Arbelaez, P. A., Aviles-Arnaut, H., Alatorre-Cobos, F., Tiessen, A., & Delano-Frier, J. P. (2015). The novel and taxonomically restricted Ah24 gene from grain amaranth (*Amaranthus hypochondriacus*) has a dual role in development and defense. *Frontiers in Plant Science*, *6*, 602.

Mota, C., Santos, M., Mauro, R., Samman, N., Matos, A. S., Torres, D., & Castanheira, I. (2016). Protein content and amino acids profile of pseudocereals. *Food Chemistry*, *193*, 55–61. https://doi.org/10.1016/j.foodchem.2014.11.043

Nadarajah, K. K. (2020). ROS homeostasis in abiotic stress tolerance in plants. *International Journal of Molecular Sciences*, *21*(15), 5208.

Netshimbupfe, M. H., Berner, J., & Gouws, C. (2022). The interactive effects of drought and heat stress on photosynthetic efficiency and biochemical defense mechanisms of Amaranthus species. *Plant-Environment Interactions*, *3*(5), 212–225.

Nezami, A., Khazaei, H. R., Boroumand, R. Z., & Hosseini, A. (2008). Effects of drought stress and defoliation on sunflower (*Helianthus annuus*) in controlled conditions. *Desert*, *12*, 99–104.

Nishiyama, Y., Allakhverdiev, S. I., & Murata, N. (2011). Protein synthesis is the primary target of reactive oxygen species in the photoinhibition of photosystem II. *Physiologia Plantarum*, *142*(1), 35–46.

Ogbaga, C. C., Stepien, P., Dyson, B. C., Rattray, N. J., Ellis, D. I., Goodacre, R., & Johnson, G. N. (2016). Biochemical analyses of sorghum varieties reveal differential responses to drought. *PLoS ONE*, *11*(5), e0154423.

Pulvento, C., Sellami, M. H., & Lavini, A. (2022). Yield and quality of *Amaranthus hypochondriacus* grain amaranth under drought and salinity at various phenological stages in southern Italy. *Journal of the Science of Food and Agriculture*, *102*(12), 5022–5033.

Reyes-Rosales, A., Cabrales-Orona, G., Martínez-Gallardo, N. A., Sánchez-Segura, L., Padilla-Escamilla, J. P., Palmeros-Suárez, P. A., & Délano-Frier, J. P. (2023). Identification of genetic and biochemical mechanisms associated with heat shock and heat stress adaptation in grain amaranths. *Frontiers in Plant Science*, *14*, 1101375.

Richard, A., Gauthier, T., Wouyou, A., Kevin, K. A. O. E., Abel, M. A., Leonard, A., ... & Bernard, G. C. (2023). Salt resistance strategies of amaranth salt-resistant mutant lines. *International Journal of Plant Physiology and Biochemistry*, *15*(1), 1–12.

Roy, A., Mandal, M., Das, S., Popek, R., Rakwal, R., Agrawal, G. K., Awasthi, A., & Sarkar, A. (2024). The cellular consequences of particulate matter pollutants in plants: Safeguarding the harmonious integration of structure and function. *Science of the Total Environment*, *914*, 169763.

Sarker, U., & Oba, S. (2020). Nutritional and bioactive constituents and scavenging capacity of radicals in *Amaranthus hypochondriacus*. *Scientific Reports*, *10*(1), 19962.

Seropian, C., & Planchon, C. (1984). Physiological responses of six bread wheat and durum wheat genotypes to water stress. *Euphytica*, *33*(3), 757–767.

Singh, J., Garai, S., Das, S., Thakur, J. K., & Tripathy, B. C. (2022). Role of C4 photosynthetic enzyme isoforms in C3 plants and their potential applications in improving agronomic traits in crops. *Photosynthesis Research*, *154*(3), 233–258.

Slabbert, M. M., & Krüger, G. H. J. (2014). Antioxidant enzyme activity, proline accumulation, leaf area and cell membrane stability in water stressed Amaranthus leaves. *South African Journal of Botany*, *95*, 123–128.

Stavi, I., Thevs, N., & Priori, S. (2021). Soil salinity and sodicity in drylands: A review of causes, effects, monitoring, and restoration measures. *Frontiers in Environmental Science*, *9*, 330.

Tebini, M., Luu, D. T., Mguis, K., Ben Ahmed, H., Meddich, A., Zribi, F., & Chalh, A. (2022). Physiological exploration of intra-specific variability in salinity tolerance of amaranth. *Russian Journal of Plant Physiology*, *69*(3), 59.

Thomason, W., & Battaglia, M. (2020). Early defoliation effects on corn plant stands and grain yield. *Agronomy Journal*, *112*(6), 5024–5032.

Torres Saldaña, G., Trinidad Santos, A., Reyna Trujillo, T., Castillo Juárez, H., Escalante Estrada, A., & de León González, F. (2006). Respuesta de genotipos de amaranto a densidades de población. *Revista Fitotecnia Mexicana*, *29*(4), 307–312.

Trovato, M., Forlani, G., Signorelli, S., & Funck, D. (2019). Proline metabolism and its functions in development and stress tolerance. In M. A. Hossain, V. Kumar, D. J. Burritt, M. Fujita, & P. S. A. Mäkelä (Eds.), *Osmoprotectant-mediated abiotic stress tolerance in plants: Recent advances and future perspectives* (pp. 41–72). Cham: Springer.

Tsutsumi, N., Tohya, M. Nakashima, T., & Ueno, O. (2017). Variations in structural, biochemical, and physiological traits of photosynthesis and resource use efficiency in Amaranthus species (NAD-ME-type C4). *Plant Production Science*, *20*(3), 300–312.

Vargas-Ortiz, E., Délano-Frier, J. P., & Tiessen, A. (2015). The tolerance of grain amaranth (*Amaranthus cruentus* L.) to defoliation during vegetative growth is compromised during flowering. *Plant Physiology and Biochemistry*, *91*, 36–40.

Vargas-Ortiz, E., Espitia-Rangel, E., Tiessen, A., & Délano-Frier, J. P. (2013). Grain amaranths are defoliation tolerant crop species capable of utilizing stem and root carbohydrate reserves to sustain vegetative and reproductive growth after leaf loss. *PLoS ONE*, *8*(7), e67879.

Vargas-Ortiz, E., Ramírez-Tobias, H. M., González-Escobar, J. L., Gutierrez-Garcia, A. K., Bojorquez-Velazquez, E., Espitia-Rangel, E., & de la Rosa, A. P. B. (2021). Biomass, chlorophyll fluorescence, and osmoregulation traits let differentiation of wild and cultivated Amaranthus under water stress. *Journal of Photochemistry and Photobiology B: Biology*, *220*, 112210.

Voronin, P. Y., Maevskaya, S. N., & Nikolaeva, M. K. (2019). Physiological and molecular responses of maize (*Zea mays* L.) plants to drought and rehydration. *Photosynthetica*, *57*(3), 850–856.

Wang, X., Fan, J., Xing, Y., Xu, G., Wang, H., Deng, J., Wang, Y., Zhang, F., Li, P., & Li, Z. (2019). The effects of mulch and nitrogen fertilizer on the soil environment of crop plants. *Advances in Agronomy*, *153*, 121–173.

Waselkov, K. E., Boleda, A. S., & Olsen, K. M. (2018). A phylogeny of the genus Amaranthus (Amaranthaceae) based on several low-copy nuclear loci and chloroplast regions. *Systematic Botany*, *43*(2), 439–458.

Wei, Z., Abdelhakim, L. O. A., Fang, L., Peng, X., Liu, J., & Liu, F. (2022). Elevated CO_2 effect on the response of stomatal control and water use efficiency in amaranth and maize plants to progressive drought stress. *Agricultural Water Management*, *266*, 107609.

Wiebers, D. O., Feigin, V. L., & Winkler Andrea, S. (2022). All life protection and our collective future. *Neuroepidemiology*, *56*(3), 147–150.

Wouyou, A., Prodjinoto, H., Zanklan, A. S., Vanpee, B., Lutts, S., & Gandonou, C. B. (2019). Implication of ions and organic solutes accumulation in amaranth (*Amaranthus cruentus* L.) salinity resistance. *American Journal of Plant Sciences*, *10*(12), 2335.

Wu, C., & Guo, D. (2023). Identification of two flip-over genes in grass family as potential signature of C4 photosynthesis evolution. *International Journal of Molecular Sciences*, *24*, 14165. https://doi.org/10.3390/ijms241814165

Yoshimura, Y., Kubota, F., & Ueno, O. (2004). Structural and biochemical bases of photorespiration in C 4 plants: Quantification of organelles and glycine decarboxylase. *Planta*, *220*, 307–317.

Zhou, H., Helliker, B. R., Huber, M., Dicks, A., & Akçay, E. (2018). C4 photosynthesis and climate through the lens of optimality. *Proceedings of the National Academy of Sciences*, *115*(47), 12057–12062.

20 CRISPR/Cas Technology
A Promising Tool for Enhancing Plant Growth and Development

Puliyampatti Gunasekaran Vishnu, Gunasekaran Arthi, Jothi Dheivasikamani Abidharini, Amardeep Ray Preethi, Palanisamy Sampathkumar, Ramalingam Sivakumar, Saravanan Renuka, and Arumugam Vijaya Anand

20.1 AN OVERVIEW OF CRISPR/CAS TECHNOLOGY

Genome adapting, promoted by precise and active changes to a container's DNA utilizing DNA endonucleases, has revolutionized biotechnology and farming. This transformational technology, attainable through programmable DNA-binding and gap domains, authorizes chemists to makeover genomic sequences, generate container lines, and scrutinize human diseases. Four key technologies – transcription electrical device-like effector nucleases, mega nucleases, metallic mineral-finger nucleases, and CRISPR/Cas9 – are instrumental in deoxyribonucleic acid erasure, healing gene insert, and ailment correction (Munawar and Ahmad, 2021). Among these, CRISPR/Cas methods have gained important attention, initially discovered three decades in the past in *Escherichia coli* during the study of salty phosphatase isozyme conversion (Ishino et al., 2018). CRISPR sequences, in the direction of prokaryotic animals like archaea and microorganisms, serve as a singular protective activity against growing invasion by taking advantage of DNA fragments from past contaminations alongside CRISPR-joined (Cas) proteins to precisely discover and demolish similar viruses all the while after contaminations (Figure 20.1). This prokaryotic-acquired invulnerable arrangement plays an important role in covering external mobile-historical ingredients (Tofazzal Islam, 2019).

CRISPR/Cas genome-editing onset finishes have remodeled the qualification of DNA and RNA sequences in living cells, accompanying uses ranging from fundamental wisdom to translational cure. These tools engage miscellaneous methods, including distinct-base refining, homology-supervised

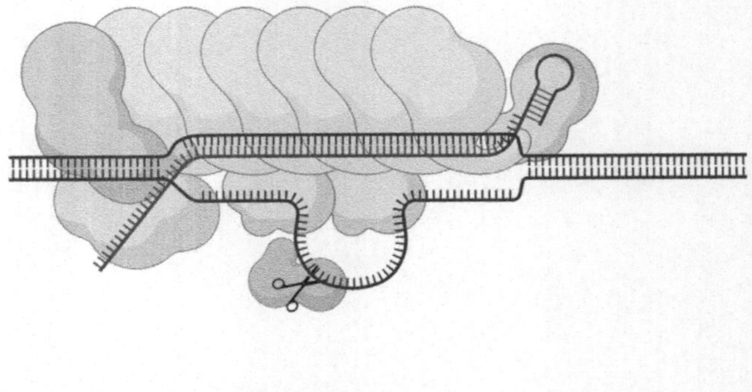

FIGURE 20.1 CRISPR cascade protein.

DOI: 10.1201/9781003457237-20

repair, and non-similar end-touching enzymes, and are increasingly employed for healing research, analysis, and scientific inquiry (Pickar-Oliver and Gersbach, 2019). Furthermore, CRISPR/Cas electronics have revolutionized the plant microscopic study of animals, permissive precise qualification of crop variety for the happening of the novel, more sustainable land structures, and the invention of germplasms with favorable looks (Zhu et al., 2020). The Cas9/sgRNA system, arising from the bacterial invulnerable structure, allows for exact genomic modifications in differing creatures and tissues, offering promise for the growth of innately edited crops to reinforce land output and resilience (Khatodia et al., 2016). Plant genome editing has enormous potential, but difficulties still exist, necessitating further research and a review of possible solutions. (Schindele et al., 2020). CRISPR/Cas science holds the key to sending all-encompassing challenges such as temperature change and snack freedom by enhancing plant abiotic stress resistance, supporting forward genetic approaches, and maximizing farming appropriateness. Its flexibility and cost-effectiveness create CRISPR/Cas a strong finish for genome manipulation, aiding mean mutagenesis, transcriptome requirement, epigenome editing, and exact deoxyribonucleic acid/allele replacement (Zhang et al., 2019). However, challenges await, specifically in realizing precise allele substitutes through similarity-supervised repair in crop plants. As research continues to advance, CRISPR/Cas electronics promise to play a more pivotal function in forming the future of biotechnology and farming, offering answers to a few of the ultimate pressing challenges braid benevolence (Bhowmik *et al.*, 2019).

20.2 AN INVESTIGATION OF PLANT DEVELOPMENT AND GROWTH

In particular, the study of arginase (ARG) and its effects on root development has made extensive usage of the CRISPR/Cas methodology for researching gene function in plant development and growth (Wang et al., 2017). When Wang and associates eliminated the ARG gene in cotton employing CRISPR/Cas9 techniques, the plant developed more roots and was able to absorb nutrients and water more effectively (Peng et al., 2021). Furthermore, CRISPR/Cas9-based genome editing has demonstrated the critical role MADS-box transcription factor genes play in endosperm cellularization and the growth of the seeds in rice, *Arabidopsis*, and oilseed rape. CRISPR/Cas9 knockouts also altered pollen production and plant architecture (Figure 20.2). CRISPR/Cas genome editing is an advantageous technique because it allows genes that negatively affect productivity to be removed (Li et al., 2018).

Investigations have shown that particular gene deletions can enhance certain traits, such as grain size, weight, quantity, density, and erect panicles. It has been shown that rice with knockouts of plant output negative regulators such as gn1a, gs3, and dep1 have larger grains, enhanced grain

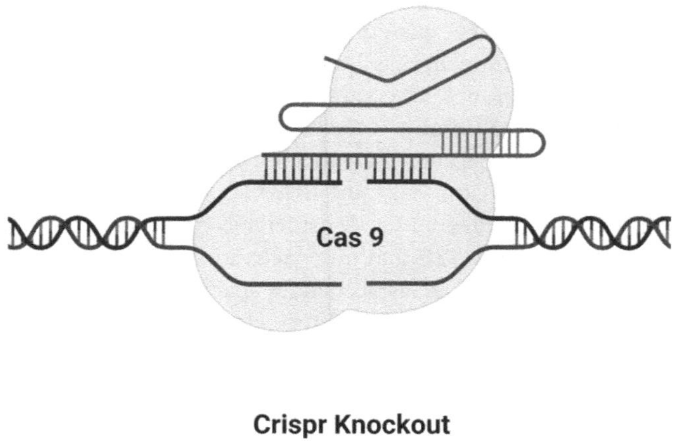

FIGURE 20.2 CRISPR knockout.

weight, and more grains and are dense (Wang et al., 2020a). Additionally, CRISPR/Cas9-edited variants of GASR7 and CLV-WUS have increased wheat grain weight and size. This indicates that CRISPR/Cas genome editing is developing as a cutting-edge exactness breeding technique to boost agricultural productivity, according to Zhang et al. (2016b).

CRISPR/Cas method has been engaged to alter plant secondary metabolism to enhance crop quality. This means boosting the biosynthesis and production of lycopene from tomato fruits, modifying the protein content of wheat and barley, and modifying the capacity of crops to produce carbohydrates (Li and associates, 2018). Additionally, CRISPR/Cas9 has been utilized to knock out genes related to high amylose levels and resistance to starch, which has good health consequences. The rice that was subjected to genome editing exhibited increased percentages of long chains in debrancher amylopectin and higher levels of amylose.

Rapeseeds with CRISPR gene knockouts had higher oil content without compromising other agronomic characteristics. Furthermore, CRISPR/Cas9 has been involved in studying agricultural domestication and unique traits in specific crops, such as making cotton fiberless (Sánchez-León et al., 2018). Both animals and plants have benefited from the use of CRISPR technologies, and researchers are always creating new instruments to solve issues. Tools for targeted alterations at the genomic, post-transcriptional, transcriptional, and epigenetic levels are included in the CRISPR toolkit. Worldwide researchers choose CRISPR approaches because they are modular and have broad-spectrum applications. For epigenetic, genomic, and post-transcriptional alterations, they can fuse with effector domains (Khan et al., 2021).

20.3 CRISPR/CAS APPROACH – STRUCTURE AND FUNCTION

CRISPR/Cas technique is a complex and diverse prokaryotic adaptive immune system. A multifaceted approach has been devised to identify CRISPR/Cas methods, wherein they are classified into three kinds and ten subtypes, all of which have signature genes. However, because Cas genes develop quickly, this classification scheme only partially reflects the complexity of CRISPR/Cas structure and evolution. Thus, additional development necessitates the integration of protein structure comparison, sensitive sequence comparison tools, and improved methods for gene neighborhood comparison (Makarova and Koonin, 2015). Class 2 systems, which rely on a single Cas effector nuclease, are employed in transcriptional control, imaging, detection, and genome editing. Class 1 systems provide the potential for DNA targeting and cleavage since they are based on a multiprotein effector complex. These systems can be reprogrammed to operate in eukaryotes as well as target the host genome for transcriptional regulation and genome editing (Hidalgo-Cantabrana and Barrangou, 2020).

CRISPR/Cas methods comprise a collection of genomic elements distinguished by their varied protein constituents. Types I, II, and III are the three main categories into which they can be divided. These systems' hallmark genes are cas3, cas9, and cas10. It is still unknown where type II CRISPR/Cas techniques originated. An extra gene, csn2, is noted in type II-A systems and is required for spacer integration but not for hindrance. In contrast, type II-B systems feature an extra cas4 gene that is a member of the PD-(D/E)xK family of nucleases but lacks csn2. It has been suggested that Cas4 proteins in CRISPR/Cas systems contribute to associated immunity rather than being directly related to any one defense mechanism. With just four Cas genes and the CRISPR repeat-spacer array, type II CRISPR/Cas is regarded as the considerable basic CRISPR/Cas system. However, other bacterial components like RNase III and trans-activating crRNA also play a role in this system's operation. The pair of Cas1 and 2 genes needed for spacer investment are present in altogether type II CRISPR/Cas modules (Chylinski et al., 2014). Homologous dsDNA regions in phages or plasmids are the target of type I CRISPR/Cas methods for destruction. The method consists of two basic steps: the surveillance complex identifies a supplemental target in foreign DNA, and Cas3, a protein containing an SF2 helicase and HD nuclease domain, cleaves the target. To recognize a target, the crRNA and the target must be complementarity and there must be a protospacer-adjacent motif (PAM) noted to inhibit self-immunity (Liu and Doudna, 2020).

20.4 CRISPR/CAS APPLICATIONS FOR CROP ADVANCEMENT

CRISPR/Cas is gaining popularity as a means of enhancing crop breeding and environmental sensitivity. It has a significant possibility for crop growth and development and has been used on important crops like rice, wheat, and cotton. Plant growth, development, and stress toleration are all controlled by CRISPR/Cas, as well as secondary metabolism (Zhang et al., 2021). The first step in using cis-regulatory elements (CREs) finding regulatory effector binding elements (EBEs) on the promoters of rice's OS SWEET genes was a necessary step in editing to increase crop resistance to disease. The resistance lines' mutations close to the TATA box eliminated the TAL effector PthXo1's capacity to bind. Resistance to Xoo strains was achieved using CRISPR/Cas9-established editing of a 149-bp promoter regulatory region of the Xa13 gene. Due to the simultaneous editing of various effectors' EBEs in separate SWEET genes, new wide-spectrum invulnerable rice types were created without impairing their normal function. The CRE editing of OsSWEET11, 13, and 14 EBEs using CRISPR/Cas9 produced wide-spectrum resistance against every tried Xoo strain (Miao et al., 2019).

This approach has been used to treat bacterial blight, citrus canker, and bacterial leaf streak (BLS). Wide-spectrum opposition to BLS and blight disease was successfully demonstrated by the transgenic rice (Saeed et al., 2022). Plant promoters' CREs are essential for enhancing crop productivity. TransFAc, PLACE, and PlantCARE databases are useful for locating CREs that have already been reported. But most CREs are either poorly defined or unknown, which means de novo discovery is required. Identification of conserved CREs can be aided by genomic and transcriptome data, bioinformatics-based strategies, and experimental techniques (Li et al., 2020a). The generation of biopharmaceuticals in plants has been changed by the CRISPR/Cas system, which modifies recombinant macromolecules. Molecular farming includes techniques like this one that modify plant metabolism without causing glycan synthesis to occur. In contrast to those in mammals, the KO of genes encoding the XylT and FucT enzymes produced favored recombinant proteins. This strategy has improved targets for genome engineering and produced significant therapeutic products (Kaul et al., 2020). Because veggies have so many health benefits, consumers are becoming more concerned about the quality of the food. Vegetable quality and shelf life can be increased with the use of biotechnological techniques like as marker-assisted selection and double haploids.

CRISPR/Cas9 genome editing can decrease undesirable chemicals in fruits and vegetables, enhance lycopene accumulation, and boost the starch and seed oil range of sweet potatoes and grapeseed. Additionally, commercial cultivars of potatoes and eggplant can be modified with this method (Das et al., 2023). The aim of the CRISPR/Cas9 approach is to maximize photosynthetic efficiency and inflorescence architecture while increasing maize production and variety. Genes that control the formation of inflorescences and stimulate the growth of meristems and flowers have been found in the system. The first commercial genome-editing crop is CRISPR/Cas9 waxy maize, which has been declared non-GMO by regulatory bodies in the United States, Argentina, Brazil, Canada, and Chile (Jiang et al., 2022).

Enhancing horticulture crop production with qualitative attributes in line with customer expectations is necessary to increase food security. Fruits without seeds, flowering and medicinal plants, and vegetables have all been edited using CRISPR/Cas technology. Reducing the length of the juvenile period can encourage early flowering and fruit development. Genome-edited tomato plants exhibit elevated levels of GABA, a chemical that promotes wellness. Additionally, this tactic has enhanced the aroma, color, and shelf life of ornamental plant flowers (Singh and Bokolia, 2021). The field of gene knockout and non-homologous end-joining (NHEJ) silencing mechanisms is the focus of study on the rapidly developing CRISPR/Cas9 technology. Various methods have been employed to achieve effective homology-directed repair processes, with positive outcomes documented in plant and mammalian cells. For effective and accurate gene knockdown, CRISPR has supplanted RNA interference. It has furthermore been utilized to generate multiplex gene knockouts and chromosomal deletions in a variety of plants (Arora and Narula, 2017).

20.5 IMPROVING PLANT RESISTANCE TO STRESS IN THE ENVIRONMENT

Abiotic stressors that affect plant development and production include salt, drought, and temperature. To survive, plants create internal defensive mechanisms, although cultivar tolerance is necessary for long-term agricultural output. Thanks to technological progress, plant genomes may now be modified through genomic engineering to enhance production, flexibility, and resistance to environmental stresses. By precisely altering genomes for desirable features, CRISPR/Cas has transformed biological research (Bhat et al., 2021). By deleting genes that impair plant machinery and make plants more vulnerable to abiotic stressors, targeted gene editing using GET systems like CRISPR/Cas has prompted the production of novel crops (Ahmad et al., 2021).

The usage of genome editing to control plant stress is becoming more and more common. Targeted gene manipulation is made easy and effective by the CRISPR-associated technology. By strengthening resistance to both PAMP-activated immunity and effectors-activated immunity, this approach strengthens plant immunity against phytopathogens (Ghosh and Dey, 2022). The probable benefits of CRISPR/Cas9 in agriculture, particularly in the evolution of crops resistant to abiotic stress, are now being investigated. Targeting important actors in sensitivity (S genes) and tolerance (T genes) can assist plants in modifying their physiological and biochemical processes to withstand stress due to abiotic conditions. A single transformation event can enhance resistance to several abiotic stressors by multiplex genome editing. Another strategy that has been suggested for enhancing resistance to abiotic stress is CRISPR/Cas9. The difficulty of converting cytosine to thymine has been solved by recent developments in nucleotide base editing, allowing for greater rates of A–T to G–C transformation. However, a stress-triggered CRISPR/Cas9 approach may be applied to overcome off-target mutations generated by Cas9 in transgenic plants, therefore reducing the incidence of off-target mutations dramatically (Zafar et al., 2020). It is hoped that the CRISPR/Cas method will help with precision plant breeding and the creation of non-transgenic grains that are resistant to climate change. However, with rice being the sole plant that is often utilized, not much research has been done on the biotic and abiotic stress toleration of cereals.

Researching these genes may boost grain productivity and growth, resulting in the creation of better cultivars for food security (Bhowmik et al., 2019). Plant development can be adversely affected by abiotic stressors, which can result in low yield and financial loss. The CRISPR/Cas gene-editing method is being used by scientists to create plant kinds that are resistant to stress. It has been demonstrated that the Cas9-based mechanism increases plants' resistance to abiotic stress. The system's effectiveness is dependent on variables like Cas9 expression, and it may be used in projects like the creation of mutant libraries and gene regulation (Biswas et al., 2021). To provide food safety, the objective is to design crops that can sustainably provide high yields under biotic and abiotic challenges.

Conventional techniques, such as random mutagenesis, take a long time and are unable to meet the growing needs. CRISPR/Cas9 and other targeted genome-editing innovations are known for their great efficiency, precision, and minimal chance of unintended consequences. Regulation and societal acceptability of genetically engineered crops, however, continue to be obstacles (Abdelrahman et al., 2018). Many nations are cultivating transgene-free genetically modified crops by eliminating selectable markers and Cas9 genes. One significant obstacle is the lack of functionally defined genes for agronomic characteristics. With multiplex genome editing enabling tolerance against several characteristics in a single transformation possibility, CRISPR is being employed more and more in field crops to create abiotic stress and heavy metal toxicity stress-acquiescent plants (Rahman et al., 2022).

20.6 APPLYING CRISPR/CAS TO IMPROVE CROP QUALITY AND YIELD IN THE FOOD ENHANCEMENT AND DEVELOPMENT

CRISPR/Cas9 technique has sparked transformative advancements in the realm of crop species (Wang et al., 2021b). Numerous research facilities worldwide have adopted this potent tool due to its vast potential. Here, we outline various applications that have been explored for enhancing both the harvest and worth of crops. CRISPR/Cas9 has been instrumental in developing cultivars with elevated nutritional content and enhanced resistance traits, marking a significant milestone in modern genome-editing technology. For instance, commercial oil derived from *Camelina sativa* seeds, renowned for their high fatty acid content, was successfully obtained. Additionally, genes such as JAZI, BRII, and GA-insensitive were targeted, resulting in mutation frequencies ranging from 26% to 84%. Manipulation of genes related to flowering regulation in the same species is directed to a high mutation frequency, particularly during the late flowering stage (Hyun et al., 2015). A multi-target CRISPR/Cas9 system, in conjunction with other tools, facilitated the influence of six modified PYL genes, achieving mutation frequencies ranging from 13% to 93% in the initial generation (Zhang et al., 2016a). Furthermore, modifications in gene expression (Gao et al., 2015), such as the alteration of the green fluorescent gene in *Nicotiana benthamiana*, were achieved using RNA-stimulated endonucleases, subsequently provided via the tobacco rattle virus to control plant gene transcriptional features (Xu et al., 2016). In rice plants, CRISPR/Cas9 has been utilized to induce heritable interpretations in the genetic makeup, leading to notable changes in gene clusters and subsequent alterations in genetic traits (Zhang et al., 2014; Xu et al., 2015). Various genes related to grain dormancy and yield modulation, like HvPM19, Gn1a, and GS3, were targeted and studied, resulting in improved crop attributes (Lawrenson et al., 2015; Shen et al., 2018; Usman et al., 2018).

Similarly, mutagenesis of CLBG1 in watermelon and WAXY allele in rice enhanced specific traits, like seed size and cooking quality (Wang et al., 2021b; Monsur et al., 2021). Similarly, investigations on MIR396f and MIR396e disruption in rice beneath low nitrogen conditions demonstrated enhanced grain yield (Zhang et al., 2020a). Notably, CRISPR/Cas9 editing of P450 and OsBADH2 homologs in rice yielded new mutants with improved yield and aroma (Usman et al., 2020). Moreover, targeted mutagenesis of transporter genes in rice cultivars, such as OsAAP6 and OsAAP10, exhibited significant enhancements in yield and quality of cooking (Wang et al., 2020b).

These results underscore the possibility for further improvements in crop traits through targeted mutagenesis of amino acid genes, promising the development of more acquiescent and best-value cultivars in various staple crops (Wang et al., 2020a; Zeng et al., 2020a). Additionally, successful applications of CRISPR/Cas9 in wheat, maize, *Brassica* species, tomatoes, apples, potatoes, citrus, grapes, and bananas have been documented, demonstrating its versatility and efficacy in enhancing harvest and worth across diverse plants (Jia and Wang, 2014; Brooks et al., 2014; Malnoy et al., 2016; Morineau et al., 2017; Klap et al., 2017; Soyk et al., 2017; Naim et al., 2018; Okuzaki et al., 2018; Lemmon et al., 2018; Zaman et al., 2019; Zhai et al., 2019; Espley et al., 2019; Toda et al., 2019; Li et al., 2020b; Ellison et al., 2020). The technique's precision and efficiency offer new avenues for crop advancement without the intro of foreign genes, ensuring food security and safety (Jia and Wang, 2014). Moreover, the development of seedless fruits and dwarf varieties through targeted gene editing further highlights the potential for CRISPR/Cas9 in shaping future agriculture practices (Klap et al., 2017).

20.7 USING CRISPR/CAS-MEDIATED BIOFORTIFIED CROPS ENHANCEMENT TO ADDRESS FOOD ENHANCEMENT AND DEVELOPMENT

The CRISPR/Cas system enables swift and precise genome modifications at specific sites within a single cell or an entire organism. It exerts control over transcription, facilitates changes in epigenomes, conducts genome-wide screenings, and enables chromosome imaging. Its versatility has led to a surge in its utilization for the development of genetically edited crops. Research teams have created crop systems to confer opposition to diseases (Schenke and Cai, 2020) and toleration to salinity, drought, and high temperatures (Chennakesavulu et al., 2022). This has been accomplished through the benefit of CRISPR/Cas. The result of climate-resilient crops and the improvement of crop quality qualities like flavor, nutritional content, and other desired traits are made possible in large part by the CRISPR/Cas method (Razzaq et al., 2021). Liu et al. (2021) primarily look into how CRISPR/Cas genome-editing devices can be utilized to enrich important nutrients in important crops.

20.7.1 Zn-Enriched Crops

Approximately 100 enzymes' catalytic activity and a host of other cellular metabolic processes depend on zinc (Sandstead, 1994). In the processes of wound healing, immunological response, protein synthesis, synthesis of DNA, and cell division, it is essential (Maggini et al., 2010). Immune system function and barrier integrity may be harmed by zinc deficiency (Sturniolo et al., 2002; Finamore et al., 2008; Maares and Haase, 2016). By damaging the inositol pentakisphosphate 2-kinase 1 (IPK1) gene and lowering the quantity of phytic acid present, the CRISPR/Cas method has been utilized to increase zinc upsurge in wheat grains (Ibrahim et al., 2022). To meet the world's nutritional demands, further work is required to use the CRISPR/Cas method to biofortify more crops with zinc.

20.7.2 Vitamin A Enhanced Crops

Carotenoids are vital isoprenoid pigments that are present in many photosynthetic species. One important carotene precursor is β-carotene (Maoka, 2020). Humans need carotenoids for vital processes like eyesight, immune system function, embryonic development, and general health, even though they cannot produce them internally (Timoneda et al., 2018). A main health concern is vitamin A deficiency, which can cause xerophthalmia and night blindness and increase the chances of morbidity and mortality (Sommer, 2008; Reddy et al., 2022). The development of carotenoid-enhanced staple plants, sometimes known as 'golden crops', through conventional genetic or breeding alteration has been attempted as a solution to this problem (Zheng et al., 2019). CRISPR-based genome editing is the most useful and extensively used genetic engineering technique available, facilitating fast, accurate, and DNA/transgene-free multiplex alterations in living things. The growth of 'golden' staple crops has been made possible by this progress (Zheng et al., 2019). To combat vitamin A deficiency, scientists have effectively created genome-edited golden crops, sometimes referred to as carotenoid biofortified crops, using CRISPR/Cas systems (Dong et al., 2020).

20.7.3 Iron-Enhanced Crops

Iron is essential for several metabolic strategies, like the transportation of oxygen and electrons. Human illnesses varying from anemia to neurodegenerative disorders are induced by disturbances in iron metabolism. One effective way to address human iron needs is by crop and dietary biofortification. Crop biofortification has been accomplished by the benefit of CRISPR/Cas9; disruption of the IPK1 gene has been shown to enrich iron concentration in wheat (Ibrahim et al., 2022).

20.7.4 Vitamin E-Enriched Crops

Vitamin E (tocopherol) is an essential lipid-soluble antioxidant that is vital to human health. Low vitamin E intake has been related to several diseases, including heart problems and several types of cancer (Rizvi et al., 2014). Humans need between 15 and 30 mg of vitamin E each day (Ungurianu et al., 2021). By precisely overexpressing *Hordeum vulgare* homogentisate phytyltransferase (HvHPT) and *H. vulgare* homogentisate geranylgeranyltransferase (HvHGGT) utilizing the CRISPR/Cas9 method, researchers have increased the concentration of tocopherol and tocotrienol (Zeng et al., 2020b). All the genes have the potential to enrich the quantity of vitamin E in a variety of plants.

20.8 CRISPR/CAS-MEDIATED ENHANCEMENT OF GENE ACTIVITY

Gene-editing devices, which as CRISPR/Cas are utilized in many fields, especially in the field of agriculture made improvements worldwide for cultivation and commercial use. Editing techniques are used in the production of food crops, and they have improved crop production worldwide.

CRISPR/Cas is one of the simplest forms. When the two pieces come together, they create the ribonucleoprotein (RNP), a complex non-coding RNA that cleaves the target site within the genome and engineers the Cas protein. The Cas protein is used for a variety of purposes. Several Cas proteins exist, and CRISPR/Cas tools are employed. CRISPR/associated genome editing is particularly helpful in agricultural biotechnology, where it can be used to recreate an essential role in gene editing and genetically modified crops. However, the method of CRISPR/Cas allows for transgene expression and genetic segregation, and genetic transformation of the plant by the RNP complex through the approach of transgene-free Cas 9 is a protein that is used in maize and wheat embryos by bombarding the particles with the proteins. Using this technique, the plant can regenerate its cargo without the need for exogenous genes, and cargo delivery is accomplished by the Cas protein miniature. This leads to optimization and new development and may facilitate the improvement of the crops. Conveyance of pre-made CRISPR/Cas RNPs is the 'cleanest' process since no exogenous DNA is forever presented; because of this, it is probably healthier and more likely to be approved by the public. Additionally, there may be a chance to get governmental approval, even though the procedure takes a long time and the editing quality is still not very excellent.

Gene expression can be regulated in many ways; one of the ways employed is CRISPR/Cas, where the target mRNA can be controlled by the small RNA molecules, which are miRNAs that are endogenous and control the mRNA in various pathways. It is possible to adjust them considerably using CRISPR/Cas to control miRNA-regulated plant activities. Mainly in polyploid species, the research is in the beginning stage, and also it has many challenges: miRNA function isn't effective because of the frameshift, and editing miRNA with sgRNA isn't a required drawback of miRNA editing. Because of this, it is advised to target a mature miRNA coding sequence using a dual sgRNA design. In the best-case scenario, this would lead to the elimination of the entire miRNA sequence or, at the very least, significantly alter the pre-miRNA structure to hinder the processing of miRNA (Deng et al., 2022). Although less effective, sgRNA-mediated miRNA regulation is also possible (Zhou et al., 2017; Chung et al., 2020). Targeting miRNA genes in many plant species has been done using various strategies. Mutations in the miRNA binding sites of controlled genes can also change the way that miRNA controls the expression of genes. Moreover, CRISPR/Cas-based genome editing can disrupt or replace CREs in the genes that need to be changed (Shi et al., 2017; Swinnen et al., 2016).

Utilizing CRISPR/Cas to break down CREs in tomatoes produced unique variation that underlies pleiotropy and quantitative trait variation, both of which regulate the phenotypic variance of beneficial qualities (Wang et al., 2021a; Hendelman et al., 2021). Lastly, dCas9 can be connected to transcriptional effectors or epigenetic modulators to modify gene expression without permanently altering the genome. Alternatively, it can be used to take advantage of the reported capability of short protospacers in conjunction with a standard nuclease to attract transcriptional activators (Pan et al., 2021, 2022; Gardiner et al., 2022). The specificity of epigenetic modification and activation/repression efficiency will be difficult to improve in future dCas-based technological research. Other obstacles include the prospect of regulating gene expression concurrently in numerous targets, tissue-specific or inducible (Li et al.,2021).

Even though first-generation CRISPR techniques were widely used with success in plants, the majority of the modifications that were introduced were knockout mutations of the targeted loci. NHEJ cannot be employed to precisely modify plant genomes to accomplish sophisticated plant breeding programs after a double-strand break (DSB) has been created (Li et al.,2016a). Furthermore, the absence of appropriate PAM sequences prevents the first generation of Cas9 nucleases from the species *Streptococcus pyogenes* (SpCas9) from targeting a large number of loci (Figure 20.3). For precise targeting by combining recently found site-directed nucleases (SDNs) or altered Cas proteins with the catalytic action of DNA-changing proteins. The formulated fusion proteins can create single-nucleotide modifications, large deletions, and in-frame insertions. Moreover, these modifications offer enhanced precision and effectiveness in DSB-free and template-free editing. Many of the latest variants have previously been favorably utilized with plants. While it will take more testing data to be widely and regularly used for agricultural progress (Wada et al., 2022),

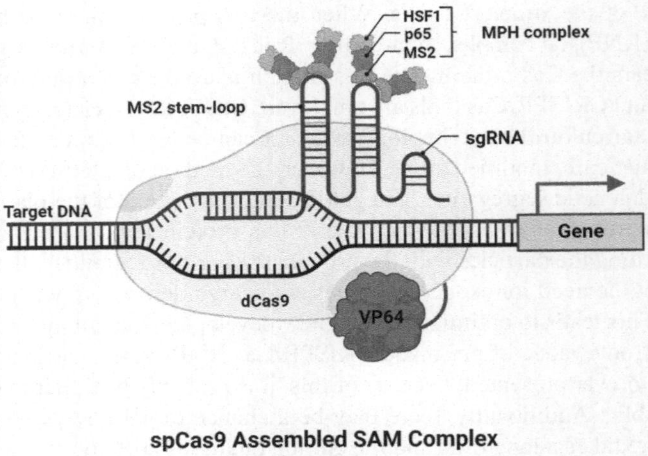

FIGURE 20.3 SpCas9 assembled SAM complex.

these novel approaches offer well-organized induction of planned edits, improving the likelihood of getting the necessary mutants consistent from plant genotypes with lower regeneration ability. Plant genome editing will be promoted by the wider use of these novel CRISPR variants in various plants and their regulation for the expression of plants. The PAM limitation of conventional SpCas9 and target specificity are already being overcome by the new CRISPR toolboxes. Smaller SDNs (Xu et al., 2021; Liu et al., 2022; Li et al., 2023) have been adopted in response to SpCas9's enormous size. These enzymes, when paired with split-Cas systems (Kaya et al., 2017; Yuan et al., 2022), will also allow for SDN expression mediated by plant viruses.

The important implication of CRISPR/Cas9 techniques in terms of support is the possibility to progress the excellence of agricultural derivatives. Miao et al. (2013) achieved the first successful CRISPR/Cas9 manipulation of rice, an important food origin for world inhabitants (Fukagawa and Ziska, 2019), demonstrating that the technique may be used for targeted modifications in rice. Since this discovery, a significant value of work has been accomplished to investigate the consequences of gene modifications in rice and comprehend the roles of particular genes to use the study findings in practical settings. Guo et al. (2020) employed CRISPR/Cas9 to overexpress and eliminate the OsProDH gene in rice. Proline dehydrogenase is found in mitochondria, is produced by the OsProDH gene, and is in charge of breaking down proline in rice. Proline is essential for protecting plants from stresses because it triggers several responses in crops and scavenges reactive oxygen species (Hayat et al., 2012). Rice with an OsProDH mutation was shown to have higher proline levels, which lowered reactive oxygen species levels (Guo et al., 2020). Thus, by altering the OsProDH gene and consequently the proline metabolism, rice can be provided with improved thermotolerance (Guo et al., 2020).

Conversely, it may be possible to produce salt-tolerant rice by modifying the OsNAC45 (Yu et al., 2018; Zhang et al., 2020b). Through the regulation of numerous additional plant stress reaction genes, OsNAC45 may recreate an important part in regulating rice's abscisic acid signal responses (OsPM1, OsCYP89G1 OsSIK1, OsDREB1F, OsERF104, OsEREBP2, OsSAMDC2). This could be essential for creating rice that can withstand more salt (Zhang et al., 2020a). In a distinct study, polygalacturonase's role in controlling the cell wall's immunological reaction via the OsPG1 gene was elucidated by the benefit of CRISPR/Cas9 (Cao et al., 2021).

This contributes to our knowledge of the function of cell wall integrity in the immune reaction of plants and emphasizes the potential of making use of the physiology of the cell wall to confer resistance to bacteria. Similar strategies to CRISPR/Cas9 were employed in other studies that sought to produce observable effects in rice by gene deletion or overexpression. These techniques have been employed to determine and successfully manipulate genes reliable for traits like anthocyanin concentration (Zheng et al., 2019; Hu et al., 2020), resistance to diseases (Zhou et al., 2015; Wang et al., 2016; Kim et al., 2019), and the length of the grains (Li et al., 2020a; Usman et al., 2021). The results reveal the undeniable efficiency of the CRISPR/Cas9 approach in modifying rice properties. It is anticipated that these techniques will be used to produce more robust, resilient, and nutrient-dense rice, which will contribute to global sustainability.

With considerable mutation efficiency, CRISPR/Cas9-established mutagenesis has also greatly helped other noteworthy plant species. Jacobs et al. demonstrated the significance of CRISPR/Cas9 gene editing in soybeans by deleting the gene for the green fluorescent protein. This landmark finding prompted various projects to use CRISPR/Cas9 gene editing in soybeans. Han et al. (2019) employed CRISPR/Cas9 to selectively mutate the E1 gene to regulate soybean flowering. They noted that, under long-day circumstances, truncating the E1 protein led to an earlier blooming time, enhanced expression of the GmFT2a/5a gene, and prevented the gene from being repressed. A photo-insensitive soybean variation was produced as a result of this transformation, and it may be appropriate to introduce soybeans to higher latitudes (Han et al., 2019).

20.9 PROSPECTS AND DIFFICULTIES

Even though the CRISPR/Cas9 technique is a simple and trustworthy way to modify a plant's DNA to boost immunity, there are a few challenges associated with it. The knowledge of focusing RNA to compare sequences with sites in the genome in addition to the target spot provides CRISPR/Cas9 the natural capability to make remote targets (Hsu et al., 2013). To facilitate off-site targeting, the Cas9's ability to target a preferred sequence should be improved more precisely (Majeed et al., 2018). In addition, altered Cas9 proteins are being made to recognize different PAMs (Agudelo et al., 2020). Furthermore, extra CRISPR/Cas9 nucleases called Cas12a and Cas13a which are aimed at DNA and single-stranded RNA have been discovered in several bacterial type II systems (Koonin et al., 2017). Despite having certain resemblances to Cas9, these two systems operate better in various circumstances, particularly when it comes to plant diseases since they employ distinct strategies for processing pre-crRNA and cleaving target nucleotides.

A possible other limit of the CRISPR/Cas9 system is the possibility of undesirable modifications due to the complex endogenous pathways that Cas nucleases use to repair the double-stranded DNA (Kosicki et al., 2018). Moreover, Cas9-influenced DSBs may be detrimental to cells by causing pathways that direct to cell death and decreasing the effectiveness of editing and transformation (Jinek et al., 2012; Roy et al., 2018; Standage-Beier et al., 2019). Using CRISPR/Cas9, plant virologists, geneticists, and molecular biologists may yield crops with increased resistance against diseases (Figure 20.4). Even though genetic engineering has experienced a revolution, there are still several issues that need to be resolved before plants may be successfully modified for human benefit. The degree of Cas9 and gRNA expression influences the specificity of CRISPR/Cas9 editing in plants. The secondary structure of the target gRNAs and the sequence composition of the target site, including the quantity of GC present, also have a significant impact on its efficacy (Arora and Narula 2017; Majeed et al., 2018). The recently created CRISPR/Cas9 RNP system eliminated the essential to depend on the target cell's capacity to interpret Cas9 and the likelihood of it coming into contact with gRNA. CRISPR/Cas9 sequence-specific nuclease editing is a useful treatment for rice blast disease (Wang et al., 2016; Li et al., 2016b).

20.10 CONCLUSION

CRISPR/Cas9 is crucial for plant development in several contexts. While CRISPR/Cas9 is essential, it also produces high yields in crop production. The country's economy will be greatly enhanced by agriculture, which is highly significant. By utilizing CRISPR/Cas 9, our crops can generate a high yield quickly and are genetically modified The genetically modified crops are pest resistance and annual crops. CRISPR/Cas9 can also develop new crop types that are adjusted for the surroundings

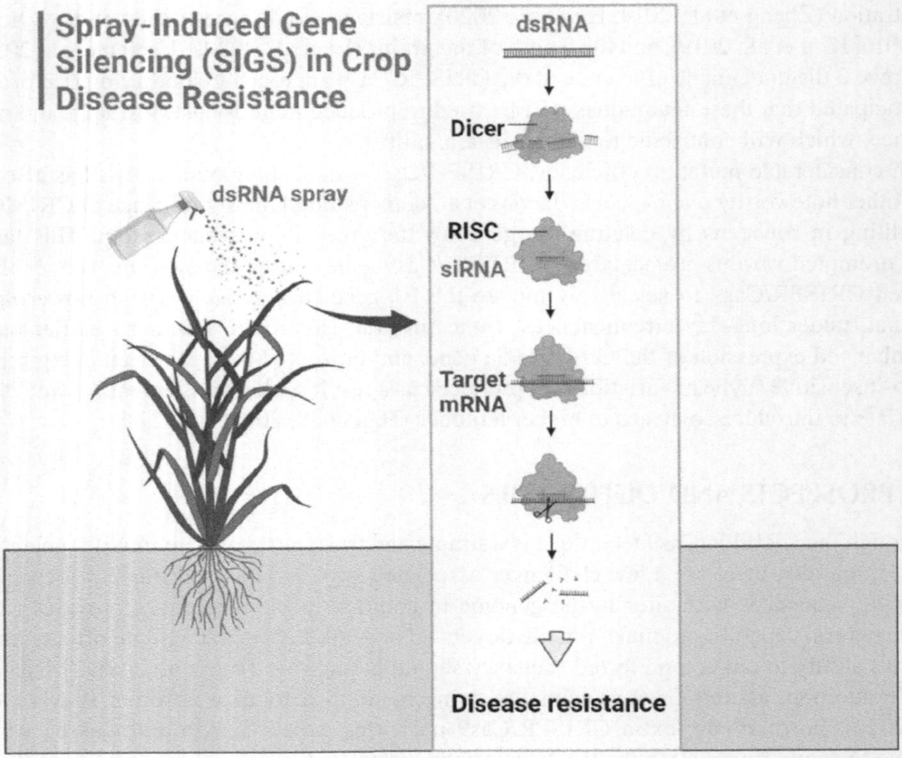

FIGURE 20.4 Spray-induced gene silencing in crop disease resistance.

and also have drought resistance and insect resistance. Many existent plant varieties have also been produced using this technique. According to the future aspects, this technology serves as an enhancing tool in the agricultural field and in plant production.

ACKNOWLEDGMENT

All the authors are thankful to their respective universities and institutions for their valuable support. The corresponding author is thankful to Skill Development Centre, BEICH, RUSA2.0 (SDC Project order No. BU/RUSA/ SDC/BEICH 2020/15; 28.12.2020), Bharathiar University, for providing financial support.

DATA AVAILABILITY

The authors confirm that the data supporting the findings of this study are available within the book/chapter.

REFERENCES

Abdelrahman, M., Al-Sadi, A. M., Pour-Aboughadareh, A., Burritt, D. J., & Tran, L. S. P. (2018). Genome editing using CRISPR/Cas9-targeted mutagenesis: An opportunity for yield improvements of crop plants grown under environmental stresses. *Plant Physiology and Biochemistry*, *131*, 31–36.

Agudelo, D., Carter, S., Velimirovic, M., Duringer, A., Rivest, J. F., Levesque, S., Loehr, J., Mouchiroud, M., Cyr, D., Waters, P. J., Laplante, M., Moineau, S., Goulet, A., & Doyon, Y. (2020). Versatile and robust genome editing with Streptococcus thermophilus CRISPR1-Cas9. *Genome Research*, *30*(1), 107–117.

Ahmad, S., Sheng, Z., Jalal, R. S., Tabassum, J., Ahmed, F. K., Hu, S., Shao, G., Wei, X., Abd-Elsalam, K. A., Hu, P., & Tang, S. (2021). CRISPR-Cas technology towards improvement of abiotic stress tolerance in plants. In K. A. Abd-Elsalam & K.-T. Lim (Eds.), *CRISPR and RNAi systems* (pp. 755–772). Elsevier.

Arora, L., & Narula, A. (2017). Gene editing and crop improvement using CRISPR-Cas9 system. *Frontiers in Plant Science*, *8*, 296675.

Bhat, M. A., Mir, R. A., Kumar, V., Shah, A. A., Zargar, S. M., Rahman, S., & Jan, A. T. (2021). Mechanistic insights of CRISPR/Cas-mediated genome editing towards enhancing abiotic stress tolerance in plants. *Physiologia Plantarum*, *172*(2), 1255–1268.

Bhowmik, P., Hassan, M. M., Molla, K., Rahman, M., & Islam, M. T. (2019). Application of CRISPR-Cas genome editing tools for the improvement of plant abiotic stress tolerance. In M. Hasanuzzaman, K. Nahar, M. Fujita, H. Oku, & T. Islam (Eds.), *Approaches for enhancing abiotic stress tolerance in plants* (pp. 459–472). CRC Press.

Biswas, D., Saha, S. C., & Dey, A. (2021). CRISPR-Cas genome-editing tool in plant abiotic stress-tolerance. *Plant Gene*, *26*, 100286.

Brooks, C., Nekrasov, V., Lippman, Z. B., & Van Eck, J. (2014). Efficient gene editing in tomato in the first generation using the clustered regularly interspaced short palindromic repeats/CRISPR-associated9 system. *Plant Physiology*, *166*(3), 1292–1297.

Cao, Y., Zhang, Y., Chen, Y., Yu, N., Liaqat, S., Wu, W., Chen, D., Cheng, S., Wei, X., Cao, L., Zhang, Y., & Liu, Q. (2021). OsPG1 encodes a polygalacturonase that determines cell wall architecture and affects resistance to bacterial blight pathogen in rice. *Rice*, *14*, 1–15.

Chennakesavulu, K., Singh, H., Trivedi, P. K., Jain, M., & Yadav, S. R. (2022). State-of-the-art in CRISPR technology and engineering drought, salinity, and thermo-tolerant crop plants. *Plant Cell Reports*, *41*(3), 815–831.

Chung, P. J., Chung, H., Oh, N., Choi, J., Bang, S. W., Jung, S. E., Jung, H., Shim, J. S., & Kim, J. K. (2020). Efficiency of recombinant CRISPR/rCas9-mediated miRNA gene editing in rice. *International Journal of Molecular Sciences*, *21*(24), 9606.

Chylinski, K., Makarova, K. S., Charpentier, E., & Koonin, E. V. (2014). Classification and evolution of type II CRISPR-Cas systems. *Nucleic Acids Research*, *42*(10), 6091–6105.

Das, T., Anand, U., Pal, T., Mandal, S., Kumar, M., Radha, Gopalakrishnan, A. V., de la Lastra, J. M. P., & Dey, A. (2023). Exploring the potential of CRISPR/Cas genome editing for vegetable crop improvement: An overview of challenges and approaches. *Biotechnology and Bioengineering*, *120*(5), 1215–1228.

Deng, F., Zeng, F., Shen, Q., Abbas, A., Cheng, J., Jiang, W., Chen, G., Shah, A. N., Holford, P., Tanveer, M., Zhang, D., & Chen, Z. H. (2022). Molecular evolution and functional modification of plant miRNAs with CRISPR. *Trends in Plant Science*, *27*(9), 890–907.

Dong, O. X., Yu, S., Jain, R., Zhang, N., Duong, P. Q., Butler, C., Li, Y., Lipzen, A., Martin, J. A., Barry, K. W., Schmutz, J., Tian, L., & Ronald, P. C. (2020). Marker-free carotenoid-enriched rice generated through targeted gene insertion using CRISPR-Cas9. *Nature Communications*, *11*(1), 1178.

Ellison, E. E., Nagalakshmi, U., Gamo, M. E., Huang, P. J., Dinesh-Kumar, S., & Voytas, D. F. (2020). Multiplexed heritable gene editing using RNA viruses and mobile single guide RNAs. *Nature Plants*, *6*(6), 620–624.

Espley, R. V., Leif, D., Plunkett, B., McGhie, T., Henry-Kirk, R., Hall, M., Johnston, J. W., Punter, M. P., Boldingh, H., Nardozza, S., Volz, R. K., O'Donnell, S., & Allan, A. C. (2019). Red to brown: An elevated anthocyanic response in apple drives ethylene to advance maturity and fruit flesh browning. *Frontiers in Plant Science*, *10*, 1248.

Finamore, A., Massimi, M., Devirgiliis, L. C., & Mengheri, E. (2008). Zinc deficiency induces membrane barrier damage and increases neutrophil transmigration in Caco-2 cells. *The Journal of Nutrition*, *138*(9), 1664–1670.

Fukagawa, N. K., & Ziska, L. H. (2019). Rice: Importance for global nutrition. *Journal of Nutritional Science and Vitaminology*, 65(Supplement), S2–S3.

Gao, J., Wang, G., Ma, S., Xie, X., Wu, X., Zhang, X., Wu, Y., Zhao, P., & Xia, Q. (2015). CRISPR/Cas9-mediated targeted mutagenesis in Nicotiana tabacum. *Plant Molecular Biology*, *87*, 99–110.

Gardiner, J., Ghoshal, B., Wang, M., & Jacobsen, S. E. (2022). CRISPR-Cas-mediated transcriptional control and epi-mutagenesis. *Plant Physiology*, *188*(4), 1811–1824.

Ghosh, S., & Dey, G. (2022). Biotic and abiotic stress tolerance through CRISPR-Cas mediated genome editing. *Journal of Plant Biochemistry and Biotechnology*, *31*(2), 227–238.

Guo, M., Zhang, X., Liu, J., Hou, L., Liu, H., & Zhao, X. (2020). OsProDH negatively regulates thermotolerance in rice by modulating proline metabolism and reactive oxygen species scavenging. *Rice*, *13*, 1–5.

Han, J., Guo, B., Guo, Y., Zhang, B., Wang, X., & Qiu, L. J. (2019). Creation of early flowering germplasm of soybean by CRISPR/Cas9 technology. *Frontiers in Plant Science*, *10*, 1446.

Hayat, S., Hayat, Q., Alyemeni, M. N., Wani, A. S., Pichtel, J., & Ahmad, A. (2012). Role of proline under changing environments: A review. *Plant Signaling & Behavior*, *7*(11), 1456–1466.

Hendelman, A., Zebell, S., Rodriguez-Leal, D., Dukler, N., Robitaille, G., Wu, X., Kostyun, J., Tal, L., Wang, P., Bartlett, M. E., Eshed, Y., Efroni, I., & Lippman, Z. B. (2021). Conserved pleiotropy of an ancient plant homeobox gene uncovered by cis-regulatory dissection. *Cell*, *184*(7), 1724–1739.

Hidalgo-Cantabrana, C., & Barrangou, R. (2020). Characterization and applications of Type I CRISPR-Cas systems. *Biochemical Society Transactions*, *48*(1), 15–23.

Hsu, P. D., Scott, D. A., Weinstein, J. A., Ran, F. A., Konermann, S., Agarwala, V., Li, Y., Fine, E. J., Wu, X., Shalem, O., Cradick, T., Marraffini, L., Bao, G., & Zhang, F. (2013). DNA targeting specificity of RNA-guided Cas9 nucleases. *Nature Biotechnology*, *31*(9), 827–832.

Hu, W., Zhou, T., Han, Z., Tan, C., & Xing, Y. (2020). Dominant complementary interaction between OsC1 and two tightly linked genes, Rb1 and Rb2, controls the purple leaf sheath in rice. *Theoretical and Applied Genetics*, *133*, 2555–2566.

Hyun, Y., Kim, J., Cho, S. W., Choi, Y., Kim, J. S., & Coupland, G. (2015). Site-directed mutagenesis in Arabidopsis thaliana using dividing tissue-targeted RGEN of the CRISPR/Cas system to generate heritable null alleles. *Planta*, *241*, 271–284.

Ibrahim, S., Saleem, B., Rehman, N., Zafar, S. A., Naeem, M. K., & Khan, M. R. (2022). CRISPR/Cas9 mediated disruption of Inositol Pentakisphosphate 2-Kinase 1 (TaIPK1) reduces phytic acid and improves iron and zinc accumulation in wheat grains. *Journal of Advanced Research*, *37*, 33–41.

Ishino, Y., Krupovic, M., & Forterre, P. (2018). History of CRISPR-Cas from encounter with a mysterious repeated sequence to genome editing technology. *Journal of Bacteriology*, *200*(7), 10–1128.

Jia, H., & Wang, N. (2014). Targeted genome editing of sweet orange using Cas9/sgRNA. *PloS One*, *9*(4), e93806.

Jiang, Y., Sun, K., & An, X. (2022). CRISPR/Cas system: Applications and prospects for maize improvement. *ACS Agricultural Science & Technology*, *2*(2), 174–183.

Jinek, M., Chylinski, K., Fonfara, I., Hauer, M., Doudna, J. A., & Charpentier, E. (2012). A programmable dual-RNA-guided DNA endonuclease in adaptive bacterial immunity. *Science*, *337*(6096), 816–821.

Kaul, T., Sony, S. K., Verma, R., Motelb, K. F. A., Prakash, A. T., Eswaran, M., Bharti, Y., Nehra, M., & Kaul, R. (2020). Revisiting CRISPR/Cas-mediated crop improvement: Special focus on nutrition. *Journal of Biosciences*, *45*, 1–37.

Kaya, H., Ishibashi, K., & Toki, S. (2017). A split Staphylococcus aureus Cas9 as a compact genome-editing tool in plants. *Plant and Cell Physiology*, *58*(4), 643–649.

Khan, Z., Binyameen, B., Waheed, U., Mubarik, M. S., Ghouri, M. Z., Sattar, T., & Razzaq, A. (2021). *CRISPR/Cas-based techniques in plants*. In A. Ahmad (Ed.), *CRISPR crops: the future of food security* (pp. 37–61). Springer.

Khatodia, S., Bhatotia, K., Passricha, N., Khurana, S. M. P., & Tuteja, N. (2016). The CRISPR/Cas genome-editing tool: Application in improvement of crops. *Frontiers in Plant Science*, *7*, 135062.

Kim, Y. A., Moon, H., & Park, C. J. (2019). CRISPR/Cas9-targeted mutagenesis of Os8N3 in rice to confer resistance to Xanthomonas oryzae pv. Oryza. *Rice*, *12*, 1–13.

Klap, C., Yeshayahou, E., Bolger, A. M., Arazi, T., Gupta, S. K., Shabtai, S., Usadel, B., Salts, Y., & Barg, R. (2017). Tomato facultative parthenocarpy results from Sl AGAMOUS-LIKE 6 loss of function. *Plant Biotechnology Journal*, *15*(5), 634–647.

Koonin, E. V., Makarova, K. S., & Zhang, F. (2017). Diversity, classification and evolution of CRISPR-Cas systems. *Current Opinion in Microbiology*, *37*, 67–78.

Kosicki, M., Tomberg, K., & Bradley, A. (2018). Repair of double-strand breaks induced by CRISPR-Cas9 leads to large deletions and complex rearrangements. *Nature Biotechnology*, *36*(8), 765–771.

Lawrenson, T., Shorinola, O., Stacey, N., Li, C., Østergaard, L., Patron, N., Uauy, C., & Harwood, W. (2015). Induction of targeted, heritable mutations in barley and Brassica oleracea using RNA-guided Cas9 nuclease. *Genome Biology*, *16*, 1–13.

Lemmon, Z. H., Reem, N. T., Dalrymple, J., Soyk, S., Swartwood, K. E., Rodriguez-Leal, D., Van Eck, J., & Lippman, Z. B. (2018). Rapid improvement of domestication traits in an orphan crop by genome editing. *Nature Plants*, *4*(10), 766–770.

Li, C., Hao, M., Wang, W., Wang, H., Chen, F., Chu, W., Zhang, B., Mei, D., Cheng, H., & Hu, Q. (2018). An efficient CRISPR/Cas9 platform for rapidly generating simultaneous mutagenesis of multiple gene homoeologs in allotetraploid oilseed rape. *Frontiers in Plant Science*, *9*, 442.

Li, J., Jiao, G., Sun, Y., Chen, J., Zhong, Y., Yan, L., Jiang, D., Ma, Y., & Xia, L. (2021). Modification of starch composition, structure and properties through editing of TaSBEIIa in both winter and spring wheat varieties by CRISPR/Cas9. *Plant Biotechnology Journal*, *19*(5), 937–951.

Li, L., Luo, Y., Chen, B., Xu, K., Zhang, F., Li, H., Huang, Q., Xiao, X., Zhang, T., Hu, J., Li, F., & Wu, X. (2016a). A genome-wide association study reveals new loci for resistance to clubroot disease in Brassica napus. *Frontiers in Plant Science*, *7*, 1483.

Li, M., Li, X., Zhou, Z., Wu, P., Fang, M., Pan, X., Lin, Q., Luo, W., Wu, G., & Li, H. (2016b). Reassessment of the four yield-related genes Gn1a, DEP1, GS3, and IPA1 in rice using a CRISPR/Cas9 system. *Frontiers in Plant Science*, *7*, 377.

Li, Q., Lu, L., Liu, H., Bai, X., Zhou, X., Wu, B., Yuan, M., Yang, L., & Xing, Y. (2020a). A minor QTL, SG3, encoding an R2R3-MYB protein, negatively controls grain length in rice. *Theoretical and Applied Genetics*, *133*, 2387–2399.

Li, Q., Sapkota, M., & van der Knaap, E. (2020b). Perspectives of CRISPR/Cas-mediated cis-engineering in horticulture: Unlocking the neglected potential for crop improvement. *Horticulture Research*, *7*, 30.

Li, Z., Zhong, Z., Wu, Z., Pausch, P., Al-Shayeb, B., Amerasekera, J., Doudna, J. A., & Jacobsen, S. E. (2023). Genome editing in plants using the compact editor CasΦ. *Proceedings of the National Academy of Sciences*, *120*(4), e2216822120.

Liu, Q., Yang, F., Zhang, J., Liu, H., Rahman, S., Islam, S., Ma, W., & She, M. (2021). Application of CRISPR/Cas9 in crop quality improvement. *International Journal of Molecular Sciences*, *22*(8), 4206.

Liu, S., Sretenovic, S., Fan, T., Cheng, Y., Li, G., Qi, A., Tang, X., Xu, Y., Guo, W., Zhong, Z., He, Y., Liang, Y., Han, Q., Zheng, X., Gu, X., Qi, Y., & Zhang, Y. (2022). Hypercompact CRISPR-Cas12j2 (CasΦ) enables genome editing, gene activation, and epigenome editing in plants. *Plant Communications*, *3*(6), 100453.

Liu, T. Y., & Doudna, J. A. (2020). Chemistry of Class 1 CRISPR-Cas effectors: Binding, editing, and regulation. *Journal of Biological Chemistry*, *295*(42), 14473–14487.

Maares, M., & Haase, H. (2016). Zinc and immunity: An essential interrelation. *Archives of Biochemistry and Biophysics*, *611*, 58–65.

Maggini, S., Wenzlaff, S., & Hornig, D. (2010). Essential role of vitamin C and zinc in child immunity and health. *Journal of International Medical Research*, *38*(2), 386–414.

Majeed, U., Yaqoob, U., Qazi, H. A., Ahmad, S., & John, R. (2018). CRISPR/Cas system as an emerging technology to enhance plant viral immunity. *Physiological and Molecular Plant Pathology*, *103*, 107–113.

Makarova, K. S., & Koonin, E. V. (2015). Annotation and classification of CRISPR-Cas systems. *Methods in Molecular Bioloogy*, *1311*, 47–75.

Malnoy, M., Viola, R., Jung, M. H., Koo, O. J., Kim, S., Kim, J. S., Velasco, R., & Nagamangala Kanchiswamy, C. (2016). DNA-free genetically edited grapevine and apple protoplast using CRISPR/Cas9 ribonucleoproteins. *Frontiers in Plant Science*, *7*, 1904.

Maoka, T. (2020). Carotenoids as natural functional pigments. *Journal of Natural Medicines*, *74*(1), 1–16.

Miao, C., Wang, Z., Zhang, L., Yao, J., Hua, K., Liu, X., Shi, H., & Zhu, J. K. (2019). The grain yield modulator miR156 regulates seed dormancy through the gibberellin pathway in rice. *Nature Communications*, *10*(1), 3822.

Miao, J., Guo, D., Zhang, J., Huang, Q., Qin, G., Zhang, X., Wan, J., Gu, H., & Qu, L. J. (2013). Targeted mutagenesis in rice using CRISPR-Cas system. *Cell Research*, *23*(10), 1233–1236.

Monsur, M. B., Ni, C., Xiangjin, W., Lihong, X., Guiai, J., Shaoqing, T., Sreenivasulu, N., Gaoneng, S., & Peisong, H. (2021). Improved eating and cooking quality of indica rice cultivar YK17 via adenine base editing of Wx allele of granule-bound starch synthase I (GBSS I). *Rice Science*, *28*(5), 427–430.

Morineau, C., Bellec, Y., Tellier, F., Gissot, L., Kelemen, Z., Nogué, F., & Faure, J. D. (2017). Selective gene dosage by CRISPR-Cas9 genome editing in hexaploid Camelina sativa. *Plant Biotechnology Journal*, *15*(6), 729–739.

Munawar, N., & Ahmad, A. (2021). CRISPR/Cas system: An introduction. In A. Ahmad (Ed.), *CRISPR crops: The future of food security* (pp. 1–35). Springer.

Naim, F., Dugdale, B., Kleidon, J., Brinin, A., Shand, K., Waterhouse, P., & Dale, J. (2018). Gene editing the phytoene desaturase alleles of Cavendish banana using CRISPR/Cas9. *Transgenic Research*, *27*, 451–460.

Okuzaki, A., Ogawa, T., Koizuka, C., Kaneko, K., Inaba, M., Imamura, J., & Koizuka, N. (2018). CRISPR/Cas9-mediated genome editing of the fatty acid desaturase 2 gene in Brassica napus. *Plant Physiology and Biochemistry*, *131*, 63–69.

Pan, C., Li, G., Malzahn, A. A., Cheng, Y., Leyson, B., Sretenovic, S., Gurel, F., Coleman, G. D., & Qi, Y. (2022). Boosting plant genome editing with a versatile CRISPR-Combo system. *Nature Plants*, *8*(5), 513–525.

Pan, C., Sretenovic, S., & Qi, Y. (2021). CRISPR/dCas-mediated transcriptional and epigenetic regulation in plants. *Current Opinion in Plant Biology*, *60*, 101980.

Peng, R., Jones, D. C., Liu, F., & Zhang, B. (2021). From sequencing to genome editing for cotton improvement. *Trends in Biotechnology*, *39*(3), 221–224.

Pickar-Oliver, A., & Gersbach, C. A. (2019). The next generation of CRISPR-Cas technologies and applications. *Nature Reviews Molecular Cell Biology*, *20*(8), 490–507.

Rahman, M. U., Zulfiqar, S., Raza, M. A., Ahmad, N., & Zhang, B. (2022). Engineering abiotic stress tolerance in crop plants through CRISPR genome editing. *Cells*, *11*(22), 3590.

Razzaq, A., Kaur, P., Akhter, N., Wani, S. H., & Saleem, F. (2021). Next-generation breeding strategies for climate-ready crops. *Frontiers in Plant Science*, *12*, 620420.

Reddy, G. B., Shalini, T., Ghosh, S., Pullakhandam, R., Kumar, B. N., Kulkarni, B., Laxmaiah, A., Kapil, U., Deb, S., Hemalatha, R., Kurpad, A. V., & Sachdev, H. S. (2022). Prevalence of vitamin A deficiency and dietary inadequacy in Indian school-age children and adolescents. *European Journal of Nutrition*, *61*(1), 197–209.

Rizvi, S., Raza, S. T., Ahmed, F., Ahmad, A., Abbas, S., & Mahdi, F. (2014). The role of vitamin E in human health and some diseases. *Sultan Qaboos University Medical Journal*, *14*(2), e157.

Rodríguez-Leal, D., Lemmon, Z. H., Man, J., Bartlett, M. E., & Lippman, Z. B. (2017). Engineering quantitative trait variation for crop improvement by genome editing. *Cell*, *171*(2), 470–480.

Roy, K. R., Smith, J. D., Vonesch, S. C., Lin, G., Tu, C. S., Lederer, A. R., Chu, A., Suresh, S., Nguyen, M., Horecka, J., Tripathi, A., Burnett, W. T., Morgan, M. A., Schulz, J., Orsley, K. M., Wei, W., Aiyar, R. S., Davis, R. W., Bankaitis, V. A., & Steinmetz, L. M. (2018). Multiplexed precision genome editing with trackable genomic barcodes in yeast. *Nature Biotechnology*, *36*(6), 512–520.

Saeed, S., Usman, B., Shim, S. H., Khan, S. U., Nizamuddin, S., Saeed, S., Yasira Shoaib, Jong- Jeon, S., & Jung, K. H. (2022). CRISPR/Cas-mediated editing of cis-regulatory elements for crop improvement. *Plant Science*, *324*, 111435.

Sánchez-León, S., Gil-Humanes, J., Ozuna, C. V., Giménez, M. J., Sousa, C., Voytas, D. F., & Barro, F. (2018). Low-gluten, nontransgenic wheat engineered with CRISPR/Cas9. *Plant Biotechnology Journal*, *16*(4), 902–910.

Sandstead, H. H. (1994). Understanding zinc: Recent observations and interpretations. *The Journal of Laboratory and Clinical Medicine*, *124*(3), 322–327.

Schenke, D., & Cai, D. (2020). Applications of CRISPR/Cas to improve crop disease resistance: Beyond inactivation of susceptibility factors. *iScience*, *23*(9), 101478.

Schindele, A., Dorn, A., & Puchta, H. (2020). CRISPR/Cas brings plant biology and breeding into the fast lane. *Current Opinion in Biotechnology*, *61*, 7–14.

Shen, L., Wang, C., Fu, Y., Wang, J., Liu, Q., Zhang, X., Yan, C., Qian, Q., & Wang, K. (2018). QTL editing confers opposing yield performance in different rice varieties. *Journal of Integrative Plant Biology*, *60*(2), 89–93.

Shi, J., Gao, H., Wang, H., Lafitte, H. R., Archibald, R. L., Yang, M., Hakimi, S. M., Mo, H., & Habben, J. E. (2017). ARGOS 8 variants generated by CRISPR-Cas9 improve maize grain yield under field drought stress conditions. *Plant Biotechnology Journal*, *15*(2), 207–216.

Singh, A., & Bokolia, M. (2021). CRISPR/Cas for crop improvement: A brief review. *Resonance*. *26*(2), 227–240.

Sommer, A. (2008). Vitamin A deficiency and clinical disease: An historical overview. *The Journal of Nutrition*, *138*(10), 1835–1839.

Soyk, S., Müller, N. A., Park, S. J., Schmalenbach, I., Jiang, K., Hayama, R., Zhang, L., Van Eck, J., Jiménez-Gómez, J. M., & Lippman, Z. B. (2017). Variation in the flowering gene SELF PRUNING 5G promotes day-neutrality and early yield in tomato. *Nature Genetics*, *49*(1), 162–168.

Standage-Beier, K., Brookhouser, N., Balachandran, P., Zhang, Q., Brafman, D. A., & Wang, X. (2019). RNA-guided recombinase-Cas9 fusion targets genomic DNA deletion and integration. *The CRISPR Journal*, *2*(4), 209–222.

Sturniolo, G. C., Fries, W., Mazzon, E., Di Leo, V., Barollo, M., & D'inca, R. (2002). Effect of zinc supplementation on intestinal permeability in experimental colitis. *Journal of Laboratory and Clinical Medicine*, *139*(5), 311–315.

Swinnen, G., Goossens, A., & Pauwels, L. (2016). Lessons from domestication: Targeting cis-regulatory elements for crop improvement. *Trends in Plant Science*, *21*(6), 506–515.

Timoneda, J., Rodríguez-Fernández, L., Zaragozá, R., Marín, M. P., Cabezuelo, M. T., Torres, L., Viña, J. R., & Barber, T. (2018). Vitamin A deficiency and the lung. *Nutrients*, *10*(9), 1132.

Toda, E., Koiso, N., Takebayashi, A., Ichikawa, M., Kiba, T., Osakabe, K., Osakabe, Y., Sakakibara, H., Kato, N., & Okamoto, T. (2019). An efficient DNA-and selectable-marker-free genome-editing system using zygotes in rice. *Nature Plants*, *5*(4), 363–368.

Tofazzal Islam, T. I. (2019). CRISPR-Cas technology in modifying food crops. *CABI Reviews*, *14*, 1–16.

Ungurianu, A., Zanfirescu, A., Niţulescu, G., & Margină, D. (2021). Vitamin E beyond its antioxidant label. *Antioxidants*, *10*(5), 634.

Usman, B., Nawaz, G., Zhao, N., Liao, S., Qin, B., Liu, F., & Li, R. (2020). Programmed editing of rice (Oryza sativa L.) OsSPL16 gene using CRISPR/Cas9 improves grain yield by modulating the expression of pyruvate enzymes and cell cycle proteins. *International Journal of Molecular Sciences*, *22*(1), 249.

Usman, B., Nawaz, G., Zhao, N., Liu, Y., & Li, R. (2018). Generation of high yielding and fragrant rice (Oryza sativa L.) lines by CRISPR/Cas9 targeted mutagenesis of three homoeologs of cytochrome P450 gene family and OsBADH2 and transcriptome and proteome profiling of revealed changes triggered by mutations. *Plants*, *9*(6), 788.

Usman, B., Zhao, N., Nawaz, G., Qin, B., Liu, F., Liu, Y., & Li, R. (2021). CRISPR/Cas9 guided mutagenesis of grain size 3 confers increased rice (Oryza sativa l.) grain length by regulating cysteine proteinase inhibitor and ubiquitin-related proteins. *International Journal of Molecular Sciences*, *22*(6), 3225.

Wada, N., Osakabe, K., & Osakabe, Y. (2022). Expanding the plant genome editing toolbox with recently developed CRISPR-Cas systems. *Plant Physiology*, *188*(4), 1825–1837.

Wang, C., Wang, G., Gao, Y., Lu, G., Habben, J. E., Mao, G., Chen, G., Wang, J., Yang, F., Zhao, X., Zhang, J., Mo, H., Qu, P., Liu, J., & Greene, T. W. (2020a). A cytokinin-activation enzyme-like gene improves grain yield under various field conditions in rice. *Plant Molecular Biology*, *102*(4), 373–388.

Wang, F., Wang, C., Liu, P., Lei, C., Hao, W., Gao, Y., Liu, Y.-G., & Zhao, K. (2016). Enhanced rice blast resistance by CRISPR/Cas9-targeted mutagenesis of the ERF transcription factor gene OsERF922. *PloS One*, *11*(4), e0154027.

Wang, X., Aguirre, L., Rodríguez-Leal, D., Hendelman, A., Benoit, M., & Lippman, Z. B. (2021). Dissecting cis-regulatory control of quantitative trait variation in a plant stem cell circuit. *Nature Plants*, *7*(4), 419–427.

Wang, C., Wang, G., Gao, Y., Lu, G., Habben, J. E., Mao, G., ... & Greene, T. W. (2020). A cytokinin-activation enzyme-like gene improves grain yield under various field conditions in rice. *Plant Molecular Biology*, *102*(4), 373–388.

Wang, S., Yang, Y., Guo, M., Zhong, C., Yan, C., & Sun, S. (2020b). Targeted mutagenesis of amino acid transporter genes for rice quality improvement using the CRISPR/Cas9 system. *The Crop Journal*, *8*(3), 457–464.

Wang, X., Aguirre, L., Rodríguez-Leal, D., Hendelman, A., Benoit, M., & Lippman, Z. B. (2021a). Dissecting cis-regulatory control of quantitative trait variation in a plant stem cell circuit. *Nature Plants*, *7*(4), 419–427.

Wang, Y., Meng, Z., Liang, C., Meng, Z., Wang, Y., Sun, G., Zhu, T., Cai, Y., Guo, S., Zhang, R., & Lin, Y. (2017). Increased lateral root formation by CRISPR/Cas9-mediated editing of arginase genes in cotton. *Science China. Life Sciences*, *60*(5), 524.

Wang, Y., Wang, J., Guo, S., Tian, S., Zhang, J., Ren, Y., Li, M., Gong, G., Zhang, H., & Xu, Y. (2021b). CRISPR/Cas9-mediated mutagenesis of ClBG1 decreased seed size and promoted seed germination in watermelon. *Horticulture Research*, *8*, 70.

Xu, R., Yang, Y., Qin, R., Li, H., Qiu, C., Li, L., Wei, P., & Yang, J. (2016). Rapid improvement of grain weight via highly efficient CRISPR/Cas9-mediated multiplex genome editing in rice. *Journal of Genetics and Genomics*, *43*(8), 529–532.

Xu, R. F., Li, H., Qin, R. Y., Li, J., Qiu, C. H., Yang, Y. C., Ma, H., Li, L., Wei, P.-C., & Yang, J. B. (2015). Generation of inheritable and "transgene clean" targeted genome-modified rice in later generations using the CRISPR/Cas9 system. *Scientific Reports*, *5*(1), 11491.

Xu, X., Chemparathy, A., Zeng, L., Kempton, H. R., Shang, S., Nakamura, M., & Qi, L. S. (2021). Engineered miniature CRISPR-Cas system for mammalian genome regulation and editing. *Molecular Cell*, *81*(20), 4333–4345.

Yu, S., Huang, A., Li, J., Gao, L., Feng, Y., Pemberton, E., & Chen, C. (2018). OsNAC45 plays complex roles by mediating POD activity and the expression of development-related genes under various abiotic stresses in rice root. *Plant Growth Regulation*, *84*, 519–531.

Yuan, G., Lu, H., De, K., Hassan, M. M., Liu, Y., Li, Y., Muchero, W., Abraham, P. E., Tuskan, G. A., & Yang, X. (2022). An intein-mediated split-nCas9 system for base editing in plants. *ACS Synthetic Biology*, *11*(7), 2513–2517.

Zafar, S. A., Zaidi, S. S. E. A., Gaba, Y., Singla-Pareek, S. L., Dhankher, O. P., Li, X., Mansoor, S., & Pareek, A. (2020). Engineering abiotic stress tolerance via CRISPR/Cas-mediated genome editing. *Journal of Experimental Botany*, *71*(2), 470–479.

Zaman, Q. U., Chu, W., Hao, M., Shi, Y., Sun, M., Sang, S. F., Mei, D., Cheng, H., Liu, J., Li, C., & Hu, Q. (2019). CRISPR/Cas9-mediated multiplex genome editing of JAGGED gene in Brassica napus L. *Biomolecules*, *9*(11), 725.

Zeng, Y., Wen, J., Zhao, W., Wang, Q., & Huang, W. (2020a). Rational improvement of rice yield and cold tolerance by editing the three genes OsPIN5b, GS3, and OsMYB30 with the CRISPR-Cas9 system. *Frontiers in Plant Science*, *10*, 1663.

Zeng, Z., Han, N., Liu, C., Buerte, B., Zhou, C., Chen, J., Wang, M., Zhang, Y., Tang, Y., Zhu, M., Wang, J., Yang, Y., & Bian, H. (2020b). Functional dissection of HGGT and HPT in barley vitamin E biosynthesis via CRISPR/Cas9-enabled genome editing. *Annals of Botany*, *126*(5), 929–942.

Zhai, Y., Cai, S., Hu, L., Yang, Y., Amoo, O., Fan, C., & Zhou, Y. (2019). CRISPR/Cas9-mediated genome editing reveals differences in the contribution of INDEHISCENT homologues to pod shatter resistance in Brassica napus L. *Theoretical and Applied Genetics*, *132*, 2111–2123.

Zhang, X., Long, Y., Huang, J., & Xia, J. (2020). OsNAC45 is involved in ABA response and salt tolerance in rice. *Rice*, *13*, 1–13.

Zhang, D., Zhang, Z., Unver, T., & Zhang, B. (2021). CRISPR/Cas: A powerful tool for gene function study and crop improvement. *Journal of Advanced Research*, *29*, 207–221.

Zhang, H., Zhang, J., Wei, P., Zhang, B., Gou, F., Feng, Z., Mao, Y., Yang, L., Zhang, H., Xu, N., & Zhu, J. K. (2014). The CRISPR/Cas9 system produces specific and homozygous targeted gene editing in rice in one generation. *Plant Biotechnology Journal*, *12*(6), 797–807.

Zhang, Y., Malzahn, A. A., Sretenovic, S., & Qi, Y. (2019). The emerging and uncultivated potential of CRISPR technology in plant science. *Nature Plants*, *5*(8), 778–794.

Zheng, J., Wu, H., Zhu, H., Huang, C., Liu, C., Chang, Y., Kong, Z., Zhou, Z., Wang, G., Lin, Y., & Chen, H. (2019). Determining factors, regulation system, and domestication of anthocyanin biosynthesis in rice leaves. *New Phytologist*, *223*(2), 705–721.

Zhou, J., Deng, K., Cheng, Y., Zhong, Z., Tian, L. I., Tang, X. U., Tang, A., Zheng, X., Zhang, T., Qi, Y., & Zhang, Y. (2017). CRISPR-Cas9 based genome editing reveals new insights into microRNA function and regulation in rice. *Frontiers in Plant Science*, *8*, 1598.

Zhou, J., Peng, Z., Long, J., Sosso, D., Liu, B. O., Eom, J. S., Huang, S., Liu, S., Cruz, C. V., Frommer, W. B., White, F. F., & Yang, B. (2015). Gene targeting by the TAL effector PthXo2 reveals cryptic resistance gene for bacterial blight of rice. *The Plant Journal*, *82*(4), 632–643.

Zhu, H., Li, C., & Gao, C. (2020). Applications of CRISPR-Cas in agriculture and plant biotechnology. *Nature Reviews Molecular Cell Biology*, *21*(11), 661–677.

21 CRISPR Tools for Developing Plant Oxidative Stress Tolerance

Protocols, Current Achievements, and Future Directions

Gunasekaran Arthi, Venkatachalam Sivaprakash, Marimuthu Oviya, Kannan Vijayarani, Natarajan Pushpa, Chenniappan Anchana Devi, K. Vijayalakshmi, and Arumugam Vijaya Anand

21.1 STRESS IN PLANTS

Oxidative stress is due to an inequality of the generation of reactive oxygen species (ROS) and the capacity of living systems to eliminate these reactive molecules or repair the ensuing damage. Under normal physiological circumstances, ROS's role as a signaling molecule is concerned with various cellular functions including growth, development, and defense responses. However, excessive ROS production, often triggered by environmental stressors like UV radiation, pollution, and heavy metals, can overwhelm the cellular antioxidant defense mechanisms, leading to oxidative harm to key biomolecules. Halliwell and Gutteridge (2015) emphasized the multifaceted nature of oxidative stress-induced cellular dysfunction and its implications for human health, including the development of chronic ailments like cancer, neurodegenerative disorders, and aging.

Environmental stressors that influence the development of plants and productivity include drought, salinity, heat, and cold. These stressors can produce morphological, physiological, biochemical, and molecular alterations in the plants. Moderate results of main crops are reduced by more than 50% due to abiotic stress, which is one of the main reasons for crop loss globally. Saline land makes up around 25% of farming land worldwide, and more places are predicted to experience drought (Burke et al., 2006). A crop's ability to detect and react to several environmental conditions seems to overlap, and crops are frequently subjected to multiple stresses. The plant experiences oxidative stress when it produces ROS in retort to various stressful situations (Ahmad et al., 2008, 2010).

Stress from the environment and biotic environment can direct oxidative stress, which damages the components of the cell and causes dysfunction of the cell. The oxidative stress is caused by the production and accumulation of ROS, which are molecules that contain activated oxygen. The main causes of oxidative stress include inequality between the generation of ROS and the detoxification of the cells, as well as the de novo production of ROS as a role of the stress signaling and immune reaction required for defense and adaptation. Stress factors that directly produce ROS can also stimulate the formation of ROS by oxidative enzymes (Rao et al., 1996; Ranieri et al., 2003; Nawkar et al., 2013), which can lead to molecular deterioration and negative cellular consequences on the plant, ultimately resulting in its death (Apel and Hirt, 2004; Zhang et al., 2010). When abiotic stress is applied to a plant, a variety of genes are triggered, appearing in raised levels of a variety

DOI: 10.1201/9781003457237-21

of metabolites. Some of these proteins may recreate a role in safeguarding the plant from abiotic stresses. However, the expression of gene profiles of plants under abiotic stress showed that although different genes are differentially regulated under different stressors, they may cause a similar defensive reaction (Ozturk et al., 2002).

The oxygen that has been partially reduced is known as ROS. Since aerobic life developed from anaerobic life, ROS has been a part of biological existence (Foyer and Noctor, 2005). The degree of oxygen reduction determines the production of three primary ROS in that order:

1. The main source of ROS is superoxide (O^{2-}), which is formed when the molecular oxygen is reduced by one electron. In this reaction, the electron is supplied by NADPH, and the reaction is catalyzed by NADPH oxidase.
2. The other ROS is hydrogen peroxide (H_2O_2), which is created when the molecular oxygen is also reduced. The one-electron lessening of superoxide preferably favors peroxide (O^{2-}) and is neutralized by two protons to form H_2O_2. The incidental dismutation is improbable to occur at physiological pH when superoxide is in anionic form (O^{2-}), as superoxides are repelled by negative charge. However, when the pH is acidic, the ratio of the neutral form (HO_2) increases, and, at that point, the incidental dismutation begins to recreate an essential role in the production of H_2O_2 (Koji et al., 2009).
3. Hydroxyl radicals (OH^-) are the next, significantly harmful, ROS, which are created when H_2O_2 undergoes additional reduction.
   a. Haber–Weiss reaction: Under normal conditions, this reaction proceeds at a very slower rate and resulting in the low production of OH – ions. – – H O O OH OH O 22 2 2 +→ ++
   b Fenton reaction: It is common in biological systems. It occurs in the presence of transition metals like Fe^{2+}, Cu^{+}, etc. 2 3 H O Fe Fe OH OH

One of the planet's oldest stresses is thought to be the action of O_2 derivatives (Dowling and Simmons, 2009). For the 2.7 billion years, or since they began generating O_2 from H_2O and CO_2, plants have dealt with this. The biochemistry of contemporary plants and animals has been shaped by the steadily rising O_2 levels, which have also guided the development of species (Dowling and Simmons, 2009). It's interesting to note that plants have developed the capability to use oxidative stress, at least ROS formation, for a variety of purposes, including signaling needs, sensing different stresses, regulating growth, and death (Demidchik et al., 2003; Foreman et al., 2003; Demidchik et al., 2010), perceiving hormones and regulatory mechanisms like purines and amino acids (Murata et al., 2001; Demidchik et al., 2004; Demidchik et al., 2009; Krishnamurthy and Rathinasabapathi, 2013), producing a gravitropic reaction (Joo et al., 2001), and a host of additional functions that are not directly associated to stress or oxidation. It is commonly recognized that plants, being sessile organisms, are continuously subjected to various stimuli derived from their natural surroundings. Plants respond to abiotic stress in reaction to even minor changes in their environment, such as temperature swings, drought, excessive salinity, exposure to radiation or heavy metals, or both (Zhang et al., 2022).

Stressors in the environment cause a plant's formation and aggregation of ROS to increase, which results in a complete stress response (Hasanuzzaman et al., 2020). All cellular compartments are severely damaged and malfunctioning due to molecules containing activated oxygen. Nucleic acids, lipids, and proteins are oxidized under oxidative stress conditions and need to be neutralized right once to stop their harmful effects on cells. To lessen oxidative deterioration in plant cells, a variety of cellular antioxidant defense systems have evolved (Chmielowska-Bak et al., 2015). Finding novel elements involved in plants' attempts to preserve cellular homeostasis and fend off the impacts of stress is the main objective of contemporary plant studies (Kmiecik et al., 2016). Through advanced genetic engineering, the outcomes of extensive study on this subject can be utilized to increase agricultural output (Roy et al., 2011).

21.2 IMPACT OF OXIDATIVE STRESS

Oxidative stress in plants can maintain various harmful effects across different regions of the plant, including (Gill and Tuteja, 2010; Suzuki et al., 2012; Chmielowska-Bak et al., 2015):

- **Cellular level:** Damage to biomolecules: Oxidative stress can cause harm to proteins, lipids, and nucleic acids, which can cause structural and functional disruptions in cells.
- **Impaired metabolism:** ROS can disrupt metabolic pathways, energy production, and nutrition intake.
- **Photosynthetic tissues:** Reduced photosynthetic efficiency: Oxidative stress can affect photosystem activity, which results in lower rates of photosynthetic activity and lower biomass production.
- **Degradation of chlorophyll:** ROS can hasten the process of chlorophyll molecule breakdown, leading to decreased pigment concentration and chlorosis.
- **Root system:** Reduced nutritional uptake: Oxidative stress can impede root development and function, making it more challenging for the roots to uptake nutrients and water.
- **Increased vulnerability to pathogens:** When ROS builds up in the roots, plant defense is weakened, leaving plants more vulnerable to illnesses and root pathogens.
- **Shoots and leaves:** Greater vulnerability to environmental stressors: Oxidative stress can intensify the harmful impacts of abiotic stressors including salinity, high temperatures, and drought, resulting in physiological problems and stunted plant growth.
- **Cell death and tissue necrosis:** Extended exposure to oxidative stress can cause plant tissues to undergo necrosis, or programmed cell death (apoptosis), which causes tissue damage and loss of function.
- **Reproductive organs:** Impaired reproductive development: Reduced seed set, fruit quality, and yield might result from oxidative stress during the flowering and fruiting stages.
- **Pollen viability and seed germination:** An accumulation of ROS can reduce pollen viability and hinder seed germination, which can jeopardize the success of plant reproduction.
- **Systemic effects**: Induction of defensive responses: Plant defense systems, such as the creation of antioxidant enzymes and secondary metabolites, can be activated in retort to moderate levels of oxidative stress to reduce damage and improve stress tolerance.
- **Hormonal imbalance:** The growth, development of plants, and retorts to environmental cues can all be impacted by disruption of hormonal signaling pathways by oxidative stress.
- Overall, oxidative stress can have a significant and diverse influence on physiology, growth, and development, emphasizing how crucial it is to preserve redox equilibrium for the well-being and productivity of plants.

21.3 CRISPR/CAS SYSTEM – OVERVIEW

Since their ability to edit DNA was demonstrated in 2012, CRISPR/Cas approaches have emerged as a considerably powerful genome-editing technology (Jinek et al., 2012). In the areas of repairing harmful mutations, locating crucial genes for cancer immunotherapy, and resolving important issues with organ xenotransplantation, they have accomplished a great deal. Crop development and breeding initiatives are made possible by the exceptional precision and performance of CRISPR/Cas technologies in plants (Baena-González and Hanson, 2017). Researchers can improve plant nutritional content, raise sensitivity to environmental challenges, and increase plant opposition to pests by concentrating on particular genes or phenotypes (Zhang et al., 2019).

The CRISPR/Cas approach has transformed genetic engineering and presents vast opportunities for enhancing crops in plants. This groundbreaking tool enables meticulous and effective genome editing, empowering researchers to selectively modify, delete, or insert genes with unparalleled precision (Georges and Ray, 2017). In the realm of plant science, CRISPR/Cas innovation emerges as a potent tool for augmenting crop characteristics, amplifying yield, fortifying stress resilience, and fostering resistance against pathogens (McAnany et al., 2022). Noteworthy improvement has been made in enhancing plant genetics through the use of CRISPR-based techniques to change plant genomes. Several CRISPRs that can enhance particular plant genes have been found by researchers (Li et al., 2013), including Cas9, Cpf1, and Cas12a. Plants with desirable characteristics, such as higher productivity, better disease resistance, and better nutritional content, can now be created thanks to the development of precise gene-editing devices (Wang et al., 2017). In plant technology, the CRISPR/Cas9 approach offers several benefits, ranging from the investigation of gene activities to the enhancement of crop types (Liu et al., 2017; Jaganathan et al., 2018). Even in cases where the genome comprises closely associated paralogous genes, precise targeting is ensured by well-designed single-guide RNAs (sgRNA) (Baysal et al., 2016).

21.3.1 CRISPR as a Tool for Modifying Plant Genomes to Enhance Oxidative Stress Tolerance

The field of molecular biology has undergone a complete transformation with the finding of alterable nucleases that cause breaking of double-strand. Zinc finger nucleases (ZFNs) are considered the foremost genome-editing technology that led to an advance in genome engineering (Chandrasegaran and Carroll, 2016). Subsequently, transcription activator-like effector nucleases (TALENs), which are established on bacterial TALEs, also enhanced the potential of genome engineering. The available techniques were soon extended to ~40 diverse organisms for genome editing (Chandrasegaran and Carroll, 2016). Nevertheless, the finding of CRISPR/Cas9 caught the attention of scientists almost in the world for its clear advantages of both TALENs and ZFNs (Mao et al., 2013). These use protein motifs to select targets; CRISPR/Cas9, on the contrary, recognizes RNA in DNA to cause double-strand breaks (DSBs). CRISPR/Cas9 has other benefits over ZFNs and TALENs, such as (1) a simple target design; (2) the ability to introduce mutations efficiently through direct injection of RNAs encoding Cas9 protein and guide RNA (gRNA); and (3) the comfort of multiplexing, which enables the targeted mutation of multiple genes in a single incident (Ma et al., 2015; Malzahn et al., 2017).

Plants' ability to withstand oxidative stress can be greatly enhanced by targeting particular genes with the CRISPR/Cas9 genome-editing innovation (Figure 21.1). Unfortunately, introducing the CRISPR/Cas9 approach into plant cells can be difficult and may cause unwanted mutations at off-target locations, which could hinder the growth and development of the plant. Notwithstanding these difficulties, abiotic stress toleration in crops may be increased by the utilization of genome-editing methods such as CRISPR/Cas9. Nevertheless, there could be unforeseen repercussions if the complex regulatory networks concerned with stress reactions are altered. Moreover, obtaining regulatory permission for the utilization of genome-editing techniques for crop enhancement could be a time-consuming and expensive procedure. However, using more specific nucleases like Cpf1 or Cas12a can reduce the danger of off-target mutation, and more recent techniques like ribonucleoprotein (RNP) transporting or electroporation offer assurance for more accurate and efficient transporting of genome-editing elements. According to Kumar et al. (2023), abiotic stress toleration is a difficult trait regulated by several genes and pathways. Hence, modifying one or occasional genes might not be enough to provide the required amount of stress toleration.

However, the CRISPR/Cas9 genome-editing approach has been effectively applied to aim at particular genes associated with plant resistance to oxidative stress. For example, the CRISPR/Cas9 genome-editing approach has been employed to alter genes relevant to abiotic stress in wheat protoplasts, such as TaDREB2 and TaERF3 (Kim et al., 2018). In different investigations, CRISPR/Cas9 RNPs were utilized to alter the TaGW2 and TaGASR7 genes in bread wheat without causing any off-target consequences. Additionally, transgenic-free plants can be obtained at T0 by the DNA-free editing method employing CRISPR/Cas9 RNPs, which can eliminate time-consuming steps like backcross breeding for transgene removal (Jaganathan et al., 2018). Plant genes like OsDERF1, OsERF922, OsRMC, GmMYB118, GmDrbza, and GmDrbzb may be aimed at by CRISPR/Cas9

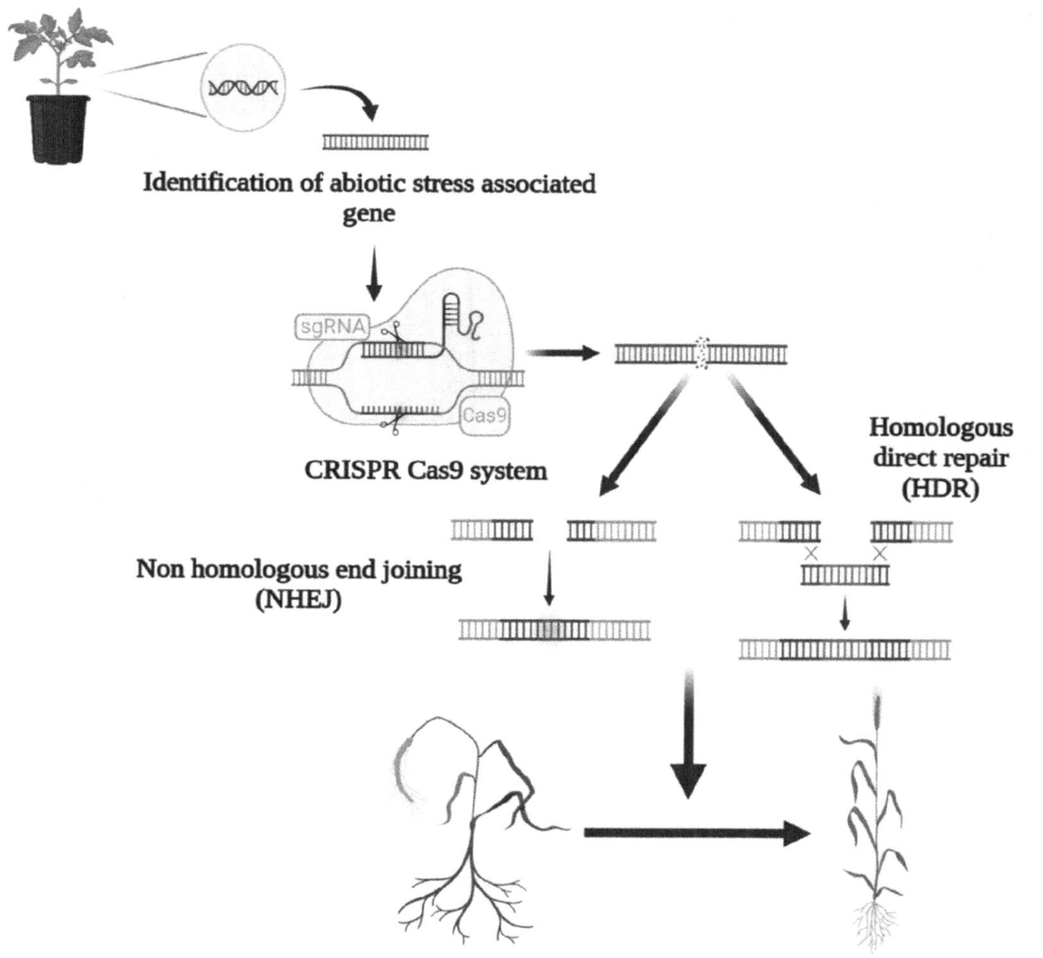

FIGURE 21.1 CRISPR tool in plant oxidative stress tolerance.

technology to produce stable lines with improved opposition to abiotic stresses, especially salt and drought (Cong et al., 2013).

Model crops and other plant species can undergo precise, site-specific changes thanks to the type 2 CRISPR/Cas9 method. Plant genetic modification using CRISPR/Cas9 has become more accessible for practical research, and TaDREB2, TaDREB3, TaHAG1, and TaALs are important targets for creating wheat cultivars that are more resilient to salinity and drought. All things considered, the CRISPR/Cas9 technique has proven reliable for genetic modification, transcriptome regulation, producing crops resistant to stress, and molecular research on the multigenic stress response in plants. The benefit of CRISPR/Cas9 technology in plant genome editing is further facilitated by the accessibility of web-based devices like CRISPR-P that assist in the building of sgRNAs in over 20 additional plants (Rai et al., 2023).

21.4 MECHANISM – OXIDATIVE STRESS MECHANISM IN PLANTS

Oxidative stress in plants arises from an imbalance between the generation of reactive oxygen species (ROS) and the plant's antioxidant defense system. ROS, comprising superoxide radicals (O_2^-), hydrogen peroxide (H2O2), and hydroxyl radicals (•OH), are natural byproducts of aerobic metabolism and environmental stressors. Elevated ROS levels result in cellular damage, including lipid peroxidation, protein oxidation, and DNA damage.

Plants have developed intricate antioxidant defense properties to maintain ROS stability and combat oxidative stress. These mechanisms encompass enzymatic antioxidants like superoxide dismutase (SOD), catalase (CAT), peroxidases (POXs), and non-enzymatic antioxidants including ascorbate (vitamin C), tocopherols (vitamin E), and glutathione. Enzymatic antioxidants directly scavenge ROS or facilitate their conversion into less harmful compounds, while non-enzymatic antioxidants neutralize ROS and replenish other antioxidants (Suzuki et al., 2012; Chmielowska-Bak et al., 2015).

Moreover, plants exploit ROS as signaling molecules in diverse physiological functions, including stress responses and defense mechanisms. ROS signaling pathways entail the actuation of mitogen-activated protein kinases, calcium ions, and plant hormones including salicylic acid, jasmonic acid, and abscisic acid (ABA). These pathways modulate gene expression, antioxidant enzyme activities, and stress-responsive genes to alleviate oxidative damage and bolster stress toleration. Comprehending the mechanisms of oxidative stress in plants is pivotal for devising strategies to enhance crop productivity and resilience to environmental stresses (Suzuki et al., 2012; Chmielowska-Bak et al., 2015).

21.4.1 CRISPR Tool Mechanism to Reduce Oxidative Stress

The CRISPR/Cas-9 genome-editing system consists of three stages: recognition, cleavage, and repair. The developed sgRNA identifies the target sequence in the target gene by utilizing a complementary base pair. The Cas-9 protein subsequently cleaves the DNA at the exact place to create a DSB. The cell's restoration machinery then uses either non-homologous end-joining (NHEJ) or homology-directed repair (HDR) to mend the break. The genomes of prokaryotic organisms contain DNA sequences belonging to the CRISPR family. These sequences originate from DNA fragments of bacteriophages that had previously infected the prokaryote. They are employed in the event of additional infections to locate and eradicate DNA from similar bacteriophages.

CRISPR/Cas9 was derived from the naturally appearing genome-editing process used by bacteria as an immune response. To create what is known as CRISPR arrays, virus-infected bacteria take little pieces of the virus's DNA and rearrange it inside their DNA in specific ways. Because of the CRISPR arrays, the bacteria can 'remember' the viruses or viruses that are closely related. Using CRISPR arrays, the bacteria create RNA segments that, if the viruses re-attack, can attach to particular DNA sequences on the viruses. The bacteria then use Cas9 or a similar enzyme to divide the DNA, rendering the virus unusable.

The CRISPR/Cas system maintains two essential elements: gRNA and the Cas nuclease. In plants, the prevalent Cas nucleases utilized are Cas9 and Cas12a (formerly known as Cpf1). The gRNA directs the Cas nuclease to a precise target DNA sequence via complementary base pairing. Upon binding, the Cas protein initiates a DSB at the designated place, prompting DNA repair mechanisms. The primary DNA repair pathways, NHEJ and HDR, are harnessed to introduce insertions, deletions, or precise sequence changes in the plant genome (Panchy et al., 2016; Puchta, 2017; Sun et al., 2021).

21.5 TARGET GENES TO ACHIEVE ABIOTIC STRESS TOLERANCE

To build plant resistance to oxidative stress, a range of genes can be targeted with CRISPR/Cas-mediated genome editing. These genes play a role in stress-responsive transcriptional regulation, ROS signaling, antioxidant defense pathways, and other oxidative stress-related functions (Bozorgmehr, 2012; Hahn and Nekrasov, 2019; Esmaeili et al., 2022). Target genes for CRISPR editing to enhance plant resistance to oxidative stress include:

- **Antioxidant enzyme genes** such as SOD, CAT, POX, GPX, APX, glutathione reductase (GR)
- **ROS signaling genes** like NADPH oxidases, respiratory burst oxidase homologs (RBOH), dual oxidases, stress-responsive transcription factors, NAC transcription factors, AP2/ERF

(APETALA2/ethylene-responsive factor) transcription factors, WRKY transcription factors, myeloblastosis transcription factors, bZIP (basic leucine zipper) transcription factors

- **Heat shock proteins and chaperones**
  - Heat shock proteins
  - Heat shock transcription factors
- **Mitochondrial and chloroplast genes**
  - Genes encoding components of the electron transport chain and synthesis of ATP
  - Genes involved in chlorophyll formation and photosystem assembly
- **Membrane transporters and ion channels**
  - Aquaporins and water transporters
  - Ion channels involved in ion homeostasis (e.g., calcium, potassium, sodium)
- **Cell wall modification enzymes**
  - POX involved in lignin biosynthesis
  - Cell wall-associated POX and polyphenol oxidases
- **Protein degradation and turnover genes**
  - Proteases and protease inhibitors involved in protein quality control and degradation pathway
- **Secondary metabolism and hormone signaling genes**
  - Genes concerned with the formation of secondary metabolites with antioxidant properties (e.g., flavonoids, phenolic)
  - Genes concerned with the biosynthesis and signaling of plant hormones
- By targeting these genes using CRISPR/Cas-mediated genome editing, researchers can modulate plant responses to oxidative stress and enhance stress tolerance in various crop species.
- **Structural genes**
  - Structural genes represent a major class of targets that can enhance distinct elements of stress toleration. ROS are important to plants because they serve as signaling molecules for the code of expression of the gene (Ribeiro et al., 2017), plant defense against viral infections (Wu et al., 2017), and symbiotic nitrogen fixation (Sinharoy et al., 2016). But excessive ROS production, a common plant reaction to abiotic and oxidative stressors, can generate growth abnormalities, decrease the rate of photosynthesis, increase cell death, and also male sterility, all of which can lower the yield of the crop (Hu et al., 2011; Zafar et al., 2019).
  - ROS molecules are scavenged by several genes that encode antioxidant enzymes, such as GRs, CAT, SOD, glutathione S-transferases, and numerous POXs. It is known that the T genes lead to abiotic stress toleration (Hu et al., 2011; Mittler, 2002). On the contrary, it has been documented that certain S genes increase programmed cell death, decrease antioxidant activity, and produce too much ROS (Mittler et al., 2004). Plants are more susceptible to abiotic stressors when some genes cause hormonal homeostasis to be disrupted (Fang et al., 2015; Liu et al., 2016; Zhao et al., 2017).
- **Regulatory genes**
  - Regulation genes are a significant class of targets for controlling the expression of many downstream genes and initiating multiple stress signals. An instance of a negative regulator of abiotic stresses in *Arabidopsis* is the ANAC069 (S-gene), along with the NAM-ATAF1/2 and CUC2 (NAC) transcription factor (TF) gene. He et al. (2017) state that this results in a high synthesis of proline and a lowered capability to scavenge ROS, increasing susceptibility to salt and osmotic stress.
  - Increased toleration to salt and osmotic stress was observed in ANAC069 knockdown mutants generated by T-DNA insertion (He et al., 2017). According to Jung et al. (2008), drought and salt tolerance are conferred via ABA-induced stomatal closure, which is enhanced by overexpressing T genes like AtMYB44. Additionally, ZmWRKY106 overexpression increases the heat and drought tolerance of transgenic

plants by controlling the indication of stress-associated genes, decreasing ROS levels, and improving the activity of antioxidant enzymes (Wang et al., 2018).

- **Cis-regulatory sequences**
  - Cis-regulatory sequences are crucial for regulating gene expression because they enable the recruitment of specific TFs. Cis-regulatory sequences are well-established to be essential for controlling abiotic stress (Liu et al., 2014). Since most of these sequences are located in the promoter area of genes, modifications to their location, identity, or sequence may have an impact on the expression of the gene, possibly leading to increased, decreased, or no expression at all. Numerous cis-regulatory sequences, including the W-box (TTGACC) and GCC box (AGCCGCC), which act as binding sites for certain TFs like GhWRKY17 and OsERF922, negatively affect the abiotic stress response and tolerance. Therefore, this may be a good target for nucleotide-level alterations using the latest genome-editing technologies, which could increase crop tolerance to abiotic stress.
  - With *Arabidopsis thaliana* as an illustration osmotic stress tolerance and salt tolerance are negatively impacted by ANAC069's ROS-scavenging abilities because they reduce the expression of numerous stress-responsive genes (T genes). By relating with cis-elements and binding particularly to the DNA sequence C[A/G]CG[T/G], ANAC069 controls the expression of these genes (Liu et al., 2015; He et al., 2017). A mutation in this core region generates stress tolerance by preventing ANAC069 from controlling the gene.

21.6 PROTOCOLS FOR CRISPR-MEDIATED GENE EDITING IN PLANTS

Plant biotechnology has undergone a revolution, thanks to CRISPR-mediated gene editing, which provides accurate and effective tools for targeted genome alterations. An outline of the procedures for CRISPR-based gene editing in plants is given in this chapter, which covers important phases such as vector manufacturing, target selection, plant transformation, and genotype analysis. This guide intends to make it easier to use CRISPR technology in plant research and agricultural improvement initiatives by providing relevant references to important papers and recent developments (Figure 21.2).

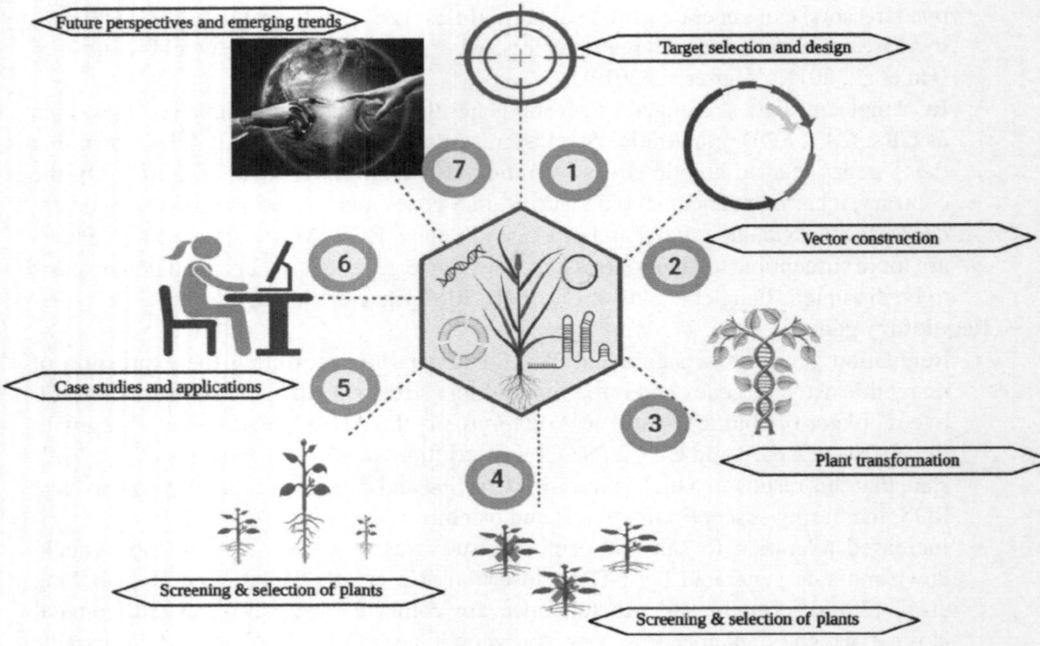

FIGURE 21.2 Protocols for CRISPR-mediated gene editing in plants.

- Target selection and design
  - Identification of target genes and genomic regions
  - Considerations for target specificity and off-target effects
  - Tools and algorithms for target prediction and design
- Vector construction
  - Selection of CRISPR/Cas systems (e.g., Cas9, Cas12a, Cas13)
  - Design of sgRNA or crRNA sequences
  - Assembly of expression cassettes and transformation vectors
- Plant transformation
  - Selection of plant species and cultivars
  - Transformation methods (e.g., *Agrobacterium*-mediated, particle bombardment)
  - Optimization of tissue culture and regeneration protocols
- Screening and selection of edited plants
  - Molecular methods for detecting mutations (e.g., PCR, T7 endonuclease I assay)
  - High-throughput genotyping techniques (e.g., amplicon sequencing, HRM analysis)
  - Phenotypic characterization of edited plants
- Validation and characterization of edited alleles
  - Sanger sequencing of target loci
  - Functional investigation of edited genes
  - Assessment of inheritance and stability of edited traits
- Case studies and applications
  - Models of successful CRISPR-associated gene editing in crop plants
  - Potential applications in crop improvement (e.g., disease resistance, stress tolerance, nutritional enhancement)
- Future viewpoints and emerging trends
  - Advances in CRISPR technology
  - Integration with other biotechnological approaches
  - Regulatory considerations and biosafety concerns

21.7 CURRENT ACHIEVEMENTS IN CRISPR-MEDIATED PLANT OXIDATIVE STRESS TOLERANCE

Several recent achievements demonstrate the successful adaptation of CRISPR tools for developing plant oxidative stress tolerance (Hossain and Dietz, 2016). Here are some examples:

Improving the expression of antioxidant enzymes: In a variety of plant species, researchers have used CRISPR/Cas9 to improve the expression of genes encoding SOD, CAT, and POX. CRISPR-mediated overexpression of these genes has been revealed to increase plant tolerance to oxidative stress by focusing on regulatory elements or TFs regulating antioxidant enzyme production.

Manipulation of ROS signaling pathways: Genes implicated in ROS signaling pathways, such as NADPH oxidases (RBOH) and ROS-scavenging enzymes, have been modified using CRISPR-based genome editing. Researchers have increased stress tolerance and plant responses to oxidative stress by adjusting ROS levels and signaling cascades.

Constructing TFs that respond to stress: CRISPR/Cas9 has been used to alter TFs that control the production of stress-inducible genes and antioxidant defense pathways, among other TFs involved in plant stress reactions. Through the manipulation of stress-responsive TFs, researchers have improved plant resistance to abiotic stressors such as oxidative stress.

Specific mutations in stress-related genes: To increase plant tolerance to stress conditions, CRISPR-mediated deletion or alteration of genes encoding proteins susceptible to oxidative stress has been used. Improved stress tolerance and increased agricultural output under harsh environmental circumstances have been attained by targeting genes involved in ROS generation, membrane integrity, or stress signaling.

Stress-tolerant traits: Precision engineering: Plants can be modified or introduced with stress-tolerant features, like disease resistance, heat tolerance, and drought tolerance, by precision genome editing made possible by CRISPR technology. Crop varieties with higher oxidative stress tolerance and better agronomic performance can be developed by researchers by merging numerous stress-tolerant characteristics or by improving gene expression networks. All things considered, these successes demonstrate how CRISPR-based genome editing can be utilized to enhance agricultural development initiatives and help plants become more resilient to oxidative stress.

Modifying genes involved in antioxidant defense pathways is one method to use the CRISPR technique to help plants gain oxidative stress toleration. For example, increasing or altering the expression of genes that code for enzymes like GPX, POX, CAT, and SOD can improve a plant's capacity to reduce oxidative damage and scavenge ROS in stressful environments. Rice is one plant that has been altered using the CRISPR technology to increase its resistance to oxidative stress. To improve plants' resistance to oxidative stress, researchers have edited genes involved in antioxidant pathways using CRISPR/Cas9 (Suzuki et al., 2012).

21.8 FUTURE DIRECTIONS IN CRISPR-BASED PLANT OXIDATIVE STRESS RESEARCH

Future research utilizing CRISPR technology to create oxidative stress tolerance in plants will examine cutting-edge tactics and targets to improve plant resistance to environmental stressors. Several possible directions for further investigation and advancement encompass:

- **Optimizing ROS signaling routes:** Examine the precise functions of regulatory elements and ROS signaling molecules in plant stress responses. Develop changes to ROS signaling pathways with CRISPR/Cas to maximize stress tolerance while maintaining growth and development.
- **Multigene editing and pathway engineering:** Target numerous genes concurrently implicated in stress-responsive signaling pathways, ROS detoxification, and antioxidant defense. Modify metabolic pathways to improve plant production and accumulation of secondary metabolites and antioxidant chemicals.
- **Enhanced CRISPR systems:** Investigate and refine new CRISPR/Cas variations (e.g., Cas12a, Cas13) with enhanced efficiency, adaptability, and targeting specificity for editing the genome of plants. Create cutting-edge CRISPR-based technologies for precise nucleotide-level changes and plant gene regulation, including prime editing and base editing.
- **Functional genomics and genome-wide screening:** Run genome-wide CRISPR screens to find new genes and regulatory components related to defense mechanisms and oxidative stress tolerance in plants. To clarify gene functions and regulatory networks behind plant stress responses, employ CRISPR/Cas-mediated knockdown and activation libraries.
- **Epigenome editing and transgenerational stress memory:** Examine how epigenetic changes affect how plants react to environmental signals and oxidative stress.
- Create CRISPR-based technologies to modify stress memory and heritable stress tolerance in plants by targeted epigenome editing.
- **Synthetic biology and genome engineering:** Use synthetic biology techniques to precisely control ROS levels and plant stress responses by designing metabolic circuits, sensors, and enzymes. Develop artificial promoters and gene regulatory networks to precisely express stress-responsive genes and pathways in agricultural plants.

- **Field trials and crop improvement:** Transfer model plant gains in oxidative stress tolerance by CRISPR to crops that are significant to agriculture. To evaluate the performance, agronomic characteristics, and stress resilience of CRISPR-edited crop types under various environmental circumstances, conduct field experiments and validation investigations.

Ultimately, utilizing cutting-edge genetic and genomic technologies to create stress-resilient crops with increased production, sustainability, and adaptation to changing climates and environmental challenges is the way that oxidative stress toleration in plants will be developed in the fortune using CRISPR technology (Jang and Joung, 2019).

21.9 APPLICATION AND ADVANTAGES OF CRISPR TOOL OVER OTHER GENE-EDITING TOOLS

There are several possible advantages to using CRISPR to help plants become more resilient to oxidative stress. For CRISPR-edited plants to have herbicide resistance, one method uses template-intervened and HDR of DSB. Furthermore, HDR-mediated gene targeting can be used to accomplish the pyramiding or stacking of several genes implicated in a stress reaction pathway. Additionally, plants can enhance their overall stress response and herbicide tolerance by editing numerous gene alleles using CRISPR. In-locus tags have been successfully inserted into rice using the tandem repeat-HDR technique (TR-HDR), with editing effectiveness varying from 3.4% to 11.4% (Jang and Joung, 2019; Kumar et al., 2023).

The possibility of CRISPR to produce unique allelic changes for the development of agricultural types resistant to drought is another possible advantage. For example, SAPK2 is a prospective gene for future crop development and is necessary for rice to be drought-tolerant. The rice crop that has a loss-of-function SAPK2 mutant is more vulnerable to drought and oxidative stressors than the wild type, which emphasizes the advantages of utilizing CRISPR to improve plant stress tolerance. In tomatoes, CRISPR/Cas9-mediated knockout can also cause parthenocarpic fruit formation in response to heat stress, providing yet another benefit for crop output in challenging circumstances. In the end, the CRISPR/Cas9 system can cause mutations in genes sensitive to abiotic stress deprivation to have any negative consequences on the target, which could increase crop yield and plant multigenic stress tolerance (Jang and Joung, 2019; Abdul Sami et al., 2021).

The CRISPR tool is superior to other gene-editing techniques in several ways, including accuracy, efficiency, adaptability, and scalability when used on plants.

- **Precision:** When corresponding to other gene-editing methods like random mutagenesis or conventional transgenic procedures, CRISPR/Cas systems allow for extremely accurate targeting of particular genomic regions, limiting off-target effects.
- **Efficiency:** Compared to more antiquated methods like ZFNs or TALENs, CRISPR-mediated gene editing enables efficient and focused changes of plant genomes, leading to better success rates and less time and resources.
- **Versatility:** By simply creating new gRNAs, CRISPR systems can be readily programmed to target distinct genes or genomic areas, giving researchers a flexible tool for a variety of plant biotechnology applications.
- **Scalability:** High-throughput applications utilizing CRISPR-based gene editing can target numerous genes at once or generate sizable mutant populations for functional genomics research or agricultural enhancement initiatives. Liu, D., Chen, M., et al. (2019).

The development of CRISPR/Cas systems, which provide incredibly effective, accurate, and adaptable tools for targeted genome changes, revolutionized gene editing. Particularly CRISPR/Cas9 became widely employed because of its comfort of use, programmability, and scalability. To facilitate gene deletion, knock-in, and other alterations, the CRISPR/Cas9 system utilizes a Cas9 nuclease guided by a sgRNA to generate DSBs at particular genomic sites. Base editing, prime editing, and upgraded CRISPR systems (such as Cas12a and Cas13) are only a few of the ongoing advancements and enhancements that have significantly increased the possibility of CRISPR techniques for specific plant genome engineering. The creation and widespread use of CRISPR/Cas systems, which have altered plant engineering and expedited efforts in crop progress and applicable genomics research, is the result of the evolution of gene-editing tools (Jang and Joung, 2019).

Although CRISPR/Cas systems have essentially replaced previous gene-editing methods, including ZFNs and TALENs, for plant gene editing, they have been created and used in the past for this purpose.

- **Complexity and design challenge:** ZFNs and TALENs need unique protein engineering for every target site, which can be laborious and time-consuming in terms of technological requirements. These proteins are complex to design and assemble, which makes it difficult for researchers who aren't experienced in protein engineering to use them widely.
- **Restricted targeting flexibility:** Because of the specificity of the DNA-binding domains, ZFNs and TALENs have limitations on the choice of target sites. This constraint may limit the scope of genomic loci that can be targeted and provide challenges for the development of effective editing tools.
- **Efficiency and off-target effects:** In contrast to ZFNs and TALENs, CRISPR/Cas systems, especially CRISPR/Cas9 offer greater specificity and efficiency. With just sgRNA, CRISPR/Cas9 may cause targeted DNA cleavage, streamlining the design process and minimizing off-target effects (Liu et al., 2019).
- **Cost and accessibility:** Because CRISPR/Cas systems rely on basic RNA molecules (sgRNAs) for target recognition rather than bespoke protein engineering, they are more affordable and accessible than ZFNs and TALENs. Because of its comfort usefulness and scalability, CRISPR technology has made genome-editing research more accessible and widely accepted.
- **Continuous innovation and advancements:** Base editing, prime editing, and improved CRISPR systems (such as Cas12a and Cas13) are just a few of the inventions and advancements that have made CRISPR/Cas technology possible. These developments have further increased the technique's potential for precise genome engineering in plants. These developments strengthen the case for CRISPR/Cas as the go-to gene-editing technique in plant science. Overall, CRISPR/Cas systems are now the strategy of preference for plant gene editing in modern research and applications, even though ZFNs and TALENs opened the door for targeted genome editing. These systems are also more affordable and easier to use.

21.10 CHALLENGES AND LIMITATIONS OF CRISPR TECHNIQUES IN PLANTS

Although the CRISPR-Cas9-gene-editing technique provides several obstacles and limits, it also presents a viable answer for improving agricultural output and stress resistance. The possibility of off-target effects, which might result in accidental mutations in non-targeted genes, is one of the main disadvantages. Moreover, a problem in crop breeding is still the effective delivery of CRISPR/Cas elements into reproductive cells, especially for resistant crops that are hard to modify. Large-scale uses of CRISPR/Cas9 in crop breeding are hindered by the intricate and time-consuming tissue culture and regeneration processes involved in genetic transformation techniques.

Nevertheless, when employing CRISPR/Cas for genome editing, the possible off-target risk cannot be disregarded. To ascertain whether phenotypic alterations are brought about by mutations in target genes, genotype–phenotype association studies should be performed on several independently edited individuals. Furthermore, some countries may not be able to use gene-edited products due to regulatory laws that label them as genetically modified organisms, thus it is important to debunk misconceptions about gene-edited crops through increased public awareness. However, the publication of criteria for the safety assessment of plants that have undergone gene editing without the introduction of foreign genes offers a foundation for the uniform development of modified crops as well as a point of reference for the future deregulation of modified crops.

According to science, after the transgenic label is removed, mutants created using CRISPR/Cas are identical to those created through natural mutation or traditional mutagenesis, making it a useful tool for crop breeding (Yang et al., 2022). Through exact, regulated genomic alterations, CRISPR technology offers a novel and revolutionary way to produce crops that are resistant to abiotic stress. It has been used to edit genes to withstand abiotic stresses like salinity, drought, and temperature. It has also been used to change genomes for biological research and crop breeding. There are certain disadvantages to the CRISPR method. The off-target impact, which is mostly brought on by the absence of designated sgRNAs and is far more common in plants (≥50%) than in animals or people, is the main drawback of CRISPR technology (Bao et al., 2019; Farhat et al., 2019).

To combat the stresses of salinity and drought, many genes are required (Wan et al., 2021). The abiotic stress results in difficult agronomic traits that are influenced by several genes (Jain et al., 2015). To successfully produce phenotypic abnormalities in response to abiotic stress, the target gene must be silenced in two or more copies (Wilson et al., 2019). Because sgRNA lacks specificity, CRISPR is unable to effectively edit a wide number of genes (Tang and Chu, 2017). CRISPR cannot work spontaneously because it needs many delivery routes to reach plant tissue, such as particle bombardment, magnetic nanoparticle non-cationic nucleic acid nanogel, cationic polymer, and viral conveyance.

Accepting CRISPR via the cell wall is a complex and time-consuming process because of a number of the constraints of these delivering tools, such as the large size of the nanoparticles and the low load-bearing capacity of viral conveyance (Boettcher and McManus, 2015; Li et al., 2022). Sometimes, when Cas9 comes from a diverse background, CRISPR needs to be adjusted for codons, which causes tissue resistance in the plant and prevents it from growing again (Rasheed et al., 2021).

21.11 ALTERATIONS TO CONFOUND THE LIMITATIONS OF CRISPR/CAS TECHNIQUES

A shortage of wild-type germplasm resources has been brought about by genetic variation and mutations, which has limited applied research on crop plants. Genome-editing technology can help solve these problems in several ways. To improve productivity, quality, and resilience to stress, hybrid breeders frequently use CRISPR/Cas9 to edit specific genes and eradicate unwanted traits. CRISPR/Cas9-mediated gene modifications, such as knock-in and replacement, also help in breeding by instructing desired alleles (Chen et al., 2019). The process of creating mutant plant CRISPR/Cas9 systems using *Agrobacterium*-interfered transformation works well, but it is costly, time-consuming, and resource-intensive. An alternative genome-editing technique that does not require tissue culture could be utilized (Manghwar et al., 2019). According to Jain et al. (2016), the creation of sgRNA libraries for the target plants may accelerate studies into the applicable genome, allow for the transformation of coding and non-coding genetic elements, and aid in the discovery of gene regulatory regions. Using several Cas9 protein variations allows CRISPR/Cas9 to be used in more applications. CRISPR interference is a technology that allows catalytically inactive Cas9 to obstruct desirable gene function.

21.12 CONCLUSION

The utilization of CRISPR/Cas techniques in enhancing plant tolerance to oxidative stress signifies a significant breakthrough with profound implications for agriculture and food security. Harnessing the accuracy and efficacy of the CRISPR/Cas system, scientists can meticulously modify plant genomes to introduce advantageous traits linked to oxidative stress resilience. An encouraging method entails pinpointing genes associated with antioxidant defense mechanisms and stress-responsive pathways. For example, editing genes responsible for crucial antioxidant enzymes like SOD, CAT, and POX could amplify their expression or activity levels, thereby strengthening the plant's capacity to neutralize ROS and alleviate oxidative harm. Moreover, CRISPR-mediated alteration or elimination of genes implicated in ROS production pathways could mitigate ROS buildup during stressful conditions McAnany, et al. (2022). By targeting genes linked to processes such as photosynthesis, respiration, or hormone signaling, which are crucial for combating oxidative stress, plants could acquire heightened resilience against environmental challenges like drought, high temperatures, and salinity.

Additionally, CRISPR technology facilitates precise manipulation of regulatory elements governing the expression of stress-responsive genes. This enables customized adjustments to enhance plant stress responses without compromising overall growth and development. Nevertheless, despite the promising potential of CRISPR in bolstering plant tolerance to oxidative stress, several obstacles persist, including concerns regarding off-target effects, regulatory frameworks, and ethical considerations. Continued research endeavors aimed at tackling these hurdles and refining CRISPR-based methodologies are imperative to fully harnessing the capabilities of this technology in sustainable agriculture, thereby protecting global food safety in the face of mounting environmental pressures.

ACKNOWLEDGMENT

All the authors are thankful to their respective universities and institutions for their valuable support. The corresponding author is thankful to Skill Development Centre, BEICH, RUSA2.0 (SDC Project order No. BU/RUSA/SDC/BEICH 2020/15; 28.12.2020), Bharathiar University, for providing financial support.

DATA AVAILABILITY

The authors confirm that the data supporting the findings of this study are available within the book chapter.

REFERENCES

Ahmad, P., Sarwat, M., & Sharma, S. (2008). Reactive oxygen species, antioxidants and signaling in plants. *Journal of Plant Biology*, *51*, 167–173.

Ahmad, P., Jaleel, C. A., Salem, M. A., Nabi, G., & Sharma, S. (2010). Roles of enzymatic and nonenzymatic antioxidants in plants during abiotic stress. *Critical Reviews in Biotechnology*, *30*(3), 161–175.

Apel, K., & Hirt, H. (2004). Reactive oxygen species: Metabolism, oxidative stress, and signal transduction. *Annual Review of Plant Biology*, *55*, 373–399.

Baena-González, E. & Hanson, J. (2017). Shaping plant development through the SnRK1-TOR metabolic regulators. *Current Opinion in Plant Biology*, *35*, 152–157.

Bao, A., Burritt, D. J., Chen, H., Zhou, X., Cao, D., & Tran, L. P. (2019). The CRISPR/Cas9 system and its applications in crop genome editing. *Critical Reviews in Biotechnology*, *39*(3), 321–336.

Baysal, C., Bortesi, L., Zhu, C., Farré, G., Schillberg, S., & Christou, P. (2016). CRISPR/Cas9 activity in the rice OsBEIIb gene does not induce off-target effects in the closely related paralog OsBEIIa. *Molecular Breeding*, *36*, 1–11.

Boettcher, M., & McManus, M. T. (2015). Choosing the right tool for the job: RNAi, TALEN, or CRISPR. *Molecular Cell*, *58*(4), 575–585.

Bozorgmehr, J. E. H. (2012). The effect of functional compensation among duplicate genes can constrain their evolutionary divergence. *Journal of Genetics*, *91*(1), 1–8.

Burke, E. J., Brown, S. J., & Christidis, N. (2006). Modeling the recent evolution of global drought and projections for the twenty-first century with the Hadley Centre climate model. *Journal of Hydrometeorology*, *7*(5), 1113–1125.

Chandrasegaran, S., & Carroll, D. (2016). Origins of programmable nucleases for genome engineering. *Journal of Molecular Biology*, *428*(5 Pt B), 963–989.

Chen, K., Wang, Y., Zhang, R., Zhang, H., & Gao, C. (2019). CRISPR/Cas genome editing and precision plant breeding in agriculture. *Annual Review of Plant Biology*, *70*, 667–697.
Chmielowska-Bąk, J., Izbiańska, K., & Deckert, J. (2015). Products of lipid, protein and RNA oxidation as signals and regulators of gene expression in plants. *Frontiers in Plant Science*, *6*, 140993.
Cong, L., Ran, F. A., Cox, D., Lin, S., Barretto, R., Habib, N., Hsu, P. D., Wu, X., Jiang, W., Marraffini, L. A., & Zhang, F. (2013). Multiplex genome engineering using CRISPR/Cas systems. *Science*, *339*(6121), 819–823.
Demidchik, V., Shabala, S. N., Coutts, K. B., Tester, M. A., & Davies, J. M. (2003). Free oxygen radicals regulate plasma membrane Ca^{2+} and K^+-permeable channels in plant root cells. *Journal of Cell Science*, *116*, 81–88.
Demidchik, V., Adobea, P., & Tester, M. A. (2004). Glutamate activates sodium and calcium currents in the plasma membrane of *Arabidopsis* root cells. *Planta*, *219*, 167–175.
Demidchik, V., Shang, Z., Shin, R., Thompson, E., Rubio, L., Laohavisit, A., Mortimer, J. C., Chivasa, S., Slabas, A. R., Glover, B. J., Schachtman, D. P., Shabala, S. N., & Davies, J. M. (2009), Plant extracellular ATP signalling by plasma membrane NADPH oxidase and Ca^{2+} channels. *The Plant Journal*, *58*, 903–913.
Demidchik, V., Cuin, T. A., Svistunenko, D., Smith, S. J., Miller, A. J., Shabala, S., Sokolik, A., & Yurin, V. (2010). Arabidopsis root K^+-efflux conductance activated by hydroxyl radicals: Single-channel properties, genetic basis and involvement in stress-induced cell death. *Journal of Cell Science*, *123*(Pt 9),1468–1479.
Dowling, D. K., & Simmons, L. W. (2009). Reactive oxygen species as universal constraints in life-history evolution. *Proceedings. Biological Sciences*, *276*(1663), 1737–1745.
Esmaeili, N., Shen, G., & Zhang, H. (2022). Genetic manipulation for abiotic stress resistance traits in crops. *Frontiers in Plant Science*, *13*, 1011985.
Fang, Y., Liao, K., Du, H., Xu, Y., Song, H., Li, X., & Xiong, L. (2015). A stress-responsive NAC transcription factor SNAC3 confers heat and drought tolerance through modulation of reactive oxygen species in rice. *Journal of Experimental Botany*, *66*(21), 6803–6817.
Farhat, S., Jain, N., Singh, N., Sreevathsa, R., Dash, P. K., Rai, R., Yadav, S., Kumar, P., Sarkar, A. K., Jain, A., Singh, N. K., & Rai, V. (2019). CRISPR-Cas9 directed genome engineering for enhancing salt stress tolerance in rice. *Seminars in Cell & Developmental Biology*, *96*, 91–99.
Foreman, J., Demidchik, V., Bothwell, J. H., Mylona, P., Miedema, H., Torres, M. A., Linstead, P., Costa, S., Brownlee, C., Jones, J. D., Davies, J. M., & Dolan, L. (2003). Reactive oxygen species produced by NADPH oxidase regulate plant cell growth. *Nature*, *422*(6930), 442–446.
Foyer, C.H., & Noctor, G. (2005). Redox homeostasis and antioxidant signaling: A metabolic interface between stress perception and physiological responses. *Plant Cell*, *17*(7), 1866–1875.
Georges, F., & Ray, H. (2017). Genome editing of crops: A renewed opportunity for food security. *GM Crops Food*, *8*(1), 1–12.
Gill, S. S., & Tuteja, N. (2010). Reactive oxygen species and antioxidant machinery in abiotic stress tolerance in crop plants. *Plant Physiology and Biochemistry*, *48*(12), 909–930.
Hahn, F., & Nekrasov, V. (2019). CRISPR/Cas precision: Do we need to worry about off-targeting in plants? *Plant Cell Reports*, *38*(4), 437–441.
Halliwell, B., & Gutteridge, J. M. (2015). *Free radicals in biology and medicine* (5th ed.). New York: Oxford University Press.
Hasanuzzaman, M., Bhuyan, M. B., Parvin, K., Bhuiyan, T. F., Anee, T. I., Nahar, K., Hossen, M. S., Zulfiqar, F., Alam, M. M., & Fujita, M. (2020). Regulation of ROS metabolism in plants under environmental stress: A review of recent experimental evidence. *International Journal of Molecular Sciences*, *21*(22), 8695.
He, L., Shi, X., Wang, Y., Guo, Y., Yang, K., & Wang, Y. (2017). Arabidopsis ANAC069 binds to C[A/G]CG[T/G] sequences to negatively regulate salt and osmotic stress tolerance. *Plant Molecular Biology*, *93*(4–5), 369–387.
Hossain, M. S., & Dietz, K. J. (2016). Tuning of redox regulatory mechanisms, reactive oxygen species and redox homeostasis under salinity stress. *Frontiers in Plant Science*, *7*, 548.
Hu, L., Liang, W., Yin, C., Cui, X., Zong, J., Wang, X., Hu, J., & Zhang, D. (2011). Rice MADS3 regulates ROS homeostasis during late anther development. *Plant Cell*, *23*, 515–533.
Jaganathan, D., Ramasamy, K., Sellamuthu, G., Jayabalan, S., & Venkataraman, G. (2018). CRISPR for crop improvement: An updated review. *Frontiers in Plant Science*, *9*, 985.
Jain, M. (2015). Function genomics of abiotic stress tolerance in plants: a CRISPR approach. *Frontiers in Plant Science*, *6*, 375.

Jain, P. K., Ramanan, V., Schepers, A. G., Dalvie, N. S., Panda, A., Fleming, H. E., & Bhatia, S. N. (2016). Development of light-activated CRISPR using guide RNAs with photocleavable protectors. *Angewandte Chemie International Edition*, *55*(40), 12440–12444.

Jang, G., & Joung, Y. H. (2019). CRISPR/Cas-mediated genome editing for crop improvement: Current applications and future prospects. *Plant Biotechnology Reports*, *13*, 1–10.

Jinek, M., Chylinski, K., Fonfara, I., Hauer, M., Doudna, J. A., & Charpentier, E. (2012). A programmable dual-RNA-guided DNA endonuclease in adaptive bacterial immunity. *Science*, *337*(6096), 816–821.

Joo, J. H., Bae, Y. S., & Lee, J. S. (2001). Role of auxin-induced reactive oxygen species in root gravitropism. *Plant Physiology*, *126*, 1055–1060.

Jung, C., Seo, J. S., Han, S. W., Koo, Y. J., Kim, C. H., Song, S. I., Nahm, B. H., Choi, Y. D., & Cheong, J. J. (2008). Overexpression of AtMYB44 enhances stomatal closure to confer abiotic stress tolerance in *Arabidopsis*. *Plant Physiology*, *146*, 623–635.

Kim, D., Alptekin, B., & Budak, H. (2018). CRISPR/Cas9 genome editing in wheat. *Functional & Integrative Genomics*, *18*, 31–41.

Kmiecik, P., Leonardelli, M., & Teige, M. (2016). Novel connections in plant organellar signalling link different stress responses and signalling pathways. *Journal of Experimental Botany*, *67*(13), 3793–3807.

Kumar, M., Prusty, M. R., Pandey, M. K., Singh, P. K., Bohra, A., Guo, B., & Varshney, R. K. (2023). Application of CRISPR/Cas9-mediated gene editing for abiotic stress management in crop plants. *Frontiers in Plant Science*, *14*, 1157678.

Koji, Y., Shiro, M., Michio, K., Mitsutaka, T., & Hiroshi, M. (2009). Antioxidant capacity and damages caused by salinity stress in apical and basal regions of rice leaf. *Plant Production Science*, *12*, 319–326.

Krishnamurthy, A., & Rathinasabapathi, B. (2013). Oxidative stress tolerance in plants: Novel interplay between auxin and reactive oxygen species signaling. *Plant Signaling & Behavior*, *8*(10), PMID: 23887492.

Li, J. F., Norville, J. E., Aach, J., McCormack, M., Zhang, D., Bush, J., Church, G. M., & Sheen, J. (2013). Multiplex and homologous recombination-mediated genome editing in Arabidopsis and Nicotiana benthamiana using guide RNA and Cas9. *Nature Biotechnology*, *31*(8), 688–691.

Li, Y., Wu, X., Zhang, Y., & Zhang, Q. (2022). CRISPR/Cas genome editing improves abiotic and biotic stress tolerance of crops. *Frontiers in Genome Editing*, *4*, 987817.

Liu, T. K., Song, X. M., Duan, W. K., Huang, Z. N., Liu, G. F., Li, Y., & Hou, X. (2014). Genome-wide analysis and expression patterns of NAC transcription factor family under different developmental stages and abiotic stresses in Chinese cabbage. *Plant Molecular Biology Reporter*, *32*(5), 1041–1056.

Liu, X. X., Fu, C., Yang, W. W., Zhang, Q., Fan, H., & Liu, J. (2016). The involvement of TsFtsH8 in *Thellungiella salsuginea* tolerance to cold and high light stresses. *Acta Physiologiae Plantarum*, *38*, 62.

Liu, D., Chen, M., Hua Cheng, B. M., Hu, R., Li, L., Trinh, C. T., Tuskan, G. A., & Yang, X. (2019). CRISPR/Cas9-mediated targeted mutagenesis for functional genomics research of crassulacean acid metabolism plants. *Journal of Experimental Botany*, *70*(22), 6621–6629.

Liu, H., Ding, Y., Zhou, Y., Jin, W., Xie, K., & Chen, L. L. (2017). CRISPR-P 2.0: An improved CRISPR-Cas9 tool for genome editing in plants. *Molecular Plant*, *10*(3), 530–532.

Liu, W., Zhu, X., Lei, M., Xia, Q., Botella, J. R., Zhu, J. K., & Mao, Y. (2015). A detailed procedure for CRISPR/Cas9-mediated gene editing in *Arabidopsis thaliana*. *Science Bulletin*, *60*, 1332–1347.

Ma, X., Zhang, Q., Zhu, Q., Liu, W., Chen, Y., Qiu, R., Wang, B., Yang, Z., Li, H., Lin, Y., Xie, Y., Shen, R., Chen, S., Wang, Z., Chen, Y., Guo, J., Chen, L., Zhao, X., Dong, Z., & Liu, Y.-G. (2015). A robust CRISPR/Cas9 system for convenient, high-efficiency multiplex genome editing in monocot and dicot plants. *Molecular Plant*, *8*, 1274–1284.

Malzahn, A., Lowder, L., & Qi, Y. (2017). Plant genome editing with TALEN and CRISPR. *Cell & Bioscience*, 7, 21.

Manghwar, H., Lindsey, K., Zhang, X., & Jin, S. (2019). CRISPR/Cas system: Recent advances and future prospects for genome editing. *Trends in Plant Science*, *24*(12), 1102–1125.

Mao, Y., Zhang, H., Xu, N., Zhang, B., Gou, F., & Zhu, J.-K. (2013). Application of the CRISPR–Cas system for efficient genome engineering in plants. *Molecular Plant*, *6*(6), 2008–2011.

McAnany, T., Rowe, D., Huang, J., & Cook, D. E. (2022). CRISPR-Cas9 ribonucleoprotein-mediated gene editing in the plant pathogenic fungus *Magnaporthe oryzae*. In *Plant Health Research and Extension Experiences for Undergraduates Poster Symposium*, Summer 2028.

Mittler, R. (2002). Oxidative stress, antioxidants and stress tolerance. *Trends in Plant Science*, *7*(9), 405–410.

Mittler, R., Vanderauwera, S., Gollery, M., & Van Breusegem, F. (2004). Reactive oxygen gene network of plants. *Trends in Plant Science*, *9*(10), 490–498.

Murata, Y., Pei, Z.-M., Mori, I. C., & Schroeder, J. (2001). Abscisic acid activation of plasma membrane Ca^{2+} channels in guard cells requires cytosolic NAD(P)H and is differentially disrupted upstream and downstream of reactive oxygen species production in abi1-1 and abi2-1 protein phosphatase 2C mutants. *The Plant Cell*, *13*, 2513–2523.

Nawkar, G. M., Maibam, P., Park, J. H., Sahi, V. P., Lee, S. Y., & Kang C. H. (2013). UV-induced cell death in plants. *International Journal of Molecular Sciences*, *14*, 1608–1628.

Ozturk, Z. N., Talame, V., Deyholos, M., Michalowski, C. B., Galbraith, D. W., Gozukirmizi, N., Tuberosa, R., & Bohnert, H. J. (2002). Monitoring large-scale changes in transcript abundance in drought- and saltstressed barley. *Plant Molecular Biology*, *48*, 551–573.

Panchy, N., Lehti-Shiu, M., & Shiu, S. H. (2016). Evolution of gene duplication in plants. *Plant Physiology*, *171*(4), 2294–2316.

Puchta, H. (2017). Applying CRISPR/Cas for genome engineering in plants: The best is yet to come. *Current Opinion in Plant Biology*, *36*, 1–8.

Rai, G. K., Khanday, D. M., Kumar, P., Magotra, I., Choudhary, S. M., Kosser, R., Kalunke, R., Giordano, M., Corrado, G., Rouphael, Y., & Pandey, S. (2023). Enhancing crop resilience to drought stress through CRISPR-Cas9 genome editing. *Plants*, *12*(12), 2306.

Ranieri, A., Castagna, A., Pacini, J., Baldan, B., Sodi, A. M., & Soldatini, G. F. (2003). Early production and scavenging of hydrogen peroxide in the apoplast of sunflower plants exposed to ozone. *Journal of Experimental Botany*, *54*, 2529–2540.

Rao, M. V., Paliyath, C., & Ormrod, D. P. (1996). Ultraviolet-B- and ozone-induced biochemical changes in antioxidant enzymes of *Arabidopsis thaliana*. *Plant Physiology*, *110*, 125–136.

Rasheed, A., Gill, R. A., Hassan, M. U., Mahmood, A., Qari, S., Zaman, Q. U., Ilyas, M., Aamer, M., Batool, M., Li, H., & Wu, Z. (2021). A critical review: Recent advancements in the use of CRISPR/Cas9 technology to enhance crops and alleviate global food crises. *Current Issues in Molecular Biology*, *43*(3), 1950–1976.

Ribeiro, T. P., Fonseca, F. L., de Carvalho, M. D., Godinho, R. M., de Almeida, F. P., Saint'Pierre, T. D., Rey, N. A., Fernandes, C., Horn, A., Jr., & Pereira, M. D. (2017). Metal-based superoxide dismutase and catalase mimics reduce oxidative stress biomarkers and extend life span of *Saccharomyces cerevisiae*. *Biochemical Journal*, *474*(2), 301–315.

Roy, B. R., Noren, S. K., Mandal, A. B., & Basu, A. K. (2011). Genetic engineering for abiotic stress tolerance in agricultural crops. *Biotechnology*, *10*(1), 1–22.

Sami, A., Xue, Z., Tazein, S., Arshad, A., He Zhu, Z., Ping Chen, Y., Hong, Y., Zhu, X. T., & Jin Zhou, K. (2021). CRISPR-Cas9-based genetic engineering for crop improvement under drought stress. *Bioengineered*, *12*(1), 5814–5829.

Sinharoy, S., Liu, C., Breakspear, A., Guan, D., Shailes, S., Nakashima, J., Zhang, S., Wen, J., Torres-Jerez, I., Oldroyd, G., Murray, J. D., & Udvardi, M. K. (2016). A *Medicago truncatula* cystathionine-β-synthase-like domain-containing protein is required for rhizobial infection and symbiotic nitrogen fixation. *Plant Physiology*, *170*(4), 2204–2217.

Sun, H., Huang, X., Fu, L., Huo, B., He, T., & Jiang, X. (2021). A powerful adaptive microbiome-based association test for microbial association signals with diverse sparsity levels. *Journal of Genetics and Genomics*, *48*(9), 851–859.

Suzuki, N., Koussevitzky, S. H. A. I., Mittler, R. O. N., & Miller, G. A. D. (2012). ROS and redox signalling in the response of plants to abiotic stress. *Plant, Cell & Environment*, *35*(2), 259–270.

Tang, J., & Chu, C. (2017). MicroRNAs in crop improvement: fine-tuners for complex traits. *Nature Plants*, *3*(7), 1–11

Wan, J., Huang, C., Li, C.-Y., Zhou, H.-X., Ren, Y.-L., Li, Z.-Y., Xing, L.-S., Zhang, B., Qiao, X., Liu, B., Liu, C.-H., Xi, Y., Liu, W.-X., Wang, W.-K., Qian, W.-Q., Mckirdy, S., & Wan, F.-H. (2021). Biology, invasion and management of the agricultural invader: Fall armyworm, *Spodoptera frugiperda* (Lepidoptera: Noctuidae). *Journal of Integrative Agriculture*, *20*(3), 646–663.

Wang, Y., Meng, Z., Liang, C., Meng, Z., Wang, Y., Sun, G., Zhu, T., Cai, Y., Guo, S., Zhang, R., & Lin, Y. (2017). Increased lateral root formation by CRISPR/Cas9-mediated editing of arginase genes in cotton. *Science China Life Sciences*, *60*, 524–527.

Wang, H., Gao, Z., Liu, X., Agarwal, P., Zhao, S., Conroy, D. W., Ji, G., Yu, J., Jaroniec, C. P., Liu, Z., Lu, X., Li, X., & He, X. (2018). Targeted production of reactive oxygen species in mitochondria to overcome cancer drug resistance. *Nature Communications*, *9*, 562.

Wilson, F. M., Harrison, K., Armitage, A. D., Simkin, A. J., & Harrison, R. J. (2019). CRISPR/Cas9-mediated mutagenesis of phytoene desaturase in diploid and octoploid strawberry. *Plant Methods*, *15*(1), 1–13.

Wu, D., Qi, T., Li, W., Tian, H., Gao, H., Wang, J., Ge, J., Yao, R., Ren, C., Wang, X., Liu, Y., Kang, L., Ding, S., & Xie, D. (2017). Viral effector protein manipulates host hormone signaling to attract insect vectors. *Cell Research*, *27*, 402–415.

Yang, N., Yan, L., Zheng, Z., Zhang, Y., Zhan, H., Tian, Y., Zhang, T., Li, R., Gong, X., Xu, M., Zhu, M., Wu, J., He, Y., & Zhao, Y. (2022). Editing gene families by CRISPR/Cas9: Accelerating the isolation of multiple transgene-free null mutant combinations with much reduced labor-intensive analysis. *Plant Biotechnology Journal*, *20*(2), 241.

Zafar, S., Li, Y. L., Li, N. N., Zhu, K. M., & Tan, X. L. (2019). Recent advances in enhancement of oil content in oilseed crops. *Journal of Biotechnology*, *301*, 35–44.

Zhang, J., Li, W., Xiang, T. T., Liu, Z. X., Laluk, K., Ding, X. J., Zou, Y., Gao, M. H., Zhang, X. J., Chen, S., Mengiste, T., Zhang, Y. L., & Zhou, J. M. (2010). Receptor-like cytoplasmic kinases integrate signaling from multiple plant immune receptors and are targeted by a *Pseudomonas syringae* effector. *Cell Host & Microbe*, *7*, 290–301.

Zhang, L., Ren, Y., Yang, T., Li, G., Chen, J., Gschwend, A. R., Yu, Y., Hou, G., Zi, J., Zhou, R., Wen, B., Zhang, J., Chougule, K., Wang, M., Copetti, D., Peng, Z., Zhang, C., Zhang, Y., Ouyang, Y., Wing, R. A., Liu, S., & Long, M. (2019). Rapid evolution of protein diversity by de novo origination in Oryza. *Nature Ecology & Evolution*, *3*, 679–690.

Zhang, H., Zhu, J., Gong, Z., & Zhu, J.-K. (2022). Abiotic stress responses in plants. *Nature Reviews Genetics*, *23*, 104–119.

Zhao, Y., Zhao, S., Cheng, Y., Luo, L., Fu, G., Guo, R., Zhao, H., Ma, H., Liu, Q., & Zhang, X. (2017). Rapeseed yield comparison between China and Canada in national trials of main producing areas. *Chinese Journal of Oil Crop Sciences*, *39*, 420–426.

22 Nanoparticle-Enhanced Strategies for Abiotic Stress Tolerance in Plants

Abhishek Singh, Shreni Agrawal, Richa Das, Vishnu D. Rajput, Tatiana Minkina, Saglara Mandzhieva, Mohamed S. Elshikh, Hassan Ragab El-Ramady, Abdel Rahman Mohammad Al Tawaha, and Karen Ghazaryan

22.1 INTRODUCTION

Nanotechnology is a rapidly expanding field of science that has garnered significant attention due to its connections with various other disciplines. Research and development in the field of nanoparticles (NPs) involves a wide range of disciplines, including chemistry, engineering, physics, and biology. These various fields come together to create innovative techniques for controlling and manufacturing NPs (Verma et al., 2022a, 2022b). The dimensions of these NPs range from 1 to 100 nm (Rajput et al., 2021). Nanotechnology involves the study and utilization of different NPs. Most NPs are produced using chemical and physical methods that involve the use of noble metals such as gold, silver, or platinum. Unfortunately, these processes are not environmentally friendly. It is crucial to develop a technique for manufacturing NPs that ensures the safety of both humans and the environment. Green synthesis methodologies for NPs have been developed that prioritize safety, ease of use, cost-effectiveness, repeatability, and scalability, all of which align with the principles of safety by design. Therefore, a variety of biological systems, including yeast, fungi, bacteria, and plant extracts, are commonly employed in green synthesis techniques for NP production. Plant-based NP green production has become the leading choice among environmentally friendly biological methods, thanks to its flexibility and ease of implementation. The enhanced chemical, physical, and biological characteristics and functions of NPs are a result of their higher surface area-to-volume ratio. By utilizing nanofertilizers, plants receive precise and targeted nutrients in the form of NPs, resulting in a significant boost in harvest yield (Dimkpa and Bindraban, 2018). Three categories of nanofertilizers cater to different scales, nano, micro, and macro, to support plant growth by providing essential nutrients (Figure 22.1).

NPs, with a diameter of 100 nm or less, can be applied in either powder or liquid form. By enhancing the accessibility of nutrients, they enhance plant growth and yield. In brief, nanofertilizers possess the following noteworthy characteristics: (1) utilizing foliar and soil treatments that effectively enhance plant growth; (2) serving as cost-effective and sustainable sources of plant nutrients; (3) offering a high rate of fertilization; and (4) playing a crucial role in reducing pollution levels (Adelere and Lateef, 2016). An additional groundbreaking fertilizer that assists in purifying contaminated water is the nanofertilizer. This review offers a comprehensive look at NPs, exploring their production and their practical applications as nanofertilizers. It also examines the effects of nanofertilizers on soil quality and plant growth, as well as their interactions with

DOI: 10.1201/9781003457237-22

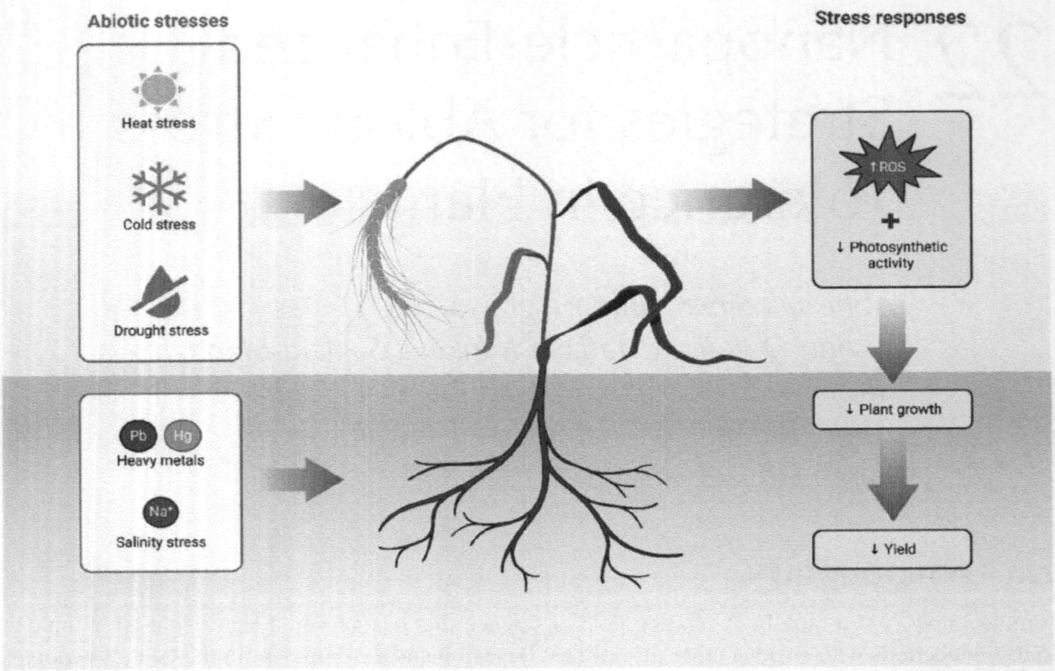

FIGURE 22.1 Different types of abiotic stress, like heat stress, cold stress, drought stress, heavy metals, and salinity stress, adversely harm the plant because of the production of reactive oxygen species, leading to decreased photosynthetic activity, plant growth, and yield.

various plant tissues. This assignment appears to be the first instance where all of these concepts have been brought together for discussion.

22.2 APPROACHES OF NP SYNTHESIS

There are two widely used methods for constructing NPs with specific shapes and sizes: the top-down approach and the bottom-up approach (Patil and Chandrasekaran, 2020). Although they follow distinct synthesis principles, both of these methods yield NPs that possess the desired properties (Gwo et al., 2016; Patil and Chandrasekaran, 2020). Using a biologist's approach, the top-down methods break down larger materials into progressively smaller pieces until they eventually form NPs (Gwo et al., 2016; Patil and Chandrasekaran, 2020). The manufacturing processes for these NPs involved various techniques such as photolithography, grinding, sputtering, and milling (Meyers et al., 2006). Every technique has its own set of pros and cons when it comes to the efficiency of producing NPs. An achievable strategy that can generate a sizable quantity of NPs is the top-down method. On the contrary, top-down approaches have some disadvantages, such as surface imperfections and the potential for damage to NPs (Pérez-de-Luque, 2017). In certain cases, the top-down synthesis of NPs has limitations due to the influence of the NPs' surface architecture on their optical, physical, and chemical functions (Khan et al., 2019). One popular technique for producing NPs at the moment is the bottom-up method. It involves the blend or assembly of individual atoms, molecules, or clusters to create a wide variety of NPs (Decarolis et al., 2020). Various methods are employed in the creation of NPs, including self-assemblage of monomer and polymer molecules, nanostructural precipitation, laser pyrolysis, flame spraying synthesis, sol–gel processing, and bio-assisted synthesis approach (Dhand et al., 2015). So, the bottom-up approach offers a versatile method for creating nanoclusters with a vast array of potential applications.

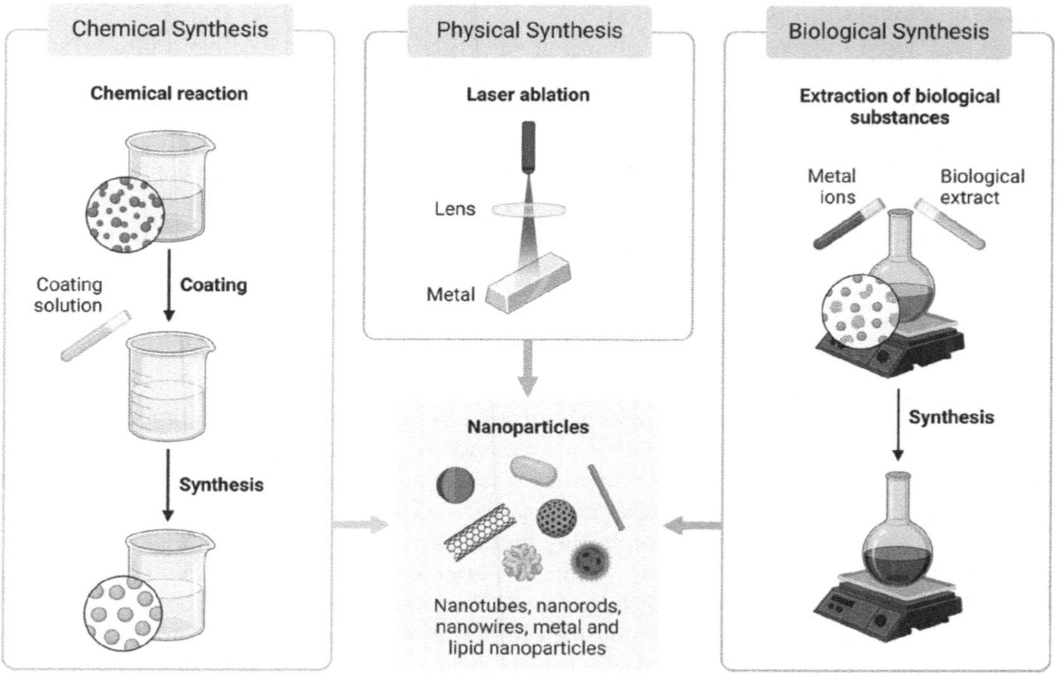

FIGURE 22.2 Nanoparticles can be synthesized from the following three methods: chemical, physical, and biological methods.

22.2.2 Methods for NP Synthesis

22.2.2.1 Physical Methods

Various physical methods can be used for NP synthesis, including plasma arcing, pyrolysis with sprays, ball milling, electron irradiation, sputter accumulation, lithography methods, layer-by-layer growth, thermal evaporation, and pulsed laser desorption. These techniques have been discussed in a recent study (Srivastava et al., 2021). Through these techniques, metal NPs are generated via evaporation condensation in a tube furnace under specific atmospheric conditions (Figure 22.2).

Various physical methodologies are used in the creation of nanomaterials, such as deposition, sputtering, ball milling, and plasma-based processes (Dhand et al., 2015). Many of these techniques exhibit a frustratingly sluggish pace in generating metal NPs. As an expert in the field, it is worth noting that in ball milling processes, the yield of nanomaterials is 50% (Seetharaman et al., 2018). According to reports, a significant portion of the sputtered substance has a size smaller than 100 nm, leading to a broad distribution of particle sizes. Methods such as laser ablation and plasma techniques consume a significant amount of energy. Physical methods tend to be quite costly due to their slow production rate, uneven size distribution, and higher energy consumption (Seetharaman et al., 2018).

22.2.2.2 Chemical Methods

Several chemical methods have been suggested for synthesizing NPs, such as pyrolysis, polyol synthesis, sol–gel method, hydrothermal synthesis, microemulsion, chemical reduction, and deposition of chemical vapor (Darroudi et al., 2013) (Figure 22.2). Furthermore, the utilization of harmful chemicals and reagents during the synthesis process results in the generation of waste that poses risks to both human health and the environment (Darroudi et al., 2013; Barzegar et al., 2023). Consequently, the utilization of these NPs is limited to the realm of biology.

22.2.2.3 Biological or Green Methods

The field of green nanotechnology is advancing quickly, aiming to produce innovative NPs with minimal harm to the environment. Using biological methods, NPs can be synthesized by harnessing the power of natural reducing and stabilizing agents. This is a more environmentally friendly and cost-effective alternative that utilizes both physical and chemical methods. It requires minimal energy and does not involve any harmful substances. The biological synthesis of NPs involves the use of various organisms, ranging from single-celled bacteria and fungi to multicellular fungi, algae, and plant constituents (Mittal et al., 2013; Panpatte et al., 2016; Singh et al., 2016; Patil and Chandrasekaran, 2020) (Figure 22.2).

22.2.2.3.1 Bacteria

Commercial biotechnology has extensively utilized bacterial species in applications such as bioremediation, bioleaching, and genetic engineering (Hoffmann et al., 1995). Biologists often find bacteria to be valuable options when it comes to preparing NPs, thanks to their ability to potentially reduce metal ions (Du et al., 2023). There is a wide variety of bacterial species that are utilized for the production of metals and NPs. Extensive research has been conducted on the synthesis of metal and metal oxide NPs using prokaryotic bacteria and actinomycetes. In the process of cadmium NP synthesis, the spherical morphology formation is mainly attributed to the presence of glutathione and cysteine desulfhydrase within *Escherichia coli* (Eroglu and Metin, 2023; Wang et al., 2023). Understanding the final size of the NP is influenced by its location within cells or outside of cells, as well as its interaction with bioactive molecules (Thakkar et al., 2010). NPs produced through extracellular synthesis by bacteria tend to be larger compared to those produced through intracellular synthesis. Biologists have utilized bacterial systems to synthesize nanomaterials such as selenium, lead sulfide, zinc sulfide, and ferrous oxide (Patil and Chandrasekaran, 2020). Many naturally occurring active molecules have been found to stabilize and decrease the size of NPs. Proteins found inside the cytosol and cell wall can stabilize and reduce bacterial NPs. Furthermore, certain sugars possess properties that can reduce and stabilize. The presence of amino acids in the cell walls and within the cells provides a protective shield, reducing the risk of toxicity toward mammalian cells (Markus et al., 2016). These dynamic compounds, present in various bacterial strains, have the ability to interact with metal ions, resulting in a reduction in their size. These interactions between the metal ions result in the formation of more complex structures, such as spherical NPs.

Bacterial NP synthesis has gained popularity due to the ease of regulating bacteria (Adelere and Lateef, 2016). Various bacteria are commonly used for the biosynthesis of different NPs. These include *Aeromonas* sp. *SH10*, *Arthrobacter gangotriensis*, *Bacillus cereus*, *Bacillus amyloliquefaciens*, *Bacillus cecembensis*, *Bacillus indicus*, *Corynebacterium* sp. *SH09*, *Enterobacter cloacae*, *E. coli*, *Geobacter* spp., *Phaeocystis antarctica*, *Pseudomonas proteolytica*, *Lactobacillus casei*, *and Shewanella oneidensis* (Singh et al., 2018).

22.2.2.3.2 Fungi

Biological processes involving fungi can be utilized to produce NPs made of metal or metal oxides. This method is highly effective in creating NPs that are uniform in size and have distinct morphological characteristics. Multiple intracellular enzymes give them an edge as biological agents for producing metal and metal oxide NPs (Hoffmann et al., 1995). It has been observed that fungi are capable of synthesizing a greater number of NPs compared to bacteria (Hoffmann et al., 1995). Biologists have discovered that fungi have a wide range of applications due to the presence of proteins, reducing constituents, and enzymes on the surface of their cells (Narayanan and Sakthivel, 2011). Enzyme reductase found in the fungal cell wall and cells plays a crucial role in the formation of metallic NPs through enzymatic reduction (Mohanpuria et al., 2008). Various fungal species are utilized to synthesize NPs such as gold, silver, zinc oxide, and titanium dioxide.

It has been shown that proteins inside cells can contribute to the production of NPs. Biologists have observed that certain enzymes found within *Trichothecium* and *Verticillium luteoalbum* have the ability to produce gold nanospheres and nanorods (Ahmad et al., 2006; Gericke and Pinches, 2006). Just like other environmentally friendly methods, fungi can produce NPs with a wide range of potential applications, including medicine and optoelectronics (Birla et al., 2009). The potential of NPs in the medical and therapeutic fields is captivating and calls for additional research. According to a study conducted by Philip in 2009, it was found that NPs derived from edible mushroom extracts possess chemotherapeutic properties that are comparable to the original extracts. Some fungi have amino acid residues, such as cysteine, that can assist in the formation of NPs (Philip, 2009). Similar to a biologist's perspective, fungi utilize amino acids found within their cell wall for stabilization and capping, much like bacteria. In addition, these NPs have been found to be non-toxic in therapeutic applications, unlike traditionally synthesized NPs (Philip, 2009).

22.2.2.3.3 Yeast

Yeasts are microorganisms that consist of a single cell and belong to the eukaryotic domain. There have been around 1,500 yeast species identified, according to Yurkov et al. (2011). Several laboratories have achieved successful NP synthesis by utilizing yeast.

Yeast that can tolerate silver, along with a broth made from *Saccharomyces cerevisiae*, can produce silver and gold NPs (Patil and Chandrasekaran, 2020). Various species are utilized to generate a vast array of metallic NPs.

Using yeast cells instead of bacteria alone allows for the synthesis of a wide range of nanosystems. Various yeast species play a crucial role in the synthesis of NPs, including gold, silver, ferrous oxide, cadmium sulfide, lead sulfide, antimony, and selenium. Biologists often use live cells or cell extracts to synthesize NPs, which can include common nanomaterial composites such as silver and gold. Proteins isolated from commercial yeast were used to successfully synthesize silver chloride NPs (Abdel-Hadi et al., 2014). The NPs were created by incubating commercial yeast extracts with precursor solutions for a period of 24 hours (Abdel-Hadi et al., 2014). Once the incubation period was complete, the solution containing the NPs was carefully collected and filtered using sterile techniques to eliminate any potential contaminants. Furthermore, it was shown that the reduction of silver chloride into NPs is a result of primary amines found in specific proteins that are present in the yeast cell wall. Furthermore, it was found that this NP exhibited efficacy against mycobacteria. In a study, it was found that yeast cells known as MKY-3 have the ability to synthesize NPs, specifically those that are silver-tolerant (Philip, 2009). These nanomaterials come in different shapes and sizes, which are influenced by various production factors such as the concentration of silver chloride and pH levels. As per the findings of this study, the reduction of extracellular silver chloride was attributed to the presence of biochemical reducing agents that were excreted (Philip, 2009). Although yeast has the ability to produce NPs through various methods, extracellular synthesis is not among them. Most research teams stated that their models of synthesis occurred within cells, with existing enzymes in the cell being responsible for constructing the nanosystem. According to previous research conducted by Kowshik et al. (2002), it was found that *Schizosaccharomyces pombe* and *Torulopsis* sp. have the ability to synthesize cadmium sulfide and lead sulfide NPs within their cells. In contrast to their previous findings, Kowshik et al. (2002) showed that a particular species of phytochelatin synthase could be responsible for the intracellular production of NPs. It is worth noting that the majority of the articles reviewed stated that the synthesized NPs showed impressive performance in various biomedical applications (Kowshik et al., 2002). Given the wide range of potential uses, NPs have proven to be incredibly versatile in various applications. The silver NPs produced by *S. cerevisiae* have been found to effectively kill mycobacteria in culture (Sivaraj et al., 2020). Yeast has the ability to produce proteins with essential amino acids, which helps to reduce and maintain the stability of the NP, similar to how living organisms function. Quinones, which are organic molecules derived from aromatic compounds, have been found to have a distinct impact on yeast and their ability to produce NPs. When the pH inside the cell becomes more alkaline, the oxidoreductases are activated and can start reducing the metal ions. Certain substances with strong nucleophilic properties and redox capabilities, along with quinones, play a valuable role in facilitating the conversion of disordered metal ions into intricate NPs (Jha et al., 2009; Faramarzi et al., 2020).

22.2.2.3.4 Algae

Algae, a fascinating group of eukaryotic organisms, have the remarkable ability to carry out photosynthesis, even though they are not classified as plants. These algal species, whether single-celled or multicellular, thrive in aquatic environments and produce chlorophyll. However, they do not possess the usual plant features such as stems, leaves, and vascular tissues. Some species, such as Anabaena, can be dangerous if their cells and toxins are consumed, while others, like *Spirulina*, have therapeutic potential due to their high concentration of natural nutrients (Carmichael et al., 1990; Khan et al., 2005). In recent years, researchers have discovered that various types of algae have the ability to catalyze the production of nanomaterials (Khan et al., 2005). When creating nanomaterials from various types of algae, the samples undergo a thorough drying process before being finely ground into powder. This powder is then mixed with water. After approximately 24 hours, the mixture is allowed to settle before undergoing the filtration process. After the material has been filtered, it is combined with the precursor to the nanomaterial and allowed to incubate at a precise room temperature. There is a change in color that indicates the formation of nanomaterial. Some algal species, such as *Tetraselmis kochinensis*, have the ability to facilitate the formation of gold NPs within their cells using enzymes found in their cytoplasm and cell wall (Senapati et al., 2012). On the contrary, when examining two different types of algae, there are numerous possibilities to consider in terms of NP structure.

According to a recent study by Mukherjee et al. (2021), certain species such as *Cystophora moniliformis*, *Leptolyngbya valderianum*, and *Scenedesmus* sp. have been found to produce NPs with a spherical shape. This finding aligns with the role of fungi and bacteria in catalyzing NP formation (Mukherjee et al., 2021). Sinha et al. (2015) also showed that *Pithophora oedogonia* has the ability to produce both cubical and hexagonal silver NPs, in addition to nanospheres (Sinha et al., 2015). Green agents that contain material similar to algae not only have a shared aesthetic, but their bioactive molecules also exhibit striking similarities, as noted by Mukherjee et al. In the realm of biology, numerous algal species have been found to be capable of synthesizing NPs. Interestingly, most of these NPs are regulated and reduced by enzymes and proteins present in the cytosol or on the cell membrane (Mukherjee et al., 2021). Lastly, the bactericidal properties of NPs synthesized by algal species are quite remarkable. Algae share some similarities with other classes of bioactive molecules in terms of NP production, but they also have their own unique characteristics. Algae possess the remarkable capability of reducing and stabilizing NPs, owing to the presence of polysaccharides and protein residues. Algae possess a remarkable range of phytochemical diversity, offering numerous advantages. Certain types of algae, like *Sargassum tenerrimum*, have a wide variety of bioactive compounds in them. These compounds include phenolic compounds, alkaloids, amino acids, carbohydrates, sterols, flavonoids, tannins, and saponins. Through purification, these compounds have the ability to alter the characteristics of nanomaterials, such as their shape, size, and function, in a distinct manner. Studying and understanding algae is an important part of exploring their potential applications in nanomaterial synthesis (Saleh, 2020).

22.2.2.3.5 Plants

Plants have the ability to accumulate different amounts of heavy metals inside their tissues. As a result, the use of plant extracts in biosynthesis processes has gained significant attention due to their higher efficiency, simplicity, cost-effectiveness, and practical feasibility compared to traditional NP manufacturing methods (Akhtar et al., 2013). According to Singh et al. (2018), plants may be employed in a 'one-pot' synthesis method to both decrease and stabilize metallic NPs concurrently. Scientists have utilized plant leaf extracts in a sustainable method called green synthesis to create

NPs made of metal or metal oxide. These NPs were then studied to explore their possible applications (Singh et al., 2018).

Previous studies have already documented the presence of NPs composed of metals such as gold, selenium, and copper. Various plant materials have been shown to possess stabilizing, capping, and reducing agents that can be utilized for the production of silver NPs (Singh et al., 2018; Huston et al., 2021). Vegetable oil, tea polyphenols, *Carpesium cernuum*, *Cannabis sativa*, and black currant, among other substances, have the potential to function as stabilizing, capping, and reducing agents (Huston et al., 2021). Moulton et al. (2010) described the production of colloidal silver NPs using tea leaves that are rich in polyphenols. Their synthesis technique is analogous to previously published methods. The group was provided with tea powder, which was derived from dehydrated and pulverized tea leaves. The tea powder was subjected to boiling and filtration, followed by the addition of silver nitrate to synthesize silver NPs. NP production was confirmed using transmission electron microscopy (TEM). Subsequently, membrane integrity and cell viability studies were employed to ascertain the toxicity of the NPs toward live beings. The results were promising, since the NPs were determined to be non-toxic and potentially capable of being compatible with living organisms (Moulton et al., 2010). Vegetable oil is a commonly used plant-derived substance that is present in most households. In their study, Kumar et al. (2008) utilized free radicals present in home paints derived from certain vegetable oils to achieve a little variation in the environmentally friendly production of metallic NPs (Kumar et al., 2008). Silver benzoate is a well-known silver compound that is commonly employed in the production of silver NPs. In addition, the research team utilized the exchange of free radicals during oxidative oil drying to transform them into silver NPs.

The group utilized the aldehydes and fatty acids obtained from the oils as stabilizing agents for the reactions, while employing alkyd resin as a protective agent. Their response prompted the creation of antimicrobial paints that incorporate silver NPs (Kumar et al., 2008). *Aloe vera* is a botanical substance that can be employed in the production of NPs. *Aloe vera* has been utilized for years in alternative medicine to address a diverse array of ailments, and it is now prevalent in the treatment of sunburns. Fardsadegh and Jafarizadeh-Malmiri (2019) utilized *Aloe vera* to effectively synthesize selenium NPs possessing antifungal and antibacterial characteristics (Fardsadegh and Jafarizadeh-Malmiri, 2019). The Fardsadegh group employed the hydrothermal method, which is widely recognized as a traditional synthesis process that is also very environmentally benign (Fardsadegh and Jafarizadeh-Malmiri, 2019). The bioactivity of chemicals present in spices can be harnessed similarly as it is harnessed in plant leaves. The aromatic fruit of the *Myristica fragrans* tree is indigenous to Indonesia. When the fruit is dehydrated and pulverized, it produces the spices nutmeg and mace. The pericarp, which is the non-seed component of the fruit, may be utilized for the synthesis of metallic NPs. The dried substances were pulverized and subsequently combined with water, then heated with either silver nitrate or cupric oxide. Sasidharan et al. (2020) conducted another study that showed the importance of flavonoids, quercetin, and phenols from *M. fragrans* in stabilizing and reducing the NPs. Furthermore, the silver NPs exhibited exceptional efficacy in dismantling the bacterial cell walls. Furthermore, Sasidharan et al. (2020) observed that copper NPs exhibited effective catalytic activity in the production of triazole rings (Sasidharan et al., 2020).

22.3 CHARACTERIZATION PARAMETERS OF NPs

22.3.1 Size

Particle size is the fundamental and crucial parameter for characterizing NPs. It determines the size and particle distribution of a substance and indicates if it falls inside the nano or micro scale. Electron microscopy is the conventional technique used to determine the size and distribution of particles. Scanning electron microscopy (SEM) and TEM pictures are used to measure particles and clusters, whereas laser diffraction methods are used to measure bulk samples in the solid phase (Wagner Ferreira Sabará et al., 2017). The measurement of light scattering from liquid particles is conducted using centrifugation and photon correlation spectroscopy. Scanning mobility particle sizers (SMPSs) are preferred due to their ability to offer rapid and precise measurements, especially when compared to other methods that struggle to image particles in the gaseous phase (Marsalek, 2014).

22.3.2 Surface Area

The surface area of NPs can also serve as a means of characterization. The surface area-to-volume ratio of an NP has a significant impact on its performance and attributes (Marsalek, 2014). The BET analysis is a commonly employed method for quantifying surface area. Particle surface area analysis in the liquid state may be achieved by a straightforward titration method, however it is a laborious and time-consuming process. NMR spectroscopy is utilized for this purpose. The surface area of NPs in the gas phase is quantified by the utilization of a differential mobility analyzer and a modified SMPS (Wagner Ferreira Sabará et al., 2017).

22.3.3 Surface Morphology

The various geometries and surface topographies of the NPs play a crucial role in harnessing their unique characteristics. Surfaces can exhibit many characteristics such as crystalline, amorphous, uniform, or uneven shapes including spheres, flat discs, cylinders, tubes, cones, and other irregular forms. Electron microscopy imaging techniques, such as SEM and TEM, are commonly employed for surface analysis (Wiechers and Musee, 2010). Electron microscopy imaging involves the electrostatic capture or filtering of liquid-phase particles, whereas gas-phase particles are first gathered on the surface and subsequently examined.

22.3.4 Composition

The purity and usefulness of the NP depend on its chemical or elemental makeup. The presence of elevated amounts of secondary and undesired components inside the NP can lead to reduced efficiency, as well as the potential for secondary reactions and contamination. X-ray photoelectron spectroscopy is commonly employed to determine the composition (Sharma and Rao, 2013). Wet chemical analysis includes atomic emission spectroscopy, ion chromatography, and mass spectrometry, which are essential components of some methodologies. The collection of gas-phase particles is achieved by the use of filtration or electrostatic methods, while analysis is conducted using spectrometric and wet chemical techniques (Bzdek et al., 2011).

22.3.5 Crystallography

Crystallography is the scientific investigation of the many components, such as atoms and molecules, that make up crystals. NP crystallography is achieved via electron or neutron diffraction, as well as powder X-ray techniques (Yano et al., 1996).

22.4 CHARACTERIZATION TOOLS AND TECHNIQUES OF NPs

X-ray diffraction (XRD), energy-dispersive spectroscopy (EDS), UV–Visible spectroscopy (UV–Vis), SEM, TEM, Fourier transform infrared (FTIR) spectroscopy, atomic force microscopy (AFM), dynamic light scattering, extended X-ray absorption fine structure spectroscopy, and nanoscale infrared (IR) spectroscopy are commonly employed analytical techniques for characterizing newly synthesized NPs (Sepeur, 2008; Strasser et al., 2010).

22.4.1 FTIR Spectroscopy

The FTIR technique of IR spectroscopy is very efficient and user-friendly. Context is crucial for evaluating chemical concentration, distinct functional groups, atomic composition of biological NP samples, and their surface chemistry in biological systems (Cao, 2004). FTIR can determine if biomolecules are involved in NP production. Additionally, it can provide insight into the specific biomolecules present in the sample (Zhang et al., 2022). The FTIR approach is based on the distinct kinds and frequency of bond vibrations in molecules (Eid, 2022). FTIR stands out due to its ability to be customized for different workspaces, its high-quality spectral images, its enhanced sensitivity, and its advanced cube corner interferometer (Eid, 2022).

22.4.2 UV–Visible Spectroscopy

The UV–visible spectroscopy method is frequently used to confirm the stability and production of metallic NPs. Absorbance at wavelengths ranging from 230 to 800 nm may be utilized to verify the production of NPs with sizes ranging from 1 to 100 nm (Awasthi et al., 2017). The absorption band of surface plasmon resonance is a result of the presence of unbound electrons. UV absorption spectroscopy may also be employed to accurately measure the quantity of particles in colloidal solutions, in addition to confirming their production (Tomaszewska et al., 2013). The measurement and description of particle size are achieved by the examination of absorption spectra, which can also provide qualitative insights. Spectroscopy involves the measurement and interpretation of electromagnetic radiation to analyze the changes in energy states of atoms, ions, and molecules in a sample (Hu et al., 2006).

The absorption of ultraviolet light by the molecule results in the excitation of electrons from the ground state to a higher energy state, which facilitates the examination of NP samples. UV–Vis spectroscopy is notable for its minimal upkeep and inclusion of double choppers, sealed optics, and adjustable spectral bandwidth that may be as low as 0.2 nm (Amendola and Meneghetti, 2009).

22.4.3 IR Spectroscopy

IR spectroscopy employs reproducible tools to analyze the molecular structure of substances by detecting and measuring their many resonant and vibrational modes (Kiefer et al., 2015). It possesses the capacity to unveil the chemical composition and bonding configuration. A distinctive spectrograph is employed to compare the strength of radiation and the frequency of the spectrum with global benchmarks for the identification of molecules (López-Lorente and Mizaikoff, 2016). IR spectroscopy may be used to both characterize and identify different chemical species. Nevertheless, it is capable of resolving features with dimensions ranging from 5 to 10 m. The NanoIR probe is a tool for chemical investigation at the nanoscale, utilizing nano-IR spectroscopy (López-Lorente and Mizaikoff, 2016). By integrating key elements of AFM with IR spectroscopy, this technique enables the capture of IR spectra at spatial resolutions ranging from 50 to 200 nm (Casalini et al., 2017). NanoIR is known for its significant property of multifunctional measurements (Baudot et al., 2010). This device provides a precise and clear representation of samples by integrating their mechanical, thermal, spectroscopic, and topographic properties.

22.4.4 X-Ray Diffraction

XRD is a highly energetic technique used to analyze the crystalline structure of both organic and inorganic substances. XRD is considered user-friendly due to its non-destructive nature, ease of use, high sensitivity, ability to provide detailed profiles, and straightforward construction (Vorontsov and Tsybulya, 2018). Furthermore, XRD has been employed in a wide range of fields such as forensic sciences, pharmaceuticals, geology, and polymer science. It is used to analyze the chemical composition by measuring qualitative and quantitative factors like the degree of crystallinity. This provides accurate information about the atomic arrangement at the interface (Stefanos Mourdikoudis et al., 2018). The crystal structures provide a description of the atomic arrangement, position, and strength of the diffraction peaks. X-rays possess a wavelength that is the size of individual atoms (Holder and Schaak, 2019). Consequently, it is commonly used to study the structure of nanomaterials. Every individual powder sample scatters a solitary X-ray beam. The dispersion of beams reflected by any crystal can generate a variety of diffraction patterns (Jemal et al., 2017). The highest level of intensity of the peak is measured at a certain angle where X-rays are dispersed. This peak is indicative of the crystal's structure and physicochemical qualities. The essential concept of XRD is based on Bragg's Law (Waseda et al., 2011). The Joint Committee, which is responsible for the Powder Diffraction Standards, maintains a database known as JCPDS. This database is used to cross-reference the patterns of individual diffraction beams. Elemental analysis of metal NPs is commonly performed using EDS (Strasser et al., 2010).

22.4.5 SEM

SEM is an adaptable and non-invasive analytical technique that employs a microscope equipped with a spacious specimen chamber and an 8.5 mm working distance. This extended working distance is achieved by the employment of angled detectors and a sharp conical objective lens (Vladár and Hodoroaba, 2020).

SEM is a method used to measure the dimensions and morphology of nano or microstructures present on the surface of materials (Batool et al., 2016). It's a great instrument for figuring out how things are put together. It is an excellent tool for analyzing the structure of objects. The tool is efficient, uncomplicated, and direct to utilize, yielding precise outcomes. Moreover, SEM is employed in several fields such as the investigation of biological molecules, biomedicine, and materials research (López Gutiérrez et al., 2022). Biomedical sciences often utilize surface characterization of cells, organs, and tissues from 3D pictures. It enhances our comprehension of cellular membranes. SEM functions are based on the premise of generating electron beams that are magnetically charged. The detection of secondary and backscattered electrons is achieved by the interaction of the magnetic field with the material (Qu and Liu, 2021). Transmission electron detection is an effective method for studying nanomaterials. SEM-EDX is a technique that may be employed to ascertain the spatial distribution of elements and the concentration of samples (Kozikowski and Sobiech, 2022).

22.4.6 Transmission Electron Microscopy

TEM is a widely used and indispensable method for characterizing NPs. The primary advantage is in the capacity to qualitatively and quantitatively examine the properties of generated NPs and their behavior within cells or tissues, including their shape, crystal structure, and size (Michen et al., 2015; Pudovkin et al., 2021; Tan, 2021). Conventional TEM techniques are employed for the investigation of all samples. Nevertheless, TEM has limitations, including its incapacity to analyze NPs smaller than 10 nm, its limited magnification for particles bigger than 100 nm, and the destructive nature of sample preparation (Mühlfeld et al., 2007; Lee et al., 2020; Neogy et al., 2020; Malatesta, 2021). TEM categorization is facilitated by applications such as immunological and energy-filtered electron microscopy (Casciardi et al., 2013).

22.5 UPTAKE AND TRANSLOCATION OF NPs

Although there have been several recent reviews on the subject of NP absorption and transport, there has been little analysis of the uptake and transportation of NPs in different plant species (Rani et al., 2022). There are three intermediate steps in the interaction between NPs and plants: NPs are absorbed and converted in the root or shoot system; NPs are absorbed through the cuticle, epidermis, and stomata; NPs are moved or converted within the plant through the xylem and phloem (Figure 22.1). Seed treatment, trunk feeding, foliar spraying, and root application are all effective methods for introducing NP into plants (Su et al., 2019; Morin et al., 2023). Among these techniques, both foliar and root exposure have been recognized as reasonably straightforward for execution, making them favorable choices for delivering NPs. Conducting a penetration investigation of NPs is essential for comprehending the interaction between NPs and plants. The absorption and translocation of NPs into plants are influenced by the concentration, stability, size, and chemical conformation of the NPs (Hong et al., 2021).

22.5.1 Foliar Application Method of NPs Uptake cum Translocation of NPs

NPs given by foliar application are absorbed by plants through the epidermis or stomata. Both the apoplastic and symplastic routes result in the transportation of substances to the vascular tissue following absorption (Hong et al., 2021). Plants generally store NPs in vacuoles and the cell wall, but they are also transported through the plant's vascular tissues in both upward and downward directions to other plant sections. Abiotic variables, such as light, temperature, and humidity, as well as plant physiological traits and properties of NPs, are crucial in determining the penetration, transfer, and accumulation of NPs. These parameters include the size, shape, and charge of the NPs (Pan et al., 2022). Studies have demonstrated that cucumber plants can take in CeO_2 NPs measuring 8–1 nm through their leaves, whereas lettuce plants can absorb $Cu(OH)_2$ NPs ranging from 50 to 100 nm in size (Zhao et al., 2016; Hong et al., 2021). Compared to exposing the roots or soil, supplementing with NP by foliar spray is more effective and sustainable (Fan et al., 2020; Zhang et al., 2020).

In agriculture, NPs are commonly used by spraying them onto the surface of leaves. Once sprayed, the NPs settle and are absorbed by plants through either the cuticle or stomata. The waxy cuticle of the leaf epidermis mostly consists of wax, cutin, and pectin. It functions as a vital natural defense mechanism by preventing water loss in developing leaves and acting as the principal barrier against the infiltration of NPs (Pérez-de-Luque, 2017). The waxy stratum corneum is composed of two different channels: one is hydrophilic, meaning it attracts water, and the other is lipophilic, meaning it attracts fats or oils (Pérez-de-Luque, 2017). The widths of both hydrophilic and lipophilic channels can range from 0.6 to 4.8 nm (Eichert et al., 2008). According to Banerjee et al., NPs that are hydrophilic and have a diameter of less than 4.8 nm are capable of diffusing through channels (2019). The cuticle surface of leaves has lipophilic channels that enable the absorption of lipophilic NPs by diffusion and infiltration (Bussières, 2014). Hu et al. (2020) have recently shown that carbon dots smaller than 2 nm may penetrate cotton leaves through the cuticular route, as shown by high-resolution confocal fluorescence microscopy analysis (Hu et al., 2020). Nevertheless, the plant epidermis has a limited capacity to absorb NPs as a result of the narrow dimensions of the pore channels in the cuticle. Studies have indicated that when NPs are administered to the surface of a leaf, they have the potential to gather in both the epidermal and vascular tissue (Le et al., 2022). Several studies have documented the translocation of NPs to different plant tissues, indicating the existence of novel channels for their entry. Researchers have suggested a novel method for the absorption of NPs, which they have named the stomatal pathway (Figure 22.1).

22.5.2 Soil Application Method of NPs, Uptake, and Translocation of NPs

The metal ions included inside the NPs are used in soil and exhibit distinct interactions based on the aging process and soil properties, persisting in the sediment for an extended period of time (Rajput et al., 2022). Research has been undertaken to analyze the mechanism of NP transformation in soil. The composition of inorganic and organic components, as well as the texture of the soil, has an impact on the transformation of NPs in the soil (Rani et al., 2022). The concentration of NPs is reduced as a result of the reduction in air pore space in clay soil (Naderi and Danesh-Shahraki, 2013; Rani et al., 2022). The presence of an Ag_2O layer surrounding Ag-NPs leads to their disintegration and subsequent release of Ag^+ in sandy soil, which is caused by the greater oxygen content of the soil (Li et al., 2017). The accumulation of copper in lettuce following exposure of its roots to copper oxide NPs (CuO NPs) was influenced by the concentration of CuO NPs, soil characteristics, and soil texture. Additionally, the accumulation of copper in lettuce increased with time. The degradation of CuO NPs in the soil has an impact on the availability of Cu, the absorption of Cu by lettuce plants, and the movement of Cu inside the lettuce plants (Servin et al., 2017). The aggregation of NPs in the soil system decreases their bioavailability due to the occurrence of both homo and hetero NP aggregation in the soil. The organic and inorganic constituents of soil interact with NPs and influence their adhesion to the ground. When ZnO-NPs formed hetero-aggregates with soil components, they exhibited reduced absorption. The presence of water in the soil might facilitate the dissolution of NPs, therefore releasing the metal ions contained inside them. The dissolving process enhanced the bioavailability and absorption of NPs. The solubility of NPs in the soil is influenced by several parameters, including soil type, texture, physicochemical qualities of the NPs, and the technique of NP administration (powder or solution) (Wu et al., 2021).

The impact of NPs on plant roots has been investigated more thoroughly compared to their utilization on plant leaves. The hair cells of the plant root absorb the NPs upon interaction with the root. They penetrate the endothelial layer from the epidermis by either symplast or exoplast and are then conveyed to the aerial portion of the plant through the xylem arteries (Tripathi et al., 2017). The structure of a plant's root consists of an inner pericycle and vascular column, surrounded by an outer cortex and epidermis. The vascular column, which consists of the pericycle and vascular tissue, is located near the midway of the root (Su et al., 2019). The endothelium is connected to the pericycle in the cortex, which also encompasses the exodermis (Figure 22.1). NPs can be delivered in the root by either the exoplast or the eutectic pathway. The Casparian strip functions as a barrier to impede the movement of NPs from the epidermis to the endothelium through the exoplast route. Certain NPs will adhere to the endothelium, while others will be carried in a radial manner along the Casparian strip but will not pass through it. Furthermore, the Casparian strip is not fully formed at the site where the lateral root connects. The circulatory system provides a convenient route for the transportation of NPs (Wang et al., 2012, 2013; Ali et al., 2021a). NPs move from one cell to another by passing via plasmodesmata, which are channels connecting the cytoplasm of adjacent cells. NPs enter the cytoplasm by crossing the plasma membrane using the symplast route.

22.6 FUNCTION OF NPS FOR MITIGATION OF MULTIPLE ENVIRONMENTAL STRESSES MANAGEMENT UNDER CLIMATE CHANGE

NPs, with their extensive surface areas and minuscule dimensions, may effectively mitigate the impacts of abiotic and biotic stressors on farmed plants through several mechanisms. Nanofertilizers in agriculture can enhance various biochemical, morphological, and physiological factors such as photosynthesis rate, nutrient uptake efficiency, regulation of phytohormones, and plant defense system strength (Figure 22.2) (Saini et al., 2021; Rani et al., 2022). The impact of nanofertilizers on crop productivity under stress conditions can be attributed to their ability to (1) mitigate oxidative stress, thereby enhancing plants' stress tolerance; (2) enhance various biochemical processes within stressed plants; (3) regulate salt toxicity; (4) reduce the accumulation of malondialdehyde and hydrogen peroxide; and (5) improve the uptake of essential nutrients for the growth and metabolism of cultivated plants. This is achieved through the use of nano-active constituents with increased surface areas, which can lead to favorable reactions. 'Smart' nanofertilizers can contain nutrients that are enclosed in nanomaterials, distributed as NPs or emulsions, or released in a controlled manner.

Global crop productivity is facing substantial threats, which require immediate research on environmentally stressed plants. This research should focus on replicating field conditions as closely as possible, specifically by studying combinations of stress factors rather than individual ones (Pandey et al., 2015; Van Zelm et al., 2020). New technologies are now available to enhance the resilience of the agricultural industry to mitigate the detrimental impacts of drought and heat stress on crop output. Biofertilizers can be effectively utilized as a way to enhance plant tolerance to many stress conditions (Duc et al., 2018). Additionally, more advanced cultivars of the crop can be utilized to enhance agricultural practices. The impact of various stresses on plants varies based on factors such as the plant's age, its inherent resistance or sensitivity to stress, and the severity of the pressures. The simultaneous exposure of plants to numerous stressors results in a diverse range of morphological, physiological, and molecular traits, as observed by Liu et al. (2021).

Osmotic effect induction occurs in plants under stress, leading to the buildup of osmoprotectants and subsequent relaxation in response to the stress. Reactive oxygen species (ROS) generation is conceivable in the majority of abiotic stress conditions. Moreover, it is widely recognized that stress induced by high temperatures and salt can affect the segregation and movement of ions inside plants, whereas situations such as drought and salinity stress lead to water deficiency in plants. This leads to a decrease in the diffusion of CO_2 inside the chloroplasts due to reduced stomatal opening, resulting in a decrease in carbon metabolism (Liu et al., 2021). Tea plants have the ability to endure environmental pressures such as drought and high temperatures due to their capacity to accumulate metabolites like as caffeine, catechins, and theanine (Liu et al., 2021). Drought, submergence, floods, chilling, freezing, and heat stress are among the stressors now under investigation in connection to plants and their ability to withstand them (Verma et al., 2021; Zahra et al., 2021).

The role of plant natriuretic peptides in regulating water and salt balance inside plants has been a recent area of discussion in relation to the various stressors experienced by plants (Liu et al., 2020).

During periods of numerous abiotic stressors, the combination of natural fluctuations and transcriptome analysis has a role in plant boron deficit and toxicity (Taylor et al., 2014; Ruotolo et al., 2018; Liu et al., 2020). Several studies, such as those conducted by Lamaoui et al. and Ashraf et al., have found that several components have been identified for their substantial role in reducing the impact of diverse stressors that contribute to enhanced plant production (Lamaoui et al., 2018; Ashraf et al., 2022). Currently, the use of nanofertilizers under diverse stress conditions is still in its early stages. Managing this process is highly complex for the agricultural industry since it relies on several soil and environmental factors. Figure 22.3 presents a concise summary of the different pressures that farmed plants encounter while utilizing nanofertilizers. The graphic displays three primary categories of stresses: single, coupled, and multiple stresses.

The presence of nutrient deficiency, salinity, water stress, drought, and heavy metals are outcomes of the pressure that can be reduced by adding nanonutrients such as copper oxide (CuO), selenium oxide (SeO_2), zinc oxide (ZnO), silicon dioxide (SiO_2), iron oxide (FeO_3), and sulfur (S) to the soil (Bhatla and Lal, 2018). The simultaneous presence of stressors such as dryness, drought, heat, salt, and heavy metals exacerbates severe environmental conditions (Van Zelm et al., 2020). Applied nanotechnology versions, such as nSe for salinity and heat stress, nSi for salinity and heavy metals, as well as salinity and drought, and nZn for heat cum stress and heavy metals cum drought are examples of nanonutrients that can assist with the second type of problem (Pinedo-Guerrero et al., 2020; Wang et al., 2019). Currently, there is a lack of sufficient published literature on the utilization of nanofertilizers in relation to various stress variables (Pinedo-Guerrero et al., 2020; Wang et al., 2019).

Agricultural areas are susceptible to nutrient loss through leaching or gaseous emissions, which can result in pollution and contribute to global warming. Nanofertilizers have the potential to mitigate some environmental issues when used during periods of stress (Vishwakarma et al., 2018). A limited number of published research (Kim et al., 2017) have examined the impact of nanofertilizers on reducing atmospheric gas emissions. NPs are the most prevalent adsorbents in soil. They have the ability to modify the transport of nutrients and the fixation of organic matter by initiating the formation of a new mineral phase. The use of NPs inside intact soil systems will grow in

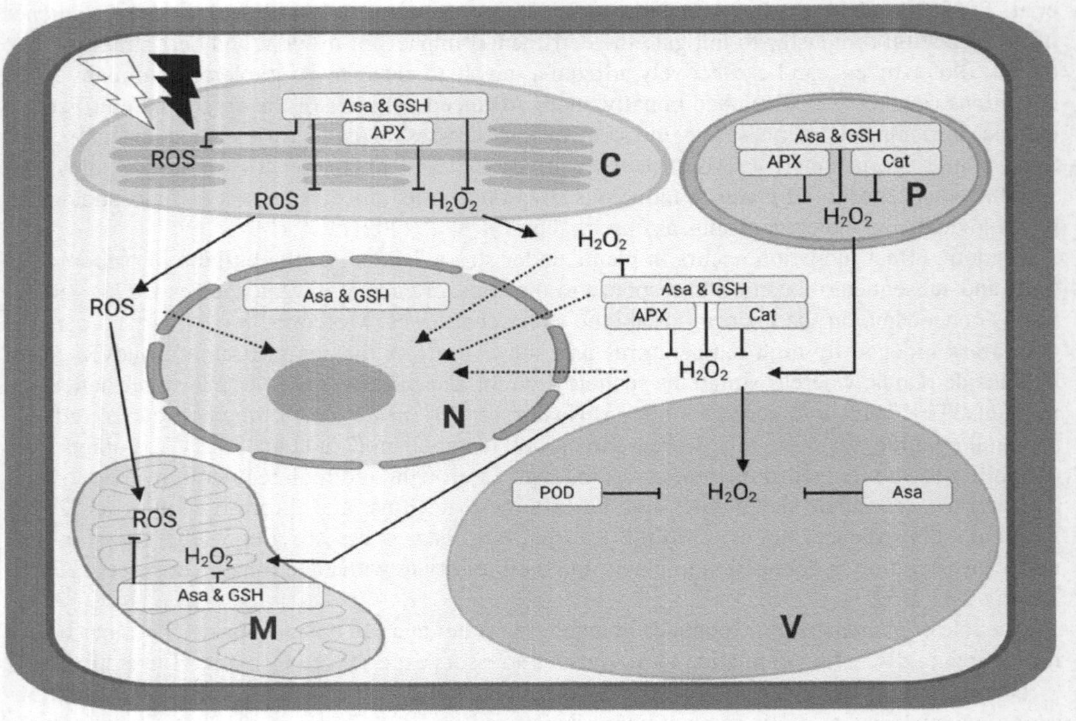

FIGURE 22.3 Diagram representing different cell organelles of plants, chloroplast, peroxisome, nucleus, mitochondria, and vacuole, which are sources of recative oxygen species (ROS).

significance in the future. The presence of NPs has been found to have a positive impact on plants by enhancing enzymatic activity, promoting seed germination, improving plant tolerance to unfavorable weather conditions, increasing carbon sequestration and nitrogen fixation, and enhancing respiratory and photosynthetic processes. These effects have been observed in various studies conducted by Hedderich and Whitman (2006) and Aronson et al. (2013). Consequently, the issue of how nanofertilizers might contribute to mitigating global warming has become a prominent concern on the world's agenda. Queries such as 'What is the role of nanofertilizers in improving plants' resilience to climate change?' and 'How do nanofertilizers facilitate crop production in highly stressful climate conditions?' require more investigation for a comprehensive response. How may the efficacy of nanofertilizers be impacted by an increase in temperature leading to heat stress? Does a universal case exist that is applicable to all climates and all types of nanofertilizers?

22.7 ACCUMULATION, TOXICITY, AND RISK ASSESSMENT OF NPs

The toxicological investigations on the impacts of nanofertilizers on crops are still in the early stages, but their utilization is already being acclaimed as a pioneering implementation of agricultural nanotechnology. The utilization of metallic NPs in plants has been demonstrated to induce an increase in oxidative stress as a result of the disruption of the electron transport chain (Khan et al., 2021). The interference results in a reduction in the plant's capacity to detoxify ROS, leading to potential genotoxic consequences. Although there is evidence suggesting that nano-nutrient components might be used in global fertilizer businesses, the study findings have not motivated these sectors to boost their investments in nanofertilizers. Nevertheless, the documentation of nanofertilizer products is limited to a tiny quantity. Certain things in this collection have originated from academic research, while others have been produced in pilot plants on a modest scale specifically for utilization in greenhouse experiments. The exorbitant expense and probable toxicity of the existing synthesis methods might be exacerbating the already limited productivity of this study. Consequently, there is a notable change in focus toward biological/green synthesis as a method of producing these NPs. This has the potential to reduce the cost and risk associated with the use of nanofertilizers. Nevertheless, several Asian nations such as Myanmar are now implementing measures to integrate nanofertilizers into their national fertilizer policy (Dimkpa and Bindraban, 2018). It is important to consider the possible disadvantages of NPs in comparison to their potential advantages. It is extremely challenging to find ways that are both highly efficient and completely free of unforeseen effects. Nanotechnology encompasses a wide range of disciplines and presents several potential hazards as a result. NPs, being of such small size, may readily penetrate the body's natural defensive mechanisms and induce injury. Nevertheless, when utilized in controlled amounts, NPs possess the capability to be both efficient and secure, without any adverse repercussions for the surrounding ecology. Zinc and iron are examples of metallic NPs that serve as nanonutrients, playing a crucial role in the nutrition of both plants and humans (Shang et al., 2019). Applying an appropriate dosage of nanofertilizers to plants has the potential to enhance their productivity and quality, while also benefiting the well-being of consumers. Precise and regulated discharge of nanofertilizers based on the specific nutritional requirements of the crops guarantees minimum leakage into the soil. The purpose of coatings on controlled-release fertilizers is to impede the interaction between the soluble material and water, as well as to hinder the diffusion of the nutrient solution out of the fertilizer. This inhibits the process of soil penetration caused by leaching. Research conducted by Zhang et al. (2014) and Pan et al. (2022) has demonstrated that the application of nano-coated materials may effectively decrease the leaching of NO_3 in a wheat-maize crop rotation (Zhang et al., 2014; Pan et al., 2022). Utilizing porous nanomaterials such as zeolites, clay, or chitosan can effectively minimize nitrogen losses by regulating the release rate according to the specific requirements of plants (Shang et al., 2019).

22.7.1 NPs-Based Oxidative Stress Damage in Plants

ROS, along with other signaling molecules, stimulate cellular responses in aerobic organisms, particularly in plants (Sharma et al., 2012). When NPs infiltrate plant cells, they have the ability to modify levels of ROS, hinder cellular metabolism, impair the antioxidant system, and eventually result in the inhibition of plant development. ROSs, such as hydrogen peroxide, hydroxyl radical, oxygen singlet, and superoxide anion, are produced by the organelles of a cell as a regular component of their metabolic processes (Keunen et al., 2011). At low levels of concentration, ROS initiates a defensive signal. However, at high levels of concentration, they can induce oxidative damage to amino acids, nucleic acids, and lipids (Mahajan and Tuteja, 2005). The extent of lipid peroxidation in a cell's membrane serves as a valuable measure of the membrane's overall health (Porter et al., 1995). ROS generation leads to lipid peroxidation, resulting in membrane damage, ion leakage, metabolic disturbance, and ultimately cell death. Plants possess antioxidant enzymes and low-molecular-weight antioxidants that can protect against cellular and subcellular harm produced by ROS (Hussain et al., 2016). Research on the frequency of oxidative stress in plants caused by NPs usually involves quantifying ROS, antioxidants, and the activity of antioxidant enzymes. The presence of silver NPs in bean leaves stimulated the generation of ROS. According to the research, there is a positive correlation between the reduction in particle diameter and the increase in the generation of ROS. This is due to the fact that smaller NPs have a larger specific surface area, which results in higher toxicity toward the cells (Yamauchi et al., 2008). The presence of TiO_2 NPs under highlight circumstances led to an increase in both tocopherol content and superoxide dismutase (SOD) activity, but catalase activity decreased (Das and Roychoudhury, 2014). Elevated concentrations of NPs induce the oxidation of membrane lipids, a process that may be associated with the activation of particles by light and an augmented formation of ROS (Abid et al., 2020). The oxidative stress in lettuce was induced by the presence of polymethyl methacrylate nanoplastics, increasing defensive mechanisms such as antioxidant enzyme activity (Zhao et al., 2016). However, it exhibited a greater concentration of active oxygen compared to the control. The recent confirmation of heightened activity of many antioxidant enzymes found in rice roots, as the concentration of plastic NPs rose, indicates that plants have the potential to induce various defensive responses and remove excessive buildup of ROS.

22.7.2 NP-Based Genotoxic Damage in Plants

NPs accumulate at different concentrations in the environment following their release at different dietary amounts. Plants readily absorb NPs and are very susceptible to nanotoxicity (Miralles et al., 2012). Plants are regarded as excellent genetic models for screening and monitoring environmentally hazardous chemicals (Zhang et al., 2015). NPs can decrease the rate of cell division (mitotic index), leading to an increase in the formation of small nuclei (micronucleus) and abnormalities in the structure of chromosomes (chromosome aberration index) in plants. These effects are a result of interactions between the NPs and the large molecules found in living organisms. The study conducted by Abdelsalam et al. (2018) revealed that wheat root tip cells had a high propensity to uptake Ag-NPs. These NPs have the potential to disrupt the normal functioning of c-myc and the segregation of chromosomes, as shown by their impact on chromosomal abnormalities and cell division. The identified aberrations included plate deformation and alterations in orientation during metaphase, chromosomal breakage, spindle malfunction, and the presence of lagging chromosomes (Abdelsalam et al., 2018).

After subjecting onion root cells to Cr_2O_3 NPs for cytogenetic investigation, a significant decrease in the mitotic index was observed. In addition, the researchers discovered chromosomal abnormalities at different dosages (Abdelsalam et al., 2018). Prior research has demonstrated that Cr_2O_3 NPs induce an elevation in SOD production as a reaction to oxidative stress, leading to subsequent damage to DNA (Pakrashi et al., 2014). The mitotic index of onions was dramatically reduced following exposure to PS-NPs. Additionally, different forms of chromosome aberrations, including chromosomal aggregation, lagging chromosomes, nuclear abnormalities, and spindle deformation, were seen, which compromised the integrity of the genome (Maity et al., 2020). Direct genotoxicity and indirect genotoxicity are two overarching classifications that are employed to delineate the underlying mechanism of genotoxicity (Jogaiah et al., 2021). The direct genotoxicity of NPs arises from their physical contact with DNA, whereas the indirect genotoxicity may occur as a consequence of a reduction in.

22.8 FUTURE PROSPECTS

The use of agrochemicals in sustainable agricultural methods should be reduced to maintain an environmentally friendly plant nutrition system. Tropical and subtropical soils are often characterized by high acidity, frequent P deficiency, and a significant propensity for phosphate sorption (Batista and Singh, 2021). Hence, cutting-edge approaches are being formulated utilizing nanotechnology and nanoengineering to enhance agricultural productivity (Verma et al., 2022a, 2022b). Phosphate rocks have significant value as sources of phosphorus, and their industrial-scale processing can contribute to global development (Majumdar et al., 2015). It is really accurate that several affluent countries possess significant and indigenous reserves of phosphate rock. Implementing strategies to optimize land use patterns, minimize soil erosion, maintain soil quality, employ effective fertilizer replacement techniques, enhance fertilizer and crop genotype selection, and utilize manure export are all viable sustainable options for regulating phosphorus consumption. It is essential to have instruments that can monitor the needs of crops and assess the demand within and between crops to alter the supply of products to the crops or individual plants. Due to the limited reserves, there is an anticipated significant increase in the costs of agricultural inputs such as fertilizers and insecticides. The subject of using nanotechnology-based delivery systems to provide critical nutrients to crops is a developing area that requires further research (Mahakham et al., 2017). Possible approaches for organizing future research on nano-based fertilizers might be explored. Due to their extensive usage and possible limited availability, N- and P-nanofertilizers should be given greater attention. Complete research may be done to investigate the varying impact of field exposure on the availability of micronutrients. By conducting a comparative analysis between calcium NPs (Ca-NPs) and calcium carbonate ($CaCO_3$) derived from alternative calcium sources such as calcium sulfate ($CaSO_4$) or calcium chloride ($CaCl_2$) or by comparing several nutrients such as calcium (Ca) and nitrogen (N) NPs, while utilizing soluble calcium nitrate ($Ca(NO_3)_2$) as a control. Commercially accessible micronutrient fertilizers may be compared to traditional fertilizers in terms of their ability to enhance biofortification and their application methods, such as fertigation. Currently, there is no evidence or conclusion to support the effectiveness of nanocarrier-dependent fertilizers in increasing fertilizer use efficiency (FUE) by improving nutrient transfer into plant cells and tissues or in reducing the harmful environmental consequences of conventional fertilizers. Additionally, it is important to determine the optimal dose for each crop to avoid any negative effects. They have the potential to discover more cost-effective and efficient carriers, such as SiO_2-NPs, Fe_2O_3-NPs, and CNTs, compared to previously accessible options, such as zeolite. These developments can help slowly incorporate active materials while decreasing the requirement and amount of wasted inputs. The architecture of nanocarriers also has the potential to serve as anchors in the organic matter and soil structure around plants' roots. To do this, it is vital to possess a comprehension of the chemical and conformational interactions among the delivery nanoscale structures, in addition to the desired structure of the soil and components. Extensive research has shown, with a high level of statistical significance, that different NPs/nanomaterials, based on their particle size, have a harmful effect on plant development when present in high concentrations. To understand the mechanisms underlying the increased growth and crop yield resulting from the use of nanofertilizers, it would be beneficial to establish a checkpoint for future research. In contrast, nanotechnology has the potential to provide controlled and sustained release of agrochemicals, as well as targeted delivery of fertilizers and pesticides to specific areas of plants. This can enhance disease resistance, increase nutrient use, and promote plant development. In addition, satellite-based navigation systems utilize geographic information systems, remote sensing technologies, global positioning systems, and other information techniques to gather data on the optimal amount of nutrients reaching plants and soils, thereby maximizing crop yield (Zulfiqar and Ashraf, 2021).

22.9 CONCLUSION

The worldwide sustainability of green plants is being questioned due to abiotic stress produced by urbanization, severe weather conditions, habitat loss, and widespread pollution, all of which impose enormous environmental burdens. The present population increase has resulted in a catastrophe, as abiotic factors such as salinity, severe temperatures, heavy metals, and water scarcity are causing substantial worries for global food security. Nanotechnology is a contemporary technology that has been widely employed to augment plants' capacity to withstand abiotic stress. Due to their small size and high reactivity, NPs can be easily absorbed by plants in several forms, such as nanofertilizers, nanoherbicides, and nanopesticides. NPs can be produced using physical, chemical, and biological synthesis processes. After the synthesis process, NPs undergo characterization using several techniques such as XRD, FTIR, UV–Vis, SEM, TEM, and IR spectroscopy. NPs interact with plants in several ways, which depend on their chemical composition, size, surface area, and vulnerability. Plants have the ability to uptake NPs via their leaves, roots, or stomata. The apoplastic and symplastic pathways facilitate the movement of chemicals to the vascular tissue after absorption. Natural products efficiently alleviate the effects of non-living stresses on cultivated plants in various ways. They have the ability to improve multiple biochemical, morphological, and physiological aspects, including the rate of photosynthesis, the efficiency of nutrient uptake, the regulation of phytohormones, and the strength of the plant's defense system. Although NPs offer certain benefits, they are also linked to several drawbacks. When NPs enter plant cells, they can alter the amounts of ROS, disrupt cellular metabolism, weaken the antioxidant system, and ultimately impede plant growth. NPs gather in varying concentrations in the environment after being released in diverse dietary quantities. NPs can inhibit cell division (mitotic index), resulting in an elevation in the occurrence of tiny nuclei (micronucleus) and abnormalities in chromosomal structure (chromosome aberration index) in plants. Hence, further inquiries should delve into the following matters, which will aid in mitigating the adverse impacts of NPs on ecosystems.

ACKNOWLEDGMENTS

VDR acknowledges the support of the Strategic Academic Leadership Program of the Southern Federal University ('Priority 2030'). Higher education and science committee MESCS Republic of Armenia provided the 23PostDoc-4D007 grant to support AS. Higher education and science committee MESCS Republic of Armenia has provided grant numbers 21AG-4C075 to support KG.

REFERENCES

Abdel-Hadi, A. M., Awad, M. F., Abo-Dahab, N. F., & Elkady, M. F. (2014). Extracellular synthesis of silver NPs by aspergillus terreus: Biosynthesis, characterization and biological activity. *Biosciences Biotechnology Research Asia*, *11*(3), 1179–1186.

Abdelsalam, N. R., Abdel-Megeed, A., Ali, H. M., Salem, M. Z. M., Al-Hayali, M. F. A., & Elshikh, M. S. (2018). Genotoxicity effects of silver NPs on wheat (Triticum aestivum L.) root tip cells. *Ecotoxicology and Environmental Safety*, *155*, 76–85.

Abid, M., Zhang, Y. J., Li, Z., Bai, D. F., Zhong, Y. P., & Fang, J. B. (2020). Effect of Salt stress on growth, physiological and biochemical characters of four kiwifruit genotypes. *Scientia Horticulturae (Amsterdam)*, *271*, 109473.

Adelere, I. A., & Lateef, A. (2016). A novel approach to the green synthesis of metallic NPs: The use of agro-wastes, enzymes, and pigments. *Nanotechnology Reviews*, *5*(6), 567–587.

Ahmad, A., Senapati, S., Khan, M. I., Kumar, R., & Sastry, M. (2006). Extra-/intracellular biosynthesis of gold NPs by an alkalotolerant fungus, *Trichothecium* sp. *Journal of Biomedical Nanotechnology*, *1*(1), 47–53.

Akhtar, M. S., Panwar, J., & Yun, Y. S. (2013). Biogenic synthesis of metallic NPs by plant extracts. *ACS Sustainable Chemistry & Engineering*, *1*(6), 591–602.

Ali, S., Mehmood, A., & Khan, N. (2021a). Uptake, translocation, and consequences of nanomaterials on plant growth and stress adaptation. *Journal of Nanomaterials*, 2021, 6677616.

Amendola, V., & Meneghetti, M. (2009). Size evaluation of gold NPs by UV-vis spectroscopy. *The Journal of Physical Chemistry C*, *113*(11), 4277–4285.

Aronson, E. L., Allison, S. D., & Helliker, B. R. (2013). Environmental impacts on the diversity of methane-cycling microbes and their resultant function. *Frontiers in Microbiology*, *4*(AUG), 225.

Ashraf, M. A., Rasheed, R., Hussain, I., Iqbal, M., Farooq, M. U., Saleem, M. H., & Ali, S. (2022). Taurine modulates dynamics of oxidative defense, secondary metabolism, and nutrient relation to mitigate boron and chromium toxicity in Triticum aestivum L. plants. *Environmental Science and Pollution Research*, *29*(30), 45527–45548.

Awasthi, A., Bansal, S., Jangir, L. K., Awasthi, G., Awasthi, K. K., & Awasthi, K. (2017). Effect of ZnO NPs on germination of Triticum aestivum seeds. *Macromolecular Symposia*, *376*(1), 1700043.

Banerjee, K., Pramanik, P., Maity, A., Joshi, D. C., Wani, S. H., & Krishnan, P. (2019). Methods of using nanomaterials to plant systems and their delivery to plants (mode of entry, uptake, translocation, accumulation, biotransformation and barriers). In M. Ghorbanpour & S. H. Wani (Eds.), *Advances in phytonanotechnology: From synthesis to application* (pp. 123–152). Academic Press.

Barzegar, M., Ahmadvand, D., Sabouri, Z., & Darroudi, M. (2023). Green synthesis of magnesium oxide NPs by chitosan polymer and assessment of their photocatalytic activity and cytotoxicity influences. *Materials Chemistry and Physics*, *301*, 127649.

Batista, B. D., & Singh, B. K. (2021). Realities and hopes in the application of microbial tools in agriculture. *Microbial Biotechnology*, *14*(4), 1258–1268.

Baudot, C., Tan, C. M., & Kong, J. C. (2010). FTIR spectroscopy as a tool for nano-material characterization. *Infrared Physics & Technology*, *53*(6), 434–438.

Bhatla, S. C., & Lal, M. A. (2018). *Plant physiology, development and metabolism*. Springer.

Birla, S. S., Tiwari, V. V., Gade, A. K., Ingle, A. P., Yadav, A. P., & Rai, M. K. (2009). Fabrication of silver NPs by Phoma glomerata and its combined effect against Escherichia coli, Pseudomonas aeruginosa and Staphylococcus aureus. *Letters in Applied Microbiology*, *48*(2), 173–179.

Bussières, P. (2014). Estimating the number and size of phloem sieve plate pores using longitudinal views and geometric reconstruction. *Scientific Reports*, *4*(1), 1–11.

Bzdek, B. R., Zordan, C. A., Luther, G. W., & Johnston, M. V. (2011). NP chemical composition during new particle formation. *Aerosol Science and Technology*, *45*(8), 1041–1048.

Cao, G. (2004). *Nanostructures and nanomaterials*. World Scientific Publishing Company.

Carmichael, W. W., Mahmood, N. A., & Hyde, E. G. (1990). Natural toxins from cyanobacteria (blue-green algae). In S. Hall & G. Strichartz (Eds.), *Marine toxins: Origin, structure, and molecular pharmacology* (pp. 87–106). Elsevier Science.

Casalini, S., Bortolotti, C. A., Leonardi, F., & Biscarini, F. (2017). Self-assembled monolayers in organic electronics. *Chemical Society Reviews*, *46*(1), 40–71.

Casciardi, S., Sisto, R., & Diociaiuti, M. (2013). The analytical transmission electron microscopy: A powerful tool for the investigation of low-dimensional carbon nanomaterials. *Journal of Nanomaterials*, 2013, 506815.

Darroudi, M., Sabouri, Z., Kazemi Oskuee, R., Khorsand Zak, A., Kargar, H., & Hamid, M. H. N. A. (2013). Sol–gel synthesis, characterization, and neurotoxicity effect of zinc oxide NPs using gum tragacanth. *Ceramics International*, *39*(8), 9195–9199.

Das, K., & Roychoudhury, A. (2014). Reactive oxygen species (ROS) and response of antioxidants as ROS-scavengers during environmental stress in plants. *Frontiers in Environmental Sciences*, *2*(DEC), 53.

Decarolis, D., Odarchenko, Y., Herbert, J. J., Qiu, C., Longo, A., & Beale, A. M. (2020). Identification of the key steps in the self-assembly of homogeneous gold metal NPs produced using inverse micelles. *Physical Chemistry Chemical Physics*, *22*(34), 18824–18834.

Dhand, C., Dwivedi, N., Loh, X. J., Jie Ying, A. N., Verma, N. K., Beuerman, R. W., Lakshminarayanan, R., & Ramakrishna, S. (2015). Methods and strategies for the synthesis of diverse NPs and their applications: A comprehensive overview. *RSC Advances*, *5*(127), 105003–105037.

Dimkpa, C. O., & Bindraban, P. S. (2018). Nanofertilizers: New products for the industry? *Journal of Agricultural and Food Chemistry*, *66*(26), 6462–6473.

Du, P., Si, R., Li, Z., & Zou, Z. (2023). Applications of X-ray absorption fine structure spectroscopy in the photocatalytic conversion of small molecules. *ACS Catalysis*, *13*(10), 6690–6703.

Duc, N. H., Csintalan, Z., & Posta, K. (2018). Arbuscular mycorrhizal fungi mitigate negative effects of combined drought and heat stress on tomato plants. *Plant Physiology and Biochemistry*, *132*, 297–307.

Eichert, T., Kurtz, A., Steiner, U., & Goldbach, H. E. (2008). Size exclusion limits and lateral heterogeneity of the stomatal foliar uptake pathway for aqueous solutes and water-suspended NPs. *Physiologia Plantarum*, *134*(1), 151–160.

Eid, M. M. (2022). Characterization of NPs by FTIR and FTIR-microscopy. In S. E. Mallakpour & C. M. Hussain (Eds.), *Handbook of consumer nanoproducts* (pp. 645–673). Springer.

Eroglu, Z., & Metin, O. (2023). Internal interactions within the complex type-II heterojunction of a graphitic carbon nitride/black phosphorus hybrid decorated with graphene quantum dots: Implications for photooxidation performance. *ACS Applied Nano Materials Journal*, *6*, 7960–7974.

Fan, X., Cao, X., Zhou, H., Hao, L., Dong, W., He, C., Xu, M., Wu, H., Wang, L., Chang, Z., & Zheng, Y. (2020). Carbon dioxide fertilization effect on plant growth under soil water stress associates with changes in stomatal traits, leaf photosynthesis, and foliar nitrogen of bell pepper (Capsicum annuum L.). *Environmental and Experimental Botany*, *179*, 104203.

Faramarzi, S., Anzabi, Y., & Jafarizadeh-Malmiri, H. (2020). Nanobiotechnology approach in intracellular selenium NP synthesis using Saccharomyces cerevisiae—fabrication and characterization. *Archives of Microbiology*, *202*(5), 1203–1209.

Fardsadegh, B., & Jafarizadeh-Malmiri, H. (2019). Aloe vera leaf extract mediated green synthesis of selenium NPs and assessment of their in vitro antimicrobial activity against spoilage fungi and pathogenic bacteria strains. *Green Processing and Synthesis*, *8*(1), 399–407.

Gericke, M., & Pinches, A. (2006). Microbial production of gold NPs. *Gold Bulletin*, *39*(1), 22–28.

Gwo, S., Chen, H. Y., Lin, M. H., Sun, L., & Li, X. (2016). Nanomanipulation and controlled self-assembly of metal NPs and nanocrystals for plasmonics. *Chemical Society Reviews*, *45*(20), 5672–5716.

Hedderich, R., & Whitman, W. B. (2006). Physiology and biochemistry of the methane-producing archaea. In M. Dworkin, S. Falkow, E. Rosenberg, K. H. Schleifer, & E. Stackebrandt (Eds.), *The prokaryotes: A handbook on the biology of bacteria. Volume 2: Ecophysiology and biochemistry* (pp. 1050–1079). Springer.

Hodoroaba, V., Rades, S., Salge, T., Mielke, J., Ortel, E., & Schmidt, R. (2016). Characterisation of NPs by means of high-resolution SEM/EDS in transmission mode. *IOP Conference Series: Materials Science and Engineering*, *109*(1), 012006.

Hoffmann, M. R., Martin, S. T., Choi, W., & Bahnemann, D. W. (1995). Environmental applications of semiconductor photocatalysis. *Chemical Reviews*, *95*(1), 69–96.

Holder, C. F., & Schaak, R. E. (2019). Tutorial on powder X-ray diffraction for characterizing nanoscale materials. *ACS Nano*, *13*(7), 7359–7365.

Hong, J., Wang, C., Wagner, D. C., Gardea-Torresdey, J. L., He, F., & Rico, C. M. (2021). Foliar application of NPs: Mechanisms of absorption, transfer, and multiple impacts. *Environmental Science: Nano*, *8*(5), 1196–1210.

Hu, M., Chen, J., Li, Z. Y., Au, L., Hartland, G. V., Li, X., Marquez, M., & Xia, Y. (2006). Gold nanostructures: Engineering their plasmonic properties for biomedical applications. *Chemical Society Reviews*, *35*(11), 1084–1094.

Hu, P., An, J., Faulkner, M. M., Wu, H., Li, Z., Tian, X., & Giraldo, J. P. (2020). NP charge and size control foliar delivery efficiency to plant cells and organelles. *ACS Nano*, *14*(7), 7970–7986.

Hussain, S., Khan, F., Hussain, H. A., & Nie, L. (2016). Physiological and biochemical mechanisms of seed priming-induced chilling tolerance in rice cultivars. *Frontiers in Plant Science*, *7*(FEB2016), 116.

Huston, M., Debella, M., Dibella, M., & Gupta, A. (2021). Green synthesis of nanomaterials. *Nanomater*, 11(8), 2130.

Jemal, K., Sandeep, B. V., & Pola, S. (2017). Synthesis, characterization, and evaluation of the antibacterial activity of allophylus serratus leaf and leaf derived callus extracts mediated silver NPs. *Journal of Nanomaterials*, 2017, 1–11.

Jha, A. K., Prasad, K., & Prasad, K. (2009). A green low-cost biosynthesis of Sb_2O_3 NPs. *Biochemical Engineering Journal*, *43*(3), 303–306.

Jogaiah, S., Paidi, M. K., Venugopal, K., Geetha, N., Mujtaba, M., Udikeri, S. S., & Govarthanan, M. (2021). Phytotoxicological effects of engineered NPs: An emerging nanotoxicology. *Science of The Total Environment*, *801*, 149809.

Keunen, E., Remans, T., Bohler, S., Vangronsveld, J., & Cuypers, A. (2011). Metal-induced oxidative stress and plant mitochondria. *International Journal of Molecular Sciences*, *12*(10), 6894–6918.

Khan, M., Khan, M. S. A., Borah, K. K., Goswami, Y., Hakeem, K. R., & Chakrabartty, I. (2021). The potential exposure and hazards of metal-based NPs on plants and environment, with special emphasis on ZnO NPs, TiO_2 NPs, and AgNPs: A review. *Environmental Advances*, *6*, 100128.

Khan, M. R., Adam, V., Rizvi, T. F., Zhang, B., Ahamad, F., Jośko, I., Zhu, Y., Yang, M., & Mao, C. (2019). NP-plant interactions: A two-way traffic. *Small*, *15*(37), e1901794.

Khan, Z., Bhadouria, P., & Bisen, P. (2005). Nutritional and therapeutic potential of spirulina. *Current Pharmaceutical Biotechnology*, *6*(5), 373–379.

Kiefer, J., Grabow, J., Kurland, H. D., & Müller, F. A. (2015). Characterization of NPs by solvent infrared spectroscopy. *Analytical Chemistry*, *87*(24), 12313–12317.

Kim, J., Yoo, G., Kim, D., Ding, W., & Kang, H. (2017). Combined application of biochar and slow-release fertilizer reduces methane emission but enhances rice yield by different mechanisms. *Applied Soil Ecology*, *117–118*, 57–62.

Kowshik, M., Deshmukh, N., Vogel, W., Urban, J., Kulkarni, S. K., & Paknikar, K. M. (2002). Microbial synthesis of semiconductor CdS NPs, their characterization, and their use in the fabrication of an ideal diode. *Biotechnology and Bioengineering*, *78*(5), 583–588.

Kozikowski, P., & Sobiech, P. (2022). Comparison of NPs' characteristic parameters derived from SEM and SMPS analyses. *The Journal for Nurse Practitioners*, *24*(6), 1–12.

Kumar, A., Vemula, P. K., Ajayan, P. M., & John, G. (2008). Silver-NP-embedded antimicrobial paints based on vegetable oil. *Nature Materials*, *7*(3), 236–241.

Lamaoui, M., Jemo, M., Datla, R., & Bekkaoui, F. (2018). Heat and drought stresses in crops and approaches for their mitigation. *Frontiers in Chemistry*, *6*, 26.

Le, N., Zhang, M., & Kim, K. (2022). Quantum dots and their interaction with biological systems. *International Journal of Molcular Sciences*, 23(18), 10763.

Lee, B., Yoon, S., Lee, J. W., Kim, Y., Chang, J., Yun, J., Ro, J. C., Lee, J. S., & Lee, J. H. (2020). Statistical characterization of the morphologies of NPs through machine learning based electron microscopy image analysis. *ACS Nano*, *14*(12), 17125–17133.

Li, W. R., Sun, T. L., Zhou, S. L., Ma, Y. K., Shi, Q. S., Xie, X. B., & Huang, X. M. (2017). A comparative analysis of antibacterial activity, dynamics, and effects of silver ions and silver NPs against four bacterial strains. *International Biodeterioration & Biodegradation*, *123*, 304–310.

Liu, X., Guan, H., Wang, T., Meng, D., Yang, Y., Dai, J., Fan, N., Guo, B., Fu, Y., He, W., & Wei, Y. (2020). ScPNP-A, a plant natriuretic peptide from Stellera chamaejasme, confers multiple stress tolerances in Arabidopsis. *Plant Physiology and Biochemistry*, *149*, 132–143.

Liu, Z., Han, Y., Zhou, Y., Wang, T., Lian, S., & Yuan, H. (2021). Transcriptomic analysis of tea plant (Camellia sinensis) revealed the co-expression network of 4111 paralogous genes and biosynthesis of quality-related key metabolites under multiple stresses. *Genomics*, *113*(1), 908–918.

López-Lorente, Á. I., & Mizaikoff, B. (2016). Recent advances on the characterization of NPs using infrared spectroscopy. *TrAC Trends in Analytical Chemistry*, *84*, 97–106.

López Gutiérrez, J. D., Abundez Barrera, I. M., & Torres Gómez, N. (2022). NP detection on SEM images using a neural network and semi-synthetic training data. *Nanomaterials*, *12*(11), 1818.

Mahajan, S., & Tuteja, N. (2005). Cold, salinity and drought stresses: An overview. *Archives of Biochemistry and Biophysics*, *444*(2), 139–158.

Mahakham, W., Sarmah, A. K., Maensiri, S., & Theerakulpisut, P. (2017). Nanopriming technology for enhancing germination and starch metabolism of aged rice seeds using phytosynthesized silver NPs. *Scientific Reports*, *7*(1), 8263.

Maity, S., Chatterjee, A., Guchhait, R., De, S., & Pramanick, K. (2020). Cytogenotoxic potential of a hazardous material, polystyrene microparticles on Allium cepa L. *Journal of Hazardous Materials*, *385*, 121560.

Majumdar, S., Almeida, I. C., Arigi, E. A., Choi, H., VerBerkmoes, N. C., Trujillo-Reyes, J., Flores-Margez, J. P., White, J. C., Peralta-Videa, J. R., & Gardea-Torresdey, J. L. (2015). Environmental effects of nanoceria on seed production of common bean (Phaseolus vulgaris), A proteomic analysis. *Environmental Science & Technology*, *49*(22), 13283–13293.

Malatesta, M. (2021). Transmission electron microscopy as a powerful tool to investigate the interaction of NPs with subcellular structures. *International Journal of Molecular Sciences*, *22*(23), 22.

Markus, J., Mathiyalagan, R., Kim, Y. J., Abbai, R., Singh, P., Ahn, S., Perez, Z. E. J., Hurh, J., & Yang, D. C. (2016). Intracellular synthesis of gold NPs with antioxidant activity by probiotic Lactobacillus kimchicus DCY51T isolated from Korean kimchi. *Enzyme and Microbial Technology*, *95*, 85–93.

Marsalek, R. (2014). Particle size and zeta potential of ZnO. *APCBEE Procedia*, *9*, 13–17.

Meyers, M. A., Mishra, A., & Benson, D. J. (2006). Mechanical properties of nanocrystalline materials. *Progress in Materials Science*, *51*(4), 427–556.

Michen, B., Geers, C., Vanhecke, D., Endes, C., Rothen-Rutishauser, B., Balog, S., & Petri-Fink, A. (2015). Avoiding drying-artifacts in transmission electron microscopy: Characterizing the size and colloidal state of NPs. *Scientific Reports*, *5*(1), 1–7.

Miralles, P., Church, T. L., & Harris, A. T. (2012). Toxicity, uptake, and translocation of engineered nanomaterials in vascular plants. *Environental Science and Technology*, *46*(17), 9224–9239.

Mittal, A. K., Chisti, Y., & Banerjee, U. C. (2013). Synthesis of metallic NPs using plant extracts. *Biotechnology Advances*, *31*(2), 346–356.

Mohanpuria, P., Rana, N. K., & Yadav, S. K. (2008). Biosynthesis of NPs: Technological concepts and future applications. *Journal of Nanoparticle Research*, *10*(3), 507–517.

Morin, J., Brochard, G., Bergé, V., Rosset, A., Artous, S., Clavaguera, S., Strekowski, R. S., & Wortham, H. (2023). Uptake of *m*-xylene and VOC emissions by mineral photocatalytic paints of indoor air building interest. Environmental Science: Nano, 10, 1704–1714.

Moulton, M. C., Braydich-Stolle, L. K., Nadagouda, M. N., Kunzelman, S., Hussain, S. M., & Varma, R. S. (2010). Synthesis, characterization and biocompatibility of "green" synthesized silver NPs using tea polyphenols. *Nanoscale*, *2*(5), 763–770.

Mourdikoudis, S., Pallares, R. M., & Thanh, K. N. T. (2018). Characterization techniques for NPs: Comparison and complementarity upon studying NP properties. *Nanoscale*, *10*(27), 12871–12934.

Mühlfeld, C., Rothen-Rutishauser, B., Vanhecke, D., Blank, F., Gehr, P., & Ochs, M. (2007). Visualization and quantitative analysis of NPs in the respiratory tract by transmission electron microscopy. *Particle and Fibre Toxicology*, *4*(1), 1–17.

Mukherjee, A., Sarkar, D., & Sasmal, S. (2021). A review of green synthesis of metal NPs using algae. *Frontiers in Microbiology*, *12*, 2152.

Naderi, M. R., & Danesh-Shahraki, A. (2013). Nanofertilizers and their roles in sustainable agriculture. *International Journal of Agriculture and Crop Sciences*, *5*(19), 2229–2232.

Narayanan, K. B., & Sakthivel, N. (2011). Synthesis and characterization of nano-gold composite using Cylindrocladium floridanum and its heterogeneous catalysis in the degradation of 4-nitrophenol. *Journal of Hazardous Materials*, *189*(1–2), 519–525.

Neogy, S., Savalia, R. T., Tewari, R., Srivastava, D., & Dey, G. K. (2020). Transmission electron microscopy of nanomaterials. *Indian Journal of Pure & Applied Physics*, *44*(2), 119–124.

Pakrashi, S., Jain, N., Dalai, S., Jayakumar, J., Chandrasekaran, P. T., Raichur, A. M., Chandrasekaran, N., & Mukherjee, A. (2014). In vivo genotoxicity assessment of titanium dioxide NPs by Allium cepa root tip assay at high exposure concentrations. *PLoS One*, *9*(2), e87789.

Pan, D., Huang, G., Yi, J., Cui, J., Liu, C., Li, F., & Li, X. (2022). Foliar application of silica NPs alleviates arsenic accumulation in rice grain: Co-localization of silicon and arsenic in nodes. *Environmental Science: Nano*, *9*(4), 1271–1281.

Pandey, P., Ramegowda, V., & Senthil-Kumar, M. (2015). Shared and unique responses of plants to multiple individual stresses and stress combinations: Physiological and molecular mechanisms. *Frontiers in Plant Science*, *6*(September), 723.

Panpatte, D. G., Jhala, Y. G., Shelat, H. N., & Vyas, R. V. (2016). NPs: The next generation technology for sustainable agriculture. In D. P. Singh, H. B. Singh, & R. Prabha (Eds.), *Microbial inoculants in sustainable agricultural productivity vol. 2: Functional applications* (pp. 289–300). Springer.

Patil, S., & Chandrasekaran, R. (2020). Biogenic NPs: A comprehensive perspective in synthesis, characterization, application and its challenges. *Journal of Genetic Engineering and Biotechnology*, *18*(1), 1–23.

Pérez-de-Luque, A. (2017). Interaction of nanomaterials with plants: What do we need for real applications in agriculture? *Frontiers in Environmental Science*, *5*(APR), 12.

Philip, D. (2009). Biosynthesis of Au, Ag and Au–Ag NPs using edible mushroom extract. *Spectrochimica Acta Part A: Molecular and Biomolecular Spectroscopy*, *73*(2), 374–381.

Pinedo-Guerrero, Z. H., Cadenas-Pliego, G., Ortega-Ortiz, H., González-Morales, S., Benavides-Mendoza, A., Valdés-Reyna, J., & Juárez-Maldonado, A. (2020). Form of silica improves yield, fruit quality and antioxidant defense system of tomato plants under salt stress. *Agriculture*, 10(9), 367.

Porter, N. A., Caldwell, S. E., & Mills, K. A. (1995). Mechanisms of free radical oxidation of unsaturated lipids. *Lipids*, *30*(4), 277–290.

Pudovkin, M. S., Shamsutdinov, N. I., Zelenikhin, P. V., & Nizamutdinov, A. S. (2021). Transmission electron microscopy and flow cytometry study of cellular uptake of unmodified Pr^{3+}:LaF_3 NPs in dynamic. *Journal of Nanoparticles Research*, *23*(6), 1–18.

Qu, J., & Liu, X. (2021). Recent advances on SEM-based in situ multiphysical characterization of nanomaterials. *Scanning*, 2021, 4426254.

Rajput, V. D., Minkina, T., Kumari, A., Shende, S. S., Ranjan, A., Faizan, M., Barakvov, A., Gromovik, A., Gorbunova, N., Rajput, P., Singh, A., Khabirov, I., Nazarenko, O., Sushkova, S., & Kızılkaya, R. (2022). A review on nanobioremediation approaches for restoration of contaminated soil. *Eurasian Journal of Soil Science*, *11*(1), 43–60.

Rajput, V. D., Singh, A., Minkina, T., Rawat, S., Mandzhieva, S., Sushkova, S., Shuvaeva, V., Nazarenko, O., Rajput, P., Komariah, Verma, K. K., Singh, A. K., Rao, M., & Upadhyay, S. K. (2021). Nano-enabled products: Challenges and opportunities for sustainable agriculture. *Plants*, *10*(12), 2727.

Rani, S., Kumari, N., & Sharma, V. (2022). Uptake, translocation, transformation and physiological effects of NPs in plants. *Archives of Agronomy and Soil Science*, *69*, 1579.

Ruotolo, R., Maestri, E., Pagano, L., Marmiroli, M., White, J. C., & Marmiroli, N. (2018). Plant response to metal-containing engineered nanomaterials: An omics-based perspective. *Environmental Science & Technology*, *52*(5), 2451–2467.

Saini, S., Kumar, P., Sharma, N. C., Sharma, N., & Balachandar, D. (2021). Nano-enabled Zn fertilization against conventional Zn analogues in strawberry (Fragaria×ananassa Duch.). *Scientia Horticulturae (Amsterdam)*, *282*, 110016.

Saleh, T. A. (2020). Nanomaterials: Classification, properties, and environmental toxicities. *Environmental Technology & Innovation*, 20, 101067.

Sasidharan, D., Namitha, T. R., Johnson, S. P., Jose, V., & Mathew, P. (2020). Synthesis of silver and copper oxide NPs using Myristica fragrans fruit extract: Antimicrobial and catalytic applications. *Sustainable Chemistry and Pharmacy*, *16*, 100255.

Seetharaman, P. K., Chandrasekaran, R., Gnanasekar, S., Chandrakasan, G., Gupta, M., Manikandan, D. B., & Sivaperumal, S. (2018). Antimicrobial and larvicidal activity of eco-friendly silver NPs synthesized from endophytic fungi Phomopsis liquidambaris. *Biocatalysis and Agricultural Biotechnology*, *16*, 22–30.

Senapati, S., Syed, A., Moeez, S., Kumar, A., & Ahmad, A. (2012). Intracellular synthesis of gold NPs using alga Tetraselmis kochinensis. *Materials Letters*, *79*, 116–118.

Sepeur, S. (2008). Nanotechnology: Technical basics and applications. Vincentz Network.

Servin, A. D., Pagano, L., Castillo-Michel, H., la Torre-Roche, R. D., Hawthorne, J., Hernandez-Viezcas, J. A., Loredo-Portales, R., Majumdar, S., Gardea-Torresday, J., Dhankher, O. P., & White, J. C. (2017). Weathering in soil increases NP CuO bioaccumulation within a terrestrial food chain. *Nanotoxicology*, *11*(1), 98–111.

Shang, Y., Hasan, M. K., Ahammed, G. J., Li, M., Yin, H., & Zhou, J. (2019). Applications of nanotechnology in plant growth and crop protection: A review. *Molecules*, *24*(14), 2558.

Sharma, P., Jha, A. B., Dubey, R. S., & Pessarakli, M. (2012). Reactive oxygen species, oxidative damage, and antioxidative defense mechanism in plants under stressful conditions. *Journal of Botany*, *2012*, 1–26.

Sharma, V., & Rao, L. J. M. (2013). An overview on chemical composition, bioactivity and processing of leaves of Cinnamomum tamala. *Critical Reviews in Food Science and Nutrition*, *54*(4), 433–448.

Singh, A., Singh, N. B., Hussain, I., Singh, H., Yadav, V., & Singh, S. C. (2016). Green synthesis of nano zinc oxide and evaluation of its impact on germination and metabolic activity of Solanum lycopersicum. *Journal of Biotechnology*, *233*, 84–94.

Singh, J., Dutta, T., Kim, K. H., Rawat, M., Samddar, P., & Kumar, P. (2018). 'Green' synthesis of metals and their oxide NPs: Applications for environmental remediation. *Journal of Nanobiotechnology*, *16*(1), 1–24.

Sinha, S. N., Paul, D., Halder, N., Sengupta, D., & Patra, S. K. (2015). Green synthesis of silver NPs using fresh water green alga Pithophora oedogonia (Mont.) Wittrock and evaluation of their antibacterial activity. *Applied Nanoscience*, *5*(6), 703–709.

Sivaraj, A., Kumar, V., Sunder, R., Parthasarathy, K., & Kasivelu, G. (2020). Commercial yeast extracts mediated green synthesis of silver chloride NPs and their anti-mycobacterial activity. *Journal of Cluster Science*, *31*(1), 287–291.

Srivastava, A., Bhatt, N. M., Patel, T. P., Dadheech, N., Singh, A., & Gupta, S. (2021). Biogenic synthesis of silver NPs, characterization and their applications – A review. *Surfaces*, *5*(1), 67–90.

Strasser, P., Koh, S., Anniyev, T., Greeley, J., More, K., Yu, C., Liu, Z., Kaya, S., Nordlund, D., Ogasawara, H., Toney, M. F., & Nilsson, A. (2010). Lattice-strain control of the activity in dealloyed core–shell fuel cell catalysts. *Nature Chemistry*, *2*(6), 454–460.

Su, Y., Ashworth, V., Kim, C., Adeleye, A. S., Rolshausen, P., Roper, C., White, J., & Jassby, D. (2019). Delivery, uptake, fate, and transport of engineered NPs in plants: A critical review and data analysis. *Environmental Science: Nano*, *6*(8), 2311–2331.

Tan, S. (2021). Transmission electron microscopy: Applications in nanotechnology. *IEEE Nanotechnology Magazine*, *15*(1), 26–37.

Taylor, A. F., Rylott, E. L., Anderson, C. W. N., & Bruce, N. C. (2014). Investigating the toxicity, uptake, NP formation and genetic response of plants to gold. *PLoS One*, *9*(4), e93793.

Thakkar, K. N., Mhatre, S. S., & Parikh, R. Y. (2010). Biological synthesis of metallic NPs. *Nanomedicine: Nanotechnology, Biology and Medicine*, *6*(2), 257–262.

Tomaszewska, E., Soliwoda, K., Kadziola, K., Tkacz-Szczesna, B., Celichowski, G., Cichomski, M., Szmaja, W., & Grobelny, J. (2013). Detection limits of DLS and UV-Vis spectroscopy in characterization of polydisperse NPs colloids. *Journal of Nanomaterials*, 2013, 313081.

Tripathi, D. K., Shweta, Singh, S., Singh, S., Pandey, R., Singh, V. P., Sharma, N. C., Prasad, S. M., Dubey, N. K., & Chauhan, D. K. (2017). An overview on manufactured NPs in plants: Uptake, translocation, accumulation and phytotoxicity. *Plant Physiology and Biochemistry PPB*, *110*, 2–12.

Van Zelm, E., Zhang, Y., & Testerink, C. (2020). Salt tolerance mechanisms of plants. *Annual Review of Plant Biology*, *71*, 403–433.

Verma, K. K., Song, X.-P., Joshi, A., Tian, D.-D., Rajput, V. D., Singh, M., Arora, J., Minkina, T., & Li, Y.-R. (2022a). Recent trends in nano-fertilizers for sustainable agriculture under climate change for global food security. *Nanomaterials (Basel, Switzerland)*, *12*(1), 173.

Verma, K. K., Song, X.-P., Joshi, A., Rajput, V. D., Singh, M., Sharma, A., Singh, R. K., Li, D.-M., Arora, J., Minkina, T., & Li, Y.-R. (2022b). Nanofertilizer possibilities for healthy soil, water, and food in future: An overview. *Frontiers in Plant Science*, *13*, 865048.

Verma, N., Sao, P., Srivastava, A., & Singh, S. (2021). Physiological and molecular responses to drought, submergence and excessive watering in plants. In A. Husen (Ed.), *Harsh environment and plant resilience* (pp. 305–321). Springer.

Vishwakarma, K., Upadhyay, N., Kumar, N., Tripathi, D. K., Chauhan, D. K., Sharma, S., & Sahi, S. (2018). Potential applications and avenues of nanotechnology in sustainable agriculture. *Nanomaterials in Plants, Algae and Microorganisms*, *1*, 473–500.

Vladár, A. E., & Hodoroaba, V. D. (2020). Characterization of NPs by scanning electron microscopy. In V.-D. Hodoroaba, W. E. S. Unger, & A. G. Shard (Eds.), *Characterization of nanoparticles: Measurement processes for nanoparticles* (pp. 7–27). Elsevier.

Vorontsov, A. V., & Tsybulya, S. V. (2018). Influence of NPs size on XRD patterns for small monodisperse NPs of Cu0 and TiO_2 anatase. *Industrial & Engineering Chemistry Research*, *57*(7), 2526–2536.

Wagner Ferreira Sabará, E., Pereira, V., Luiz Molisani, A., Mary Ealias, A., Saravanakumar, M. P. (2017). A review on the classification, characterisation, synthesis of NPs and their application. *IOP Conference Series Materials Science and Engineering*, *263*(3), 032019.

Wang, C., Mirzaei, A., Wang, Y., Chaker, M., Zhang, Q., & Ma, D. (2023). $BiVO_4$ microspheres coated with nanometer-thick porous TiO_2 shells for photocatalytic water treatment under visible-light irradiation. *ACS Applied Nano Materials Journal*, *6*, 5545–5556.

Wang, W., Tarafdar, J., & Biswas, P. (2013). NP synthesis and delivery by an aerosol route for watermelon plant foliar uptake. *Journal of Nanoparticle*, *15*(1), 1417.

Wang, Z., Li, H., Li, X., Xin, C., Si, J., Li, S., Li, Y., Zheng, X., Li, H., Wei, X., Zhang, Z., Kong, L., & Wang, F. (2019). Nano-ZnO priming induces salt tolerance by promoting photosynthetic carbon assimilation in wheat. *Archives of Agronomy and Soil Science*, *66*(9), 1259–1273.

Wang, Z., Xie, X., Zhao, J., Liu, X., Feng, W., White, J. C., & Xing, B. (2012). Xylem- and phloem-based transport of CuO NPs in maize (Zea mays L.). *Environmental Science & Technology*, *46*(8), 4434–4441.

Waseda, Y., Matsubara, E., & Shinoda, K. (2011). *X-ray diffraction crystallography: Introduction, examples and solved problems*. Springer-Verlag.

Wiechers, J. W., & Musee, N. (2010). Engineered inorganic NPs and cosmetics: Facts, issues, knowledge gaps and challenges. *Journal of Biomedical Nanotechnology*, *6*(5), 408–431.

Wu, P., Cui, P., Du, H., Alves, M. E., Zhou, D., & Wang, Y. (2021). Long-term dissolution and transformation of ZnO in soils: The roles of soil pH and ZnO particle size. *Journal of Hazardous Materials*, *415*, 125604.

Yamauchi, Y., Furutera, A., Seki, K., Toyoda, Y., Tanaka, K., & Sugimoto, Y. (2008). Malondialdehyde generated from peroxidized linolenic acid causes protein modification in heat-stressed plants. *Plant Physiology and Biochemistry PPB*, *46*(8–9), 786–793.

Yano, F., Hiraoka, A., Itoga, T., Kojima, H., Kanehori, K., & Mitsui, Y. (1996). Influence of ion-implantation on native oxidation of Si in a clean-room atmosphere. *Applied Surface Science*, *100–101*, 138–142.

Yurkov, A. M., Kemler, M., & Begerow, D. (2011). Species accumulation curves and incidence-based species richness estimators to appraise the diversity of cultivable yeasts from beech forest soils. *PLoS One*, *6*(8), e23671.

Zahra, N., Shaukat, K., Hafeez, M. B., Raza, A., Hussain, S., Chaudhary, M. T., Akram, M. Z., Kakavand, S. N., Saddiq, M. S., & Wahid, A. (2021). Physiological and molecular responses to high, chilling, and freezing temperature in plant growth and production: Consequences and mitigation possibilities. In A. Husen (Ed.), *Harsh environment and plant resilience* (pp. 235–290). Springer.

Zhang, P., Ma, Y., & Zhang, Z. (2015). Interactions between engineered nanomaterials and plants: Phytotoxicity, uptake, translocation, and biotransformation. In M. H. Siddiqui, M. H. Al-Whaibi, & F. Mohammad (Eds.), *Nanotechnology and plant sciences: Nanoparticles and their impact on plants* (pp. 77–99). Springer.

Zhang, Q., Ying, Y., & Ping, J. (2022). Recent advances in plant nanoscience. *Advanced science (Weinheim, Baden-Wurttemberg, Germany)*, *9*(2), e2103414.

Zhang, Y., Hu, C. X., Tan, Q. L., Zheng, C. S., Gui, H. P., Zeng, W. N., Sun, X. C., & Zhao, X. H. (2014). Plant nutrition status, yield and quality of satsuma mandarin (Citrus unshiu Marc.) under soil application of Fe-EDDHA and combination with zinc and manganese in calcareous soil. *Scientia Horticulturae (Amsterdam)*, *174*(1), 46–53.

Zhang, Y., Zhou, Y., Zhang, D., Tang, X., Li, Z., Shen, C., Han, X., Deng, W., Yin, W., & Xia, X. (2020). PtrWRKY75 overexpression reduces stomatal aperture and improves drought tolerance by salicylic acid-induced reactive oxygen species accumulation in poplar. *Environmental and Experimental Botany*, *176*, 104117.

Zhao, L., Ortiz, C., Adeleye, A. S., Hu, Q., Zhou, H., Huang, Y., & Keller, A. A. (2016). Metabolomics to detect response of lettuce (Lactuca sativa) to $Cu(OH)_2$ nanopesticides: Oxidative stress response and detoxification mechanisms. *Environmental Science & Technology*, *50*(17), 9697–9707.

Zulfiqar, F., & Ashraf, M. (2021). NPs potentially mediate salt stress tolerance in plants. *Plant Physiology and Biochemistry*, *160*, 257–268.

23 Nanobiotechnology Approaches for Enhancing Crop Resilience to Salinity Stress

Abhishek Singh, Shreni Agrawal, Shivangee Solanki, Vishnu D. Rajput, Tatiana Minkina, Saglara Mandzhieva, Mohamed S. Elshikh, Hassan Ragab El-Ramady, Abdel Rahman Mohammad Al Tawaha, and Karen Ghazaryan

23.1 INTRODUCTION

Plant growth and development, significantly influenced by climate change, is critically stunted in the face of abiotic stresses, thereby negatively impacting crop yield and quality (Jithesh et al. 2006). As a consequence, abiotic stressors compromise the long-term viability and yield of the global agricultural sector. With respect to crop growth and productivity, salt stress has garnered immense concern as one of the most critical abiotic stresses (Hasanuzzaman et al. 2020). With respect to land area, salinity afflicts an estimated 53% of Asia and Australia, 13% of Latin America, 10% of Europe, 9% of Africa, and 2% of North America (Singh et al. 2020). When the salinity of the soil substrate increases, the plants' own internal supplies of sodium (Na^+) and chloride (Cl) are observed to be elevated. While both Na^+ and Cl ions are deemed important for plant metabolism, beyond their threshold concentrations, they greatly limit plant life, and Cl is documented to inflict a greater threat (Wang and Huang 2019). Osmotic stress, ionic stress, and nutrient stress are negative aftermaths of salinity. An overabundance of Na^+ disturbs cell proliferation and development by promoting osmotic and ionic imbalance, membrane dysfunction, and an increase in the production of reactive oxygen species (ROS) (Sharma et al. 2012). Plant cells generate and accumulate ROS when exposed to salinity, which manifests as ionic stress, osmotic stress, and nutritional stress. Superoxide anion radicals ($O_2^{\bullet-}$), hydroxyl radicals ($OH^{\bullet}$), hydrogen peroxide (H_2O_2), and singlet oxygen ($^1O_2$) are common oxidative stress-inducing species generated within chloroplasts, mitochondria, peroxisomes, and the apoplast of plant cells, respectively. The rate of generation of chloroplastic ROS is 10× higher than that of mitochondrial ROS under light conditions. ROS are paradoxical chemicals which damage cells at lethal concentrations and serve as essential secondary messengers that facilitate fundamental processes at the cellular level. In plants, ROSs sustain several crucial processes, for instance, cell cycle regulation and programmed cell death, and they may crosstalk with other signaling molecules like salicylic acid, jasmonic acid (JA), ethylene, nitric oxide, and calcium. Under salt stress, rice has been found to produce both ROS and phytohormones (particularly, ethylene).

Potent ROSs are produced as a result of high-energy exposure reactions, such as the electron transfer chain, that gradually lowers the molecular oxygen (O_2) (Figure 23.1). Plants consistently generate ROS as a byproduct of metabolism occurring across all cellular compartments or as a

DOI: 10.1201/9781003457237-23

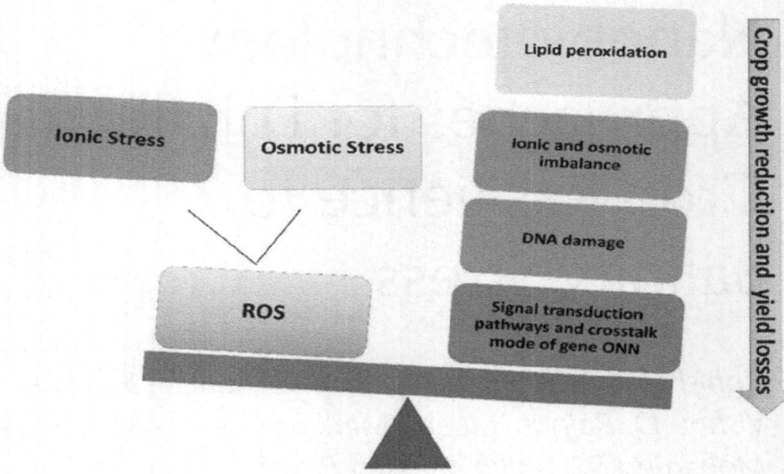

FIGURE 23.1 The role of reactive oxygen species (ROS) in damaging plant growth, yield, and underlying mechanisms.

result of the inevitable leaking of electrons onto O_2 from the electron transport processes of the plasma membrane, mitochondria, and chloroplasts. The plant cellular homeostasis is thrown into chaos when under salt stress resulting in exaggerated production of ROS (Figure 23.1) (Van Zelm et al. 2020). Beyond the safety-threshold level, ROSs are deadly and environment-unfriendly. The physiological condition wherein overabundance of ROS impairs all other cellular processes is identified as oxidative stress, and it eventually develops into critical oxidative injuries (Chawla et al. 2013). Normal cells are put in jeopardy of protein oxidation, lipid peroxidation (LPO), nucleic acid degradation, initiation of programmed cell death or apoptosis, enzyme suppression, and enzyme inactivation due to the overabundance of ROS induced by extrinsic environmental variables, ultimately resulting in cell cytotoxicity (Chawla et al. 2013).

ROS accumulation within plant tissues is recognized as the benchmark symptom of hyperosmotic disparity and/or ionic misproportion that can potentially inflict subcellular injuries (Hasanuzzaman et al. 2020). One of the byproducts of the lipid oxidation pathway, malondialdehyde (MDA), is produced in significant amounts when the cell is under oxidative stress, and it negatively impacts metabolism and other physiological activities (Figure 23.1). Irreversible loss of membrane integrity causes a state of crisis within the cell involving disruption of functions of bio-membranes, cellular homeostasis, molecule transport, cell signal transduction, and process compartmentalization (Hasanuzzaman et al. 2019). Besides losing to hyperosmotic stress and ionic disturbance, the nutritional imbalance thus caused diminished photosynthetic proficiency due to altered levels of photosynthetic pigments thereby compromising the survivability of the plant (Ali et al. 2013). Furthermore, researchers have revealed an intricate correlation linking modification in gene expression and the unrestrained ROS production in organelles, namely, chloroplasts, mitochondria, and peroxisomes of plants. Expression of antioxidant enzyme coding gene expression is a general stress response extensively evaluated in plants.

Immoderate levels of free radicals and ROS/RNS are managed by cells' intrinsic antioxidative machinery consisting of both enzymatic and non-enzymatic antioxidants (Gill and Tuteja 2010). The ascorbate–glutathione (AsA–GSH) cycle is mediated by the enzymes monodehydroascorbate reductase (MDHAR), dehydroascorbate reductase (DHAR), glutathione reductase (GR), and ascorbate peroxidase (APX) (Gill and Tuteja 2010) and non-enzymatic antioxidant compounds carotenoids, phenolics, ascorbate (AsA), glutathione, and tocopherols. The degree of advantage or harm inflicted by ROS is strictly a function of its intracellular concentration (Wang and Huang 2019).

In light of the same, a myriad of plant morphological, physiological, biochemical, and molecular processes, namely, seed germination and development, root formation and elongation, shoot development, lateral growth, flower induction and maturation, tropic movements, water and nutrient uptake, internal translocation, stomatal conductance, number of trichomes, rate of photosynthesis, leaf water potential and turgor pressure, synthesis of osmoprotectants, activity index of antioxidant enzymes, production of phytohormones (abscisic acid [ABA], JA, CH_4), and ionic equilibrium (Na^+ and Cl^-), are affected by ROS-induced oxidative and ionic stress (Ahmad et al. 2019). With the advent of technological advancements, plant's ability to tolerate lethal salt concentrations is augmented in innovative ways such as those involving the exogenous application of antioxidants and genetic engineering in conjugation with nanoparticles (NPs). Of all approaches, nanotechnological interventions have exhibited particularly promising and transformative potential to revolutionize the domain of crop improvement. Fostering plant salt tolerance with the help of nanotechnology provides a versatile and feasible alternative to conventional crop improvement programs. NPs' potency to protect photosynthesis, promote ROS detoxification, and reduce osmotic and ionic stress has rendered them propitious candidates to serve as frontiers of plant enhancement initiatives (Etesami et al. 2021; Parihar et al. 2015; Zulfiqar and Ashraf 2021). Owing to their unique morphology (size 1–100 nm, shapes and structure) and high surface reactivity, boosted efficacy, minimal eco-toxicity, and resource-effective fabrication process, NPs have emerged as viable replacements for traditional ways of crop protection. Studies evaluating the effects of the application of Ag-NPs and C-NPs indicated a distinct increase in enzymatic activity of CAT, POD, and APX, which subsequently restored the content of carbohydrates and osmoprotectants, re-equilibrated redox potential, and re-established ion homeostasis (Mohamed et al. 2017; Shafiq et al. 2019). Similar observations were recorded upon Fe-NP application, which enhances CAT and POX enzymatic activities along with enhanced soluble carbohydrates, and reduced MDA content, thereby improving the dry weight of shoots and amplifying seed yield (Torabian et al. 2018).

23.2 PHYSIO-BIOCHEMICAL RESPONSE IN PLANT UNDER SALINITY STRESS

Salinity stress leads to alteration of multiple metabolic and biochemical processes, the degree of damage of which depends on the magnitude and time duration of the salt stress that impairs plant expansion and growth. Large amounts of soluble salts affect the biochemical traits of the crops by raising the osmotic and ionic potential of the soil mix. Different biochemical traits like photosynthetic pigments content, antioxidant enzymatic activity, MDA, H_2O_2 level, ions, and intercellular proline accumulation are negatively affected by salt toxicity.

23.2.1 Osmotic Stress

Some plants cannot adapt to dry or salty terrains, while others can wilt (dehydrate) under normal conditions. A normally functioning plant root establishes a water potential gradient to draw water up from the soil and into the root system (ideally, the water potential is lower in the roots than in the surrounding soil to facilitate diffusive uptake). Numerous drought-resistant plants adjust their solute potential (Ψs) to survive short-term or chronic water shortages. Osmotic adjustments occur during a net influx of solute into the plant cell. In contrast to the osmolalities attained through passive concentration of solutes through dehydration, those attained through osmotic adjustment are significantly higher. Osmotic alterations allow water to migrate from the soil and into the plant vascular system along a potential gradient by lowering root Ψw value below that of soil Ψw. Plants are thought to rely heavily on osmotic adjustment to survive in environments with low water availability or high salt concentration. Osmotic stress is triggered by salt stress due to ionic imbalance and a severe paucity of water in the plant cell. Some of the short-lived biophysical effects of this process include a drop in cell turgor pressure, plasma membrane shrinkage, and structural changes to the cell wall (Munns and Tester 2008). Plants employ osmotic signaling pathways, which control molecular phenomena like gene expression and the activation of enzymes involved in osmolyte biosynthesis and water transport to deal with osmotic stress (Yang and Guo 2018). Under salt stress, osmolytes like proline, polyols, and sugars build up. By decreasing the osmotic potential in the cytosolic compartment, these osmolytes contribute to the regulation of osmotic pressure. In addition, they have been evaluated to regulate plant growth in the face of salt stress by inducing ABA accumulation, influencing related gene expression, and affecting ABA biosynthesis (Marusig and Tombesi 2020). Protein kinases are a hub for both fast osmoregulation and salt stress signaling (Chen et al. 2021). Histidine kinases, mitogen-activated protein kinases (MAPKs), and mitogen-activated protein kinases are all upregulated in response to osmotic treatments, which in turn increases osmolyte synthesis and accumulation (Moustafa et al. 2014; Zhou et al. 2016). Research indicates that MAPKs play a pivotal role in maintaining a healthy level of ROS in the body (Pan et al. 2012; Zhang et al. 2014a). For instance, ZmMPKs are induced by salt stress, and activation of oxidative stress regulation confers salt stress tolerance (Pan et al. 2012; Zhang et al. 2014a; Moustafa et al. 2014). MPK3 and MPK6 activate the receptor-like kinase salt intolerance 1, which actively participates in the oxidative stress response triggered by salt stress (Li et al. 2014). Elevation of cytosolic Ca^{2+} is a general response to osmotic and salt stresses. Osmotic stress signaling relies heavily on calcium, and a copine protein called Ca^{2+} responsive phospholipid-binding BONZAI1 (BON1) is a key upstream regulator of calcium signaling (Li et al. 2014; Chen et al. 2020). If BON1 is damaged, the osmotic stress response of the cytosolic Ca^{2+} signal is reduced.

Plant oxidative damage is triggered by metabolic imbalance and the accretion of toxic ROS, both of which are triggered via ion stress and osmotic stress (Chen et al. 2020). Plant organelles like the chloroplast, peroxisome, mitochondria, and apoplast all produce ROS when subjected to salt stress. As ROS builds up in a plant cell, the cell quickly responds by activating a series of adaptive responses by scavenging the ROS (Van Zelm et al. 2020).

At physiologically non-lethal concentrations, ROS serve as fundamental signaling molecules. In other words, ROS production and ROS scavenging are finely tuned to a delicate balance through the use of stringent control mechanisms. The activation of ROS scavengers and the mediation of the gene expression of ROS-responsive genes were both found to play principal roles in oxidative stress regulation under salt stress (Nadarajah 2020). An assortment of evaluations have documented that salt stress acts as a stimulus for the activation of ROS-scavenging enzymes and antioxidants (Sofo et al. 2015). For instance, APX and catalase (CAT) are both triggered by salt stress, which results in bolstered defense responses to both salinity and oxidative stresses (Lu et al. 2007). Increased salt tolerance in rice has been linked to overexpression of the OsAPXa and OsAPXb. OsGSTU4 (glutathione S-transferase) constitutive expression in *Arabidopsis* also improves salt tolerance (Sharma et al. 2014). OsMADS25, a MADS-box transcription factor (TF) in rice, is essential for salt tolerance due to its function in ROS homeostasis (Xu et al. 2018). Detoxification processes in response to salt stress stimuli are mediated by senescence-associated genes (Seo et al. 2011). When SAG29 is overexpressed in *Arabidopsis*, the plant is extremely susceptible to salt treatment, while knockout seedling mutants are resistant to the adversities of salt toxicity. To regulate plant growth and response to stress, the RBOH gene, which encodes a NADPH oxidase homolog, has been shown to mediate ROS synthesis and ROS scavenging (Chapman et al. 2019).

23.2.2 Ionic Stress

The structural constituents of cells are built up of minerals that typically exist in the form of ions. Mineral elements are vital sub-parts of enzymes and coenzymes essentially regulating enzyme activity. Electrochemically, they contribute to osmotic regulation, colloidal stabilization, and charge neutralization. Maintenance of cell membrane stability, physiological integrity, and promotion of healthy growth and development in plants relies, in large part, on the appropriate regulation of ion metabolism. Damage from salt is typically caused by an excess of Na^+ and Cl and a paucity of other functionally important ions like Ca^{2+} and K^+. Plants store excess sodium because of the high concentration of sodium in the soil. A chemical gradient favors the uptake of Cl, which can be facilitated by a decrease in membrane potential caused by an excess of Na^+. A cell's metabolism and some enzymes can be damaged by an excess of Na^+ (Flowers and Colmer 2015). Osmotic imbalance, increased production of ROS, and membrane dysfunction are all repercussions of the presence of excess Na^+, which in turn impacts cell proliferation and development (Yang and Guo 2018). High levels of Cl inhibit plant growth by reducing chlorophyll content and disrupting the cell membrane system and organelle structure (Munns and Tester 2008). Additionally, Na^+ exhibits a clear competitive inhibition of K^+ due to their shared ionic radius and hydration energy. To keep functioning normally, the cytoplasm of most cells keeps a high concentration of K^+ and a low concentration of Na^+. As a result, plant damage from K^+ deficiency will occur if there is an excessive amount of external Na^+ influx. The Ca^{2+} level in plant cells is lowered because of the competitive inhibition of Na^+, which is an essential element in signal transduction across a wide range of plant physiologies. Zhou et al. (2016) found that as the NaCl concentration in flowering *Schizonepeta tenuifolia* increased, the Ca^{2+}/Na^+ value decreased. Ottow et al. (2005) state that plants' osmotic pressure is partially controlled by a compensatory decrease in Ca^{2+} when exposed to high concentrations of Na^+. The uptake of NH_4^+ and NO_3 by plants is also inhibited under salt stress (Song et al. 2006). In a nutshell, salt stress hampers plant development by upsetting the delicate ionic imbalance established within plant cells.

23.3 GENERATION OF ROS IN PLANTS UNDER SALINITY STRESS

Under salinity stress, biochemical changes occur and produce ROS (Zhu 2016). ROS are overly reactive, and their oxidative damage to proteins, lipids, and nucleic acids can seriously harm metabolism in the absence of any protective mechanisms (Apel and Hirt 2004; Rengasamy and Rengasamy 2010). In salt-stressed plants, the electron acceptor is O_2, which increases the accumulation of ROS, including ($^1O_2$), (OH^*), and (O_2^{*-}), and H_2O_2 is a strong oxidizing agent and therefore is potentially harmful to the integrity of cell membranes and organelles (Santos et al. 2018). The primary target of salt stress is cell membranes, because the resulting ROS may oxidize the membrane lipids and destroy the metabolic activity (Caverzan et al. 2019). Alteration of the lipid–protein integrity of the membranes by the dint of peroxidation of lipids further denatures diverse biomolecules like carbohydrates and nucleic acids. The intrinsic amalgamation of ROS causes the production of MDA through lipid oxidation that imposes a negative impact on cellular metabolism and physiological activities in conjugation with inhibition of enzyme activities and damage to prominent cellular components, thereby impeding plant growth and development (Zhang et al. 2007).

23.3.1 Lipids

Autoxidation of lipids in living systems occurs via a series of sequential reactions, beginning from initiation, followed by propagation and termination. The first step is peroxidation of lipids, which begins when a highly reactive carbon-centered lipid radical (•L) is formed when an activated free radical, like ($^1O_2$, $O_2^•$, or •OH), reacts with a lipid substrate (LH). Lipid peroxyl radical (LOO•) is produced during the second step of LPO when O_2 reacts with lipids at a high rate. Following the initial peroxidation, LOO• removes an atom of hydrogen from another LH, thus creating lipid hydroperoxide (LOOH) and an extremely unstable, chemically reactive carbon-centered radical (L•), which then extends the chain reaction; in the subsequent and final step, LPO is terminated by the coupling of any two radicals to form non-radical products (NRP). When it comes to LPO reactions, NRPs are stable but unable to propagate them (Porter et al. 1995). In LPO, transition metal ions like Cu and Fe are crucial. By degrading LOOH, ferrous (Fe^{2+}) and ferric (Fe^{3+}) ions catalyze the elongation of the LPO chain. By reacting with other lipid molecules, the resulting alkoxyl (LO•) and peroxyl (LOO•) radicals can trigger the formation of new radical chains. Cell membrane LPO has been linked to a variety of physiological processes, including those that affect membrane rigidity, cellular deformability, and erythrocyte membrane lipid fluidity

(Matkovics et al. 1997). In membranes of cells and organelles, LPO takes place spontaneously when ROS formation exceeds the equilibrium level, disrupting normal cellular functioning. Damage to proteins and DNA is caused by LPO, which in turn generates lipid free radicals. Increased lipid degradation and ROS production have both been observed in saline-stressed plants (Mishra et al. 2011). Distortion of cellular bio-membranes and loss of structural integrity are majorly credited to the production of MAD (malonaldehyde) (Foyer and Halliwell 1976). Phospholipids are largely constituted of polyunsaturated fatty acids that are highly susceptible to ROS.

22.3.2 Protein

Alterations to protein biochemistry and rebarbative modifications to their functional efficiencies can be induced by ROS in a variety of ways, both directly and indirectly. Nitrosylation, carboxylation, disulfide bond formation, and glutathionylation cause direct modifications to protein activity, while conjugation with simplified products of fatty acid peroxidation is an indirect modification (Yamauchi et al. 2008). Proteins become more susceptible to proteolysis when exposed to high levels of ROS and altered electric charge. Tissues are damaged by oxidative stress because they typically have a high concentration of carbonylated proteins and are used as characteristic markers for the detection of protein oxidation (Møller and Kristensen 2004). Protein modification is increased in stressed plants (Tanou et al. 2009). Iron-sulfur center oxidation by $O_2^{\bullet-}$ that inhibits enzyme activity (Gardner and Fridovich 1991). Metal treatment improved the degradation of oxidized proteins and increased proteolytic activity by 20% (Romero-Puertas et al. 2002). Multiple studies have reported that an increase in damage causes protein enzymes to become inactive and lose their catalytic functionality (Bechtold et al. 2009).

23.3.3 DNA

DNA damage is largely credited to ROS (Imlay and Linn 1988). DNA is the cell-derived genetic library, which when hampered sets off a domino effect across fundamental components of the cell such as proteins. Nuclear, mitochondrial, and chloroplastic DNA are all vulnerable to ROS-induced oxidative damage. Nicks in strands, oxidized sugars, loss of some nucleotides, broken phosphodiester bonds within DNA, and severed bonds between DNA and proteins are a few molecular manifestations of ROS stress onto the DNA. Further, mutations arise from changes in both strands' nucleotides. Nucleic acid degradation in plants is accelerated and exaggerated in the face of environmental adversities, including salinity (Liu et al. 2000). When exposed to ROS, DNA's sugar and base components are damaged. ROS can damage DNA in ways that can lead to mutations (Fink et al. 1997). DNA strand breakage is caused by ROS attacking DNA sugar (Evans et al. 2004). Interestingly, nuclear DNA was evaluated to be more tolerant to oxidative injuries than mitochondrial and chloroplastic DNA due to the presence of shielding proteins and histones (Gill and Tuteja 2010). Despite the presence of an advanced and well-established repair system to mend DNA damage, acute alterations caused by ROS result in irreversible changes to the DNA and irreparable impairment to the cell.

23.4 PHOTOSYNTHETIC CONTENTS UNDER SALT STRESS

The prominent and functionally pivotal photosynthetic pigment of the plant system is chlorophyll, and the rate of photochemical reaction ATP, NADPH, sucrose, and starch production depends on the activity of chlorophyll molecules. A decrease in crop production during salt stress conditions is directly related to reducing the level of chlorophyll content of the plants (Ghosh et al. 2011; Kibria et al. 2017).

Hu et al. (2019) observed that the first observable symptom of a plant under salt stress is a decline in the amount of chlorophyll in the leaves. Change in the content of chlorophyll in leaves depends on the genotypes of plants; i.e., salt-sensitive genotypes show more loss of chlorophyll compared to salt-resistant genotypes. Because chlorophyll pigment production is inhibited by Na^+ deposition under salt stress due to the degradation of the membrane of the chloroplast, plants experience reduced chlorophyll concentration in their leaves. Elsheery and Cao (2008) elucidated that chlorophyll a and b content decreased under salinity stress because of the inhibition of photo-protectants and reducing the absorption of light due to a decrease in the chlorophyll quantity. Plants subjected to salt stress for a prolonged period of time tend to produce potent ROS, especially $^1O_2$ species, due to the critical inactivation of photosystem II (PSII) and the diversion of electrons to O_2 (Ling and Jarvis 2015; Gulzar et al. 2020). Chloroplasts in photosynthetically active leaves are the primary sites of ROS production during the day (Wrzaczek et al. 2013). Furthermore, photorespiration in C3 plants is a source of ROS generation in peroxisomes (Gulzar et al. 2020). This process is triggered by the oxygenase activity of Rubisco in chloroplasts. Several plant species have been reported to have increased electron flow to O_2 in response to salinity (Stepien and Johnson 2009; Gulzar et al. 2020). ($^1O_2$), the ($O_2^{\bullet}$), H_2O_2, and $^{\bullet}OH$ are the primary ROS produced by chloroplasts (Stepien and Johnson 2009; Bose et al. 2014). As H_2O_2 (a non-radical ROS) formation after exposure to salinity is easier to detect, it is the more common focus of studies (Stepien and Johnson 2009). Chloroplasts have a wide variety of enzymatic and non-enzymatic antioxidants to neutralize ROS and prevent oxidative damage (Mittler 2002). AsA and GSH are principal non-enzymatic antioxidants originating in chloroplasts, while superoxide dismutase (SOD) and enzymes of the Foyer–Halliwell–Asada pathway (also identified as the (AsA–GSH) are key members of the enzymatic antioxidant pool (Sharma et al. 2012). Various combinations of antioxidants play a crucial role in maintaining ROS levels within 'functionally useful' ranges for signaling a wide variety of plant processes and stress responses. Salt tolerance is heavily dependent on the water–water cycle, which is also one of the fundamental processes involved in the sustenance of a healthy ROS balance in chloroplasts (Suo et al. 2017). As a consequence of the Mehler reaction, ground-state O_2 evolved during the photolysis of water at PSII readily accepts electrons from excited photosystems, in particular the thylakoid membrane-bound primary electron acceptor of PSI (Asada 2006). This results in the formation of free radical species such as $O_2^{\bullet}$. The production of $O_2^{\bullet}$ by PSII from electron leakage to O_2 may also involve the acceptor side of the electron transport chain. The 'water–water cycle' is completed when the thylakoid membrane-bound enzyme Cu/Zn SOD converts $O_2^{\bullet}$ to H_2O_2, which is then reduced into water by the thylakoid membrane-bound enzyme tAPX (Edreva 2005). The (AsA–GSH) is a functional extension of the water–water cycle and is responsible for quenching free radicals and oxidative elements in chloroplasts by consuming NADPH (Noctor and Foyer 2003). This pathway further aids in restricting the over-reduction of photosystems by providing NADP (the final electron acceptor of PSI), which reduces the likelihood of further ROS generation. Dismutation of $O_2^{\bullet}$ by SOD results in the production of H_2O_2, which is then neutralized by stromal APX using AsA as an electron donor. MDHAR and/or DHAR are responsible for the recycling of oxidized AsA-DHAR. By utilizing NADPH as an electron donor, GR recycles GSH (Noctor and Foyer 2003). Under environmentally unfavorable conditions, chloroplasts are documented to upregulate enzymes involved in antioxidant processes, thereby bolstering resistance (Ozgur et al. 2013). For instance, the halophytic wild tomato, *Lycopersicon pennellii*, exhibited remarkable SOD, APX, and MDHAR activities in their chloroplasts when exposed to salt stressor than the common tomato *L. esculentum* (Ozgur et al. 2013). Moreover, it has been reported that thioredoxin/peroxiredoxin (Trx/Prx) and GPX quenched the excess H_2O_2 detected in the primary photosynthetic apparatus found in plants (Suo et al. 2017).

23.5 TREND IN ALTERATION OF NA^+/K^+ RATIO IN PLANTS UNDER SALINITY STRESS

Salinity stress inflicts both osmotic and ionic injuries onto plants that eventually develop into nutrient disorders. Salt stress disrupts plant cytoplasmic Na^+/K^+ homeostasis, which further increases the cytosolic Na^+ to K^+ ratio (Tavakkoli et al. 2010). Salt tolerance is closely linked to osmotic regulation that competes with Na^+ absorption (Gul et al. 2019). Salt tolerance mechanisms aimed to alleviate Na^+ toxicity via diminishing osmotic and ionic stress also increase K^+ concentrations (Gul et al. 2019). Salinity tolerance genotypes of crops essentially target intrinsic ionic ratios and, therefore, reduce the Na^+/K^+ ratio and increase the K^+/Na^+ ratio simultaneously in both roots and shoots; meanwhile, salt-sensitive crop genotypes have higher Na^+/K^+ ratio that decreases the K^+ concentration and, hence, leads to an undesirable decline in crop productivity and yield (Zhang et al. 2010; Nakhoda et al. 2012; Zheng et al. 2012).

23.6 NPS: A SUSTAINABLE APPROACH FOR COMBATING AGAINST SALINITY STRESS

After several imperative evaluations, it has been verified that NPs boost the operation of the antioxidant enzyme pool in response to abiotic adversities. Researchers and investigators have ascribed improvement of salt tolerance and stress acclimatization in plants to similarities shared between NPs and naturally occurring antioxidant enzymes. With respect to their experimental observations, few NPs assist the plant in combating oxidative injuries and associated anomalies by exhibiting properties like some antioxidant enzymes. For instance, the activity of CAT and POD enzymes overlaps with those of iron, cesium, and cobalt-derived NPs (Rico et al. 2015). Furthermore, supplementation of pre-treated wheat seeds with zeolite and zinc enhanced the effects of Ag-NPs application. The plantlets exhibited enhanced plant growth, proline content, soluble sugar content, and POD activity (Mohamed et al. 2017). Si and B NPs illustrated substantial upgradation in water and nutrient retention and utilization, energetic efficiency of photo autotrophy, and antioxidant enzymatic activity (Mahmoud et al. 2019; Zulfiqar and Ashraf 2021).

Cu is an indispensable metallic micronutrient necessary for the basic metabolic processes of the cell. Several proteins participating in the electron transfer chain in the thylakoid lumen require Cu as a functional cofactor. For instance, plastocyanin is the most abundant ETC protein and a significant Cu user (Yokotani et al. 2011). The importance of Cu for ethylene production, photosynthesis, respiration, and the metabolism of C and N has been exhaustively studied by several investigators (Iqbal et al. 2018). Furthermore, enzymes catalyzing redox reactions are limited by a conformational requirement of Cu for their activity (Lwalaba et al. 2020). One of the many isoforms of SOD utilizes Cu in its active form, indicating that Cu indirectly aids in alleviating oxidative distress in plants (Lwalaba et al. 2020). Reports claim that the application of Cu-NPs to maize plants meliorates the amino acid pool and osmo-protectant content and strengthens the defense responses of antioxidants, which in turn aids plants in escaping the negative ramifications of salt stress on nutrition, photosynthesis, and water translocation (Iqbal et al. 2018). Fruits treated with engineered Cu-NPs possessed high levels of vitamin C (increment by 80%), phenols (16%), and GSH (increased by 81%) against controls. Other enhancements were perceived as a 140% increase in APX activity in leaves, 26% in GPX, an 8% increase in SOD, and a 93% increase in CAT (Pérez-Labrada et al. 2019).

A wide array of research analyses has testified to the potent facilitator role of Fe-NPs in fostering plant defense systems against abiotic stressors. Fe_2O_3 in the form of NPs provided favorable results upon application on the plants of *Dracocephalum moldavica* species stressed by salt (Moradbeygi et al. 2020). A group of investigators revealed that plants treated with 60 mg/L of nano Fe_2O_3 resulted in a direct increment of antioxidant enzymatic activity. While under salt stress, treatment of Fe_2O_3 (3 mM) in conjugation with SA on *Trachyspermum ammi* curbed salt toxicity by increasing the K^+/Na^+ ratio and proficiency of the antioxidant system (Abdoli et al. 2020). Additionally, foliar application of Fe-NPs on *Helianthus annuus* grown in a saline soil regimen enhanced the enzymatic efficiency of CAT, POD, and polyphenol oxidase (Torabian et al. 2018). Low concentrations of multi-walled carbon nanotubes (MWCNTs) successfully enriched photosynthetic pigments, triggering antioxidant defense system enzymes (such as APX, CAT, and GP) and non-enzymatic antioxidants (such as the amount of phenol) in *Ocimum basilicum* compromised with salt toxicity (Gohari et al. 2020). A fairly recent study conducted on *D. moldavica* plants exposed to 100 mg/L TiO_2 NPs exemplified enhanced plant growth and a concomitant enhancement in the functioning profiles of various enzymes (Gohari et al. 2020).

Photosynthesis is a distinctive energy-generating process restricted to only photoautotrophs. It is the absolute source of energy available directly to autotrophs and indirectly to heterotrophs. Starter materials, namely, chlorophyll, water, and carbon dioxide are utilized during the conversion of sunlight into usable, chemical energy, excess of which can be stored in sugar molecules in chloroplasts for future use. An assortment of experimental evidence has validated that chlorophyll complexes and photosynthesis are primary targets of salinity toxicity however, the application of NPs substantially precludes major compromising consequences, thereby enabling plants to increase their photosynthetic quantum yield and create additional light acquisition compounds (Aghdam et al. 2016). Recently, an extensive experiment recorded a remarkable improvement in plant profile in response to the application of TiO_2-NPs (Ullah et al. 2020). Preconditioning of plants with ZnO-NPs resulted in optimization of the ROS-scavenging system, improvement of photosynthetic productivity, and proficiency electron transfer chain via upregulation of SOD, APX, and CAT activities which mitigated salt stress-induced ROS. Under salt stress, foliar treatment of ZnO-NPs at a concentration of 20 mg/L reduced ion release inside roots and enhanced density of stomatal cells, Hill's response, and relative water content (RWC) (Hezaveh et al. 2019). Foliar application of Se-NPs (10 and 20 mg/L) favorably impacted strawberry plants that were made to grow under variable saline soils (0, 25, 50, and 75 mM NaCl). Se-NPs promoted growth by targeting photosynthetic pigments and shielding them from the degrading tendencies of ROS. Additionally, resistance was credited to a decline in LPO due to an increase in the activities of antioxidant enzymes like SOD and POD (Zahedi et al. 2019).

Ionic imbalance and toxicity are direct manifestations of salt stress which indirectly put the plant cell in jeopardy of disruption of normal life cycle. Since, the presence of salt greatly influences the bioavailability, uptake, and translocation of nutrients from the supporting substratum (soil or artificial growing medium), it is hypothesized that nutritional deficiency and imbalance are additional primary benchmark parameters of exposure to salinity. Salt concentration beyond permissible level results in depolarization of the plasma membrane of root cells and induces outward-rectifying K channels, thus increasing Na^+ and decreasing K^+ levels in plant cells. *S. lycopersicum*, *Vigna radiata*, and *Lens culinaris* seedlings exposed to high concentrations of Na^+ possessed a greater Na^+/K^+ ratio and lower Ca^{2+} and Mg^{2+} content in the roots and shoots. Overabundance of Na^+ in soil displaces the potassium present in enzymes, thereby obstructing cellular processes mandatory for cell survival. In addition, mineral homeostatic aberrations and ionic stress develop due to a decline in the uptake of K^+, Ca^{2+}, and Mg^{2+} and an excessive accumulation of Na^+. Rahman et al., through their investigation, verified that a peak in Na^+ and the Na^+/K^+ ratio distorted ion homeostasis in *O. sativa* seedlings. He further claimed that Na^+ concentration in the shoots was greater than that in other plant organs for reasons unidentified. Moreover, zinc (Zn) content in roots and shoots is also altered in the face of salinity. Kopittke et al. inferred the possibility of improvement of nutrient uptake, absorption, translocation, and allocation by the action of NPs to reverse the adversities of salt stress. Rectifying activity of cerium oxide NPs (nanoceria) includes specific targeting of ROS, OH, in plants lacking enzymatic scavenging pathways. Application of 50 mg/L of cerium-NPs on leaves of *Arabidopsis thaliana* renders OH^- and other free radical species inactive, which subsequently reduces its accumulation. Ce-NPs accomplish so by stimulating the K^+-selective (GORK) channels situated in the plasma membrane primarily responsible for potassium influx within the cell (Wu et al. 2018). Fe is yet another pivotal micronutrient with respect to plant development and growth due to its active participation in enzymatic reactions, photosynthesis, and the operation of photosystems. Iron predominantly exists as insoluble Fe^{3+} hydroxides and/or complexed with organic compounds. Nonetheless, very low concentration (0.01–1 nM) of soluble inorganic Fe^{2+} is also detected in exceptional soil types. Plants resort to two primary Fe mobilization mechanisms, phyto-siderophores (Fe-mobilizing complexes) and chelation, for intrinsic fixing of Fe from the substrate. Under ionic stress, soil pH and ROS levels dramatically rise, hence undermining the bioavailability of Fe. Engineered iron NPs are commonly employed to confer plant resistance against abiotic stressors. At the cellular level, Fe sets off a series of chain reactions, similar to a domino effect, which supports plant nutrient uptake under unfavorable environmental conditions. Growth indices, antioxidant enzyme activities, and chlorophyll content were remarkably improved after exposure to FeO-NPs (Manzoor et al. 2021). Therefore, NPs thwart intracellular salt accumulation to reduce plant susceptibility to fatal salt concentrations.

23.7 APPROACHES AND METHODS FOR NANOPARTICLE SYNTHESIS

Producing NPs with the appropriate size and form may be achieved by the widely used top-down and bottom-up methodologies, accordingly (Patil and Chandrasekaran 2020). Despite having different synthesis principles, the two approaches result in NPs possessing favorable traits (Gwo et al. 2016; Patil and Chandrasekaran 2020). Top-down methods involve consistent breaking of large materials into significantly smaller fragments until they finally form particles of size ranging between 1 and 100 nm (Gwo et al. 2016; Patil and Chandrasekaran 2020). Various manufacturing process options available for the fabrication of NPs include photolithographic methods, grinding, sputtering, and milling (Meyers et al. 2006). Regarding the accuracy of producing NPs, each approach offers benefits and drawbacks. The top-down approach is a feasible strategy that can produce a significant number of NPs. NPs' surface flaws and potential for impairment, however, are the main obstacles to top-down methods (Pérez-de-Luque 2017). In other situations, top-down production of NPs is restricted given that the exterior structure of the NPs affects their visual and physio-chemical characteristics (Khan et al. 2019). The bottom-up method, inclusive of coalescence or assembly of individual atoms, molecules, or clusters to produce a wide variety of NPs, is another popular technique for NP production in the present day and age (Decarolis et al. 2020). NPs are fabricated using conventional methods like self-assembly of monomer or polymer molecules, chemical or electrochemical nano structural precipitation, sol–gel processing, laser pyrolysis, chemical vapor deposition, plasma or flame spraying synthesis, and bio-assisted synthesis (Dhand et al. 2015). Therefore, bottom-up is a versatile technique for fabricating nanoclusters with broad-spectrum pragmatic utility.

23.7.1 Physical Methods

Physical processes include plasma arcing, ball milling, thermal evaporation, spray pyrolysis, ultrathin films, electron irradiation, pulsed laser desorption, lithographic techniques, sputter deposition, layer-by-layer growth, and the diffusion flame synthesis of NPs (Srivastava et al. 2021). In these techniques, metal NPs are produced through evaporation condensation in a tube furnace at atmospheric pressure.

Physical methods for creating NMs include deposition, sputtering, ball milling, and plasma-based processes (Dhand et al. 2015). Nevertheless, physical approaches are deemed to be inadequate due to the very slow rate of production of NPs. For instance, in ball milling processes, the yield of NMs is 50% (Seetharaman et al. 2018). It has been reported that only 6–8% of sputtered material is less than 100 nm in size, resulting in a wide particle size distribution. Laser ablation and plasma methods are high-energy consumers. A large number of redundancies are prohibitively expensive because of their extensive size distribution, slow production rate, waste byproducts, and high energy consumption (Seetharaman et al. 2018).

23.7.2 Chemical Methods

Chemical reduction, pyrolysis, the sol–gel method, microemulsion, polyol synthesis, hydrothermal synthesis, and CVP are a few popular chemical methods proposed for synthesizing NPs (Darroudi et al. 2013). However, from the viewpoint of ecological safety, the incorporation of toxic chemicals and reagents in the manufacturing process and the generation of waste products are harmful to both people and the environment (Darroudi et al. 2013; Barzegar et al. 2023). As a result, the use of these NPs is restricted in the biological sphere.

23.7.3 Biological or Green Methods

The domain of green nanotechnology has garnered significant intrigue in the scientific community as a means to create novel NPs with minimal environmental impact and maximum ecological compatibility. Biological methods of NP synthesis enable the synthesis of NPs utilizing natural reducing and stabilizing agents. It is a greener and cheaper alternative to chemical and physical methods, requiring no energy or harmful substances. Bottom-up biological synthesis of NPs utilizes single-celled organisms like bacteria and fungi to multicellular ones like actinomycetes and yeast, algae, and plant materials (Mittal et al. 2013; Panpatte et al. 2016; Singh et al. 2016; Patil and Chandrasekaran 2020).

23.7.3.1 Bacteria

Commercial biotechnological applications, such as bioremediation, genetic engineering, and bioleaching, have made extensive use of bacterial species (Hoffmann et al. 1995). Bacteria are smart and feasible choices in NP preparation due to their ability to reduce metal ions (Du et al. 2023). Many different types of bacteria are used in the production of metal and other novel NPs. Metal and metal oxide NPs synthesis using prokaryotic bacteria and actinomycetes has been widely used. However, GSH and cysteine desulfhydrase within *Escherichia coli* were primarily involved in the formation of spherical morphology during cadmium NP synthesis (Eroglu and Metin 2023; Wang et al. 2023). The NP's final size may be associated with its intracellular versus extracellular location and its interaction with bioactive molecules (Thakkar et al. 2010). The NPs produced by bacteria in extracellular synthesis are typically larger than those produced by intracellular synthesis. Bacterial systems have been used in the synthesis of NMs made of Se, FeO, $ZnSO_4$, and $PbSO_4$ (Patil and Chandrasekaran 2020). Numerous naturally occurring active molecules have been shown to stabilize and reduce NPs. Bacterial NPs can be stabilized and reduced by amino acids of proteins found in the cell wall and cytosol, such as tyrosine and tryptophan. In addition, some sugars, like aldose and ketose, have reducing and stabilizing properties. In particular, the amino acids in the cell walls and inside the cells act as a protective layer that keeps them from being harmful to mammalian cells (Markus et al. 2016). These active molecules, which are found in different types of bacteria, react with the metal ions and make them smaller. This facilitates intricate interplay among metal ions, thus forming more complex structures like spherical NPs.

Owing to the ease of regulation, bacterial synthesis is largely favored over other fabrication approaches (Adelere and Lateef 2016). *E. coli*, *Arthrobacter gangotriensis*, *Lactobacillus casei*, *Bacillus cereus*, *Aeromonas* sp. *SH10 Phaeocystis antarctica*, *Bacillus cecembensis*, *Pseudomonas proteolytica*, *Bacillus amyloliquefaciens*, *Enterobacter cloacae*, *Geobacter* spp., *Corynebacterium* sp. *SH09*, *Bacillus indicus*, and *Shewanella oneidensis* are some propitious bacterial candidates for the biosynthesis of different types of NPs (Singh et al. 2018).

23.7.3.2 Fungi

A highly efficient method for producing monodispersed NPs with well-defined morphologies is the fungi-mediated biosynthesis of metal/metal oxide NPs. By virtue of multiple intracellular enzymes, fungi prove to be superior biological agents in the production of metal and metal oxide NPs (Hoffmann et al. 1995). Comparatively more NPs can be synthesized by competent fungi than bacteria (Hoffmann et al. 1995). Fungi have many advantages over other organisms because they have enzymes, proteins, and reducing components on the surface of their cells (Narayanan and Sakthivel 2011). Enzymatic reduction (reductase) in the cell wall or within the fungal cell is a likely mechanism for the formation of metallic NPs (Mohanpuria et al. 2008). Metal and metal oxide NPs, such as silver, gold, titanium dioxide, and zinc oxide, are synthesized using a wide variety of fungal species.

Experimental studies have revealed that intrinsic proteins can aid in the production of NPs. The intracellular enzymes of *Trichothecium* and *Verticillium luteoalbum*, respectively, have been shown to actively participate in and facilitate the formation of gold nanospheres and nanorods (Gericke and Pinches 2006; Ahmad et al. 2006). NPs produced by fungi, like those produced by other environmentally friendly methods, have a wide spectrum of prospective applications, from medicine to optoelectronics (Birla et al. 2009). NPs' potential in the medical and therapeutic domains is intriguing and warrants further research. NPs synthesized from edible mushroom extracts have been reported to have similar chemotherapeutic properties to those of the extracts themselves (Philip 2009). Fungi contain amino acid residues, such as cysteine, that can aid in the formation of NPs (Philip 2009). Fungi use amino acids in their cell wall for the same stabilizing and capping purposes as bacteria. Furthermore, unlike conventionally synthesized NPs, these are non-toxic and safe for consumption when used in therapeutic applications and do not have unwarranted off-targets (Philip 2009).

23.7.3.3 Yeast

Yeasts are single-celled microorganisms by cellular composition, and eukaryotes possess well-defined nuclei and intact membranes for organelles by cellular complexity. There are approximately 1,500 recognized species of yeast (Yurkov et al. 2011). Several laboratories have reported the successful synthesis of NPs/NMs using yeast. Silver-tolerant yeast and *Saccharomyces cerevisiae* broth have been reported to biosynthesize silver and gold NPs (Patil and Chandrasekaran 2020). A myriad of yeast species harbors immense commercial potential and, hence, is utilized to produce an enormous number of metallic NPs.

Yeast cells can be used to synthesize a variety of nanosystems that would be impossible using bacteria alone. Yeast species have been used in the synthesis of NPs of silver, gold, cadmium sulfide, lead sulfide, ferrous oxide, selenium, and antimony. Nanosystems can be synthesized using either living cells or cell extracts with more common nanomaterial (NM) composites like silver and gold. Silver chloride NPs were successfully synthesized using proteins isolated from commercial yeast (Abdel-Hadi et al. 2014). The NPs were initially produced by incubating commercial yeast extracts with precursor solutions for 24 hours (Abdel-Hadi et al. 2014). After incubation, the solution containing the NPs was collected and filtered sterilely to remove any contaminants. In addition, they demonstrated that the reduction of silver chloride into NPs was caused by primary amines of specific proteins embedded in the yeast cell wall. It was also discovered that this NP was effective against mycobacteria. Silver-tolerant MKY-3 yeast cells could synthesize NPs (Philip 2009). The size, shape, and other morphological attributes of these NPs are governed by the parameters of their production process (concentration of silver chloride, pH, etc.). According to the results of this study, extracellular silver chloride reduction was caused by biochemical reducing agents that were excreted (Philip 2009). Extracellular synthesis of NPs is not attainable when using yeast as the starter raw material. The vast majority of research teams validated the same and stated that their models of synthesis took place within cells and that enzymes already present in the cell were responsible for constructing the nanosystem. Based on their previously undertaken research, *Schizosaccharomyces* pombe and *Torulopsis* sp. could intracellularly synthesize cadmium sulfide and lead sulfide NPs, respectively (Kowshik et al. 2002). In contrast to their earlier work, Kowshik et al. (2002) demonstrated that a specific species of phytochelatin synthase was responsible for the NPs' intracellular synthesis. Interestingly, nearly all the reviewed articles claimed that the synthesized NPs performed admirably in a variety of biomedical settings (Kowshik et al. 2002). Due to the versatility of NPs, the term 'biological applications' covers a wide range of possible uses. Mycobacteria in culture are killed by the silver NPs produced by *S. cerevisiae* (Sivaraj et al. 2020). To reduce and stabilize the NP, yeast, like other living organisms, can generate proteins containing the necessary amino acids. Organic molecules derived from aromatic compounds, quinones are unique to yeast and have been shown to aid in NP production. When the pH inside the cell becomes more basic, the oxidoreductases are activated and can begin reducing the metal ions. Strong nucleophiles with redox properties, quinones are useful for catalyzing the transformation of unorganized metal ions into more complex NPs (Jha et al. 2009; Faramarzi et al. 2020).

23.7.3.4 Algae

Algae are eukaryotic organisms exhibiting vast diversity that perform photosynthesis, yet they are not identified as plants. These single- or multicellular (depending on the species) chlorophyll producers thrive in aquatic environments but lack the typical plant characteristics of stems, leaves, and vascular tissues. Species like *Anabaena* are fatal if their cells or toxins are ingested, while others like *Spirulina* can be used therapeutically due to the high concentration of natural nutrients they contain (Carmichael et al. 1990; Khan et al. 2005). Many different kinds of algae have been discovered to harbor the potential of production of NMs in recent years (Khan et al. 2005). To fabricate NMs from different types of algae, the samples are first completely dried, followed by grounding the dried algae into a fine powder and mixing with water, leaving it to sit for 24 hours, and subsequent filtering of the mixture. After the biomass has been filtered, it is mixed with the NM precursors and left to incubate at room temperature until the solution's color changes, indicating the formation of the NM. Some algal species, such as *Tetraselmis kochinensis*, facilitate the intracellular formation of gold NPs, via enzymes mobilized on the cell wall and free-floating in the cytoplasmic compartment (Senapati et al. 2012). However, when comparing the two groups of agents, algal species typically have a wider range of options for NP morphology. In line with fungi and bacteria catalyzing NP formation, species like *Cystophora moniliformis*, *Scenedesmus* sp., and *Leptolyngbya valderianum* all form spherical NPs (Mukherjee et al. 2021). In addition to nanospheres, Sinha et al. (2015) successfully demonstrated that *Pithophora oedogonia* can synthesize cubical and hexagonal silver NPs (Sinha et al. 2015). Algae and other green agents share more than a common aesthetic; their bioactive molecules are strikingly similar (Mukherjee et al. 2021). The vast majority of algal-derived NPs are reduced/stabilized by enzymes/proteins in the cytosol or on the membrane (Mukherjee et al. 2021). Lastly, most NPs synthesized by algal species are effective bactericides. Algae use similar but also slightly different bioactive molecules from other classes in the production of NPs. Algae's exceptional ability to reduce and stabilize NPs is dominantly credited to its polysaccharides and protein residues. Phytochemical diversity is a key feature of algae exploited by many to attain utility benefits. Some algae, such as *Sargassum tenerrimum*, contain an array of bioactive compounds, including amino acids, alkaloids, carbohydrates, flavonoids, saponins, sterols, tannins, and phenolic compounds. Once the compounds have been purified, they can each contribute in their own unique way to modifying the morphological characteristics of NPs such as size, shape, and active properties. As much remains uncharted concerning the integration of algae in NP fabrication processes, the characterization of algae has been proven to be a fundamental, rudimentary step toward further study and potential use in NM synthesis (Saleh 2020).

23.7.3.5 Plants

Plant tissues serve as natural repositories of heavy metals. Therefore, biosynthesis techniques using plant extracts have received more attention as a straightforward, efficient, cost-effective, and practically viable alternative to standard preparation procedures for NP production (Akhtar et al. 2013). According to Singh et al. (2018), metallic NPs may be concurrently reduced and stabilized by using plants in a 'one-pot' synthesis method. Scientists developed metal/metal oxide NPs and explored their potential applications using plant leaf extracts in an environmentally sustainable manufacturing procedure (Singh et al. 2018).

NPs of other metals, such as Cu, Au, and Se, have also been reported. Several plant materials have been found to contain reducing, stabilizing, and capping agents that can be used to synthesize Ag-NPs (Singh et al. 2018; Huston et al. 2021). Synthesis of colloidal Ag-NPs from polyphenol-rich tea leaves was reported by Moulton et al. (2010). Their synthesis strategy shares similarities with other previously reported strategies. The group utilized tea powder prepared from dried and ground tea leaves. They boiled, filtered, and added $AgNO_3$ to the tea powder to fabricate Ag-NPs, which was confirmed by transmission electron microscope (TEM). They then used cell viability and membrane integrity assays to determine the toxicity of the NPs to living organisms. Their findings were encouraging, as NPs were found to be non-toxic and highly biocompatible (Moulton et al. 2010). Vegetable oil is yet another well-known, essential household ingredient and commonly utilized plant product. Silver benzoate is a widely utilized silver salt for Ag-NP synthesis, and the team used free radical exchange during oxidative drying of oils to convert it into silver NPs. The group also used fatty acids and aldehydes from the oils as reaction-stabilizing agents, and alkyl resin as a protective agent. Their reaction led to the development of antimicrobial paints containing Ag-NPs (Kumar et al. 2008). *Aloe vera* is yet another plant agent that can be utilized during the creation of NPs. *A. vera* has been used for centuries in alternative medicine to treat a wide range of conditions, and it is now also commonly used to treat sunburns. *A. vera* was used by Fardsadegh and Jafarizadeh-Malmiri (2019) to successfully create selenium NPs with antifungal and antibacterial properties. The hydrothermal approach was used by the Fardsadegh group, and while it is lauded as a classic synthesis method, it is actually more eco-friendly than is commonly believed (Fardsadegh and Jafarizadeh-Malmiri 2019). The bioactivity of molecules in spices can be exploited in the same way that it is exploited in plant leaves. The fragrant fruit of the *Myristica fragrans* tree is native to Indonesia. When the fruit is dried and ground, it yields the spices nutmeg and mace. The non-seed portion of the fruit, known as the pericarp, can be used to produce metallic NPs by being dried, crushed, added to water, and boiled with CuO or $AgNO_3$. Further research by Sasidharan et al. (2020) demonstrated that *Myristica fragrans* flavonoids, quercetin, and phenols were crucial to the NPs' stabilization and reduction. In addition, the Ag-NPs were found to be especially efficient at destroying the bacterial cell walls. Furthermore, Sasidharan et al. (2020) discovered that the Cu-NPs functioned well as catalysts in the synthesis of triazole rings.

23.8 CHARACTERIZATION PARAMETERS OF NPS

23.8.1 Size

NP size is fundamentally basic and the most significant characteristic trait while describing nanotechnological aspects. It is the very trait that distinguishes materials fabricated on the nanoscale from those produced on the macro-scale. Electron microscopy is the standard method for determining particle size and distribution. Scanning electron microscope (SEM) and TEM images are used to measure particles and clusters, while laser diffraction techniques are used to measure bulk samples in the solid phase (Wagner Ferreira Sabará et al. 2017). Light scattering from liquid particles is measured by centrifugation and photon correlation spectroscopy. Scanning mobility particle sizers (SMPSs) are widely favored, owning to their rapidity and ability to produce precise measurements compared to other methods, and gaseous phase particles are notoriously difficult and challenging to image (Marsalek 2014).

23.8.2 Surface Area

NPs can also be characterized by their surface area. A NP's performance and properties are profoundly affected by its surface area-to-volume ratio (Marsalek 2014). BET analysis is widely used to quantify surface area. Surface area analysis of particles in the liquid phase can be accomplished with a simple titration; however, it is deemed highly redundant due to its time ineffectiveness. NMR spectroscopy is employed to address the aforementioned time constraints. The NP surface area in the gas phase is measured using a differential mobility analyzer and a modified SMPS (Wagner Ferreira Sabará et al. 2017).

23.8.3 Surface Morphology

The versatility and functional activity of NMs are accredited to their surface morphology, shape, and three-dimensional conformations. Surfaces can be crystalline, amorphous, uniform, or have irregularities, and some of the shapes include spheres, flat discs, cylinders, tubes, cones, and irregular shapes. Electron microscopy imaging methods like SEM and TEM are typically used to determine surface characteristics (Wiechers and Musee 2010). For electron microscopy imaging, liquid-phase particles are captured electrostatically or through filtration, while gas-phase particles are deposited on a surface and analyzed.

23.8.4 Composition

The chemical constitution and elemental composition of NPs govern essential traits such as purity and functionality. Lower efficiency and the possibility of secondary reaction and contamination can result from the presence of higher levels of secondary or undesirable elements in the NP. In most cases, X-ray photoelectron spectroscopy is used to determine the composition (Sharma and Rao 2013). Wet chemical analysis, including mass spectrometry, atomic emission spectroscopy, and ion chromatography, is an integral part of some methods. Filtration or electrostatic collection is used to collect gas-phase particles, and spectrometric or wet chemical techniques are used for analysis (Bzdek et al. 2011).

23.8.5 Crystallography

Crystallography is the scientific study of the order of constituent atoms and molecules in crystals. Nanoparticle crystallography is accomplished through powder X-ray, electron, or neutron diffraction (Yano et al. 1996).

23.9 CHARACTERIZATION TOOLS AND TECHNIQUES OF NPS

Analytical techniques such as UV–Visible spectroscopy, Fourier transform infrared (FTIR) spectroscopy, X-ray diffraction (XRD), energy-dispersive spectroscopy (EDS), dynamic light scattering, SEM, TEM, atomic force microscopy (AFM), nanoscale infrared spectroscopy (IR), and extended X-ray absorption fine structure spectroscopy are typically used to the characterization of de novo synthesized NPs (Sepeur 2008; Strasser et al. 2010).

23.9.1 Fourier Transform Infrared Spectroscopy

The FTIR method of IR is both effective and easy to use. It is a vital analytical element in biological systems for measuring unknown concentrations of chemicals, determining accurate surface chemistry, and identifying unique functional groups and atomic configurations of biological NP samples (Cao 2004). FTIR differentiates and determines the presence of biomolecules in the synthesis of NP. Furthermore, it reveals the types of biomolecules found in the sample (Zhang et al. 2022). The FTIR method relies on the different types and frequencies of bond vibrations in molecules (Eid 2022). By virtue of high sensitivity, high cube corner interferometer, customizable workspaces, and hyperspectral imaging, detection laboratories harbor a biased inclination toward FTIR (Eid 2022).

23.9.2 UV–Visible Spectroscopy

UV–Visible spectroscopy is commonly used to validate the synthesis and stability of metal NPs/colloidal particles. Using NPs between 1 and 100 nm in size, one can confirm synthesis using absorbance at wavelengths between 230 and 800 nm (Awasthi et al. 2017). The surface plasmon resonance absorption band is caused by the free electrons. In addition to verifying synthesis, UV absorption spectroscopy can be used to precisely quantify particles in colloidal solutions (Tomaszewska et al. 2013). Particle size is quantified and characterized via absorption spectra, which can also be used for analysis on a qualitative level. By principle, electromagnetic radiation is measured and interpreted as molecules, atoms, or ions of a sample transitioning from one energy state to another

(Hu et al. 2006). Absorption of ultraviolet light (200–400 nm) by specific molecules causes the excitation of constituent electrons from their ground state to a higher energy state, allowing for their analysis in NP samples. UV-Vis spectroscopy stands out due to its low maintenance, sealed optics, double choppers, and spectral bandwidth, which can be adjusted as low as 0.2 nm (Amendola and Meneghetti 2009).

23.9.3 Infrared Spectroscopy

Reproducible instruments are used in IR to examine the molecular structure of matter by measuring unique and characteristic resonant vibration modes of molecules (Kiefer et al. 2015). It is designed in a way to delineate the chemical makeup and bonding structure of unknown, uncharacterized samples. A characteristic spectrograph is employed to compare the intensity of the radiation and the frequency of the spectrum with international standards and to identify the molecules (López-Lorente and Mizaikoff 2016). It is possible to characterize and identify chemical species with the help of IR. However, it can only resolve features between 5 and 10 m in size. The nano-infrared spectroscopy (NanoIR) probe is a nanoscale chemical analysis tool (López-Lorente and Mizaikoff 2016). It combines essential features of AFM and IR to allow for the acquisition of infrared spectra at spatial resolutions of 50–200 nm (Casalini et al. 2017). Multifunctional measurements are one of the prominent characteristics of nanoIR (Baudot et al. 2010). This analytical instrument provides a complete picture of samples by compiling their topographic, spectroscopic, mechanical, and thermal properties.

23.9.4 X-Ray Diffraction

XRD is a powerful tool for studying the crystalline structure of both organic and inorganic substances. XRD is user-friendly because it is not destructive, easy to use, sensitive, deep profile, and simple to construct (Vorontsov and Tsybulya 2018). In addition to scoping applications in pharmaceutical, glass, polymer, geological, and forensic industries, it has been employed to analyze undetermined chemical compositions based on quantitative and qualitative data, and values of degree of crystallinity, thus providing precise information on atomic arrangements at interfaces (Stefanos Mourdikoudis et al. 2018). The atomic arrangement, location, and intensity of the diffraction peaks are all described by the crystal structure. X-rays have an atomic-scale wavelength (Holder and Schaak 2019). As a result, it is typically employed for investigating NM architecture. Each powder sample atom scatters a single X-ray beam. Many different diffraction patterns can be created by the scattered beams reflected by any crystal (Jemal et al. 2017). The maximum intensity of the peak is sampled at a specific angle where the X-rays are scattered. The crystal's structure and physicochemical properties are reflected in this peak. XRD operates principally on Bragg's Law (Waseda et al. 2011). The Joint Committee on Powder Diffraction Standards maintains a database that is used to cross-reference the individual diffraction beam's pattern. Analysis of metal NPs for elemental content typically involves EDS (Strasser et al. 2010).

23.9.5 Scanning Electron Microscope

SEM is a flexible, non-destructive method of analysis that uses a microscope with a large specimen chamber and a working distance of 8.5 mm, which is made possible by angled detectors and a sharp conical objective lens (Vladár and Hodoroaba 2020). SEM is a type of imaging technique used to measure the size and shape of nano and microstructures on the surface of samples (Batool et al. 2016). It's a great instrument for figuring out how things are put together. It's quick, simple, and straightforward to use, and it gives accurate results. Furthermore, SEM is used in a wide variety of disciplines, including the study of biological molecules, biomedicine, and the study of materials (López Gutiérrez et al. 2022). Surface characterization of cells, organs, and tissues from 3D images is a common application in biomedical sciences. It adds to our understanding of cellular membranes. SEM operates on the principle of producing magnetically charged electron beams. Secondary and backscattered electrons generated by their magnetic field's interaction with the sample are detected (Qu and Liu 2021). Transmission electron detection is a powerful tool possessing high reproducibility for investigating NMs. SEM-EDX is also employed to determine the elemental makeup and concentration of samples (Kozikowski and Sobiech 2022).

23.9.6 Transmission Electron Microscope

TEM is a common method and an essential tool for characterizing NPs. The main benefit is the ability to analyze prepared NPs and internalized NPs in cells or tissues qualitatively and quantitatively with respect to their morphology, crystal structure, and size (Michen et al. 2015; Tan 2021; Pudovkin et al. 2021). Standard transmission electron microscopy methods are used for all of the sample analyses. However, TEM has drawbacks, such as its inability to characterize NPs with sizes 10 nm or smaller, its low magnification for particles larger than 100 nm, and the destructive nature of sample preparation (Mühlfeld et al. 2007; Neogy et al. 2020; Lee et al. 2020; Malatesta 2021). Applications such as immune and energy-filtered electron microscopy help categorize TEM (Casciardi et al. 2013).

23.10 UPTAKE AND TRANSLOCATION OF NANOFERTILIZERS

The literature database is enriched with review papers concerning NP uptake and bio-migration. Unfortunately, there exists a severe paucity of literature elucidating precise mechanisms of NP uptake and translocation within plants (Rani et al. 2022). There are three stages in the interaction between NPs and plants: NPs absorbed and transformed at the leaf, root, or stem surface; NPs absorbed through the cuticle, epidermis, and stomata; and NPs transferred and transformed within the plant via the xylem and phloem (Figure 23.1). Seed treatment, foliar spraying, trunk feeding, and even root application are all viable routes for NP introduction into plants (Su et al. 2019; Morin et al. 2023). Among these, foliar and root exposure is relatively simple to implement, making them good options for NPs delivery. Penetration analysis of NPs is crucial for understanding the intricate crosstalk between NPs and diverse elements of a plant system. The size, stability, chemical conformation, and concentration of the NPs all affect the uptake and translocation into plants (Hong et al. 2021).

23.10.1 Foliar Application Method of NPs Uptake and Translocation of Nanofertilizers

The foliar-applied NPs are taken up by the plants via the epidermis or the stomata. Both apoplastic and symplastic pathways converge in the vascular tissue after uptake (Hong et al. 2021). Plants accumulate NPs primarily in vacuoles and the cell wall, but the vascular tissue (xylem and phloem) transports them both upward and downward to other parts of the plant. Atmospheric abiotic variables (humidity, light, and temperature), plant physiological characteristics, and properties of NPs all play a role in the penetration, transfer, and accumulation of NPs (size, shape, and charge) (Pan et al. 2022). Cucumber plants have been shown to absorb CeO_2 NPs with a size of 8.1 nm when applied to their leaves and lettuce plants have been shown to absorb $Cu(OH)_2$ NPs with a size of 50–100 nm (Zhao et al. 2016; Hong et al. 2021). When compared to root or soil exposure, foliar spray NP supplementation is more effective and sustainable (Zhang et al. 2020; Fan et al. 2020).

NPs designed for direct agriculture application are predominantly sprayed onto leaves. NP depositions are subsequently internalized by plants via the cuticle or stomatal openings. The outermost, waxy cuticle of the leaf epidermis is dominantly constituted of wax, cutin, and pectin, which essentially prevents unwarranted water loss in growing leaves and functions as a frontline, physical barrier against the entry of foreign elements such as NPs (Pérez-de-Luque 2017). However, the waxy stratum corneum is littered with two chemically distinct types of channels: hydrophilic and lipophilic (Pérez-de-Luque 2017). Both hydrophilic and lipophilic channels possess diameters ranging between 0.6 and 4.8 nm (Eichert et al. 2008). Hydrophilic NPs with a diameter of less than 4.8 nm can diffuse through the channels (Banerjee et al. 2019). Lipophilic NPs can be absorbed by leaves via diffusion and infiltration due to the presence of lipophilic channels in the cuticle surface (Bussières 2014). Hu et al., based on inferences from high-resolution confocal fluorescence microscopy analysis published that carbon dots smaller than 2 nm can easily penetrate cotton leaves via the cuticular pathway (2020). However, the plant epidermis can only absorb minute amounts of NPs due to the small size of the pore channels in the cuticle. NPs applied to the leaf surface may accumulate in the epidermis and vascular tissue, according to some studies (Le et al. 2022). The translocation of NPs to other plant tissues has been observed in many studies that suggested some new pathways for entering NPs. Scientists have proposed a new route for NP absorption; they call it the stomatal pathway (Figure 23.1).

23.10.2 Soil Application Method of NPs, Uptake, and Translocation of NPs

The metal ions of the NPs used in soil interact differently depending on the aging process and soil characteristics, and they stay in sediment for a longer time (Rajput et al. 2022). Numerous studies have been conducted to analyze the mechanism of NP transformation in soil. The presence of inorganic and organic components, as well as the texture of the soil, influence the transformation of NPs in the soil (Rani et al. 2022). Due to the decrease in air pore space in clay soil, the concentration of NPs is decreased (Naderi and Danesh-Shahraki 2013; Rani et al. 2022). For instance, the formation of an Ag_2O layer around Ag-NPs results in the dissolution and subsequent release of Ag^+ in sandy soil due to the soil's higher oxygen content (Li et al. 2017). The accumulation of Cu in lettuce after root exposure to CuO NPs was time-dependent and affected by the concentration of CuO NPs, soil property, and soil texture. The weathering of CuO NPs in soil affects soil Cu availability, lettuce plant Cu uptake, and lettuce plant Cu translocation (Servin et al. 2017). Aggregation of NPs in the soil system reduces their bioavailability because homo- and hetero-aggregation of NPs is common in soil. Soil's organic and inorganic components interact with NPs and affect how well they stick to the ground. For instance, ZnO-NPs absorbed less when they formed hetero-aggregates with soil moieties. Soil water can dissolve NPs, freeing the metal ions inside. The bioavailability and absorption of NPs were improved through dissolution. How well NPs dissolve in the soil is affected by factors such as soil type, texture, NP physicochemical properties, and NP application method (powder vs. solution) (Wu et al. 2021).

Comparative analyses have revealed that NP effects on roots are extensively evaluated however, the response of plant shoots to NP application remains uncharted territory. The hair cells in the plant root take in the NPs when they interact with the root. They enter the endothelial layer from the epidermis via the symplast or exoplast and are transported to the above-ground part of the plant via the xylem vessels (Tripathi et al. 2017). The root of a plant is made up of a pericycle and vascular column on the inside, followed by the cortex and the epidermis on the outside. The vascular column, composed of pericycle and vascular tissue, can be found at the root's midpoint (Su et al. 2019). The endothelium is linked to the pericycle in the cortex, which also includes the exodermis (Figure 23.1). NPs can be transported in the root via either the exoplast or the eutectic. The Casparian strip acts as a barrier to prevent NPs from passing from the epidermis to the endothelium along the exoplast pathway. Some NPs will eventually deposit in the endothelium, while others will be transported radially along the Casparian strip but will not penetrate the Casparian strip. However, the Casparian strip is incomplete at the lateral root junction. The vascular system is an accessible pathway for NPs (Wang et al. 2012, 2013; Ali et al. 2021). NPs travel between cells via plasmodesmata after entering the cytoplasm through the plasma membrane via the symplastic pathway.

23.11 NANOPARTICLE-BASED-NANOOMICS RESPONSE TO PLANTS UNDER SALINITY

Plant biological functions are determined by their molecular activity. Each cell has a sophisticated and stringently regulated process regulating gene expression with replication, transcription, and translation being the fundamental events. Plants use diverse mechanisms to upregulate or downregulate target gene expression in the face of salt stress. An underlying strategy of conferring salinity tolerance involves gene regulation which involves altering components of the central dogma, from initiation factors to the TFs participating in the production of mRNA and eventually, proteins. Plants are exposed to saline in two ways: at first, it reduces the water-holding capacity of the soil, which interferes with water intake and causes osmotic stress; secondly, it causes an ion accumulation that leads to ionic stress. The salt content can be regulated at the biochemical and physiological levels with NPs; nevertheless, this is limited by a lack of knowledge about the NPs' physio-biochemical activity and molecular hindrance processes. Thorough comprehension of NP interaction at the molecular level is a crucial prerequisite for exploiting the efficiency of NPs against salinity stress. A transformative step toward curbing the adversities of salinity tolerance via comprehending the intricacies of adaptation in plants is the incorporation of multiple 'omics' approaches, namely, genomics, transcriptomics, proteomics, and metabolomics. Meticulous experiments designed to study salt-stressed plants using 'omics' technologies have enabled close observation of traits, genes, proteins, and metabolomes modulating different defense responses.

23.11.1 Nanoomics-Based Genomics and Transcriptomics Response

Catastrophic salt concentration, as was previously said, physiologically presents as ionic stress (produced by catastrophic ionic accumulation in the soil) and osmotic stress (induced by a drastic decrease in water intake owing to salinity) (Nongpiur et al. 2016). Plants retain and exclude Na^+ to lessen the negative effects of ionic stress. Many ion transporters linked to salt resistance in crops have been identified in the past 20 years. Both the high-affinity K^+ transporter (HKT) and the Na^+/H^+ antiporter (NHX) on the vacuolar membrane are well-known transporters that are crucial for the absorption, extended transportation, and redistribution of Na^+ and K^+. The degree of salt resistance in plants is highly dependent on the presence of HKTs, which include Na^+ or Na^+–K^+ transporters. They decrease sodium ion concentration in the shoot and maintain stable potassium ion concentration in living organisms. HKT proteins can be generically classified into two groups, one involved in the transport of Na^+ and the other in that of K^+. HKT1 regulates Na^+ uptake by functioning as a Na^+ transporter when external K^+ is low, whereas HKT2 functions as a K^+–Na^+ cotransporter. In the vacuolar membrane, sodium ion transport is facilitated by proton gradient-driven (established by H^+-ATPase and H^+-pyrophosphatase) NHXs. Overexpression of VP and NHX putatively increases the strength of the proton gradient generated by H^+-ATPase and H^+-PPase, thereby augmenting the efficiency of sodium ion shuttles. Applying MWCNTs to *Brassica napus* L. may improve the regulation of the antioxidant genes APX and SOD in combination with their associated transcripts, BnAPX, BnCu/ZnSOD, and BnMnSOD. This would lead to a notable decrease in ROS and the Na^+/K^+ ratio, which would return to normal ion and redox discrepancies (Zhao et al. 2019). This beneficial effect was also related to the MWCNTs-induced alteration in Na^+/H^+ exchanger 1 (NHX1) and K^+ transporter 1 (KT1) transcripts (Zhao et al. 2019). In *B. oleracea*, the application of Ag-NPs synthesized from *Vitex negundo* plant extract resulted in the overexpression of anthocyanin regulatory genes, which ultimately led to anthocyanin accumulation that correlated with the hyper-catalytic activity of SOD, CAT, and APX (Baskar et al. 2015). Fe-NPs and ZnO-NPs have been shown in different studies to be able to promote the activity of genes encoding antioxidant enzymes such as Cu/Zn SOD, APX, and CAT2, which participate in H_2O_2 purification and the sustaining of Na^+/K^+ homeostasis in tomato (*Solanum lycopersicon* L.) plants (Machado et al. 2023).

Reduction of ion leakage and enhanced Hill reaction were observed post-exogenous application of ZnO-NPs to rapeseed grown under salinity stress. At the molecular level, ZnO-NPs interfered with the expression of stress proteins. These studies validated the promising potential of ZnO-NPs in averting physiological distress caused by salinity (Zulfiqar and Ashraf 2021).

Osmo sensors have emerged as important contributors in reducing the harmful effects of salt by controlling osmotic potential. *Arabidopsis* spp. possess an osmo sensor enriched with histidine kinase (AHK1) which functions to reduce hyperosmolality-induced Ca^{2+} increase by targeting linked (OSCA1) genes (Zorb et al. 2019). Multiple kinase families influence the activity of genes that respond to stress. Maintaining osmotic equilibrium and protecting the structure of cells by providing enough water is the main goal of osmolytes. As an attempt to evade the toxic effects of salt, plants resort to intrinsic accumulation of proline, an essential osmolyte that scavenges ROS. Proline preserves the structural integrity of bio-macromolecules by maintaining the cytoplasmic water content of cells. Proline is documented to aid in stress recovery by serving as a repository of organic nitrogen, and in stress response via elevating the production of salt-reactive proteins. Plant adaptation mechanisms fundamentally involve two phenomena: elimination of excess ROS and stimulation of activity of the antioxidant enzyme pool. Salt causes the P5CS gene, which regulates proline biosynthesis, to become overexpressed.

Several studies have verified that the application of NPs can significantly trigger overexpression of *P5CS* gene and ultimately increase cellular proline levels for improved acclimatization to osmotic stress (Liu et al. 2020; Aazami et al. 2021; Kumari et al. 2022; Samrot et al. 2022; Machado et al. 2023).

Crop resistance to salt is one of the direct products of enhanced antioxidant activity. Enhanced AsA levels in modified *Arabidopsis* are linked to higher seed sprouting and seedling development, suggesting that DHAR is essential for preserving cellular AsA resources (Ushimaru et al., 2006). Transgenic *Arabidopsis* spp. that express OsAPXb or OsAPXa coding for cytosolic POD have demonstrated bolstered salt resistance (Lu et al. 2007). As inferred from extensive experimental data, OsAPXb confers far more potent tolerance to salt stress than OsAPXa because it directly targets H_2O_2 metabolism. The overexpression of antioxidative genes such as OsAPX7, OsMIOX (Duan et al. 2012), OsGR2,3 (Hong et al. 2009), cCuZn-SOD2 (Kim et al. 2005), OsSRO1c (Teixeira et al. 2006), and OsAPx8 (Hong et al. 2007) are modulated by of salt stress.

23.11.2 Nanoomics-Based Proteomics Response

The molecular aspects of mechanistic strategies employed by plants to adapt to abiotic stressors have never ceased to cause intrigue across the scientific community. With the advent of proteomics, the study of plant responses to salt stress has unlocked greater depths of comprehension. As the plant prepares for the launch of stress responses, all systemic operations beginning from carbohydrate metabolism, protein expression, energy production, defense mechanism, secondary metabolic synthesis, cell wall maintenance, photosynthesis, enzymatic activity, ROS scavenging to osmolyte generation, are all amplified (Deinlein et al. 2014; Van Zelm et al. 2020). Dramatic elevation of molecular chaperones has been documented in salt-stressed plants (Wang et al. 2019). Similarly, expression of participating enzymes of the glycolytic and TCA cycle such as pyruvate kinase isozyme A (PK), 6-phosphofructokinase 3 (PFK), succinate dehydrogenase (SDH), malate dehydrogenase (MDH), isocitrate dehydrogenase (IDH), pyruvate decarboxylase 2 (PDC), and aconitate was indicated to be upregulated during salinity stress. As an obvious consequence, the plant was able to optimize major metabolic processes afflicted during salt stress, thereby restoring normal glycolysis and TCA cycle to provide adequate energy for normal plant growth (Xing et al. 2019). Antioxidant enzymes form a remarkable, cellular, enzymatic system that directly or indirectly enables the alleviation of physiological and biochemical liabilities caused by salinity stress. The catalytic proficiency of enzymes constituting the antioxidant system, namely, CAT, SOD, and glutaredoxin sulfhydryl oxidoreductase (GPx), is stimulated against salinity, which aids in constraining H_2O_2 accumulation and promoting ROS detoxification by scavengers. A multitude of experimental studies have unanimously claimed that the application of NPs significantly enhances the concentration and enzymatic efficiency of the antioxidant enzyme pool in the face of salt stress (Zulfiqar and Ashraf 2021). The Rubisco enzyme is a key player in the Calvin cycle and a dominant contributor to carbon dioxide fixation in plants. A study examined the catalytic activity of a small subunit of Rubisco post-TiO_2-NP topical application to find it exceptionally elevated in response to salt stress (Xuming et al. 2008). Yet another study conducted in four cultivars of tomatoes demonstrated that the application of Fe-NPs and ZnO-NPs can induce the expression of SOD, APX, and GR genes, which consequently enhanced the formation of antioxidant enzymes under varying NaCl stress levels. As a result which relative growth rate, dry matter percentage, osmotic potential, proline content, callus formation, and shoot formation in tomatoes were significantly improved (Aazami et al. 2021).

Outside of the enzymatic antioxidant pool, a non-enzymatic antioxidant recognized as phytoene synthase, capable of scavenging ROS and RNS has been indicated to be upregulated. Under regulated conditions, the application of 300 mg/L ZnO-NPs positively impacted the expression of phytoene synthase in *Arabidopsis* (Wang et al. 2016). The cumulative effect of activation of transporter channels and antioxidant enzymes in response to NP application offers a conducive physiological environment for plants to compete for survival in saline soil regimens (Pérez-Labrada et al. 2019; Das et al. 2019; Liu et al. 2019; Zafar et al. 2021; Aazami et al. 2021; Soliman et al. 2022; Alam et al. 2022; Moradi et al. 2022; Machado et al. 2023).

23.11.3 Nanoomics-Based Metabolomics Response

The study of metabolisms or, metabolomics, is an ingenious approach to precisely explicate how the plant system molds itself to adapt to unfavorable conditions and evade toxicity of abiotic stressors. Metabolomics provides extensive insights and in-depth comprehension regarding plant cellular responses to salt stress. The resulting metabolome harbors immense potential to serve as an effector, facilitating stress responses like osmotic balance maintenance, membrane stabilization, elimination of ROS, and antioxidant defense (Jiang et al. 2014). Metabolites are biomolecular outcomes of biosynthetic pathways and regulatory mechanisms and are essentially synthesized by organisms to withstand external stimuli. More than 200,000 metabolites are known to be secreted by plants, and they may be broadly divided into two categories: primary and secondary metabolites. Structural properties of primary metabolites are highly conserved across all plant species however, the elemental makeup of these secondary compounds is species-specific and highly susceptible to modifications driven by time, evolution, or other variables. The molecular database of a cell keeps track of all modifications in plant physiology generated by engineered NPs. These tracks also record the metabolites that participate in biological pathways, such as the citric acid cycle, glycolysis, gluconeogenesis, synthesis of amino acid, nitrogen fixation, fatty acid metabolism, and more. Initial stress responses in plants involve the secretion of metabolites which function as signaling molecules and alteration of soil chemistry and biochemical pathways to improve nutrient uptake. It's interesting to note that certain secondary metabolites, including carotenoids, phenolics, flavonoids, and anthocyanins, act as non-antioxidant enzymes that aid in eliminating ROS (Jithesh et al. 2006; Abid et al. 2020). Several imperative evaluations suggest that the application of NPs has a direct, positive impact on the cellular concentration of secondary metabolites (Aazami et al. 2021; Kumari et al. 2022). These compounds are classified as non-antioxidant enzymes due to their potent ROS detoxification activity and ability to scavenge free radicals such as $^1O_2$ and $^{\bullet}OH$ (Demidchik 2010, 2015; Akhavan Hezaveh et al. 2020). A novel study demonstrated that flavonoids, phenolics, and protein content in *Momordica charantia* and *Prunella vulgaris* were augmented by Cu, Ag, and Au-derived NPs (Fazal et al. 2016). Characteristic increase in the content of secondary metabolites validates that free radical generation is positively correlated with the accretion of non-antioxidant enzyme compounds (Jiang et al. 2014; Da Costa and Sharma 2016). Cellular levels of 6-benzylaminopurine, 1-naphthalene acetic acid, and indole-3-butyric acid were documented to rise in wheat upon treatment with Ag-NPs, while those of ABA drop under similar conditions. Hence, it can be inferred that Ag-NPs stimulate seed germination during salinity. All biotic elements of an ecosystem, including plants, demand iron a crucial inorganic element to ensure competent metabolic performance (Tripathi et al. 2018). Moreover, the formation of various essential proteins that regulate cellular respiration, photosynthesis, oxygen transport, metabolic rate, and DNA repair requires Fe. Therefore, iron deficiency can severely impair crop growth and yield (Chen et al. 2017). It has been observed that Fe-NPs particularly target antioxidant enzymes to confer salt resistance (Singh and Bhatla 2016). Exposure to lethal soil concentrations (200 and 150 mM NaCl) can be successfully attenuated by CNPs (0.3%) as evaluated in lettuce (Baz et al. 2020). The negative effects of salinity in wheat plantlets were curbed by preconditioning the seeds with polyhydroxy fullerene NPs, which resulted in an improvement in antioxidant activity (Shafiq et al. 2019). The addition of TiO_2 during salinity facilitated the enhancement of essential oils (Gohari et al. 2020). Varying concentrations of TiO_2 NPs sprayed onto broad bean crop afflicted with salt toxicity demonstrated a significant increase in the content of soluble sugars, proline, other amino acids along with enzymatic functions (Abdelaal et al. 2020).

23.12 ACCUMULATION, TOXICITY, AND RISK ASSESSMENT OF NANOFERTILIZERS

Toxicology studies on nanofertilizers' effects on crops are in their infancy, but their use is already being heralded as a groundbreaking application of nanotechnology in agriculture. It has been demonstrated that applying metallic NPs to plants can induce an oxidative spike by interfering with the electron transport chain (Khan et al. 2021). Additionally, the interference lessens the plant's capacity to eliminate ROS, which might have genetically toxic consequences.

Despite evidence that nano-nutrient elements can be used as fertilizers, global fertilizer industries are plagued by technical caveats and, hence, do not appear to have been encouraged by the research outcomes to increase their investments in nanofertilizers. However, only a small number of nanofertilizer products have been documented so far. Some of these items have originated from university research, while others have been made in pilot plants on a small scale for use in greenhouse trials. The high cost and potential toxicity of the current synthesis methods are a major pitfall restricting the widespread utilization of nanofertilizers and hindering the realization of their full potential. Therefore, there is a significant shift in emphasis toward biological/green synthesis as a means of manufacturing such NPs. This has the potential to lessen the expense and danger of using nanofertilizers. However, initiatives are under way in some Asian countries like Myanmar to incorporate nanofertilizers into their national fertilizer strategy (Dimkpa and Bindraban 2018). The potential downsides of NPs must be weighed against the potential upsides. Highly efficient methods free of unintended consequences are practically impossible to discover. Nanotechnology is a broad field with many potential dangers because of this. NPs are so small that they can easily pass through a human body's defenses and cause harm. However, if used in measured quantities, NPs have the potential to be both effective and safe, with no negative consequences for the surrounding ecosystem. Metal NPs, like zinc and iron, are examples of nanonutrients that are essential for both plants and humans (Shang et al. 2019). Giving plants the right amount of nanofertilizers could improve their yield and quality, as well as the health of the people who eat them. Controlled and targeted release of nanofertilizers according to the nutritional needs of the crops ensures minimal losses into the soil. Coatings on controlled-release fertilizers prevent the soluble material from coming into contact with water and/or prevent the nutrient solution from diffusing out of the fertilizer. This prevents the leaching-induced soil penetration. The use of nano-coated materials has been shown to reduce NO_3 leaching during a wheat-maize crop rotation (Zhang et al. 2014b; Pan et al. 2022). In addition, using a porous NM like zeolites, clay, or chitosan can greatly reduce nitrogen losses by controlling the rate of release based on plant need (Shang et al. 2019).

23.12.1 NPs-Based Oxidative Stress Damage in Plants

ROS, like other signal molecules used by aerobic organisms to trigger cellular responses to environmental fluctuations, is of particular interest to plants (Sharma et al. 2012). When NPs enter plant cells, they can alter ROS levels, impede cell metabolism, damage the antioxidant system, and ultimately retard plant growth. ROS, which includes OH, $^1O_2$, $O_2^{\bullet -}$, and H_2O_2, are generated naturally and constantly by the organelles of a cell as part of their metabolism (Keunen et al. 2011). At low concentrations, ROS activates a defense signal, but at high concentrations, it causes oxidative damage to amino acids, lipids, and nucleic acids (Mahajan and Tuteja 2005). The level of LPO in a cell's membrane is an undebatable indicator of membrane integrity and health (Porter et al. 1995). ROS production causes LPO, which in turn causes membrane damage, ion leakage, metabolic disruption, and eventually cell death. Antioxidant enzymes and low-molecular-weight antioxidants in plants can prevent cellular and subcellular damage caused by ROS (Hussain et al. 2016). Studies on oxidative stress in plants due to NPs typically involve measuring ROS or antioxidants, as well as the activity of antioxidant enzymes. Bean leaves containing Ag-NPs triggered the production of ROS. Published data has established an inversely proportional trend between NP diameter and ROS generation. This is largely credited to the fact that smaller NPs possess a high surface-to-volume ratio, thus inflicting greater toxicity (Yamauchi et al. 2008). Tocopherol content and SOD activity both increased in the presence of TiO_2 NPs under high light conditions, while CAT activity decreased (Das and Roychoudhury 2014). Higher concentrations of NPs cause membrane LPO, which may be related to particle photoactivation and increased ROS production (Abid et al. 2020). Lettuce's oxidative stress was triggered by exposure to polymethyl methacrylate nanoplastics (PMMA-NPs), which also led to an uptick in defense mechanisms like antioxidant enzyme activity (Zhao et al. 2016). Still, it had a higher concentration of active oxygen than the control. The recent, unequivocal verification of the fact that increased activities of several antioxidant enzymes in rice roots in the face of increased concentration of plastic NPs suggests that plants possess the potential to trigger defense responses, eliminate ROS and restrict intrinsic accumulation of ROS.

23.12.2 NPS-Based Genotoxic Damage in Plants

NPs accumulate at almost all levels of a food chain after being released into the environment via the ecological phenomenon of biomagnification. NPs can be absorbed by plants, and they are very sensitive to nanotoxicity (Miralles et al. 2012). Thus, plants are considered exceptional genetic models for screening and monitoring environmental toxic compounds (Zhang et al. 2015). NPs can reduce mitotic index and increase micronucleus and chromosome aberration index in plants through interactions with biological macromolecules (nucleus, cytoplasmic components, or lipids). Wheat root tip cells were found to readily internalize Ag-NPs, and the particles could interfere with normal *c-myc* function and chromosome segregation when examined for their effects on chromosome aberrations and cell division (Abdelsalam et al. 2018). Aberrations observed included multicore, incorrect orientation of sister chromatids at metaphase, chromosome breakage, metaphase plate deformation, spindle dysfunction, and lagging chromosomes (Abdelsalam et al. 2018). The mitotic index was found to significantly decline in cells taken from onion roots after being exposed to Cr_2O_3 NPs for cytogenetic analysis. Additionally, chromosomal aberrations were observed at various doses (Abdelsalam et al. 2018). Previous studies have shown that Cr_2O_3 NPs cause an increase in SOD production in response to oxidative stress and subsequent DNA damage (Pakrashi et al. 2014). Onion mitotic index was significantly decreased after being exposed to PS-NPs, and multiple types of chromosome aberrations (such as chromosome aggregation, lagging chromosomes, etc.), spindle deformation, and nuclear abnormality were observed, all of which undermined the integrity of the genome (Maity et al. 2020). Direct genotoxicity and indirect genotoxicity are two broad categories used to elucidate the mechanism of genotoxicity (Jogaiah et al. 2021). Direct genotoxicity of NPs arises from the physical interaction of NPs with DNA, while indirect genotoxicity may be the result of a decrease in.

23.13 FUTURE PROSPECTS

Reducing the widespread use of agrochemicals in the agrarian sector is the primary prerequisite to achieving sustainability of bioresources and ensuring the development of eco-friendly plant nutrient systems. Tropical and subtropical soils are acidic, substantially P-deficient, and exhibit high potential for phosphate sorption, and therefore, are ideal sources of agrochemicals (Batista and Singh 2021). Therefore, novel and state-of-the-art strategies are being developed using nanotechnology and nanoengineering to boost agricultural output and maximize crop yield (Verma et al. 2022). The optimal utilization of phosphate rocks as P-sources, and the processing of phosphate rocks at an industrial scale is promising for global development and crop improvement (Majumdar et al. 2015). The statement holds true, particularly, for nations in possession of abundant native phosphate rock resources. Optimizing the usage of available land resources, reducing erosion, preserving and restoring soil quality, exploring novel alternatives to fertilizer, optimizing fertilizer recommendations, choosing the most appropriate crop genotypes, and exporting manure are all sustainable alternatives for managing P consumption. Instruments that closely monitor crop relations and requirements to evaluate the inter- and intra-crop demand are vital for adjusting the amount of product supplied to the crops or individual plants. Regrettably, bio-limitations of natural reserves undoubtedly result in a dramatic spike in commercial prices of agrochemicals such as fertilizers and pesticides. Delivery systems based on nanotechnology to supply crops with essential nutrients are an emerging field that demands extensive research and comprehensive evaluations (Mahakham et al. 2017). Future research on nano-based fertilizers could be organized in a few different ways. Because of their widespread use and potential scarcity, N- and P-nanofertilizers might be given more consideration. In the case of micronutrients, a comprehensive study may be conducted to elucidate the impact of the variables influencing their availability in field conditions by comparing Ca-NPs from $CaCO_3$ with those from other Ca sources, like $CaSO_4$ or $CaCl_2$, or by comparing multiple nutrients, like Ca- and N-NPs, using soluble Ca $(NO_3)_2$ as a control. Micronutrient fertilizers are available commercially and can be compared to conventional fertilizers in terms of biofortification and application methods (such as fertigation). However, (1) there exists a dearth of conclusive evidence to validate the ability of nanocarrier-based fertilizers to increase FUE via improved nutrient transfer into the plant tissues or cells or to reduce the environmental hazards related to the use of conventional fertilizers, and (2) the optimal dosage for each crop (without any adverse effects) is not precisely determined. There is the potential for further exploration of economically feasible and functionally proficient carriers, such as SiO_2-NPs, Fe_2O_3-NPs, and CNTs, than what has been available up to this point, such as zeolite. These advancements would aid in the gradual absorption of active substances and reduce the need for, and amount of, wasteful inputs. Nanocarriers' design also has the potential to make them useful as anchors in the organic matter and soil structure that surround plants' roots. To achieve so, it is necessary to have an expansive comprehension of the chemical and conformational interactions between the delivery nanoscale structures and the target soil structures and materials. (3) Rigorous statistical analyses and toxicity profiles have strongly claimed that various NPs/NMs, depending on their particle size and dosage of application, are detrimental to plant growth at lethal concentrations. To expand the current knowledge pool concerning mechanisms of enhanced growth, productivity, and performance of crop plants in response to the application of nanofertilizers, the development of a checkpoint might prove to be beneficial. Nanotechnology, on the contrary, may allow for the regulated, long-term release of agrochemicals and the site-specific delivery of macromolecules like fertilizers and pesticides, all of which, synergistically, can improve plant disease resistance, nutrient uptake efficiency, and plant growth. Moreover, satellite-based navigation systems, such as geographic information systems, remote sensing, global positioning systems, and information technologies, can be integrated into agricultural systems to provide an optimally adequate amount of nutrients to plants and soils for maximum yield, thereby ensuring global food security and promoting sustainability of bioresources (Zulfiqar and Ashraf 2021).

23.14 CONCLUDING REMARKS

The immense potential of nanotechnology as a transformative tool in advancing crop salinity tolerance is evident enough. Through a meticulous examination of recent developments, it can be concluded that nanotechnology serves as a multifaceted approach to mitigating the detrimental effects of salt stress on food crops. By leveraging the unique properties of NMs, researchers have devised novel techniques for bolstering crop resilience to salt at molecular, cellular, and physiological levels. Nanoparticle-mediated delivery of essential nutrients, antioxidants, and stress-responsive molecules has shown promising results in ameliorating salinity-induced damage and improving crop performance under saline conditions. Additionally, nano-formulations designed for targeted delivery of bioactive compounds offer precise and efficient means of augmenting stress tolerance in crops while minimizing unwarranted environmental impacts. Furthermore, the blend of nanotechnology with modern biotechnological tools, such as gene editing and RNA interference, holds immense prospective for the development of salt-tolerant crop varieties with improved yields and nutritional profiles. Nonetheless, it is imperative to proceed with caution and prioritize the principles of sustainability, environmental integrity, and biosafety. Robust risk assessment frameworks ought to be established to evaluate the potential ecological impacts of NMs, and measures must be implemented to minimize off-target effects on soil profile, water quality, and human health. Continued interdisciplinary research, collaboration, and innovation must be undertaken to realize the full potential of nanotechnology in troubleshooting intricate challenges plaguing modern agriculture. By harnessing the power of nanotechnology, it is possible to forge a resilient and sustainable agricultural system, wherein production equals consumption while safeguarding our planet's bioresources for generations to follow.

ACKNOWLEDGMENT

VDR acknowledges the support of the Strategic Academic Leadership Program of the Southern Federal University ('Priority 2030'). Higher education and science committee MESCS Republic of Armenia provided the 23PostDoc-4D007 grant to support AS. Higher education and science committee MESCS Republic of Armenia has provided grant numbers 21AG-4C075 to support KG.

REFERENCES

Aazami, M. A., Rasouli, F., & Ebrahimzadeh, A. (2021). Oxidative damage, antioxidant mechanism and gene expression in tomato responding to salinity stress under in vitro conditions and application of iron and zinc oxide NPs on callus induction and plant regeneration. *BMC Plant Biology*, *21*, 1–23.

Abdel-Hadi, A. M., Awad, M. F., Abo-Dahab, N. F., & Elkady, M. F. (2014). Extracellular synthesis of silver NPs by aspergillus terreus: Biosynthesis, characterization and biological activity. *Biosciences Biotechnology Research Asia*, *11*, 1179–1186.

Abdelaal, K. A. A., Mazrou, Y. S. A., & Hafez, Y. M. (2020). Silicon foliar application mitigates salt stress in sweet pepper plants by enhancing water status, photosynthesis, antioxidant enzyme activity and fruit yield. *Plants*, 9, 733. https://doi.org/10.3390/PLANTS9060733

Abdelsalam, N. R., Abdel-Megeed, A., Ali, H. M., Salem, M. Z. M., Al-Hayali, M. F. A., & Elshikh, M. S. (2018). Genotoxicity effects of silver NPs on wheat (Triticum aestivum L.) root tip cells. *Ecotoxicology and Environmental Safety*, *155*, 76–85. https://doi.org/10.1016/J.ECOENV.2018.02.069

Abdoli, S., Ghassemi-Golezani, K., & Alizadeh-Salteh, S. (2020). Responses of ajowan (Trachyspermum ammi L.) to exogenous salicylic acid and iron oxide NPs under salt stress. *Environmental Science and Pollution Research*, *27*, 36939–36953. https://doi.org/10.1007/S11356-020-09453-1

Abid, M., Zhang, Y. J., Li, Z., Bai, D. F., Zhong, Y. P., & Fang, J. B. (2020). Effect of Salt stress on growth, physiological and biochemical characters of Four kiwifruit genotypes. *Scientia Horticulturae (Amsterdam)*, *271*, 109473. https://doi.org/10.1016/J.SCIENTA.2020.109473

Adelere, I. A., & Lateef, A. (2016). A novel approach to the green synthesis of metallic NPs: The use of agro-wastes, enzymes, and pigments. *Nanotechnology Reviews*, *5*, 567–587.

Aghdam, M. T. B., Mohammadi, H., & Ghorbanpour, M. (2016). Effects of nanoparticulate anatase titanium dioxide on physiological and biochemical performance of Linum usitatissimum (Linaceae) under well-watered and drought stress conditions. *Revista Brasileira de Botânica*, *39*, 139–146.

Ahmad, A., Senapati, S., Khan, M. I., Rajiv, K., & Murali, S. (2006). Extra-/intracellular biosynthesis of gold NPs by an alkalotolerant fungus, *Trichothecium* sp. *Journal of Biomedical Nanotechnology*, *1*, 47–53. https://doi.org/10.1166/JBN.2005.012

Ahmad, R., Hussain, S., Anjum, M. A., Khalid, M. F., Saqib, M., Hassan, A., Fahad, S., & Ahmad, S. (2019). Oxidative stress and antioxidant defense mechanisms in plants under salt stress. In M. Hasanuzzaman, K. Nahar, H. F. Alharby, & K. R. Hakeem (Eds.), Plant abiotic stress tolerance: Agronomic, molecular, and biotechnological approaches (pp. 191–205). Springer.

Akhavan Hezaveh, T., Pourakbar, L., Rahmani, F., & Alipour, H. (2020). Effects of ZnO NPs on phenolic compounds of rapeseed seeds under salinity stress. *The Journal of Plant Process and Function*, *8*, 11–18.

Akhtar, M. S., Panwar, J., & Yun, Y. S. (2013). Biogenic synthesis of metallic NPs by plant extracts. *ACS Sustainable Chemistry & Engineering*, *1*, 591–602.

Alam, P., Arshad, M., Al-Kheraif, A. A., Azzam, M. A., & Balawi, T. A. (2022). Silicon nanoparticle-induced regulation of carbohydrate metabolism, photosynthesis, and ROS homeostasis in Solanum lycopersicum subjected to salinity stress. *ACS Omega*, *7*, 31834–31844.

Ali, S., Mehmood, A., & Khan, N. (2021). Uptake, translocation, and consequences of NMs on plant growth and stress adaptation. *Journal of Nanomaterials*, 2021, 6677616. https://doi.org/10.1155/2021/6677616

Ali, Y., Aslam, Z., Ashraf, M. Y., & Tahir, G. R. (2013). Effect of salinity on chlorophyll concentration, leaf area, yield and yield components of rice genotypes grown under saline environment. *International Journal of Environmental Science and Technology*, 1, 221–225. https://doi.org/10.1007/BF03325836

Amendola, V., & Meneghetti, M. (2009). Size evaluation of gold NPs by UV-vis spectroscopy. *The Journal of Physical Chemistry C*, *113*, 4277–4285.

Apel, K., & Hirt, H. (2004). ROS: Metabolism, oxidative stress, and signal transduction. Annual Review of Plant Biology, *55*, 373–399. https://doi.org/10.1146/ANNUREV.ARPLANT.55.031903.141701

Asada, K. (2006). Production and scavenging of ROS in chloroplasts and their functions. *Plant Physiology*, *141*, 391–396. https://doi.org/10.1104/PP.106.082040

Awasthi, A., Bansal, S., Jangir, L. K., Awasthi, G., Awasthi, K. K., & Awasthi, K. (2017). Effect of ZnO NPs on germination of Triticum aestivum seeds. *Macromolecular Symposia*, *376*, 1700043. https://doi.org/10.1002/MASY.201700043

Banerjee, K., Pramanik, P., Maity, A., Joshi, D. C., Wani, S. H., & Krishnan, P. (2019). Methods of using NMs to plant systems and their delivery to plants (mode of entry, uptake, translocation, accumulation, biotransformation and barriers). In M. Ghorbanpour, & S. H. Wani (Eds.), *Advances in phytonanotechnology: From synthesis to application* (pp. 123–152). Academic Press. https://doi.org/10.1016/B978-0-12-815322-2.00005-5

Barzegar, M., Ahmadvand, D., Sabouri, Z., & Darroudi, M. (2023). Green synthesis of magnesium oxide NPs by chitosan polymer and assessment of their photocatalytic activity and cytotoxicity influences. *Materials Chemistry and Physics*, *301*, 127649. https://doi.org/10.1016/J.MATCHEMPHYS.2023.127649

Baskar, V., Venkatesh, J., & Park, S. W. (2015). Impact of biologically synthesized silver NPs on the growth and physiological responses in Brassica rapa ssp. pekinensis. *Environmental Science and Pollution Research*, *22*, 17672–17682.

Batista, B. D., & Singh, B. K. (2021). Realities and hopes in the application of microbial tools in agriculture. *Microbial Biotechnology*, *14*, 1258–1268. https://doi.org/10.1111/1751-7915.13866

Batool, M., Godoy, A. O., Birnbach, M., Hodoroaba, V-D., Rades, S., Salge, T., Mielke, J., Ortel E. & Schmidt R. (2016). Characterisation of NPs by means of high-resolution SEM/EDS in transmission mode. *IOP Conference Series: Materials Science and Engineering*, *109*, 012006. https://doi.org/10.1088/1757-899X/109/1/012006

Baudot, C., Tan, C. M., & Kong, J. C. (2010). FTIR spectroscopy as a tool for nano-material characterization. *Infrared Physics & Technology*, *53*, 434–438. https://doi.org/10.1016/J.INFRARED.2010.09.002

Baz, H., Creech, M., Chen, J., Gong, H., Bradford, K., & Huo, H. (2020). Water-soluble carbon NPs improve seed germination and post-germination growth of lettuce under salinity stress. *Agronomy*, *10*, 1192. https://doi.org/10.3390/AGRONOMY10081192

Bechtold, U., Rabbani, N., Mullineaux, P. M., & Thornalley, P. J. (2009). Quantitative measurement of specific biomarkers for protein oxidation, nitration and glycation in Arabidopsis leaves. *The Plant Journal*, *59*, 661–671. https://doi.org/10.1111/J.1365-313X.2009.03898.X

Birla, S. S., Tiwari, V. V., Gade, A. K., Ingle, A. P., Yadav, A. P., & Rai, M. K. (2009). Fabrication of silver NPs by Phoma glomerata and its combined effect against Escherichia coli, Pseudomonas aeruginosa and Staphylococcus aureus. *Letters in Applied Microbiology*, *48*, 173–179. https://doi.org/10.1111/J.1472-765X.2008.02510.X

Bose, J., Rodrigo-Moreno, A., & Shabala, S. (2014). ROS homeostasis in halophytes in the context of salinity stress tolerance. *Journal of Experimental Botany*, *65*, 1241–1257. https://doi.org/10.1093/JXB/ERT430

Bussières, P. (2014). Estimating the number and size of phloem sieve plate pores using longitudinal views and geometric reconstruction. *Scientific Reports*, 4, 1–11. https://doi.org/10.1038/srep04929

Bzdek, B. R., Zordan, C. A., Luther, G. W., & Johnston, M. V. (2011). Nanoparticle chemical composition during new particle formation. Aerosol Science and Technology, *45*, 1041–1048. https://doi.org/10.1080/02786826.2011.580392

Cao, G. (2004). *Nanostructures and nanomaterials. Synthesis, properties, and applications*. Imperial College Press. https://doi.org/10.1142/P305

Carmichael, W. W., Mahmood, N. A., & Hyde, E. G. (1990). Natural toxins from cyanobacteria (blue-green algae). In S. Hall, & G. Strichartz (Eds.), *Marine toxins: Origin, structure, and molecular pharmacology* (pp. 87–106). American Chemical Society. https://doi.org/10.1021/BK-1990-0418.CH006

Casalini, S., Bortolotti, C. A., Leonardi, F., & Biscarini, F. (2017). Self-assembled monolayers in organic electronics. *Chemical Society Reviews*, *46*, 40–71. https://doi.org/10.1039/C6CS00509H

Casciardi, S., Sisto, R., & Diociaiuti, M. (2013). The analytical transmission electron microscopy: A powerful tool for the investigation of low-dimensional carbon NMs. *Journal of Nanomaterials*, 2013. https://doi.org/10.1155/2013/506815

Caverzan, A., Piasecki, C., Chavarria, G., Stewart, Jr., C. N., & Vargas, L. (2019). Defenses against ROS in crops and weeds: The effects of interference and herbicides. *International Journal of Molecular Science*, *20*, 1086. https://doi.org/10.3390/IJMS20051086

Chapman, J. M., Muhlemann, J. K., Gayomba, S. R., & Muday, G. K. (2019). RBOH-dependent ROS synthesis and ROS scavenging by plant specialized metabolites to modulate plant development and stress responses. *Chemical Research in Toxicology*, *32*, 370–396. https://doi.org/10.1021/ACS.CHEMRESTOX.9B00028

Chawla, S., Jain, S., & Jain, V. (2013). Salinity induced oxidative stress and antioxidant system in salt-tolerant and salt-sensitive cultivars of rice (Oryza sativa L.). *Journal of Plant Biochemistry and Biotechnology*, *22*, 27–34.

Chen, K., Gao, J., Sun, S., Zhang, Z., Yu, B., Li, J., Xie, C., Li, G., Wang, P., Song, C.-P., Bressan, R. A., Hua, J., Zhu, J.-K., Zhao, Y. (2020). The calcium-responsive phospholipid-binding BONZAI proteins control global osmotic stress responses in plants through repression of immune signaling. *SSRN Electron J*. https://doi.org/10.2139/SSRN.3569555

Chen, M., Wang, D., Yang, F., Xu, X., Xu, N, & Cao, X. (2017). Transport and retention of biochar NPs in a paddy soil under environmentally-relevant solution chemistry conditions. *Environmental Pollution*, *230*, 540–549. https://doi.org/10.1016/J.ENVPOL.2017.06.101

Chen, X., Ding, Y., Yang, Y., Song, C., Wang, B., Yang, S., Guo, Y., & Gong, Z. (2021). Protein kinases in plant responses to drought, salt, and cold stress. *Journal of Integrative Plant Biology*, *63*, 53–78. https://doi.org/10.1111/JIPB.13061

Da Costa, M. V. J., & Sharma, P. K. (2016). Effect of Cu oxide NPs on growth, morphology, photosynthesis, and antioxidant response in Oryza sativa. *Photosynthetica International Journal for Photosynthesis Research*, *54*, 110–119. https://doi.org/10.1007/S11099-015-0167-5

Darroudi, M., Sabouri, Z., Kazemi Oskuee, R., Zak, A. K., Kargar, H., Hamid, M. H. N. A. (2013). Sol–gel synthesis, characterization, and neurotoxicity effect of zinc oxide NPs using gum tragacanth. *Ceramics International*, *39*, 9195–9199. https://doi.org/10.1016/J.CERAMINT.2013.05.021

Das, A., Das, B., Das, A., & Das, B. (2019). Nanotechnology a potential tool to mitigate abiotic stress in crop plants. In A. K. Shanker, & C. Shanker (Eds.), *Abiotic and biotic stress plants*. IntechOpen. https://doi.org/10.5772/INTECHOPEN.83562

Das, K., & Roychoudhury, A. (2014). ROS and response of antioxidants as ROS-scavengers during environmental stress in plants. *Frontiers in Environmental Science*, *2*, 53.

Decarolis, D., Odarchenko, Y., Herbert, J. J., Qiu, C., Longo, A., & Beale, A. M. (2020). Identification of the key steps in the self-assembly of homogeneous gold metal NPs produced using inverse micelles. *Physical Chemistry Chemical Physics*, *22*, 18824–18834. https://doi.org/10.1039/C9CP03473K

Deinlein, U., Stephan, A. B., Horie, T., Horie, T., Luo, W., Xu, G., & Schroeder, J. I. (2014). Plant salt-tolerance mechanisms. *Trends in Plant Science*, *19*, 371–379. https://doi.org/10.1016/J.TPLANTS.2014.02.001

Demidchik, V. (2010). ROS, oxidative stress and plant ion channels. In V. Demidchik, & F. Maathuis (Eds.), *Ion channels and plant stress responses* (pp. 207–232). Springer. https://doi.org/10.1007/978-3-642-10494-7_11

Demidchik, V. (2015). Mechanisms of oxidative stress in plants: From classical chemistry to cell biology. *Environmental and Experimental Botany*, *109*, 212–228. https://doi.org/10.1016/J.ENVEXPBOT.2014.06.021

Dhand, C., Dwivedi, N., Loh, X. J., Ying, A. N. J., Verma, N. K., Beuerman, R. W., Lakshminarayanan, R., & Ramakrishna, S. (2015). Methods and strategies for the synthesis of diverse NPs and their applications: A comprehensive overview. *RSC Advances*, *5*, 105003–105037. https://doi.org/10.1039/C5RA19388E

Dimkpa, C. O., & Bindraban, P. S. (2018). Nanofertilizers: New products for the industry? *Journal of Agricultural and Food Chemistry*, *66*, 6462–6473. https://doi.org/10.1021/ACS.JAFC.7B02150

Du, P., Si, R., Li, Z., & Zou, Z. (2023). Applications of X-ray absorption fine structure spectroscopy in the photocatalytic conversion of small molecules. *ACS Catalysis*, *13*, 6690–6703. https://doi.org/10.1021/ACSCATAL.2C06413

Duan, J., Zhang, M., Zhang, H., Zhang, H., Xiong, H., Liu, P., Ali, J., Li, J., & Li, Z. (2012). OsMIOX, a myo-inositol oxygenase gene, improves drought tolerance through scavenging of ROS in rice (Oryza sativa L.). *Plant Science*, *196*, 143–151. https://doi.org/10.1016/J.PLANTSCI.2012.08.003

Edreva, A. (2005). Generation and scavenging of ROS in chloroplasts: A submolecular approach. *Agriculture, Ecosystems & Environment, 106*, 119–133. https://doi.org/10.1016/J.AGEE.2004.10.022

Eichert, T., Kurtz, A., Steiner, U., & Goldbach, H. E. (2008). Size exclusion limits and lateral heterogeneity of the stomatal foliar uptake pathway for aqueous solutes and water-suspended NPs. *Physiologia Plantarum*, *134*, 151–160. https://doi.org/10.1111/J.1399-3054.2008.01135.X

Eid, M. M. (2022). Characterization of NPs by FTIR and FTIR-microscopy. In S. E. Mallakpour & C. M. Hussain (Eds.), Handbook of consumer nanoproducts (pp. 645–673). Springer Nature. https://doi.org/10.1007/978-981-16-8698-6_89

Elsheery, N. I., & Cao, K. F. (2008). Gas exchange, chlorophyll fluorescence, and osmotic adjustment in two mango cultivars under drought stress. *Acta Physiologiae Plantarum*, *30*, 769–777.

Eroglu, Z., & Metin, O. (2023). Internal interactions within the complex type-II heterojunction of a graphitic carbon nitride/black phosphorus hybrid decorated with graphene quantum dots: Implications for photooxidation performance. *ACS Applied Nano Materials*, *6*, 7960–7974.

Etesami, H., Fatemi, H., & Rizwan, M. (2021). Interactions of NPs and salinity stress at physiological, biochemical and molecular levels in plants: A review. *Ecotoxicology and Environmental Safety*, *225*, 112769. https://doi.org/10.1016/J.ECOENV.2021.112769

Evans, M. D., Dizdaroglu, M., & Cooke, M. S. (2004). Oxidative DNA damage and disease: Induction, repair and significance. *Mutation Research - Reviews in Mutation Research*, *567*, 1–61. https://doi.org/10.1016/j.mrrev.2003.11.001

Fan, X., Cao, X., Zhou, H., Hao, L., Dong, W., He, C., Xu, M., Wu, H., Wang, L., Chang, Z., & Zheng, Y. (2020). Carbon dioxide fertilization effect on plant growth under soil water stress associates with changes in stomatal traits, leaf photosynthesis, and foliar nitrogen of bell pepper (Capsicum annuum L.). *Environmental and Experimental Botany*, *179*, 104203. https://doi.org/10.1016/J.ENVEXPBOT.2020.104203

Faramarzi, S., Anzabi, Y., & Jafarizadeh-Malmiri, H. (2020). Nanobiotechnology approach in intracellular selenium nanoparticle synthesis using Saccharomyces cerevisiae – Fabrication and characterization. *Archives of Microbiology*, *202*, 1203–1209.

Fardsadegh, B., & Jafarizadeh-Malmiri, H. (2019). Aloe vera leaf extract mediated green synthesis of selenium NPs and assessment of their in vitro antimicrobial activity against spoilage fungi and pathogenic bacteria strains. *Green Processing and Synthesis*, *8*, 399–407.

Fazal, H., Abbasi, B. H., Ahmad, N., & Ali, M. (2016). Elicitation of medicinally important antioxidant secondary metabolites with silver and gold NPs in callus cultures of Prunella vulgaris L. *Applied Biochemistry and Biotechnology*, *180*, 1076–1092. https://doi.org/10.1007/S12010-016-2153-1

Fink, S. P., Reddy, G. R., & Marnett, L. J. (1997). Mutagenicity in Escherichia coli of the major DNA adduct derived from the endogenous mutagen malondialdehyde. *Proc Natl Acad Sci U S A*, *94*, 8652. https://doi.org/10.1073/PNAS.94.16.8652

Flowers, T. J., & Colmer, T. D. (2015). Plant salt tolerance: Adaptations in halophytes. *Annals of Botany*, *115*, 327–331. https://doi.org/10.1093/AOB/MCU267

Foyer, C. H., & Halliwell, B. (1976). The presence of glutathione and glutathione reductase in chloroplasts: A proposed role in ascorbic acid metabolism. *Planta*, *133*, 21–25. https://doi.org/10.1007/BF00386001

Gardner, P. R., & Fridovich, I. (1991). Superoxide sensitivity of the Escherichia coli 6-phosphogluconate dehydratase. *Journal of Biological Chemistry*, *266*, 1478–1483. https://doi.org/10.1016/S0021-9258(18)52319-X

Gericke, M., & Pinches, A. (2006). Microbial production of gold NPs. *Gold Bulletion*, *39*, 22–28.

Ghosh, N., Adak, M. K., Ghosh, P. D., Gupta, S., Gupta, D. N., & Mandal, C. (2011). Differential responses of two rice varieties to salt stress. *Plant Biotechnology Reports*, *5*, 89–103.

Gill, S. S., & Tuteja, N. (2010). ROS and antioxidant machinery in abiotic stress tolerance in crop plants. *Plant Physiology and Biochemistry*, *48*, 909–930. https://doi.org/10.1016/J.PLAPHY.2010.08.016

Gohari, G., Mohammadi, A., Akbari, A., Panahirad, S., Dadpour, M. R., Fotopoulos, V., & Kimura, S. (2020). Titanium dioxide NPs (TiO_2 NPs) promote growth and ameliorate salinity stress effects on essential oil profile and biochemical attributes of Dracocephalum moldavica. *Scientific Reports*, 10, 1–14. https://doi.org/10.1038/s41598-020-57794-1

Gul, M., Wakeel, A., Steffens, D., & Lindberg, S. (2019). Potassium-induced decrease in cytosolic Na^+ alleviates deleterious effects of salt stress on wheat (Triticum aestivum L.). *Plant Biology*, *21*, 825–831. https://doi.org/10.1111/PLB.12999

Gulzar, S., Hussain, T., Gul, B., & Hameed, A. (2020). Photosynthetic adaptations and oxidative stress tolerance in halophytes from warm subtropical region. In M.-N. Grigore (Ed.), *Handbook of halophytes* (pp. 1–31). Springer. https://doi.org/10.1007/978-3-030-17854-3_52-1

Gwo, S., Chen, H. Y., Lin, M. H., Sun, L., & Li, X. (2016). Nanomanipulation and controlled self-assembly of metal NPs and nanocrystals for plasmonics. *Chemial Society Reviews*, *45*, 5672–5716. https://doi.org/10.1039/C6CS00450D

Hasanuzzaman, M., Bhuyan, M. H. M. B., Zulfiqar, F., Raza, A., Mohsin, S. M., Mahmud, J. A., Fujita, M., & Fotopoulos, V. (2020). ROS and antioxidant defense in plants under abiotic stress: Revisiting the crucial role of a universal defense regulator. *Antioxidants*, *9*, 1–52. https://doi.org/10.3390/ANTIOX9080681

Hasanuzzaman, M., Borhannuddin Bhuyan, M. H. M., Anee, T. I., Parvin, K., Nahar, K., Mahmud, J. A., & Fujita, M. (2019). Regulation of ascorbate-glutathione pathway in mitigating oxidative damage in plants under abiotic stress. *Antioxidants*, 8, 384. https://doi.org/10.3390/ANTIOX8090384

Hezaveh, T. A., Pourakbar, L., Rahmani, F., & Alipour, H. (2019). Interactive effects of salinity and zno nps on physiological and molecular parameters of rapeseed (Brassica napus L.). Communications in Soil Science and Plant Analysis, *50*, 698–715. https://doi.org/10.1080/00103624.2019.1589481

Hoffmann, M. R., Martin, S. T., Choi, W., & Bahnemann, D. W. (1995). Environmental applications of semiconductor photocatalysis. *Chemical Reviews*, *95*, 69–96.

Holder, C. F., & Schaak, R. E. (2019). Tutorial on powder X-ray diffraction for characterizing nanoscale materials. *ACS Nano*, *13*, 7359–7365.

Hong, C. Y., Chao, Y. Y., Yang, M. Y., Cheng, S. Y., Cho, S.-C., & Kao, C.-H. (2009). NaCl-induced expression of glutathione reductase in roots of rice (Oryza sativa L.) seedlings is mediated through hydrogen peroxide but not abscisic acid. *Plant Soil*, *320*, 103–115. https://doi.org/10.1007/S11104-008-9874-Z

Hong, C. Y., Hsu, Y. T., Tsai, Y. C., & Kao, C. H. (2007). Expression of ASCORBATE PEROXIDASE 8 in roots of rice (Oryza sativa L.) seedlings in response to NaCl. *Journal of Experimental Botany*, *58*, 3273–3283. https://doi.org/10.1093/JXB/ERM174

Hong, J., Wang, C., Wagner, D. C., Gardea-Torresdey, J. L., He, F., & Rico, C. M. (2021). Foliar application of NPs: Mechanisms of absorption, transfer, and multiple impacts. *Environmental Science: Nano*, *8*, 1196–1210. https://doi.org/10.1039/D0EN01129K

Hu, M., Chen, J., Li, Z. Y., Au, L., Hartland, G. V., Li, X., Marquez, M., Xia, Y. (2006). Gold nanostructures: Engineering their plasmonic properties for biomedical applications. *Chemical Society Reviews*, *35*, 1084–1094. https://doi.org/10.1039/B517615H

Hu, P., An, J., Faulkner, M. M., Wu, H., Li, Z., Tian, X., & Giraldo, J. P. (2020). Nanoparticle charge and size control foliar delivery efficiency to plant cells and organelles. *ACS Nano*, *14*, 7970–7986.

Hu, Q., Liu, R., & Liu, J. (2019). Effects of Bacillus subtilis QM3 on germination and antioxidant enzymes activities of wheat seeds under salt stress. *OALib*, *6*, 1–9. https://doi.org/10.4236/OALIB.1105218

Hussain, S., Khan, F., Hussain, H. A., & Nie, L. (2016). Physiological and biochemical mechanisms of seed priming-induced chilling tolerance in rice cultivars. *Frontiers in Plant Sciences*, *7*, 116.

Huston, M., Debella, M., Dibella, M., & Gupta, A. (2021). Green synthesis of NMs. *Nanomater*, 11, 2130. https://doi.org/10.3390/NANO11082130

Imlay, J. A., & Linn, S. (1988). DNA damage and oxygen radical toxicity. *Science*, *240*, 1302–1309. https://doi.org/10.1126/SCIENCE.3287616

Iqbal, M. N., Rasheed, R., Ashraf, M. Y., Ashraf, M. A., & Hussain, I. (2018). Exogenously applied zinc and Cu mitigate salinity effect in maize (Zea mays L.) by improving key physiological and biochemical attributes. *Environmental Science and Pollution Research*, *25*, 23883–23896.

Jemal, K., Sandeep, B. V., & Pola, S. (2017). Synthesis, characterization, and evaluation of the antibacterial activity of allophylus serratus leaf and leaf derived callus extracts mediated silver NPs. *Journal of Nanomaterials*, 2017, 4213275. https://doi.org/10.1155/2017/4213275

Jha, A. K., Prasad, K., & Prasad, K. (2009). A green low-cost biosynthesis of Sb_2O_3 NPs. *Biochemical Engineering Journal*, *43*, 303–306. https://doi.org/10.1016/J.BEJ.2008.10.016

Jiang, H. S., Qiu, X. N., Li, G. B., Li, W., & Yin, L.-Y. (2014). Silver NPs induced accumulation of ROS and alteration of antioxidant systems in the aquatic plant Spirodela polyrhiza. *Environmental Toxicology and Chemistry*, *33*, 1398–1405. https://doi.org/10.1002/ETC.2577

Jithesh, M. N., Prashanth, S. R., Sivaprakash, K. R., & Parida, A. K. (2006). Antioxidative response mechanisms in halophytes: Their role in stress defence. *Journal of Genetics*, 85, 237–254. https://doi.org/10.1007/BF02935340

Jogaiah, S., Paidi, M. K., Venugopal, K., Geetha, N., Mujtaba, M., Udikeri, S. S., & Govarthanan, M. (2021). Phytotoxicological effects of engineered NPs: An emerging nanotoxicology. *The Science of the Total Environment*, *801*, 149809. https://doi.org/10.1016/J.SCITOTENV.2021.149809

Keunen, E., Remans, T., Bohler, S., Vangronsveld, J., & Cuypers, A. (2011). Metal-induced oxidative stress and plant mitochondria. *International Journal of Molecular Science*, *12*, 6894–6918. https://doi.org/10.3390/IJMS12106894

Khan, M., Khan, M. S. A., Borah, K. K., Goswami, Y., Hakeem, K. R., & Chakrabartty, I. (2021). The potential exposure and hazards of metal-based NPs on plants and environment, with special emphasis on ZnO NPs, TiO_2 NPs, and AgNPs: A review. *Environmental Advances*, *6*, 100128. https://doi.org/10.1016/J.ENVADV.2021.100128

Khan, M. R., Adam, V., Rizvi, T. F., Zhang, B., Ahamad, F., Jośko, I., Zhu, Y., Yang, M., & Mao, C. (2019). Nanoparticle-plant interactions: A two-way traffic. *Small*, *15*, e1901794. https://doi.org/10.1002/SMLL.201901794

Khan, Z., Bhadouria, P., & Bisen, P. (2005). Nutritional and therapeutic potential of Spirulina. *Current Pharmaceutical Biotechnology*, *6*, 373–379. https://doi.org/10.2174/138920105774370607

Kibria, M. G., Hossain, M., Murata, Y., & Hoque, M. A. (2017). Antioxidant defense mechanisms of salinity tolerance in rice genotypes. *Rice Science*, *24*, 155–162. https://doi.org/10.1016/J.RSCI.2017.05.001

Kiefer, J., Grabow, J., Kurland, H. D., & Müller, F. A. (2015). Characterization of NPs by solvent infrared spectroscopy. *Analytical Chemistry*, *87*, 12313–12317.

Kim, J.-H., Yeoup Chung, B., Baek, M.-H., Wi, S., Yang, D., Lee, M.-C., & Kim, J.-S. (2005). Expression of antioxidant isoenzyme genes in rice under salt stress and effects of jasmonic acid and γ-radiation. *Journal of Applied Biological Chemistry*, *48*, 1–6

Kowshik, M., Deshmukh, N., Vogel, W., Urban, J., Kulkarni, S. K., & Paknikar, K. M. (2002). Microbial synthesis of semiconductor CdS NPs, their characterization, and their use in the fabrication of an ideal diode. *Biotechnology and Bioengineering*, *78*, 583–588. https://doi.org/10.1002/BIT.10233

Kozikowski, P., & Sobiech, P. (2022). Comparison of NPs' characteristic parameters derived from SEM and SMPS analyses. *Journal of Nanoparticle Research*, 24, 1–12. https://doi.org/10.1007/S11051-022-05480-W

Kumar, A., Vemula, P. K., Ajayan, P. M., & John, G. (2008). Silver-nanoparticle-embedded antimicrobial paints based on vegetable oil. *Nature Materials*, 7, 236–241. https://doi.org/10.1038/nmat2099

Kumari, S., Khanna, R. R., Nazir, F., Nazir, F., Albaqami, M., Chhillar, H., Wahid, I., & Khan, M. I. R. (2022). Bio-synthesized NPs in developing plant abiotic stress resilience: A new boon for sustainable approach. *International Journal of Molecular Science*, 23, 4452. https://doi.org/10.3390/IJMS23084452

Le, N., Zhang, M., & Kim, K. (2022). Quantum dots and their interaction with biological systems. *International Journal of Molecular Science*, 23, 10763. https://doi.org/10.3390/IJMS231810763

Lee, B., Yoon, S., Lee, J. W., Kim, Y., Chang, J., Yun, J., Ro, J. C., Lee, J.-S., & Lee, J. H. (2020). Statistical characterization of the morphologies of NPs through machine learning based electron microscopy image analysis. *ACS Nano*, *14*, 17125–17133.

Li, C. H., Wang, G., Zhao, J. L., Zhang, L.-Q., Ai, L.-F., Han, Y.-F., Sun, D.-Y., Zhang, S.-W., & Sun, Y. (2014). The receptor-like kinase SIT1 mediates salt sensitivity by activating MAPK3/6 and regulating ethylene homeostasis in rice. *Plant Cell*, *26*, 2538–2553. https://doi.org/10.1105/TPC.114.125187

Li, W. R., Sun, T. L., Zhou, S. L., Ma, Y.-K., Shi, Q.-S., Xie, X.-B., Huang, X.-M. (2017). A comparative analysis of antibacterial activity, dynamics, and effects of silver ions and silver NPs against four bacterial strains. *International Biodeterioration & Biodegradation*, *123*, 304–310. https://doi.org/10.1016/J.IBIOD.2017.07.015

Ling, Q., & Jarvis, P. (2015). Regulation of chloroplast protein import by the ubiquitin E3 ligase SP1 is important for stress tolerance in plants. *Current Biology*, *25*, 2527–2534. https://doi.org/10.1016/j.cub.2015.08.015

Liu, J., Shabala, S., Shabala, L., Zhou, M., Meinke, H., Venkataraman, G., Chen, Z., Zeng, F., & Zhao, Q. (2019). Tissue-specific regulation of Na^+ and K^+ transporters explains genotypic differences in salinity stress tolerance in rice. *Frontiers in Plant Science*, *10*, 1361.

Liu, T., Van Staden, J., & Cress, W. A. (2000). Salinity induced nuclear and DNA degradation in meristematic cells of soybean (Glycine max (L.)) roots. *Plant Growth Regulation*, 30, 49–54. https://doi.org/10.1023/A:1006311619937

Liu, W., Worms, I., & Slaveykova, V. I. (2020). Interaction of silver NPs with antioxidant enzymes. *Environmental Science: Nano*, *7*, 1507–1517. https://doi.org/10.1039/C9EN01284B

López Gutiérrez, J. D., Abundez Barrera, I. M., & Torres Gómez, N. (2022). Nanoparticle detection on SEM images using a neural network and semi-synthetic training data. *Nanomaterials*, *12*, 1818. https://doi.org/10.3390/NANO12111818

López-Lorente, Á. I., & Mizaikoff, B. (2016). Recent advances on the characterization of NPs using infrared spectroscopy. *TrAC Trends in Analytics Chemistry*, *84*, 97–106. https://doi.org/10.1016/J.TRAC.2016.01.012

Lu, Z., Liu, D., & Liu, S. (2007). Two rice cytosolic ascorbate peroxidases differentially improve salt tolerance in transgenic Arabidopsis. *Plant Cell Reports*, *26*, 1909–1917. https://doi.org/10.1007/S00299-007-0395-7

Lwalaba, J. L. W., Louis, L. T., Zvobgo, G., Richmond, M. E. A., Fu, L., Naz, S., Mwamba, M., Mundende, R. P. M., & Zhang, G. (2020). Physiological and molecular mechanisms of cobalt and Cu interaction in causing phyto-toxicity to two barley genotypes differing in Co tolerance. *Ecotoxicology and Environmental Safety*, *187*, 109866. https://doi.org/10.1016/J.ECOENV.2019.109866

Machado, J., Vasconcelos, M. W., Soares, C., Fidalgo, F., Heuvelink, E. & Carvalho, S. M. P. (2023). Enzymatic and non-enzymatic antioxidant responses of young tomato plants (cv. micro-tom) to single and combined mild nitrogen and water deficit: Not the sum of the parts. *Antioxidants*, 12, 375. https://doi.org/10.3390/ANTIOX12020375

Mahajan, S., & Tuteja, N. (2005). Cold, salinity and drought stresses: An overview. *Archives of Biochemistry and Biophysics*, *444*, 139–158. https://doi.org/10.1016/j.abb.2005.10.018

Mahakham, W., Sarmah, A. K., Maensiri, S., & Theerakulpisut, P. (2017). Nanopriming technology for enhancing germination and starch metabolism of aged rice seeds using phytosynthesized silver NPs. *Scientific Reports*, *7*, 8263. https://doi.org/10.1038/S41598-017-08669-5

Mahmoud, A. W. M., Abdelaziz, S. M., El-Mogy, M. M., & Abdeldaym, E. A. (2019). Effect of foliar Zno and Feo NPs application on growth and nutritional quality of red radish and assessment of their accumulation on human health. *Agriculture*, *65*, 16–29. https://doi.org/10.2478/AGRI-2019-0002

Maity, S., Chatterjee, A., Guchhait, R., De, S., & Pramanick, K. (2020). Cytogenotoxic potential of a hazardous material, polystyrene microparticles on Allium cepa L. *Journal of Hazardous Materials*, *385*, 121560. https://doi.org/10.1016/J.JHAZMAT.2019.121560

Majumdar, S., Almeida, I. C., Arigi, E. A., Choi, H., VerBerkmoes, N. C., Trujillo-Reyes, J., Flores-Margez, J. P., White, J. C., Peralta-Videa, J. R., & Gardea-Torresdey, J. L. (2015). Environmental effects of nanoceria on seed production of common bean (Phaseolus vulgaris): A proteomic analysis. *Environmental Science & Technology*, *49*, 13283–13293.

Malatesta, M. (2021). Transmission electron microscopy as a powerful tool to investigate the interaction of NPs with subcellular structures. *International Journal of Molecular Science*, *22*, 12789. https://doi.org/10.3390/IJMS222312789

Manzoor, N., Ahmed, T., Noman, M., Shahid, M., Nazir, M. M., Ali, L., Alnusaire, T. S., Li, B., Schulin, R., & Wang, G. (2021). Iron oxide NPs ameliorated the cadmium and salinity stresses in wheat plants, facilitating photosynthetic pigments and restricting cadmium uptake. *Science of the Total Environment*, *769*, 145221. https://doi.org/10.1016/J.SCITOTENV.2021.145221

Markus, J., Mathiyalagan, R., Kim, Y. J., Abbai, R., Singh, P., Ahn, S., Perez, Z. E. J., Hurh, J., Yang, D. C. (2016). Intracellular synthesis of gold NPs with antioxidant activity by probiotic Lactobacillus kimchicus DCY51T isolated from Korean kimchi. *Enzyme and Microbial Technology*, *95*, 85–93. https://doi.org/10.1016/J.ENZMICTEC.2016.08.018

Marsalek, R. (2014). Particle size and zeta potential of ZnO. *APCBEE Procedia*, *9*, 13–17. https://doi.org/10.1016/J.APCBEE.2014.01.003

Marusig, D., & Tombesi, S. (2020). Abscisic acid mediates drought and salt stress responses in Vitis vinifera – A review. *International Journal of Molecular Science*, 21, 8648. https://doi.org/10.3390/IJMS21228648

Matkovics, B., Kotorman, M., Varga, I. S., Hai, D. Q., & Varga, C. (1997). Oxidative stress in experimental diabetes induced by streptozotocin. *Acta Physiologica Hungarica*, *85*, 29–38

Meyers, M. A., Mishra, A., & Benson, D. J. (2006). Mechanical properties of nanocrystalline materials. *Progress in Materials Science*, *51*, 427–556. https://doi.org/10.1016/J.PMATSCI.2005.08.003

Michen, B., Geers, C., Vanhecke, D., Endes, C., Rothen-Rutishauser, B., Balog, S., & Petri-Fink, A. (2015). Avoiding drying-artifacts in transmission electron microscopy: Characterizing the size and colloidal state of NPs. *Scientific Reports*, 5, 1–7. https://doi.org/10.1038/srep09793

Miralles, P., Church, T. L., & Harris, A. T. (2012). Toxicity, uptake, and translocation of engineered NMs in vascular plants. *Environmental Science & Technology*, *46*, 9224–9239. https://doi.org/10.1021/ES202995D

Mishra, S., Jha, A. B., & Dubey, R. S. (2011). Arsenite treatment induces oxidative stress, upregulates antioxidant system, and causes phytochelatin synthesis in rice seedlings. *Protoplasma*, *248*, 565–577. https://doi.org/10.1007/S00709-010-0210-0

Mittal, A. K., Chisti, Y., & Banerjee, U. C. (2013). Synthesis of metallic NPs using plant extracts. *Biotechnology Advances*, *31*, 346–356. https://doi.org/10.1016/J.BIOTECHADV.2013.01.003

Mittler, R. (2002). Oxidative stress, antioxidants and stress tolerance. *Trends in Plant Science*, *7*, 405–410. https://doi.org/10.1016/S1360-1385(02)02312-9

Mohamed, A. K. S. H., Qayyum, M. F., Abdel-Hadi, A. M., Rehman, R. A., Ali, S., & Rizwan, M. (2017). Interactive effect of salinity and silver NPs on photosynthetic and biochemical parameters of wheat. Archives of Agronomy and Soil Science, *63*, 1736–1747. https://doi.org/10.1080/03650340.2017.1300256

Mohanpuria, P., Rana, N. K., & Yadav, S. K. (2008). Biosynthesis of NPs: Technological concepts and future applications. *Journal of Nanoparticle Research*, *10*, 507–517.

Møller, I. M., & Kristensen, B. K. (2004). Protein oxidation in plant mitochondria as a stress indicator. *Photochemical & Photobiological Sciences*, *3*, 730–735. https://doi.org/10.1039/B315561G

Moradbeygi, H., Jamei, R., Heidari, R., & Darvishzadeh, R. (2020). Investigating the enzymatic and non-enzymatic antioxidant defense by applying iron oxide NPs in Dracocephalum moldavica L. plant under salinity stress. *Scientia Horticulturae (Amsterdam)*, *272*, 109537. https://doi.org/10.1016/J.SCIENTA.2020.109537

Moradi, P., Vafaee, Y., Mozafari, A. A., & Tahir, N. A.-R. (2022). Silicon NPs and methyl jasmonate improve physiological response and increase expression of stress-related genes in strawberry cv. paros under salinity stress. *Silicon*, *14*, 10559–10569.

Morin, J., Brochard, G., Bergé, V., Rosset, A., Artous, S., Clavaguera, S., Strekowskiad, R. S., & Wortham, H. (2023). Uptake of *m*-xylene and VOC emissions by mineral photocatalytic paints of indoor air building interest. *Environmental Science: Nano*, 179, 106979. https://doi.org/10.1039/D3EN00084B

Moulton, M. C., Braydich-Stolle, L. K., Nadagouda, M. N., Kunzelman, S., Hussain, S. M., & Varma, R. S. (2010). Synthesis, characterization and biocompatibility of "green" synthesized silver NPs using tea polyphenols. *Nanoscale*, *2*, 763–770. https://doi.org/10.1039/C0NR00046A

Mourdikoudis, S., Pallares, R. M., & Thanh, N. T. K. (2018). Characterization techniques for NPs: Comparison and complementarity upon studying nanoparticle properties. *Nanoscale*, *10*, 12871–12934. https://doi.org/10.1039/C8NR02278J

Moustafa, K., AbuQamar, S., Jarrar, M., Al-Rajab, A. J., & Trémouillaux-Guiller, J. (2014). MAPK cascades and major abiotic stresses. *Plant Cell Reports*, *33*, 1217–1225.

Mühlfeld, C., Rothen-Rutishauser, B., Vanhecke, D., Blank, F., Gehr, P., & Ochs, M. (2007). Visualization and quantitative analysis of NPs in the respiratory tract by transmission electron microscopy. *Particle and Fibre Toxicology*, 4, 1–17. https://doi.org/10.1186/1743-8977-4-11

Mukherjee, A., Sarkar, D., & Sasmal, S. (2021). A review of green synthesis of metal NPs using algae. *Frontiers in Microbiology*, *12*, 2152.

Munns, R., & Tester, M. (2008). Mechanisms of salinity tolerance. *Annual Review of Plant Biology*, *59*, 651–681. https://doi.org/10.1146/ANNUREV.ARPLANT.59.032607.092911

Nadarajah, K. K. (2020). ROS homeostasis in abiotic stress tolerance in plants. *International Journal of Molecular Science*, 21, 5208. https://doi.org/10.3390/IJMS21155208

Naderi, M. R., & Danesh-Shahraki, A. (2013). Nanofertilizers and their roles in sustainable agriculture. *International Journal of Agricutural and Crop Science*, *5*, 2229–2232

Nakhoda, B., Leung, H., Mendioro, M. S., Mohammadi-nejad, G., & Ismail, A. M. (2012). Isolation, characterization, and field evaluation of rice (Oryza sativa L., Var. IR64) mutants with altered responses to salt stress. *Field Crops Research*, *127*, 191–202. https://doi.org/10.1016/J.FCR.2011.11.004

Narayanan, K. B., & Sakthivel, N. (2011). Synthesis and characterization of nano-gold composite using Cylindrocladium floridanum and its heterogeneous catalysis in the degradation of 4-nitrophenol. *Journal of Hazardous Materials*, *189*, 519–525. https://doi.org/10.1016/J.JHAZMAT.2011.02.069

Neogy, S., Savalia, R. T., Tewari, R., Srivastava, D., & Dey, G. K. (2020). Transmission electron microscopy of NMs. *Indian Journal of Pure & Applied Physics*, *44*, 119–124. https://doi.org/10.5772/INTECHOPEN.92212

Noctor, G., & Foyer, C. H. (2003). Ascorbate and glutathione: Keeping active oxygen under control. Annual Review of Plant Physiology and Plant Molecular Biology, *49*, 249–279. https://doi.org/10.1146/ANNUREV.ARPLANT.49.1.249

Ottow, E. A., Brinker, M., Teichmann, T., Fritz, E., Kaiser, W., Brosché, M., Kangasjärvi, J., Jiang, X., & Polle, A. (2005). Populus euphratica displays apoplastic sodium accumulation, osmotic adjustment by decreases in calcium and soluble carbohydrates, and develops leaf succulence under salt stress. *Plant Physiology*, *139*, 1762–1772. https://doi.org/10.1104/PP.105.069971

Ozgur, R., Uzilday, B., Sekmen, A. H., & Turkan, I. (2013). ROS regulation and antioxidant defence in halophytes. *Functional Plant Biology*, *40*, 832–847. https://doi.org/10.1071/FP12389

Pakrashi, S., Jain, N., Dalai, S., Jayakumar, J., Chandrasekaran, P. T., Raichur, A. M., Chandrasekaran, N., & Mukherjee, A. (2014). In vivo genotoxicity assessment of titanium dioxide NPs by Allium cepa root tip assay at high exposure concentrations. *PLoS One*, *9*, e87789. https://doi.org/10.1371/JOURNAL.PONE.0087789

Pan, D., Huang, G., Yi, J., Cui, J., Liu, C., Li, F., & Li, X. (2022). Foliar application of silica NPs alleviates arsenic accumulation in rice grain: Co-localization of silicon and arsenic in nodes. *Environmental Science: Nano*, *9*, 1271–1281. https://doi.org/10.1039/D1EN01132D

Pan, J., Zhang, M., Kong, X., Xing, X., Liu, Y., Zhou, Y., Liu, Y., Sun, L., & Li, D. (2012). ZmMPK17, a novel maize group D MAP kinase gene, is involved in multiple stress responses. *Planta*, *235*, 661–676.

Panpatte, D. G., Jhala, Y. G., Shelat, H. N., & Vyas, R. V. (2016). NPs: The next generation technology for sustainable agriculture. In D. P. Singh, H. B. Singh, & R. Prabha (Eds.), *Microbial inoculants in sustainable agricultural productivity vol., 2, functional applications* (pp. 289–300). Springer. https://doi.org/10.1007/978-81-322-2644-4_18

Parihar, P., Singh, S., Singh, R., Singh, V. P., & Prasad, S. M. (2015). Effect of salinity stress on plants and its tolerance strategies: A review. *Environmental Science and Pollution Research International*, *22*, 4056–4075.

Patil, S., & Chandrasekaran, R. (2020). Biogenic NPs: A comprehensive perspective in synthesis, characterization, application and its challenges. *Journal of Genetic Engineering and Biotechnology*, 18, 1–23. https://doi.org/10.1186/S43141-020-00081-3

Pérez-de-Luque, A. (2017). Interaction of NMs with plants: What do we need for real applications in agriculture? *Frontiers in Environmentaal Science*, *5*, 12.

Pérez-Labrada, F., López-Vargas, E. R., Ortega-Ortiz, H., Cadenas-Pliego, G., Benavides-Mendoza, A., & Juárez-Maldonado, A. (2019a). Responses of tomato plants under saline stress to foliar application of Cu NPs. *Plants*, 8, 151. https://doi.org/10.3390/PLANTS8060151

Philip, D. (2009). Biosynthesis of Au, Ag and Au–Ag NPs using edible mushroom extract. *Spectrochimica Acta, Part A: Molecular and Biomolecular Spectroscopy*, *73*, 374–381. https://doi.org/10.1016/J.SAA.2009.02.037

Porter, N. A., Caldwell, S. E., & Mills, K. A. (1995). Mechanisms of free radical oxidation of unsaturated lipids. *Lipids*, *30*, 277–290. https://doi.org/10.1007/BF02536034

Pudovkin, M. S., Shamsutdinov, N. I., Zelenikhin, P. V., & Nizamutdinov, A. S. (2021). Transmission electron microscopy and flow cytometry study of cellular uptake of unmodified Pr^{3+}:LaF_3 NPs in dynamic. *Journal of Nanoparticle Research*, *23*, 1–18.

Qu, J., & Liu, X. (2021). Recent advances on SEM-based in situ multiphysical characterization of NMs. *Scanning*, 2021, 4426254. https://doi.org/10.1155/2021/4426254

Rahman A, Nahar K, Hasanuzzaman M, Fujita M. Calcium Supplementation Improves $Na^{(+)}/K^{(+)}$ Ratio, Antioxidant Defense and Glyoxalase Systems in Salt-Stressed Rice Seedlings. *Front Plant Sci.* 2016 May 12;7:609.

Raj, S., Trivedi, R., & Soni, V. (2022). Biogenic synthesis of silver NPs, characterization and their applications – A review. *Surfaces*, *5*, 67–90. https://doi.org/10.3390/SURFACES5010003

Rajput, V. D., Minkina, T., Kumari, A., Shende, S. S., Ranjan, A., Faizan, M., Barakvov, A., Gromovik, A., Gorbunova, N., Rajput, P., Singh, A., Khabirov, I., Nazarenko, O., Sushkova, S., & Kızılkaya, R. (2022). A review on nanobioremediation approaches for restoration of contaminated soil. *Eurasian Journal of Soil Science*, *11*, 43–60. https://doi.org/10.18393/EJSS.990605

Rani, S., Kumari, N., & Sharma, V. (2022). Uptake, translocation, transformation and physiological effects of NPs in plants. Archives of Agronomy and Soil Science, 69, 1579–1599. https://doi.org/10.1080/03650340.2022.2103549

Rengasamy, P., & Rengasamy, P. (2010). Soil processes affecting crop production in salt-affected soils. *Functional Plant Biology*, *37*, 613–620. https://doi.org/10.1071/FP09249

Rico, C. M., Peralta-Videa, J. R., & Gardea-Torresdey, J. L. (2015). Chemistry, biochemistry of NPs, and their role in antioxidant defense system in plants. In M. H. Siddiqui, M. H. Al-Whaibi, & F. Mohammad (Eds.), *Nanotechnology and plant sciences: Nanoparticles and their impact on plants* (pp. 1–17). Springer.

Romero-Puertas, M. C., Palma, J. M., Gómez, M., Del Río, L. A., & Sandalio, L. M. (2002). Cadmium causes the oxidative modification of proteins in pea plants. *Plant, Cell & Environment*, *25*, 677–686. https://doi.org/10.1046/J.1365-3040.2002.00850.X

Saleh, T. A. (2020). NMs: Classification, properties, and environmental toxicities. *Environmental Technology & Innovation*, 20, 101067. https://doi.org/10.1016/J.ETI.2020.101067

Samrot, A. V., Singh, S. P. R., Deenadhayalan, R., Rajesh, V. V., Padmanaban, S. & Radhakrishnan, K. (2022). NPs, a double-edged sword with oxidant as well as antioxidant properties – A review. *Oxygen*, *2*, 591–604. https://doi.org/10.3390/OXYGEN2040039

Santos, A. L., Sinha, S., & Lindner, A. B. (2018). The good, the bad, and the ugly of ROS: New insights on aging and aging-related diseases from eukaryotic and prokaryotic model organisms. *Oxidative Medicine and Cellular Longevity*, *2018*, 1941285. https://doi.org/10.1155/2018/1941285

Sasidharan, D., Namitha, T. R., Johnson, S. P., Jose, V., & Mathew, P. (2020). Synthesis of silver and Cu oxide NPs using Myristica fragrans fruit extract: Antimicrobial and catalytic applications. *Sustainable Chemistry and Pharmacy*, *16*, 100255. https://doi.org/10.1016/J.SCP.2020.100255

Seetharaman, P. K., Chandrasekaran, R., Gnanasekar, S., Chandrakasan, G., Gupta, M., Manikandan, D. B., Sivaperumal, S. (2018). Antimicrobial and larvicidal activity of eco-friendly silver NPs synthesized from endophytic fungi Phomopsis liquidambaris. *Biocatalysis and Agricultural Biotechnology*, *16*, 22–30. https://doi.org/10.1016/J.BCAB.2018.07.006

Senapati, S., Syed, A., Moeez, S., Kumar, A., & Ahmad, A. (2012). Intracellular synthesis of gold NPs using alga Tetraselmis kochinensis. *Materials Letters*, *79*, 116–118. https://doi.org/10.1016/J.MATLET.2012.04.009

Seo, P. J., Park, J. M., Kang, S. K., Kim, S.-G., & Park, C.-M. (2011). An Arabidopsis senescence-associated protein SAG29 regulates cell viability under high salinity. *Planta*, *233*, 189–200.

Sepeur, S. (2008). Nanotechnology: Technical basics and applications. Vincentz Network.

Servin, A. D., Pagano, L., Castillo-Michel, H., la Torre-Roche, R. D., Hawthorne, J., Hernandez-Viezcas, J. A., Loredo-Portales, R., Majumdar, S., Gardea-Torresday, J., Dhankher, O. P., & White, J. C. (2017). Weathering in soil increases nanoparticle CuO bioaccumulation within a terrestrial food chain. Nanotoxicology, *11*, 98–111. https://doi.org/10.1080/17435390.2016.1277274

Shafiq, F., Iqbal, M., Ali, M., & Ashraf, M. A. (2019). Seed pre-treatment with polyhydroxy fullerene NPs confer salt tolerance in wheat through upregulation of H_2O_2 neutralizing enzymes and phosphorus uptake. *Journal of Soil Science and Plant Nutrition*, *19*, 734–742.

Shang, Y., Hasan, M. K., Ahammed, G. J., Li, M., Yin, H., & Zhou, J. (2019). Applications of nanotechnology in plant growth and crop protection: A review. *Molecules*, *24*, 2558. https://doi.org/10.3390/molecules24142558

Sharma, P., Jha, A. B., Dubey, R. S., & Pessarakli, M. (2012). ROS, oxidative damage, and antioxidative defense mechanism in plants under stressful conditions. *Journal of Botany*, *2012*, 1–26. https://doi.org/10.1155/2012/217037

Sharma, R., Sahoo, A., Devendran, R., & Jain, M. (2014). Over-expression of a rice tau class glutathione s-transferase gene improves tolerance to salinity and oxidative stresses in Arabidopsis. *PLoS One*, *9*, e92900. https://doi.org/10.1371/JOURNAL.PONE.0092900

Sharma, V., & Rao, L. J. M. (2013). An overview on chemical composition, bioactivity and processing of leaves of Cinnamomum tamala. Critical Reviews in Food Science and Nutrition, *54*, 433–448. https://doi.org/10.1080/10408398.2011.587615

Singh, A., Singh, N. B., Hussain, I., Singh, H., Yadav, V., & Singh, S. C. (2016). Green synthesis of nano zinc oxide and evaluation of its impact on germination and metabolic activity of Solanum lycopersicum. *Journal of Biotechnology*, *233*, 84–94. https://doi.org/10.1016/J.JBIOTEC.2016.07.010

Singh, J., Dutta, T., Kim, K. H., Rawat, M., Samddar, P., & Kumar, P. (2018). 'Green' synthesis of metals and their oxide NPs: Applications for environmental remediation. *Journal of Nanobiotechnology*, 16, 1–24. https://doi.org/10.1186/S12951-018-0408-4

Singh, N., & Bhatla, S. C. (2016). Nitric oxide and iron modulate heme oxygenase activity as a long distance signaling response to salt stress in sunflower seedling cotyledons. *Nitric Oxide*, *53*, 54–64. https://doi.org/10.1016/J.NIOX.2016.01.003

Singh, V. K., Singh, R., Tripathi, S., Devi, R. S., Srivastava, P., Singh, P., Kumar, A., Bhadouria, R. (2020). Seed priming: State of the art and new perspectives in the era of climate change. In M. N. V. Prasad & M. Pietrzykowski (Eds.), *Climate change and soil interactions* (pp. 143–170). Elsevier. https://doi.org/10.1016/B978-0-12-818032-7.00006-0

Sinha, S. N., Paul, D., Halder, N., Sengupta, D., & Patra, S. K. (2015). Green synthesis of silver NPs using fresh water green alga Pithophora oedogonia (Mont.) Wittrock and evaluation of their antibacterial activity. *Applied Nanoscience*, *5*, 703–709.

Sivaraj, A., Kumar, V., Sunder, R., Parthasarathy, K., & Kasivelu, G. (2020). Commercial yeast extracts mediated green synthesis of silver chloride NPs and their anti-mycobacterial activity. *Journal of Cluster Science*, *31*, 287–291.

Sofo, A., Scopa, A., Nuzzaci, M., & Vitti, A. (2015). Ascorbate peroxidase and catalase activities and their genetic regulation in plants subjected to drought and salinity stresses. *International Journal of Molecular Sciences*, *16*, 13561–13578. https://doi.org/10.3390/IJMS160613561

Soliman, M. H., Alnusairi, G. S. H., Khan, A. A., Alnusaire, T. S., Fakhr, M. A., Abdulmajeed, A. M., Aldesuquy, H. S., Yahya, M., & Najeeb, U. (2022). Biochar and selenium NPs induce water transporter genes for sustaining carbon assimilation and grain production in salt-stressed wheat. *Journal of Plant Growth Regulation*, *42*, 1522–1543.

Song, J., Ding, X., Feng, G., & Zhang, F. (2006). Nutritional and osmotic roles of nitrate in a euhalophyte and a xerophyte in saline conditions. *New Phytologist*, *171*, 357–366. https://doi.org/10.1111/J.1469-8137.2006.01748.X

Stepien, P., & Johnson, G. N. (2009). Contrasting responses of photosynthesis to salt stress in the glycophyte Arabidopsis and the halophyte thellungiella: Role of the plastid terminal oxidase as an alternative electron sink. *Plant Physiology*, *149*, 1154–1165. https://doi.org/10.1104/PP.108.132407

Strasser, P., Koh, S., Anniyev, T., Greeley, J., More, K., Yu, C., Liu, Z., Kaya, S., Nordlund, D., Ogasawara, H., Toney, M. F., & Nilsson, A. (2010). Lattice-strain control of the activity in dealloyed core–shell fuel cell catalysts. *Nature Chemistry*, 2, 454–460. https://doi.org/10.1038/nchem.623

Su, Y., Ashworth, V., Kim, C., Adeleye, A. S., Rolshausen, P., Roper, C., Jason White, J., & Jassby, D. (2019). Delivery, uptake, fate, and transport of engineered NPs in plants: A critical review and data analysis. *Environmental Science: Nano*, *6*, 2311–2331. https://doi.org/10.1039/C9EN00461K

Suo, J., Zhao, Q., David, L., Chen, S., & Dai, S. (2017). Salinity response in chloroplasts: Insights from gene characterization. *International Journal of Molecular Science*, *18*, 1011. https://doi.org/10.3390/IJMS18051011

Tan, S. (2021). Transmission electron microscopy: Applications in nanotechnology. *IEEE Nanotechnology Magazine*, *15*, 26–37. https://doi.org/10.1109/MNANO.2020.3037432

Tanou, G., Molassiotis, A., & Diamantidis, G. (2009). Induction of ROS and necrotic death-like destruction in strawberry leaves by salinity. *Environmental and Experimental Botany*, *65*, 270–281. https://doi.org/10.1016/J.ENVEXPBOT.2008.09.005

Tavakkoli, E., Rengasamy, P., & McDonald, G. K. (2010). High concentrations of Na^+ and Cl^- ions in soil solution have simultaneous detrimental effects on growth of faba bean under salinity stress. *Journal of Experimental Botany*, *61*, 4449–4459. https://doi.org/10.1093/JXB/ERQ251

Teixeira, F. K., Menezes-Benavente, L., Galvão, V. C., Margis, R., & Margis-Pinheiro, M. (2006). Rice ascorbate peroxidase gene family encodes functionally diverse isoforms localized in different subcellular compartments. *Planta*, *224*, 300–314. https://doi.org/10.1007/S00425-005-0214-8

Thakkar, K. N., Mhatre, S. S., & Parikh, R. Y. (2010). Biological synthesis of metallic NPs. *Nanomedicine: Nanotechnology, Biology and Medicine*, *6*, 257–262. https://doi.org/10.1016/J.NANO.2009.07.002

Tomaszewska, E., Soliwoda, K., Kadziola, K., Tkacz-Szczesna, B., Celichowski, G., Cichomski, M., Szmaja, W., & Grobelny, J. (2013). Detection limits of DLS and UV-Vis spectroscopy in characterization of polydisperse NPs colloids. *Journal of Nanomaterials*, 2013, 1. https://doi.org/10.1155/2013/313081

Torabian, S., Farhangi-Abriz, S., & Zahedi, M. (2018). Efficacy of $FeSO_4$ nano formulations on osmolytes and antioxidative enzymes of sunflower under salt stress. *Indian Journal of Plant Physiology*, *23*, 305–315.

Tripathi, D. K., Shweta, Singh, S., Singh, S., Pandey, R., Singh, V. P., Sharma, N. C., Prasad, S. M., Dubey, N. K., & Chauhan, D. K. (2017). An overview on manufactured NPs in plants: Uptake, translocation, accumulation and phytotoxicity. *Plant Physiology and Biochemistry*, *110*, 2–12. https://doi.org/10.1016/J.PLAPHY.2016.07.030

Tripathi, D. K., Singh, S., Gaur, S., Singh, S., Yadav, V., Liu, S., Singh, V. P., Sharma, S., Srivastava, P., Prasad, S. M., Dubey, N. K., Chauhan, D. K., & Sahi, S. (2018). Acquisition and homeostasis of iron in higher plants and their probable role in abiotic stress tolerance. *Frontiers in Environmental Science*, *5*, 86.

Ullah, S., Adeel, M., Zain, M., Rizwan, M., Irshad, M. K., Jilani, G., Hameed, A., Khan, A., Arshad, M., Raza, A., Baluch, M. A., & Rui, Y. (2020). Physiological and biochemical response of wheat (Triticum aestivum) to TiO_2 NPs in phosphorous amended soil: A full life cycle study. *Journal of Environmental Management*, *263*, 110365. https://doi.org/10.1016/J.JENVMAN.2020.110365

Ushimaru, T., Nakagawa, T., Fujioka, Y., Daicho, K., Naito, M., Yamauchi, Y., Nonaka, H., Amako, K., Yamawaki, K., & Murata, N. (2006). Transgenic Arabidopsis plants expressing the rice dehydroascorbate reductase gene are resistant to salt stress. *Journal Plant Physiology*, *163*, 1179–1184. https://doi.org/10.1016/J.JPLPH.2005.10.002

Van Zelm, E., Zhang, Y., & Testerink, C. (2020). Salt tolerance mechanisms of plants. Annual Review of Plant Biology, *71*, 403–433. https://doi.org/10.1146/ANNUREV-ARPLANT-050718-100005

Verma, K. K., Song, X.-P., Joshi, A., Tian, D.-D., Rajput, V. D., Singh, M., Arora, J., Minkina, T., & Li, Y.-R. (2022). Recent trends in nano-fertilizers for sustainable agriculture under climate change for global food security. *Nanomater*, 12, 173. https://doi.org/10.3390/NANO12010173

Vladár, A. E., & Hodoroaba, V. D. (2020). Characterization of NPs by scanning electron microscopy. In V. D. Hodoroaba, W. E. S. Unger, & A. G. Shard (Eds.). *Characterization of nanoparticles* (pp. 7–27). Elsevier. https://doi.org/10.1016/B978-0-12-814182-3.00002-X

Vorontsov, A. V., & Tsybulya, S. V. (2018). Influence of NPs size on XRD patterns for small monodisperse NPs of Cu0 and TiO_2 anatase. *Industrial & Engineering Chemistry Research*, *57*, 2526–2536.

Wagner Ferreira Sabará, E., Pereira, V., Luiz Molisani, A., et al. (2017). A review on the classification, characterisation, synthesis of NPs and their application. *IOP Conference Series: Materials Science and Engineering*, *263*, 032019.

Wang, C., Mirzaei, A., Wang, Y., Chaker, M., Zhang, Q., & Ma, D. (2023). $BiVO_4$ microspheres coated with nanometer-thick porous TiO_2 shells for photocatalytic water treatment under visible-light irradiation. *ACS Applied Nano Materials*, *2023*, 132579.

Wang, J., & Huang, R. (2019). Modulation of ethylene and ascorbic acid on ROS scavenging in plant salt response. *Frontiers in Plant Science*, *10*, 319.

Wang, W., Tarafdar, J., & Biswas, P. (2013). Nanoparticle synthesis and delivery by an aerosol route for watermelon plant foliar uptake. Journal of Nanoparticle Research, *15*, 1. https://doi.org/10.1007/s11051-013-1417-8

Wang, X., Yang, X., Chen, S., Li, Q., Wang, W., Hou, C., Gao, X., Wang, L., & Wang, S. (2016). Zinc oxide NPs affect biomass accumulation and photosynthesis in Arabidopsis. *Frontiers in Plant Science*, *6*, 1243.

Wang, Z., Li, H., Li, X., Xin, C., Si, J., Li, S., Li, Y., Zheng, X., Li, H., Wei, X., Zhang, Z., Kong, L. & Wang, F. (2019). Nano-ZnO priming induces salt tolerance by promoting photosynthetic carbon assimilation in wheat. Archives of Agronomy and Soil Science, *66*, 1259–1273. https://doi.org/10.1080/03650340.2019.1663508

Wang, Z., Xie, X., Zhao, J., Liu, X., Feng, W., White, J. C., & Xing, B. (2012). Xylem- and phloem-based transport of CuO NPs in maize (Zea mays L.). *Environmental Science & Technology*, *46*, 4434–4441.

Waseda, Y., Matsubara, E., & Shinoda, K. (2011). X-Ray diffraction crystallography: Introduction, examples and solved problems. Springer.

Wiechers, J. W., & Musee, N. (2010). Engineered inorganic NPs and cosmetics: Facts, issues, knowledge gaps and challenges. *Journal of Biomedical Nanotechnology*, *6*, 408–431. https://doi.org/10.1166/JBN.2010.1143

Wrzaczek, M., Brosché, M., & Kangasjärvi, J. (2013). ROS signaling loops – Production, perception, regulation. *Current Opinion in Plant Biology*, *16*, 575–582. https://doi.org/10.1016/J.PBI.2013.07.002

Wu, H., Shabala, L., Shabala, S., & Giraldo, J. P. (2018). Hydroxyl radical scavenging by cerium oxide NPs improves Arabidopsis salinity tolerance by enhancing leaf mesophyll potassium retention. *Environmental Science: Nano*, *5*, 1567–1583. https://doi.org/10.1039/C8EN00323H

Wu, P., Cui, P., Du, H., Alves, M. E., Zhou, D., & Wang, Y. (2021). Long-term dissolution and transformation of ZnO in soils: The roles of soil pH and ZnO particle size. *Journal of Hazardous Materials*, *415*, 125604. https://doi.org/10.1016/J.JHAZMAT.2021.125604

Xu, N., Chu, Y., Chen, H., Li, X., Wu, Q., Jin, L., Wang, G., & Huang, J. (2018). Rice transcription factor OsMADS25 modulates root growth and confers salinity tolerance via the ABA–mediated regulatory pathway and ROS scavenging. *PLOS Genetics*, *14*, e1007662. https://doi.org/10.1371/JOURNAL.PGEN.1007662

Xuming, W., Fengqing, G., Linglan, M., Jie, L., Sitao, Y., Ping, Y., & Fashui, H. (2008). Effects of nano-anatase on ribulose-1, 5-bisphosphate carboxylase/oxygenase mRNA expression in spinach. *Biological Trace Element Research*, *126*, 280–289.

Yamauchi, Y., Furutera, A., Seki, K., Toyoda, Y., Tanaka, K., & Sugimoto, Y. (2008). Malondialdehyde generated from peroxidized linolenic acid causes protein modification in heat-stressed plants. *Plant Physiology and Biochemistry*, *46*, 786–793. https://doi.org/10.1016/J.PLAPHY.2008.04.018

Yang, Y., & Guo, Y. (2018). Elucidating the molecular mechanisms mediating plant salt-stress responses. *New Phytologist*, *217*, 523–539. https://doi.org/10.1111/NPH.14920

Yano, F., Hiraoka, A., Itoga, T., Kojima, H., Kanehori, K., & Mitsui, Y. (1996). Influence of ion-implantation on native oxidation of Si in a clean-room atmosphere. *Applied Surface Science*, *100–101*, 138–142. https://doi.org/10.1016/0169-4332(96)00274-7

Yokotani, N., Higuchi, M., Kondou, Y., Ichikawa, T., Iwabuchi, M., Hirochika, H., Matsui, M., & Oda, K. (2011). A novel chloroplast protein, CEST induces tolerance to multiple environmental stresses and reduces photooxidative damage in transgenic Arabidopsis. *Journal of Experimental Botany*, *62*, 557–569. https://doi.org/10.1093/JXB/ERQ290

Yurkov, A. M., Kemler, M., & Begerow, D. (2011). Species accumulation curves and incidence-based species richness estimators to appraise the diversity of cultivable yeasts from beech forest soils. *PLoS One*, *6*, e23671. https://doi.org/10.1371/JOURNAL.PONE.0023671

Zafar, S., Hasnain, Z., Aslam, N., Mumtaz, S., Jaafar, H. Z. E., Wahab, P. E. M., Qayum, M. & Ormenisan, A. N. (2021). Impact of Zn NPs synthesized via green and chemical approach on Okra (Abelmoschus esculentus L.) growth under salt stress. *Sustainablity*, 13, 3694. https://doi.org/10.3390/SU13073694

Zahedi, S. M., Abdelrahman, M., Hosseini, M. S., Hoveizeh, N. F., & Phan Tran, L.-S. (2019). Alleviation of the effect of salinity on growth and yield of strawberry by foliar spray of selenium-NPs. *Environmental Pollution*, *253*, 246–258. https://doi.org/10.1016/J.ENVPOL.2019.04.078

Zhang, D., Jiang, S., Pan, J., Kong, X., Zhou, Y., Liu, Y., & Li, D. (2014a). The overexpression of a maize mitogen-activated protein kinase gene (ZmMPK5) confers salt stress tolerance and induces defence responses in tobacco. *Plant Biology (Stuttg, Germany)*, *16*, 558–570. https://doi.org/10.1111/PLB.12084

Zhang, J. L., Flowers, T. J., & Wang, S. M. (2010). Mechanisms of sodium uptake by roots of higher plants. *Plant and Soil*, *326*, 45–60.

Zhang, L. X., Li, S. X., Zhang, H., & Liang, Z. S. (2007). Nitrogen rates and water stress effects on production, lipid peroxidation and antioxidative enzyme activities in two maize (Zea mays L.) genotypes. *Journal of Agronomy and Crop Science*, *193*, 387–397. https://doi.org/10.1111/J.1439-037X.2007.00276.X

Zhang, P., Ma, Y., & Zhang, Z. (2015). Interactions between engineered NMs and plants: Phytotoxicity, uptake, translocation, and biotransformation. In M. H. Siddiqui, M. Al-whaibi, & F. Mohammad (Eds.), *Nanotechnology and plant sciences: Nanoparticles and their impact on plants* (pp. 77–99). Springer.

Zhang, Q., Ying, Y., & Ping, J. (2022). Recent advances in plant nanoscience. *Advanced Science (*Weinheim, Baden-Wurttemberg, Germany), 9, e2103414. https://doi.org/10.1002/ADVS.202103414

Zhang, Y., Hu CX, Tan, Q. L., Zheng, C.-S., Gui, H.-P., Zeng, W.-N., Sun, X.-C., & Zhao, X.-H. (2014b). Plant nutrition status, yield and quality of satsuma mandarin (Citrus unshiu Marc.) under soil application of Fe-EDDHA and combination with zinc and manganese in calcareous soil. *Scientia Horticulturae (Amsterdam)*, *174*, 46–53. https://doi.org/10.1016/J.SCIENTA.2014.05.005

Zhang, Y., Zhou, Y., Zhang, D., Tang, X., Li, Z., Shen, C., Han, X., Deng, W., Yin, W., & Xia, X. (2020). PtrWRKY75 overexpression reduces stomatal aperture and improves drought tolerance by salicylic acid-induced ROS accumulation in poplar. *Environmental and Experimental Botany*, *176*, 104117. https://doi.org/10.1016/J.ENVEXPBOT.2020.104117

Zhao, G., Zhao, Y., Lou, W., Su, J., Wei, S., Yang, X., Wang, R., Guan, R., Pu, H., & Shen, W. (2019). Nitrate reductase-dependent nitric oxide is crucial for multi-walled carbon nanotube-induced plant tolerance against salinity. *Nanoscale*, *11*, 10511–10523. https://doi.org/10.1039/C8NR10514F

Zhao, L., Ortiz, C., Adeleye, A. S., Hu, Q., Zhou, H., Huang, Y., & Keller, A. A. (2016). Metabolomics to detect response of lettuce (Lactuca sativa) to $Cu(OH)_2$ nanopesticides: Oxidative stress response and detoxification mechanisms. *Environmental Science & Technology*, *50*, 9697–9707.

Zheng, Y. H., Li, X., Li, Y. G., Miao, B. H., Xu, H., Simmons, M., & Yang, X. H. (2012). Contrasting responses of salinity-stressed salt-tolerant and intolerant winter wheat (Triticum aestivum L.) cultivars to ozone pollution. *Plant Physiology and Biochemistry*, *52*, 169–178. https://doi.org/10.1016/J.PLAPHY.2012.01.007

Zhou, X., Naguro, I., Ichijo, H., & Watanabe, K. (2016). Mitogen-activated protein kinases as key players in osmotic stress signaling. *Biochimica et Biophysica Acta (BBA) - General Subjects*, *1860*, 2037–2052. https://doi.org/10.1016/J.BBAGEN.2016.05.032

Zhu, J. K. (2016). Abiotic stress signaling and responses in plants. *Cell*, *167*, 313–324. https://doi.org/10.1016/J.CELL.2016.08.029

Zulfiqar, F., & Ashraf, M. (2021). NPs potentially mediate salt stress tolerance in plants. *Plant Physiol Biochem*, *160*, 257–268. https://doi.org/10.1016/J.PLAPHY.2021.01.028

[illegible]

[illegible]

[illegible]

[illegible]

[illegible]

[illegible]

[illegible]

[illegible]

[illegible]

[illegible]

[illegible]

[illegible]

[illegible]

[illegible]

[illegible]

[illegible]

Index

Note: **Bold** page numbers refer to tables and *italic* page numbers refer to figures.

[illegible]